Cryptography and Secure Communication

Today's pervasive computing and communications networks have created an intense need for secure and reliable cryptographic systems. Bringing together a fascinating mixture of topics in engineering, mathematics, computer science, and informatics, this book presents the timeless mathematical theory underpinning cryptosystems both old and new.

Major branches of classical and modern cryptography are discussed in detail, from basic block and stream cyphers through to systems based on elliptic and hyperelliptic curves, accompanied by concise summaries of the necessary mathematical background. Practical aspects such as implementation, authentication, and protocol-sharing are also covered, as are the possible pitfalls surrounding cryptographic methods.

Written specifically with engineers in mind, and providing a solid grounding in the relevant algorithms, protocols, and techniques, this insightful introduction to the foundations of modern cryptography is ideal for graduate students and researchers in engineering and computer science, and practitioners involved in the design of security systems for communications networks.

Richard E. Blahut is the Henry Magnuski Professor of Electrical and Computer Engineering at the University of Illinois, Urbana–Champaign. He is a Fellow of the Institute of Electrical and Electronics Engineers and the recipient of many awards including the IEEE Alexander Graham Bell Medal (1998), the IEEE Claude E. Shannon Award, the Tau Beta Pi Daniel C. Drucker Eminent Faculty Award, and the IEEE Millennium Medal. He was named a Fellow of IBM Corporation in 1980 (where he worked for over 30 years) and was elected to the US National Academy of Engineering in 1990.

Cryptography and Secure Communication

Richard E. Blahut

Henry Magnuski Professor of Electrical and Computer Engineering,
University of Illinois, Urbana–Champaign

Shaftesbury Road, Cambridge CB2 8EA, United Kingdom

One Liberty Plaza, 20th Floor, New York, NY 10006, USA

477 Williamstown Road, Port Melbourne, VIC 3207, Australia

314–321, 3rd Floor, Plot 3, Splendor Forum, Jasola District Centre, New Delhi – 110025, India

103 Penang Road, #05–06/07, Visioncrest Commercial, Singapore 238467

Cambridge University Press is part of Cambridge University Press & Assessment, a department of the University of Cambridge.

We share the University's mission to contribute to society through the pursuit of education, learning and research at the highest international levels of excellence.

www.cambridge.org
Information on this title: www.cambridge.org/9781107014275

First published 2014

A catalogue record for this publication is available from the British Library

ISBN 978-1-107-01427-5 Hardback

Yet it may be roundly asserted that human
ingenuity cannot concoct a cipher which
human ingenuity cannot resolve.

– Edgar Allen Poe
The Gold Bug

A hundred ounces of silver spent for information
may save ten thousand spent on war.

– Sun-Tzu
4th century AD

Contents

Preface

Information transmission and information protection are two sides of the same tapestry, but with the information-protection side having more tangled and multitextured threads. At the core of the subject of information protection is the more specific subject of classical cryptography, which protects the content of a message from being understood by unauthorized receivers, but does not protect the message in other ways. Much of this book is concerned with cryptography in this classical sense, but treated in its modern sophisticated form. The modern subject of cryptography, and of information protection in general, is a fascinating mixture of mathematics, engineering, informatics, and computer science, and the same mixture is found in this book.

The subject of information protection is rapidly evolving into a subject that goes well beyond the classical notions of point-to-point cryptography. Now there is an intense need for secrecy and security in large public networks. Within this larger setting of public networked communication, many other issues are important, including issues of authorization, certification, and authentication, that bring many subtle considerations into the discussion. While the emphasis of the book is cryptography, it touches on these other topics as well. My goal, as in my other books, is to concentrate on the formal, and presumably timeless, aspects of the subject rather than on the details of systems in current use. Although this book is not designed to serve as a handbook describing the current standard cryptosystems, some topics are best described by discussing practical systems that are now in use.

Modern cryptography uses a great deal of rather advanced mathematical material from the subjects of number theory, abstract algebra, and algebraic geometry, and I believe that one cannot be an expert in the subject of cryptography without having some understanding of this material. Accordingly, this book provides a formal and rigorous development of all relevant mathematical topics, but abridged to suit the needs of the moment.

This book was written by an engineer, a noncryptographer, for those – especially engineers – who want to learn the subject of information protection in some depth. While I readily admit to the dangers of this recipe, I also hope that there will be positive pedagogical consequences. As an outsider to the subject, I can more easily see when points that are obvious to the expert can be opaque to the novice, and so require more careful treatment. But, at the same time, I also believe strongly that the engineering student of cryptography must not be shortchanged. Though the starting background may be different than that of a mathematician, the engineer can and should follow the main flow of the mathematics to the core, not taking any of the fundamentals for granted.

In writing this book, I sometimes had to find my own way to a result that is beyond my formal training. For this reason, the development is goal-focused and direct, but without sacrificing rigor. My hope is that such a book written by a nonspecialist in a specialized subject will be accessible to the general technically educated reader.

Of course, the soft underbelly of much of modern cryptography is the subject of complexity, a subject that is not formally addressed in this book. Secrets are protected by the apparent intractability of the computational problem that is presented to the adversary. Evidence for intractability is often anecdotal. Formal statements, when known, are qualified, and often may apply only obliquely. Statements regarding the complexity usually refer to the asymptotic complexity, which is of theoretical interest, but can be very different from the practical complexity of real instances of the problem. Because our preference is to try to avoid unsupported assertions in this book, many statements regarding complexity often appear only in general terms, or in the end-of-chapter notes.

Most major notions of classical and modern cryptography are discussed in this book. Even some techniques that are out of date or discredited are discussed if they are important to the history and culture of the subject. Such ideas contribute to understanding, and may lead to future developments.

Many of the various topics of mathematics that underlie the subject of cryptography are gathered midway through the book, not appearing until Chapter 9, although the relevant elements of number theory do appear earlier in Chapter 2. The deferred placement of background material in Chapter 9 helps to shape the character of the book, but it necessitates the occasional forward-reference to the definitions and theorems of Chapter 9. The first half of the book – Chapters 1 through 8 – discusses classical cryptography and the basic earlier methods of public-key cryptography, mostly those based on number theory. The mathematics required in this half of the book is primarily number theory, which is developed in Chapter 2, and elementary notions of group theory. Public-key cryptography is studied in Chapters 3 and 4. Information-theoretic issues are studied in Chapter 5, conventional block and stream ciphers are studied in Chapters 6 and 7, and message authentication is covered in Chapter 8.

At the midpoint of the book, in Chapter 9, a concise summary is given of the mathematics that is needed throughout the latter chapters and occasionally in the early chapters. The latter half of the book also requires other advanced topics of mathematics, especially notions of algebraic geometry. For the most part, these topics are developed in place, as needed. In particular, cryptography based on elliptic and hyperelliptic curves, including pairing methods, is presented in Chapters 10, 11, and 12. The last three chapters round out the book. Chapter 13 discusses practical issues of implementation. Chapter 14 discusses identification, and Chapter 15 discusses lattice-based and code-based cryptography. Most of the treatment throughout is self-contained, or so it is intended.

The mathematics that is developed, beautiful and elegant, is in some ways related to the engineering mathematics of signal processing, though far more advanced and expressed in its own language. Perhaps some of this theory will one day pass into the engineer's workaday toolbox.

Acknowledgments

This book began as an assortment of unedited lecture notes from a course on cryptography that I taught in 1999 with Professor Nigel Boston, and repeated in 2003 and 2005 with Professor Iwan Duursma. Those early lecture notes were only intended to clarify the lectures as an aid to the class participants, and to help me with my own understanding of the mathematical material. Because of the many rough edges at that time, those notes were not intended for general distribution. The notes continued to evolve into the current book from 2009 to 2011 when I taught the course alone to mostly engineering students.

I owe my understanding of the deeper mathematical topics to my shared time with Boston both in the classroom and out of the classroom, as well as my interactions with Duursma. Without the closeness of this association, I could not have developed my understanding of this material. Although I do thank them for giving me this new interest, I also blame them for burdening me with a new addiction. My long friendships with Ian Blake and Jim Massey must also be mentioned as two early feathers tickling the skin of my curiosity. This book resulted from scratching that itch. And, of course, the stimulation and challenge of the many attentive and questioning students in the ECE Illinois classroom is invaluable in preparing a book such as this.

Expert criticism of the manuscript was kindly provided by Professor Nigel Boston, Professor Alfred Menezes, and Professor Ian F. Blake. Their help was invaluable and saved me from many errors. Early conversations with Negar Kiyavash, Sam Spencer, Patricio Parada, Figen Oktem, Sara Bahramian, and Leila Fuladi also helped me with the evolution of the manuscript. The quality of the book has much to do with the composition skills of Ms. Frances Bridges who provided that and so much more, and with the editing skills of Ms. Debra Rosenblum. And, as always, Barbara made it possible.

1 Introduction

Information and the communication of that information comprise the nerve system of civilization, and civilization depends on the availability of reliable methods for the protection of information from intruders and adversaries. There are many ways the collection and communication of information needs to be protected and made trustworthy. The requirements are central to the orderly functioning of society and may include secrecy, integrity, nonrepudiation, authentication, covertness, copy resistance, certification, authorization, and ownership protection. These various topics can be regarded as more or less distinct requirements, although of course there are considerable overlaps. Together they form the topic of secure communications. At the center of these various topics, as well as at the heart of this book, is the classical topic of cryptography.

Communication and cryptography are closely related topics in the general field of telecommunication. Communication is the process of exchanging data and messages. By itself, the term communication carries an active, positive tone and suggests cooperation and openness. Yet the process of communication does have its competitive, defensive side. The nature of social and economic interaction can impose a great variety of subtle requirements on the structure of a communication system to ensure various forms of security, privacy, and trustworthiness.

Secrecy and authentication are complementary functions in a communication system. Secrecy is the function that ensures that a message cannot be understood by an eavesdropper. Authentication is the function that ensures that the message originated with the indicated source of that message. The purpose of authentication is to verify the source of the message. It does not verify the identity of the transmitter of a message, a function known as identification. Authentication can be provided by a digital signature that must be impervious to forgery. This requirement leads to connections with cryptography. Cryptography consists of the study, development, and implementation of methods for protecting data; cryptanalysis is the study, development, and implementation of methods for attacking and breaking cryptosystems. Taken together, the two subjects of cryptography and cryptanalysis form the topic of cryptology.

The elementary notion of a classical *cryptosystem* is familiar to many puzzle solvers who enjoy recreational cryptograms. Recreational cryptograms are a widespread feature of newspapers and are popular with puzzle solvers. Because each letter of the alphabet is represented in a cryptogram by another letter, a solution amounts to finding which of 26! possible permutations is the "key." These recreational cryptograms usually appear with the word spaces intact, which gives useful information to the solver and allows the puzzles to be solved quite simply. However, even without the word spaces, the cryptograms are still rather easy to solve. We conclude from this observation that one does not actually try all 26! possible permutations. This would be a formidable task. Instead, the puzzle solver uses some kind of hierarchical structure in which the permutation is deduced by a sequence of inferences. In the language of this book, the recreational *cryptanalyst* attacks the cryptosystem using prior knowledge of both the structure of the cryptosystem and the linguistic structure and content of the likely encrypted messages.

1.1 Classical cryptography

A message $(x_1, x_2, \ldots, x_n)$, herein called a *plaintext message*, consists of a sequence of n symbols from a given finite alphabet $\mathcal{A}$ of size $\#\mathcal{A}$. For convenience, we may often regard the length n of the plaintext message to be fixed in advance in order to avoid certain uninteresting distractions that can arise when treating variable-length messages. Then the message $(x_1, x_2, \ldots, x_n)$ is an element of $\mathcal{A}^n$, but not every element of $\mathcal{A}^n$ need be a legitimate plaintext message of the given application. Let $\mathcal{M} \subset \mathcal{A}^n$ denote the set of legitimate plaintext messages. When appropriate, the set $\mathcal{M}$ is called a *natural language* or, more simply, a *language*. This statement implies that, in general, not every element of $\mathcal{A}^n$ is a plaintext message. This is denoted by $\mathcal{M} \neq \mathcal{A}^n$. Indeed, the cardinality of $\mathcal{M}$ is usually much smaller than the cardinality of $\mathcal{A}^n$, which we write as $\#\mathcal{M} \ll \#\mathcal{A}^n$. This observation plays an important role in the information-theoretic approach to cryptography.

For the simplest model of the English language, $m = 26$, all letters are upper case, and there is no symbol for a space. Whenever we require messages to have a fixed length n, we may pad a shorter message at the end with copies of a filler symbol, such as the symbol Z (or a blank space or other alternative null symbol), to make the message have the standard length n. For this simple model of English, $\mathcal{A}^n = 26^n$, but the legitimate plaintext messages, as determined by $\mathcal{M}$, are far fewer.

Ciphers may be classified as either of two types: *block ciphers* and *stream ciphers*. A block cipher of blocklength n first segments a longer plaintext message $\boldsymbol{x}$, say of length N, into N/n blocks (or segments); each block has length n, and the message is now written $\{\boldsymbol{x}_1, \boldsymbol{x}_2, \ldots, \boldsymbol{x}_{N/n}\}$, where $\boldsymbol{x}_\ell$ is the ℓth block of the message. For this purpose, we require that n divides N. Each plaintext block $\boldsymbol{x}_\ell$, for $\ell = 1, \ldots, N/n$, consists of n sequential symbols of the message, with each symbol from the given alphabet $\mathcal{A}$. Thus $\boldsymbol{x}_\ell \in \mathcal{A}^n$.

For many elementary block ciphers, $n = 1$. This means that symbols are encrypted independently, although by a common encryption rule. Such ciphers are far too trivial to be used in serious applications, although they are popular in recreational applications, and will be described here.

A *block encryption function* is a map, denoted $e(\boldsymbol{x})$, that maps a message block $\boldsymbol{x}$ of length n into another block, often also of length n, called a *ciphertext block*, and denoted $\boldsymbol{y} = e(\boldsymbol{x})$. Thus $\boldsymbol{y} \in \mathcal{A}^n$. A *key* is an element k from a set $\mathcal{K}$ called the *keyspace*. The function of a key $k \in \mathcal{K}$ is to specify an encryption function, now denoted $e_k(\boldsymbol{x})$, from a predefined set of encryption functions indexed by k. A *block encryptor* is a collection of maps, denoted $\{e_k(\boldsymbol{x}), k \in \mathcal{K}\}$, forming the set of encryption functions. A *block decryptor* is the collection of inverse maps $\{d_k(\boldsymbol{y}) = e_k^{-1}(\boldsymbol{y})\}$, also indexed by k. Thus for each k, $d_k(e_k(\boldsymbol{x})) = \boldsymbol{x}$. This arrangement, in which both the encryptor and the decryptor have access to the same key k, is called a *symmetric-key cryptosystem*. We will see eventually, and perhaps surprisingly, that some cryptosystems have a different key at the encryptor and the decryptor. This is called an *asymmetric-key cryptosystem*. In this case, each decryption key, which itself must be kept secret, is associated with a corresponding encryption key, which can be made public. This encryption key needs to be suitably published only once by that decryptor or by a trusted proxy. Thereafter, the decryptor or the proxy can be completely passive. Indeed, the proxy can be dissolved. It is no longer needed for this purpose. In contrast, in order to form a key using a symmetric cryptosystem, the decryptor must actively interact with every transmitter that intends to send one or more encrypted messages. For practical reasons, this interaction needed to create a secret symmetric key is usually public, and so it is referred to as a *public key exchange*. A public key exchange is different from a *public encryption key*, though both are public.

Another advantage of an asymmetric-key cryptosystem is that the leaking of the encryption key does not compromise the system because, in fact, the encryption key is public. Of course, this consideration requires that neither the decryption key nor the plaintext can be deduced from the public encryption key and the ciphertext. In particular, this system is vulnerable if it is used with a message space so small that it is possible to simply encrypt every possible message with the public encryption key until the given ciphertext is observed, thereby revealing the actual plaintext corresponding to that ciphertext. This possibility is countered by padding all short messages with randomly generated bits. In this way, inadequate entropy in the message space is supplemented with the additional entropy of the random padding.

The *cipherspace*, or *cryptspace*, of the key k is the set $e_k(\mathcal{M})$. The cipherspace is contained in $\mathcal{A}^N$ whenever the alphabet and blocklength of the ciphertext are the same as the alphabet and blocklength of the plaintext. It is clear that $d_k(e_k(\mathcal{M})) = \mathcal{M}$. However, in general, $d_{k'}(e_k(\mathcal{M})) \neq \mathcal{M}$ for $k' \neq k$. Indeed, it may be true that $d_{k'}(e_k(\mathcal{M})) \cap \mathcal{M} = \phi$ for every $k' \neq k$. In that case, in principle, the encryption is not secure. It is vulnerable to a *direct attack*. Simply compute $d_{k'}(\boldsymbol{y})$ for every k' and choose that k' for which $d_{k'}(\boldsymbol{y}) \in \mathcal{M}$. Of course, if $\#\mathcal{K}$, the cardinality of the keyspace $\mathcal{K}$, is large enough, a direct attack may not be feasible because of excessive computations. Thus from a practical point of view, secrecy may depend on computational resources.

In the contrary case, it may be that $d_{k'}(e_k(\mathcal{M})) = \mathcal{M}$ for all $k' \neq k$ and for all k. This is an example of perfect secrecy. For any ciphertext $\boldsymbol{y}$ and any plaintext $\boldsymbol{x}$ there is a key k for which $e_k(\boldsymbol{x}) = \boldsymbol{y}$. *Any* plaintext message could be the right one. Only if the key is known or partially known can the plaintext be deduced, or partially deduced, from the ciphertext.

The only cryptosystem proven to have perfect secrecy is the one-time pad. Suppose that both the encryptor and the decryptor have an identical copy of a long, random, binary sequence $\boldsymbol{u} = (u_\ell, \ell = 0, \ldots)$, called a one-time pad. Let $\boldsymbol{x} = (x_\ell, \ell = 0, \ldots,)$ be the plaintext represented in the form of a binary sequence. The encryptor transmits $\boldsymbol{y} = \boldsymbol{x} + \boldsymbol{u}$ (where $+$ denotes componentwise modulo-two addition) and the decryptor computes $\boldsymbol{y} + \boldsymbol{u}$. Because $\boldsymbol{u} + \boldsymbol{u} = \boldsymbol{0}$ modulo two, the decryptor has recovered the plaintext $\boldsymbol{x}$. The adversary, or eavesdropper, sees only $\boldsymbol{y}$, which is an unintelligible ciphertext with its alphabet and blocklength the same as that of the plaintext. It is impossible to recover $\boldsymbol{x}$, in whole or in part, without knowing $\boldsymbol{u}$.

The one-time pad will not be fully secure if the same random sequence $\boldsymbol{u}$ of the pad is used twice. This is because, $\boldsymbol{u} + \boldsymbol{u} = \boldsymbol{0}$ modulo two, so if the two messages $\boldsymbol{x}_1$ and $\boldsymbol{x}_2$ are encrypted with the same binary key, then $\boldsymbol{y}_1 = \boldsymbol{x}_1 + \boldsymbol{u}$ and $\boldsymbol{y}_2 = \boldsymbol{x}_2 + \boldsymbol{u}$, from which the cryptanalyst can compute that $\boldsymbol{y}_1 + \boldsymbol{y}_2 = \boldsymbol{x}_1 + \boldsymbol{x}_2$. Because $\boldsymbol{x}_1, \boldsymbol{x}_2 \in \mathcal{M}$, and $\mathcal{M}$ is presumably sparse in $\mathcal{A}^n$, it is possible with sufficient computational resources to list all $\boldsymbol{x}_1$ and $\boldsymbol{x}_2$ for which $\boldsymbol{x}_1 + \boldsymbol{x}_2$ equals $\boldsymbol{y}_1 + \boldsymbol{y}_2$. We may expect this list of $\boldsymbol{x}_1$ and $\boldsymbol{x}_2$ to be much smaller than $|\mathcal{M}|^2$. In this way, the two ciphertexts have been partially decrypted by reducing them to a list of possible pairs of plaintexts.

Although the one-time pad is secure at the cryptographic level, it may still fail against *side-channel attacks*. Its weakness is that it requires both the encryptor and the decryptor to have an identical copy of the one-time pad. In practice, this means that the problem of cryptographic security is replaced by the problem of the distribution and protection of the one-time pad. This would be an equivalent problem unless the circumstances for the distribution of the one-time pad could be made different from the circumstances for the distribution of the ciphertext. For example, a trusted courier can physically carry the one-time pad from one user to the other, or the one-time pad could be distributed at an earlier time, in a different situation, and over a channel known to be secure.

Classical methods of key distribution are not practical on a large public network. Accordingly, new methods of public key exchange or public key agreement have been developed in recent years both for message secrecy and for message signatures. Surprisingly, such methods are in use for creating secret keys between two strangers in full view of adversarial third parties. These modern methods depend on the computational intractability of the adversarial cryptanalyst's task.

The practice of cryptography requires that formal procedures for using a given cryptographic technique be in place so that the key is not divulged by improper use. These methods are referred to as *cryptographic protocols*, which often can be understood without understanding the cryptographic techniques themselves. The protocols are variously studied under the name *information security*, rather than *information secrecy*, which more usually refers to the cryptography. The subject of this book falls primarily under the heading of information secrecy, but the methods of information security are close at hand, and also require discussion. Indeed, it is often not possible to draw a clear distinction between the tasks of secrecy and security.

1.2 Notions of cryptographic secrecy

Classical cryptography relies on the distribution of a secret key over a secure channel. This is unacceptable in modern applications because communication routinely takes place over large public networks between strangers, with no opportunity to prearrange the distribution of a secret key. Secure communication between two such parties will require that the two parties set up their key over a public channel and in the clear. This is both possible and routinely done. More precisely, it is generally believed that the methods now in common use for doing this are secure. It is possible – or so we believe – for two parties to arrange a secret key, one that only they will know, using only communications over a public channel in full view of a sophisticated adversary. Such methods can only give computational or practical secrecy, never perfect secrecy. They can be broken by an adversary with infinite computational resources. Much of this book is devoted to the study of methods of doing this, and to ways of attacking such methods.

Early attempts to define cryptographic security were based on the notion of *perfect secrecy* or *perfect security*, a notion that is only assured by a one-time pad and is usually impossible to meet in everyday situations. This notion has now been supplemented by alternative and more practical definitions of cryptography motivated by the question "When is a cryptographic system computationally secure?" The notion of computational security underlies the modern methods of public-key cryptography, but the notion of perfect secrecy remains as a stronger and more desirable requirement that is rarely met. A computationally secure system may be perfectly insecure when judged by a more rigid standard.

We will define four notions of secrecy against an attack in which the cryptanalyst has only the ciphertext, but always has full knowledge of the method of encryption. Only the key is unknown to the cryptanalyst.

1. Perfect or information-theoretic secrecy: The ciphertext gives no information about the plaintext.

2. Unconditional secrecy: It is impossible to break the ciphertext with infinite computational resources, even with the best *possible* algorithm.
3. Computational secrecy: It is intractable to break the ciphertext with "practical" computational resources, even with the best *possible* algorithm.
4. Practical secrecy: It is intractable to break the ciphertext with "practical" computational resources using the best *known* algorithm.

These four notions of secrecy are listed in the order of their decreasing apparent weakness. We would like the statements we make about cryptographic systems to be as strong as possible. However, we are usually unable to make a statement that a given cryptosystem is perfectly secure – because most are not. We can state only what we know, which is that for many cryptography systems, the system is perfectly insecure from a theoretical point of view, meaning that each of these systems can be broken in unlimited computational time or with unlimited computational resources. Moreover, we do not often know the complexity of the best possible algorithm for a given task. Usually we do not know the best possible algorithm. Statements about the best possible algorithm are either not known or are heavily qualified. Usually, we are content with practical secrecy even though this notion is informal and imprecise. The set of known algorithms changes with time, and the notion of a practical computation changes with time as well. Thus, to claim practical secrecy is to claim nothing of permanent truth.

Any discussion of practical secrecy should include some consideration of the desired longevity of the secrecy. One will use a stronger system for a message that must remain secret for decades than for a message that must remain secret for months. One should assume that, except for the one-time pad, every cryptography system can be broken eventually, though perhaps not for centuries.

A cryptography system also may be vulnerable to various kinds of side-channel attacks. A side-channel attack uses a variety of "externalities" to gain information about a message or a key through the "backdoor." The perceived relationship between the source of the message and the user of the message, or the time and circumstances of message transmission, may be useful to the cryptanalyst. Any measurement regarding the amount of time taken for an encryption or decryption computation may give some information about the key. Such information, even when it is meager, may be useful to the cryptanalyst when combined with other sources of information. It is possible that a faint echo of the plaintext message will be present on one or more wires leaving the encryption device, even perhaps unintentionally leaked in this way on the power cord. Even the electromagnetic or infrared signature of the encryption device may be scrutinized to gain information about the encryption key, however slight. This side-channel information might be used to supplement a direct attack.

Recurring message patterns of the user, such as starting every message with the date or a formal salutation, have been used to successfully attack a cryptosystem. For such reasons, some systems alter the key for each encrypted message with an unencrypted but randomly chosen number to further randomize the message space. The secret key is modified by adding the random number to it, then appending this random number unencrypted as an attachment to the ciphertext. That random number is added to the secret key for encrypting and for decrypting that specific message. In this way, although a new key is not exchanged, each message is effectively encrypted with a different key in order to further hide recurring user patterns that may be present. Of course, the secrecy resides only in the original key, but certain attacks are foiled in this way.

In the end, it may be that the best way for an adversary to circumvent a modern cryptosystem is to find a way to enter the encryptor before the encryption takes place or to enter the decryptor after the decryption takes place, and to so read the plaintext directly. A totally secure lock on the front door does no real good if the windows are unprotected. This kind of vulnerability to an intruder is outside of the scope of this book.

1.3 Block ciphers

A *block cipher* segments a message into plaintext blocks of a fixed length, encrypts each plaintext block into a corresponding ciphertext block of a fixed length, then concatenates the encrypted blocks to form the encrypted message. Generally, the plaintext blocks and the encrypted blocks have the same length. We will usually assume that this is the case, though it is not always so. The decryptor reverses this procedure, breaking the concatenated stream of ciphertext blocks into individual blocks and decrypting these blocks one by one. The decrypted plaintext blocks are then concatenated to recover the original message.

We will describe elementary block ciphers in this section, mostly by way of examples. These elementary block ciphers are much too weak to be useful in any serious application. Some elementary ciphers may be useful for providing rudimentary privacy, but are easily broken by straightforward computational methods. They serve here only as a way to introduce the subject of cryptography and its terminology, to suggest its history, and to motivate the search for secure methods.

The simpler of the elementary block ciphers have a blocklength n equal to one. Each block x_ℓ of the plaintext is only a single symbol of the alphabet $\mathcal{A}$, and each block $y_\ell = e_k(x_\ell)$ of the ciphertext is also a single symbol of the alphabet $\mathcal{A}$. The message $(x_1, \ldots, x_n)$ then is encrypted, one symbol at a time, to produce the ciphertext $(y_1, \ldots, y_n) = (e_k(x_1), \ldots, e_k(x_n))$. In the simplest case, the key k and the encryption function e_k remain the same for all symbols of the message.

For an elementary block cipher of blocklength n equal to one, an encryption function will only be a permutation of the alphabet $\mathcal{A}$, i.e., the $q = \#\mathcal{A}$ symbols of the alphabet $\mathcal{A}$ are encrypted by a permutation of the alphabet, denoted by $\sigma_k : \mathcal{A} \to \mathcal{A}$, or $x \mapsto \sigma_k(x)$. Each key k specifies one permutation. Thus the keyspace $\mathcal{K}$ is a subset of $\mathcal{S}_q$, where $\mathcal{S}_q$ is the group of permutations on a set of q elements.

Some elementary ciphers have standard names. A *substitution cipher*[1] of blocklength one (not to be confused with a permutation cipher to be discussed later) is a cipher in which the set of keys is the set of all permutations of an alphabet of size $q = \#\mathcal{A}$. Thus $\mathcal{K} = \mathcal{S}_q$, and $\#\mathcal{K} = q!$. It is common to refer to $q!$ as the cardinality of the key space, and the size of the equivalent binary key needed to specify a permutation is $\log_2(q!)$ bits. For example, if $q = 26$, the size of the equivalent binary key is $\log_2(26!) \approx 90$, so the equivalent binary key size is about 90 bits. A single key of the recreational cryptogram is the permutation

$$\begin{array}{l}(A\ B\ C\ D\ E\ F\ G\ H\ I\ J\ K\ L\ M\ N\ O\ P\ Q\ R\ S\ T\ U\ V\ W\ X\ Y\ Z)\\(F\ G\ Q\ P\ N\ A\ D\ E\ O\ B\ U\ J\ V\ H\ Y\ I\ T\ W\ K\ X\ R\ C\ L\ S\ Z\ M).\end{array}$$

There are approximately 2^{90} such substitution keys on an alphabet of 26 letters. The set of recreational cryptograms is an example of a substitution cipher on an alphabet with 26 letters.

For a larger example of a substitution cipher, a standard keyboard is often regarded as having 256 characters. Then the alphabet has size 256. Thus there are 256! permutations on this set, so a substitution cipher that used all possible keys would have an equivalent key size of $\log_2(256!)$ bits. This is a very large number, and it would require a binary word with more than a thousand bits to specify one key from this set. Accordingly, one will usually choose to restrict the keyspace in some way, as determined by the details of the encryption algorithm. The actual key will be much smaller than $\log_2(256!)$ bits, and so the set of allowable permutations will be much smaller as well.

All block ciphers of blocklength one are actually substitution ciphers. The named block ciphers of blocklength one are defined by restricting the set of permutations to those that can be described by some specialized simpler rule. A *shift cipher* is a special case of a substitution cipher. A shift cipher consists only of permutations that are cyclic shifts of alphabet $\mathcal{A}$. Let $\mathcal{A}$ be represented by the set of integers with addition modulo q, denoted $\mathbf{Z}_q$. If $\mathcal{A}$ is the roman alphabet, then $q = 26$ and the letters of $\mathcal{A}$ can be represented by the integers from 1 to 26. Then the encryption function is $e_k(x_i) = x_i + k \pmod{26}$. The keyspace of the shift cipher has size 26 and the equivalent binary key size is $\log_2 26$, which is less than five bits. Because the keyspace is so small, the shift cipher is not secure, though it may be used as a privacy cipher.

An *affine cipher* of blocklength one is given by $e_k(x_i) = ax_i + b \pmod{26}$. So that this function can be inverted, $a^{-1} \pmod{26}$ must exist, but b can be any element of $\mathbf{Z}_{26}$. Elementary number theory, which is reviewed in the next chapter, states that for the inverse a^{-1} to exist, a must satisfy $\mathrm{GCD}(a, q) = 1$. The set of a that have inverses under modulo-q arithmetic is denoted $\mathbf{Z}_q^*$. For the roman alphabet, $q = 26$ and $\mathbf{Z}_{26}^* = 12$.

[1] A substitution cipher permutes the letters of the alphabet. A permutation cipher permutes the letters of the message block.

Therefore the affine cipher $e_k(x_i) = ax_i + b \pmod{26}$ has $12 \cdot 26$ keys (12 choices for a and 26 choices for b), and the equivalent binary key size is less than nine bits. Of these, the trivial key with $a = 1$ and $b = 0$, and perhaps other trivial keys, should be avoided.

Block ciphers with a blocklength n larger than one can be defined similarly. If each symbol is from an alphabet $\mathcal{A}$ of size q, then each block of length n is from an alphabet $\mathcal{A}^n$ of size q^n. Then $\mathcal{K}$ is a subset of $\mathcal{S}_{q^n}$, the group of permutations on q^n elements. It may even be that $\mathcal{K}$ is equal to $\mathcal{S}_{q^n}$. Then the cipher is called a *block substitution cipher*.

For example, the simplest model of the English language has $q = 26$. If $n = 2$, then there are $26^2 = 676$ pairs of letters. A block substitution cipher on blocks of length two replaces each pair of letters with a substitute pair of letters. This block cipher is a set of simple look-up tables, each with 676 entries, one entry for each pair of letters, and one such look-up table for each key. To break this blocklength-two substitution cipher is much harder than it is to break the elementary recreational cryptogram of blocklength one described earlier. Because there are $(26^2)!$ permutations on pairs of letters, there are $(26^2)!$ distinct blocklength-two substitution ciphers on an alphabet of size 26. This is an immense number, not to be confused with $(26!)^2$. If all of these permutations are indexed by keys in a keyspace $\mathcal{K}$, then the size of the keyspace is $(26^2)!$. One cannot attack this cipher by exhaustively trying all keys; there are far too many. Because $(26)^2!$ is an immense number – too large to index the keys – one should expect that only a subset of these permutations would be used as keys in a given practical cipher, and that restricted keyspace would be defined by a tractable rule for defining the key. A cryptanalyst would use knowledge of this restricted keyspace, were it known, but this knowledge will presumably be useless in a well-designed cipher.

There are elementary block ciphers of this kind that have standard names: A *Vigenère cipher* is a componentwise additive cipher on blocks of length n symbols using an additive key of length n symbols. For example, let $SNOW$ denote a Vigenère key with blocklength $n = 4$. This key has the numerical equivalent $(19, 15, 16, 23)$. To encrypt the word "ball," add componentwise modulo 26 the numerical equivalents $(2, 1, 12, 12) + (19, 15, 16, 23) = (21, 16, 2, 9)$, which becomes "UPBI." To decrypt, again using the key "SNOW," write $(21, 16, 2, 9) - (19, 15, 16, 23) = (2, 1, 12, 12)$, which recovers the word "ball."

A *Hill cipher* of blocklength two has the form

$$\begin{bmatrix} y_1 \\ y_2 \end{bmatrix} = \boldsymbol{M} \begin{bmatrix} x_1 \\ x_2 \end{bmatrix},$$

under modulo-q arithmetic, where $\boldsymbol{M}$ is an invertible matrix modulo q. The matrix will be invertible if $(\det \boldsymbol{M})^{-1}$ exists in $\boldsymbol{Z}_q$, and this inverse exists if the greatest common divisor of $\det \boldsymbol{M}$ and q is equal to one. The Hill cipher provides our first example of a cipher that uses number theory in a nontrivial way. For example, to specify a Hill cipher of blocklength two requires a two-by-two matrix that is invertible modulo 26. The matrix

$$M = \begin{bmatrix} 1 & 2 \\ 3 & 4 \end{bmatrix}$$

satisfies GCD(det M, 26) = 2, so this M cannot be used to key a Hill cipher. The matrix

$$M = \begin{bmatrix} 1 & 2 \\ 3 & 5 \end{bmatrix}$$

is suitable because now GCD(det M, 26) = 1. Therefore

$$M^{-1} = \begin{bmatrix} 21 & 2 \\ 3 & 25 \end{bmatrix}$$

modulo 26. To encrypt the word "ball" with this Hill cipher, the numerical equivalent, (2, 1, 12, 12), is encrypted two letters at a time as

$$\begin{bmatrix} y_1 \\ y_2 \end{bmatrix} = \begin{bmatrix} 1 & 2 \\ 3 & 5 \end{bmatrix} \begin{bmatrix} 2 \\ 1 \end{bmatrix}$$

$$\begin{bmatrix} y_3 \\ y_4 \end{bmatrix} = \begin{bmatrix} 1 & 2 \\ 3 & 5 \end{bmatrix} \begin{bmatrix} 12 \\ 12 \end{bmatrix},$$

modulo 26. Then $y = (4, 11, 10, 18)$, which becomes "DKJR" in the ciphertext. In this way, the plaintext "ball" is represented by the ciphertext "DKJR."

To decrypt, write

$$\begin{bmatrix} x_1 \\ x_2 \end{bmatrix} = \begin{bmatrix} 21 & 2 \\ 3 & 25 \end{bmatrix} \begin{bmatrix} 4 \\ 11 \end{bmatrix}$$

$$\begin{bmatrix} x_3 \\ x_4 \end{bmatrix} = \begin{bmatrix} 21 & 2 \\ 3 & 25 \end{bmatrix} \begin{bmatrix} 10 \\ 18 \end{bmatrix},$$

modulo 26. Then $x = (2, 1, 12, 12)$, which reduces to the word "ball," thereby recovering the plaintext message.

A *permutation cipher* of blocklength N is given by

$$\begin{bmatrix} y_1 \\ y_2 \\ \vdots \\ y_N \end{bmatrix} = M \begin{bmatrix} x_1 \\ x_2 \\ \vdots \\ x_N \end{bmatrix},$$

where M is an N-by-N matrix with a single one in every column and a single one in every row and all other elements are equal to zero. Such a matrix is called a *permutation*

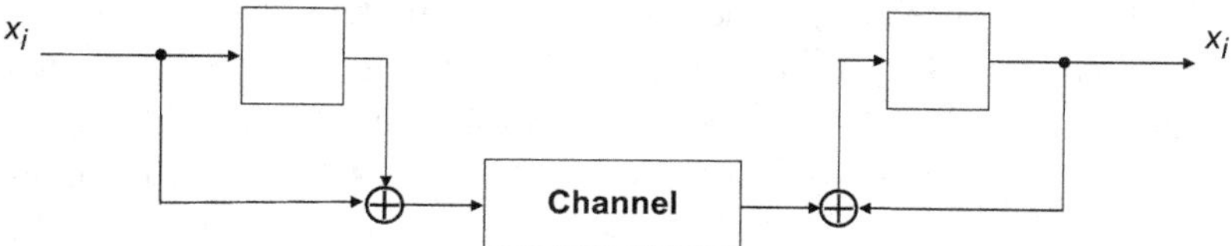

Figure 1.1 An autokey cipher

matrix. There are $N!$ permutation matrices on blocks of N letters. This means that the size of the keyspace of the permutation cipher is $\log_2(N!)$ bits.

An example of a permutation matrix of blocklength four is

$$\boldsymbol{M} = \begin{bmatrix} 0 & 0 & 1 & 0 \\ 1 & 0 & 0 & 0 \\ 0 & 0 & 0 & 1 \\ 0 & 1 & 0 & 0 \end{bmatrix}.$$

When this matrix is used to form a permutation cipher, the word "ball" becomes the ciphertext "ALBL." A permutation cipher simply scrambles the letters of the word. The encryption key is a designation of the permutation matrix.

1.4 Stream ciphers

A block cipher encrypts a long message by breaking it into blocks, each of a fixed blocklength. The blocks are independently encrypted and then reassembled into a long ciphertext. In contrast, a nonblock cipher, such as a *stream cipher*, uses a different structure to encrypt a message. The message is regarded to have an unspecified length, and it can even be modeled as infinitely long. The encryptor, however, must be practical, so it has a finite memory. This will normally mean that the encryptor can only observe, or retain, a limited segment of the message at one time. As new message symbols enter the encryptor, old message symbols are discarded. At the same time that new message symbols are entering, cipher symbols are formed and shifted out of that encryptor. Accordingly, the plaintext is serially shifted into the stream-cipher encryptor and the ciphertext is serially shifted out of the encryptor.

An example of an encryption technique that is neither a block cipher nor a proper stream cipher is one known as the *autokey cipher*. The autokey cipher uses portions of the message itself to disguise other portions of the message. There is no true key nor does the autokey provide true cryptographic secrecy. A simple form of an autokey cipher is shown in Figure 1.1, with the letters of the alphabet represented as the integers from 1 to 26, and addition modulo 26. This autokey cipher, given by

$$y_i = x_i + x_{i-1} \pmod{26},$$

is a simple example of a feedforward cipher with no key.

The autokey cipher used to encode English text with the usual numerical representation would encode "DEED" into "DIJID" as follows. The numerical representation of the plaintext "deed" is 4, 5, 5, 4. Regarding this to be padded with zeros, and adding each symbol to its predecessor symbol, gives 4, 9, 10, 8, 4, which is the numerical equivalent of "DIJID."

This autokey cipher, of course, is merely a privacy cipher. It has no secrecy. It is easily inverted by the computation

$$x_i = y_i - x_{i-1} \pmod{26}.$$

The autokey cipher, which as such is a keyless privacy cipher, can be made into a keyed cipher by the more general rule

$$y = x_i + \sum_{\ell=1}^{n} k_\ell x_{i-\ell} \quad \pmod{26},$$

where k_ℓ is a zero or a one, and $k_\ell x_{i-\ell}$ has the obvious meaning. Now, the n-bit number $(k_1, k_2, \ldots, k_n)$ is the key. The encryption is inverted by

$$x_i = y_i - \sum_{\ell=1}^{n} k_\ell x_{i-\ell} \quad \pmod{26},$$

which provides the decryption.

A more popular stream cipher is one known as an *additive stream cipher*. This is an encryption technique in which an unending *keystream* is generated by a finite-state machine using a secret key as a seed. A binary keystream should appear to be an infinite stream of independent random symbols resembling random equiprobable binary symbols, though of course, the keystream is a deterministic function of the key that both the encryptor and the decryptor can compute. For encryption, a binary keystream is added, modulo two, bit by bit to the binary *datastream* to produce the *cipherstream*. For decryption, the keystream is again added, modulo two, bit by bit to the binary cipherstream. Because adding the keystream, modulo two, to the cipherstream is the same as subtracting, it removes the keystream, and the binary datastream is recovered.

A simple method of generating a keystream is by means of a *linear recursion* as by using a *linear-feedback shift register*. For example, define

$$x_i = x_{i-3} + x_{i-4}$$

with a 4-bit seed as the initialization. This linear recursion produces a periodic sequence of period fifteen. To see that the period of this sequence cannot be larger than fifteen, notice that to compute each x_i, it suffices to retain only the previous four bits. These four bits cannot take on more than the fifteen possible nonzero values because, if these four bits ever become all zeros, then the sequence is zero thereafter. Thus, the period cannot be more than fifteen. A deeper analysis will show that the period is indeed fifteen.

Because the keystream is generated by a finite-state machine from a random seed given by the key, which consists of a binary block of finite length, the keystream is not maximally random. It is completely predictable from its initialization seed. Hence it is called a *pseudorandom sequence*, meaning that it only superficially appears to be maximally random. Only the key is actually random, but it is of finite blocklength. Thus, there are only a limited number of such keystreams as determined by the specific value of the key. The task is to construct these keystreams from the key in such a way that a short known segment of the keystream does not divulge the entire keystream. We will study the security of stream ciphers in Chapter 7.

1.5 Public-key cryptography

Classical cryptography deals with a small community of users exchanging messages only occasionally. The issue of establishing keys was not a pressing one early on because keys could be exchanged well in advance of their use, and only a few users in any community would be involved. The issue of secret communications has completely changed. A huge number of secret messages are exchanged every day and the community of users is very large, with no opportunity for prior key exchange in a secure setting. Cryptography now must take place in full public view. Such methods involve the use of one-way functions. These are functions that are feasible to compute, but not computationally feasible to invert. This requirement has led to the modern era in cryptography in which public ciphers depend utterly on deep and difficult topics of modern mathematics, topics that occupy much of this book. These methods are primarily used to create the keys that are then to be used by other ciphers, either block ciphers or stream ciphers.

Figure 1.2 shows a very stylized and abbreviated depiction of the role of abstract mathematics in public-key cryptography, and gives a glimpse of some topics that will be studied in this book. Starting at the base of the stack with the familiar properties of the integers, a sequence of mathematics structures are built one on top of the other: prime fields from number theory, extension fields from prime fields, elliptic and hyperelliptic curves over finite fields from extension fields, and bilinear pairings from elliptic curves. At each level, new methods of cryptography have emerged, have rapidly developed, and have passed into common use. It is interesting to observe that the newer topics have often been introduced into cryptography by the cryptanalyst as a method of attack, but are then quickly adopted by the cryptographer as a method of protection.

The outline of the theory depicted in Figure 1.2 encompasses the major emphasis of modern cryptography, but the outline is not exhaustive. Other methods, such as methods based on lattices or codes, are also available. These alternative methods of cryptography are important as potential alternatives to the standard methods, if the standard methods become compromised.

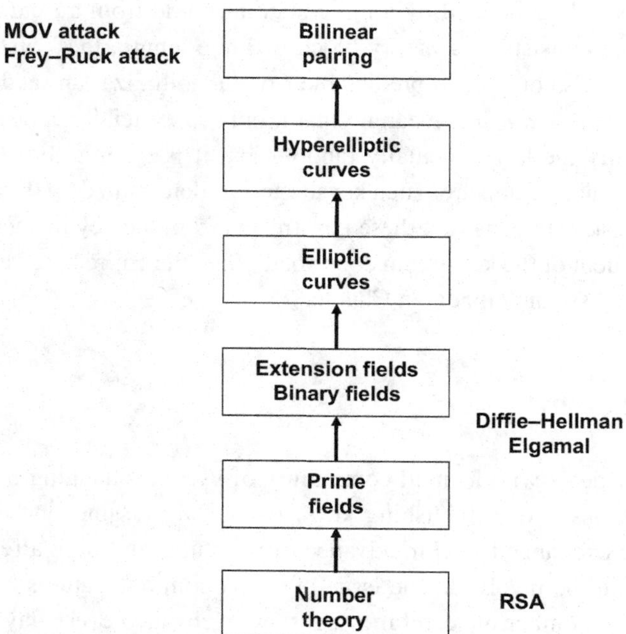

Figure 1.2 Mathematics in cryptography

Finally, it must be remarked that public-key cryptography does not eliminate the need for the more traditional methods of secret-key cryptography. The reason for using both the public-key and the secret-key cryptosystems in many applications is that the public-key system is generally rather slow and so it is not well suited to bulk encryption. Therefore it is common practice to use the public-key cryptography system only to create a key that is then used by the secret-key cryptography system.

1.6 Iterated and cascade ciphers

Given a keyspace $\mathcal{K}$, a set of encryption functions $\{e_k(x)\}$, and a set of decryption functions $\{d_k(x)\}$, the cryptographic security is limited by the cardinality of the keyspace, denoted $\#\mathcal{K}$ (or $\|\mathcal{K}\|$). If the keys are m-bit binary numbers, then $\#\mathcal{K} = 2^m$, provided all bits are effectively used. One way of trying to improve the security is to encrypt twice, using two keys. Thus, encrypt by using $y = e_{k'}(e_k(x))$ and decrypt by using $x = d_k(d_{k'}(y))$. This is called an *iterated cipher* or, if the two ciphers are of different kinds, a *cascade cipher*. The product keyspace $\mathcal{K}^2$ then has the apparent size $(\#\mathcal{K})^2$ or $(\#\mathcal{K}_1)(\#\mathcal{K}_2)$. If the keys are each an m-bit binary number, then $(\#\mathcal{K})^2 = 2^{2m}$. However, ensuring that these double keys are all different might not be enough. It may be that two different pairs of keys produce the same encrypted data, then the two pairs of keys are equivalent. For this reason, the effective keyspace could be much smaller than $(\#\mathcal{K})^2$.

Thus although the use of an iterated cipher may superficially appear to improve security, in many cases it does not. For example, it is clear that the cascade of an additive cipher, given by $y = x + k$, with another additive cipher, given by $y' = y + k'$, is actually the additive cipher $y' = x + k + k'$, which is another additive cipher with key $k + k'$. Thus, the keyspace for the iterated additive cipher actually contains only $\#\mathcal{K}$ distinct keys, not $(\#\mathcal{K})^2$. A similar fact may be true for other iterative ciphers, but in some ciphers this situation may be much harder to recognize. In general, it may be difficult to evaluate the true security of an iterated cipher or a cascade cipher.

The so-called *meet-in-the-middle* plaintext attack can be used against any iterated or cascade cipher. This attack is a conceptual attack. One does not claim it to be practical. The meet-in-the-middle attack is an attack that requires the cryptanalyst to have both a ciphertext and a corresponding plaintext. The known plaintext is repeatedly encrypted with the first encryptor using every possible key of $\mathcal{K}_1$ to form a list of possible "middle-texts." The known ciphertext is decrypted with the second decryptor using every possible key of $\mathcal{K}_2$ repeatedly to form a second list of possible middle-texts. Then the meet-in-the-middle attack examines the two lists to find an entry that is common to both. Of course, these two lists would be immense, and it would be intractable to store or to search such lists in the way described.

The computational work of a meet-in-the-middle attack is on the order of $\#\mathcal{K}_1 + \#\mathcal{K}_2$. Thus the effective keyspace and the secrecy have not been significantly increased by the iterated cipher. Any cipher that can be mathematically modeled as a cascade cipher is potentially vulnerable to a hypothetical improved version of a meet-in-the-middle attack. Assertions about the effective size of the keyspace of a cascade cipher must be adjusted accordingly.

Although this simple argument does not assert that the meet-in-the-middle attack is practical, it does assert that the effective keyspace of the iterated cipher is something closer to $(\#\mathcal{K}_1) + (\#\mathcal{K}_2)$ than to $(\#\mathcal{K})^2$. Because of the meet-in-the-middle attack, one cannot justify the use of an iterated cipher by simply citing the increased size of the keyspace.

1.7 Cryptanalysis

The standard assumption of cryptology is that a cryptanalyst will know the cryptosystem being used, but will not know the key. This conservative assumption means that the system should remain secure even if the cryptanalyst steals an encryptor and decryptor, or otherwise has a complete description of these devices. Indeed, the system should be secure even if the designer of the cryptosystem becomes the cryptanalyst and lacks only the decryption key. (An even more conservative requirement is that at the time of the design, the designer of the cryptosystem already intends to become the cryptanalyst in the future.) Of course, if the structure of the cryptosystem is not known to the cryptanalyst, the task of breaking it will be more difficult – perhaps much more difficult – but we do not want to base the security of a cryptosystem on the premise that the cryptanalyst does not know the method of encryption being employed. Conventional practice assumes that the cryptanalyst has full knowledge of the structure of the cryptosystem: only the decryption key is unknown.

Direct attacks can be regarded according to the following classification.

Known public encryption key: The cryptanalyst possesses only the public encryption key, if there is one, and the principles of operation and must determine the secret decryption key k.

Known ciphertext attack: The cryptanalyst knows the principles of operation and possesses only the ciphertext $\boldsymbol{y}$ and must determine the plaintext $\boldsymbol{x}$ or the secret decryption key k.

Known plaintext attack: The cryptanalyst knows the principles of operation and possesses both a plaintext $\boldsymbol{x}$ and the corresponding ciphertext $\boldsymbol{y}$ and must determine the secret decryption key k.

Chosen plaintext attack: The cryptanalyst knows the principles of operation and has temporary access to the encryptor containing the key k. Hence, the cryptanalyst can choose any plaintext $\boldsymbol{x}$ and construct the corresponding ciphertext $\boldsymbol{y}$ and must determine the secret decryption key k.

Chosen ciphertext attack: The cryptanalyst knows the principles of operation and has temporary access to the decryptor containing the key k and has access to one or more ciphertexts $\boldsymbol{y}$ corresponding to this key. Hence the cryptanalyst can deduce the plaintext $\boldsymbol{x}$ corresponding to ciphertext $\boldsymbol{y}$ and must determine the secret decryption key k.

Protocol attack: The cryptanalyst infers the message or partial information about the message based on weaknesses in the use of the system, without attacking the system itself.

Backdoor attack: The adversary sends an agent into the transmitter or receiver to observe the plaintext, or an indication of the plaintext, before it is encrypted or after it is decrypted.

Externalities attack: The cryptanalyst observes various physical aspects of the encryptor or decryptor such as patterns of use, temperature fluctuations, radiation, or battery-power fluctuations in order to determine partial or full knowledge of the key.

The first of these levels of attack pertains to asymmetric-key systems in which the encryption key and the decryption key are different, and the encryption key, in whole or in part, is public. It also pertains to various key-exchange systems. The next four of these five levels of attack are listed, in part, in increasing order of vulnerability. A known ciphertext attack is always a threat to a public-key cryptosystem, which will be discussed in later chapters.

A simple example of a protocol attack was mentioned in Section 1.1 for a cryptosystem that uses an elementary modulo-two (exclusive-or) addition of a binary key to a binary dataword. If the same key k is used twice to form the two ciphertexts $\boldsymbol{y}_1 = \boldsymbol{x}_1 + k$ and $\boldsymbol{y}_2 = \boldsymbol{x}_2 + k$, then the cryptanalyst can add the two ciphertexts to form $\boldsymbol{y}_1 + \boldsymbol{y}_2 = \boldsymbol{x}_1 + \boldsymbol{x}_2$. Because $\boldsymbol{x}_2$ is presumably a message with a known structure, it will be easier to recover $\boldsymbol{x}_1$ corrupted by $\boldsymbol{x}_2$ than to recover $\boldsymbol{x}_1$ hidden by k.

A recreational cryptogram is intended to be solved by the weakest type of attack, namely a known ciphertext attack. In this case, the plaintext message is ordinary English text, usually with the word spaces left in place and possibly the punctuation left in place as well. The known ciphertext attack is based on the fact that the message space is so constrained by the structure of the language that there is only one combination of a proper message and a key that is consistent with the ciphertext.

Some classical techniques of cryptanalysis use the statistical properties of the English language. The relative frequencies of the 26 letters have been estimated many times by compiling statistics from numerous novels, magazines, and newspapers. Equally important, the relative frequencies of doublets and triplets of letters appearing in English text have also been estimated. Similarly, the relative frequencies of standard English words and sequences of English words have been estimated many times as well. If word boundaries can be identified, then a statistical analysis of words may be attempted by the cryptanalyst.

A ciphertext-only attack attempts to break a cryptographic system when only an instance of a ciphertext is known, or perhaps several such instances. The usual goal of a ciphertext attack is either to find the plaintext message or to find the key by processing only the ciphertext and using full knowledge of the method of encryption. In most systems, learning the key also reveals the plaintext message, but learning the plaintext message does not necessarily reveal the key.

Sometimes a ciphertext attack might have different goals. For example, the goal may be to alter the message without decrypting it. Even when the ciphertext cannot be broken, there may be ways to modify a legitimate ciphertext, without deciphering it, for the purpose of interfering with the proper function of the user of that encrypted message.

In one form of public-key cryptography, the encryption is public, consisting of group exponentiation having the form $e_k(x) = x^a \pmod{n}$. For simplicity, suppose that we could have a decryptor $d_k(y) = y^b \pmod{n}$, where $ab = 1$ in the appropriate arithmetic system. Then $d_k(e_k(c)e_k(x)) = (c^a x^a)^b = (cx)^{ab} \pmod{n}$ which gives the incorrect message cx. In this way, it is possible to convert any such plaintext x into a multiple of x by multiplying the ciphertext $e_k(x)$ by any randomly chosen constant, which we can write as $c^a \pmod{n}$. This property can be used to alter a ciphertext for the purpose of confusing or disrupting the decryptor. Even though this may be a meaningless message, it may be a disruptive nuisance. This possibility is mentioned not simply to present this particular vulnerability, but more importantly to illustrate that it is sometimes possible to cause damage to a cryptographic system without actually breaking it. A thorough validation of a cryptographic system must anticipate such vulnerabilities.

The choice of a cryptosystem may depend in part on the application. One may be satisfied with reasonable privacy, or one may demand extreme secrecy. If the information is not of lasting value, then it is not important if the cipher will eventually be broken after a long time. If the information has lasting value, then it may even be desired to protect it for many decades, though probably not for centuries. The level of protection depends on the application. Nevertheless, this specification is ultimately subjective because we do not know how secure the known cryptography methods really are.

We will spend a good part of our time describing attacks on cryptosystems to uncover weaknesses. For example, a common attack on a stream cipher is a *linear-complexity attack*. This attempts to generate the entirety of a binary keystream from a small exposed segment of the keystream by means of a linear recursion that is computed from the exposed segment of the keystream. If an additive-keystream cryptosystem is vulnerable to a linear-complexity attack, then the effective secrecy may be as if that cryptosystem had a shorter key. Every stream cipher must be analyzed to show that it is not vulnerable to a linear-complexity attack.

1.8 Implementation attacks

This book discusses the mathematical aspects of cryptography, dealing only with protection against direct attacks. There are other attacks to which modern cryptography is also vulnerable, perhaps even more vulnerable. These are generally known as implementation attacks and consist of protocol attacks, backdoor attacks, and externalities attacks. The most obvious implementation attack is to penetrate the encryptor or decryptor to read the plaintext before it is encrypted or after it is decrypted. In addition, it may be possible to make inferences by observing external observables associated with the encryptor or decryptor. If an implementation has statistical information leaking out unintentionally in some way, then statistical analysis of this information may be used to attack the system. A sequence of computations always consumes power, and each individual computation in the sequence uses energy. The amount of energy depends on both the operation and the operands. The energy comes from a power supply, such as a battery, and is dissipated as heat. This means that the fluctuations in power used by the processor can divulge information about the key. If one could make precise detailed measurements of time-varying or spatially varying power use or temperature fluctuations as the computation proceeds, one could gain information and make inferences about the key or the plaintext. Perhaps surprisingly, such power attacks, though delicate, have sometimes been successful.

The cryptanalyst also wants to use other side information whenever it is available. For example, the time interval needed to decrypt, in most software implementations, depends on both the key and the message, if only weakly. This is because the number of ones or zeros in a key may determine the sequence of branches taken by the program, and the branches might not take the same amount of time to execute. The cryptologist wants to be assured that such information is impractical to adequately measure, or is so meager as to be inconsequential. Again, timing attacks, though delicate, have been successful.

1.9 Complexity theory

Public-key cryptography depends on the asymmetry of certain mathematical relationships. These are relationships in which the *forward problem* of $y = f(x)$ is tractable to compute, but the *inverse problem* $x = f^{-1}(y)$ is intractable to compute. Our acceptance and satisfaction with various public-key cryptography techniques now in use comes down to an interpretation of the word "intractable" in the previous sentence. Since much of the assets of civilization are protected by cryptographic techniques, it seems essential to have a precise, formal, and applicable meaning of the word "intractable." Unfortunately, this is not to be, at least not for the present generation.

However, even with that being said, there are two notions of intractability that are in play, and are each accepted in appropriate settings. These notions might be called "practical intractability" and "formal intractability." The two notions are generally aligned, but not completely so. A practical notion of intractability is what ultimately matters to the users of a cryptographic system, although it may not be known how to quantify practical intractability. In contrast, the notion of formal intractability deals with the perceived asymptotics of an algorithm in order to shed light on the practical cases. The limitation of such formal statements is that the asymptotic complexity need not be a satisfactory indicator of the complexity of a practical problem. Indeed, the asymptotic complexity might only be relevant in very large, and completely unreasonable, instances of the problem.

The goal of formal complexity theory is to demonstrate, or at least make plausible, that a certain class of computational problems has computational complexity that either is, or is not, exponential in the size of the problem, or is some other function of the size of the problem. This can give enormous insight as well as confidence in the security of a cryptosystem. However, such a statement must not be read as saying more than it says. For one thing, the statement is usually made about *typical* instances of a problem – and sometimes about *worst-case* instances of a problem – within a class of problems, and not about *every* instance of the problem or if there are subtle alterations. Rather innocent-looking practical constraints imposed on the statement of a particular problem may actually exclude all of the typical instances of the problem, unintentionally leaving on the table only those instances of the problem to which the complexity statement does not apply (as happened in the infamous knapsack cryptosystem).

Computational algorithms are of two kinds: deterministic algorithms and probabilistic algorithms. A deterministic algorithm always gives an answer, at least in principle. A randomized algorithm requires a randomly chosen parameter as an input, and it may give an answer or it may fail to give an answer, usually with a probability that is known. If the randomized algorithm fails to give an answer, it can be repeated. Multiple repetitions with random chosen parameters will make the probability of failure arbitrarily small. Although a deterministic algorithm may appear to be preferable, this need not be the case. A randomized algorithm may be superior. Moreover, a deterministic algorithm could have an unbounded and unpredictable stopping time, in which case it behaves like a randomized algorithm.

Theoretical computational complexity studies the asymptotic complexity of an algorithm and attempts to judge the algorithm by comparing this asymptotic complexity to the asymptotic complexity of the best possible algorithm for that task – even though the best possible algorithm might not be known. Algorithms are described as of polynomial, subexponential, or exponential complexity, always referring to the asymptotic growth of complexity with respect to some characterizing parameter describing the size of the problem. This, of course, depends on what is meant by the size of the problem. The size of a problem that depends on an integer n might be the integer n itself or it might be the number of digits needed to state the integer, which is proportional to the log of the integer n.

For a large number of difficult computational problems, the best algorithm is not known, and the asymptotic complexity of the best algorithm is not known either. Many of these computational problems are collected to form large equivalence classes of problems called the class of *nondeterministic polynomial problems* and the class of *NP-complete problems* for which large instances are widely believed to be intractable. The problems in this latter class are equivalent in the sense that a good algorithm that solves one of these problems can be converted into a good algorithm that solves any other problem in the class. We will not explore these subtle notions of complexity theory in this book, nor do we often make formal statements about particular problems from this point of view. We may refer to such problems only as *formally intractable*, but without clarifying this term further.

The classification of algorithms by their asymptotic complexity is an important guide in the development of practical algorithms, but it is only part of the story. We are ultimately concerned with computational problems of practical size and not with asymptotic complexity. An algorithm with an exponential asymptotic complexity may be quite satisfactory for a practical computational problem even for large instances of a problem, while an algorithm with polynomial asymptotic complexity may be unusable in practice. Accordingly, the notion of intractability will usually be left a little vague. By "intractable," we generally mean practical intractability, meaning that under the necessary conditions on the size of the problem, the computation is beyond reasonable, or believed to be so.

Even having said all these things, one must acknowledge that the asymptotic complexity of an algorithm is a compelling indicator and is hard to ignore. Because we may not know the practical complexity of a problem, we may take some comfort in statements about the asymptotic complexity of that problem.

Because most modern cryptography depends on the computational intractability of some underlying problems, the complexity of algorithms is of fundamental importance to the subject. Nevertheless, just as we do not study implementation in this book beyond general qualitative statements, we do not study complexity in this book beyond general qualitative statements.

1.10 Authentication and identification

In addition to the topic of *secrecy*, the broad subject of data protection also includes the topics of *authentication* and *identification*. The notions of authenticity and identity are quite different from the notion of secrecy. The notion of secrecy deals with the control of who can receive a message. The notions of authenticity and identity deal with the control of who can prepare or send a message. Although these notions are closely related, authentication usually refers to the validation of the source of a message, while identification refers to the validation of the identity of the source of a message.

Authentication is a more subtle concept than secrecy, and it is more difficult to achieve. Identification is also a subtle concept, and can even be a somewhat elusive notion. It is especially troublesome to define exactly what is meant by the verification of the identity of a total stranger. An authenticated message contains a *signature* that presumably cannot be forged, at least not with available resources. Only that stranger can send that signature. But if the stranger has no identity, then to what extent is this authentication useful? For this reason, the task of authentication may become entangled with questions of identification and certification, although it is a distinct task. We regard authentication and identification as more difficult problems than secrecy, although even this is a rather imprecise statement because practical cryptography methods provide neither perfect secrecy nor perfect authentication.

A signature can be applied to a plaintext or to a ciphertext. A plaintext can be encrypted after a signature is attached and can include the signature within the encryption. This is because the functions of signature and secrecy are independent. Moreover, a signed document can be inserted as a section into a larger document and that larger document can, in turn, be signed separately and independently, without taking note of internal signatures within various sections of the document. In another kind of signature system, a chain of signatures within the same document can be prepared under a unified protocol that permits an *aggregation* of signatures, whereby the final signature gives assurance that all internal signatures are valid.

A signature is not itself a document, but it is associated with a document. The signature is usually an attachment to that document. It must be producible and verifiable by a machine. Thus it usually takes the form of a digital file. It must not be possible for an unauthorized agent to modify a signature or to transfer that signature to an alternative document, though the signature may be separated from its host document and communicated or stored independently. The purpose of authentication is to establish the origin of a message, not the ownership of the message. The signature, unless encrypted itself, or otherwise secured, can be examined in conjunction with its host document at any time and by anyone, to verify that it is a valid signature of that specific document.

A signature can have another use. Instead of authenticating the source of a document, a signature can be used later by the receiver to verify that the document already existed at the time of transmission of the signature. The source prepares a cryptographically secure digest of the message, the digest is then signed and the signed digest transmitted. The document itself can be transmitted or revealed at a later time. At this later time, the receiver computes the message digest from the received message and verifies it using the signature received earlier, proving that the document existed at the time the signature was received. This prevents an archived and unreleased document from being rewritten prior to release. This protects the receiver. It is also possible to protect the transmitter with a timestamp on the document, but this requires a somewhat different protocol.

An authentication system, no matter how secure at the cryptographic level, suffers from vulnerability at a higher level. This is *identity theft*, which is addressed within the topic of identification. Even though the message contains a legitimate signature and that signature cannot be forged, it may be that the receiver has been deceived into believing that a signature generated by an imposter is the legitimate signature. If an imposter can establish a false identity with a fake signature, then that imposter can impersonate the legitimate source. To protect against this, a trusted certification authority will be consulted to certify that the signature does belong to the individual. In effect, the certification authority re-signs the signature with its own signature. This verification binds the signature to an individual and precludes identity theft. This then leads to the question of how the certification authority certifies the identity of that source and also leads to the deeper questions of what constitutes the identity of an individual.

Of course, the introduction of a trusted certification authority does not end the discussion. One must now consider identity theft of the trusted certification authority. In practice, the certification authorities will be organized into a hierarchy, with higher levels delegating power to certification subauthorities. Ultimately, the protection against identity fraud and identity theft must come from the continued scrutiny of the broad community.

Authentication is closely related to the subject of *hashing*, and it creates the need for cryptographically secure hash functions, which will be studied in Chapter 8. A hash function creates a message digest. The need for hashing arises because an authentication signature must authenticate the entire document as a unit. Altering any part of the digital document in any way must invalidate the signature. But the signature is to be an attachment to a document, and is not otherwise an alteration of the document. Although the document is of arbitrary length, it is desirable to have a short signature of fixed length – a 160-bit signature is a frequent standard. A hash function, then, is an extreme compression of a file of arbitrary length into a short abbreviation called a *message hash* or a *message digest*. The message digest is then signed rather than signing the message itself. To authenticate the message, the message digest is recovered by removing the signature from the signed message digest using an appropriate key. At the same time, the received message itself is rehashed to reform a new message digest that is compared to the recovered message digest. If the results of these two operations do not agree, the authentication is rejected.

In practice, the signed message digest is appended to the digital document as an attachment. However, as mentioned earlier, one can also send the signature in advance of the message or deposit it in a vault. This is appropriate in applications in which the message is not yet to be released, but must be validated when released at some future time to prove that it did exist at the time that the signature was produced, and that the document has not been changed after that signature was produced. Since a date is only defined by reference to an external time standard, it is clear that dating and timestamping a document requires inserting a reference to an external standard. A timestamp is a verifiable dated item that is attached to the message and hashed with the message. The dated item could be a page from the day's newspaper or could be a long binary sequence posted by a trusted timestamping service. The signed timestamped message digest is then posted in a suitable trusted archiving service such as the next-day's newspaper. In this way, the time of the message signing can be verified as occurring after the timestamp is created and before the archive is established.

1.11 Ownership protection

Another need of secure communications is ownership protection. Many large digital documents contain valuable intellectual property that is expensive to create, but easy to copy. The information contained in these documents is an asset that must be protected for the orderly conduct of the business of society. *Data watermarking* and *data fingerprinting* are two topics dealing with methods that are used to establish or protect ownership of a document. Watermarking and fingerprinting are similar topics, but they do have different specific goals and different methods of solution. These are forms of data protection that are tangentially related to cryptography and the other topics of this book, but the known methods of watermarking and fingerprinting, though quite useful, are not yet fully satisfactory from a formal or theoretical perspective.

The goal of *digital watermarking* is to establish ownership of a digital document by embedding a digital signature into the document in a way that cannot be removed without destroying or invalidating the document. The owner can later prove ownership by demonstrating the presence of the watermark within the document.

The goal and task of ownership protection is different from the goal and task of authentication. Authentication allows the recipient of a file to verify the source of that file. Ownership protection allows the source of a file to prove that it is the origin of the file.

The goal of *digital fingerprinting* is to detect unauthorized copying by embedding a *unique* digital signature into each authorized copy of a digital document in such a way that the fingerprint cannot be removed without destroying or invalidating the document, and such that the digital document cannot be copied without copying the signature as well. All unauthorized copies can then be traced back to the mother copy, thereby establishing from which fingerprinted copy the unauthorized copies were made.

In normal practice, it is not necessary that a watermark or a fingerprint be invisible – only that it be indelible. In some applications, however, it may be required that the watermark or fingerprint also be invisible and secret. In current practice, the methods of fingerprinting and watermarking are in use, but the formal goals have not been fully achieved. Indeed, in many applications, the intellectual property that needs protection does not ultimately reside in the digital file itself, but in the function performed by or described in that digital file. One can always subvert a method of digital ownership protection by copying the function performed by or described in a file, rather than by copying the file itself.

1.12 Covert communications

There are numerous situations in which the existence of a communication between two parties is to be hidden. It is not enough that the message has been encrypted and cannot be read. Even to know that this communication took place can provide information to an adversary. One instance in which this may be true is the situation in which there is a clique of nodes that are communicating. Then the sorting of the nodes, as well as the timing, sequence, and origin of a sequence of messages can provide useful information about the relationships between the nodes, or can suggest their relative status. Even though the content of messages may be hidden, it may be possible to make useful inferences. Even though the content of the message may be unreadable, the transmitters may also want to hide the existence of the message or the sequence of messages. This leads to the topic of covert communications.

One example of covertness is a message constructed so that it has the appearance of the background noise that is always present in every environment. Another example is the modulation of the precise timing of packet transmissions in a packetized communication network so that the sequence of interpacket time intervals conveys the covert information.

More generally, the requirement of covertness leads to the topic of *steganography*, which is the process of hiding one message inside another message. Steganography is obliquely related to the topic of cryptography, but it is not the same. Whereas encryption provides confidentiality by concealing the content of a message, steganography provides confidentiality by concealing the existence of a message. For an example of steganography, notice that each sentence of any popular novel is composed of either an odd number or an even number of keyboard characters. The author of the novel can carefully choose the words in each sentence to control the parity of the number of characters in that sentence. In this way, the parity of the number of characters in each sentence can be chosen to represent either a zero or a one. This sequence of zeros and ones corresponding to the sequence of sentences can be selected by the author of the novel to carry a hidden message expressed in a binary representation, one bit per sentence. It is unlikely that the existence of a hidden message of this kind could be discovered. Indeed, such a hidden message could be present in this book!

Steganography is different from cryptography because the goal is that the message be invisible rather than unreadable. Even the existence of the message is to be hidden. That message may, of course, be encrypted as well as hidden, but this is a separate function. We do not deal with covert communications such as steganography in this book. Or do we?

1.13 History of information protection

In the early times in the history of civilization, reading and writing were for the privileged few and printing was nonexistent. The need for secret writing was limited and simple ciphers seem to have sufficed. Caesar is said to have used a trivial substitution cipher illustrated by the form $y_i = x_i + 3 \pmod{26}$. More generally, the simple shift cipher of the form $y_i = x_i + k \pmod{26}$ has been in use for thousands of years.

Although the methods for the protection of information have been in use for a long time, the serious organized development of these methods has emerged in cohesive form only recently. The formal and scholarly study of the general theory of such methods is correspondingly recent. Indeed, it is only in the past several decades that the emergence and importance of widespread digital communication has created an intense need for information protection, especially in banking and commerce, and the subject of information protection has now developed a vast and open literature. Prior to these recent developments, the study of cryptography was primarily in the hands of military and government entities, and scholarly research was actively discouraged.

The modern development of the subject of cryptography owes much to the 1949 mathematical paper by Shannon and the 1967 historical book by Kahn. The paper by Shannon laid the early mathematical framework for understanding the nature of cryptography. It also mentioned the possibility of using hard mathematical problems as the core of a cryptosystem, though not mentioning public-key cryptography. The book by Kahn stimulated broader interest in this subject, advocated the notion of a formal subject of cryptography, and helped to provoke widespread challenges to governmental restrictions to cryptographic research of that time. Later, the needs of electronic commerce ultimately led to relaxation of government regulation and freedom of cryptographic research. In particular, Feistel and others at IBM Research developed the Data Encryption Standard (DES), which soon became an accepted standard and was widely adopted for commercial purposes. Accordingly, the security and the theory of the Data Encryption Standard attracted a great deal of interest and scrutiny by the many users, and by stimulating this interest and activity, DES also helped to form an open community of cryptographers in academia and industry.

At about the same time, the notion of public-key cryptography was advocated in 1976 by Diffie and Hellman, and by others, though initially without providing a practical embodiment of their idea. This concept was soon made practical both by the announcement of the Rivest, Shamir, and Adleman (RSA) method of encryption and by the Diffie–Hellman method of key exchange. Thus public-key cryptography was soon in widespread use. In this way, the mathematical topic of number theory became central to the field of cryptography. This once lauded example of "pure mathematics" lost the luster of the ivory tower, and entered the realm of "applied mathematics."

The next major milepost in the history of cryptography was the introduction of elliptic curves into the subject by Miller in 1985 and Koblitz in 1987, influenced by other developments in cryptology at the time. This development is important for its own sake in that an elliptic-curve cryptosystem is generally regarded as a stronger form of cryptography, but it is also important because it brought the mathematical subject of algebraic geometry into cryptography as a full partner. Elliptic curves had earlier entered the field of cryptology as a tool to attack those public-key cryptography systems that are based on number theory. It is worth underscoring the fact that elliptic curves were turned from a means of attack to a means of protection by the work of Miller and Koblitz.

Some fifteen years later, the pattern of turning a means of attack into a means of protection was repeated when, in 1993, the topic of bilinear pairings was brought to the field of cryptology, initially as a tool to attack public-key cryptography based on elliptic curves. Joux (2000) soon turned this tool around and devised a tripartite key-exchange scheme based on bilinear pairing. This was followed by the work of Boneh and Franklin (2001), which led to several decades of exciting research and rapid progress on the use of pairings in other ways both for secrecy and for authentication.

It is interesting, and perhaps informative, that this pattern was repeated yet once again by the introduction of lattice theory into the subject of cryptography. Lattice theory was first introduced as a method of cryptanalysis to attack existing cryptosystems. Only later was it turned into a method of devising cryptosystems.

The practical role of cryptography and cryptology in shaping the major historical events of the twentieth century is as immense as it is obscure and invisible. The outcome of every modern war as well as many peacetime political events have, in large part, been determined by successful attacks on presumed unbreakable ciphers. Perhaps more importantly, the modern world of electronic commerce could not exist as such without the protection of business information, financial transactions, intellectual property, and many kinds of personal information provided by the methods of modern cryptography.

Problems for Chapter 1

1.1 Prove that there are $N!$ distinct ways to arrange a sequence of N distinct symbols. How many nontrivial recreational cipher keys on an alphabet of size 26 are there? How many recreational cipher keys are there satisfying the requirement that no letter is represented by itself?

1.2 Estimate the value of $\log_2(256!)$. How many bits does it take to express this number? Estimate the value of $\log_2(26^2!)$. How many bits does it take to express this number?

1.3 Prove that $a \pmod{n} = b \pmod{n}$ if, and only if, $a - b = 0 \pmod{n}$.

1.4 If 10-digit phone numbers, by themselves, are encrypted by a certain asymmetric encryption device that requires 1 microsecond per encryption, how long would it take, on average, to simply encrypt every 10-digit number until a given ciphertext is obtained. How long would it take if it requires 1 millisecond per encryption?

1.5 **a** How many integers in the set $\mathbf{Z}_{26} = \{0, \ldots, 25\}$ have inverses under multiplication modulo 26?

b How many two-by-two matrices over $\mathbf{Z}_{23}$ have inverses modulo 23?

c How many two-by-two matrices over $\mathbf{Z}_{26}$ have inverses modulo 26?

d In $\mathbf{Z}_{26}$, does the matrix

$$\boldsymbol{M} = \begin{bmatrix} 4 & 5 \\ 5 & 19 \end{bmatrix}$$

have an inverse modulo 26? If so, find it.

1.6 **a** Determine the number of keys in the affine cipher $y = ax + b \pmod{28}$.

b Determine the number of keys in the affine cipher $y = ax + b \pmod{29}$.

1.7 By squaring all elements of $\mathbf{Z}_{15}$, find all square roots of 1 modulo 15. How many zeros does the polynomial $x^2 - 1$ have in $\mathbf{Z}_{15}$? Repeat this analysis for $\mathbf{Z}_{17}$. In what way are the results different? Is there an explanation?

1.8 **a** Prove that

$$\sum_{i=1}^{n} i^2 = \frac{n(n+1)(2n+1)}{6}.$$

b Is there any value of n for which this sum is a square?

c Express the condition that $\sum_{i=1}^{n} i^2$ is a square in terms of the (rational) zeros of a bivariate polynomial $p(x, y)$ of degree three.

d Ponder but don't answer the following question: Are there an infinite number of n for which $\sum_{i=1}^{n} i^2$ is a square? Pose this query in terms of a diophantine equation.

1.9 Prove that there are an infinite number of primes.

1.10 List the first ten primes of the form $4k + 1$ with k an integer. List the first ten primes of the form $4k + 3$ with k an integer. Are there any primes that are not of either of these forms? Comment on whether or not there are an infinite number of primes of each of these forms?

1.11 Prove that if n is an integer and the sum of the digits of the decimal representation of n is divisible by 3, then n is divisible by 3. This problem is a curiosity in that the mathematical notation becomes part of the mathematics.

1.12 The algebraic field with three elements, denoted $\boldsymbol{F}_3 = \{0, 1, 2\}$, is defined by the addition and multiplication tables

+	0	1	2
0	0	1	2
1	1	2	0
2	2	0	1

×	0	1	2
0	0	0	0
1	0	1	2
2	0	2	1

Calculate the determinant of the following matrix over the field $\boldsymbol{F}_3$, and show that its rank is three

$$\boldsymbol{M} = \begin{bmatrix} 2 & 1 & 0 \\ 1 & 0 & 1 \\ 2 & 1 & 2 \end{bmatrix}$$

1.13 **a** What is the period of the binary keystream sequence produced by the linear recursion

$$x_i = x_{i-3} + x_{i-4},$$

where the plus sign denotes modulo-two addition (exclusive-or)?

b What is the largest possible value of the period of a binary linear recursion of length L?

1.14 Read brief biographies of Carl Frederick Gauss, Bernard Riemann, Alan Turing, Whitfield Diffie, and Claude Shannon and write a paragraph stating whom you most admire and why. Comment on their contributions to cryptography.

1.15 The Bézout theorem says that the two bivariate polynomials, $p(x, y)$ and $q(x, y)$ of degree m and n, respectively, over the complex number system have exactly mn common zeros in the plane $\boldsymbol{C}^2$ provided multiple zeros are counted as such and so-called points at infinity are counted whenever appropriate. Restate the Bézout theorem in a simpler form for the special case in which $q(x, y) = y$.

1.16 **(Three-pass protocols)** Paul and Paula use the following protocol for transmitting a binary message $\boldsymbol{m}$ of blocklength n. Paul privately chooses a random bit pattern $\boldsymbol{r}$ of blocklength n and adds it componentwise (bit by bit exclusive-or) to $\boldsymbol{m}$ to obtain the encrypted message $\boldsymbol{e} = \boldsymbol{m} + \boldsymbol{r}$, which he sends to Paula. Paula privately chooses another random bit pattern $\boldsymbol{s}$ and adds it to $\boldsymbol{e}$ to obtain another encrypted message $\boldsymbol{e}' = \boldsymbol{e} + \boldsymbol{s}$, which she sends to Paul. Paul subtracts $\boldsymbol{r}$ and returns $\boldsymbol{e}'' = \boldsymbol{m} + \boldsymbol{s}$ to Paula, who then subtracts $\boldsymbol{s}$ to recover $\boldsymbol{m}$. Comment on the security of this procedure.

1.17 The warden of a small prison with 100 prisoners has devised a game as a diversion for the prisoners. One hundred otherwise identical cards have been prepared with the name of one prisoner on one side of each card. Every Monday morning the cards are shuffled and randomly placed face down in a row on a long table. The prisoners enter the room one at a time wherein each prisoner, acting alone, is allowed to turn over any 50 cards of the 100 cards. If the name of that prisoner is not exposed, the game ends and all prisoners are returned to their cells. If that prisoner's name is exposed, that prisoner is sent through a side door to a waiting room. Then each card is turned face down in the same place, and the next prisoner enters the room to repeat the process, turning over any 50 cards. If all 100 prisoners expose their own names during their individual turns, all 100 prisoners are freed from prison. The game ends, however, as soon as one prisoner fails to expose that prisoner's own name among the 50 cards exposed, and all prisoners are returned to their cells. The warden reasons that each prisoner has a probability 1/2 of success and that the trials are independent. Therefore the probability that all 100 prisoners will succeed is 2^{-100}, which is far, far smaller than the probability that the prison door will be accidently left unlocked. The warden concludes that this game is a harmless diversion that will amuse the prisoners.

Is the warden right? What is the prisoners' best strategy? What is the corresponding probability that the prisoners will be freed in a given week? The prisoners are allowed to freely discuss their weekly game between games and to prepare a joint strategy, but cannot communicate while the game is in process. (The moral of this problem is that a "secure" cryptography system might not be what it appears to be.)

Notes for Chapter 1

The conventional requirement that a cryptosystem should be secure even when the cryptanalyst knows all the details of the cryptosystem except for the secret key is known as *Kerckhoff's principle*. The one-time pad was first described in 1917 by Gilbert Vernam for applications in telegraphy. The one-time pad was proved by Shannon (1949) to be secure in the sense of perfect secrecy, and, moreover, Shannon proved that it is the only way to obtain perfect secrecy.

The Vigenère cipher has been in use since the sixteenth century. It is vulnerable to multitrack frequency analysis. Hill (1929) proposed a protection against a straightforward frequency-analysis attack by introducing a matrix operation into the encryption, thereby introducing deeper levels of number theory into the subject. The autokey cipher, though not a true cipher, was patented by Guanella as early as 1946.

Many books are available that deal with the various topics of secure communication, differing in the choice of topics and the background assumed of the reader. The book by Barr (2002) is written for the reader without a technical background. It has an expository style with many explicit and detailed examples. The book by Schneier (1996) is also written for the general reader and takes the form of a broad technical survey rather than an exposition. The book by Garrett (2001) is a cryptography textbook, written for the general reader who does have a general technical background, and the book is quite accessible at this level. The book of Stinson (1995, 2006) is a popular treatment of cryptography at a medium level of difficulty written from the point of view of computer science. Koblitz (1998) is a more advanced book written with flair by a mathematician, but restricted in its selection of topics. The book by Menezes, van Oorschot, and Vanstone (1997) is a highly regarded source of information but does assume some mathematics background. Other books coming from the direction of algebraic geometry are the specialized books on elliptic curves by Silverman (1986), and the specialized book on elliptic curves in cryptography by Washington (2008).

Other recent books on cryptography are those by Hoffstein, Pipher, and Silverman (2008), which focuses on public-key cryptography; by Galbraith (2012), which is intended for mathematicians; by Trappe and Washington (2006), which is intended for undergraduate mathematics students; by Katz and Lindell (2007), which is intended for undergraduates in theoretical computer science; and Paar and Pelzl (2009), which is intended for an audience with less mathematics background.

The book by Kahn (1967, 1996) discusses the history and impact of cryptography on society and is written for a general audience. This book is credited with popularizing the subject and turning the interest of the many future cryptographers toward the then secluded world of cryptology. More recent books for a general audience include those by Singh (1999) and Levy (2001). These books provide an excellent introduction to the recent history and the widespread impact of modern cryptography.

2 The integers

Number theory, the oldest branch of mathematics, can be found in the early history of cryptography and number theory continues to have an important role in the subject. Since these first days, the integers have been used to represent the symbols of a message, and the operations of arithmetic have been used to combine these numbers with a cryptographic key to hide the information that the numbers represent. Modern cryptographic systems depend on number theory in a much deeper way by using difficult or unsolved problems of number theory, and other branches of mathematics, to try to hide information. In turn, the adversarial cryptanalyst often attacks those cryptosystems by using deep theorems of mathematics to try to break a cryptosystem and recover the hidden information.

2.1 Basic number theory

The set of positive and negative integers $\{0, \pm 1, \pm 2, \ldots\}$, denoted $\mathbf{Z}$, is closed under the operation of addition, which is an operation that is familiar and has many familiar properties. Integer addition is *commutative*, meaning that $a + b = b + a$. Integer addition is *associative*, meaning that the sum $a + b + c$ can be executed from either side. There is an *identity element* under integer addition, namely the special integer called *zero*, and the operation of addition has an inverse operation called *subtraction*.

The set of integers $\mathbf{Z}$ is an early example of a structure called a *group*. A group is a set G together with an operation $*$ on pairs of elements of the set such that under the operation $*$ the set is closed, and the operation $*$ is associative, has an *identity element*, denoted e, and an *inverse* of every element. An inverse for the element a is an element b such that $a * b = b * a = e$. Groups will be studied briefly now and more formally in Section 9.1.

The set of integers $\mathbf{Z}$ is a group with ordinary addition, denoted $+$, as the group operation, zero as the identity element, and $-a$ as the inverse of a.

A *finite group* is a group with a finite number of elements. For any element β of the finite group G, let $\beta^2 = \beta * \beta$, $\beta^3 = \beta * \beta * \beta$, and so forth in the obvious way. Thus define $\beta^{i+1} = \beta * \beta^i$. In such a sequence $\beta, \beta^2, \beta^3, \beta^4, \ldots,$ of consecutive powers of β, eventually an element of the finite group G must be repeated because there are only a finite number of elements in a finite group. The first element repeated must be β itself because if $\beta^i = \beta^j$, then $\beta^{i-1} = \beta^{j-1}$, so β^{i-1} is repeated if β^i is repeated. The element of the sequence before the first repetition of β must be the identity element e because if $\gamma * \beta = \beta$ then $\gamma = e$. The set of distinct powers of β is called the *orbit* of β. The number of elements in the orbit of β is called the *order* of the element β. The number of elements in the finite group G is called the *order* of the group G. Any subset of G that itself is a group under the inherited operation $*$ is called a *subgroup* of G. The orbit of any element of G is a subgroup of G, called a *cyclic subgroup* because all of its elements lie in a single cycle. If the orbit of any element of G is equal to G, then G is called a *cyclic group*.

A discussion of the structure of finite groups begins with the following theorem.

Theorem 2.1.1 (Lagrange) *The order of any element β of a finite group divides the order of the group.*

Proof Let s be the order of β. Form the elements of the finite group G into a two-dimensional array as follows. The first row of the array is the orbit of the element β written as

$$\beta^0, \beta^1, \beta^2, \beta^3, \cdots, \beta^{s-1},$$

where $\beta^0 = e$, the identity element of G, and $\beta^s = \beta^0$. If some elements of G are not used in the array, choose any unused element of G, call it g_1, and multiply each element β^i of the first row by g_1 to produce the second row, having elements $g_1 * \beta^i$. If again, at this point, some elements of G are not yet used, choose another unused element of the group, call it g_2, and multiply each element of the first row by g_2 to form the third row. Continue in this way forming new rows by choosing an unused element and multiplying each element of the first row by this element. Stop after a row is completed if at this point there is no element that is unused. The process must stop because the group is finite.

We now argue that every element of G must appear exactly once in the final array. To this end, notice first that no element can appear twice in the same row because if $g_k * \beta^i = g_k * \beta^j$, then the two elements β^i and β^j are equal, contrary to the nature of the first row. Notice second that no element can appear twice in two different rows because if $g_k * \beta^i = g_\ell * \beta^j$ with $k < \ell$, then $g_\ell = g_k * \beta^{i-j}$, contradicting the fact that $g_k * \beta^{i-j}$ has not been used previously. Therefore every element of G appears exactly once in the array, so the number of elements in the array is the product of the number of rows and the number of columns. In particular, the number of elements in G is a multiple of the order of β. □

The set of integers $\mathbf{Z}$ is closed under the operation of addition. It is a group under the operation of addition. The set of integers is also closed under the operation of multiplication. Multiplication of integers is commutative, meaning that $a \cdot b = b \cdot a$. Multiplication of integers is associative, meaning that $a \cdot b \cdot c$ can be executed from either side. The set of integers has an identity element under multiplication, called *one* and denoted 1. The operation of integer multiplication, in general, does not have an inverse. In some instances, however, as when $c = ab$, multiplication will have an inverse, $a = c/b$, called *division*, and a and b are called *factors* of c. (Thus $6/3 = 2$.) In other instances, when b is not a factor of c, division of c by b does not exist in the set of integers. (Thus $6/4$ is undefined in the set of integers $\mathbf{Z}$.) Instead, in the set of integers, there always exists the operation of *division with remainder*. The subset consisting of those integers that have the same remainder r under division by n is denoted $\langle r \rangle_n$, and is called an *equivalence class*. Such an equivalence class $\langle r \rangle_n$ exists for each r from 0 to $n-1$, and the set of such equivalence classes $\{\langle r \rangle_n\}$ is denoted $\mathbf{Z}/\langle n \rangle$ or $\mathbf{Z}_n$.

The elements of $\mathbf{Z}_n$ may be regarded as (or represented by) the integers from zero to $n-1$. Thus $\mathbf{Z}_n = \{0, \ldots, n-1\}$. These integers are the *canonical representatives* of $\mathbf{Z}/\langle n \rangle$. The set $\mathbf{Z}_n$ inherits an addition operation and a multiplication operation from $\mathbf{Z}$. Thus addition in $\mathbf{Z}_n$ is defined[1] as addition of integers reduced modulo n. This makes $\mathbf{Z}_n$ into a finite group under addition because the required properties are inherited from $\mathbf{Z}$, as can easily be checked. Similarly, multiplication in $\mathbf{Z}_n$ is defined as multiplication of integers reduced modulo n. The set $\mathbf{Z}_n$, however, is not a finite group under multiplication because multiplication need not have an inverse in $\mathbf{Z}_n$. Indeed, the element zero never has an inverse under multiplication in $\mathbf{Z}_n$ and, for example, 3 does not have an inverse under multiplication in $\mathbf{Z}_6$.

The sets $\mathbf{Z}$ and $\mathbf{Z}_n$ have similar abstract structures. Each has two operations, called addition and multiplication, such that the set is a group under addition, the distributivity property, $a(b+c) = ab + ac$ and $(a+b)c = ac + bc$ is satisfied, and the associativity property $a(bc) = (ab)c$ is satisfied. Any algebraic system with this abstract structure is called a *ring*. Algebraic rings will be discussed more formally in Section 9.2.

The set of elements of $\mathbf{Z}_n$ that have an inverse under multiplication is denoted $\mathbf{Z}_n^*$. This is the set of those integers b of $\mathbf{Z}_n$ for which b and n have no common integer factor, meaning no nontrivial positive integer that divides both b and n. For example, $\mathbf{Z}_{10}^* = \{1, 3, 7, 9\}$ and $\mathbf{Z}_{12}^* = \{1, 5, 7, 11\}$. All elements of $\mathbf{Z}_n^*$ have an inverse, and the identity element of $\mathbf{Z}_n^*$ is the element one. Thus the set $\mathbf{Z}_n^*$ is a group under the operation of multiplication. In the finite group $\mathbf{Z}_{10}^*$ under multiplication, $1^{-1} = 1$, $3^{-1} = 7$, $7^{-1} = 3$, and $9^{-1} = 9$; while in the finite group $\mathbf{Z}_{12}^*$ under multiplication, $1^{-1} = 1$, $5^{-1} = 5$, $7^{-1} = 7$, and $11^{-1} = 11$.

[1] When the context of a discussion is clearly $\mathbf{Z}_n$, the modulo-n reduction is implicit in the context and need not be stated. Then $a + b$ is understood to be addition in $\mathbf{Z}_n$. When the larger context of the discussion is $\mathbf{Z}$, then the modulo-n reduction is explicit and must be stated whenever it occurs.

The set of integers has many important properties, many of which follow from the notion of a prime. A *composite integer* c is an integer that can be written as $c = a \cdot b$, where neither a nor b is ± 1. A *prime integer* is a positive integer, larger than one, that is not composite. The integer one is not a prime. Two integers a and b are *coprime* if no integer other than ± 1 divides both a and b. The integers ± 1 are coprime with every integer. If a and b are coprime, then we may also say that a is coprime to b.

The *greatest common divisor* of two integers r and s, denoted GCD(r, s), is the largest positive integer that divides both r and s. Two integers r and s are coprime if their greatest common divisor equals one. The elements of $\mathbf{Z}_n^*$ are the elements a of $\mathbf{Z}_n$ such that GCD$(a, n) = 1$. The *least common multiple* of two integers r and s, denoted LCM(r, s), is the smallest integer divisible by both r and s.

Definition 2.1.2 *The totient function, denoted $\phi(n)$, is the number of positive integers less than or equal to n and coprime to n.*

The set of positive integers less than n and coprime to n is the set denoted $\mathbf{Z}_n^*$. Thus $\phi(n) = \#\mathbf{Z}_n^*$, where $\#S$ denotes the cardinality of the set S. The elements of $\mathbf{Z}_n^*$ are called the *units* of $\mathbf{Z}_n$. For example, $\phi(10) = 4$ because $\mathbf{Z}_{10}^* = \{1, 3, 7, 9\}$, and $\phi(12) = 4$ because $\mathbf{Z}_{12}^* = \{1, 5, 7, 11\}$. If n is a prime p, then every nonzero element of $\mathbf{Z}_n$ is an element of $\mathbf{Z}_n^*$, which means that $\#\mathbf{Z}_p^* = p - 1$ if p is a prime. Later, when we study the extended euclidean algorithm in Corollary 2.2.2, we will state a corollary which asserts that for any $a \in \mathbf{Z}_n^*$, there will always exist integers A and N such that

$$Aa + Nn = 1.$$

Therefore $Aa = 1$ modulo n, so $A \pmod{n}$, denoted a^{-1}, is the inverse of a in $\mathbf{Z}_n^*$, and such an A exists for every element of $\mathbf{Z}_n^*$. This means that $\mathbf{Z}_n^*$ is a group under multiplication.

Theorem 2.1.3 *For primes p and q, the totient function satisfies $\phi(p) = p - 1$, $\phi(q) = q - 1$, $\phi(pq) = (p - 1)(q - 1)$, and $\phi(p^e) = p^{e-1}(p - 1)$.*

Proof If n is equal to a prime p, then every nonzero element of $\mathbf{Z}_p$ is coprime with p, so $\#\mathbf{Z}_p^* = p - 1$. Thus $\phi(p) = p - 1$ if p is a prime.

Furthermore, if $n = pq$, where p and q are both primes, arrange the pq elements of $\mathbf{Z}_{pq}$ in a p by q array by writing element i in row i modulo p and column i modulo q. Every i smaller than pq appears exactly once in such an array. All elements in the last column are divisible by p and can be struck out. All elements in the last row are divisible by q and can be struck out. The reduced array now has $p - 1$ columns and $q - 1$ rows. These are the $(p - 1)(q - 1)$ elements not divisible by p or q. This means that $\phi(pq) = (p - 1)(q - 1)$.

Finally, in $\mathbf{Z}_{p^e}$, every pth element is divisible by p. There are p^{e-1} such multiples. Thus $p^e - p^{e-1}$ elements are coprime with p, and so $\phi(p^e) = p^{e-1}(p-1)$. □

Corollary 2.1.4 *If $n = \prod_{i=1}^{\ell} p_i^{e_i}$, then $\phi(n) = \prod_{i=1}^{\ell}(p_i - 1)p_i^{e_i-1}$.*

Proof Exercise. □

Theorem 2.1.5 (Euler's theorem) *If β is an element of $\mathbf{Z}_n^*$, then $\beta^{\phi(n)} = 1 \pmod{n}$.*

Proof By the definition of the totient function, $\#\mathbf{Z}_n^* = \phi(n)$. Let k be the order of the element β. Because $\mathbf{Z}_n^*$ is a group with $\phi(n)$ elements and the set of powers of β is a subgroup with k elements, the theorem of Lagrange (Theorem 2.1.1) requires that $\phi(n) = rk$ for some integer r. Because $\beta^k \equiv 1 \pmod{n}$, this means that $\beta^{\phi(n)} \equiv 1 \pmod{n}$. □

This theorem asserts that the inverse of $x \in \mathbf{Z}_n^*$ under multiplication modulo n can be written $x^{\phi(n)-1}$ because $xx^{\phi(n)-1} = 1 \pmod{n}$. For example, $\phi(10) = 4$, and $\mathbf{Z}_{10}^* = \{1, 3, 7, 9\}$, so the theorem tells us that $1^{-1} = 1^3 = 1$, $3^{-1} = 3^3 = 7$, $7^{-1} = 7^3 = 3$, and $9^{-1} = 9^3 = 9$, as noted earlier.

Corollary 2.1.6 (Fermat's little theorem) *If p is a prime and $\beta \in \mathbf{Z}_p$, then $\beta^p = \beta \pmod{p}$.*

Proof This statement is obvious if β equals zero. If p is a prime, then $\phi(p) = p - 1$, and Euler's theorem gives $\beta^{p-1} = 1 \pmod{p}$. Therefore $\beta^p = \beta \pmod{p}$. □

A prime p is divisible only by p and by one. Moreover,

$$\phi(p) + \phi(1) = p.$$

A *biprime* $n = pq$, where p and q are both primes, is divisible only by n, p, q, and 1. Moreover,

$$\phi(n) + \phi(p) + \phi(q) + \phi(1) = n.$$

These two statements are special cases of the following theorem.

Theorem 2.1.7 *The totient function satisfies $\sum_{d|n} \phi(d) = n$, where the indicated sum is over the set of d dividing n including n and one.*

Proof Let $n = pq$, where p and q are distinct primes. Then the divisors of n are n, p, q, and 1, so

$$\phi(pq) + \phi(p) + \phi(q) + 1 = (p-1)(q-1) + (p-1) + (q-1) + 1 = pq.$$

Let $n = p^e$, where p is a prime. Then p^ℓ for $\ell = 0, \ldots, e$ are the divisors of n, so

$$\sum_{\ell=0}^{e} \phi(p^\ell) = \sum_{\ell=1}^{e} p^{\ell-1}(p-1) + 1$$
$$= (p-1)\sum_{\ell=0}^{e-1} p^\ell + 1 = (p-1)\frac{p^e - 1}{p-1} + 1 = p^e.$$

Let $n = p^e r$, where r does not have p as a factor. The integers d that divide n are those of the form $d = p^e s, p^{e-1}s, p^{e-2}s, \ldots, ps, s$ for every integer s that divides r including $s = 1$. Suppose we are given that $\sum_{s|r} \phi(s) = r$. We will show that then $\sum_{d|p^\ell s} \phi(d) = p^\ell r$. Thus

$$\sum_{d|p^\ell s} \phi(d) = \sum_{s|r}\sum_{\ell=0}^{e} \phi(p^\ell s) = \sum_{s|r}\left[\sum_{\ell=1}^{e} p^{\ell-1}(p-1) + 1\right]\phi(s)$$
$$= (p-1)\left[\sum_{\ell=0}^{e-1} p^{\ell-1} + 1\right]\sum_{s|r}\phi(s) = r(p-1)\frac{p^e - 1}{p-1} + r$$
$$= p^e r.$$

We conclude that the theorem holds for all n. □

We close this introductory section by stating a theorem whose proof is more difficult than the other theorems of this section and depends on topics that are not given until later chapters.

Theorem 2.1.8 *If p is a prime, then $\mathbf{Z}_p^*$ contains an element of order $\phi(p) = p - 1$ under multiplication, and so is a cyclic group under multiplication.*

Proof Because $\mathbf{Z}_p^*$ is a finite group with $p - 1$ elements, every element of $\mathbf{Z}_p^*$ has order dividing $p - 1$. If all factors of $p - 1$ are distinct primes, then every element has a prime order and the chinese remainder theorem for integers (Theorem 2.2.5) gives an element of order $p - 1$, so the group is cyclic.

It is more difficult to show that if some factors of $p - 1$ are repeated primes, then there is a cyclic subgroup with order equal to that prime power. Instead, we refer to Theorem 9.8.1, which is an alternative statement of the theorem. □

More generally, it is also true that if p^m is a prime power, then $\mathbf{Z}_{p^m}^*$ is a cyclic group under multiplication and so contains an element of order $\phi(p^m) = p^{m-1}(p - 1)$. We will not prove this statement at this time because we will not have a use for it. The proof is lengthy.

2.2 The euclidean algorithm

A central algorithm of computational number theory is the *euclidean algorithm*, which is given in the following theorem. The euclidean algorithm rests on the *division algorithm*, which states that for any two nonnegative integers, a and b, with b not larger than a, there exist two nonnegative integers Q and r, called the *quotient* and the *remainder*, such that

$$a = Qb + r,$$

and r is smaller than b. Elementary algorithms for computing the quotient and remainder are well known. If r is equal to zero, then b is said to *divide* a, which is written $b|a$.

The division algorithm, applied repeatedly, leads to the following procedure for computing the greatest common divisor.

Theorem 2.2.1 (Euclidean algorithm) *Given two distinct positive integers, r and s, with s larger than r, their greatest common divisor can be computed by the iteration*

$$\begin{aligned}
s &= Q_1 r + r_1\\
r &= Q_2 r_1 + r_2\\
r_1 &= Q_3 r_2 + r_3\\
&\vdots\\
r_{n-2} &= Q_n r_{n-1} + r_n\\
r_{n-1} &= Q_{n+1} r_n,
\end{aligned}$$

where the process stops when a remainder of zero is obtained. The last nonzero remainder, r_n, is the greatest common divisor.

Proof The process must stop because the remainders are nonnegative and decreasing. We will show that r_n divides GCD(r, s) and GCD(r, s) divides r_n, and so they are equal.

To this point, observe that the first equality of the theorem shows that, because GCD(r, s) divides both r and s, GCD(r, s) also divides r_1. Then the second equality shows that GCD(r, s) divides r_2. Continuing, we see that GCD(r, s) divides every r_i, so it also divides r_n.

The last equation shows that r_n divides r_{n-1}. Because r_n divides both itself and r_{n-1}, the second to last equation shows that r_n divides r_{n-2}. Continuing this way, we conclude that r_n divides GCD(r, s).

Because r_n both divides GCD(r, s) and is divisible by GCD(r, s), we conclude that r_n and GCD(r, s) are equal. □

The following corollary, which is called the *extended euclidean algorithm*, is an important observation of number theory. The equation that appears in the corollary is analogous to an equation in the ring of polynomials known as the *Bézout identity*.

Corollary 2.2.2 (Extended euclidean algorithm) *For any integers r and s, there exist integers R and S such that*

$$\text{GCD}(r, s) = Rr + Ss.$$

Proof The last nonzero remainder r_n in the statement of the theorem is the greatest common divisor. It satisfies $r_{n-2} = Q_n r_{n-1} + r_n$. Working backwards through the set of equations in Theorem 2.2.1 to successively eliminate all other remainders by back-substitution gives r_n as a linear combination of r and s with integer coefficients, as was to be proved. □

An efficient way to compute two integers R and S satisfying

$$\text{GCD}(r, s) = Rr + Ss,$$

is contained in the proof of Corollary 2.2.2. Simply execute the euclidean algorithm, then back-substitute as described in that proof.

It is easy to rewrite the expression of the theorem as

$$\text{GCD}(r, s) = r(R - \ell s) + s(S + \ell r),$$

which makes it obvious that R can be decremented by any multiple of s if S is incremented by the same multiple of r.

Corollary 2.2.3 *If integers r and s are coprime, then there exist integers R and S such that*

$$Rr + Ss = 1.$$

Proof Because by assumption $\text{GCD}(r, s) = 1$, the corollary follows immediately from the extended euclidean algorithm. □

In particular, the equation

$$Rr = 1 \pmod{s}$$

always can be solved for a nonnegative R smaller than s if s is coprime to r.

This line of thought is followed further in the next theorem to conclude, as indicated earlier, that $\mathbf{Z}_n^*$ forms a group under multiplication because every element of the set has an inverse. Moreover, the extended euclidean algorithm can be used to find the inverse of each element.

Corollary 2.2.4 *The set $\mathbf{Z}_n^*$ is a group under multiplication.*

Proof The set $\mathbf{Z}_n^*$ contains the identity element one, and is closed under multiplication, an operation that is clearly commutative and associative. Every element x of $\mathbf{Z}_n^*$ satisfies $\text{GCD}(x, n) = 1$, so by Corollary 2.2.3 there exists an integer X such that $xX = 1 \pmod{n}$. Then $x^{-1} = X \pmod{n}$, so every element x has an inverse under multiplication. □

Another important and useful consequence of the euclidean algorithm is a condition under which the set of remainders (or the *residues*) under division by each of a set of integers uniquely specifies an integer.

Theorem 2.2.5 (Chinese remainder theorem) *For any set of pairwise coprime positive integers $\{m_0, m_1, \ldots, m_{K-1}\}$, the set of congruences*

$$c \equiv c_k \pmod{m_k}, \qquad k = 0, \ldots, K-1$$

has exactly one nonnegative solution smaller than the product $M = \prod_k m_k$, which is given by

$$c = \sum_{k=0}^{K-1} c_k N_k M_k \pmod{M},$$

where $M_k = M/m_k$, and N_k is the integer that satisfies

$$N_k M_k + n_k m_k = 1.$$

Proof Corollary 2.2.2 asserts that the specified N_k must exist because $\text{GCD}(M_k, m_k) = 1$.

To show that the solution is valid, it is necessary to show that $c \pmod{m_k} = c_k$ where

$$c = \sum_{k=0}^{K-1} c_k N_k M_k \pmod{M}.$$

But $M_\ell = 0 \pmod{m_k}$ if $\ell \neq k$, so for $k = 0, \ldots, K-1$, this equation becomes

$$\begin{aligned} c \pmod{m_k} &= \textstyle\sum_{\ell=0}^{K-1} c_\ell N_\ell M_\ell \pmod{m_k} \\ &= c_k N_k M_k \pmod{m_k} \\ &= c_k(1 - n_k m_k) \pmod{m_k} \\ &= c_k, \end{aligned}$$

so the solution is valid.

To show that the solution is unique, suppose that c and c' are solutions to the theorem, both smaller than M. Then the residues c_k and c'_k are equal for every k, so $c - c'$ is a multiple of m_k for every k. Consequently, $c - c'$ must be zero because it is a multiple of M and lies between $-M$ and M. □

2.3 Prime fields

The ring of integers modulo n can be regarded as a ring consisting of n equivalence classes. Each equivalence class consisting of all elements of $\mathbf{Z}$ that have the same remainder under division by n. With this interpretation, we usually call the ring $\mathbf{Z}/\langle n\rangle$. Each equivalence class contains a unique integer between zero and $n - 1$, called the *canonical representative* of that equivalence class. Accordingly, the ring can be regarded simply as the set $\{0, 1, \ldots, n - 1\}$ with modulo n addition and modulo n multiplication. With this interpretation, we usually call the ring $\mathbf{Z}_n$.

Whenever $c = ab \pmod{n}$, we can define *modular division* in $\mathbf{Z}$, or *division* in $\mathbf{Z}_n$, as $a = c/b \pmod{n}$, or $a = b^{-1}c \pmod{n}$. Corollary 2.2.3 asserts that b^{-1} exists whenever $\text{GCD}(b, n) = 1$. If n is prime, b^{-1} will exist for all nonzero values of b. If n is composite, b^{-1} will not exist for some nonzero values of b, those not in $\mathbf{Z}_n^*$.

If every nonzero element of the ring $\mathbf{Z}_n$ has a unique multiplicative inverse, then for every nonzero element of the ring, the operation of division is defined as multiplication by the multiplicative inverse of that element. Thus $a/b = b^{-1}a$. In this case, the ring is called a *field*, which is a term of algebra that refers to any arithmetic system that has the operations of addition, subtraction, multiplication, and division by all nonzero elements, provided that these operations are well behaved with respect to certain standard axioms of behavior. Algebraic fields will be defined and discussed in more detail in Section 9.3.

We have already seen that for prime p, the extended euclidean algorithm asserts that $a^{-1} \pmod{p}$ exists for every positive integer smaller than p. The following theorem establishes this fact without direct reference to the extended euclidean algorithm.

Theorem 2.3.1 *If p is a prime, then every nonzero element of $\mathbf{Z}_p$ has a multiplicative inverse.*

Proof We know that $\mathbf{Z}_p$ is a ring under the natural modulo p addition and modulo p multiplication. To show that $\mathbf{Z}_p$ is a field, it is only necessary to show that multiplication by a nonzero element has an inverse. Multiplication by a nonzero a will have an inverse if the integers $a, 2a, 3a, \ldots, (p - 1)a$ modulo p are distinct because this means that one of these $p - 1$ distinct positive integers smaller than p must be one. To show that these integers are distinct, observe that if $ax = ay \pmod{p}$, then $a(x - y) = 0 \pmod{p}$ so $a(x - y) = cp$ for some value of c. But neither a nor $x - y$ is divisible by p, so both c and $x - y$ must equal zero. Therefore for each x, $ax = 1$ is satisfied by one and only one element a. This means that the operation of multiplication is unique and so has an inverse. □

Because every nonzero element of $\boldsymbol{Z}_p$ has an inverse under multiplication whenever p is a prime, $\boldsymbol{Z}_p$ has the structure of a field as defined in Section 9.3. For this reason, $\boldsymbol{Z}_p$ will also be denoted $\boldsymbol{F}_p$. It is called the *prime field* with p elements, and p is called the *characteristic* of the field $\boldsymbol{F}_p$. We shall refer to it either as the ring $\boldsymbol{Z}_p$ or as the field $\boldsymbol{F}_p$ according to which label best suits the situation under discussion. The set of nonzero elements of $\boldsymbol{F}_p$ will be denoted $\boldsymbol{F}_p^*$.

A consequence of Theorem 2.3.1 is the following theorem.

Theorem 2.3.2 (Wilson's theorem) *An integer p greater than 1 is prime if, and only if, $(p-1)! = -1 \pmod{p}$.*

Proof This is trivial when p equals 2 or 3, so we consider p larger than 3. In any field, the only square roots of one are ± 1, so referring to Theorem 2.3.1, each of the other elements of the field has an inverse element different from itself. Thus the elements of the field other than ± 1 come in pairs. Each pair has a pairwise product that is one modulo p, so $\prod_{\ell=1}^{p-2} \ell = 1 \pmod{p}$. Therefore $(p-1)! = (p-1)\prod_{i=1}^{p-2} \ell = -1 \pmod{p}$ provided p is a prime, as asserted in the statement of the theorem. On the other hand, if p is composite, then the factors of p must divide $(p-1)!$ which means that the expression $(p-1)! = -1 + cp$ cannot hold for any integer value of c, and so $(p-1)! \neq -1 \pmod{p}$. □

2.4 Quadratic residues

Suppose that p is an odd prime and that x is any nonzero integer smaller than p. Call x a *quadratic residue* modulo p of $\boldsymbol{Z}$, or a *square*[2] of $\boldsymbol{F}_p$ if $y^2 \equiv x \pmod{p}$ has a solution for y. Otherwise, call x a *quadratic nonresidue* modulo p of $\boldsymbol{Z}$, or a *nonsquare* of $\boldsymbol{F}_p$. The quadratic residues are the squares in the field $\boldsymbol{F}_p$, and so each quadratic residue has a square root. Because p is an odd prime, γ and $-\gamma$ are distinct elements of $\boldsymbol{F}_p$ and both have the same square in $\boldsymbol{F}_p$. Thus, evidently, of the $p-1$ nonzero elements of $\boldsymbol{F}_p$, only $(p-1)/2$ of them can be squares. To show this formally, suppose that γ and β have the same squares with $\beta < \gamma$. Then $\gamma^2 - \beta^2 = 0 \pmod{p}$. But $\gamma^2 - \beta^2 = (\gamma - \beta)(\gamma + \beta)$, so either $\gamma - \beta = 0$ or $\gamma + \beta = 0 \pmod{p}$. Thus

[2] Because p is a prime, the ring $\boldsymbol{Z}_p$ is the same as the field $\boldsymbol{F}_p$. In the language of fields, the quadratic residues are called the *squares* of $\boldsymbol{F}_p$, and the quadratic nonresidues are called the *nonsquares* of $\boldsymbol{F}_p$. The terms "square" and "nonsquare" of the finite field $\boldsymbol{F}_p$ are both clearer and more concise than "quadratic residue" and "quadratic nonresidue" (mod p) of the integer ring $\boldsymbol{Z}$. However, the latter terms are traditional and appropriate to the language of the ring $\boldsymbol{Z}$. The terms are firmly established and hard to avoid.

$\beta = \pm\gamma$, so the only square roots of γ^2 are $\pm\gamma$ (mod p). Thus x has two square roots in $\boldsymbol{F}_p$ for exactly $(p-1)/2$ nonzero values of γ and has no square roots for the other $(p-1)/2$ nonzero values of x. Accordingly, the number of quadratic residues modulo p is equal to the number of quadratic nonresidues.

For example, with $p = 11$, we have

$$\begin{aligned}
1^2 &\equiv 1,\\
2^2 &\equiv 4,\\
3^2 &\equiv 9,\\
4^2 &\equiv 5,\\
5^2 &\equiv 3,\\
6^2 &\equiv 3,\\
7^2 &\equiv 5,\\
8^2 &\equiv 9,\\
9^2 &\equiv 4,\\
10^2 &\equiv 1,
\end{aligned}$$

so the quadratic residues modulo 11 are $\{1, 3, 4, 5, 9\}$. The quadratic nonresidues modulo 11 are $\{2, 6, 7, 8, 10\}$.

Theorem 2.4.1 (Euler's criterion) *The element β is a quadratic residue modulo the odd prime p if, and only if,*

$$\beta^{(p-1)/2} = 1 \pmod{p}.$$

Proof The element β is a quadratic residue in $\boldsymbol{F}_p$ if, and only if, $\beta \equiv \gamma^2 \pmod{p}$ for some element γ of $\boldsymbol{F}_p$. The set $\boldsymbol{F}_p^*$ is a group, as asserted by Corollary 2.2.4, so every nonzero element of $\boldsymbol{F}_p$ satisfies $\gamma^{p-1} = 1 \pmod{p}$. If $\beta = \gamma^2$, this can be written $1 = (\gamma^2)^{(p-1)/2} = \beta^{(p-1)/2} \pmod{p}$. If β cannot be so written, then it is not a quadratic residue. □

Each of the nonzero integers from 0 to $p-1$ is either a quadratic residue or a quadratic nonresidue. We may further extend this definition to regard *any* integer coprime with p as a quadratic residue or a quadratic nonresidue according to whether its value modulo p is a quadratic residue or a quadratic nonresidue. Now assign the symbol 0 to each multiple of p, assign the symbol $+1$ to each quadratic residue, and assign the symbol -1 to each quadratic nonresidue, then we have a sequence of symbols 0, $+1$, and -1 marking the quadratic residues modulo p and the quadratic nonresidues modulo p. The elements of the sequence are described by a rather old-fashioned appearing notation (apparently going back to Legendre or Gauss) as follows.

Definition 2.4.2 *For any integer a and any odd prime p, the Legendre symbol is defined as*

$$(a \mid p) = \begin{cases} 0 & \text{if } p \text{ divides } a \\ 1 & \text{if } a \text{ is a quadratic residue modulo } p \\ -1 & \text{if } a \text{ is a quadratic nonresidue modulo } p. \end{cases}$$

The Legendre symbol is simply a label identifying which integers are quadratic residues. The Legendre symbol is not defined for $p = 2$.

For example, the squares of $\boldsymbol{F}_{13}$ are 1, 3, 4, 9, 10, and 12. The Legrendre symbol $(x \mid 13)$ as a periodic function of x, (with -1 denoted as *not one*), is given by

$$\begin{array}{rcccccccccccccccccccl} x = & 0, & 1, & 2, & 3, & 4, & 5, & 6, & 7, & 8, & 9, & 10, & 11, & 12, & 13, & 14, & 15, & 16, & 17, & \ldots \\ (x|13) = & 0 & 1 & \not1 & 1 & 1 & \not1 & \not1 & \not1 & \not1 & 1 & 1 & \not1 & 1 & 0 & 1 & \not1 & 1 & 1 & \end{array}$$

Theorem 2.4.3 *The Legendre symbol satisfies*

$$(x \mid p) = x^{(p-1)/2} \pmod{p}.$$

Proof If p divides x, then the left side equals zero by the definition of the Legendre symbol, and the term $x^{(p-1)/2}$ equals zero modulo p, so both sides are zero. If x is a quadratic residue, then we know that the left side equals one by the definition of the Legendre symbol. We know that the right side equals one by the Euler criterion. If x is a quadratic nonresidue, the Euler criterion says $x^{(p-1)/2} \not\equiv 1$, and we know $x^{p-1} \equiv 1$. But $(x^{\frac{p-1}{2}} - 1)(x^{\frac{p-1}{2}} + 1) = x^{p-1} - 1 = 0 \pmod{p}$. By the Euler criterion, p does not divide $x^{(p-1)/2} - 1$, so p must divide $x^{(p-1)/2} + 1$. That is, $x^{(p-1)/2} \equiv -1 \pmod{p} \equiv (x \mid p)$, as was to be proved. □

Theorem 2.4.4 *The integer 2 is a quadratic residue modulo p if, and only if, $p = \pm 1 \pmod{8}$. Moreover,*

$$(2 \mid p) = (-1)^{(p^2-1)/8}.$$

Proof This proof uses the notion of an extension field $\boldsymbol{F}_{p^m}$, which is explained in Section 9.7. (By creating and appending new elements, every finite field can be extended to a larger finite field, called an *extension field*, that contains elements of a chosen order, provided only that the order is coprime with the field characteristic p.)

Let β be an element of order 8, possibly in an extension field. Then $\beta^8 = 1$, and $\beta^4 = -1$, which means that $\beta^2 = -\beta^{-2}$. Consider the element $\gamma = \beta + \beta^{-1}$. Then $\gamma^2 = \beta^2 + 2 + \beta^{-2} = 2$, so γ is the square root of 2. It remains to show that $\gamma = \beta + \beta^{-1}$ is an element of the field $\boldsymbol{F}_p$ if, and only if, $p = 8k \pm 1$. Every odd prime p

can be written as $p = 8k \pm e$ where e is either 1 or 3. Then

$$\begin{aligned}(\beta + \beta^{-1})^p &= \beta^p + [\text{terms involving } p] + \beta^{-p} \\ &= \beta^p + \beta^{-p} \qquad \pmod p \\ &= \beta^{8k\pm e} + \frac{1}{\beta^{8k\pm e}} \\ &= \beta^{\pm e} + \beta^{\mp e} \qquad (\beta^8 = 1) \\ &= \beta^e + \beta^{-e}.\end{aligned}$$

We conclude that if $e = 1$, $\gamma^p = \gamma$ so γ, the square root of 2, is an element of $\boldsymbol{F}_p$ because $\boldsymbol{F}_p$ consists of those field elements satisfying $\gamma^p = \gamma$. On the other hand, by noting that $\beta^3 = -\beta^{-1}$ (because $\beta^2 = -\beta^{-2}$), we conclude that if $e = 3$, $\gamma^p = -\gamma$, which means that γ, the square root of 2, is not an element of $\boldsymbol{F}_p$.

To show, moreover, that this fact can be expressed as

$$(2|p) = (-1)^{(p^2-1)/8},$$

write $p = 8k \pm e$ where e is either 1 or 3. Then

$$(p^2 - 1)/8 = (e^2 - 1)/8 + 2ke + 8k^2.$$

The last two terms on the right are even. This means that

$$(-1)^{(p^2-1)/8} = (-1)^{(e^2-1)/8}.$$

If 2 has a square root in $\boldsymbol{F}_p$, then e is ± 1 and the term $(e^2 - 1)/8$ in the exponent on the right is zero. If 2 does not have a square root in $\boldsymbol{F}_p$, then e is ± 3 and the term in the exponent is one. This means that $(2|p) = (-1)^{(p^2-1)/8}$, as was to be proved. □

The following theorem is a helpful compendium that gives many useful properties of the Legendre symbol.

Theorem 2.4.5 *The Legendre symbol satisfies the following properties:*

(i) $(ab|p) = (a|p)(b|p)$
(ii) If $a = b \pmod p$, *then* $(a|p) = (b|p)$
(iii) $(a^2|p) = 1$ *if* p *does not divide* a
(iv) $(-1|p) = (-1)^{(p-1)/2}$
(v) $(1|p) = 1$.

Proof The proof consists of five parts.

(i) Theorem 2.4.3 asserts that $(x|p) = x^{(p-1)/2} \pmod p$ for any integer x. This means that we can write

$$\begin{aligned}(ab|p) &= (ab)^{(p-1)/2} = a^{(p-1)/2}b^{(p-1)/2} \pmod p \\ &= (a|p)(b|p) \pmod p.\end{aligned}$$

Finally, because the Legendre symbol can only be 0 or ± 1, the modulo-p operation is superfluous in the last line and that condition can be removed.

(ii) This follows immediately from the fact that

$$x^{(p-1)/2} = (x+p)^{(p-1)/2} \pmod{p}.$$

(iii) From (i), $(a^2|p) = (a|p)^2 = 1$ because $(a|p) = \pm 1$.

(iv) This is a restatement of the forthcoming Theorem 2.4.6, which says that if $p = 4k+1$ for some k, then $(-1|p) = 1$. Moreover, for such p,

$$(-1)^{(p-1)/2} = (-1)^{2k} = 1.$$

On the other hand, if $p = 4k+3$, then both sides are equal to -1.

(v) This property is trivial because the element one is always a square in $\boldsymbol{F}_p$.

This completes the proof of the five parts. □

The next theorem gives a condition on the existence or nonexistence of $\sqrt{-1}$ in the prime field $\boldsymbol{F}_p$.

Theorem 2.4.6 *The integer -1 is a quadratic residue modulo p if, and only if, $p = 4k+1$ for some integer k.*

Proof Let $\boldsymbol{F}_p^*$ denote the nonzero elements of the finite field $\boldsymbol{F}_p$. The elements of $\boldsymbol{F}_p^*$ form a cyclic group, as will be shown in Section 9.8. Let π denote a generator of this cyclic group. The generator π is itself a nonsquare because if it were a square, then every power of π would be a square and there would be no nonsquares in $\boldsymbol{F}_p^*$. The orbit of π can be written $\pi^1, \pi^2, \pi^3, \ldots,$ in which nonsquares and squares alternate (with π^i a square if i is even). Because $\pi^{p-1} = 1$ and $(-1)^2 = 1$, it is clear that the middle term of the sequence satisfies $\pi^{(p-1)/2} = -1$. If $(p-1)/2$ is even, (which means that $p = 4k+1$ for some integer k), then -1 appears in the above sequence in the position of a square. Otherwise, -1 appears in the position of a nonsquare. □

Theorem 2.4.6 has an intriguing consequence. Because the theorem states that $\sqrt{-1}$ exists in every prime field of the form $\boldsymbol{F}_{4k+1}$, we know that $1^2 + \sqrt{-1}^2 = 0$ in $\boldsymbol{F}_{4k+1}$, where $\sqrt{-1}$ is an element of that prime field, and so an integer. By regarding the integer $\sqrt{-1}$ as an element of $\boldsymbol{Z}$, this becomes $1^2 + \sqrt{-1}^2 = 0 \pmod{p}$, where $\sqrt{-1}$ is some integer of $\boldsymbol{Z}$. Likewise, for every element x of the prime field $\boldsymbol{F}_p$, $x\sqrt{-1}$ is also an element of $\boldsymbol{F}_p$, and so, as integers, $x^2 + (x\sqrt{-1})^2 = 0 \pmod{p}$, which can be written as $x^2 + y^2 = 0 \pmod{p}$ by replacing $x\sqrt{-1}$ by the integer notation y. Therefore for every positive integer x smaller than p, we have $x^2 + y^2 = \ell p$ is always satisfied for some integer y and for some value of ℓ that depends on x. If there is one x for which this is satisfied with $\ell = 1$, then for that x and some integer y, we have $x^2 + y^2 = p$, where x and y are integers. The following theorem, often regarded as one of the great theorems of classical number theory, states that for $p = 4k + 1$, there is always such an x with $\ell = 1$.

Theorem 2.4.7 (Fermat two-squares theorem) *An odd prime p is the sum of two squared integers x^2 and y^2 if, and only if, $p = 4k + 1$ for some positive integer k.*

Proof (Necessary) The square of an even integer is always equal to zero modulo 4. The square of an odd integer is always equal to 1 modulo 4 because every odd integer has the form $2k + 1$. Therefore the sum of two squares is either 0, 1, or 2 modulo 4. This means that an odd prime of the form $4k + 3$ cannot be the sum of two squares.

(Sufficient) Suppose that $p = 4k + 1$ for some positive integer k. By Theorem 2.4.6, $\sqrt{-1}$ exists as an element in $\boldsymbol{F}_p$. Let $i = \sqrt{-1}$, and consider elements of the form $c = a - bi \pmod{p}$, where a and b are elements of $\boldsymbol{F}_p$ satisfying $0 \le a, b \le \lfloor\sqrt{p}\rfloor$ as integers. Then a and b each take on $\lfloor\sqrt{p}\rfloor + 1$ values so there are $(\lfloor\sqrt{p}\rfloor + 1)^2$ such pairs, which is a number larger than p. Therefore, some such element c of $\boldsymbol{F}_p$ is produced by two distinct pairs (a_1, b_1) and (a_2, b_2). Then $a_1 - b_1 i = a_2 - b_2 i \pmod{p}$. Consequently,

$$(a_1 - a_2) = \sqrt{-1}(b_1 - b_2)$$

in $\boldsymbol{F}_p$. Let $x = (a_1 - a_2)$ and $y = (b_1 - b_2)$, so $x = \sqrt{-1}y$. Therefore $x^2 = -y^2$ in $\boldsymbol{F}_p$. Thus, as integers, $x^2 + y^2 = \ell p$ for some nonzero ℓ. On the other hand, as integers $|a_1 - a_2|$ and $|b_1 - b_2|$ are both smaller than $\sqrt{p}$, so $x^2 + y^2 < (\sqrt{p})^2 + (\sqrt{p})^2 = 2p$. This can hold only if $\ell = 1$. The proof is complete. □

As a result of this theorem, we see that the prime p can be factored in the ring[3] $\mathbf{Z}[i]$ if, and only if, $p = 4k + 1$ for some integer k. For example, $13 = (2 + 3i)(2 - 3i) = 2^2 + 3^2$, $29 = (2 + 5i)(2 - 5i) = 2^2 + 5^2$, $53 = (2 + 7i)(2 - 7i) = 2^2 + 7^2$, and $113 = (7 + 8i)(7 - 8i) = 7^2 + 8^2$. Accordingly, a prime of the form $4k + 1$ is said to *split* or *ramify* in $\mathbf{Z}[i]$. In contrast, a prime of the form $p = 4k + 3$ is said to be *inert* in $\mathbf{Z}[i]$.

2.5 Quadratic reciprocity

For the proof of the following theorems, we shall use the *signed representation* of $\boldsymbol{Z}_p$, which is defined to be the set of integers $\{-\frac{1}{2}(p - 1), -\frac{1}{2}(p - 1) + 1, \ldots, \frac{1}{2}(p - 1) - 1, \frac{1}{2}(p - 1)\}$. These integers are the representatives for $\boldsymbol{Z}_p$ that are smallest in absolute value. For example, $\boldsymbol{Z}_7 = \{-3, -2, -1, 0, 1, 2, 3\}$ in the signed representation of $\boldsymbol{Z}_7$, whereas $\boldsymbol{Z}_7 = \{0, 1, 2, 3, 4, 5, 6\}$ in the canonical representation.

[3] The ring $\mathbf{Z}[i]$ is defined as the set of complex numbers with integers for both the real and imaginary part.

The positive integers in the signed representation are, of course, distinct. A powerful property of the signed representation is that the signed representatives of the positive integers remain distinct in absolute value when multiplied by any integer. Thus, for example, in $\boldsymbol{Z}_7$, $\{3, 6, 9\} = \{3, -1, 2\}$, $\{4, 8, 12\} = \{-3, 1, -2\}$, and $\{5, 10, 15\} = \{-2, 3, 1\}$. This is the content of the following theorem.

Theorem 2.5.1 *For any odd prime p of $\boldsymbol{F}_p$ and any positive integer a, the $\frac{1}{2}(p-1)$ elements of the set $\{a, 2a, 3a, \ldots, \frac{1}{2}(p-1)a\}$ of $\boldsymbol{Z}_p$, when expressed in the signed representation, are distinct in absolute value, and so the absolute values of these signed representatives are the integers $1, 2, \ldots, \frac{1}{2}(p-1)$.*

Proof The subset of positive signed representatives $\{1, 2, 3, \ldots, \frac{1}{2}(p-1)\}$ can be multiplied by any integer a to form the subset $\{a, 2a, 3a, \ldots, \frac{1}{2}(p-1)a\}$ modulo p. Because $\boldsymbol{Z}_p$ is the field $\boldsymbol{F}_p$, the $(p-1)/2$ elements of this subset must all be distinct and each integer $\ell a \pmod p$ must have a unique inverse under addition. Replace each element by its signed representative. These are all distinct. We will show moreover that even the absolute values of all signed representatives of this set are distinct. That is, dropping the minus sign does not produce any duplicate of a number already there. To show this, label the positive entries by $r_1, r_2, \ldots,$ and label the negative entries by $-s_1, -s_2, \ldots,$ with the sign shown explicitly, so that the s_j are positive. We will show that no r_i is equal to an s_j (even though the sign of $-s_j$ is removed).

Let $m_i a = r_i$ and $m_j a = -s_j$ be any two distinct elements of the set. Then $r_i = s_j$ would imply that $a(m_i + m_j) = 0 \pmod p$, which implies that $m_i + m_j = 0 \pmod p$. This cannot be true because m_i and m_j are both nonnegative and both smaller than $p/2$. This means that r_i and s_j are different. Hence together there are $\frac{1}{2}(p-1)$ distinct integers taken on by the r_i and s_j. All are positive and each is not larger than $\frac{1}{2}(p-1)$, and so they must consist of all of the distinct integers between 1 and $\frac{1}{2}(p-1)$. Therefore they comprise exactly the integers from 1 to $\frac{1}{2}(p-1)$ in some order. □

Theorem 2.5.2 (Gauss's lemma) *Let p be a prime. The Legendre symbol satisfies*

$$(a \mid p) = (-1)^{\mu},$$

where μ is the number of negative integers appearing among the signed representatives of the subset $\{a, 2a, 3a, \ldots, \frac{1}{2}(p-1)a\}$ of $\boldsymbol{Z}_p$.

Proof A consequence of Theorem 2.5.1 is that the product of all of the $\frac{1}{2}(p-1)$ signed representatives in the subset $\{a, 2a, 3a, \ldots, \frac{1}{2}(p-1)a\}$ (modulo p) is

$$(a)(2a)\cdots(\tfrac{p-1}{2}a) = (-1)^{\mu}\tfrac{p-1}{2}! \pmod p.$$

By canceling $\frac{p-1}{2}!$ from both sides, this simplifies to

$$a^{(p-1)/2} = (-1)^{\mu}.$$

Combining this with

$$(a\,|\,p) = a^{(p-1)/2},$$

as given in Theorem 2.4.3 completes the proof of the theorem. □

For example, with $a = 3$ and $p = 31$

$$\begin{aligned}&\{3, 6, 9, 12, \ldots, 45\} \pmod{31}\\ &\quad= \{3, 6, 9, 12, 15, -13, -10, -7, -4, -1, 2, 5, 8, 11, 14\}.\end{aligned}$$

Hence $\mu = 5$, so by Gauss's lemma, $(3|31) = -1$. We conclude that 3 is not a square modulo 31.

Our next theorem is a celebrated and important theorem of number theory. It says that for two distinct odd primes, p and q, the Legendre symbol satisfies

$$(p|q)(q|p) = \pm 1,$$

the minus sign occurring if, and only if, $\frac{1}{2}(p-1)$ and $\frac{1}{2}(q-1)$ are both odd. This requires that both p and q are of the form $4k+3$.

Theorem 2.5.3 (Quadratic reciprocity) *For distinct odd primes, p and q, the Legendre symbol satisfies*

$$(p|q)(q|p) = (-1)^{\frac{(p-1)}{2}\frac{(q-1)}{2}}.$$

Proof Because p and q are distinct odd primes, the two Legendre symbols, $(p|q)$ and $(q|p)$, are each equal to either $+1$ or -1. This means that the left side of the statement of the theorem is either plus one or minus one. By Gauss's lemma, $(q|p) = (-1)^{\nu}$ and $(p|q) = (-1)^{\mu}$ where the numbers ν and μ are the number of negative terms in the signed representations of the sets $\mathcal{S}_x = \{xq \pmod{p} : x = 1, \ldots, \frac{1}{2}(p-1)\}$ and $\mathcal{S}_y = \{yp \pmod{q} : y = 1, \ldots, \frac{1}{2}(q-1)\}$, respectively. Thus

$$(p|q)(q|p) = (-1)^{\nu+\mu}.$$

Accordingly, we must show that

$$\nu + \mu = \frac{(p-1)}{2} \cdot \frac{(q-1)}{2} \pmod{2}$$

where the left side is the sum of the number of negative terms in S_x and the number of negative terms in S_y. The raw product on the right side is the parity of the total number of x, y pairs such that $1 \le x \le \frac{1}{2}(p-1)$ and $1 \le y \le \frac{1}{2}(q-1)$.

Because there are p integers between $\frac{-p}{2}$ and $\frac{p}{2}$ and the integers $p\ell$ are spaced by p in the signed representation, there must be a unique nonnegative integer y for each x satisfying $-\frac{p}{2} < qx - py < \frac{p}{2}$. Thus for each value of x, there is one and only one nonnegative integer y for which $qx - py$ is a signed representative of $\mathbf{Z}_p$. Because $x < \frac{1}{2}p$ and $qx < \frac{1}{2}qp$, it follows that $0 \leq py \leq \frac{1}{2}qp$, and so $0 \leq y < \frac{1}{2}q$. Thus $x \in \{1, 2, \ldots, \frac{1}{2}(p-1)\}$ and $y \in \{1, 2, \ldots, \frac{1}{2}(q-1)\}$. The value of zero for y has been excluded from further consideration because if $y = 0$, then $qx - py$ cannot be negative and does not contribute to μ. We conclude that the integer y is an element of $\{1, \ldots, \frac{1}{2}(q-1)\}$. Accordingly, we can change the definition of $\mathcal{S}_x$ to a set of lattice points of $\mathbf{Z}^2$ of the form

$$\mathcal{S}_x = \{(x, y) \,|\, x = 1, \ldots, \tfrac{1}{2}(p-1);\, y = 1, \ldots, \tfrac{1}{2}(q-1); -\tfrac{1}{2}p < qx - py < \tfrac{1}{2}p\}.$$

Similar statements apply as well to S_y, but with p and q interchanged, and x and y also interchanged. Accordingly,

$$\mathcal{S}_y = \{(x, y) \,|\, y = 1, \ldots, \tfrac{1}{2}(q-1);\, x = 1, \ldots, \tfrac{1}{2}(p-1); -\tfrac{1}{2}q < py - qx < \tfrac{1}{2}q\}.$$

Now recall that $\mu + \nu$ is equal to the number of pairs of such x and y for which either $-\frac{1}{2}p < qx - py < 0$, or $-\frac{1}{2}q < py - qx < 0$. We must count only those points for which the signed representation is negative, so we can now reduce the two sets of lattice points to the sets

$$\mathcal{S}_x^- = \{(x, y) \,|\, x \in \{1, \ldots, \tfrac{1}{2}(p-1)\};\, y \in \{1, \ldots, \tfrac{1}{2}(q-1)\}; -\tfrac{1}{2}p < qx - py < 0\}$$
$$\mathcal{S}_y^- = \{(x, y) \,|\, y \in \{1, \ldots, \tfrac{1}{2}(q-1)\};\, x \in \{1, \ldots, \tfrac{1}{2}(p-1)\}; -\tfrac{1}{2}q < py - qx < 0\}.$$

These two sets can be illustrated by the sketch

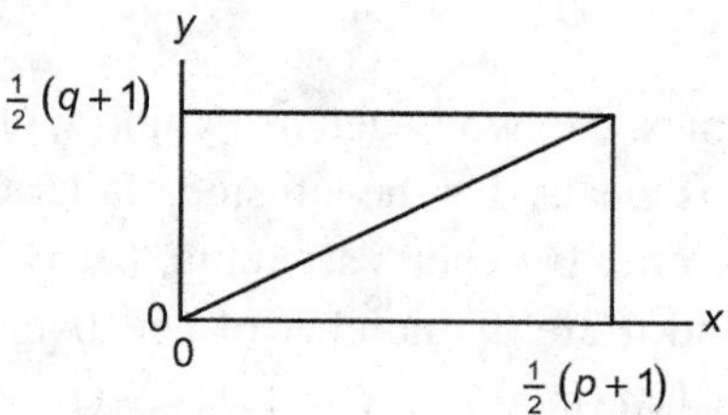

which is meant to suggest a $\frac{1}{2}(p+1)$ by $\frac{1}{2}(q+1)$ array of lattice points, although the lattice points themselves are not shown in the sketch. The diagonal line $qx = py$ separates the sets S_x^- and S_y^-. The set S_x^- consists of lattice points slightly above, or on, the diagonal. The set S_y^- consists of lattice points slightly below, or on, the diagonal. Points on the outside rectangular boundary of the array are not included in the set $\mathcal{S}_x^-$ and $\mathcal{S}_y^-$.

We will show that the set of lattice points of the upper triangle that are not in $S_x^- \cup S_y^-$ is congruent to the set of points in the lower triangle that are not in $S_x^- \cup S_y^-$. Let λ denote the number of (x, y) pairs of the array excluded because $qx - py < -p/2$. We will show that the number of (x, y) pairs of the array excluded because $qx - py > q/2$ is also equal to λ.

The two triangles in the figure are congruent, and can be made identical by a change of variables. Thus let $x = \frac{1}{2}(p+1) - x'$ and $y = \frac{1}{2}(q+1) - y'$. The variables x' and y' index the same sets, but in the opposite direction. Under this change of variables, the inequality $qx - py < -\frac{1}{2}p$ becomes the inequality $qx' - py' > \frac{1}{2}q$, which is the same as the condition on the previous set of excluded pairs (x, y). Hence the total number of excluded pairs (x, y) is 2λ, of which λ is the number of such pairs for which $qx - py$ is larger than $\frac{1}{2}q$ and λ is also the number of such pairs for which $qx - py$ is smaller than $-\frac{1}{2}p$. Therefore, for some integer λ,

$$\begin{aligned}\frac{(p-1)}{2} \cdot \frac{(q-1)}{2} &= \mu + \nu + 2\lambda \\ &= \mu + \nu \pmod 2,\end{aligned}$$

as was to be proved. □

Quadratic reciprocity is a surprisingly useful tool for certain computations such as for primality testing. For an example of the use of quadratic reciprocity, 11 and 31 are both primes, and $(11-1)/2$ and $(31-1)/2$ are both odd. Therefore $(31|11)(11|31) = -1$. But $31 = 9 \pmod{11}$ which is a square modulo 11, so $(31|11) = 1$. Therefore $(11|31) = -1$, so we can conclude that 11 is a nonsquare modulo 31.

2.6 The Jacobi symbol

The Legendre symbol $(x|p)$ is defined only if the second integer p is a prime. Consequently, to use the properties of the Legendre symbol for a calculation, one must first know that the second integer is a prime. Thus the Legendre symbol cannot be used as such in any algorithm whose purpose is to determine primality of the second integer. For this reason, the Legendre symbol will now be generalized by replacing the prime p by an arbitrary positive integer n. The generalization is based on the fact that any odd integer n has a unique prime factorization of the form $n = p_1^{e_1} p_2^{e_2} \cdots p_k^{e_k}$.

Definition 2.6.1 *The Jacobi symbol is defined for an odd positive integer n, with prime factorization* $p_1^{e_1} p_2^{e_2} \cdots p_s^{e_s}$, *and any integer m, as*

$$(m|n) = \prod_{i=1}^{s} (m|p_i)^{e_i},$$

where, because each p_i *is a prime, each instance of the symbol* $(m|p_i)$ *on the right is the Legendre symbol.*

Whenever n is equal to a prime p, the Jacobi symbol reduces to the Legendre symbol, so it is reasonable to use the same notation for both. The properties of the Legendre symbol given in Theorem 2.4.5 can now be carried over to prove similar properties of the Jacobi symbol.

Unlike the Legendre symbol, the Jacobi symbol $(a|n)$ does not reveal whether or not a is a quadratic residue modulo n, where a quadratic residue modulo n is defined as an element of $\mathbf{Z}_n^*$ that is the square, modulo n, of another element. For example, although $(5|21) = 1$, the integer 5 is not a square of the ring $\mathbf{Z}_{21}^*$.

Theorem 2.6.2 *The Jacobi symbol satisfies the following properties:*

(0) *For any odd integer n, $(1|n) = 1$.*

(1) *If $a \equiv b \pmod{n}$, then*

$$(a|n) = (b|n).$$

(2) *If n is an odd positive integer, then*

$$(2|n) = \begin{cases} 1 & \text{if} \quad n \equiv \pm 1 \pmod{8} \\ -1 & \text{if} \quad n \equiv \pm 3 \pmod{8}. \end{cases}$$

(3) *If n is an odd positive integer, then*

$$(ab|n) = (a|n)(b|n).$$

(4) *If m and n are coprime odd positive integers, then*

$$(m|n)(n|m) = (-1)^{\left(\frac{m-1}{2}\right)\left(\frac{n-1}{2}\right)}.$$

Proof The proof consists of five parts.

(0) Because $(1|p) = 1$ for any prime, $(1|n)$ is equal to the product of ones, and so itself is one.

(1) If $a \equiv b \pmod{n}$, then for any prime factor p_i of n, $a = b \pmod{p_i}$, and so $(a|p_i) = (b|p_i)$. Then

$$(a|n) = \prod_i (a|p_i)^{e_i} = \prod_i (b|p_i)^{e_i} = (b|n).$$

(2) Let $n = \prod_{i=1}^{s} p_i^{e_i} = \prod_{\ell=1}^{L} p_\ell$ where the primes p_ℓ in the second product are not necessarily distinct. In that product, each prime is written the number of times it appears in n, and $L = \sum_{i=1}^{s} e_i$. Use the equality

$$\prod_{\ell=1}^{L}(1 + x_\ell) = 1 + \sum_i x_i + \sum_{i \neq j} \sum_j x_i x_j + \cdots + \sum_{\substack{i \neq j \\ i \neq k}} \sum_{i \neq j} \sum_k x_i x_j x_k + \cdots$$

to write

$$\begin{aligned} n^2 &= \prod_{\ell=1}^{L} p_\ell^2 = \prod_{\ell=1}^{L} (1 + p_\ell^2 - 1) \\ &= 1 + \sum_i (p_i^2 - 1) + \sum_{i \neq j} \sum_j (p_i^2 - 1)(p_j^2 - 1) + \ldots \end{aligned}$$

We now show, for every odd prime p, that $p^2 - 1$ is always a multiple of eight. Because p is odd, $p = 4k \pm 1$ for some k. Then $p^2 = 16k^2 \pm 8k + 1$, so $p^2 - 1 = 8(2k^2 \pm k)$. We can conclude that $p_i^2 - 1$ is a multiple of eight for all terms in the above expansion of n^2. This means that all terms except for the first two are multiples of 64. Therefore

$$n^2 \equiv 1 + \sum_{\ell=1}^{L} (p_\ell^2 - 1) \pmod{64}.$$

Thus

$$\frac{1}{8}(n^2 - 1) \equiv \sum_{\ell=1}^{L} \frac{1}{8}(p_\ell^2 - 1) \pmod{8}.$$

By Theorem 2.4.4 for the Legendre symbol, we have $(2|p) = (-1)^{(p^2-1)/8}$. Then because $(-1)^8 = 1$, we have

$$\begin{aligned} (2|n) &= \prod_{\ell=1}^{L} (2|p_\ell) = \prod_{\ell=1}^{L} (-1)^{(p_\ell^2-1)/8} \\ &= (-1)^{\sum_\ell (p_\ell^2-1)/8} = (-1)^{(n^2-1)/8} \\ &= \begin{cases} 1 & \text{if } n \equiv \pm 1 \pmod{8} \\ -1, & \text{if } n \equiv \pm 3 \pmod{8}. \end{cases} \end{aligned}$$

(3) If n is a prime, then this is just part (i) of Theorem 2.5.3. Otherwise, let $n = \prod_{i=1}^{s} p_i^{e_i}$. Then

$$\begin{aligned} (ab|n) &= \prod_{\ell=1}^{L} (ab|p_\ell)^{e_\ell} \\ &= \prod_{\ell=1}^{L} (a|p_\ell)^{e_\ell} (b|p_\ell)^{e_\ell} = (a|n)(b|n). \end{aligned}$$

(4) The statement that, for m and n both odd integers,

$$(m|n) = \begin{cases} -(n|m) & \text{if both } m \text{ and } n \text{ are equal to } 3 \pmod{4} \\ +(n|m) & \text{otherwise} \end{cases}$$

is a generalization of quadratic reciprocity, as given in Theorem 2.5.3, from the case in which m and n are both odd primes to the case in which m and n are any odd integers. Let $n = \prod_{i=1}^{s} p_i^{e_i} = \prod_{\ell=1}^{L} p_\ell$ where $L = \sum_{i=1}^{s} e_i$ and each distinct prime is repeated e_i times in the second expression on the right. Similarly, let $m = \prod_{j=1}^{s'} q_j^{e'_j} = \prod_{\ell'=1}^{L'} q_{\ell'}$ where $L' = \sum_{j=1}^{s} e'_j$ and each distinct prime $q_{\ell'}$ is repeated e'_j times in the second expression on the right. Then by Definition 2.6.1,

$$(m|n) = \prod_{\ell=1}^{L} (m|p_\ell).$$

Using part (3) of this theorem gives

$$(m|n) = \prod_{\ell=1}^{L} \prod_{\ell'=1}^{L'} (q_{\ell'}|p_\ell),$$

and by Theorem 2.5.3,

$$\begin{aligned}(m|n) &= \prod_{\ell=1}^{L} \prod_{\ell'=1}^{L'} \left[(-1)^{(\frac{p_\ell-1}{2})(\frac{q_{\ell'}-1}{2})}\right](p_\ell|q_{\ell'}) \\ &= (n|m) \prod_{\ell=1}^{L} \prod_{\ell'=1}^{L'} \left[(-1)^{(\frac{p_\ell-1}{2})(\frac{q_{\ell'}-1}{2})}\right].\end{aligned}$$

Each term within the product is equal to either $+1$ or -1 depending on whether the exponent is even or odd. A term is equal to -1 only if $(p_i - 1)/2$ and $(q_j - 1)/2$ are both odd. The entire product is -1 only if there are an odd number of such -1 terms in the product. This means that there are both an odd number of terms with $p_i = 3 \pmod 4$ and an odd number of terms with $q_j = 3 \pmod 4$. But the product of an odd number of terms each equal to $3 \pmod 4$ is equal to $3 \pmod 4$. Therefore $m = 3 \pmod 4$ and $n = 3 \pmod 4$ if, and only if, the sign of the product is negative. Thus $(m|n) = (n|m)(-1)^{\frac{(m-1)}{2}\frac{(n-1)}{2}}$. Recalling that $(n|m) = \pm 1$, this becomes

$$(m|n)(n|m) = (-1)^{(\frac{m-1}{2})(\frac{n-1}{2})}$$

as was to be proved. □

As an example of the use of Theorem 2.6.2, we compute $(888|1999)$ as follows:

$$\begin{aligned}(888|1999) &= (2|1999)(2|1999)(2|1999)(111|1999) && \text{Property 3}\\ &= (111|1999) && \text{Property 2}\\ &= -(1999|111) && \text{Property 4}\\ &= -(1|111) && \text{Property 1}\\ &= -1 && \text{Property 0.}\end{aligned}$$

This calculation tells us that if 1999 is a prime, then by Definition 2.4.2, 888 is a quadratic nonresidue modulo 1999. If we do not know that 1999 is a prime, then we cannot reach this conclusion. If 1999 is not a prime then the calculation is uninformative by itself.

In general, by repeating the process of flipping and reducing the top integer modulo the bottom integer as has been illustrated in this example, we can compute a Jacobi symbol in a time proportional to the number of decimal digits of the integers.

2.7 Primality testing

To find a large prime – say one with 100 digits – is not a trivial task. Suppose that an integer with a 100-digit decimal representation is randomly selected by first choosing the digits randomly and independently. One then wants to determine whether this large integer is a prime. It is certainly not a prime if the low-order digit is even or a 5. But if the low-order digit is 1, 3, 7, or 9, then it can be difficult to determine whether that integer is a prime. Some popular methods of testing for primality are described in the next three sections.

The distribution of primes among the set of integers is quite irregular, seemingly with no simple pattern. Sometimes this irregularity can be made useful, as will be the case in later chapters. The following theorem, a deep theorem of number theory, rules out the possibility of a periodic sequence of nonprimes. A proof is not given.

Theorem 2.7.1 (Dirichlet) *For any coprime positive integers a and b, the sequence $ak + b$ for $k = 1, 2, 3, \ldots,$ contains an infinite number of primes.*

The theorem immediately informs us that there are an infinite number of primes of the form $p = 4k + 1$, and also an infinite number of primes of the form $p = 4k + 3$. We say that there are an infinite number of primes congruent to one modulo four, and there are an infinite number of primes congruent to three modulo four.

The theorem can be regarded as a statement about nonprimes. Indirectly it says that certain elementary sequences of gaps in the primes do not exist. Nevertheless, there is an average behavior of primes, as described by the following deep and remarkable theorem of number theory. A proof is not given.

Theorem 2.7.2 (Prime number theorem) *The distribution of primes among the integers is such that*

$$\#\{\text{primes} \leq x\} \sim \int_2^x \frac{dt}{\log_e t},$$

meaning that the ratio of the two sides approaches one as x goes to infinity.

Loosely, the prime number theorem says that the fraction of integers that are prime in a small interval of integers near x is approximately $1/\log_e x$. For example, if x is a randomly generated 100-digit integer, the probability that x is a prime is about

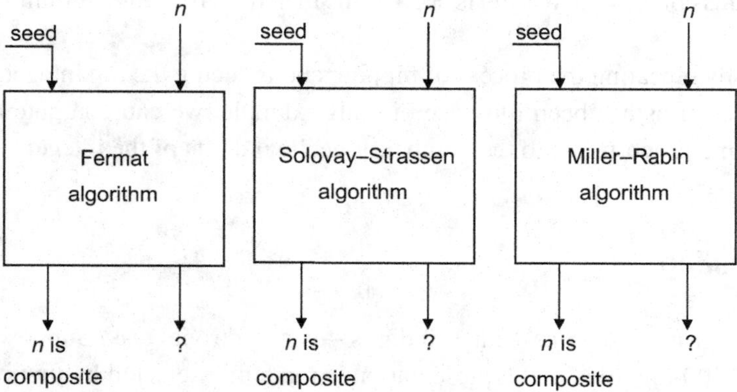

Figure 2.1 Three primality-testing algorithms

$1/\log_e 10^{100}$. Thus among integers of length 100 digits, a little more than one in 200 is a prime. Of course, a prime represented in decimal notation can only end in 1, 3, 7, or 9. If we consider only integers that end in 1, 3, 7, or 9, the situation improves in the ratio of four to ten. Of such integers of length 100 decimal digits, a little fewer than one in 100 is a prime. This means that there are approximately 10^{98} 100-digit primes, an enormous number of primes.

To find a 100-digit prime, one can randomly choose a candidate 100-digit decimal sequence, then test it to see whether that sequence specifies a prime integer. This means that we will need algorithms for primality testing. Several practical primality-testing algorithms that have a structure known as a yes-biased probability test will be described. Each test can be applied to the question "Is n composite?" Such a procedure is controlled by a large randomly chosen seed. The outcome of the procedure is either "yes" or "don't know," which means that n might be a prime. If the answer is "don't know," then another trial is needed with a new randomly chosen seed. A yes-biased probability test can show with certainty that an integer is composite, but it cannot show with certainty that an integer is prime.

The three methods that are the most well known are shown in Figure 2.1. We will describe the *Fermat algorithm*, the *Solovay–Strassen algorithm*, and the *Miller–Rabin algorithm* in the next three sections.

It is remarkable, and perhaps unsettling, that each of the three primality-testing algorithms shown in Figure 2.1 can determine that an integer n is composite and yet apparently provides no help at all toward determining the factors of n. If it were otherwise, cryptography based on number theory would not succeed.

2.8 The Fermat algorithm

The Fermat little theorem gives a necessary condition that for every prime p is satisfied for every integer b smaller than p. The Fermat little theorem leads to a simple and long-standing test for primality. This test is called the Fermat primality-testing algorithm. The Fermat algorithm is simple. To test the integer n for primality, the Fermat algorithm randomly chooses a test integer b. If $b^n \neq b \pmod{n}$, then by the Fermat little theorem, n is composite. Otherwise, if $b^n = b \pmod{n}$, the test fails and makes no decision. It can be repeated with another randomly chosen b. However, because the converse to Fermat's little theorem does not hold, the Fermat algorithm cannot show that an integer is a prime. It can only show that an integer is not a prime. Because n might be an integer for which the test almost always fails, the Fermat algorithm is unsatisfactory as the principal test of primality. The Fermat algorithm can be useful as a preliminary simple test used to reject some n quickly before turning to a stronger algorithm. However, even if $b^n = b \pmod{n}$ for *every* integer b that is coprime to n, that n still need not be a prime. This motivates the following definition.

Definition 2.8.1 *An odd composite integer n is called a Carmichael integer if $b^n = b \pmod{n}$ for every b in $\mathbf{Z}_n$.*

Carmichael integers are sometimes called *false primes* because they satisfy the condition of the Fermat little theorem but are not primes. The existence of Carmichael integers is important to us because this means that the Fermat theorem, in itself, is not strong enough to prove that an integer is a prime, even if all b smaller than n are tested. Carmichael integers satisfy certain necessary and sufficient conditions, making it possible to identify a Carmichael integer from its prime factorization, were the prime factorization known. Of course, if the primality of n is not known, the factorization of n is certainly not known either. The next theorem, however helps to show that Carmichael integers do exist.

Theorem 2.8.2 (Korselt) *An odd composite positive integer n is a Carmichael integer if, and only if, n is square-free and $p_i - 1$ divides $n - 1$ for every prime factor p_i of n.*

Proof (Sufficiency) Suppose that the odd composite positive integer n is a square-free integer and so its prime factors are odd and distinct, and that $p_i - 1$ divides $n - 1$ for every prime factor p_i of n. For any integer b, let $b_i = b \pmod{p_i}$ for each p_i. Each b_i is either coprime with p_i or is zero. If b_i is coprime with p_i, then by the Fermat little theorem, $b_i^{p_i-1} = 1 \pmod{p_i}$. Moreover, if b_i is nonzero, we have

$$b_i^{n-1} = (b_i^{p_i-1})^{(n-1)/(p_i-1)} = 1 \pmod{p_i}.$$

If $b_i = 0$, then $b_i^n = 0$. Therefore $b_i^n = b_i \pmod{p_i}$ for each i. Finally, by applying the chinese remainder theorem to each side of this equality, we conclude that $b^n = b \pmod{n}$. This holds for every b, so n is a Carmichael integer.

(Necessity) Suppose that n is a Carmichael integer. Then $b^n = b \pmod{n}$ for every $b \in \mathbf{Z}_n$. First, we observe that n must be square-free as follows. Let $n = p^2 m$ for some integer m and prime p. Because n is a Carmichael integer, $p^n = p \pmod{n}$. This means that $p^n = p + Qp^2 m$ for some Q, and so $1 = p(p^{n-2} - Qm)$. But the prime p does not divide one, so this cannot be true. Therefore n cannot have p^2 as a factor.

Let $n = p_1 \cdots p_k$, where the p_i are distinct odd primes. For each i, let a_i generate the cyclic group $\mathbf{Z}^*_{p_i}$. Then for each i, p_i and a_i are coprime, and a_i has order $\phi(p_i) = p_i - 1$. The chinese remainder theorem asserts that there exists an integer a coprime to n for which $a = a_i \pmod{p_i}$ for all i. Moreover, because n is a Carmichael integer, $a^{n-1} = 1 \pmod{n}$, which means that $a^{n-1} = 1 \pmod{p_i}$ and, in turn, $a_i^{n-1} = 1 \pmod{p_i}$. Because a_i has order $p_i - 1 \pmod{p_i}$, we can conclude that $p_i - 1$ divides $n - 1$. This holds for all i, and so the proof of necessity is complete. □

Every Carmichael integer, as defined, is odd. This condition could have been left out of Definition 2.8.1. Then it would come out that n is odd as an additional conclusion of the theorem.

Invoking Theorem 2.8.2, notice that $3 \cdot 11 \cdot 17 = 561$ and 560 is divisible by 2, 10, and 16, so 561 is a Carmichael integer. Notice also that $5 \cdot 13 \cdot 17 = 1105$ and 1104 is divisible by 4, 12, and 16, so 1105 is also a Carmichael integer. Finally, notice that $7 \cdot 13 \cdot 19 = 1729$ and that 1728 is divisible by 6, 12, and 18, so 1729 is a Carmichael integer. Indeed, the three smallest Carmichael integers are 561, 1105, and 1729. Each of these has three prime factors.

Corollary 2.8.3 *Every Carmichael integer n is the product of at least three distinct primes.*

Proof The Carmichael integer itself is not a prime. All prime factors must be distinct because n is square free. Suppose that n is the product of two primes $n = p_1 p_2$ with $p_2 > p_1$. Then the theorem requires that $p_2 - 1$ divides $p_1 p_2 - 1$. But the division algorithm gives

$$n - 1 = (p_2 - 1)p_1 + (p_1 - 1).$$

But because $p_1 - 1 \neq 0$, $p_2 - 1$ does not divide $n - 1$, contrary to the statement of the theorem. Thus n cannot be the product of two primes. □

It is known that there is an infinite number of Carmichael integers. However, their density decreases as the size of the integers increases, and eventually they become quite rare. Figure 2.2 shows the approximate density of Carmichael integers as a function of the number of decimal digits in the integer n. For example, the 18-digit integers have fewer than one Carmichael integer out of 10^{12}. Thus there is only about one chance in 10^{12} of randomly choosing an 18-digit composite integer for which the

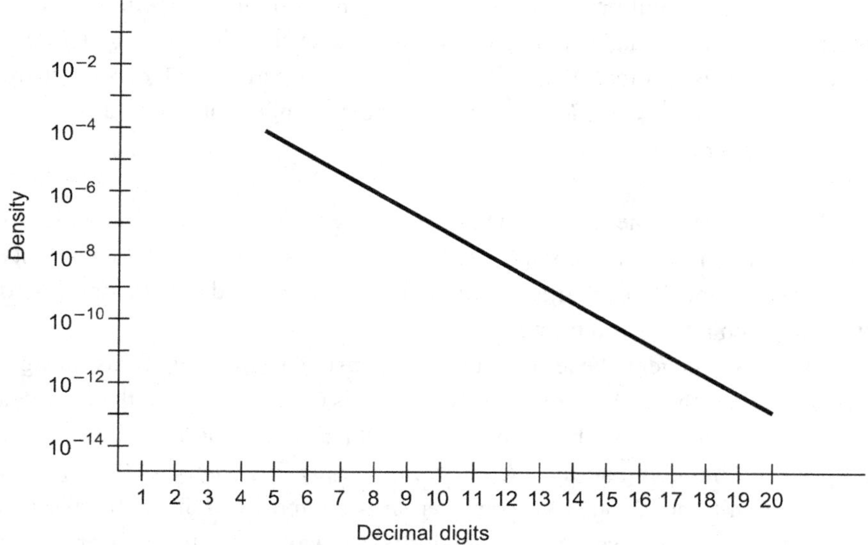

Figure 2.2 Approximate density of Carmichael integers

equality of Fermat's little theorem holds for all b less than n. This rarity, however, is not a sufficient reason to be comfortable. One presumes that there are far more n for which the equality of Fermat's little theorem holds for "many" or "most" values of b, whatever "many" or "most" might mean in this sentence. Thus, absent any statement to the contrary, we presume that "almost-Carmichael" integers need not be rare. Because of this defect, Fermat's little theorem is not considered to be a satisfactory theorem for serious checking of primality, although it does see use as a simple test of compositeness that can quickly reject some integers that are to be tested for primality.

2.9 The Solovay–Strassen algorithm

The Solovay–Strassen algorithm is a probabilistic algorithm for determining whether an integer n is a prime. It is a probabilistic algorithm because it requires a randomly chosen integer and so can be repeated many times. It does not use the Fermat little theorem as such, so it is not vulnerable to encounters with the Carmichael integers. If the algorithm calls n composite, then n is definitely composite. If, after multiple trials, it calls n likely to be prime, then n is prime with high probability, but not certainty. The Solovay–Strassen algorithm is not the only popular primality-testing algorithm. The Miller–Rabin primality-testing algorithm, given in the next section, may have a smaller failure rate, but its structure is more complex and less transparent.

Theorem 2.9.1 *If the integer n is a prime, then $(\beta|n) = \beta^{\frac{n-1}{2}} \pmod{n}$.*

Proof By the definition of the Legendre symbol, if n is a prime then $(\beta|n) = 1$ whenever β is a square, and $(\beta|n) = -1$ whenever β is a nonsquare. By Euler's criterion, if n is a prime, then $\beta^{\frac{n-1}{2}} = 1 \pmod{n}$ if and only if β is a square. Moreover, because $\beta^{n-1} = 1 \pmod{n}$ for all β, we conclude that if β is a nonsquare, then $\beta^{\frac{n-1}{2}} = -1 \pmod{n}$. □

The converse of the theorem is not true. Therefore the theorem can be used only to show that an integer is composite. It cannot be used to show with certainty that an integer is a prime. By repeating multiple trials, it can be used only to assert a probability that the given integer is a prime.

Let n be an integer to be tested for primality using the Solovay–Strassen algorithm. If the last digit of the decimal representation of n is 0, 2, 4, 5, 6, or 8, then n is composite. Such an integer need not be further tested. Otherwise, randomly choose an integer β, $1 \leq \beta < n$, and compute the Jacobi symbol $(\beta|n)$. If $(\beta|n) = \beta^{(n-1)/2} \pmod{n}$, then answer "don't know," otherwise, answer "n is composite." Theorem 2.9.1 asserts that if n is prime, the Solovay–Strassen algorithm always gives the answer "don't know." If n is composite, we may get the answer "don't know" and we may get the answer "n is composite." This is because $(\beta|n) \neq \pmod{n}$ with probability at least 1/2 if n is composite.

If the answer is "don't know," then randomly choose a new integer β independently of the first integer chosen and repeat.[4] The test is repeated many times, stopping if the answer "n is composite" is obtained, or if the answer "don't know" is observed m times consecutively for some chosen m.

The Solovay–Strassen algorithm requires an efficient method to compute both the Jacobi symbol and a large power of β. The law of quadratic reciprocity, given in Theorem 2.5.3, combined with the properties of the Jacobi symbol provides the efficient method of computing the Jacobi symbol. Fast exponentiation and modular reduction provide the efficient method of computing the power of β.

For an example of the Solovay–Strassen algorithm, suppose that we want to know whether 221 is a prime. Choose any integer smaller than 221. The integer 47 will do. Then compute (47|221), as by using quadratic reciprocity, and compute $47^{110} \pmod{221}$, as by using the powers-of-two product method of fast exponentiation. Thus

$$\begin{aligned}(47|221) &= (221|47) = (33|47)\\ &= (47|33) = (14|33)\\ &= (2|33)(7|33) = (33|7)\\ &= (5|7) = (7|5)\\ &= (2|5) = -1.\end{aligned}$$

[4] The usually careful observer will notice that most computer programs for "randomly" selecting an integer a may actually be deterministic methods. The notion of independent trials here is tainted with some degree of pragmatism.

On the other hand, using the binary expansion of the decimal integer 110, we can write

$$\begin{aligned} 47^{110} &\equiv 47^{64}47^{32}47^{8}47^{4}47^{2} \pmod{221} \\ &= -1 \pmod{221}. \end{aligned}$$

Therefore

$$(47|221) = 47^{110},$$

so the Solovay–Strassen test is inconclusive. The integer 221 may be a prime or may be composite. The test must be repeated.

Choose another integer smaller than 221. The integer 2 will do. Now, we compute $(2|221) = -1$ and compute $2^{110} = 30 \pmod{221}$. Because $-1 \neq 30$, we conclude that 221 is not a prime. Notice that the procedure, although it shows that 221 is composite, gives no information about the factors. The computational work is linear in the number of digits in the integer n. Therefore the computational work is logarithmic in n itself.

To reduce the probability of falsely declaring n to be a prime, the Solovay–Strassen algorithm is run m times, each time with a different randomly chosen value of the integer β. Let A be the event "a randomly selected odd integer n of specified size is composite." Let B be the event "the algorithm answers 'don't know' m times in succession." By Bayes' formula, we have

$$\Pr(A|B) = \frac{\Pr(B|A)\Pr(A)}{\Pr(B|A)\Pr(A) + \Pr(B|\bar{A})\Pr(\bar{A})}.$$

By the prime-number theorem, because n is odd, $1 - \Pr(A) \approx \frac{2}{\log_e n}$, where the factor of two accounts for the fact that only odd integers are tested. Further,

$$\Pr(B|A) \approx 2^{-m}.$$

Therefore, after multiplying numerator and denominator by $\log_e n$ and 2^m, the expression becomes

$$\Pr(A|B) \approx = \frac{\log_e n - 2}{\log_e n - 2 + 2^{m+1}},$$

which gives the probability that the chosen integer is composite but undetected as such.

For example, if n is a 256-bit number, then the value n is on the order of 2^{256}. Choose $m = 50$. Then $\Pr(A|B)$ is equal to about 10^{-13}, which is a very small probability of failure to detect that n is composite.

2.10 The Miller–Rabin algorithm

The Miller–Rabin primality-testing algorithm is a yes-biased probabilistic algorithm for the question "Is n composite?" If the algorithm answers "n is composite," then n is indeed composite. If the algorithm answers "don't know", then n is either prime or

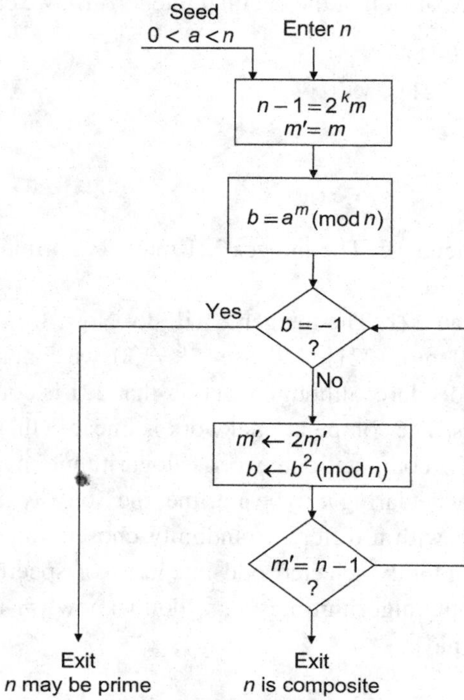

Figure 2.3 The Miller–Rabin algorithm

composite. It is a probabilistic algorithm because it requires a randomly chosen integer and so can be repeated many times. The Miller–Rabin algorithm is considered to be faster than the Solovay–Strassen algorithm, though more difficult to understand. It is summarized in Figure 2.3.

The Miller–Rabin algorithm is described as follows. We need consider only odd n, so we may write $n - 1 = 2^s t$ for some s, where t is odd. Because $n - 1$ is even, s is nonzero. Both s and t are easily computed by repeatedly dividing $n - 1$ by two until an odd integer is obtained.

Randomly choose a positive integer a smaller than n and coprime to n. The latter condition is easily checked and only fails if $\mathrm{GCD}(a, n)$ is a nontrivial factor of n, which itself means that n is composite. Set $m = t$. Then compute $b = a^t \pmod{n}$ and enter the following loop. If $m = 2^s t = n - 1$, stop and answer "n is composite." If $b \equiv -1 \pmod{n}$, answer "don't know" and exit. Otherwise, replace b by $b^2 \pmod{n}$, replace m by $2m$, and return to the start of the loop.

Proposition 2.10.1 *If n is prime, the Miller–Rabin algorithm will not answer that n is composite.*

Proof We must show that the test never falsely declares that a prime is a composite. Suppose that n is actually a prime, but we get the answer "n is composite." We will show that this leads to a contradiction, so it could not occur. We would get "n is composite" only if $a^{2^i t} \not\equiv -1 \pmod{n}$ for $0 \le i \le s-1$.

Suppose that n is a prime. Then $a^{n-1} \equiv 1 \pmod{n}$. Thus n divides $(a^{n-1} - 1)$. But

$$\begin{aligned} a^{n-1} - 1 &= a^{2^s t} - 1 \\ &= (a^{2^{s-1}t} - 1)(a^{2^{s-1}t} + 1), \end{aligned}$$

where the second factor is not divisible by n because the algorithm did not exit at the next to last step. Hence n divides $(a^{2^{s-1}t} - 1)$. Next, write

$$a^{2^{s-1}t} - 1 = (a^{2^{s-2}t} - 1)(a^{2^{s-2}t} + 1)$$

to conclude that n divides $(a^{2^{s-2}t} - 1)$. Continuing in this way, we ultimately find that n divides $(a^t - 1)$, so that $a^t = 1 \pmod{n}$. Because t divides $n-1$, this means that $a^{n-1} = 1 \pmod{n}$, contrary to the above reasoning. This contradicts the assumption that n is a prime, and n must indeed be composite. □

To illustrate the algorithm, the integer $41 = 2^3 \cdot 5 + 1$ is prime and the integer $81 = 2^4 \cdot 5 + 1$ is composite. In the first instance, choosing $a = 2$, the algorithm computes $2^5 \pmod{41} = 32$, then $32^2 \pmod{41} = -1$, so the algorithm returns "41 may be a prime." Another attempt may be made with another value of a. In the second instance, with $a = 2$, the algorithm computes $2^5 \pmod{81} = 32$, then $32^2 \pmod{81} = 52$, then $52^2 \pmod{81} = 31$, and then $31^2 \pmod{81} = 70$. The value -1 was never obtained so the algorithm returns "81 is composite."

For a slightly larger example, suppose that we want to determine whether $n = 221$ is a prime. First, write $n - 1 = 220$ as $2^2 \cdot 55$, so $s = 2$ and $t = 55$. Select any integer a smaller than 221. The integer 174 will do. Then

$$\begin{aligned} a^{2^0 t} &= 174^{55} && \pmod{221} \\ &= 47 \not= -1 && \pmod{221} \\ a^{2^1 t} &= 174^{110} && \pmod{221} \\ &= 220 = -1 && \pmod{221}. \end{aligned}$$

Because $220 \equiv -1 \bmod 221$, the test is inconclusive. Choose another integer a, this time choosing $a = 137$. Compute

$$\begin{aligned} a^{2^0 t} &= 137^{55} && \pmod{221} \\ &= 188 \not= -1 && \pmod{221} \\ a^{2^1 t} &= 137^{110} && \pmod{221} \\ &= 205 \not= -1, && \pmod{221} \\ a^{2^2 t} &= 1 && \pmod{221}. \end{aligned}$$

Hence we conclude that 221 is composite. Note that this procedure (apparently) tells us nothing about the factors of 221 (which are 13 and 17). Remarkably, although the algorithm finds that 221 is composite, it is of no help in factoring 221.

The modular exponentiation is performed by repeated squaring, so the algorithm has a running time proportional to $k \log^3 n$, where k is the number of different values of a that are tested. Thus this is an efficient polynomial-time algorithm. Asymptotically, one can accelerate the computation by using an algorithm for fast multiplication of integers. This can push the running time down to an order proportional to $k \log^2 n \log \log n \log \log \log n$, which is nearly the same as $k \log^2 n$. This last enhancement, however, is only of theoretical interest because it is meaningful only for impractically massive integers.

The more integers a we test, the better the probability of success of the Miller–Rabin test. It is known that only one-fourth of the integers that might be chosen for integer a will cause the Miller–Rabin algorithm to fail to recognize a composite n as such, but the proof of this will not be given. Therefore if the Miller–Rabin test is repeated k times, with independently selected values of the integer a, the probability of failing to recognize a composite integer as such is 4^{-k}. Thus with $n = 25$, the probability of failure is not more than $2^{-50} \sim 10^{-15}$. The Miller–Rabin test is strictly stronger than the Solovay–Strassen primality test in the sense that for every composite n, any a that fails the Miller–Rabin test will also fail the Solovay–Strassen test. If n is composite, then the Miller–Rabin primality test declares that n is probably prime with a probability of error of at most 4^{-k}. On the other hand, the Solovay–Strassen primality test declares that n is probably prime with a probability of error at most 2^{-k}. In practice, the probability that the Miller–Rabin test fails to detect a composite number as such is regarded as significantly smaller than 4^{-k} regardless of the number being tested.

Some other algorithms, not discussed here, of testing integers of unknown origin for primality should be used with caution because an adversary might try to insert an integer that is likely to fail the test. In such cases, the error bound of 4^{-k} of the Miller–Rabin algorithm still can be relied upon.

The proof of Proposition 2.10.1 is clear and concise, but the direct proof does hide an instructive detail that would help to make the procedure less mysterious. This detail is the fact that the Miller–Rabin algorithm, in effect, is testing to see if there are more than two roots of unity in $\mathbf{Z}_n$. If there are more than two roots of unity, then n must be composite because, in any field, the second-degree polynomial $x^2 - 1$ can have at most two zeros. In contrast, if n is not a prime, then the polynomial $x^2 - 1$ can have more than two zeros in $\mathbf{Z}_n$. In particular, an integer can have four square roots modulo a biprime, which is the statement of the following proposition.

Theorem 2.10.2 *If n is the product of two distinct odd primes, then $x^2 \equiv 1 \pmod{n}$ has four solutions, expressed as ± 1 and $\pm k$ for some integer k, provided* $\mathrm{GCD}(x^2, n) = 1$.

Proof Let $n = pq$ where p and q are odd primes. Then $x^2 \equiv 1 \pmod{n}$ is equivalent to

$$
\begin{aligned}
x^2 &= 1 + an \\
&= 1 + apq
\end{aligned}
$$

for some integer a. This leads, in turn, to the statements $x^2 \equiv 1 \pmod{p}$ and $x^2 \equiv 1 \pmod{q}$. Because p and q are prime, and $x^2 - 1$ has only two zeros in any field, including the fields $\boldsymbol{F}_p$ and $\boldsymbol{F}_q$, this implies that

$$x \equiv \pm 1 \pmod{p}, \quad \text{and } x \equiv \pm 1 \pmod{q}.$$

Thus there are four pairs of congruences. By the chinese remainder theorem, each pair corresponds to a different solution of the original equation. For the first pair, with $x \equiv 1 \pmod{p}$, and $x \equiv 1 \pmod{q}$, we know that the solution is $x = 1$. For the second pair, with $x \equiv -1 \pmod{p}$, and $x \equiv -1 \pmod{q}$, the solution is $x = -1$. For the third pair, with $x \equiv 1 \pmod{p}$ and $x \equiv -1 \pmod{q}$, we know by the chinese remainder theorem that a unique solution exists, which is easily computed. Call the solution k. Finally, because $-k \equiv -1 \pmod{p}$ and $-k \equiv 1 \pmod{q}$, we conclude that $-k$ solves the fourth pair of congruences. In summary, the four solutions are ± 1, $\pm k$ for some integer k. □

It follows easily from the theorem that if a has a square root $\sqrt{a}$ in $\boldsymbol{Z}_{pq}$, then $-\sqrt{a}$, $k\sqrt{a}$, and $-k\sqrt{a}$ are also square roots of a in $\boldsymbol{Z}_{pq}$. For example, in $\boldsymbol{Z}_{35}$, the square roots of 16 are 4, 11, 24, and 31.

With this theorem in the background, we can outline an alternative proof of Proposition 2.10.1. Consider the sequence of values of $b = a^{2^i t}$ tested in the Miller–Rabin algorithm. Because b is squared in each iteration to form the next value of b, we are testing the values of $a^t, a^{2t}, \ldots, a^{2^{s-1}t}$. Suppose that n is a prime, but the algorithm answers "n is composite." Then $a^{2^i t} \not\equiv -1 \pmod{n}$ for $i = 0, \ldots, s-1$. But n is a prime so, because $n - 1 = 2^s t$, Fermat's little theorem says that $a^{2^s t} = 1 \pmod{n}$. Then $a^{2^{s-1}t}$ is a square root of one in the field $\boldsymbol{F}_n$. There are only two square roots of one in any field, namely, $+1$ and -1, and we know that $a^{2^{s-1}t} \neq -1$. Therefore $a^{2^{s-1}t} = 1$, so $a^{2^{s-2}t}$ is a square root of one, and thus it is either $+1$ or -1. Again, as before, it cannot be -1. Continuing in this way, we conclude that $a^t \pmod{n} = 1$, which means that the algorithm must have halted immediately at the start. Hence the algorithm could not have answered that n is composite. The contradiction shows that the method cannot mistakenly identify a composite integer as a prime.

2.11 Factoring of integers

The factoring of large composite integers has been a task that has challenged number theory for many centuries with only limited success. Some methods of cryptography depend on the fact that known methods of factoring integers are far too weak, and stronger methods of factoring are not expected. This later statement, of course, does not rest on a formal proof. It is, however, an educated and widely held opinion, resting on the apparent lack of breakthroughs in the past.

The task of factoring a positive integer n is to find two positive integers a and b, both larger than one, such that n can be written as $n = ab$. More specifically, the task of factoring n is to find all prime factors of n. To this end, the following theorem is important. Although the theorem appears to be intuitive, it does require a proof.

Theorem 2.11.1 (Unique factorization theorem) *Every integer n larger than one can be written uniquely (up to the order of terms) as*

$$n = p_1^{e_1} p_2^{e_2} \cdots p_k^{e_k}$$

where the p_i are distinct primes and the e_i are positive integers.

Proof Suppose that the theorem is false, and let n be the smallest positive integer for which there are two different factorizations:

$$p_1^{e_1} p_2^{e_2} \cdots p_k^{e_k} = q_1^{e'_1} q_2^{e'_2} \cdots q_{k'}^{e'_{k'}}$$

where the p_i and q_i are all primes, and the terms on the two sides are not the same. All the p_i must be different from all the q_i because otherwise, the common terms could be canceled to give a smaller integer that is factored in two distinct ways. Without loss of generality, let p_1 and q_1 be the smallest primes on the left and right side, respectively, and suppose that q_1 is smaller that p_1. Then $p_1 = aq_1 + b$ for some nonnegative integers a and b. Then we can express one copy of p_1 in this way so as to obtain

$$\begin{aligned} bp_1^{e_1-1} p_2^{e_2} \cdots p_k^{e_k} k &= q_1^{e'_1} q_2^{e'_2} \cdots q_{k'}^{e'_{k'}} - aq_1 p_1^{e_1-1} p_2^{e_2} \cdots p_k^{e_k} \\ &= q_1 \left(q_1^{e'_1-1} q_2^{e'_2} \cdots q_{k'}^{e'_{k'}} - ap_1^{e_1-1} p_2^{e_2} \cdots p_k^{e_k} \right). \end{aligned}$$

The two sides are equal and smaller than n. Because the prime q_1 appears on the right side but not on the left, the two sides (after the right side is refactored) are distinct prime factorizations of an integer that is smaller than n. But by assumption, n is the smallest such integer. The contradiction shows that there is no such n, and the proof is complete. □

An elementary method of finding the factors of n, as described by the unique factorization theorem, is to divide n by each prime smaller than $\sqrt{n}$. Each successful division, meaning each division that occurs without remainder, identifies a factor of n. This method can be used for integers with about 12 digits or so, or perhaps a little more. It is not useful for larger integers because the computational burden is unacceptable.

Factoring of large integers is hard, and cryptography based on number theory depends on the fact that, despite intense efforts to devise tractable factoring algorithms, none of the known algorithms for integer factoring are tractable for very large composite integers. In this chapter, we will study only the *Pollard algorithm* for factorization. In

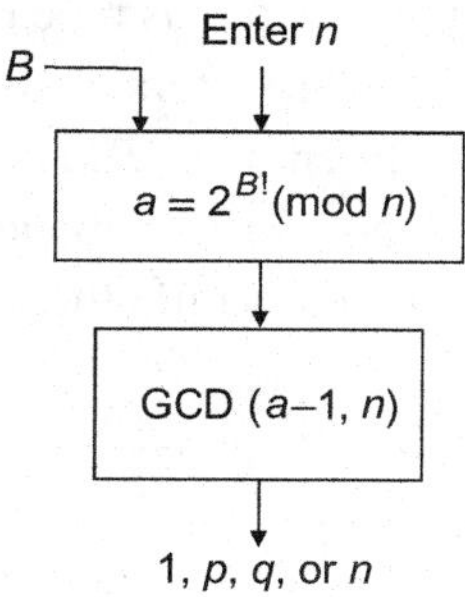

Figure 2.4 Pollard factoring algorithm

Chapter 3, we will return to the topic of integer factoring, describing stronger methods, primarily the *quadratic sieve* and the *number-field sieve*.

2.12 The Pollard algorithm for factoring

A *biprime* n is a composite integer of the form pq where p and q are each a prime. Pollard's $p - 1$ factoring algorithm, in special situations, can be used to find the factors p and q of a biprime or, indeed, of any composite integer, provided some very restrictive conditions are satisfied. We will consider factoring only biprimes. The performance of the Pollard algorithm depends on the nature of the factors of $p - 1$ or $q - 1$, which are very different from the factors of p and q. The factoring method works only if every prime power dividing $p - 1$ or $q - 1$ is small. The algorithm does not work – meaning that it is computationally intractable – if $p - 1$ or $q - 1$ each has some prime or prime-power factor that is large. The algorithm is not suitable for general factoring of large integers because virtually every sufficiently large biprime $n = pq$ has at least one large prime factor in each of $p - 1$ and $q - 1$. Large integers of the form pq that have only small prime factors in either $p - 1$ and $q - 1$ apparently are rare.

Because we have not given a formal definition of a small integer or a large integer in the context of this topic, nor could we know the factors of $p - 1$ and $q - 1$ before the factorization takes place, we cannot state a prior test on p and q under which the Pollard algorithm works. We only can say heuristically that the algorithm works in appropriate situations, but such a situation is rare and cannot be predicted by a superficial inspection of n before attempting the algorithm.

The Pollard algorithm for factoring is shown in Figure 2.4. The algorithm produces, as its output, either $1, p, q$, or n. It succeeds if it produces p or q as its output. It fails if it provides 1 or n as its output. A failure only means that the attempt at factoring has failed. It does not mean that n is a prime.

The Pollard algorithm works by picking a large integer B in hopes that $p - 1$ and $q - 1$ both divide $B!$. This will be so if the prime-power factors of $p - 1$ and $q - 1$ are sufficiently small. If the algorithm fails, B is increased and the attempt is repeated. The algorithm will succeed for the new value of B if all prime-power factors of $p-1$ and $q-1$ divide $B!$ for the new B. If not, B is increased again. Eventually $B!$ is too large for the algorithm to continue, and the computation will fail and must be abandoned.

Let n be a composite integer with a prime factor p. By Fermat's little theorem, any positive integer a smaller than p satisfies

$$a^{p-1} = 1 \pmod{p}.$$

Hence

$$a^{k(p-1)} = 1 \pmod{p}$$

for any positive integer k.

The algorithm starts with an odd integer n to be factored, and chooses an arbitrary integer B. If the value of B is chosen too small, the algorithm is likely to fail. If the value of B is chosen too large, the amount of computational work will be unacceptable. We first compute $a = 2^{B!} \pmod{n}$. One way to do this is by computing $2^2, 2^{3!}, 2^{4!}, 2^{5!}, \ldots,$ all modulo n, by the expression $2^{\ell!} = (2^{(\ell-1)!})^{\ell}$, stopping at $2^{B!}$. Then by computing $\mathrm{GCD}(a-1, n)$, there is a chance that the algorithm has found a nontrivial factor of n.

The reasoning behind the algorithm is as follows. First note that if n has a prime factor p for which every prime power dividing $p-1$ is not larger than $B!$, then $p-1$ divides $B!$. This means that if $p-1 = p_1^{e_1} \cdots p_k^{e_k}$ and $p_i^{e_i} \leq B$ for every i, then $p_i^{e_i}$ must divide $B!$ for every i, and so $p-1$ divides $B!$. In this case, $B! = k(p-1)$ for some k.

Now recall that a is defined as

$$a = 2^{B!} \pmod{n},$$

which means that $a = 2^{B!} + Qpq$ for some Q. Therefore it is also true that

$$\begin{aligned} a &= 2^{B!} \pmod{p} \\ &= (2^{p-1})^k \pmod{p}, \end{aligned}$$

provided $B! = k(p-1)$. But p is a prime, so $2^{p-1} = 1 \pmod{p}$. Therefore

$$a = 1 \pmod{p}.$$

This means that p divides $a-1$. Because p divides both n and $a-1$, we conclude that it divides $\mathrm{GCD}(a-1, n)$. Therefore $\mathrm{GCD}(a-1, n)$ is either equal to n or is a factor of n. If it is equal to n, the algorithm has failed to find a factor of n because B is too small.

It is easy to counter the Pollard factoring algorithm by choosing primes p and q in such a way that the product $n = pq$ is not vulnerable to the Pollard factoring algorithm. An easy way to choose such a p is to first choose a smaller prime p'.

Then set $p = 2p' + 1$ and test p for primality. If p is a prime, it avoids the condition needed by the Pollard algorithm because $p - 1$ has the large prime p' as a factor. If p is not a prime, choose a new p' and repeat. Then choose q by the same process. This leads to the biprime $n = pq$ for which $p - 1$ and $q - 1$ each have a large prime factor.

2.13 Square roots in a prime field

The computation of the two square roots of an integer modulo p, where p is a prime, is a problem that is difficult enough, and important enough, to have attracted a serious search to find special methods for its solution. It is important for us because computing square roots in $\boldsymbol{F}_p$ is a subtask in the quadratic-sieve algorithm for factoring composite integers, which we will study later, as well as in other related algorithms. We will describe three methods of finding a square root modulo p, provided a square root modulo p exists.

Let a be a quadratic residue modulo p. This means that a is a square in the field $\boldsymbol{F}_p$. The task of computing a square root of a is to find $\pm r$ in $\boldsymbol{F}_p$ such that $(\pm r)^2 = a \pmod p$. It is enough to find either of these two square roots, because if r is a square root of a in $\boldsymbol{F}_p$, then $p - r$ is the other square root of a in $\boldsymbol{F}_p$.

To test whether a square root of x modulo a prime p exists, it suffices to look at the Legendre symbol $(x\,|\,p)$, which is easily computed. If the Legendre symbol is not equal to one, then x has no square root, and a square-root algorithm is not needed. Whenever the Legendre symbol is equal to one, the square root exists. Then if p is an odd prime of the form $4k + 3$, again a square-root algorithm is trivial, because in such a case it is simple to use the identity $\sqrt{x} = x^{(p+1)/4}$. This identity is a consequence of the Euler's criterion given in Theorem 2.4.1, which says that β is a square if, and only if, $\beta^{(p+1)/2} = \beta \pmod p$. Therefore $\sqrt{\beta} = \beta^{(p+1)/4} = \beta^{k+1}$.

We conclude that an algorithm for computing the square root of x in $\boldsymbol{F}_p$ can presume that $(x\,|\,p) = 1$ and that p has the form $4k + 1$. Otherwise, an algorithm is not needed.

The first two methods that we will describe for finding the square roots are probabilistic methods and have a complexity that grows as a polynomial in p for any given square a. The third method is a deterministic method and has a complexity that grows as a polynomial in p for a fixed a, and grows exponentially in a for a fixed p.

An algorithm for computing square roots modulo a prime p cannot be used to compute square roots modulo a biprime. Were such an algorithm for computing square roots modulo a biprime known, it would be easy to convert that algorithm to an algorithm for integer factorization. This conversion is based on the statement given in Theorem 2.10.2 that there are four square roots modulo a biprime.

Proposition 2.13.1 *The task of solving for the four roots of $x^2 = a$ modulo the biprime pq for any value of the integer a is at least as hard as the task of factoring pq.*

Proof It is necessary to show that if $x^2 = a \pmod{pq}$ can be solved for its four roots, then the equation $n = pq$ can be factored. Choose any positive integer x and, using the hypothetical square-root algorithm, find the four roots of the equation $x^2 = a \pmod{n}$, where n is the biprime pq. Let the four roots be $\pm x$, $\pm y$, where y is a positive integer. Then $x^2 - y^2 = 0 \pmod{pq}$. But, as integers, x^2 is not equal to y^2 and both are smaller than n^2. Thus

$$(x - y)(x + y) = Qpq \neq 0.$$

Because x and y are each larger than zero and smaller than n, either GCD$[(x + y), n]$ or GCD$[|x - y|, n]$ must be either p or q. □

If the hypothetical algorithm for computing square roots is replaced by an algorithm that only computes one square root, then the theorem still applies but in a probabilistic sense. Again randomly choose x and compute a square root of $x^2 \pmod{pq}$, hoping to find a square root not equal to $\pm x$. The algorithm will fail if it returns either x or $-x$. This happens with probability one-half. When it fails, randomly pick a new x and repeat. The probability of failure m times in succession is 2^{-m}. Otherwise, when the algorithm does not return $\pm x$, it returns $\pm y$. But when it returns a square root $\pm y$ that is not equal to $\pm x$, we can write, as before,

$$x^2 - y^2 = 0 \quad \pmod{pq}.$$

Thus

$$(x - y)(x + y) = Qpq \neq 0.$$

Because x and y are each larger than zero and smaller than n, one of these greatest common divisors of $|x - y|$ and $(x + y)$ with n must be either p or q.

Therefore any algorithm for computing a square root modulo a biprime can be converted to an algorithm for factoring a biprime. If we accept the premise that factoring biprimes is intractable, we must also accept the premise that computing a square root modulo a biprime is also intractable.

The Tonelli–Shanks algorithm

An algorithm for computing square roots in $\boldsymbol{F}_p$ is needed only for the case in which p is an odd prime of the form $p = 4k + 1$. Suppose that p is an odd prime of this form. The *Tonelli–Shanks algorithm* solves an equation of the form $x^2 = n \pmod{p}$, where

n is a quadratic residue modulo p (that is, n is a square of the field $\boldsymbol{F}_p$). It should first be determined whether the equation has a solution for the specified n, as by observing whether the Jacobi symbol $(n|p)$ is equal to one.

The Tonelli–Shanks algorithm is a probabilistic algorithm because it requires the use of a nonsquare in the prime field $\boldsymbol{F}_p$. Surprisingly, no deterministic algorithm is known for choosing a nonsquare in $\boldsymbol{F}_p$, even though half of the nonzero elements of $\boldsymbol{F}_p$ are nonsquares. The best way known to find a nonsquare is to choose an arbitrary nonzero integer and test it, as by computing the Jacobi symbol. If it is not a nonsquare, choose another integer and repeat the test. Each trial will be a success with probability one-half, so a nonsquare will usually be found rather quickly. Occasionally, however, the process can go on for a long time before a nonsquare is found. The probability of failing to find a nonsquare after k independent attempts is not larger than 2^{-k}.

Any integer can be readily factored as $2^s t$ with t an odd integer. Simply repeatedly divide by two for as long as possible. In particular, $p-1$ can be so factored to obtain 2^s and t. (If $p = 4k+3$, s will equal one. Otherwise, s will be larger than one.) Because $\boldsymbol{F}_p^*$ under multiplication is a cyclic group of order $p-1 = 2^s t$, any subgroup of $\boldsymbol{F}_p^*$ under multiplication is a cyclic subgroup whose order divides $2^s t$ as is shown in Section 9.8. In particular there is a unique cyclic subgroup of $\boldsymbol{F}_p^*$ of order 2^s. We need a generator π of this cyclic subgroup, which we obtain as follows. Choose any nonsquare b of $\boldsymbol{F}_p^*$, and let $\pi = b^t$. Then $\pi^{2^s} = b^{p-1} = 1$, whereas by Theorem 2.4.1 (Euler's criterion), $\pi^{2^{s-1}} = (\sqrt{b})^{p-1}$ can equal one only if b is a square in $\boldsymbol{F}_p$. Because b was chosen as a nonsquare, we conclude that $\pi = b^t$ has order 2^s. Hence, to find a generator π of the unique cyclic subgroup of order 2^s in $\boldsymbol{F}_p^*$, it suffices to find any nonsquare of $\boldsymbol{F}_p$ and to raise it to the power t. Then for every a, because a^t has order dividing 2^s, the power a^t is in the orbit of π, an element of order 2^s.

To compute the square root of a in $\boldsymbol{F}_p$, with $p-1 = 2^s t$, iteratively compute the sequence of pairs (π_i, r_i) as follows. Let $\pi_1 = \pi = b^t$, where b is any nonsquare of $\boldsymbol{F}_p$, and let $r_1 = a^{(t+1)/2}$ noting that $(t+1)/2$ is an integer because t is odd. Then $r_1^2 = a \cdot a^t$. But a^t is in the orbit of π_1, which means that $(r_1)^2 a^{-1}$ is also in the orbit of π_1.

Now set $r_2 = r_1$ if $(r_1^2 a^{-1})^{2^{s-2}} = 1$, and set $r_2 = r_1\pi_1$ otherwise. Then set $\pi_2 = \pi_1^2$. This means that $(r_2)^2 a^{-1}$ is in the orbit of π_2, where π_2 has order 2^{s-1}. Continued iteration of this procedure gives pairs (π_i, r_i), where $(r_i)^2 a^{-1}$ is in the orbit of π_i for each i, and with π_i an element of order 2^{s-i+1}. The process stops when $i = s$, $r_s^2 = a$, and $(r_s)^2 a^{-1} = 1$. Thus $r_s = \sqrt{a}$.

A flow chart of the Tonelli–Shanks algorithm is depicted in Figure 2.5.

For an example of the Tonelli–Shanks algorithm, let $p = 1009$ and $a = 7$. The Tonelli–Shanks algorithm first computes the factorization $p-1 = 2^s t$, which is

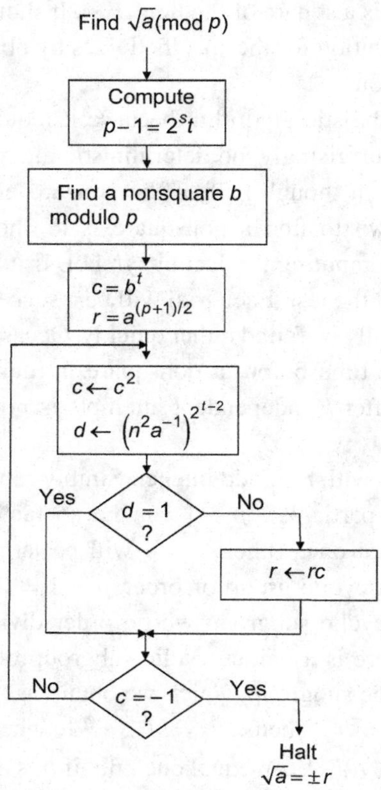

Figure 2.5 The Tonelli–Shanks algorithm

$1008 = 2^4 \cdot 63$. It next finds a nonsquare. The integers $1, 2, 3, \ldots, 10$ are tested and found to be squares in $\boldsymbol{F}_{1009}$, but $b = 11$ is not a square in $\boldsymbol{F}_{1009}$. Therefore choose $b = 11$. Let $r_1 = a^{(t-1)/2} = a^{32} = 993$, and $\pi_1 = b^t = 11^{63} = 179$. Then

$$
\begin{aligned}
((r_1)^2 a^{-1})^4 &= 1 r_2 = r_1 = 993, & \pi_2 &= \pi_1^2 = 762. \\
((r_2)^2 a^{-1})^2 &= -1 r_3 = r_2\pi_2 = 925, & \pi_3 &= \pi_2^2 = 469. \\
((r_3)^2 a^{-1}) &= -1 r_4 = r_3\pi_3 = 964, & (\pi_4 &= \pi_3^2 = -1). \\
(r_4)^2 a^{-1} &= 1.
\end{aligned}
$$

Therefore $\sqrt{7} = 964$ and its negative is $1009 - 964 = 45$. Thus the two values of $\sqrt{7}$ in $\boldsymbol{F}_{1009}$ are 45 and 964.

Berlekamp square-root algorithm

The *Berlekamp square-root algorithm* is an algorithm for computing square roots in the prime field $\boldsymbol{F}_p$, where p has the form $p = 4k + 1$. It can be required that the element

a is known to have a square root, which can be established by first computing the Jacobi symbol. The Berlekamp algorithm executes the euclidean algorithm for certain polynomials in a sequence of trial problems in hopes that a suitable solution for one of them will be obtained eventually, thereby resulting in a probabilistic algorithm for computing square roots.

To find the square root of a in $\boldsymbol{F}_p$, one can factor the polynomial $x^2 - a$ in the field $\boldsymbol{F}_p$, recalling that

$$x^2 - a = (x - \sqrt{a})(x + \sqrt{a})$$

whenever $\sqrt{a}$ exists in the given field. Thus $\pm\sqrt{a}$ are the two zeros of the polynomial $x^2 - a$. Moreover, Euler's criterion (Theorem 2.4.1) states that an element β is a square root modulo p if, and only if, β is a zero of the polynomial $x^{(p-1)/2} - 1$. We conclude that any factor of the form $x - \beta$ that divides both polynomials, $x^2 - a$ and $x^{(p-1)/2} - 1$ yields $\beta = \sqrt{a}$.

We might choose to use the euclidean algorithm for polynomials to compute $\text{GCD}(x^2 - a, x^{(p-1)/2} - 1)$. However, because both zeros of $x^2 - a$ are square roots, this computation will always result in $x^2 - a$, which is of no help. Instead, we must alter the polynomial $x^2 - a$ so that it has only one factor of the form $x \pm \sqrt{a}$. The hope is that if β is a square root of a, there may be some ℓ for which $\beta + \ell$ is a quadratic residue but $\beta - \ell$ is not a quadratic residue. Specifically, the algorithm requires that for some ℓ, only one of the two terms $\pm\sqrt{a} - \ell$ is a square of $\boldsymbol{F}_p$. Although it is not obvious, such an ℓ always does exist.

For any arbitrarily chosen integer ℓ, compute $\text{GCD}((x + \ell)^2 - a, x^{(p-1)/2} - 1)$ in hopes of obtaining a factor of degree one. Repeat this computation for other values of ℓ until the algorithm produces a factor of degree one. We may simply choose $\ell = 0, 1, \ldots,$ in order, for as long as necessary. The greatest common divisor is 1 if both $\sqrt{a} - \ell$, and $-\sqrt{a} - \ell$ are quadratic nonresidues, and the greatest common divisor is $(x + \ell)^2 - a$ if both $\sqrt{a} - \ell$, and $-\sqrt{a} - \ell$ are quadratic residues. As soon as precisely one of the pair, $\sqrt{a} - \ell$ and $-\sqrt{a} - \ell$, is a quadratic residue, the greatest common divisor yields a single linear factor $(x - s)$ with a zero at $x = s$ such that $(s + \ell)^2 = a$. Then $\sqrt{a} = \pm(s + \ell)$.

The algorithm is a probabilistic algorithm in the sense that the number of values of ℓ that must be tested is not specified. The sequence of trial values of ℓ might be chosen arbitrarily just as well as sequentially. Either way, if p is very large, it may be that the waiting time to find a suitable ℓ is very large and the algorithm will fail because the waiting time is unacceptable.

For an example of the Berlekamp square-root algorithm, let $p = 1009$ and $a = 7$. Compute $\text{GCD}[(x + \ell)^2 - 7, x^{504} - 1]$ for successive values of ℓ, looking for a nontrivial factor. After failures for $\ell = 0, 1, \ldots, 7$, setting $\ell = 8$ results in the computation $\text{GCD}[(x + 8)^2 - 7, x^{504} - 1] = x + 972$. Therefore $s = -972 = 37$ and

$r = s + \ell = 45$. We conclude that $45^2 = 7 \pmod{1009}$, which is easily checked. Therefore the two values of $\sqrt{7}$ in $\boldsymbol{F}_{1009}$ are 45 and 964.

Schoof square-root algorithm

The *Schoof square-root algorithm* is a mathematically sophisticated algorithm for computing square roots modulo p that is too complicated to be useful. Given the simplicity of other square-root algorithms and the difficulty of the Schoof square-root algorithm, its study is not warranted based on practical considerations. However, one should always be concerned that in the future new algorithms for some important problems may be found based on connections that are not yet recognized, thereby threatening the relevant methods of cryptography. Making this point is the primary reason for describing the Schoof square-root algorithm in this section, even though the necessary mathematical background is scattered throughout the latter parts of the book.

The *Schoof point-counting algorithm*, as is described in Chapter 10, is an algorithm that computes the number of points on an elliptic curve over a large finite field. That task is still the immediate purpose of the Schoof algorithm. Surprisingly, knowing the number of points on a suitable elliptic curve can be used to compute a square root of an integer modulo p. Therefore, the Schoof algorithm also has an application, in principle, for computing square roots in a prime field, a task that on its face has nothing to do with elliptic curves.

In its current form, this method of computing square roots based on point counting on elliptic curves appears not to be competitive with other methods of computing square roots. It is included here because it does illustrate how an algorithm can come from an unexpected or unnoticed connection. Because modern cryptography is largely based on the presumed, but unproven, nonexistence of good algorithms for certain computational problems, it is important to develop a sense of caution.

The development of the method of computing square roots based on the Schoof algorithm uses a number of advanced topics that are not discussed until later in the book, namely the topics of elliptic curves and number fields. Elliptic curves are defined and studied in Chapter 10, and number fields are defined and studied in Chapter 9. These topics are essential to the square-root algorithm described here. A complete development of the Schoof point-counting algorithm itself will appear within the discussion of elliptic curves in Chapter 10. In this section, we do not discuss the Schoof algorithm, but only discuss its use in computing square roots of integers modulo a prime p. For this purpose, the Schoof point-counting algorithm is a black box that gives the number of rational points on an elliptic curve over a finite field. Indeed, any good algorithm for counting the number of rational points could be used instead, were another good algorithm available.

We shall make use of the properties of elliptic curves, viewing the same elliptic curve both on the rational field $\boldsymbol{Q}$ and on the prime field $\boldsymbol{F}_p$. To apply this method, we use the same polynomial $p(x, y)$ with integer coefficients to form both an elliptic curve on $\mathcal{X}(\boldsymbol{Q})$ on the rational field $\boldsymbol{Q}$ and an elliptic curve $\mathcal{X}(\boldsymbol{F}_p)$ on the prime field $\boldsymbol{F}_p$. Indeed, we may choose to regard these two curves simply as two manifestations of the same abstracted curve.

To compute $\sqrt{d}$ modulo p, we require that the elliptic curve over the rational field $\boldsymbol{Q}$ has a *complex multiplication* by $\sqrt{-d}$. As described in Chapter 10, the term refers to a notion of multiplying the curve by a complex number, meaning that, in some sense, each point of the curve $\mathcal{X}(\boldsymbol{Q})$ is "multiplied" by that complex number, and the result is another point of an elliptic curve in an imaginary quadratic extension field of $\boldsymbol{Q}$, a field known as an imaginary quadratic number field.

The simplest example of complex multiplication is multiplication of the elliptic curve $\mathcal{X} : y^2 = x^3 + x$ by $i = \sqrt{-1}$, which consists of replacing each point P by the point iP defined by $(x, y) \mapsto (-x, iy)$. If (x, y) is a point of $\mathcal{X}(\boldsymbol{Q})$, then as is easily checked $(-x, iy)$ is a point of $\mathcal{X}(\boldsymbol{Q}(i))$. This is dubbed multiplication by $\sqrt{-1}$ because "multiplication" of the curve $\mathcal{X}(\boldsymbol{Q})$ by $\sqrt{-1}$ four times in this way maps the point $P = (x, y)$ onto itself, which can be regarded as multiplication by one.

Complex multiplication of elliptic curves is a concept that is explained more fully in Chapter 10. Some elliptic curves allow such a complex multiplication for a specified square-free integer d. Most curves do not allow complex multiplication for any value of d. Methods for finding an elliptic curve with complex multiplication for a given d are available, but still the curves are hard to find, the difficulty growing rapidly as d becomes large.

The Frobenius trace t can be found from the number of rational points on the elliptic curve $\mathcal{X}(\boldsymbol{F}_p)$ by $t = p + 1 - \#\mathcal{X}(\boldsymbol{F}_p)$, and it is known that $|t| \leq 2\sqrt{p}$. The Schoof square-root algorithm is based on a remarkable relationship between the number of rational points on the curve $\mathcal{X}(\boldsymbol{F}_p)$ and the order of complex multiplication on the curve $\mathcal{X}(\boldsymbol{Q})$, which relationship we now describe, but do not prove. Specifically, it is known that if d is a square-free positive integer in the prime field $\boldsymbol{F}_p$ and $\mathcal{X}(\boldsymbol{Q})$ has complex multiplication in $\boldsymbol{Q}(\sqrt{-d})$, then the polynomial $x^2 - tx + p$ has its two zeros, denoted α and $\overline{\alpha}$, in the imaginary quadratic number field $\boldsymbol{Q}(\sqrt{-d})$, written as $\alpha = u + \sqrt{-d}v$ and $\overline{\alpha} = u - \sqrt{-d}v$, where $2u$ and $2v$ are both integers. But, $(x - \alpha)(x - \overline{\alpha}) = x^2 - tx + p$, so we know that

$$\begin{aligned} p &= \alpha\overline{\alpha} \\ &= u^2 + dv^2 \end{aligned}$$

and

$$\begin{aligned} t &= \alpha + \overline{\alpha} \\ &= 2u. \end{aligned}$$

Combining these gives $4p - t^2 = 4dv^2$. Therefore $2v$ can be computed as $2v = \sqrt{(4p - t^2)/d}$. Although this expression also involves a square root, it is a square root in $\boldsymbol{Z}$, not in $\boldsymbol{F}_p$. Algorithms for computing square roots in $\boldsymbol{Z}$ are well known. Finally, because $u^2 + dv^2 = p$, we know that $(u/v)^2 = -d \pmod{p}$. We conclude that $\sqrt{-d} = \pm 2u/2v$ modulo p, which is straightforward to compute once $2u$ and $2v$ are known. Finally, $\sqrt{d} = \sqrt{-1}\sqrt{-d}$.

Some examples of suitable elliptic curves, selected from the table given in Section 10.12, are the following:

(1) The curve $\mathcal{X}$ defined by the polynomial $y^2 = x^3 + x$ has complex multiplication by $\sqrt{-1}$. The endomorphism of order four that corresponds to $i = \sqrt{-1}$ is given by $(x, y) \longmapsto (-x, iy)$.
(2) The curve $\mathcal{X}$ defined by the polynomial $y^2 = x^3 + 1$ has complex multiplication by $\sqrt{-3}$.
(3) The curve $\mathcal{X}$ defined by the polynomial $y^2 + 5xy = x^3 - x^2 + 7x$ has complex multiplication by $\sqrt{-7}$.

Such polynomials can be found for any square-free integer d, but the search becomes more difficult as d becomes larger. For this reason, we might regard this algorithm to be of limited usefulness. This indeed is the natural conclusion if one regards the problem of computing $\sqrt{d} \pmod{p}$ as a task with p fixed and d ranging over a large number of values. The algorithm may be more attractive if one views the computation $\sqrt{d} \pmod{p}$ as a collection of computations with d fixed and p ranging over a number of values. Thus for example, one can view the algorithm as a method to compute $\sqrt{-1} \pmod{p}$ or $\sqrt{2} \pmod{p}$ for many values of the prime p at little additional cost for each new value of p beyond the cost of computing the number of points on the curve in the field $\boldsymbol{F}_p$.

An example of computing a square root using this method is the following. Let $p = 1009$ and $a = 7$. To compute $\sqrt{7}$ in $\boldsymbol{F}_{1009}$, noting that $\sqrt{7} = \sqrt{-1}\sqrt{-7}$, we choose the two elliptic curves based on polynomials $y^2 = x^3 + x$ and $y^2 + 5xy = x^3 - x^2 + 7x$.

The curve $\mathcal{X} : y^2 = x^3 + x$ has 1040 rational points in the field $\boldsymbol{F}_p$. Thus for this curve, $t = -30$ and $4p = 30^2 + 56^2$. Therefore $\sqrt{-1} = 30/56 \pmod{p}$.

The curve $\mathcal{X} : y^2 + 5xy = x^3 - x^2 + 7x$ has 1008 rational points in the field $\boldsymbol{F}_p$. Thus for this curve, $t = 2$ and $4p = 2^2 + 7 \cdot 24^2$. Therefore $\sqrt{-7} = 2/24 \pmod{p}$.

Now it is trivial to compute that

$$\sqrt{7} = \sqrt{-1}\sqrt{-7} = (30/56) \cdot (2/24) = 5/112.$$

But $112 \cdot 9 = -1 \pmod{1009}$ so that $5/112 = 5 \cdot (-9)$. Then

$$\sqrt{7} = 5 \cdot (-9) = -45.$$

Therefore $\sqrt{7}$ in $\boldsymbol{F}_{1009}$ is equal to ± 45. Because $-45 = 964$ modulo 1009, we can conclude that the two values of $\sqrt{7}$ in $\boldsymbol{F}_{1009}$ are 45 and 964.

Problems for Chapter 2

2.1 **a** Show that if integer a divides integer b and integer b divides integer c, then integer a divides integer c.

b Show that if integer a divides both integer b and integer c, then integer a divides $bx + cy$ for all integers x and y.

2.2 **a** Find all square roots of 12 modulo 37. How many are there?

b Find all square roots of 11 modulo 35. How many are there?

2.3 Prove that $\mathrm{LCM}(a, b)\mathrm{GCD}(a, b) = ab$.

2.4 **a** Prove that if p and q are distinct primes, then $\phi(pq) = (p - 1)(q - 1)$ and $\phi(p^e) = p^{e-1}(p - 1)$.

b Prove that $\phi(p^e m) = p^{e-1}\phi(pm)$, where p does not divide m.

c Prove that $\phi(\Pi_{i=1}^{\ell} p_i^{e_i}) = \Pi_{i=1}^{\ell}(p_i - 1)p_i^{e_i-1}$.

d Prove that $\phi(n) = n \prod_{p_i|n} \left(1 - \frac{1}{p_i}\right)$ where the product is over all distinct primes that divide n.

e When is it true that $\phi(mn) = \phi(m)\phi(n)$?

2.5 Prove that if q is a power of a prime, then $\phi(q^m - 1)/m$ is an integer.

2.6 Prove that, in any group, the identity element is unique and the inverse of any element is unique.

2.7 Prove that if α is a generator of a cyclic group of order n, then α^i has order $n/\mathrm{GCD}(n, i)$.

2.8 Prove that there are $\phi(n)$ generators in a cyclic group of order n.

2.9 Prove that all primes other than 2 and 3 are of the form $6k \pm 1$. Devise a primality test using this fact.

2.10 The following primality test is proposed. Suppose that n is an odd integer larger than one. Choose a random integer a, $1 \leq a \leq n - 1$. If $a^{n-1} \equiv 1 \pmod{n}$, then answer "$n$ might be prime." Otherwise, answer "n is composite." What is the justification of this test? Use this test, if possible, to determine whether 561 is a prime. Does this show that the test may sometimes fail to give a useful output?

2.11 Let $\mathcal{S} = \{1, 2, \ldots, 100\}$, the first 100 integers. Prove that any permutation on $\mathcal{S}$ can be regarded as a partition of $\mathcal{S}$ into ordered subsets with the permutation within each ordered subset consisting of a cyclic shift.

2.12 **(Sieve of Eratosthenes)**

a Prove that if m has no prime divisor smaller than $\sqrt{m} + 1$, then m is a prime.

b Remove all multiples of all primes smaller than $n + 1$ from the set $\{n + 1, n + 2, \ldots, n^2\}$. Prove that all integers remaining in the set are primes.

c Sketch a flow diagram for a procedure that will compute the first 1000 primes using the sieve of Eratosthenes.

2.13 Fermat's little theorem states that if p is a prime, then every nonzero element of $\boldsymbol{Z}_p$ is a zero of the polynomial $x^{p-1} - 1$.

a Show that the factorization

$$(x^{p-1} - 1) = \left(x^{\frac{p-1}{2}} - 1\right)\left(x^{\frac{p-1}{2}} + 1\right)$$

implies that $\frac{p-1}{2}$ elements of $\boldsymbol{Z}_p$ are squares and $\frac{p-1}{2}$ elements $\boldsymbol{Z}_p$ are nonsquares.

b Show that the Legendre symbol satisfies $(\beta|p) = \beta^{(p-1)/2}$.

2.14 Prove that if a and b are in the same equivalence class of $\boldsymbol{Z}/\langle p\rangle$, that a is a quadratic residue modulo p if, and only if, b is a quadratic residue modulo p.

2.15 Let $p(x, y) = y^2 - x^3 - x - 6$ be a bivariate polynomial over the field $\boldsymbol{F}_{11}$. By determining whether $x^3 + x + 6$ is a quadratic residue for each value of x, determine the number of zeros of the polynomial $p(x, y)$. Find them. Repeat for the bivariate polynomial $p(x, y) = y^2 - x^5 - x - 6$.

2.16 Find all the squares and nonsquares in $\boldsymbol{F}_{13}$. Demonstrate that the cube of any nonsquare generates a cyclic group of order four. State and prove the generalization of this to $\boldsymbol{F}_p$.

2.17 Prove Fibonacci's identity, that the product of two integers, each expressible as the sum of two squares, can itself be expressed as the sum of two squares. Thus the set of sums of two squares of $\boldsymbol{Z}$ is closed under multiplication. Does this fact extend to $\boldsymbol{Q}$?

2.18 Let $n = 9676489$ be a biprime that is to be factored. Then $3000^2 < n < 3200^2$. Determine the number of m with $3000 < m < 3200$ such that all prime divisors of $m^2 - n$ are less than 20. For N as small as possible, find $3000 < m_1 < \cdots < m_N < 3200$ such that $(m_1^2 - n) \cdots (m_N^2 - n) = s^2$ with all prime divisors of s less than 20. Factor n.

2.19 Let $n = 9676489$ be a biprime that is to be factored. Let $f(x) = x^2 + 1 \pmod{n}$. Using the iteration $x_0 = 1$, $x_i = x_i^2 + 1 \pmod{n} = f(x_{i-1})$, and $x_{2j} = f(f(x_{2j-2}))$, find a collision such that $x_i = x_{2j} \pmod{p}$. Then using $\mathrm{GCD}(x_j - x_{2j}, n)$, factor n.

2.20 Identify the squares and the nonsquares in $\boldsymbol{F}_3$, $\boldsymbol{F}_5$, $\boldsymbol{F}_7$, $\boldsymbol{F}_{11}$, and $\boldsymbol{F}_{13}$. Verify quadratic reciprocity when the primes p and q are equal to 3, 5, 7, 11, and 13.

2.21 Show that 5 is a quadratic residue (modulo p) if, and only if, $p = 5k \pm 1$.

2.22 Prove that if n has k distinct prime factors, then $x^2 = 1$ has exactly 2^k solutions in $\boldsymbol{Z}_n$.

2.23 **(Stein's recursion)**

a Prove that if n_1 and n_2 are both odd positive integers with $n_2 < n_1$, then

$$\mathrm{GCD}(n_1, n_2) = \mathrm{GCD}\left(\frac{n_1 - n_2}{2}, n_2\right).$$

b Using this recursion, describe a procedure for computing the greatest common denominator.

2.24 For the two primes $p = 11$ and $q = 19$, construct a $\frac{1}{2}(p-1)$ by $\frac{1}{2}(q-1)$ array with entries $qx - py$. How does this array illustrate Theorem 2.5.1? How does this array relate to the proof of Theorem 2.5.3?

2.25 **(Fermat factoring algorithm)** To factor n, choose the largest integer s for which $s^2 - n$ is positive. Determine if $s^2 - n$ is a square. If it is a square, denote that square as t^2 and set $n = (s-t)(s+t)$. Otherwise, replace s by $s+1$ and repeat. Either prove that this procedure always halts and factors n, or give an example where it fails to do so. Factor 3811 by this method. Compare the complexity with trial factoring of integers.

2.26 Given a black box that computes $\sqrt{x}\,(\text{mod } pq)$ for any x, provided p and q are prime, devise a procedure for factoring pq.

2.27 Suppose that n is an odd composite integer of the form $4k+1$. Can n be the sum of two squares? Can a primality testing algorithm be constructed based on the Fermat two-squares theorem?

2.28 **(Wilson primality test)**

a Show that $x - \beta$ is a factor of $x^{p-1} - 1$ for every nonzero element β of the field $\boldsymbol{F}_p$.

b Prove the *Wilson primality testing algorithm*, which is based on Wilson's theorem, that $(p-1)! = -1(\text{mod } p)$ if, and only if, p is prime. Is this a practical test for primality?

2.29 A probabilistic algorithm for computing square roots in $\boldsymbol{F}_p$ where $p = 5 \ (\text{mod } 8)$ is developed as follows. Let a be a square and let $d = a^{(p-1)/4}$. Show that if $d = 1$, then $\sqrt{a} = \pm a^{(p+3)/8}\ (\text{mod } p)$, and if $d = p-1$, then $\sqrt{a} = \pm 2a(4a)^{(p-5)/8}\ (\text{mod } p)$. For other values of d, the method fails.

2.30 Prove that the chinese remainder theorem can be executed in a pairwise fashion. That is, if m_1, m_2, and m_3 are coprime, then one can nest the expression of the theorem using coprime factors m_2 and m_3 into that same expression using coprime factors m_1 and m_2m_3. Is there a computational advantage or disadvantage?

2.31 Show that every prime field $\boldsymbol{F}_p$, for p odd, contains an element ℓ for which $\gamma_1^2 - \gamma_2^2 \neq \ell$ for every distinct pair γ_1 and γ_2 of $\boldsymbol{F}_p$. How does this relate to the Berlekamp square-root algorithm?

2.32 Two basic conjectures of number theory that remain open are the following:

(Goldbach conjecture) Every even integer larger than 2 can be expressed as the sum of two primes.

(Twin-prime conjecture) There are an infinite number of primes p such that $p+2$ is also a prime.

Write a brief paragraph expressing your reactions and opinions regarding these two conjectures.

Notes for Chapter 2

Many of the topics of number theory that are used in cryptography are classical topics, often breathtaking, that were largely developed by the following mathematicians: Fermat (1601–65), Euler (1707–83), Lagrange (1736–1813), Legendre (1752–1833), Gauss (1777–1855), Jacobi (1804–51), Dirichlet (1805–59), Galois (1811–32), Riemann (1826–66), and Frobenius (1849–1917). Many textbooks on number theory are available, including the books by Niven, Zuckerman, and Montgomery (1991), and LeVeque (1996).

The law of quadratic reciprocity, which has been called the start of modern number theory, was stated both by Euler (1783) and by Legendre (1785), and apparently first proved by Gauss (1808). The celebrated theorem on the density of prime numbers is a deep theorem of number theory whose proof was developed over time and owes much to the work of Riemann (1859). The theorem on the irregularity of prime numbers is another deep theorem of number theory and is due to Dirichlet (1835).

The euclidean algorithm comes down to us from Euclid's *Elements* (*c.* 300 BC), though the algorithm probably predates Euclid. Much of number theory and abstract algebra is built upon the extended euclidean algorithm. The identity that we have referred to as the Bézout identity for integers was known prior to Bézout's development of the corresponding identity for polynomial rings and other rings.

Carmichael (1912) first noticed the existence of certain composite integers that always satisfy the Fermat congruence, a fact that was anticipated by Korselt (1899). The proof that there is an infinite number of Carmichael integers was given by Alford, Granville, and Pomerance (1994). The Fermat two-squares conjecture of 1640 was proved by Euler in 1747, with modern proofs coming much later.

The success of number-theoretic cryptography depends critically upon two fundamental observations of computational number theory. On the one hand, tractable (probabilistic) algorithms for finding large (100-digit) primes do exist, particularly the Solovay–Strassen (1977) algorithm and the Miller (1976)–Rabin (1979) algorithm. On the other hand, no tractable algorithm for finding the two factors of a large integer known to be the product of two large primes is (apparently) known, despite intense effort for several centuries to find such an algorithm. Known methods of factoring large integers by machine are countered simply by choosing larger integers.

The algorithms of Solovay–Strassen, Miller–Rabin, and Fermat are general primality-testing algorithms, suitable for embedded applications, that will decide primality with reasonable certainty and with no restrictions on the nature of the integers tested. Other algorithms such as the Goldwasser–Kilian primality-testing algorithm can test much larger integers, even those with more than 1000 digits, but generally with certain restrictive conditions on the integers tested. The Pocklington–Lehmer primality test is of a different character. When it can be applied, it answers with certainty that an integer is prime, but it has limitations on when it can be applied.

The subject of cryptography has repaid its debt to the subject of number theory, at least in part, by some new insights and some related theorems that it has given to number theory. The Pollard algorithm for factoring integers (1975) is a contribution of cryptography back to number theory. An improvement of Pollard's factoring method due to H. W. Lenstra, Jr. (1983) is known as the *elliptic curve method* of factoring integers. The Tonelli–Shanks algorithm was first published by Tonelli (1891) and refined by Shanks (1970). The Berlekamp (1967) factoring algorithm is a specialization of a more general algorithm for factoring polynomials over finite fields.

3 Cryptography based on the integer ring

All cryptographic systems require some form of key exchange. If there are only a few users engaged in a long-term relationship, then it may be possible to exchange keys over a secure private channel, such as by the use of a trusted courier. Then the security of the system is no better than the security of the private channel. If there are a great number of users in a network, however, and their relationships are unpredictable, unforeseen, and brief, then it is inappropriate to have a single shared key, and it is unreasonable to have an individual private key for each pair of users. Consequently, public-key cryptosystems are unavoidable. The keys must be exchanged, or created, over a public channel, but in a way that cannot be reproduced or broken by an adversary.

The earliest public-key cryptographers used the fact that the factoring of large composite integers is apparently very hard. The historical evidence for this premise was compelling and remains so; mathematicians had been searching for suitable integer-factoring algorithms for hundreds of years with very limited success. Since the introduction of such public-key cryptosystems, there has been a continuing intense effort to find methods of factoring large composite integers. Accordingly, improved methods of factoring integers have been found, but these improvements are modest. For large n, even these improved methods of factoring are still impractical. To counter these improved algorithms and the modern fast computers, cryptosystems simply use larger composite integers, relying on the fact that the complexity of any known algorithm grows faster than any polynomial in the size of the integer being factored. This countermeasure of simply using larger integers, of course, only works to counter modest improvements in algorithms for factoring. It will fail to counter a revolutionary improvement in factoring, if such an improvement were to be found. We have no evidence that such fast algorithms do not exist, nor that they do exist. Nor can we assert, if such algorithms do exist, that they are not known to anyone. Such algorithms may be held in confidence by individuals or by groups. It is a matter of collective faith that this is not the case.

3.1 Biprime cryptography

The earliest widespread public-key cryptosystem, known as the RSA (Rivest, Shamir, and Adleman) *cryptosystem*, is an asymmetric key system based on modular exponentiation of integers and on the difficulty of factoring large integers. The general technique is often called by the generic term *biprime cryptography* to distinguish it from specific implementations of RSA, although the essential idea remains the same. The RSA encryption technique is based on elementary number theory. Let n equal the product pq where p and q are large primes, perhaps more than 100 digits each. Such a composite integer is called a *biprime*. A composite integer $n = pq$, perhaps having 140 digits or perhaps 200 digits, is regarded as very difficult, or perhaps intractable, to factor for the primes p and q. This intractability of factoring is essential to the security of RSA. There is no proof, however, that integer factoring is difficult. The only evidence is that centuries of effort have apparently failed to find a simple method of factoring.

We will work in the ring $\mathbf{Z}_n$, where $n = pq$ is the product of two distinct large primes. The integer n is made public but the primes p and q are kept secret by the decryptor, even from the encryptor. Because $n = pq$ is the product of two primes, the totient function of n is given by $\phi(n) = (p-1)(q-1)$, which can be easily computed from p and q, but (apparently) not easily computed from n. Next, randomly choose the encryption exponent $b \in \mathbf{Z}_n$ and check that $\text{GCD}(b, \phi(n)) = 1$ as by using the euclidean algorithm. If not, choose another b. Then compute $a = b^{-1} \pmod{\phi(n)}$ by using the extended euclidean algorithm. The integer a does exist because, by choice of b, the integers b and $\phi(n)$ are coprime. An alternative procedure is to first randomly choose a coprime to $\phi(n)$, then compute b.

The encryption key is b, which is made public, as is n. The decryption key is a, which is kept secret. The decryption key a cannot be computed from b and n. This computation is apparently intractable. The decryption key could be computed from $\phi(n)$, but $\phi(n)$ is not known to those not knowing p and q, or $\phi(n)$ itself. It is believed that $\phi(n)$ cannot be computed from n because, as we shall see, determining $\phi(n)$ from n is computationally equivalent to factoring n.

A plaintext message is an element of $\mathbf{Z}_n$. This means that there is some agreed way to represent user messages by elements of $\mathbf{Z}_n$. This is straightforward because each element of $\mathbf{Z}_n$ can be represented as a binary number and every message can be expressed as a binary sequence. Longer binary messages can be broken into blocks, which are represented as elements of $\mathbf{Z}_n$, and the blocks encrypted separately, either with the same key or with a sequence of keys. Alternatively, the encrypted message could be simply another key to be used by an associated bulk encryption system.

Encryption is as follows: for any $x \in \mathbf{Z}_n$, $y = e_k(x) = x^b \pmod{n}$. Thus encryption is a map from $\mathbf{Z}_n$ into $\mathbf{Z}_n$. Decryption is as follows: for $y \in \mathbf{Z}_n$, compute $d_k(y) = y^a \pmod{n}$ where $a = b^{-1} \pmod{\phi(n)}$.

First, we check that decryption gets us back to the original message.

Proposition 3.1.1 *For the biprime cryptosystem,* $d_k(e_k(x)) = x$.

Proof Because $a = b^{-1} \pmod{\phi(n)}$, we can write $ab = 1 + r\phi(n)$ for some integer r. Then $d_k(e_k(x)) = (x^b)^a \pmod{n} = x^{1+r\phi(n)} \pmod{n} = x(x^{\phi(n)})^r \pmod{n}$. We will show that this is equal to x by considering two cases separately. These cases are $x \in \mathbf{Z}_n^*$ and $x \notin \mathbf{Z}_n^*$. If $x \in \mathbf{Z}_n^*$, then by Euler's theorem, given in Theorem 2.1.5, $x^{\phi(n)} \equiv 1 \pmod{n}$, so $d_k(e_k(x)) = x$ for $x \in \mathbf{Z}_n^*$.

Otherwise, if $x \notin \mathbf{Z}_n^*$, then x and n are not coprime. Either $\text{GCD}(x, p) = p$ or $\text{GCD}(x, q) = q$. Without loss of generality, assume that $\text{GCD}(x, p) = p$ and $\text{GCD}(x, q) = 1$. Then $x = 0 \pmod{p}$, which means that $x^{ab} = 0 = x \pmod{p}$. Moreover, because $x^{\phi(q)} = 1 \pmod{q}$ and $\phi(n) = \phi(q)\phi(q)$, we can also write $x^{ab} = x^{1+r\phi(p)\phi(q)} = x(x^{\phi(q)})^{r\phi(p)} = x \pmod{q}$. Thus both $x^{ab} = x \pmod{p}$ and $x^{ab} = x \pmod{q}$, so the chinese remainder theorem states that $x^{ab} = x \pmod{n}$. Thus $d_k(e_k(x)) = x$ for all $x \notin \mathbf{Z}_n^*$ as well as for all $x \in \mathbf{Z}_n^*$. Therefore, $d_k(e_k(x)) = x$ for all x, as was to be proved. □

The security of RSA depends on the fact that $e_k(x)$ is effectively a one-way function. However, we will see that it is a one-way function with a hidden trapdoor. The trapdoor is the factorization of n. A cryptanalyst who knows this factorization can compute the totient function $\phi(n) = (p-1)(q-1)$ and then find a by using the extended euclidean algorithm. This means that p and q must be so large that it is unreasonable to factor n by known algorithms and modern computers. The integers p and q could be on the order of 500 bits each, and possibly as large as 2000 bits each. Such a pair of primes gives a biprime on the order of 1000 to 4000 bits.

An example of biprime cryptography, but one that is too small to be secure, is as follows. Choose $p = 1171$ and $q = 1019$. Both are primes. Then $n = 1193249$ and $\phi(n) = 1191060$. The decryptor randomly chooses the decryption exponent $a = 1076531$ and computes the encryption exponent $b = a^{-1} \pmod{\phi(n)} = 120251$. Then message x is encrypted as $y = x^{120251} \pmod{1193249}$, and the ciphertext y is decrypted as $x = y^{1076531} \pmod{1193249}$.

The method of biprime cryptography can be used in reverse to become a signature scheme. The signer randomly chooses p and q, and forms $n = pq$ as before, then selects a signature key a and computes a verification key $b = a^{-1} \pmod{\phi(n)}$. The verification key b and the biprime n are made public. The signer forms a signature on the message x by

$$y = x^a \pmod{n}.$$

The signed message is (x, y). To verify the signature, the verifier only needs to employ the signer's public key b to first recover the message

$$\widehat{x} = y^b \pmod{n},$$

that corresponds to the signature. The message x that accompanies the signature is valid if the message $\widehat{x}$ under the signature matches the actual message x. Thus, the message x is valid if $x = y^b \pmod{n}$. Such a signature cannot be placed on message x without knowing a, which presumably ensures the security of the signature. Deeper issues regarding the security of signatures are discussed in Chapter 8.

3.2 Implementing biprime cryptography

As was described in the previous section, the decryptor selects two large primes, p and q, which are kept secret, and computes $n = pq$ and $\phi(n) = (p-1)(q-1)$. Next, the decryptor randomly chooses a positive integer b smaller than $\phi(n)$ such that $\text{GCD}(b, \phi(n)) = 1$, and computes $a = b^{-1} \pmod{\phi(n)}$ using the extended euclidean algorithm. The integers n and b comprise the public encryption key k and are openly published. All else remains private. The encryptor knows only the published integers, n and b.

To encrypt, the integer arithmetic of $\mathbf{Z}_n$ is used. If n is represented as a k-bit binary number, then addition can be done in $O(k)$ binary operations, multiplication can be done in $O(k^2)$ binary operations, and reduction modulo n can be done in $O(k^2)$ binary operations. This means that modular multiplication can be done in $O(k^2)$ operations and modular exponentiation (computing $x^c \pmod{n}$) in $O(k^3)$ operations by the square-and-multiply method. For instance, to compute $x^{13} \pmod{n}$, one might compute x, x^2, x^4, $x^8 \pmod{n}$ by repeated squaring and then multiply x, x^4, and $x^8 \pmod{n}$ to get $x^{13} \pmod{n}$.

To send the message x, the encryptor encodes x as follows:

$$e_k(x) = x^b \pmod{n}.$$

To decode the ciphertext $y = e_k(x)$, the decryptor simply makes use of the private decryption exponent a to compute:

$$\begin{aligned} d_k(y) &= y^a \pmod{n} \\ &\equiv x^{ab} \pmod{n} \\ &\equiv x. \end{aligned}$$

The security of the RSA cryptosystem rests on several assumptions:

(1) *The inverse*[1] *of exponentiation in* $\mathbf{Z}_n^*$ *is hard to compute.* Specifically, given the integers b, n, and $x^b \pmod n$, it is difficult, or intractable, to compute x. If one could determine x easily from these integers, then any cryptanalyst could decode an encrypted transmission.

(2) *Factoring n into its prime factors is hard.* Specifically, given the biprime pq, it is difficult to compute the factors p and q. If one could compute p and q easily, then one could easily compute $\phi(n) = (p-1)(q-1)$ and use the euclidean algorithm to compute the private decryption exponent a, which then could be used to decrypt a transmitted message x.

(3) *Computing* $\phi(n)$ *from the biprime n is hard.* If one could compute $\phi(n)$ from n, then one could compute the private decryption exponent a from the public encryption exponent b.

(4) *Computing the private decryption exponent a is hard.* Specifically, given b and n, it is difficult to compute $a = b^{-1} \pmod{\phi(n)}$. If one could compute a easily, then one could determine $x^{ab} = x$, which is the message sent by the encryptor.

The last two assumptions are similar, and perhaps equivalent. Statement (4) is included in the list as a reminder that careful questions such as this one must be raised. It is obvious that the ability to factor n lets one compute a easily by using the extended euclidean algorithm. It is less obvious, but true, that knowledge of a allows us to factor n easily with high probability. This means that if the private decryption exponent a is somehow divulged, it is not enough to choose a new a. It is also necessary to choose a new p and q as well.

The following proposition establishes the fact that computing a from b and n is equivalent to factoring pq. The equivalence is established by a probabilistic algorithm in this proposition, meaning that it can fail. The probability of failure can be made arbitrarily small by repeating the algorithm enough times.

Proposition 3.2.1 *Given n, a, and b, where* $a = b^{-1} \pmod{\phi(n)}$, *one can probabilistically factor the biprime* $n = pq$ *efficiently.*

Proof To find p and q, the factors of n, one first computes $r = ab - 1$. Because we know that $ab \equiv 1 \pmod{\phi(n)}$, which can be written $ab - 1 = 0 \pmod{\phi(x)}$, we know that $\phi(n)$ divides r. By Lagrange's theorem, $g^{\phi(n)} = 1 \pmod n$ which implies that $g^r \equiv 1 \pmod n$ for all nonzero integers g. But, by Theorem 2.10.2, the biprime n has four square roots of 1. These are $1, -1, k, -k$ for some $k \in \mathbf{Z}_n$. One may attempt to determine k by the following procedure. Let $r = ab - 1$, and randomly choose an integer $g \in \mathbf{Z}_n^*$. Then compute $g^r, g^{\frac{r}{2}}, g^{\frac{r}{4}}, \ldots$, stopping when the exponent of g is an odd number. Because $g^r \equiv 1 \pmod n$, $g^{\frac{r}{2}}$ is a square root of 1 and must be an element

[1] Although this requires computing x from x^b with b known, it is not the usual discrete log problem which requires computing b from x^b with x known.

of the set $\{1, -1, k, -k\}$. If $q^{\frac{r}{2}}$ is neither 1 nor -1, then compute GCD($x - 1, n$) to get a factor of n. Otherwise, repeat for $g^{r/4}$ in the same way. The test fails for that g if we eventually reach the last term without success. Another g can then be chosen, and the proof is repeated. □

A closer analysis of this test shows that it succeeds with probability one-half. By repeating the test m times with other randomly chosen values of g, this probability of continued failure becomes 2^{-m}, which can be made as small as desired. The implication of Proposition 3.2.1 is that if we are confident that factoring of biprimes is intractable, then we must be confident that computing $\phi(n)$ from the biprime n is also intractable. This is because if $\phi(n)$ is known, then n, a, and b are also known, and so n can be factored.

3.3 Protocol attacks on biprime cryptography

Direct attacks on biprime cryptography are attacks on the mathematics of the process of encryption and decryption. In contrast, protocol attacks are attacks on the way in which biprime cryptography is used to provide security. We will study protocol attacks in this section and direct attacks in the next section.

The RSA encryption system has a long history and apparently has resisted all direct attacks. However, if it is used improperly, successful attacks on the protocol are possible. In this section, we look at vulnerabilities due to improper use.

One improper protocol is a shared modulus protocol. In order to obtain a perceived efficiency, a small community of users decides to share a common modulus n. A naive central authority chooses the biprime $n = pq$ in common for all members of the community. The central authority also chooses the individual encryption and decryption exponents a_i and b_i for each member. However, the ith member then knows a_i, b_i, and n and, by Proposition 3.2.1, can compute p and q. This means that each member can decrypt any message sent to any other member of the community.

Another elementary protocol attack is a brute-force attack against a small message space using the fact that the encryption exponent is public. Given an application in which the message space is small, such as the set of all 10-digit phone numbers encrypted with RSA, a brute-force protocol attack is tractable. Given the ciphertext y, known to be an encrypted phone number, and the encryption exponent b, simply compute $x^b \pmod{n}$ for all possible 10-digit phone numbers, (of which there are fewer than 10 billion) until the known ciphertext y is obtained. The corresponding plaintext x is then exposed. While this direct search is burdensome, it is not intractable. To prevent this attack, all short messages should be padded prior to encryption with additional superfluous symbols by using a random-number generator. The superfluous symbols are ignored by the user, but do foil the elementary protocol attack.

There is also a small vulnerability in encrypting a message x if p or q divides x. This is because computing the greatest common denominator of x and n would reveal the factors p and q. However, it is not unreasonable to ignore this slight vulnerability if p and q are large integers because choosing a message x that is a multiple of p or of q will be exceedingly unlikely, occurring in the ratio of $p+q$ to pq. If p and q are 100-digit integers, such an x occurs with probability on the order of 10^{-100}. It is as unlikely to have such a message as it is to simply guess the factorization of n.

The cryptanalyst will try to infer the primes p and q by any means. The digits of p and q must not contain any sequence known to have significance to the decryptor. These should be chosen randomly. The method of randomly choosing p and q must be fully unpredictable even if the cryptanalyst has access to an exact replica of the encryptor.

Because the user will normally choose a new decryption exponent a for each new message even though new values of p and q are not chosen, it may seem that a careless exposure of a for one message may expose only that individual message. In fact, knowing the decryption exponent $a = b^{-1}$ for one value of b compromises the security of n, as demonstrated by Proposition 3.2.1.

The decryption exponent a in RSA should be kept secret even after the messages it has encrypted are no longer secret. This is not an option because it is computationally tractable to factor n if a is known – at least in a probabilistic sense – as was shown in Proposition 3.2.1. This would also compromise all other messages encrypted using that value of n, either in the past or in the future.

The cryptanalyst, knowing that n is the product of two primes and knowing both a and b, could also proceed as follows. Pick any w smaller than n, and compute GCD(w, n) using the euclidean algorithm. If GCD(w, n) is larger than one, then it is a proper factor of n, so it must be p or q, and the factorization of n is complete. Otherwise, GCD$(w, n) = 1$. Write $ab - 1 = 2^k m$ with m odd, and as in the Miller–Rabin algorithm, proceed as follows. Compute $w^m \pmod{n}$. If this is $1 \pmod{n}$, it is of no use. In that case, start over with another w. Otherwise, calculate $w^{2m}, w^{4m}, \ldots$ until reaching a term equal to $1 \pmod{n}$. Let $w^{2^t m} \equiv 1 \pmod{n}$ for the first such t. Then $w^{2^{t-1}m}$ is a square root of 1 modulo the biprime n. Hopefully, it is a nontrivial square root, denoted k, because then either $k = 1 \pmod{p}$ and $k = -1 \pmod{q}$, or $k = 1 \pmod{q}$ and $k = -1 \pmod{p}$.

For example, let $n = 89855713$ and $b = 34986517$. Let $w = 5$. Suppose that the cryptanalyst learns that $a = 82330933$, and so computes $ab - 1 = 2^3 \cdot 360059073378795$. In our example, $w^m \pmod{n}$ is nontrivial. It is 85877701. But then $w^{2^{t-1}m} + 1 \equiv 2 \pmod{p}$, and $w^{2^{t-1}m} + 1 \equiv 0 \pmod{q}$. This means that $w^{2^{t-1}m} + 1$ is divisible by q, but not p. To get q, calculate the greatest common divisor with n. Finally, because $w^{2m} \equiv 1 \pmod{n}$ and $w^m \not\equiv \pm 1 \pmod{n}$, GCD$(1 + w^m, n) = 9103$, which is a factor of n. Thus $89855713 = 9103 \cdot 9871$.

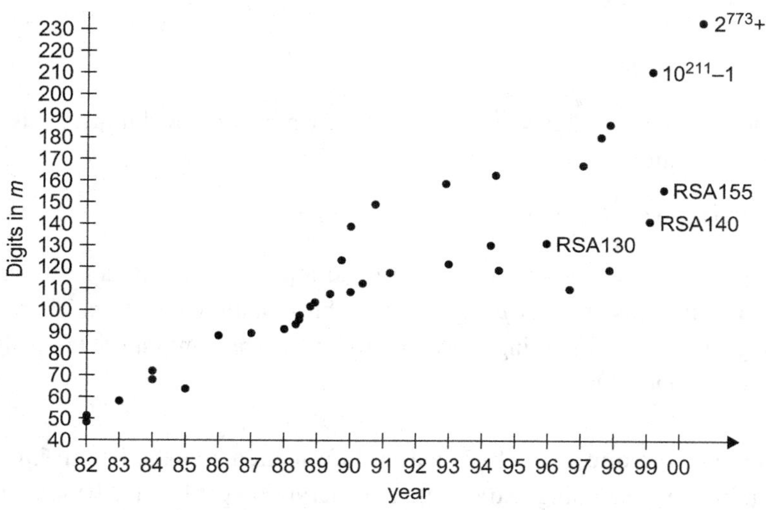

Figure 3.1 The history of integer factoring

3.4 Direct attacks on biprime encryption

The most obvious way to attack RSA is to attempt to factor the public integer n. The integer n is known to be the product of two primes, usually each with approximately the same number of digits. For this reason, there has been intense study of methods of factoring biprimes. Such methods are discussed in the latter part of this chapter.

To assess the security of RSA in past years, successively larger biprimes have been declared to be public challenges, such as those biprimes denoted RSA130, RSA140, RSA155, RSA160, and RSA768 where the 3-digit integer indicates the number of decimal digits in the challenge biprime except for the last instance in which it indicates the number of bits. The early history of factoring these and other large integers is shown in Figure 3.1. Some of the integers, such as $2^{773}+1$ have a special form that allows the use of special factoring methods, and so their successful factoring apparently says little about factoring arbitrary biprimes.

One can also attack RSA by attempting to compute the totient function $\phi(n)$ from n because the decryption exponent a is easily computed from b and $\phi(n)$. The following proposition shows, however, that this is not essentially different from a factoring attack.

Proposition 3.4.1 *Computing the totient function $\phi(n)$ when n is the product of two primes $n = pq$ is computationally equivalent to factoring $n = pq$.*

Proof Let $n = pq$. It is trivial that $\phi(n)$ can be computed from p and q, so it only is necessary to show that p and q can be computed from $\phi(n)$ and n. Suppose that $\phi(n)$ is known. Because $\phi(n) = (p-1)(q-1) = pq - (p+q) + 1$, we can compute $p + q$

by

$$p + q = n + 1 - \phi(n).$$

Now, define $c = p + q$ and $d = pq = n$, both of which are now known. This leads to the quadratic equation

$$(x + p)(x + q) = x^2 + cx + d = 0,$$

which has p and q as its roots. Factoring quadratic equations is not difficult, so if $\phi(n)$ is known, it is easy to find p by factoring this quadratic equation. Therefore any tractable algorithm for computing $\phi(n)$ directly from n becomes a tractable algorithm for computing p and q from n. □

A variation of the direct attacks is partial information attacks. Partial information attacks require that something extra about the encryption parameters be known. If, for example, it is learned that the decryption key must lie in a restricted set of integers because the procedure used to randomly select that key is not uniformly random, then this knowledge can be used to attack the cryptosystem.

3.5 Factoring biprimes

The security of biprime cryptography, such as RSA, rests on the apparent fact that it is difficult to factor integers of the form pq where both p and q are prime. If it is known that n has the form $n = pq$, then one may simply attempt to divide n by every prime that has not more than $(b + 1)/2$ decimal digits where b is the number of decimal digits in n. But the prime number theorem tells us that there are about $10^{(b+1)/2}/(b \log_e 10)$ such trial primes. If $b = 200$, this is about 10^{98} trials, were a list of trial primes known. The trial primes, in turn, can be found in the same way. These have only 100 digits, but there are a great many of them. Clearly, this is an intractable approach. Something better is needed.

The *Dixon method of factoring* is a better approach to the factoring of composite integers. The core idea of the Dixon method is at the heart of the best modern methods of integer factorization. Let n be a composite, for our purposes a biprime. Suppose that we can find two integers, x and y smaller than n, such that $y \neq \pm x \pmod{n}$, but $y^2 = x^2 \pmod{n}$. Then $x^2 - y^2 = 0 \pmod{n}$, which means that n divides $x^2 - y^2$. Therefore n divides $(x - y)(x + y)$. But n does not divide $(x - y)$ or $(x + y)$ because $x \pm y$ is not a multiple of n. This means that $\mathrm{GCD}(x \pm y, n) \neq 1$, so the two instances of $\mathrm{GCD}(x \pm y, n)$ must each be a nontrivial factor of n. In the case where $n = pq$, we conclude that the two primes, p and q, are given by $\mathrm{GCD}(x \pm y, n)$.

For example, suppose that the factors of $n = 84923$ are required. By starting at $\lfloor \sqrt{n} \rfloor = 292$ and testing squares, one eventually finds that $505^2 \pmod{84923} = 256$, which is also the square of 16. Accordingly, let $x = 505$ and $y = 16$. Then use the extended euclidean algorithm to compute $\text{GCD}(505 \pm 16, n) = 163; 521$. Thus, in conclusion, $84923 = 163 \cdot 521$.

Computing the greatest common divisor, even for very large integers, is computationally tractable and does not impose a computational burden on the Dixon method. However, finding the integers x and y with the required equal squares is difficult, and apparently intractable. Indeed, we must believe that finding these integers is very difficult for large n if we accept the premise that the integer factoring problem itself is very difficult.

The task of finding an integer $y \neq \pm x \pmod{n}$ such that $y^2 = x^2 \pmod{n}$ is a difficult task as it stands because there are a great many numbers to be considered. Dixon's method must be used with a systematic search procedure. Such search procedures are known if the factors of n are not too large. The best search procedures of this kind currently known are those known as the *quadratic sieve* and the *number-field sieve*. The quadratic sieve works with a list of all the primes smaller than a chosen integer B. If B is chosen too small, the process may fail. If B is chosen too large, the amount of computation required by the quadratic sieve may be excessive. The number-field sieve is based on a similar approach, but softens the computational function by introducing a mathematical structure richer than the integers and a larger class of numbers to work with.

3.6 The quadratic sieve

The most successful attacks on RSA have been by the methods known as the *quadratic sieve* and the *number-field sieve*. These are algorithms for factoring large integers that have evolved from Dixon's method. The quadratic sieve has been used successfully to factor a biprime that is more than 129 decimal digits long – although with a massive amount of computation.

Let $n = pq$ where n is known and p and q are unknown. To find the factors p and q, first find $\lceil \sqrt{n} \rceil$ and compute the sequence of integers $(\lceil \sqrt{n} \rceil)^2 - n$, $(\lceil \sqrt{n} \rceil + 1)^2 - n$, $(\lceil \sqrt{n} \rceil + 2)^2 - n, \ldots$, stopping at some term that is yet to be specified. Factor each of these terms into a product of primes and prime powers to form the following list:

$$\begin{aligned}
(\lceil \sqrt{n} \rceil)^2 - n &= 2^{a_2^{(0)}} \cdot 3^{a_3^{(0)}} \cdot 5^{a_5^{(0)}} \cdot 7^{a_7^{(0)}} \cdot 11^{a_{11}^{(0)}} \ldots \\
(\lceil \sqrt{n} \rceil + 1)^2 - n &= 2^{a_2^{(1)}} \cdot 3^{a_3^{(1)}} \cdot 5^{a_5^{(1)}} \cdot 7^{a_7^{(1)}} \cdot 11^{a_{11}^{(1)}} \ldots \\
(\lceil \sqrt{n} \rceil + 2)^2 - n &= 2^{a_2^{(2)}} \cdot 3^{a_3^{(2)}} \cdot 5^{a_5^{(2)}} \cdot 7^{a_7^{(2)}} \cdot 11^{a_{11}^{(2)}} \ldots
\end{aligned}$$

and so on. Thus to factor an integer n that has m digits, we must factor many integers that each has about $m/2$ digits. This is a burdensome subtask of the quadratic sieve. Now, choose an arbitrary small prime (but not too small), and discard all entries of the list that have one or more prime factors larger than that prime.

The next step is to select from the remaining set of equations a subset of M equations involving only M primes. This requires a search through the list of equations. Finally, select a subset of the equations so that each of the M primes appears, in total, an even number of times, counting multiplicity appropriately. The product of all terms on the left of these equations is then a square modulo n, and the product of all terms on the right is a square. If this process is successful in finding such a set of polynomials, we will have found integers A and B satisfying the expression

$$A^2 + Qn = B^2$$

where A and B are known. Hence

$$Qpq = (B - A)(B + A).$$

If neither term on the right is divisible by n, then set $p = \text{GCD}[n, B - A]$ and $q = \text{GCD}[n, B + A]$ to obtain the desired factorization. If either of the terms is divisible by n, then the factorization attempt by this method has failed. It can be repeated, making different choices whenever there is an option. This may require using larger prime factors. If repeated attempts continue to fail, then eventually the process must be abandoned because the computational burden has become too great.

Example 3.1 Find the two prime factors of $n = 1649$. To begin, compute that $\lceil \sqrt{n}\, \rceil = 41$. The values of $(41 + i)^2 - n$ are

$$\begin{aligned}41^2 - n &= 32 = 2^5\\ 42^2 - n &= 115 = 5 \cdot 23\\ 43^2 - n &= 200 = 2^3 \cdot 5^2.\end{aligned}$$

Therefore

$$\begin{aligned}41^2 &= 2^5 && \pmod{n}\\ 42^2 &= 5 \cdot 23 && \pmod{n}\\ 43^2 &= 2^3 \cdot 5^2 && \pmod{n}.\end{aligned}$$

By inspection of these equations looking to accumulate an even power of each small prime factor on the right, we conclude that

$$41^2 \cdot 43^2 = 2^8 \cdot 5^2 \pmod{n}$$

in which each prime factor on the right appears with an even power. This means that taking the square root of each side of this expression leads to

$$41 \cdot 43 = \pm 2^4 \cdot 5 \pmod{n}.$$

Because $41 \cdot 43 = 1763 = 114 \pmod{n}$, this reduces to

$$114 = \pm 80 \pmod{n}$$

or

$$114 \pm 80 = 0 \pmod{n}$$

Finally, because $114 \pm 80 = 34, 194$, we compute

$$\text{GCD}(1649, 34) = 17$$
$$\text{GCD}(1649, 194) = 97,$$

as can easily be computed by using the euclidean algorithm. Hence we conclude that $p = 17$ and $q = 97$ are the two factors of n.

Example 3.2 Given that $n = pq$, find the two prime factors, p and q, of $n = 89855713$. To begin, compute that $\lceil \sqrt{n} \rceil = 9480$. All k in the sequence for $k = 9480, \ldots, 10639$ for which $k^2 - n$ has only prime factors smaller than, or equal to, 103 – an arbitrary choice of upper limit on the size of prime factors – are given in the following list.

k	primes in $k^2 - n$
9482	3, 13, 19, 71
9485	2, 3, 13
9487	2,3
9521	2, 3, 13, 53
9553	2, 3, 71, 103
9575	2, 3, 19, 23, 29
9607	2, 3, 23, 47
9615	2, 13, 19, 41
9656	3, 41, 89, 103
9710	3, 19, 29, 47
9745	2, 3, 13, 23, 89
9759	2, 23, 71, 103
9837	2, 23, 71
9862	3, 13, 19, 97, 103
9901	2, 3, 13, 41, 71
9983	2, 3, 41, 47, 53
10012	3, 19, 23, 89
10145	2, 3, 13, 19, 29
10171	2, 3, 13, 47, 103
10223	2, 3, 13, 19, 103
10265	2, 3, 13, 23, 47
10274	3, 23, 47, 103
10369	2, 3, 13, 53, 89
10435	2, 3, 23, 29, 41
10457	2, 3, 13, 89
10547	2, 3, 47, 71, 89
10603	2, 3, 13, 19, 47
10619	2, 3, 23, 29, 53
10639	2, 3, 13, 19, 41 .

From this list we must choose, for any integer M, a set of M entries that among them involves only M primes. Some searching is required to find such a set. The following set will do

$$\begin{aligned}
9485^2 - n &= 2^3 \cdot 3^4 \cdot 13^2\\
9487^2 - n &= 2^{14} \cdot 3^2\\
9607^2 - n &= 2^4 \cdot 3 \cdot 23 \cdot 47^2\\
10171^2 - n &= 2^3 \cdot 3^3 \cdot 13 \cdot 47 \cdot 103\\
10265^2 - n &= 2^4 \cdot 3 \cdot 13 \cdot 23^2 \cdot 47\\
10274^2 - n &= 3 \cdot 23 \cdot 47^2 \cdot 103.
\end{aligned}$$

We must now choose a product of some of these equations so that the right side is a square. In this example, this is trivially accomplished by choosing only the second equation,

$$9487^2 - n = (2^7 \cdot 3)^2 = 384^2.$$

Hence

$$\begin{aligned}
Qn &= (9487 - 384)(9487 + 384),\\
&= 9103 \cdot 9871
\end{aligned}$$

(which, in this case, happens to be the desired factorization). In general, the algorithm is completed by the computation

$$\begin{aligned}
&\text{GCD}(9103, 89855713) = 9103\\
&\text{GCD}(9871, 89855713) = 9871.
\end{aligned}$$

In this instance, the algorithm is a success. It has given us the factorization, $89855713 = 9103 \cdot 9871$.

If, however, we fail to notice that the second equation on the list is a square, we may notice instead that the product of all other equations is a square. Thus

$$(9485 \cdot 9607 \cdot 10171 \cdot 10265 \cdot 10274)^2 = 2^{14} \cdot 3^{10} \cdot 13^4 \cdot 23^4 \cdot 47^6 \cdot 103^2 \pmod{n},$$

which becomes

$$32989949^2 = 32989949^2 \pmod{n}.$$

The conclusion then is uninformative, and the procedure cannot continue. Because this attempt at factorization has failed, another attempt is needed.

In general, whenever an attempt at factoring using the quadratic sieve fails, one may make a second attempt. To do so, choose another set of M equations that involves only M prime factors, possibly enlarging the list of equations, or else choose a larger limit on the largest prime factor. Eventually, if every reasonable attempt fails, the computational resources will be exhausted and the factoring problem will remain unsolved.

Finding a set of equations whose right sides multiply to form a square can be organized by writing down the array of exponents modulo two. Thus

$$9485^2 - n = 2^3 \cdot 3^4 \cdot 13^2 \cdot 23^0 \cdot 47^0 \cdot 103^0$$

becomes the row of modulo-two exponents

1 0 0 0 0 0

in the array designating whether each exponent is odd or even. The full array of exponents modulo two for this example is

	2	3	13	23	47	103
9485	1	0	0	0	0	0
9487	0	0	0	0	0	0
9607	0	1	0	1	0	0
10171	1	1	1	0	1	1
10265	0	1	1	0	1	0
10274	0	1	0	1	0	1.

From this array, simply choose any set of rows that sum to zero modulo two. This amounts to the task of solving the matrix equation

$$[a_0\ a_1\ a_2\ a_3\ a_4\ a_5] \begin{bmatrix} 1 & 0 & 0 & 0 & 0 & 0 \\ 0 & 0 & 0 & 0 & 0 & 0 \\ 0 & 1 & 0 & 1 & 0 & 0 \\ 1 & 1 & 1 & 0 & 1 & 1 \\ 0 & 1 & 1 & 0 & 1 & 0 \\ 0 & 1 & 0 & 1 & 0 & 1 \end{bmatrix} = [0\ 0\ 0\ 0\ 0\ 0] \pmod 2)$$

for a nonzero vector $[a_0\, a_1\, a_2\, a_3\, a_4\, a_5]$. In this example, there are only three nonzero solutions [010000], [101111] (both of which we have already found), and [111111].

This matrix formulation applies to any instance of the quadratic sieve. In general, to follow the procedure, we must compute the null space of a matrix over the field with two elements $\boldsymbol{F}_2$.

3.7 The number-field sieve

The task of factoring integers in the ring $\boldsymbol{Z}$ can be attacked with a sharper set of tools when the ring $\boldsymbol{Z}$ is embedded into a larger arithmetic system known as a number field because of the additional algebraic structure that then becomes available. The smallest field that contains $\boldsymbol{Z}$ is the field of rational numbers $\boldsymbol{Q}$, but we will need fields larger than Q. Accordingly, we shall make use of certain extension fields of the rationals known as *number fields*. A general discussion of fields, extension fields, and methods to construct them is given later in Section 9.8. Our discussion here of the number-field sieve assumes some familiarity with those topics of Chapter 9.

The *number-field sieve* is a generalization of the quadratic sieve. It requires considerable skill and massive computation to use. The number-field sieve replaces the integer ring Z used in the quadratic sieve with a number field. A number field is an extension of the rational field of finite degree. A number field is defined as the quotient ring $Q[x]/\langle p(x)\rangle$, where $p(x)$ is an irreducible polynomial over Q. The polynomial $p(x)$ has only rational coefficients, which can be taken to be integers by clearing denominators if $p(x)$ is not required to be a monic polynomial. The number field $Q[x]/\langle p(x)\rangle$ also goes by the alternative label $Q(\xi)$ where ξ is any zero of the irreducible polynomial $p(x)$. Thus $Q[x]/\langle x^2-2\rangle = Q(\sqrt{2}) = \{a + b\sqrt{2} \mid a, b \in Q\}$. The irreducible polynomial $x^2 - 2$ over Q is uniquely determined by its zero $\sqrt{2}$, so the alternative notations $Q[x]/\langle x^2-2\rangle$ and $Q(\sqrt{2})$ are equivalent.

The *degree* of the number field $Q[x]/\langle p(x)\rangle$ is equal to the degree of the polynomial $p(x)$. The degree of a number field is finite. The real field R is not a number field even though it is an extension of the rationals because it is not an extension of finite degree.

The number-field sieve is conceptually more difficult than the quadratic sieve, but is more efficient computationally for very large biprimes. It has been observed by some that the number-field sieve works better than the quadratic sieve if the biprime to be factored has more than about 120 to 130 decimal digits. For smaller biprimes, the quadratic sieve evidently works better. The number-field sieve has been used to factor RSA-140, a 140-digit composite integer that had been once posed as a challenge to the community. Since then the number-field sieve has been used to factor even larger integers.

Like the quadratic sieve, the number-field sieve finds congruent squares $x^2 = y^2 \pmod{n}$, where x and y are distinct positive integers. It then calculates GCD$(x - y, n)$ and GCD$(x + y, n)$. If either is a proper factor of n, the process is a success. If not, another attempt is made. Like the quadratic sieve, the number-field sieve finds x and y by setting up a matrix, although the procedure for setting up this matrix is somewhat different. The resulting matrix equation is then solved by any suitable method, preferably by a good method for solving sparse matrix equations.

Number fields are introduced in this section because of their use in integer factoring by means of the number-field sieve. The purpose is to deflect some of the computational burden of the quadratic sieve into the more powerful structure of a number field where it can be absorbed by the efficiency of the field structure. This requires the notion of a *prime ideal*, which is a generalization of the notion of a prime of the integer ring to a more general ring. A prime ideal of any commutative ring R is a proper ideal I of R such that whenever the product ab of two elements a and b of the ring R is in I, then a and b are both in I as well. The number-field sieve depends on the generalizations of the notion of a prime integer of the ring $\boldsymbol{Z}$ to the notion of a prime ideal of the ring $\boldsymbol{Z}[\xi]$, where the *number ring* $\boldsymbol{Z}[\xi]$ is the subset of the number field $\boldsymbol{Q}(\xi)$ consisting of those elements of $\boldsymbol{Q}(\xi)$ with only integer coefficients.

In general, the number field $\boldsymbol{Q}[x]/\langle p(x)\rangle$, where $p(x)$ is an irreducible polynomial over $\boldsymbol{Q}$, is defined as the set of all polynomials over $\boldsymbol{Q}$ with degree smaller than the degree of $p(x)$. Addition in $\boldsymbol{Q}[x]/\langle p(x)\rangle$ is defined as elementary addition of polynomials. Multiplication in $\boldsymbol{Q}[x]/\langle p(x)\rangle$ is defined as elementary multiplication of polynomials reduced modulo $p(x)$.

The exemplar form of a number field is the field of *gaussian rationals* $\boldsymbol{Q}(i) = \boldsymbol{Q}[x]/\langle x^2+1\rangle$. The integers of $\boldsymbol{Q}(i)$ are the elements of $\boldsymbol{Z}[i]$ and are known as *gaussian integers*. Theorem 2.4.7 implies that every prime integer of the form $p = 4k+1$ can be factored in $\boldsymbol{Q}(i)$ as $p = (a+ib)(a-ib)$ where a and b are positive integers and $i^2 = -1$. Thus every prime p of $\boldsymbol{Z}$ of this form can be replaced by two "primes," $a+ib$ and $a-ib$ of $\boldsymbol{Z}[i]$, thereby increasing the size of the factor base. This observation is expressed by saying that $\boldsymbol{Z}[i]$ is *smoother* than $\boldsymbol{Z}$, meaning that an irreducible factorization of an integer has more factors in $\boldsymbol{Z}[i]$ than in $\boldsymbol{Z}$. Thus, $377 = 13 \cdot 29$ in $\boldsymbol{Z}$, but $377 = (3+2i)(3-2i)(5+2i)(5-2i)$ in $\boldsymbol{Z}[i]$. However, $403 = 13 \cdot 31$ in $\boldsymbol{Z}$, but $403 = (3+2i)(3-2i)\cdot 31$ is the prime factorization in $\boldsymbol{Z}[i]$. The term 31 is not of the form $4k+1$ and so does not factor in $\boldsymbol{Z}[i]$.

Recall that the quadratic sieve first constructed an appropriate set of linear equations in the integer ring $\boldsymbol{Z}$, then solves that set of equations to form an appropriate difference of squares: $B^2 - A^2 = Qn$ in hopes that $B-A$ or $B+A$ is then a factor of n. The number-field sieve is helpful only for the first task: constructing a set of linear equations. The number-field sieve reduces the computational complexity by embedding the original task into the richer structure of a number field. The second part of the process – solving the system of equations – will proceed as before.

Before introducing the number-field sieve, the quadratic sieve will be restated more abstractly in terms of a polynomial, $f(x) = x^2 - n$, together with a collection of small primes $\{p_1, p_2, \ldots, p_m\}$. Under this description, that sieving process finds a set of integers x_i such that for each x_i, $f(x_i) = p_1^{e_{1i}} p_2^{e_{2i}} \ldots p_m^{e_{mi}} = \prod_{\ell=1}^m p_\ell^{e_\ell}$. This leads to the desired difference of two squares by selecting a subset $\mathcal{U}$ of the x_i such that for each ℓ, the sum $\sum_{i:x_i\in\mathcal{U}} e_{\ell i}$ is even. The next step is to write $x^2 = \prod_{x_i\in\mathcal{U}} x_i^2$ and $y^2 = \prod_{i:x_i\in\mathcal{U}} p_1^{e_{1i}} p_2^{e_{2i}} \ldots p_m^{e_{mi}}$. These are equal modulo n because then

$$x^2 = \prod_{x_i\in\mathcal{U}} x_i^2 \equiv \prod_{x_i\in\mathcal{U}} (x_i^2 - n) \equiv \prod_{x_i\in\mathcal{U}} f(x_i) = y^2 \pmod{n}.$$

Thus $x^2 - y^2 = (x-y)(x+y) = Qn$. One then computes $\mathrm{GCD}(x-y, n)$ and $\mathrm{GCD}(x+y, n)$ in the hope of finding a factor of n.

The number-field sieve mimics this description of the quadratic sieve by using a different polynomial for $f(x)$. Because the polynomial used by the number-field sieve need not have the form of a square modulo n, it will be necessary to sieve on both the set of x_i and the set of $f(x_i)$. Thus one searches for a subset $\mathcal{U}$ of the x_i such that both $\prod_{x_i \in \mathcal{U}} x_i^2$ and $\prod_{x_i \in \mathcal{U}} f(x_i)$ are squares modulo n. This is where the number field plays a role.

The more general form of the number-field sieve chooses two polynomials $f(x)$ and $g(x)$, both irreducible over the rational field with integer coefficients and small degrees, denoted d and e, and that have a common zero modulo n. Elementary methods to find suitable polynomials for this task by a straightforward and orderly search have been developed. It is not known whether there are fast methods for finding these polynomials that are significantly better than a straightforward and orderly search.

Once the two polynomials are obtained, the number-field sieve sets up a factor base in the number rings $\mathbf{Z}[x]/\langle f(x)\rangle$ and $\mathbf{Z}[x]/\langle g(x)\rangle$. These number rings are denoted $\mathbf{Z}[\theta]$ and $\mathbf{Z}[\gamma]$, respectively, where θ and γ are zeros of $f(x)$ and of $g(x)$, respectively. A simple instance of the number-field sieve first finds an irreducible polynomial $f(x)$ with an integer, say m, as a zero, and then chooses $g(x) = x - m$ as the second polynomial.

Define the function $\phi(x)$ taking elements of the number ring $\mathbf{Z}[\theta]$ to $\mathbf{Z}_n$ in the natural way by defining $\phi(1) = 1 \pmod{n}$ and $\phi(\theta) = m$ so that $\phi(a + b\theta) = a + bm \pmod{n}$. Next, let $\mathcal{U}$ be a set of pairs of integers (a, b) that satisfy both of the following two statements. In the ring $\mathbf{Z}[\theta]$,

$$\prod_{(a,b)\in\mathcal{U}} (a + b\theta) = \beta^2$$

with $\beta \in \mathbf{Z}[\theta]$. In the ring $\mathbf{Z}$

$$\prod_{(a,b)\in\mathcal{U}} (a + bm) = y^2$$

with $y \in \mathbf{Z}$. Finding such a set involves simultaneously sieving against two requirements. This is the demanding part of the algorithm. Assuming that a suitable set $\mathcal{U}$ can be found, a difference of two squares is now straightforward to compute. Simply apply the homomorphism $x = \phi(\beta)$, where $x \in \mathbf{Z}_n$, as follows

$$\begin{aligned} x^2 = \phi(\beta)^2 = \phi(\beta^2) &= \phi\Big(\prod_{(a,b)\in\mathcal{U}} (a + b\theta)\Big) \\ &= \prod_{(a,b)\in\mathcal{U}} \phi(a + b\theta) = \prod_{(a,b)\in\mathcal{U}} (a + bm) = y^2 \pmod{n}. \end{aligned}$$

This gives the desired difference of squares: $x^2 - y^2 = 0 \pmod{n}$. Then $(x - y)(x + y) = Qn$. If either GCD$(x - y, n)$ or GCD$(x + y, n)$ is nontrivial, it is a factor of n.

Thus the main task is to find the set $\mathcal{U}$. This is done by forming a factor base in $\mathbf{Z}[\theta]$ and another factor base in $\mathbf{Z}$. For each of many integer pairs (a, b), the quantity $a + b\theta$ is expressed in $\boldsymbol{Q}(\theta)$ in terms of the first factor base and the quantity $a + bn$ is expressed in the ring $\boldsymbol{Z}$ in terms of the second factor base. Then having a list of such (a, b) pairs, the next task is to choose a subset $\mathcal{U}$ such that the product of $(a + b\theta)$ over

$\mathcal{U}$ is a square in $\mathbf{Z}[\theta]$ and the product of $(a + bm)$ over $\mathcal{U}$ is a square in $\mathbf{Z}$. This search for a suitable subset is similar to the corresponding task in the quadratic sieve except that two such conditions must be satisfied rather than one.

This completes our outline of the number-field sieve. The various computational tasks of the number-field sieve that were outlined here are quite demanding and require careful implementation of the many details that are needed. Massive amounts of computation are required and the method is far beyond the resources of the casual user.

3.8 The Rabin cryptosystem

Another cryptosystem that depends on the intractability of the integer factoring problem is the *Rabin cryptosystem*. Indeed, it is known that breaking the Rabin cryptosystem by a direct attack is as hard as factoring, which is a stronger statement than can be said for RSA because other direct attacks on RSA appear viable. Thus the Rabin cryptosystem may appear superior to RSA, which is not known to be as hard as factoring. However, the Rabin cryptosystem is quite vulnerable to a chosen plaintext attack – and RSA is not. This is an attack that attempts to find the key when given the opportunity to encrypt and decrypt a chosen plaintext.

Let n equal pq, where p and q are distinct odd primes, both large. For now, we put no other conditions on p and q. Later, we will give several reasons to choose both p and q of the form $4k + 3$ for some value of k. The decryptor picks the two primes p and q and computes the biprime $n = pq$, which is made public.

The encryptor picks an integer B, makes it public, and encrypts message x by using $y = e(x) = x(x + B) \pmod{n}$. The integers n and B are public; the integers p and q are kept secret, known only to the decryptor.

To recover the plaintext x, the decryptor must solve the quadratic equation $x^2 + Bx \equiv y \pmod{n}$, where y is the received ciphertext. Complete the square by letting $z = x + B/2$. If B is even, this is immediate. If B is odd, $\frac{B}{2}$ is found by using $\frac{1}{2} \equiv \frac{n+1}{2} \pmod{n}$. Then

$$\begin{aligned} z^2 &= x^2 + Bx + \frac{B^2}{4} \\ &= y + \frac{B^2}{4} \pmod{n}. \end{aligned}$$

The right side is an integer that is easily computed from y without knowing the factorization of n. We will call this integer C.

The task of decryption now is to solve the equation $z^2 = C = y + \frac{B^2}{4} \pmod{n}$ for the unknown z. The message x is then found as $x = z - B/2$. The task, then, is to compute $z = \sqrt{C}$. Theorem 2.10.2 asserts that there are four solutions to $z^2 = C$ modulo pq.

The decryptor knows p and q, so it simply solves $z^2 \equiv C \pmod{p}$ and $z^2 \equiv C \pmod{q}$ by taking square roots in the two fields $\boldsymbol{F}_p$ and $\boldsymbol{F}_q$. This gives both z modulo p and z modulo q. The decryptor then uses the chinese remainder theorem to recover z modulo n. The adversary, however, knows only n, not the factors p and q, and so, as discussed in Section 2.13, it cannot solve the quadratic equation.

By the definition of the encryption, there is a solution z, and so there are two solutions to $z^2 \equiv C \pmod{p}$ and two solutions to $z^2 \equiv C \pmod{q}$. The chinese remainder theorem then gives four possible solutions for $z \pmod{n}$, and so results in four possible plaintext messages. To disambiguate, the decryptor may simply pick that solution corresponding to a legitimate message. This method is usually satisfactory because the false messages will likely be nonsense. If necessary, the message x can have a brief preamble that distinguishes it as a valid message and thereby resolves the ambiguity. Alternatively, the four solutions can be simply tested against the original equations modulo n. Moreover, at the end of this section, we will show that if both p and q have the form $4k + 3$ for some k, then only one of the four solutions will survive this test.

For example, let $p = 11$ and $q = 19$. Then $n = 209$. A message consists of any integer between 1 and 208 inclusively, although some messages are easy to decrypt and should be avoided. If the message x is 173 and $B = 118$, then the encryption is $y = 173(173 + 118) \pmod{209} = 183$. The integer 183 is the ciphertext corresponding to the plaintext 173. To recover the plaintext, the decryptor computes $183 + \frac{B^2}{4} \pmod{209} = 111$, then computes $111 \pmod{11} = 1$ and $111 \pmod{19} = 16$, then solves two equations,

$$z_1^2 = 1 \pmod{11}$$
$$z_2^2 = 16 \pmod{19},$$

using any preferred algorithm for computing square roots modulo a prime. The square roots are $z_1 = \pm 1 \pmod{11}$ and $z_2 = \pm 4 \pmod{19}$. Thus $z_1 = 1$ or 10, and $z_2 = 4$ or 15. The chinese remainder theorem uses the Bézout identity $1 = 7 \cdot 11 - 4 \cdot 19$. Then $z = -76z_1 + 77z_2 \pmod{209} = 23, 36, 175$, or 186. Finally, $x = z - B/2$. Therefore $x = 173, 186, 116$, or 127. This completes the main decryption process, but the correct message must be selected from these four candidate messages by a subsequent process.

To compute a square root modulo a prime p is easy when p is of the form $4k + 3$. Euler's criterion, given in Theorem 2.3.1, says that the square root $\beta^{(p+1)/2} = \beta \pmod{p}$. This is the first reason for choosing p and q of the form $4k + 3$. The second reason is that it is easy to mark the message so as to disambiguate for the true solution, as shown in the next theorem.

Theorem 3.8.1 *If p and q are both primes of the form $4k + 3$, then the equation $z^2 = A \pmod{pq}$ has only one integer solution for z that itself is a square* mod pq.

Proof The two equations $u^2 = A \pmod{p}$ and $v^2 = A \pmod{q}$ each have two solutions, $\pm u$ and $\pm v$. By the chinese remainder theorem, there are four candidate solutions for z. These four solutions are $\pm au \pm bv$ where

$$a = \begin{cases} 1 & \pmod{p} \\ 0 & \pmod{q} \end{cases} \qquad b = \begin{cases} 0 & \pmod{p} \\ 1 & \pmod{q} \end{cases}.$$

Step 1 This step will show that $au + bv$ and $au - bv$ cannot both be squares mod pq if p and q both have the form $4k + 3$. The Jacobi symbols satisfy

$$\begin{aligned}(au \pm bv|pq) &= (au \pm bv|p)(au \pm bv|q) \\ &= (au|p)(\pm bv|q)\end{aligned}$$

where the second line follows because $b = 0 \pmod{p}$ and $a = 0 \pmod{q}$. Moreover $a = 1 \pmod{p}$ and $b = 1 \pmod{q}$, so this can be further manipulated as

$$\begin{aligned}(au \pm bv|pq) &= (u|p)(\pm 1|q)(v|q) \\ &= (\pm 1)^{(q-1)/2}(u|p)(v|q).\end{aligned}$$

Because p and q are both equal to three (mod 4), it follows that $(q - 1)/2$ is odd, so the right side has the sign of ± 1. This leads to the statement

$$(au + bv|pq) = -(au - bv|pq),$$

so because the Jacobi symbols $(au + bv|pq)$ and $(au - bv|pq)$ cannot both be positive, either $au + bv$ or $au - bv$ is not a square modulo pq.

Step 2 This step will show that z and $-z$ are not both squares modulo pq if p and q are equal to $3 \pmod{4}$. Suppose that $z = x^2 \pmod{pq}$ and $-z = y^2 \pmod{pq}$. Then $z = x^2 + Qpq$, and $-z = y^2 + Q'pq$. Then $z = x^2 \pmod{p}$ and $-z = y^2 \pmod{p}$, so $y^2 = -x^2$ in $\boldsymbol{F}_{4k+3}$. Then $y = \sqrt{-1}x$. But $\sqrt{-1}$ does not exist in $\boldsymbol{F}_{4k+3}$. This means that x^2 and $-x^2$ cannot both be squares modulo $4k + 3$. Thus z and $-z$ are not both squares modulo pq.

Combining Step 1 and Step 2, we conclude that of the four terms $\pm au \pm bv$, only one can be a square, so only one has a square root. □

The earlier example with $p = 11$ and $q = 19$ has four solutions for z. These are 23, 36, 175, and 186. Of these, only 36 is a square (mod 209), and the corresponding plaintext message x is $36 - 59 = 186$. Because the actual message 173 is not a square, it is not identified by this method. To make this feature into a useful method of disambiguation, each message should be extended to make it correspond to a square.

As a consequence of Proposition 2.13.1, we can conclude that the Rabin cryptosystem is at least as strong as RSA with respect to a direct attack. However, the Rabin cryptosystem is vulnerable to a chosen plaintext attack whenever the adversary can encrypt a known plaintext that is not a square and insert that ciphertext into the decryptor, then observe the decrypted output. Because the decryptor must choose one of four candidate solutions based on the values of p and q, it may be that the decrypted output is not consistent with the plaintext message. This information may help the adversary determine p and q.

3.9 The rise and fall of knapsack cryptosystems

A now discredited public-key cryptosystem is one known as the *Merkle–Hellman knapsack cryptosystem.* This cryptosystem was proposed in 1978 and was broken in less than ten years. It is still of pedagogical interest as an example of an apparently secure cryptosystem that was eventually undone by a successful attack. The lesson to be learned is that an innocent-looking side condition, when attached to an apparently hard problem, may make the problem significantly easier. A statement about intractability may fail to be true when embedded in a context, however innocuous the context might appear.

The original attraction of knapsack cryptosystems was due to the fact that a knapsack cryptosystem is apparently protected by a computational problem that is provably intractable in the sense of complexity theory, the so-called *subset-sum problem.*[2] This was seen as an exciting contrast to other cryptosystems that are only protected by the presumed, but unproved intractability of tasks such as factoring biprimes or the discrete-log problem, each an apparent intractability that is only supported anecdotally.

Before defining the Merkle–Hellman knapsack cryptosystem, we remark that the subset-sum problem is known to be hard. This is the computational task of choosing a subset of the general set of integers $\{s_1, s_2, \ldots, s_n\}$ whose sum is equal to a given integer y. There are 2^n subsets of this set to consider, so any trial-and-error procedure would be exponentially complicated in n and, indeed, the problem is known to be formally intractable. This would appear to state that the Merkle–Hellman cryptosystem is secure because it is based on solving the subset-sum problem. However, the subset-sum problem is only hard in the general case. It is not hard in every instance of the problem. Indeed, it turns out that it is not hard in each of those instances in which we are interested. Namely, for those instances with the hidden Merkle–Hellman trapdoor, the problem is not as hard as the general case.

The term "knapsack" is a metaphor suggesting that a knapsack can be packed with certain objects from a given set in only one way. In the case of the knapsack cryptosystem, the objects are a set of integers $\{s_1, s_2, \ldots, s_n\}$. A binary number x, with

[2] This problem is in the class of problems known as the class of NP-complete problems.

bits denoted $b_1, b_2, \ldots, b_n$, is packed into a single integer $\boldsymbol{y}$ by writing $\boldsymbol{y} = \sum_i b_i s_i$. Then $\boldsymbol{y}$ is the ciphertext. To make this representation of x unique, we need to define a special property for the set $\{s_1, s_2, \ldots, s_n\}$.

Definition 3.9.1 *An ordered set of nonnegative integers is superincreasing if every integer in the set is strictly larger than the sum of all smaller integers in the set.*

For example, the ordered set $\{3, 5, 11, 20, 41, 81, 167, 339\}$ is a superincreasing set of integers because each integer is larger than the sum of the integers that precede it. In general, a set of ordered integers $\{s_i\}$ is superincreasing if $s_j > \sum_{i=1}^{j-1} s_i$.

Now, using the superincreasing set $\{s_i\}$, it is evident that the binary message $b_1, b_2, \ldots, b_n$, with $b_i \in \{0, 1\}$, is uniquely represented by the integer

$$\boldsymbol{y} = \sum_i b_i s_i,$$

where the sum is regarded as a sum in $\boldsymbol{Z}$. It is easy to recover the bits of the message from the integer $\boldsymbol{y}$. Simply subtract the terms of the superincreasing set one by one, if possible, starting with the largest, with $i = n$. Thus if $\boldsymbol{y} - s_i \geq 0$ at step i, then bit b_i is a one and $\boldsymbol{y}$ is replaced by $\boldsymbol{y} - s_i$. Otherwise, bit b_i is a zero and $\boldsymbol{y}$ is unchanged. Then decrease i by one and repeat this process.

As defined so far, this knapsack is not secure, nor is it yet intended to be secure. To make this process into a cryptosystem, the superincreasing set must be hidden behind a trapdoor, yet it must have a public counterpart that is used for encoding. In its public form, the encryption key appears as a set $\{t_i\}$, which itself is not superincreasing, but does have some secret relationship to the superincreasing set $\{s_i\}$. The decryptor announces the set t_i and announces that it can decrypt any binary message $b_1, b_2, \ldots, b_n$ packed into a single integer $y = \sum_i b_i t_i$, using the published set $\{t_i\}$.

Choose any superincreasing set $\{s_1, s_2, \ldots, s_n\}$. The trapdoor consists of any prime p larger than $\sum_i s_i$ and any integer a smaller than p, both of which are kept secret, as is the set $\{s_1, s_2, \ldots, s_n\}$. The public set of integers is $\{t_1, t_2, \ldots, t_n\}$ where $t_i = as_i \pmod{p}$. The encryption of the binary plaintext $(b_1, b_2, \ldots, b_n)$ is $y = \sum_i b_i t_i$. The integer y is the ciphertext.

To decrypt, find $a^{-1} \pmod{p}$ which must exist because p is a prime. Compute $a^{-1}y \pmod{p}$. This gives

$$\begin{aligned} a^{-1}y &= a^{-1} \sum_i b_i t_i \pmod{p} \\ &= \sum_i b_i (a^{-1} a s_i) \pmod{p} \\ &= \sum_i b_i s_i. \end{aligned}$$

Because the set $\{s_1, s_2, \ldots, s_n\}$ is a superincreasing set, it is now easy to find the plaintext bits.

The task of the cryptanalyst is to recover the binary data when given y and the public set of integers $\{t_1, t_2, \ldots, t_n\}$. Surprisingly, at least to those who believed in the security of the knapsack cryptosystem, there are tractable ways to do this.

The system is not protected by the intractability of the subset-sum problem because the set $\{t_1, t_2, \ldots, t_n\}$ is not an arbitrary set of nonnegative integers. It was constructed by a restrictive rule, which required that the sequence be a disguised superincreasing sequence. Thus, within the class of subset-sum problems, which are known to be intractable in general, the Merkle–Hellman knapsack imposes a condition that restricts the problem to a special instance of the subset-sum problem that is not intractable. The intractability of the class of subset-sum problems, which was intended to assure cryptographic security, is inadvertently destroyed by the need to provide a trapdoor.

Problems for Chapter 3

3.1 Verify that 274177 divides $2^{64} + 1$. Check if the quotient is likely to be a prime. Show your work.

3.2 Let $n = pq$ where $p = 1019$ and $q = 1171$. Using the extended euclidean algorithm, find $1076531^{-1} \pmod{\phi(n)}$.

3.3 Let n be the product of k distinct primes. How many solutions does the equation $x^2 = c \pmod{n}$ have? Give an example with $k = 3$.

3.4 Factor 87463 by the quadratic sieve method. Show your work.

3.5 Suppose that p and q are both odd primes equal to three modulo four. Prove that the Jacobi symbol satisfies $(x \mid pq) = (x \mid p)(x \mid q)$. Is this true for any odd primes p and q?

3.6 Given the superincreasing set, $\{3, 5, 11, 20, 41, 81, 167, 339\}$, the prime 701, and the integer $a = 223$, construct a Merkle–Hellman encryption set $\{t_i\}$. Encrypt the binary message (10011101). Decrypt the ciphertext.

3.7 **a** Show that every prime integer of the form $p = 4k + 1$ can be expressed as the sum of two squares.

b Show that every prime integer of the form $p = 4k + 3$ can be expressed as the difference of two squares.

3.8 The integer 6 has the unique factorization $6 = 2 \cdot 3$ in the ring $\mathbf{Z}$. Does the integer 6 have a unique factorization in the ring $\mathbf{Z}[x]/\langle x^2 + 5\rangle$?

3.9 Using RSA with public $N = 22879$ and encryption exponent 259, the following message is intercepted by the adversary: 08864 00235 20699 08186 16629 01277 16675 19828 12791.

Help the adversary decrypt the message, given that letters are converted to numbers as follows:

$a = 187$	$f = 800$	$k = 313$	$p = 937$	$u = 123$
$b = 133$	$g = 215$	$l = 015$	$q = 788$	$v = 321$
$c = 051$	$h = 665$	$m = 114$	$r = 621$	$w = 434$
$d = 412$	$i = 432$	$n = 601$	$s = 430$	$x = 264$
$e = 982$	$j = 732$	$o = 072$	$t = 199$	$y = 246$
				$z = 380.$

3.10 Suppose that x is a quadratic residue modulo $n = pq$ with p and q primes that are both congruent to 3 (mod 4). The Rabin cryptosystem requires the decryptor to solve $y^2 = x \pmod{n}$, and this equation has four solutions. Show that exactly one of these four solutions is itself a quadratic residue modulo n. (So this can uniquely determine the message.)

3.11 Show that the Legendre symbol satisfies $(au + pv|p) = (au|p)$.

3.12 Prove that the ideal I of the ring $\boldsymbol{Z}$ is a prime ideal if, and only if, I consists of the set of all integer multiples of a prime p.

3.13 Because $x^2 - 5$ is an irreducible polynomial over $\boldsymbol{Q}$, the set $\boldsymbol{Q}(\sqrt{5})$ is a number field. Because $(1 + \sqrt{5})/2$ is a zero of $x^2 - x - 1$, it is an algebraic integer that is also an element of the number field $\boldsymbol{Q}(\sqrt{5})$. Show that $\boldsymbol{Z}[\sqrt{5}]$ does not contain $(1 + \sqrt{5})/2$. In general, does the number ring $\boldsymbol{Z}[\xi]$ contain all algebraic integers of $\boldsymbol{Q}(\xi)$?

Notes for Chapter 3

RSA cryptography, introduced in 1978 by Rivest, Shamir, and Adleman, was evidently the first public-key cryptographic system. A similar method was apparently suggested by Cocks in 1973, but never published. Even earlier, the use of one-way functions in cryptography was described by Jevons (1874), using the factorization problem to create a trapdoor function.

Any efficient algorithm for factoring large composite integers will break RSA. However, it was shown by Boneh and Venkatesan (1998) that the converse is not true. Breaking RSA is not equivalent to factoring large composite integers. This means that there may be an efficient way of breaking RSA even if an efficient algorithm for factoring the biprime n does not exist or is not yet known. Moore (1992) discusses the many dangers of making innocuous and apparently harmless variations in the RSA protocol.

The knapsack cryptosystem was first described by Merkle and Hellman (1978). The knapsack cryptosystem was broken by Shamir (1984). There followed an intense effort, as by Brickell (1985), Odlyzko (1990), and others to repair the flaw in the Merkle–Hellman knapsack cryptosystem, but these efforts failed. The method is now considered unrepairable. It remains important as an example and a reminder that other public-key cryptosystems may also have a fatal flaw, as yet undiscovered, or at least unreported.

Morrison and Brillhart (1975) introduced the idea of a factor base and the notion of a B-smooth integer as an integer none of whose prime factors is larger than B. The use of random squares to search for the factors of biprimes was introduced by Dixon (1981). The quadratic sieve was developed by Pomerance (1985) as an improvement to Dixon's factorization method. A method of factoring using elliptic curves was introduced by Lenstra (1987). The use of number fields was introduced into the subject of integer factoring by Pollard in 1988, although it was not published at that time. The general technique is rooted in work by Coppersmith, Odlyzko, and Schroeppel (1986), who used the field of gaussian integers to compute discrete logarithms in $\boldsymbol{F}_p^*$ in subexponential time. Pollard generalized this to a number field. The number-field sieve was later described in detail by Lenstra and Lenstra (1993). Modifications to the number-field sieve are due to Coppersmith (1993). Murphy and Brent (1998) and Kleinjung (2006) have studied the selection of polynomials for the number-field sieve.

No polynomial-time algorithm is known for integer factoring, and it is not known whether such an algorithm is possible. A 512-bit "challenge" biprime called RSA155 was factored by a massive international cooperation in August 1999 and another, called RSA160, was factored in the same manner in April 2003. More recently after two years of a massive effort culminating in 2009, a 232-digit biprime called RSA768 was factored. These improved factoring methods are easily countered, at least temporarily, simply by using larger integers. In practice, however, this may be unacceptable.

4 Cryptography based on the discrete logarithm

Many cryptographic systems use exponentiation in an appropriate finite group as the critical part of the method of encryption. The security of any such system depends on the mathematical intractability of the computational problem of inverting the operation of exponentiation in that group, a problem known as the *discrete-log problem*. The difficulty or intractability of the discrete-log problem may – and will – depend on the specific group. The discrete-log problem is difficult in some groups and is easy in other groups. The belief in the intractability of this computational problem in many groups is based on anecdotal evidence rather than on mathematical proof. This means that a public-key cryptosystem based on exponentiation always entails the risk that the underlying inverse computational problem is actually easy. Indeed, that computational problem may already have been solved by a secluded and secretive cryptanalyst.

4.1 Diffie–Hellman key exchange

The *Diffie–Hellman key exchange* (or, more descriptively, *the Diffie–Hellman key agreement or key creation*) is a method by which two parties with no prior communication establish a secret key by messages sent over a clear public two-way channel. No key is actually exchanged despite the terminology. More correctly, a key is created by both parties working together. The two parties, between them, establish a common secret key even though all of their communication takes place on a public channel. It is believed that no eavesdropper is able to determine the secret key even though the eavesdropper has access to all communications. The Diffie–Hellman key exchange is believed to be computationally secure. However, it does not provide perfect secrecy. It can, in principle, always be broken by a brute-force attack with sufficiently large computational resources. The assumption, or intent, is that it is far beyond current computational resources to carry out a brute-force attack on the discrete-log problem.

Party A and party B agree on a suitable large finite group, G, and an element $P \in G$ that has order p where p is a large prime. Typically G is cyclic or it has a large cyclic subgroup. In the usual applications, both the group G and the element P will be a public standard shared by the entire community. Party A chooses a random integer, a, which will be kept secret, and party B chooses a random integer, b, which also will be kept secret. Party A exponentiates P, computing $P^a \in G$ and sends it to party B over the public channel. Party B exponentiates P, computing $P^b \in G$ and sends it to party A over the public channel. Party A, who knows a, but not b, computes $(P^b)^a \in G$. Party B, who knows b, but not a, computes $(P^a)^b \in G$. Thus both party A and party B have the same element P^{ab} of group G. This group element is converted to a common key that both parties share. At this point, the integers a and b can be destroyed. They will not be used again. Because both parties have agreed on a common group element P^{ab}, they have agreed on a key, which can be expressed as a binary number representing that group element. The key can then be used in any symmetric cryptography system for bulk encryption.

For example, if G is the cyclic group F_p^* under multiplication, then the group elements are nonnegative integers smaller than p, and P is an element of the group of large order, usually a generator of F_p^*. The group elements can be expressed as binary numbers of length $\log_2 p$ bits.

Because all communication takes place on a public channel, an adversary can observe P, P^a, and P^b, but not P^{ab}. The key exchange is based on the premise that there is no tractable computational method for computing P^{ab} from P^a, P^b, and P. This computational task clearly is not harder than the task of computing a from P and P^a. It could possibly be easier because both P^a and P^b are known. However, it is believed that the discrete-log problem is intractable. A cryptanalyst, knowing P and observing P^a and P^b, cannot compute P^{ab} using current computers and algorithms.

Although it is generally believed that the discrete-log problem in certain groups is intractable, no proof of this assertion is apparently known. We can only say that a general computationally tractable solution to the discrete-log problem has not been publicly reported despite intense and widespread efforts to find a good algorithm.

Solving the discrete-log problem breaks the Diffie–Hellman key exchange. A required solution to the discrete-log problem would be a tractable method of computing a when given P^a. However, it is not true – or at least not known to be true – that breaking the Diffie–Hellman key exchange is equivalent to solving the discrete-log problem. It may be that one can compute P^{ab} from P^a, P^b, and P without ever computing a or b. No algorithm for this computation is known, and there is widespread belief that such an algorithm will not be found. Nevertheless, the evidence for this belief is anecdotal, not theoretical.

To make this distinction explicit, we state two widely accepted premises:

(1) For many groups, it is hard to compute a from P^a.
(2) For many groups, it is hard to compute P^{ab} from P, P^a, and P^b.

These two statements are not known to be equivalent. If the first statement is false, then it is obvious that the second statement must be false as well. If the second statement is false, the first statement may still be true.

4.2 Discrete logarithms

Let G be any cyclic group, and represent the group operation as multiplication. The discrete logarithm of group element $y = \alpha^x$ to the base α is $x = \log_\alpha y = \log_\alpha \alpha^x$. The discrete-log problem is to compute x from α^x and α (not to be confused with computing α from α^x and x).

For many very large finite groups, it is believed to be intractable to compute discrete logarithms, yet in these same groups, it is easy to exponentiate. This apparent computational asymmetry is the basis of many cryptography systems, such as the Diffie–Hellman key exchange. However, the intractability of computing a logarithm in any of these large finite groups has never been proven, so any such cryptosystem has not been proven to be secure. The problem of finding such an algorithm is known as the discrete-log problem. The belief in the security of these systems rests on the apparent difficulty of the discrete-log problem. This belief is largely based on anecdotal and historical evidence regarding the longstanding failure to find good algorithms. There is no known satisfactory theoretical statement regarding the difficulty of the discrete-log problem.

To compute a discrete exponentiation in the group $\boldsymbol{G}$ is to compute $y = \alpha^x$. To compute α^x efficiently, one can express the integer x in a binary representation of the form:

$$x = \sum_{i=0}^{m-1} x_i 2^i.$$

The integer m is the number of bits in the binary representation of x, and each x_i of the binary expression is either a zero or a one. To compute α^x, first precompute the sequence of successive squares, $\alpha, \alpha^2, \alpha^4, \alpha^8, \ldots, \alpha^{m-1}$. This requires $m - 2$ successive squarings. Then compute α^x by

$$\alpha^x = \prod_{i:x_i=1} \alpha^{2^i}.$$

Thus to compute α^x requires only $m - 2$ squarings in the group G and at most $m - 1$ multiplications in the same group. The number of operations at most is proportional to the logarithm of the order of the group G.

The inverse of exponentiation is the discrete logarithm which is believed to be computationally intractable for some large groups. Although one might be able to do somewhat better than the obvious procedure in those groups – computing consecutive powers of α until y is found – one cannot do sufficiently better. This is not the case for all groups. For example, finding a discrete logarithm in the group $\boldsymbol{Z}_n$ under addition amounts to solving the equation $y = ax$ for a. This is an easy calculation by using the extended euclidean algorithm. On the other hand, to compute a discrete logarithm in the finite field $\boldsymbol{F}_p^*$ is to compute the integer x that satisfies $\alpha^x = y$ when given the element α and the field element y. This computation of $x = \log_\alpha y$ is evidently a hard problem.

For example, the integer equation

$$3^x = 2 \pmod{9871},$$

despite its apparent simplicity, is not readily solved for x. To find x we can compute the powers of 3 modulo 9871 in sequence until 2 is obtained. Although there are somewhat better ways to find x, it appears that there are no ways that are substantially better.

More to the point, we will see that the complexity of the discrete-log problem in a group of order n is not substantially greater than the complexity of the discrete-log problem in a group of order equal to the largest prime factor of n. The reason is that if n can be factored, then the problem can be decomposed into smaller problems, one for each prime factor of n. These are solved individually and combined using the chinese remainder theorem. To avoid this decomposition, it is common practice in cryptographic applications to choose n to be a prime. If the order of the group n is composite, then there is little to be gained in security by choosing anything other than the largest prime factor of n, and choosing an element of this prime order as the logarithm base.

Furthermore, in view of the formula

$$\log_\alpha y = \log_\alpha \beta \log_\beta y,$$

the complexity of computing the discrete logarithm in a group G does not significantly depend on the logarithm base.

4.3 The Elgamal cryptosystem

The Elgamal cryptosystem is a cryptosystem that is based on the discrete-log problem. The Elgamal ciphertext is longer than the plaintext, usually twice as long, and this can be regarded as a disadvantage. On the other hand, the Elgamal cryptosystem is an asymmetric cryptosystem, and this can be regarded as an advantage. This means that the encryption key and the decryption key are different. Only one key need be known to the encryptor, and only one key need be known to the decryptor. A cryptanalyst who has learned only the encryption key cannot decipher the ciphertext. Even the encryptor itself cannot recover the plaintext from the ciphertext that it had created were the plaintext to be lost. The encryption key can even be made public and published in a directory.

The Elgamal cryptosystem can be formed using any cyclic group in which the discrete-log problem is intractable, or believed to be intractable. It will be broken if the discrete-log problem is solved. The cyclic group G can be $\boldsymbol{Z}_p^*$, or can arise as a cyclic subgroup of some other large group, such as the group of point addition on an elliptic curve, as will be discussed in Chapter 10.

The cyclic group G is made public and a chosen generator α of G is also made public. These may be regarded as the fixed part of the system. To prepare to receive a ciphertext, an integer a is randomly chosen secretly by the decryptor, and the encryption key $A = \alpha^a$ is computed and sent to the encryptor over a clear public channel. This key $A = \alpha^a$ can even be a publicly published encryption key for all messages sent by anyone to that decryptor. To encrypt message x, which must be represented as an element of G, the encryptor randomly and secretly chooses any integer b and computes α^b and xA^b in the group G. The encrypted message is $\boldsymbol{y} = (y_1, y_2) = e(x) = (\alpha^b, xA^b)$. The decryption computation is $d(y_1, y_2) = y_2/y_1^a$, which is easily seen to be equal to x.

To ensure cryptographic security, the exponent b must be used only once by the encryptor, which protects against a certain attack. This need not be a disadvantage because it is straightforward to randomly choose a new integer b for every plaintext message to be encrypted. This is because the random integer b is chosen by the encryptor and is not revealed to the decryptor, nor to any third party, so there is no communication burden due to using a new value of b for every message.

The Elgamal cryptosystem does not encrypt a fixed message into a fixed ciphertext. The actual ciphertext is random because of the random choice of b. Encrypting the same message twice will result in different ciphertexts. The Elgamal cryptosystem is not symmetric, and it is not a key exchange algorithm in the sense of the Diffie–Hellman key exchange. Of course, because the message x is arbitrary, the message to be encrypted could actually be a key for another cryptosystem that is to be transmitted by the Elgamal cryptosystem for future use with some other method of bulk encryption.

Example Let the group G be the group $\boldsymbol{F}_{2357}^*$, noting that 2357 is a prime. This 4-digit prime is, of course, too small for an actual system. Practical systems will use a prime of more than 50 digits, and perhaps even 100 digits or more. The element $\alpha = 2$ has order 2356 in $\boldsymbol{F}_{2357}$. In general, such an element α of order $q - 1$ of the finite field $\boldsymbol{F}_q$ is called a *primitive element* of $\boldsymbol{F}_q$. The group $\boldsymbol{F}_q$ and the primitive element α might be unique to a single decryptor or they might be a standard in a large community. In either case, they are public.

To initialize, the decryptor randomly chooses the integer $a = 1751$ and computes $A = \alpha^a$ as

$$\begin{aligned} A &= 2^{1751} \pmod{2357} \\ &= 1185, \end{aligned}$$

then sends this integer to the encryptor over a clear public channel. This is the public encryption key for that decryptor.

To encrypt the plaintext $x = 2035$, the encryptor randomly chooses the integer $b = 1520$ and computes

$$2^{1520} \pmod{2357} = 1430$$

and

$$2035 \cdot 1185^{1520} \pmod{2357} = 697.$$

The ciphertext is (1430, 697).

To decrypt the ciphertext (1430, 697), the decryptor first computes

$$1430^{-1751} = 1430^{605} = 872 \pmod{2357}.$$

The decryptor next computes

$$\begin{aligned} x &= 697 \cdot 872 \pmod{2357} \\ &= 2035 \end{aligned}$$

which recovers the plaintext x.

4.4 Trapdoor one-way functions

A *bijective function*, $y = f(x)$, meaning that each point of the input space is mapped to one and only one point of the output space, is called a *one-way function* if it is computationally tractable to compute y from x, but is computationally intractable to compute x from y according to the inverse function $x = f^{-1}(y)$. We will not try to define the notion of computationally intractable, preferring to leave it as an intuitive notion. Indeed, it is difficult to give a definition that is both useful and precise.

The one-way function $f(x)$ is called a *trapdoor one-way function* if the inverse map becomes computationally tractable when some pertinent side information is given. This side information offsets the computationally intractable part of the computation. The side information may be regarded as provided by a genie through a mathematical trapdoor.

To compute square roots modulo a biprime $n = pq$ is hard. This is the standard example of a trapdoor one-way function, and perhaps the only interesting example. Given that $n = pq$ where p and q are both prime, but not known, the task is to find x satisfying

$$x^2 = y \pmod{n}$$

when y is given. This is easy to solve if the factors p and q are known. If p and q are not known, then the computational problem is essentially as hard as the computational problem of factoring n to find p and q, as discussed in Section 2.13. This is an example of a trapdoor one-way function, the trapdoor being the values of p and q. In this problem, although computing y from x is easy, computing x from y is apparently computationally intractable. However, computing x from y becomes computationally easy if a genie divulges the two prime factors p and q.

To compute square roots modulo a prime p was discussed extensively in Section 2.13. This task is especially easy if $p \equiv 3 \pmod 4$. Recall that x is a square of $\boldsymbol{F}_p$ if, and only if, $x^{\frac{p-1}{2}} = 1 \pmod p$. We are given that x is a square, so if p has the form $4k + 3$, we can write

$$\begin{aligned} x &= x \cdot x^{\frac{p-1}{2}} = x^{\frac{p+1}{2}} \\ &= x^{\frac{4k+4}{2}} = (x^{k+1})^2. \end{aligned}$$

Hence the square roots of x (modulo p) are $\pm x^{k+1}$. If the prime p has the form $4k + 1$, then it is more involved to compute the square root, but not intractable. Algorithms for this case are given in Section 2.13.

To compute the square root of x modulo the biprime pq, when given the two prime factors p and q, one must first compute the square root modulo p and the square root modulo q and combine these using the chinese remainder theorem, which gives four square roots modulo pq. However, the task of the cryptanalyst is to compute square roots of $x \pmod n$ without knowing p and q. The security of a cryptographic system rests on the premise that this is computationally intractable when the trapdoor is closed.

4.5 The Massey–Omura cryptosystem

The *Massey–Omura cryptosystem* is a three-pass double-lock public cryptosystem that does not involve a public key or a prior secret key. It is not regarded as a practical cryptosystem. It is described here, not because of any presumed practicality, but because it illustrates another approach to cryptography. The Massey–Omura cryptosystem is based on the intractability of the discrete-log problem. The system uses a large cyclic group G of order n with a designated generator α. The group G, its order n, and the designated generator α are made public. The plaintext message x to be transmitted is represented by an element of G. The transmitter randomly chooses an integer $a \in Z_n^*$ and computes $a^{-1} \pmod{n}$, as by using the euclidean algorithm. The receiver randomly chooses an integer b and computes $b^{-1} \pmod{n}$, again as by using the euclidean algorithm. The four-step protocol for encryption of message m is as follows:

(1) The transmitter sends x^a, an element of G.
(2) The receiver observes x^a, computes $(x^a)^b$, and returns x^{ab}.
(3) The transmitter observes x^{ab}, computes $(x^{ab})^{a^{-1}}$, and now sends x^b.
(4) The receiver observes x^b, computes $(x^b)^{b^{-1}}$, and now has the message x.

The process is repeated for each message x or for each packet x of a packetized message. There is no need for a and b to remain the same from packet to packet. Only the transmitter of the message knows a, and so the transmitter can change it at will. Only the receiver of the message knows b, and so the receiver can change it at will.

The advantage of the Massey–Omura cryptosystem is that the distribution or exchange of keys between the two parties is avoided as such. Instead, this function is merged with message transmission. A minor disadvantage is that three passes are required: two transmissions from the message source and one from the destination. A more serious disadvantage of the Massey–Omura cryptosystem is that the system, as such, is vulnerable to an impersonation attack. A third party can choose a random integer c and respond to the initial message with $(x^a)^c = x^{ac}$. If the transmitter fails to recognize this as an impersonation, the response $(x^{ac})^{a^{-1}} = x^c$ is returned to the impersonating party, from which the message x is recovered. Therefore a method to authenticate the source of messages is essential, not only for its own sake, but also to protect the secrecy of the message. But then the need for authentication suggests that an alternative method of encryption consistent with the method of authentication should be used instead of the Massey–Omura cryptosystem.

Two comments are appropriate. We have asserted that $(x^{ab})^{a^{-1}} = x^b$. But this holds in every cyclic group because every cyclic group is abelian and so $x^{ab} = x^{ba}$. We have also asserted that the group element x represents the message, so as in any such cryptosystem, a map from messages into group elements must be specified. It is trivial if the group G is $F_{2^m}^*$. In this case, the message can be a nonzero binary message of length m, which is regarded as an m-bit field element.

4.6 The Pohlig–Hellman algorithm

The task of computing a discrete logarithm in a finite cyclic group G of order n generated by α is to find $x \in Z_n$ satisfying $y = \alpha^x$ in that group for the integer x when given α and y. An important instance of the discrete logarithm is in the cyclic group $G = F_p^*$, in which the group order n is equal to $p - 1$. Now the expression to be solved takes the form $y = \alpha^x \pmod{p}$.

The *Pohlig–Hellman algorithm* is a method in any cyclic group whose order n is composite to break down a discrete logarithm into discrete logarithms in smaller groups. The group $\boldsymbol{F}_p^*$ of order $n = \phi(p) = p - 1$ is an important example. Even the group $\boldsymbol{Z}_\ell^*$ with ℓ composite is an example. Because $\phi(p)$ and $\phi(\ell)$ are, in general, composite, the computation of the discrete logarithm can be replaced by the computation of multiple smaller discrete logarithms. Each smaller discrete logarithm is in one of those smaller cyclic subgroups whose order is a prime or a prime power that divides n. By means of the chinese remainder theorem, the solution to the larger discrete-log problem can be obtained from the set of solutions to the smaller discrete-log problems. To continue the computation, the discrete logs modulo a prime power can be broken into even smaller pieces, but in a different way, now recognizing the prime power as a product of multiple copies of the same prime.

The Pohlig–Hellman algorithm requires that not only the order n of the group be known, but also that the prime factorization of n be known. Because computing the prime factorization of n may be intractable, this requirement may already make the algorithm intractable. However, it is common practice during discussions of the Pohlig–Hellman algorithm to assume that the factorization is known. Accepting this assumption, the Pohlig–Hellman algorithm reduces the difficulty of the discrete-log problem to the order of the difficulty of the discrete-log problem in the largest subgroup of prime order. For this reason, a cryptographic scheme may be restricted to the largest cyclic subgroup of prime order of the given group. As a consequence of the Pohlig–Hellman algorithm, one prefers to use a group that has a large prime order so that there are no proper subgroups.

The Pohlig–Hellman algorithm computes the discrete logarithm for any cyclic group G of composite order. When the order of the group G is written in terms of its prime factorization, the group order n is expressed as a factorization into prime powers as given by

$$\begin{aligned} n &= \prod_{i=1}^{I} p_i^{k_i} \\ &= p_1^{k_1} p_2^{k_2} \cdots p_I^{k_I}, \end{aligned}$$

where $k_i \geq 1$. To compute $x = \log_\alpha y$, where α has order n, the computation will be organized into two loops, an outer loop and an inner loop. The outer loop will be indexed by i, which runs from 1 to I, and the ith iteration of the outer loop computes $x_i = \log_\alpha y \pmod{p_i^{k_i}}$. It does this by calling the inner loop for each i, which regards $p_i^{k_i}$ as the product of k_i copies of p_i. The inner loop will be indexed by ℓ, which runs from 1 to k_i for each value of i. The k_i iterations of the inner loop will be executed for each value of i.

The Pohlig–Hellman algorithm requires on the order of $\sum_{i=1}^{k} k_i(\sqrt{p_i}\log p_i + \log n)$ operations. One concludes that the complexity of solving a discrete-log problem of order n is dominated by the complexity of the discrete-log problem of order equal to the largest prime factor of n. This is why one usually prefers to use as the cryptographic group only the cyclic subgroup whose order is the largest prime factor of the order of some larger group because, in view of the Pohlig–Hellman algorithm, the smaller prime factors contribute little to the security if the factorization of n is known.

For example, the integer $p = 5^2 \cdot 2^{448} + 1$ is a very large prime, yet a discrete logarithm in $\boldsymbol{F}_p^*$ can be computed using the Pohlig–Hellman algorithm with only $2 + 448$ inner-loop iterations followed by an application of the chinese remainder theorem. This is because $p - 1 = 5^2 \cdot 2^{448}$. Moreover, in this instance, factorization of $p - 1$ is straightforward simply by repeated division by 2 until an odd remainder is obtained.

To explain the central idea of the outer loop, it is enough to first decompose a composite n into only two factors, given by $n = qr$. The factors q and r need not be coprime for the initial discussion. Indeed, the inner loop will have use for the case in which q and r are powers of the same prime p. However, when later we invoke the chinese remainder theorem, we must require the factors to be coprime.

To compute $x = \log_\alpha y$, suppose that $y = \alpha^x$ in a cyclic group of order $n = qr$. Write $y^r = (\alpha^x)^r$, and write $x = Qq + x_0$ with quotient Q and remainder x_0, so that

$$
\begin{aligned}
y^r &= (\alpha^{Qq+x_0})^r \\
&= (\alpha^n)^Q \alpha^{x_0 r} \\
&= (\alpha^r)^{x_0}.
\end{aligned}
$$

Now let $y_0 = y^r$ and $\beta = \alpha^r$ so that this expression becomes $y_0 = \beta^{x_0}$, and $x_0 = \log_\beta y_0$ with $x_0 = x \pmod q$. This has the same form as the original discrete-log problem, but has a smaller size. By the same reasoning, with r and q interchanged, we can write $y_0' = \gamma^{x_0'}$ and $x_0' = \log_\gamma y_0'$ where $y_0' = y^q$, $\gamma = \alpha^q$, and $x_0' = x \pmod r$. Thus the original task of solving $x = \log_\alpha y \pmod n$ has been replaced by the easier task of solving the two smaller discrete-log problems given by

$$
\begin{aligned}
x_0 &= \log_\beta y_0 \\
x_0' &= \log_\gamma y_0'.
\end{aligned}
$$

The solutions x_0 and x_0' to these two equations are equal to $x \pmod q$ and $x \pmod r$, respectively. If q and r are coprime, the two solutions can be combined using the chinese remainder theorem to obtain $x \pmod{qr}$.

The two smaller discrete-log problems can be computed by any suitable algorithm. In particular, if either q or r is itself composite, then that smaller problem can be decomposed in the same way as before. By repeating this process sufficiently many times, the original computation of a discrete logarithm can be decomposed into a collection of computations of smaller discrete logarithms, each of the form

$$
x_i = \log_{\beta_i} y_i.
$$

The outer loop of the Pohlig–Hellman algorithm decomposes n into all of the coprime factors $p_i^{k_i}$, where β_i has order $p_i^{k_i}$. Because the prime-power factors are coprime, the set of solutions x_i can be combined using the chinese remainder theorem to obtain x. This is the outer loop. It remains to describe the inner loop, which calculates x_i.

Before continuing, we begin a detailed running example, first breaking the problem into subproblems based on the prime powers comprising the outer loop. Computation of each prime-power discrete logarithm comprising the inner loop will be described after introducing the example. Consider the prime field $\boldsymbol{F}_{8101}$ and the generator $\alpha = 6$. The element 6 has order 8100. To compute $x = \log_6 7531$ in $\boldsymbol{F}_{8101}$, first observe the prime-power factorization

$$\begin{aligned} 8101 - 1 &= 2^2 \cdot 3^4 \cdot 5^2 \\ &= 4 \cdot 81 \cdot 25. \end{aligned}$$

Thus we must compute $x_1 = x \pmod 4$, $x_2 = x \pmod{81}$, and $x_3 = x \pmod{25}$ satisfying

$$\begin{aligned} y_1 &= \beta_1^{x_1} \pmod{8101} \\ y_2 &= \beta_2^{x_2} \pmod{8101} \\ y_3 &= \beta_3^{x_3} \pmod{8101} \end{aligned}$$

respectively, where the new terms are

$$\begin{aligned} y_1 &= y \pmod{8100/4} \text{ and } \beta_1 = \alpha^{8100/4}, \\ y_2 &= y \pmod{8100/81} \text{ and } \beta_2 = \alpha^{8100/81}, \\ y_3 &= y \pmod{8100/25} \text{ and } \beta_3 = \alpha^{8100/25}, \end{aligned}$$

all in $\boldsymbol{F}_{8101}^*$. For the moment, we do not specify how these three discrete logarithms

$$\begin{aligned} x_1 &= \log_{\beta_1} y_1 \\ x_2 &= \log_{\beta_2} y_2 \\ x_3 &= \log_{\beta_3} y_3 \end{aligned}$$

are to be computed. Any method of computing x_1, x_2, and x_3 will do for now, including direct search. No matter how they are computed, the individual discrete logs are found to be $x_1 = 1$, $x_2 = 47$, and $x_3 = 14$. The chinese remainder theorem then gives

$$\begin{aligned} x &= 1 \cdot 2025 x_1 + 64 \cdot 100 x_2 + 24 \cdot 324 x_3 \pmod{8100} \\ &= 6689, \end{aligned}$$

which is the desired solution.

Before continuing this running example, we must develop the rest of the algorithm. This is the inner loop, which computes a discrete log in a group of order n of the form p^k where p is a prime. Thus we must solve $y = \alpha^x \pmod{p^k}$ for x, which is written as $x = \log_\alpha y \pmod{p^k}$. The computation of the discrete log will be decomposed into

k passes through the inner loop. To this purpose, represent x as a base p integer, which to emphasize the notation, will now be denoted $x(p)$,

$$x(p) = \sum_{j=0}^{k-1} x_j p^j,$$

where $0 \le x_j < p$. This representation is more suggestive if x is expressed using *Horner's rule* as follows:

$$\begin{aligned} x(p) &= (\cdots((x_{n-1}p + x_{n-2})p + x_{n-3})p + \cdots + x_1)p + x_0 \\ &= x'(p)p + x_0, \end{aligned}$$

where $x'(p)$ denotes the outer parenthesized term in the previous line. Notice that if we remove the end term x_0 and then divide by p, the form of the expression is reproduced. Let $x_1(p) = (x(p) - x_0)/p$. Thus α^x can be written as

$$\alpha^{x(p)} = \alpha^{x_1(p)p}\alpha^{x_0},$$

which can be manipulated into the expression

$$\alpha^{x(p)}/\alpha^{x_0} = (\alpha^p)^{x_1(p)}.$$

This leads to the iterative procedure to compute $x = \log_\alpha y$, where α has order p^k. First compute x_0 by writing the decomposition $n = pp^{k-1}$ and using $\beta = \alpha^{n/p^{k-1}}$. Then in order to compute x_1, compute $\alpha^{x(p)}/\alpha^{x_0}$ and replace α by α^p. Then repeat the same process.

This completes the development of the Pohlig–Hellman algorithm, which has reduced the discrete log to multiple discrete logs, each of prime order. The algorithm depends on the availability of suitable methods of computing discrete logs of prime order. A flow diagram of the logic of the Pohlig–Hellman algorithm is shown in Figure 4.1.

We will now finish the running example of the discrete logarithm of $y = 7531$ that was started earlier in the prime field $\boldsymbol{F}_{8101}$, as generated by $\alpha = 6$. To compute $x = \log_6 7531$ in $\boldsymbol{F}_{8101}$, recall the factorization $8101 - 1 = 2^2 \cdot 3^4 \cdot 5^2$. The computation was replaced earlier by three smaller computations of the three residues

$$x_1 = \log_{\beta_1} y_1 \pmod{4},$$

where $\beta_1 = 6^{8100/4} \pmod{8101}$ has order 4, and

$$x_2 = \log_{\beta_2} y_2 \pmod{81},$$

where $\beta_2 = 6^{8100/81} \pmod{8101}$ has order 81, and

$$x_3 = \log_{\beta_3} y_3 \pmod{25}$$

where $\beta_3 = 6^{8100/25} \pmod{8101}$ has order 25.

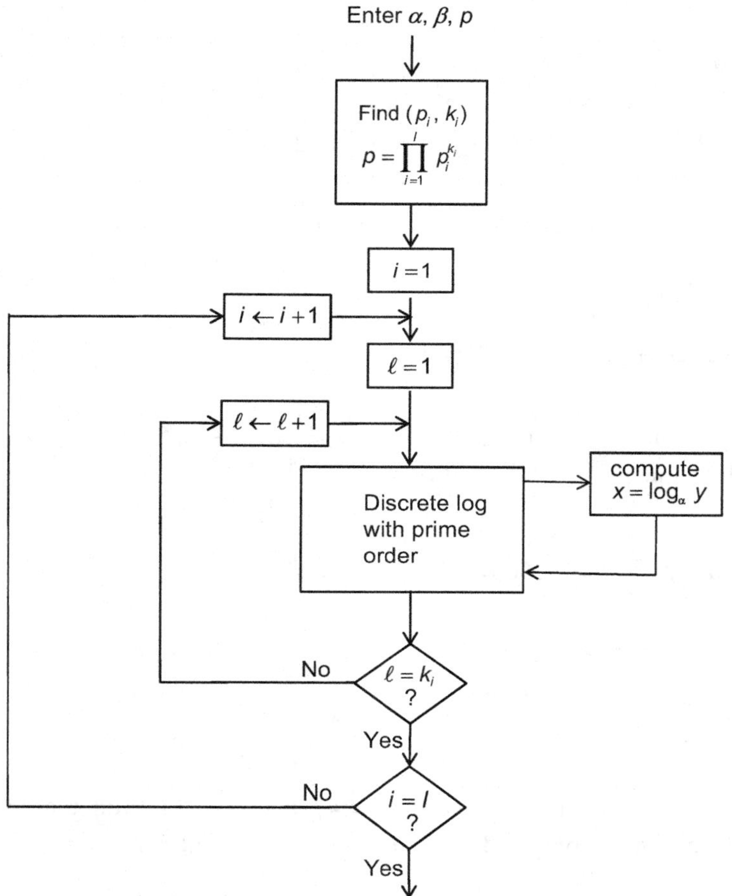

Figure 4.1 The Pohlig–Hellman algorithm

Step 1 To compute $x_1 = x \pmod 4$, let $x_1 = x_{11}2^1 + x_{10}2^0$, where $x_{11}, x_{10} \in \{0, 1\}$. Define $\beta_0 = \alpha^{8100/2}$ and $\beta_1 = \alpha^{8100/4}$. Then

$$\begin{aligned} y_{10} &= 7531^{8100/2} &&= -1 = \beta_0^1, \text{ so } x_{10} = 1 \\ y_{11} &= (7531/\alpha^{x_{10}})^{8100/4} &&= \quad 1 = \beta_1^0, \text{ so } x_{11} = 0. \end{aligned}$$

Therefore

$$\begin{aligned} x_1 &= x_{11}2 + x_{10} \\ &= 0 \cdot 2 + 1 = 1. \end{aligned}$$

Step 2 To compute $x_2 = x \pmod{81}$, let $x_2 = x_{23}3^3 + x_{22}3^2 + x_{21}3^1 + x_{20}3^0$, where $x_{23}, \ldots, x_{20} \in \{0, 1, 2\}$. Define $\beta_0 = \alpha^{8100/3} = 5883$, $\beta_1 = \alpha^{8100/9}$, $\beta_2 = \alpha^{8100/27}$, and

$\beta_3 = \alpha^{8100/81}$. Then

$$\begin{aligned}
y_{20} &= 7531^{8100/3} &&= 2217 = \beta_0^2, \text{ so } x_{20} = 2,\\
y_{21} &= (7531/\alpha^{x_{20}})^{8100/9} &&= 1 = \beta_1^0, \text{ so } x_{21} = 0,\\
y_{22} &= (7531/\alpha^{3x_{21}})^{8100/27} &&= 6735 = \beta_2^2, \text{ so } x_{22} = 2,\\
y_{23} &= (7531/\alpha^{9x_{22}})^{8100/81} &&= 5883 = \beta_3^1, \text{ so } x_{23} = 1.
\end{aligned}$$

Therefore

$$\begin{aligned}
x_2 &= x_{23}3^3 + x_{22}3^2 + x_{21}3^1 + x_{20}3^0\\
&= 1 \cdot 3^3 + 2 \cdot 3^2 + 0 \cdot 3^1 + 2 \cdot 3^0 = 47.
\end{aligned}$$

Step 3 To compute $x_3 = x \pmod{25}$, let $x_3 = x_{31}5^1 + x_{30}5^0$, where $x_{31}, x_{30} \in \{0, 1, 2, 3, 4\}$. Define $\beta_0 = \alpha^{8100/5}$ and $\beta_1 = \alpha^{8100/25}$. Then

$$\begin{aligned}
y_{30} &= 7531^{8100/5} &&= 5221 = \beta_0^4, \text{ so } x_{30} = 4,\\
y_{31} &= (7531/\alpha^{x_{30}})^{8100/25} &&= 356 = \beta_1^2, \text{ so } x_{31} = 2.
\end{aligned}$$

Therefore

$$\begin{aligned}
x_3 &= x_{31}5^1 + x_{30}5^0\\
&= 2 \cdot 5 + 4 = 14.
\end{aligned}$$

This completes the calculation of $x_1 = 1$, $x_2 = 47$, and $x_3 = 14$. From these residues, the discrete logarithm is computed using the chinese remainder theorem as described earlier.

For a larger example of the power of the Pohlig–Hellman algorithm in a prime field, consider the 107-digit prime

$$\begin{aligned}
p = {}&22708823198678103974314518195029102158525052496759285596453\\
&2691897983114274751597764112766422771396508339 37.
\end{aligned}$$

In this case, because all its factors are moderate, it is possible to factor $p - 1$ as

$$p - 1 = 2^4 \cdot 104729^8 \cdot 224737^8 \cdot 350377^4,$$

by repeated trial division by all primes smaller than 1 million. Accordingly, for this prime p, it is easy to compute discrete logarithms in $\boldsymbol{F}_p^*$ using the Pohlig–Hellman algorithm. This is only because $p - 1$ has acceptably small factors for this p. For a more typical p, the discrete logarithm will be protected by the intractability of the factoring problem, as well as by the intractability of the discrete logarithm of the largest prime factor.

4.7 The Shanks algorithm

The task of computing a discrete logarithm in a finite cyclic group G of order n generated by the element α is to find $x \in \mathbf{Z}_n$ satisfying $y = \alpha^x$ when given α and y. The integer x is written as $x = \log_\alpha y$.

The *Shanks algorithm* is a method of computing a discrete logarithm based on a trade between arithmetic requirements and memory requirements. The Shanks algorithm is also called the *Shanks baby-step, giant-step method* and has some similarity to a meet-in-the-middle algorithm. The Shanks algorithm is the fastest known algorithm for finding a discrete logarithm that works in an arbitrary cyclic group. Algorithms that work only for certain groups or only for certain values of the group order may be faster in appropriate situations because such algorithms rely on more specific properties of the particular group.

Let $x = \log_\alpha y$ be the inverse of $y = \alpha^x$ in a cyclic group G, where the logarithm base α generates the cyclic group G and so α has order n where n is the order of the group G. The task is to compute $x = \log_\alpha y$, which means to compute the integer x from y and α. The solution is the integer x for which $y = \alpha^x$. The Shanks algorithm reduces the computation time at the expense of a large memory. The amount of computation is proportional to $n^{1/2}$, where $n = p - 1$. The amount of required storage is also proportional to $n^{1/2}$.

Observe that any integer x can be written in terms of a fine index j and a coarse index i as $x = mi + j$ where m is a fixed integer and both i and j are smaller than m. Then the expression $y = \alpha^x$ becomes

$$y = \alpha^{mi+j},$$

which means that

$$(\alpha^{-m})^i y = \alpha^j.$$

We will solve the equation in this form by a combination of computation and stored precomputations. To get a good balance between these two, m should be chosen approximately equal to $\sqrt{p-1}$. The integer

$$m = \lfloor \sqrt{p-1} \rfloor$$

will do. Then i and j satisfy the conditions $0 \le i \le m-1$ and $0 \le j \le m-1$.

The Shanks algorithm to compute $\log_\alpha y$ consists of four steps. The first two steps provide an initialization that only needs to be done once and stored.

Step 1 Compute α^j for $j = 0, \ldots, m-1$ and store $\gamma_j = \alpha^j$. This gives a table of γ_j versus j.

Step 2 Sort the entries of this table by the values of γ_j to get a table of j versus γ_j. This gives a look-up table of $\log_\alpha \gamma$ that gives j for each computed value of γ.

Step 3 Compute $\alpha^{-m} = \alpha^{n-m}$, as by the square-and-multiply method.

Step 4 For $i = 0, 1, 2, \ldots$, compute $\alpha^j = (\alpha^{-m})^i y$ and look up α^j in the table to find j.

Because the first two steps can be regarded as a precomputation, the table computed in Step 2 can be saved and reused for many values of y. To compute the discrete log of a given y, the fourth step may be first executed for all i to form a second table. The two tables are then searched for a common entry.

As an example of the Shanks algorithm, we will compute $x = \log_5 13$ in $\boldsymbol{F}^*_{23}$, which we know to be a cyclic group because $\boldsymbol{F}^*_q$ is a cyclic group for every finite field $\boldsymbol{F}_q$. Because 23 is prime, the multiplicative group $\boldsymbol{F}^*_{23}$ has order $n = 22$, and $\alpha = 5$ is an element of order 22.

In a preliminary step, choose $m = \lfloor\sqrt{n}\rfloor = 4$, so that $x = 4i + j$. In Step 1, compute the first m powers of 5 in $\boldsymbol{F}_{23}$. This gives the following table of 5^j in $\boldsymbol{F}_{23}$ for $j = 0, \ldots, 3$.

$$5^0 = 1$$
$$5^1 = 5$$
$$5^2 = 2$$
$$5^3 = 10.$$

In Step 2, sort these entries in terms of the entries on the right side in order to invert the list. The result is the table

$$\log_5 1 = 0$$
$$\log_5 2 = 2$$
$$\log_5 5 = 1$$
$$\log_5 10 = 3.$$

Both Step 1 and Step 2 can be regarded as precomputations. We will also need to know $\alpha^{-m} = 5^{-4}$, which can be computed as follows:

$$5^{-1} = 14$$
$$5^{-2} = 14^2 = 12$$
$$5^{-4} = 12^2 = 6.$$

To compute $\log_5 y$ for $y = 13$, first check the precomputed table to see that $\log_5 13$ is not already an entry, in which case the algorithm would halt. Because $\log_5 13$ is not an entry in the table, the algorithm continues as follows: compute $(5^{-4})^i y = 6^i y$ for

$i = 0, 1, 2, 3, 4, 5$ in $\boldsymbol{F}_{23}$. This gives the list of $6^i y = 13, 9, 8, 2, 12, 3$

$$6^0 y = 13$$
$$6^1 y = 9$$
$$6^2 y = 8$$
$$6^3 y = 2$$
$$6^4 y = 12$$
$$6^5 y = 3.$$

Because 2 is the common entry on this list and the precomputed list and $2 = 5^2$, this leads to

$$6^3 y = 5^2.$$

Because $6 = 5^{-4}$, this can be written

$$y = 5^{4 \cdot 3 + 2} = 13.$$

Thus, $5^{14} = 13$, so $\log_5 13 = 14$.

4.8 The Pollard algorithm for discrete logarithms

The task of computing a discrete logarithm in a finite cyclic group G of order n generated by α is to find $x \in \boldsymbol{Z}_n$ satisfying $y = \alpha^x$ when given α and y. The integer x is written as $x = \log_\alpha y$.

The *Pollard algorithm* for computing discrete logarithms is regarded as a probabilistic algorithm because its computational burden varies widely with y. Its expected running time is proportional to $\sqrt{n}$ which is an enormous improvement compared to direct search, but is still inadequate for most applications. The expected running time of the Pollard algorithm is the same as the expected running time of the Shanks algorithm. The Pollard algorithm only requires a modest amount of storage. For this reason, in practice, the Pollard algorithm is often preferred to the Shanks algorithm. However, the running time of the Pollard algorithm is more unpredictable, and the algorithm may fail because of an unacceptable waiting time for the answer. The Pollard algorithm for computing discrete logarithms does not make use of any properties that are unique to the group G, so it is valid for any group.

In the simplest case, we require that the finite group G has an order that is a prime. This means that G has no proper subgroups. Because the group order is a prime, the group G must be a cyclic group, and so it is abelian. The group G is isomorphic to $\boldsymbol{Z}_n$, where n is the order of G.

The plan of the algorithm is to choose an appropriate function f and to compute the sequence $\gamma_i = f(\gamma_{i-1})$ for $i = 1, \ldots,$ in hopes of finding a collision, meaning a repetition $\gamma_k = \gamma_j$ for some k and some j smaller than k. This collision must eventually occur because the group is finite, but if the group is very large, the waiting time may be unacceptable, in which case the algorithm fails.

After the first collision, the next term must satisfy $\gamma_{k+1} = \gamma_{j+1}$, so the sequence must repeat as before. Therefore, the sequence has entered a cycle where it remains thereafter periodically returning to the value of γ_j. For this reason, the Pollard algorithm for computing discrete logarithms is called a *rho algorithm*, because the symbol "rho" has the appearance of a circle with an entering tail attached. The cycle-finding structure of the Pollard discrete-log algorithm can also be used to form an algorithm for integer factoring, called the Pollard rho factoring algorithm.

To define the function f, the group G is partitioned into three sets $\mathcal{S}_1$, $\mathcal{S}_2$, and $\mathcal{S}_3$ of approximately equal size, and such that it is convenient to determine to which set an element of G belongs. The only other condition on the partition is that $\mathcal{S}_2$ not contain the group identity element. To compute $\log_\alpha \beta$, the function f is defined as

$$f(\gamma) = \begin{cases} \beta\gamma & \text{if} \quad \gamma \in \mathcal{S}_1 \\ \gamma\gamma & \text{if} \quad \gamma \in \mathcal{S}_2 \\ \alpha\gamma & \text{if} \quad \gamma \in \mathcal{S}_3 \end{cases}$$

with $\gamma_0 = 1$ and $\gamma_{i+1} = f(\gamma_i)$. The reason that $\mathcal{S}_2$ must not contain the identity element e is now evident. If it did, then $f(e)$ would be the same as e, and the algorithm would stagnate. At the first collision, which is defined as $\gamma_k = \gamma_j$, we have $\alpha^{a_k}\beta^{b_k} = \alpha^{a_j}\beta^{b_j}$ where a_i and b_i at iteration i count the number of times that the algorithm has passed through $\mathcal{S}_1$ and $\mathcal{S}_3$, respectively, with adjustments to account for passages by the algorithm through $\mathcal{S}_2$.

For example, if G is equal to the group $\mathbf{Z}_n^*$, then an easy and common way to define the partition $\{\mathcal{S}_1, \mathcal{S}_2, \mathcal{S}_3\}$ is by the modulo-three value of the integers of G. Because $\mathcal{S}_2$ must not contain zero, we define these sets as

$$\mathcal{S}_1 = \{x : x = 0 \pmod{3}\}$$
$$\mathcal{S}_2 = \{x : x = 1 \pmod{3}\}$$
$$\mathcal{S}_3 = \{x : x = 2 \pmod{3}\}.$$

With the specification that $\gamma_0 = 1$ and $\gamma_i = f(\gamma_{i-1})$, the function produces the sequence $\gamma_1, \gamma_2, \gamma_3, \ldots$. We will want to keep track of how many copies of α or β have been multiplied together in order to compute γ, so we introduce the two integers a and b to count these multiples. For this reason, γ_i is replaced by the triplet (γ_i, a_i, b_i), and the function f is redefined on the triplet as

$$f(\gamma, a, b) = \begin{cases} (\beta\gamma, a, b+1) & \text{if} \quad \gamma \in \mathcal{S}_1 \\ (\gamma\gamma, 2a, 2b) & \text{if} \quad \gamma \in \mathcal{S}_2 \\ (\alpha\gamma, a+1, b) & \text{if} \quad \gamma \in \mathcal{S}_3. \end{cases}$$

Each time the algorithm passes through $\mathcal{S}_1$ or $\mathcal{S}_3$, respectively, γ is multiplied by β or α and the count b or a is increased by one. Each time the algorithm passes through $\mathcal{S}_2$, γ is squared and the current values of a and b are both doubled.

In this way, a specifies the number of copies of α that have appeared in γ, and b specifies the number of copies of β that have appeared in γ. Then the sequence element (γ_i, a_i, b_i) allows us to write immediately that $\gamma_i = \alpha^{a_i}\beta^{b_i}$.

The algorithm depends on a collision. In particular, there is a collision whenever $\gamma_i = \gamma_k$. Then

$$\alpha^{a_i}\beta^{b_i} = \alpha^{a_k}\beta^{b_k},$$

which can be rewritten

$$\alpha^{a_i + b_i \log_\alpha \beta} = \alpha^{a_k + b_k \log_\alpha \beta}.$$

Therefore

$$a_i + b_i \log_\alpha \beta = a_k + b_k \log_\alpha \beta \pmod{n}.$$

This means that

$$\log_\alpha \beta = -\frac{a_i - a_k}{b_i - b_k} \pmod{n}$$

is the desired solution to the discrete-log problem.

As described up to this point, the computation has a significant and unacceptable limitation: the entire sequence of γ_i for $i = 1, 2, \ldots,$ must be retained and examined to recognize the first collision. To avoid this, instead of finding the first collision, the algorithm is modified to find any later collision that can be found without storing the entire sequence. To this end, notice that it is not essential that the first collision be the one observed. Any collision will do. If the first collision occurs when $\gamma_{\mu+\lambda} = \gamma_\mu$, then $\gamma_{\mu+\lambda+i} = \gamma_{\mu+i}$ for all nonnegative i, which provides a continuing stream of collisions. In particular, when $i = \lambda - \mu$, we have the collision $\gamma_{2\lambda} = \gamma_\lambda$. The advantage of waiting for this collision is that for each i only γ_i and γ_{2i} need to be compared. There is no need to compare γ_i to all prior terms in the sequence, so there is no need to store the terms of the sequence. The minor disadvantage is that 2λ terms of the sequence must be computed rather than $\lambda + \mu$ terms.

This collision is detected by simultaneously computing the sequence twice, once on each of two tracks, with the computation of the fast track executing two iterations for every one iteration of the computations on the slow track. Thus, the fast track computes

Table 4.1 *Example of Pollard algorithm*

i	γ_i	a_i	b_i	γ_{2i}	a_{2i}	b_{2i}
1	228	0	1	279	0	2
2	279	0	2	184	1	4
3	92	0	3	14	1	6
4	184	1	4	256	2	7
5	205	1	5	304	3	8
6	14	1	6	121	6	18
7	28	2	6	144	12	38
8	256	2	7	235	48	152
9	152	2	8	72	48	154
10	304	3	8	14	96	309
11	372	3	9	256	97	310
12	121	6	18	304	98	311
13	12	6	19	121	196	624
14	144	12	38	144	392	1250

γ_{2i} at the same time that the slow track computes γ_i. These two values are compared when they are computed. If they are the same, computation of the sequence stops. At this time, all quantities for the final computation of the discrete logarithm according to the formula

$$\log_\alpha \beta = -\frac{a_i - a_{2i}}{b_i - b_{2i}} \pmod n$$

are known, and the computation can be completed.

For an example of the Pollard algorithm, we will compute $\log_2 228$ in the prime field $\boldsymbol{F}_{383}$, as is summarized in Table 4.1. Because the integer 2 has order 191 (mod 383), not every nonzero element of $\boldsymbol{F}^*_{383}$ has a base-two logarithm. In the language of fields, 2 is an element of order 191 in the field $\boldsymbol{F}_{383}$, so it generates a cycle with 191 elements. Only elements in this cycle have a base-two logarithm in $\boldsymbol{F}_{383}$. To compute $\log_2 228$, let G be the cyclic subgroup of $\boldsymbol{F}^*_{383}$ of order $n = 191$ generated by 2 (mod 383). Let G be partitioned as $\{\mathcal{S}_1, \mathcal{S}_2, \mathcal{S}_3\}$, as was described earlier. With the values $\alpha = 2$ and $\beta = 228$, start the iteration with the initialization $\gamma_0 = 1$, $a_0 = 0$, and $b_0 = 0$. The subsequent values of γ_i, a_i, and b_i are shown in Table 4.1, the slow-track computations are on the left of Table 4.1 and the fast-track computations are on the right. Because γ_i is an element of $\boldsymbol{F}_{383}$, it is reduced modulo 383. Because G is a group of order 191, the exponents a_i and b_i are reduced modulo 191. To keep the iterates small, all entries would be so reduced as they are computed, but for ease of understanding, a_i and b_i have not been reduced in this table. The first repetition of γ_i is at $\gamma_{14} = \gamma_{28}$ because both equal 144. To complete the computation, note that

$b_{28} - b_{14}\,(\text{mod } 191) = 125$ and $a_{14} - a_{28}\,(\text{mod } 191) = 189$. Then the result of the computation is $\log_2 228 = -189/125\,(\text{mod } 191) = 110$.

4.9 The method of index calculus

Index calculus is a subexponential method of computing a discrete logarithm $\log_\alpha x$ in F_p^* or F_2^{*m} by breaking the computation into small pieces. It is a deterministic algorithm for any instance, but the computational burden varies widely and unpredictably with x so, in this sense, it has the appearance of a probabilistic algorithm. It is the algorithm of choice for very large discrete-log problems.

Suppose that we want to compute x such that $3^x \equiv 2\,(\text{mod } 9871)$. This is expressed as the computation of the discrete logarithm given by $x = \log_3 2$ in F_{9871}. We first work out small powers of $3\,(\text{mod } 9871)$ and save those "smooth" integers that have only small primes as factors. For this purpose, an arbitrary limit on the size of the set of small primes is specified based on the amount of computation that is deemed to be acceptable. The specified set of small primes is called a *factor base* for the computation. The specified number of small primes is arbitrary, and can be increased when necessary. For a given instance of the problem, the algorithm may fail if the specified factor base is too small. However, the computation may be too formidable if the specified factor base is too large.

For example, to compute $\log_3 2$ in F_{9871}, compute and factor all powers of 3 (mod 9871) up to 3^{86} (mod 9871), and list those that as integers have no prime factors larger than 31, the prime 31 being chosen as an arbitrary limit on the amount of computation. The list of such powers of 3, starting at 3^{19}, is

$$\begin{aligned}
3^{19} &\equiv 2^2 \cdot 11 \cdot 13,\\
3^{45} &\equiv 2 \cdot 7 \cdot 11 \cdot 17,\\
3^{62} &\equiv 5^2 \cdot 31,\\
3^{65} &\equiv 7 \cdot 13^2,\\
3^{70} &\equiv 2 \cdot 5 \cdot 11^2,\\
3^{80} &\equiv 2^4 \cdot 11 \cdot 17,\\
3^{86} &\equiv 2^2 \cdot 7 \cdot 11 \cdot 31.
\end{aligned}$$

The prime 3 is omitted from the factor base because 3 is the logarithm base.

Now, for any choice of the parameters, $a, b, c, \ldots,$ and g, multiply powers of all terms on the left, and the corresponding powers of all terms on the right as follows:

$$(3^{19})^a(3^{45})^b \ldots (3^{86})^g \equiv 2^{2a+b+e+4f+2g} \cdot 5^{2c+e} \cdot 7^{b+d+g} \cdot 11^{a+b+2e+f+g} \cdot 13^{a+2d} \cdot 17^{b+f} \cdot 31^{c+g}.$$

Next, because we are looking for the solution of $3^x = 2$, set each exponent to zero except the exponent of 2. This gives the set of equations

$$\begin{aligned} 2c + e &= 0 \\ b + d + g &= 0 \\ a + b + 2e + f + g &= 0 \\ a + 2d &= 0 \\ b + f &= 0 \\ c + g &= 0. \end{aligned}$$

If this set of equations is satisfied, the right side of the earlier equation reduces to a power of 2. Solving this set of linear equations is elementary linear algebra. There are six equations and seven unknowns, so six of the parameters can be expressed in terms of any one of them. We see immediately that $a = -2d$, $e = -2c$, $f = -b$, and $g = -c$, so the remaining two equations reduce to

$$\begin{aligned} b + d - c &= 0 \\ 2d + 5c &= 0. \end{aligned}$$

To reduce these equations to a single unconstrained parameter, let $c = -2x$ where x is arbitrary. Then $d = 5x$, $b = -7x$, $a = -10x$, $e = 4x$, $f = 7x$, and $g = 2x$. Choosing $x = 1$, one solution of the system of equations is $a = -10$, $b = -7$, $c = -2$, $d = 5$, $e = 4$, $f = 7$, and $g = 2$.

This solution allows the term appearing as the exponent of 2, which is $2a + b + e + 4f + 2g$, to be evaluated. It is equal to 9. The situation now reduces to

$$\begin{aligned} 2^9 \pmod{9871} &= (3^{19})^a (3^{45})^b \ldots (3^{86})^g \\ &= 3^{708}. \end{aligned}$$

Then $2^9 = 3^{708}$ in $\boldsymbol{F}_{9871}$. This means that $2 = 3^{\frac{708+9870k}{9}}$ for some k, and it is enough to consider only k in the interval $0 \le k \le 8$. Finally, test each k, for $k = 0, \ldots, 8$, to find a k for which the exponent of 3 is an integer. Indeed, with $k = 2$, this becomes $2 = 3^{2272}$. We can conclude that $\log_3 2 = 2272$ in $\boldsymbol{F}_{9871}$.

If x is not a product of primes in the factor base, the attempt is not yet finished. Test in turn the integer $2^k x \pmod{9871}$ for each k to see if it factors totally, as an integer, in the factor base as $3^k x = 2^{i_2} 5^{i_5} 7^{i_7} 11^{i_{11}} 13^{i_{13}} 17^{i_{17}} 31^{i_{31}}$. Upon finding such a k, the discrete logarithm is given by

$$\log_3 x = i_2 \log_3 2 + i_3 \log_3 5 + \cdots + i_{31} \log_3 31 - k.$$

For a field $\boldsymbol{F}_p$ with very large p, it will be intractable to test most values of k, so the procedure may still fail. If the search over k is unsuccessful, the index calculus attempt has failed. It may be attempted again with a larger factor base, or abandoned.

To compute discrete logarithms in very large fields using index calculus, the number of linear equations may be many thousands, even millions. Thus computing a discrete logarithm with index calculus is tractable only for moderate-sized problems in $\boldsymbol{F}_p^*$. It can be countered by increasing the size of the group $\boldsymbol{F}_p^*$ by increasing p. It can also be countered by replacing $\boldsymbol{F}_p^*$ with an alternative group in which the method of index calculus is not available.

4.10 Complexity of the discrete-log problem

The security of many public-key cryptosystems depends on the intractability of the discrete-log problem in the particular group used by that cryptosystem. Accordingly, one wants to make formal statements about the computational intractability of the discrete-log problem. However, formal statements of an adequate form are not known, so we cannot make definite statements that are fully satisfying. Only anecdotal evidence is known to support the view that the discrete-log problem is intractable. Much of this extensive and rather compelling evidence about intractability has been garnered by searching for algorithmic methods to compute discrete logarithms. Upper bounds on complexity arise from known algorithms and the complexity statements are usually asymptotic without extensive discussion of practical instances. Lower bounds on complexity are generally absent.

In principle, the discrete-log problem always can be solved by computing a table of x versus α^x, then sorting this table into an inverse table of α^x versus x. This, however, is exponentially expensive in storage and also requires an exponentially large amount of precomputation. For these reasons, this straightforward method is not feasible for large groups.

Many approaches to solve the discrete-log problem have been explored. They may be categorized as follows:

(i) Algorithms that work in any cyclic group.
(ii) Algorithms that work in any cyclic group that has many small subgroups.
(iii) Index calculus methods that depend on the specifics of a particular group.
(iv) Methods that exploit isomorphisms between groups.

Not surprisingly, a method that can be used for an arbitrary cyclic group is typically not the best method for a specific group because it will not take advantage of special properties of that group. Index calculus is not a method that can be used for an arbitrary group. It works in the multiplicative group of $\boldsymbol{F}_p$ and can be generalized in some closely related situations, but with some possible exceptions, it has not been successfully applied to other groups that lack the structure of a field, and apparently cannot be so applied. Index calculus requires the notion of smoothness of group elements so that some sort of factor base can be provided. Such has been shown to be possible, for example, for groups based on hyperelliptic curves of large genus.

When given a collection of computational problems parameterized by n, one may express the complexity of that problem by describing the asymptotic growth with n of the amount of computation. One attempts to find such a description by deriving a lower bound and an upper bound. If these bounds agree asymptotically, the asymptotic complexity is known. One way that this might be done is by finding the asymptotic complexity of a known algorithm for that problem, then proving that no other algorithm can be asymptotically better. However, even though an analysis of asymptotic complexity gives valuable insight into the difficulty of a problem, the asymptotic complexity need not describe realistic instances of a problem. It may be that the asymptotic complexity is meaningful only for outrageously large instances of a problem. Nevertheless, asymptotic statements are instructive and accepted as useful guides.

To compare the asymptotic behavior of two nonnegative real functions, one defines the notation $o(n)$ and $O(n)$. These are not functions of n in the usual sense. Instead they are descriptions of asymptotic behavior defined as follows. The expression $f(n) = O(g(n))$, comparing two functions $f(n)$ and $g(n)$, means that for some constant c and for all sufficiently large n, $f(n) \leq cg(n)$. The expression $f(n) = o(g(n))$ means that $\lim_{n\to\infty} f(n)/g(n) = 0$. In particular, any function that grows no faster than linearly with n is an $O(n)$ function. An $o(n)$ function is one that goes to zero as n goes to infinity. Every $o(n)$ function is, of course, an $O(n)$ function.

To describe the computational complexity of a problem or an algorithm asymptotically, define the general complexity classes of *polynomial complexity*, *subexponential complexity*, and *exponential complexity*. To this end, one may use the expression

$$L_n[\alpha, c] = O(e^{(c+o(1))(\log n)^\alpha (\log\log n)^{1-\alpha}})$$

where $0 \leq \alpha \leq 1$ and c is a constant. The term $\exp(c(\log n)^\alpha (\log\log n)^{1-\alpha})$ describes the dominant asymptotic behavior. The $o(n)$ term captures the remainder. If $\alpha = 0$, then the expression $L(n, c)$ describes a function that is polynomial in $\log n$. Thus

$$L_n[0, c] = (\log n)^{c+o(1)}.$$

If $\alpha = 1$, then $L(n, \alpha)$ describes a function that is exponential in $\log n$. Thus,

$$L_n[1, c] = n^{c+o(1)}.$$

If $0 < \alpha < 1$, then $L(n, \alpha)$ describes a function that is subexponential in $\log n$. A subexponential function in $\log n$ grows more quickly than any polynomial in $\log n$, but more slowly than an exponential in $\log n$.

A compressed summary of the asymptotic complexity of known attacks in various groups (some of which are not defined until later in the book) is shown in Table 4.2. The first column states the specific algebraic system, and the other three columns summarize what is evidently known about complexity of algorithms for the discrete-log problem in each system. These conclusions, however, must be understood with

Table 4.2 *Complexity of known attacks on the discrete-log problem*

	Complexity		
	Exponential	Subexponential	Polynomial
$\boldsymbol{Z}/p\boldsymbol{Z}$			euclidean
$\boldsymbol{Z}_p^*$	Pollard	index calculus	none expected
$\mathcal{X}(\boldsymbol{F}_q)$	Pollard	none expected	none expected
jac($\mathcal{X}$)	Pollard	index calculus (?)	none expected

appropriate caution. The statement that no subexponential algorithms for computing discrete logarithms in $\mathcal{X}(\boldsymbol{F}_q)$ are expected, is a statement that refers to the class of all such elliptic curves. It may be, and is true, that specific subclasses of elliptic curves do have a subexponential discrete-log algorithm.

In the multiplicative group $\boldsymbol{F}_p^*$ of the finite field $\boldsymbol{F}_p$, the complexity of the discrete-log problem is subexponential, though evidently still very large. Nevertheless, this observation has caused some concern about the security of a cryptosystem based on the group $\boldsymbol{F}_p^*$. That concern is based on the asymptotic complexity of the discrete-log problem in this group, rather than the difficulty of any particular instance of practical intractability.

The complexity of the discrete-log problem in the multiplicative group $\boldsymbol{F}_p^*$ of the finite field $\boldsymbol{F}_p$ is known to grow asymptotically not faster than

$$e^{c(\log P)^{\frac{1}{3}}(\log\log p)^{\frac{2}{3}}},$$

where P is the largest prime divisor of $p-1$, and the constant parameter c is about 1.4. This asymptotic statement follows from a complexity analysis of the index-calculus attack.

The last two groups listed in Table 4.2, $\mathcal{X}(\boldsymbol{F}_q)$ and jac$\mathcal{X}(\boldsymbol{F}_q)$, are groups defined on elliptic curves and hyperelliptic curves, respectively. These groups are discussed in Chapters 10 and 11. The absence of a known subexponential attack on these two groups for small genus is an important motivation for introducing these two groups into cryptography. In particular, because there is no notion of smoothness in the group of points of an elliptic curve, the method of index calculus does not have a successful generalization. This statement also holds for hyperelliptic curves of small genus, and, to a lesser extent, for hyperelliptic curves of larger genus. The intractability of the discrete-log problem in any group leads to its possible use both for cryptographic key exchange and for digital signatures.

In the finite field $\boldsymbol{F}_q$, the asymptotic complexity of the number-field sieve is described heuristically as follows. For $q=2^m$, the running time is $L\,(2^m, q/3)$, with $c\approx 1.4$. For

q prime, the running time is $L\left(p, \frac{1}{3}\right)$, with $c = 3^{\frac{2}{3}}$. For $q = p^m$ and m fixed, the running time of the number-field sieve is $L\left(p^m, \frac{1}{3}\right)$, with c depending on m.

Finally, for the purpose of comparison, we provide some general statements regarding the complexity of factoring biprimes. The complexities of biprime factoring algorithms are as follows:

Quadratic sieve	$L\left(n, \frac{1}{2}\right), c = 1$
Elliptic curve methods	$L\left(p, \frac{1}{2}\right), c = \sqrt{2}$
Number field sieve	$L\left(n, \frac{1}{3}\right), c = 1.92$ (p smallest prime factor of n).

The asymptotic complexity of the discrete-log problem might be compared to these statements.

Because many cryptography systems depend on the intractability of the discrete-log problem, it is important to reflect on why the discrete-log problem is more vulnerable in some groups than in others. Every cyclic group can be written in terms of a generator π as $G = \{\pi^i \mid i = 0, 1, \ldots, \#G - 1\}$. In every such group, i is an integer, but π is an element of the group. In $\boldsymbol{F}_p$, π is an integer and in the term π^i, both π and i are integers. The set of integers has considerable structure. It has both the operation of addition and the operation of multiplication, as well as the distinction between prime and composite integers. The cryptosystem uses only the operation of multiplication, yet it presents the cryptanalyst with a group that has other operations and properties that might be used by the cryptanalyst to devise an attack. It may be wise to use a group that provides only the operation that is used by the cryptosystem and does not have other structure, such as the rich structure of primes and composites, so as to deny the cryptanalyst opportunities to use this structure. Accordingly, other groups with less accompanying structure, especially the group of point addition on an elliptic curve, have become very important.

Problems for Chapter 4

4.1 Show that discrete logarithms are easy to compute in the additive group $\boldsymbol{Z}_p$ for a prime p. In this group, the task is to compute x from a and b, given that $ax = b \pmod{p}$ and a is nonzero. This example shows that the discrete-log problem is easy in some groups. What is special about this group that makes the problem easy?

4.2 Let $n = pq$ where the integers p and q are both prime. Suppose that p and q are known integers. Give a procedure, when given a, for computing an x that satisfies $x^2 = a \pmod{n}$.

4.3 Let G be a cyclic group, and let α and β each be a generator for this group. Prove that any algorithm that can compute $\log_\alpha \gamma$ for any value of γ in G can be used to compute $\log_\beta \gamma$ for any value of γ in G. Conclude that the difficulty of the discrete-log problem is not affected by the choice of logarithm base.

4.4 Verify that 809 is a prime and that 3 is a primitive element of $\boldsymbol{F}_{809}$. Use the Shanks algorithm to compute $\log_3 525$ in the field $\boldsymbol{F}_{809}$.

4.5 **(Tripartite key exchange)** Devise a protocol based on the discrete-log problem by which three parties can establish a single common key by each making multiple transmissions. How many transmissions are necessary? Can your protocol be generalized so that four users can establish a common key? Can your protocol be arranged so that each party transmits to only one other party?

4.6 Show that the Pollard algorithm for computing a discrete log is likely to enter a loop after on the order of $\sqrt{n}$ steps, where n is the order of the group. (The birthday surprise, which is discussed in Section 8.6, suggests that the expected lengths of both the tail and the circle are $\sqrt{\pi n/8}$.)

4.7 The Pollard rho factoring algorithm iteratively computes $x \leftarrow f(x)$ and $y \leftarrow f(f(x))$ and $d = \text{GCD}(|x-y|, n)$ for any appropriate function $f(x)$ (such as x^2+1) and appropriate initial conditions. The algorithm continues while $d = 1$, fails if $d = n$, and otherwise halts with a factor of n. Fill in the details and explain the algorithm. Describe the estimated running time with the aid of the birthday surprise. Write a program to factor the integer 1649 using this algorithm or, better, carry out the iterations by hand.

4.8 Verify that 809 is a prime and that the element 3 generates a cyclic group of order 808 in $\boldsymbol{F}^*_{809}$. Use the Pollard algorithm to compute $\log_3 525$ in the field $\boldsymbol{F}_{809}$.

4.9 After reading Chapter 10, use the Shanks algorithm to solve the following elliptic-curve discrete-log problem:
Let $P = (519, 681)$ and $Q = (513, 40)$ be two points on the elliptic curve

$$\mathcal{X}(\boldsymbol{F}_{719}) : y^2 = x^3 + 231x + 508,$$

a curve that has order 727, a prime. Find the unique positive integer k smaller than 727 such that $Q = kP$.

4.10 After reading Chapter 10, describe how the Massey–Omura cryptosystem can be implemented using elliptic curves. Do you think that this should be used in practice?

Notes for Chapter 4

The formal notion of a public-key cryptosystem was first presented by Diffie and Hellman (1976). Their landmark paper "New Directions in Cryptography" proposed the now widely used notion of public-key exchange – though without then providing a specific method. This suggestion was both timely and brilliant; however some slight hints of this notion can be found in earlier literature. In this direction, one might notice that the use of a one-way function to store keys in encrypted form was discussed earlier by Wilkes (1968) and by Purdy (1974). At that time, however, all cryptographically secure communication used what we might now call "private keys." While their first paper was still in press, Diffie and Hellman (1976), in a second paper, described their key-exchange protocol, and shortly thereafter, Rivest, Shamir, and Adleman (1978) announced the now well-known and widely used RSA public-key encryption process. Steiner, Tsudik, and Waidner (2000) discussed extensions of public-key exchange to groups of three or more.

The Pohlig–Hellman algorithm for computing discrete logarithms was published in 1978. Other standard methods of computing discrete logarithms in an arbitrary cyclic group were proposed by Shanks (1971) and Pollard (1978). Blake *et al.* (1982), Coppersmith (1984), and others have shown that the complexity of the discrete-log problem is subexponential in some fields of characteristic two. Shoup (1997) has shown that all meaningful algorithms for the general discrete-log problem on groups of prime order p have complexity on the order of $\sqrt{p}$.

The index calculus method takes its name from the common classical usage of the term "index" for what is now called the discrete logarithm of an integer modulo p. The central idea of the index-calculus method was described by Kraitchik (1922). The method was refined through the years and applied by others, including Adleman (1979), Merkle (1979), and Pollard (1978), as well as Odlyzko (1984), describing the extension of index calculus to fields of characteristic two. The generalization of index calculus attacks to hyperelliptic curves was studied by Gaudry (2000) and Thériault (2003). The method requires an adequate frequency of smooth numbers of a given size. This frequency is estimated by the Dickman–deBruijn formula due to Dickman (1930) and deBruijn (1951).

5 Information-theoretic methods in cryptography

The strongest notion of cryptographic secrecy is known as perfect secrecy or information-theoretic secrecy. A cryptography system has perfect secrecy if the ciphertext gives no information about either the plaintext or the key. This definition is clear qualitatively, but it does require a definition of "information" to make it precise. The notion of perfect secrecy and its method of achievement is in contrast to most notions of secrecy and the methods of its achievement that are discussed in this book. These latter methods are perfectly insecure from the point of view of perfect secrecy, but instead protect the information behind a wall of computational intractability.

The notion of perfect secrecy is due to Shannon, who sought to give a formal and broad foundation for the subject of cryptography. He studied the deep question of when a cryptography system is provably secure. He espoused the principle that perfect secrecy means that the ciphertext gives no information; neither information about the plaintext nor information about the key. To quantify this principle, he referred to his formal definition of the term "information." The Shannon formulation rests on the methods and terminology of probability theory. This requires that each message $\boldsymbol{x}$ has a prior probability, denoted $p(\boldsymbol{x})$, of being chosen as the transmitted message. After the ciphertext $\boldsymbol{y}$ is observed by an adversary, the probability that $\boldsymbol{x}$ was the transmitted message is then given by the conditional probability $p(\boldsymbol{x}\,|\,\boldsymbol{y})$. The decryptor, however, knows the key k, so, to the decryptor, the conditional probability is $p(\boldsymbol{x}\,|\,\boldsymbol{y}, k)$. This $p(\boldsymbol{x}\,|\,\boldsymbol{y}, k)$, for a proper cryptography system, equals one for the correct message $\boldsymbol{x}$ and otherwise $p(\boldsymbol{x}\,|\,\boldsymbol{y}, k)$ equals zero.

The essential notion of Shannon is that the ciphertext $\boldsymbol{y}$ gives no information about the plaintext $\boldsymbol{x}$ if, and only if, $p(\boldsymbol{x}\,|\,\boldsymbol{y}) = p(\boldsymbol{x})$ even though $p(\boldsymbol{x}\,|\,\boldsymbol{y}, k)$ is equal to one when $\boldsymbol{x}$ is the correct message. That is, the conditional probability distribution on $\boldsymbol{x}$, when given $\boldsymbol{y}$ but not k, is the same as the unconditional probability distribution on $\boldsymbol{x}$ when $\boldsymbol{y}$ is not given. This statement, however, is an equality of probability vectors on a very large space. Because $\boldsymbol{x}$ is an element in the large space $\mathcal{A}^m$, this condition can be cumbersome in this form as a statement about probability distributions. It is better to encapsulate the condition into a more manageable form by introducing the notion of entropy, which is a scalar parameter describing a probability vector. The notion of entropy will be developed in this chapter.

5.1 Probability space

The plaintext message is a sequence of symbols from a discrete alphabet $\mathcal{A}$. The sequence comes from a message source whose output is the message $\boldsymbol{x} = (x_1, \ldots, x_n)$, where x_ℓ is the ℓth symbol of the message and n is the length of the message. We model the source as random, producing message $\boldsymbol{x}$ with probability $p(\boldsymbol{x}) = p(x_1, \ldots, x_n)$, which we take to be an understood notion. We do not inquire deeply into the objective meaning of probability in this application.

The simplest model of a probabilistic source is a (stationary) *memoryless source.* The source is called a memoryless source if

$$\begin{aligned} p(\boldsymbol{x}) &= p(x_1, \ldots, x_n) \\ &= \prod_{\ell=1}^{n} p(x_\ell). \end{aligned}$$

A memoryless source randomly generates the individual symbol x_ℓ of its output message independently but with the same probability distribution on each symbol. A memoryless source provides an abstract model of a source output that is used to develop understanding of many topics, both in the subject of cryptography and elsewhere.

The block probability distribution $p(\boldsymbol{x}) = p(x_1, \ldots, x_n)$, not necessarily stationary and memoryless, is associated with *marginal* and *conditional* probability distributions. The probability distribution $p(\boldsymbol{x})$ can be *marginalized* to any subset of its components by summing out all other components. For example,

$$p(x_1, \ldots, x_\ell) = \sum_{x_{\ell+1}} \sum_{x_{\ell+2}} \cdots \sum_{x_n} p(x_1, \ldots, x_n)$$

and

$$p(x_\ell) = \sum_{x_1} \sum_{x_2} \cdots \sum_{x_{\ell-1}} \sum_{x_{\ell+1}} \cdots \sum_{x_n} p(x_1, \ldots, x_n).$$

A source is a memoryless source if, and only if, its probability distribution $p(\boldsymbol{x})$ is the product of its marginals.

The *conditional* $p(x_\ell | x_{\ell-1}, \ldots, x_1)$ is given by

$$p(x_\ell | x_{\ell-1}, \ldots, x_1) = p(x_\ell | x_1, \ldots, x_{\ell-1}) = \frac{p(x_1, \ldots, x_\ell)}{p(x_1, \ldots, x_{\ell-1})}.$$

Other conditionals are defined similarly.

In contrast to a memoryless source is a source with memory. The most useful model of a source with memory is a *first-order Markov source.* This is a source in which $p_\ell(x_\ell | x_{\ell-1}, \ldots, x_1) = p(x_\ell | x_{\ell-1})$ for $\ell = 2, \ldots, n$, and the conditional probability distribution $p(x_\ell | x_{\ell-1})$ is the same function for every value of ℓ greater than one. Then

$$p(\boldsymbol{x}) = p(x_1) \prod_{\ell=2}^{n} p(x_\ell | x_{\ell-1}).$$

Each marginal $p(x_\ell)$ on a single component is defined as described earlier. Under standard conditions on the conditionals $p(x_\ell|x_{\ell-1})$, the marginals $p(x_\ell)$ of a Markov source will converge with n to the unique eigenvector of the probability matrix $p(x_\ell|x_{\ell-1})$ corresponding to the eigenvalue equal to one. This eigenvector is always a probability vector. Such an eigenvector always exists for a Markov conditional probability matrix.

5.2 Entropy

The information content of a discrete memoryless source with probability distribution $\boldsymbol{p}$ is expressed neatly by a simple number called the *entropy* of the source. The entropy is a scalar function of a probability vector $\boldsymbol{p}$, and is defined by

$$H(\boldsymbol{p}) = -\sum_{j=0}^{J-1} p_j \log p_j.$$

To assign a meaning to the jth term when $p_j = 0$, define

$$0 \log 0 = \lim_{x \to 0}[-x \log x] = 0.$$

If a logarithm to the base two is used, then the entropy has units of *bits*. If a logarithm to the base e is used, then the entropy has units of *nats*. The logarithm to the base two is usually preferred for quantitative statements. The logarithm to the base e is usually preferred for theoretical developments.

The entropy can be regarded as a scalar that measures the amount of randomness produced by the probability distribution $\boldsymbol{p}$. The entropy of a probability vector is analogous to the magnitude of a real vector.

We shall analyze the entropy using the useful inequality $\log_e x \leq x - 1$, which holds with equality if, and only if, $x = 1$.

Theorem 5.2.1 *The entropy satisfies $H(\boldsymbol{p}) \leq \log J$ with equality if, and only if, $p_j = \frac{1}{J}$ for $j = 0, \ldots, J-1$.*

Proof

$$\begin{aligned} H(\boldsymbol{p}) - \log J &= -\sum_{j=0}^{n-1} p_j \log p_j + \sum_{j=0}^{n-1} p_j \log \frac{1}{J} \\ &= \sum_j p_j \log \frac{1}{Jp_j} \\ &\leq \sum_j p_j \left(\frac{1}{Jp_j} - 1 \right) = \sum_j \frac{1}{J} - \sum_j p_j = 0, \end{aligned}$$

which follows from the inequality $\log_e x \leq x - 1$. □

A probability vector of dimension n is a real-valued vector, all of whose components are nonnegative and that sum to one. Let $\boldsymbol{P}^n$ denote the set of all probability vectors of dimension n. The set $\boldsymbol{P}^n$ is called *probability space* of dimension n. We may regard the distribution $p_j = \frac{1}{n}$ for all j as the "origin" of the space $\boldsymbol{P}^n$ and write $\boldsymbol{0} = \left(\frac{1}{n}, \frac{1}{n}, \ldots, \frac{1}{n}\right)$. Then, motivated by Theorem 5.2.1, we can define the "distance" from the origin of any $\boldsymbol{p} \in \boldsymbol{P}^n$ as $d(\boldsymbol{p}, \boldsymbol{0}) = H(\boldsymbol{p}) - \log J$, which equals zero if $\boldsymbol{p} = \boldsymbol{0}$. This motivates a more general definition.

Definition 5.2.2 *The Kullback distance, divergence, or relative entropy between* $\boldsymbol{p}$ *and* $\boldsymbol{q}$*, both elements of* $\boldsymbol{P}^n$*, is*

$$d_K(\boldsymbol{p}, \boldsymbol{q}) = \sum_{j=1}^{n} p_j \log \frac{p_j}{q_j}.$$

The Kullback distance satisfies the *Gibbs inequality* $d_K(p, q) \geq 0$, which again follows from the inequality $\log_e x \leq x - 1$. Although the Kullback distance is nonnegative, it is not symmetric and it does not satisfy a triangle inequality. It is not a metric.

Often, when the probability vector that is associated with random variable X has not been assigned a label, it is convenient and common, although improper, to write $H(X)$ instead of $H(\boldsymbol{p})$. In such a case, it is understood that H is not really a function of X, but rather is a function of the probability vector $\boldsymbol{p}$ associated with X, even though that probability vector has not been mentioned.

Side information may be given about a random variable X. Let Y be a second random variable taking value k with probability q_k, and such that X takes value j with probability $P_{j|k}$ whenever Y takes value k. Then the entropy of X, when given that the random variable Y has value k, is $H(X|k) = -\sum_j P_{j|k} \log P_{j|k}$ and the *conditional entropy* is the expectation

$$H(X|Y) = \sum_k q_k H(X|k) = -\sum_k \sum_j q_k P_{j|k} \log P_{j|k}.$$

The following theorem says that side information never increases entropy.

Theorem 5.2.3 *The conditional entropy satisfies*

$$H(X|Y) \leq H(X).$$

Proof The proof uses the inequality $\log_e x \leq x - 1$. Because $p_j = \sum_k q_k P_{j|k}$, we can write

$$\begin{aligned} H(X|Y) - H(X) &= \sum_j \sum_k q_k P_{j|k} \log \frac{P_{j|k}}{\sum_k q_k P_{j|k}} \\ &\leq \sum_j \sum_k q_k P_{j|k} \left(\frac{P_{j|k}}{\sum_k q_k P_{j|k}} - 1 \right) = 0, \end{aligned}$$

which proves the theorem. □

The information content of a message of length n that is produced by a stationary memoryless source is $nH(\boldsymbol{p})$ because $\boldsymbol{p}$ is a product probability distribution. The information content in a message of length n that is produced by an arbitrary source, possibly with memory, is

$$H(\boldsymbol{p}^{(n)}) = -\sum_{x \in \mathcal{P}} p^{(n)}(\boldsymbol{x}) \log p^{(n)}(\boldsymbol{x}),$$

where $\boldsymbol{p}^{(n)}$ is the probability distribution on the set of plaintext messages. It is easy to see that $H(\boldsymbol{p}^{(n)}) = nH(\boldsymbol{p})$ if the source is memoryless. Moreover $H(\boldsymbol{p}^{(n)}) < nH(\boldsymbol{p})$ if the source is not memoryless. Memory in a source always reduces the entropy, which means that the information content of a source is always reduced by dependence.

5.3 Perfect secrecy

The cryptographic notion of *perfect secrecy*, or *information-theoretic secrecy*, is a very strong notion, and a demanding goal. Perfect secrecy requires that the ciphertext gives *no* information about the plaintext. This notion requires that, even with infinite computational resources, the cryptanalyst knows no more about the plaintext or the key after examining the ciphertext than it knew before. This condition can be quantified whenever a probabilistic description of the message source is appropriate. Then perfect secrecy requires that the probability distributions on the message space and on the keyspace are not changed after observing the ciphertext. Knowing the ciphertext $\boldsymbol{y}$ has no effect on the posterior probability distribution $p(\boldsymbol{x}\,|\,\boldsymbol{y})$ of the plaintext $\boldsymbol{x}$. Thus the condition for perfect secrecy, as defined by Shannon, is written $p(\boldsymbol{x}\,|\,\boldsymbol{y}) = p(\boldsymbol{x})$ for all $\boldsymbol{x}$ and $\boldsymbol{y}$. Of course, in order that the message can be decoded, it is also necessary that $p(\boldsymbol{x}\,|\,\boldsymbol{y}, k) = 1$ for the true message $\boldsymbol{x}$.

A cryptographic system that fails to have perfect secrecy, as most do, will still be expected to superficially appear to have perfect secrecy. The deterministic structure must be deeply hidden and not revealed by any standard statistical procedure or tractable algorithm. This motivates the heuristic notion of *data diffusion*, meaning that changing any bit of the plaintext should typically change about half of the bits of the ciphertext, and the notion of *data confusion*, meaning that changing one bit of the key should typically change about half of the bits of the ciphertext. Moreover, under any protocol examination, these changes should be unpredictable and appear to be random.

Theorem 5.3.1 *A cryptosystem with keyspace $\mathcal{K}$ and message space $\mathcal{M}$ cannot have perfect secrecy unless $\#\mathcal{K}$ is at least as large as $\#\mathcal{M}$.*

Proof Suppose the contrary, that $\#\mathcal{K}$ is smaller than $\#\mathcal{M}$. Then, for each ciphertext $\boldsymbol{y}$, an encryption function $\boldsymbol{y} = e_k(\boldsymbol{x})$ indexed by k satisfies $\cup_{k \in \mathcal{K}} e_k^{-1}(\boldsymbol{y}) \leq \mathcal{M}$. Then given $\boldsymbol{y}$, some $\boldsymbol{x}$ are excluded as impossible for that $\boldsymbol{y}$, so $\boldsymbol{y}$ does give information about $\boldsymbol{x}$. □

Of course, Theorem 5.3.1 gives only a necessary condition, not a sufficient condition. Even if $\#\mathcal{K} \geq \#\mathcal{M}$, one can carelessly design a cryptosystem that leaks information.

The conventional cryptosystem can be perfectly secure from a direct attack by the cryptanalyst only if the key is as long as the message. The standard and only example of a cryptosystem with perfect secrecy is a *one-time pad*. A one-time pad is a cryptographic system in which the keyspace is at least as large as the message space. The message symbols are from an alphabet of size q. The key is a randomly and independently chosen sequence of symbols of the message alphabet that is as long as the message. The key is known to both the transmitter and the receiver. It is not known to the cryptanalyst. Symbols of the key are added (modulo q) to the symbols of the message, one key symbol to one message symbol. The random keystream of a one-time pad must be as long as the entire set of messages over the life of the cryptosystem. A one-time pad is not reused.

More specifically, a binary one-time pad is a random binary sequence b_ℓ for $\ell = 0, \ldots, N-1$ that is known to both the transmitter and the receiver, but not the cryptanalyst. Let a_ℓ for $\ell = 0, \ldots, n-1$ be the binary data sequence. The ciphertext is the modulo-two sum

$$c_\ell = a_\ell + b_\ell \qquad \ell = 0, \ldots, n-1.$$

Decryption is simple. Using the fact that $b_\ell + b_\ell = 0$ for modulo-two addition, we have the decoding rule

$$a_\ell = c_\ell + b_\ell.$$

A one-time pad is a perfectly secure cryptosystem. It can be penetrated only by an attack on the secure side channel that distributes or stores the one-time pad. However, many digital communication systems transmit very long messages – perhaps sending many millions of bits per second. Because it is difficult to distribute and store long keys, most cryptosystems have short keys compared to the length of the entire message. Then there is redundancy in the encrypted message that the cryptanalyst may use to deduce the key. A cryptosystem will compensate for using a short key by relying on the complexity of the way in which the key is combined with the data. The goal is to mix the data with the key in a highly involuted and nonlinear way to produce the message $\boldsymbol{c} = f(\boldsymbol{a}, k)$ so that, even if the function f is known, the plaintext or the key k cannot be computed from the codeword $\boldsymbol{c}$ and the message $\boldsymbol{a}$ or a fragment of the message $\boldsymbol{a}$. This means that the function f is so complex that there is no practical method by which it can be inverted to recover the plaintext or the key.

Theorem 5.3.2 *A cryptosystem in which the message space, the cipher space, and the keyspace have the same size has perfect secrecy if, and only if, keys are used equiprobably, and for every plaintext and every ciphertext there is one and only one key mapping that plaintext to that ciphertext.*

Proof For any key k, $e_k(\boldsymbol{x}) = \boldsymbol{y}$ is a one-to-one map of the message space into the cipher space. Otherwise decryption would not be possible. Moreover, it is easy to see that for any fixed $\boldsymbol{x}$ and $\boldsymbol{y}$, $e_k(\boldsymbol{x}) = \boldsymbol{y}$ for exactly one key k. This means that $\boldsymbol{y}$ is uniquely determined by $\boldsymbol{x}$ and k.

Let $\boldsymbol{x}$ be a random message, and let $p_X(\boldsymbol{x})$ be the probability distribution on the space of messages. Let $p_K(k)$ be the probability distribution on the space of keys. The random variable X takes values $\boldsymbol{x}$ with probability $p_X(\boldsymbol{x})$. The random variable K takes value k with probability $p_K(k)$. The messages and the keys are chosen independently. This means that the probability that $\boldsymbol{y} = e_k(\boldsymbol{x})$ is given by $p(\boldsymbol{y}|\boldsymbol{x}) = p_K(k|e_k(\boldsymbol{x}) = \boldsymbol{y}) = p_K(k)$. Then by the Bayes formula

$$p(\boldsymbol{x}\,|\,\boldsymbol{y}) = \frac{p(\boldsymbol{x},\boldsymbol{y})}{p_Y(\boldsymbol{y})} = \frac{p(\boldsymbol{y}|\boldsymbol{x})p_X(\boldsymbol{x})}{p_Y(\boldsymbol{y})}.$$

Thus

$$p(\boldsymbol{x}\,|\,\boldsymbol{y}) = \frac{p_K(k)p_X(\boldsymbol{x})}{p_Y(\boldsymbol{y})}.$$

But perfect secrecy requires that $p(\boldsymbol{x}\,|\,\boldsymbol{y}) = p_X(\boldsymbol{x})$. It follows that $p_K(k) = p_Y(\boldsymbol{y})$ for all k and $\boldsymbol{y}$. This holds if $p_K(k) = 1/K$, and it does not hold otherwise. □

Corollary 5.3.3 *The only cryptosystem with perfect secrecy is the one-time pad.*

Proof The proof is an immediate consequence of the theorem. □

Every one-time pad can be regarded as equivalent to a binary one-time pad used with a binary message. A key consists of a random binary string of length n bits. There are 2^n such keys. A message $\boldsymbol{x}$ consists of an n-bit string. The ciphertext $\boldsymbol{y}$ is the componentwise modulo-two sum of $\boldsymbol{x}$ and k. Thus $\boldsymbol{y} = \boldsymbol{x} + k$.

The definition of perfect secrecy can be restated more concisely in terms of the entropy. A cryptographic system has perfect security if, and only if, the conditional entropy satisfies $H(X|Y) = H(X)$. This statement is equivalent to the earlier requirement that $p(\boldsymbol{x}\,|\,\boldsymbol{y}) = p(\boldsymbol{x})$ for all $\boldsymbol{x}$ and $\boldsymbol{y}$, but it makes the statement more concisely because entropy is a scalar, while a probability distribution is a vector, perhaps on a very large space. Such statements of interest are simpler when expressed in terms of the entropy.

5.4 The Shannon–McMillan theorem

Given the set of sequences $\mathcal{A}^n$ and a subset $\mathcal{M} \subset \mathcal{A}^n$, which is the set of messages, we want to estimate $\#\mathcal{M}$ so that we can analyze the effectiveness of a cryptosystem. That analysis is the purpose of this section. If the messages occur with equal probability, then each message occurs with probability $1/\#\mathcal{M}$. Accordingly, the entropy is $H(\mathcal{M}) = \log_e(\#\mathcal{M})$ and the entropy per symbol is $\frac{1}{n}\log_e(\#\mathcal{M})$. The number of possible sequences of symbols from alphabet $\mathcal{A}$ is $(\#\mathcal{A})^n = e^{n \log_e(\#\mathcal{A})}$ while the actual number of sequences is $e^{H(\mathcal{M})}$.

There are many ways to estimate the number of legitimate messages in $\mathcal{M}$, such as grammatical methods that study the syntax of the language. Instead, we will estimate $\#\mathcal{M}$ by a more amenable probabilistic analysis. We form this estimate indirectly by means of a simple probability distribution on the letters of $\mathcal{A}$. Then we replace $\mathcal{M}$ by the typical sequences of $\mathcal{A}^n$. This approach will require some background.

Theorem 5.4.1 (Chebychev's inequality) *Suppose that X is a real-valued discrete random variable with mean $\overline{x}$. Then*

$$\Pr[|x - \overline{x}| > \alpha] \le \frac{\text{var}(x)}{\alpha^2}.$$

where var(x) *is the variance of random variable $\overline{x}$.*

Proof Let

$$\phi(x) = \begin{cases} 0 & \text{if} \quad |x - \overline{x}| < \alpha \\ 1 & \text{if} \quad |x - \overline{x}| \ge \alpha. \end{cases}$$

Then because

$$\phi(x) \le \left(\frac{x - \overline{x}}{\alpha} \right)^2,$$

we can write

$$\begin{aligned} \Pr[|x - \overline{x}| > \alpha] &= \sum_j \phi(x_j) p(x_j) \\ &\le \sum_j \left(\frac{x_j - \overline{x}}{\alpha} \right)^2 p(x_j) \\ &= \frac{\text{var}(x)}{\alpha^2}, \end{aligned}$$

which is the statement of the theorem. □

Definition 5.4.2 *Let $\delta > 0$ be given. A sequence $\boldsymbol{x}$ of blocklength n generated by a memoryless source $\boldsymbol{p}$ with entropy $H(\boldsymbol{p})$ is called a typical sequence (or a δ-typical sequence) if $\left|\frac{1}{n}\log p(\boldsymbol{x}) - H(\boldsymbol{p})\right| < \delta$ where the logarithm base agrees with that used in computing the entropy.*

The set of typical sequences of blocklength n will be written as

$$\mathcal{F}(\delta) = \left\{ \boldsymbol{x} : \left| \frac{1}{n} \log p(\boldsymbol{x}) - H(\boldsymbol{p}) \right| < \delta \right\}.$$

This notion of typicality is strong enough for our needs. An alternative notion, which we do not require, is given by

$$\mathcal{L}(\delta) = \left\{ \boldsymbol{x} : \left| \frac{1}{n} n_j(\boldsymbol{x}) - p_j \right| < \frac{\delta}{J} \text{ for } j = 0, 1, \ldots, J-1 \right\},$$

where $n_j(\boldsymbol{x})$ is the number of occurrences of symbol j in the sequence $\boldsymbol{x}$.

Whereas typical sequences of Definition 5.4.2 approximately display the right entropy, the alternative typical sequences approximately display the right frequency of symbols. Loosely, both are called *typical sequences*, though with some slight ambiguity of meaning.

The following theorem, for which the weaker notion of typicality suffices, states that almost all sequences are typical, in that the sample entropy is nearly the actual entropy.

Theorem 5.4.3 (Shannon–McMillan theorem) *Given a discrete memoryless source $\boldsymbol{p}$ of entropy $H(\boldsymbol{p})$ and any δ greater than zero, we can choose n large enough so that the set of all possible sourcewords of blocklength n produced by the source can be partitioned into two sets, $\mathcal{F}(\delta)$ and $\mathcal{F}^c(\delta)$, for which the following statements hold:*

(i) The probability of a sourceword belonging to $\mathcal{F}^c(\delta)$ is less than δ.
(ii) If a sourceword $\boldsymbol{x}$ is in $\mathcal{F}(\delta)$, then its probability of occurrence is approximately $2^{-nH(\boldsymbol{p})}$ in the sense that

$$| -n^{-1} \log p(\boldsymbol{x}) - H(\boldsymbol{p}) | < \delta.$$

Proof Let $\mathcal{F}$ denote the set of typical sequences $\{\boldsymbol{x} : | -\log p(\boldsymbol{x}) - nH(\boldsymbol{p})| < n\delta\}$. Then $\mathcal{F}$ satisfies the requirements of statement (ii). By Chebychev's inequality,

$$\begin{aligned} \Pr[| -\log p(\boldsymbol{x}) - nH(\boldsymbol{p})| \geq n\delta] &\leq \frac{\text{var}[-\log p(\boldsymbol{x})]}{n^2\delta^2} \\ &\leq \frac{n\sigma^2}{n^2\delta^2} = \frac{\sigma^2}{n\delta^2}, \end{aligned}$$

where

$$\sigma^2 = \sum_j p_j (\log p_j)^2 - \left(\sum_j p_j \log p_j \right)^2$$

is a constant, independent of n. Consequently,

$$\sum_{x \in \mathcal{F}^c} p(\boldsymbol{x}) \leq \delta$$

for sufficiently large n. This completes the proof of the theorem. □

Corollary 5.4.4 *The number of elements in $\mathcal{F}$ satisfies*

$$(1-\delta)2^{n(H(p)-\delta)} \leq \#\mathcal{F} \leq 2^{n(H(p)+\delta)}.$$

Proof Let $H = H(p)$ and write

$$1 \geq \sum_{x \in \mathcal{F}} p(x) \geq \sum_{x \in \mathcal{F}} 2^{-n(H+\delta)} = \#\mathcal{F}2^{-n(H+\delta)},$$

and using statement (i),

$$1-\delta \leq \sum_{x \in \mathcal{F}} p(x) \leq \sum_{x \in \mathcal{F}} 2^{-n(H-\delta)} = \#\mathcal{F}2^{-n(H-\delta)},$$

which completes the proof of the corollary. □

The proof of the theorem contains a slightly stronger result than the statement of the theorem. We state this separately as a corollary.

Corollary 5.4.5 *The probability* $\Pr[\mathcal{F}^c]$ *that a sourceword of blocklength n belongs to the set $\mathcal{F}^c$ is bounded by*

$$\Pr[\mathcal{F}^c] \leq \frac{1}{n\delta^2}\left[\sum_j p_j(\log p_j)^2 - \left(\sum_j p_j \log p_j\right)^2\right].$$

Proof This is an explicit statement of Chebychev's inequality where the bracketed term is the variance of the random variable $-\log p_j$. □

The sequences that are the elements composing the set $\mathcal{F}$ are the typical sequences generated by the memoryless plaintext source. The theorem says that for large n there is a negligible probability that a nontypical sequence will occur. All typical sequences are nearly equally probable, and there are about 2^{nH} of them.

The Shannon–McMillan theorem is a form of the law of large numbers. It says that even if the probability distribution on the output of a memoryless source is highly skewed, when viewed on the level of large blocks, the source output is almost surely a typical block, and the probability of any individual typical block occurring is almost equiprobable. A typical block is a block with a typical frequency of symbols.

5.5 Unicity distance

The task of recovering the plaintext message is a stronger requirement than the requirement of leaking no information. The one-time pad leaks no information, and so provides perfect secrecy. Weaker systems may provide some information and yet still be regarded as adequate. If infinite computational resources can be used to state, for example, only that the possible plaintext messages are reduced to a list of size 10^{30}, then this may be regarded as adequate, even though it does not constitute perfect secrecy. How much ciphertext is needed in order to determine the message $m \in \mathcal{M}$ or the key $k \in \mathcal{K}$ by means of a known ciphertext attack? This is an attack that uses unlimited computational resources to attempt to decode by successively trying every legitimate key. If only one legitimate message is decoded after trying every key, then the message and the key are found.

The smallest ciphertext message length at which the expected number of false keys is approximately equal to zero is called the *unicity distance*. The unicity distance[1] is the shortest length of typical ciphertext from which the secret key can, in principle, be determined in a ciphertext-only attack because, on average, only one plaintext corresponds to each ciphertext. This notion underlies the construction of recreational cryptotexts, which usually are about forty letters long. We will see that the unicity distance of English text under a substitution cipher is about forty.

Let $\mathcal{M}$ be a set of plaintext messages with m a plaintext message in the set $\mathcal{M}$. The entropy of the plaintext space is denoted $H(\mathcal{M})$. The ciphertext messages have blocklength n on an alphabet $\mathcal{A}$.

Let $\mathcal{K}$ be a keyspace of size $2^{\log_2 \|\mathcal{K}\|}$. If the keys are used with equal probability, then $H(\mathcal{K})$ is equal to $\log_2 \|\mathcal{K}\|$. Then there are $2^{H(\mathcal{K})}$ keys in $\mathcal{K}$. If the keys are not used with equal probability, then $H(\mathcal{K})$ is less than $\log_2 \|\mathcal{K}\|$ and there are, effectively, $2^{H(\mathcal{K})}$ keys. For example, a 128-bit key may have $H(\mathcal{K})$ smaller than 128 bits if the key distribution system does not use all possible keys, or uses them with other than the equiprobable probability distribution, which is known to the cryptanalyst.

Given a plaintext message $m \in \mathcal{M}$ and $k \in \mathcal{K}$, define the set of false keys

$$\mathcal{K}_F = \{k' \in \mathcal{K} : d_{k'}(e_k(m)) \in \mathcal{M}, k' \neq k\}.$$

A cryptosystem is not secure if the cardinality of $\mathcal{K}_F$ is small for most choices of m and k. We want to characterize this in a simple way.

Definition 5.5.1 *The unicity distance n_u is the smallest n such that there is almost always only one key k that is consistent with a random ciphertext* $\mathbf{y}$.

The cipher space of blocklength n over an alphabet $\mathcal{A}$ of size $\#\mathcal{A}$ contains $2^{\log_2 \|\mathcal{A}\|}$ ciphertexts. A proper cryptosystem uses all ciphertexts in this space, and uses them equally often. There are $2^{H(\mathcal{K})}$ keys, about $2^{H(\mathcal{M})}$ messages, so about $2^{H(\mathcal{K})}2^{H(\mathcal{M})}$ key-message pairs. There are about $2^{H(\mathcal{K})}2^{nH(p)}/2^{n\log_2 \|A\|}$ key-message pairs corresponding

[1] The unicity distance is not a distance in the usual mathematical sense. It might better be called the unicity length.

Key entropy (bits)	Unicity distance $H(\boldsymbol{p}) = 4.03$	$H(\boldsymbol{p}) = 2.3$
40	55	10
56	76	14
64	88	16
80	110	20
88	121	22
128	176	32
256	352	64

Figure 5.1 Unicity distance of encrypted English

to each ciphertext. This equals one if $H(\mathcal{K}) + nH(\boldsymbol{p}) - n\log_2 \|A\| = 0$. Thus the unicity distance is defined as

$$n_u = \frac{H(\mathcal{K})}{\log_2 \|\mathcal{A}\| - H(\boldsymbol{p})}.$$

A message of length n_u can almost always be decrypted by a direct search, trying all possible keys until a legitimate plaintext is recovered. Of course, for most systems in use, trying all possible keys is not computationally tractable. Nevertheless, the notion of unicity distance is one of the core concepts of classical cryptography.

For example, a memoryless binary source with a probability distribution $(p_0, p_1) = (0.89, 0.11)$ has entropy $H(\boldsymbol{p}) = 0.5$. Therefore if $\|\mathcal{A}\| = 2$ and $H(\mathcal{K}) = 128$, the unicity distance is 256. Therefore a message of length 256 bits can usually be uniquely decrypted by trying all possible keys until a typical sequence is recovered. To make the unicity distance large, one can make $H(\mathcal{K})$ large or one can make $H(\boldsymbol{p})$ close to $\log_2 \|\mathcal{A}\|$. For a maximally random binary source, $H(\boldsymbol{p}) = 1$ and the unicity distance is infinite.

To determine the unicity distance of encrypted English text, it is necessary to estimate the entropy of English text, which is done in the next section. Such estimates range from 2.3 to 4.03 bits. A fair compromise estimate is an entropy of 3.0 bits per symbol. The unicity distance of various key lengths used to encrypt the English language is shown in Figure 5.1, for various values of $H(\mathcal{K})$ and $H(\boldsymbol{p})$, using $\log_2 \#\mathcal{A} = \log_2 26 = 4.76$ bits. In particular, for a simple substitution cipher, $H(\mathcal{K}) = \log_2 26! = 88.3$ bits. Thus the unicity distance of a recreational cipher is somewhere in the range from 22 to 121 symbols depending on the specific statistical model of the English language. If the entropy were 3.0 bits per letter, then the unicity distance would be fifty symbols. This is consistent with the fact that recreational cryptograms are typically about forty English letters or a little more. The word spaces provide additional information, which helps the puzzle solver.

5.6 Entropy of natural language

Determining the entropy of a natural language, such as English, takes considerable effort and requires that an idealized model of the language be formulated. This requires that the structure of the language, denoted $\mathcal{M}$, be defined in some way. One approach is to define a grammatical model of the language so that valid messages can be distinguished from invalid messages. A grammatical model would implicitly list all possible messages by stating a set of rules that a valid message must satisfy. However, this can be complicated, and for our purposes, a probabilistic model suffices. This requires that a probability be assigned to each possible message. The notion of an invalid message is replaced by the notion of an improbable message. The statement of a probabilistic model requires one to first specify the message space $\mathcal{M}$ and then to assign a probability $p(m)$ to each message m of $\mathcal{M}$ such that $p(m) \geq 0$ and $\sum_{m \in \mathcal{M}} p(m) = 1$. For each $m \in \mathcal{M}$, $p(m)$ is the probability that m is the message. The entropy on the message space is

$$H(\mathcal{M}) = -\sum_{m} p(m) \log p(m).$$

An equiprobable distribution on the message space $\mathcal{M}$ containing $\#\mathcal{M}$ messages is

$$p(m) = \frac{1}{\#\mathcal{M}} \qquad \text{for all } m \in \mathcal{M}.$$

To reconcile the probabilistic model with the grammatical model for a message consisting of English text, we could enumerate all sequences that are proper English messages, count them, and assign the probability $p(m) = (\#\mathcal{M})^{-1}$. Then $H(\mathcal{M}) = \log_2(\#\mathcal{M})$. This procedure is impractical to carry out because it is too massive a task and because a precise definition of $\mathcal{M}$ is difficult to formulate.

Because this method of computing $H(\mathcal{M})$ is impractical and perhaps a bit vague, simple probabilistic models of the message space are used. The simplest of these is a memoryless componentwise model. Each letter of an alphabet $\mathcal{A}$ occurs with a specified probability that does not change with time, and messages are blocks of fixed length n. Symbol $x \in \mathcal{A}$ occurs with probability $p(x)$. The message $\boldsymbol{x} = (x_1, \ldots, x_n)$ occurs with product probability

$$p(\boldsymbol{x}) = \prod_{\ell=1}^{n} p(\boldsymbol{x}_\ell)$$

because the symbols are independent. The entropy per message is $H(\mathcal{M}) = -\sum_{\boldsymbol{x}} p(\boldsymbol{x}) \log p(\boldsymbol{x})$ and the entropy per symbol is $H(\boldsymbol{p}) = \frac{1}{n} H(\mathcal{M}) = -\sum_j p_j \log p_j$.

The probability of any letter occurring in English text is estimated as the frequency of occurrence of that letter in a collection of sample text. A table of the frequency

Table 5.1 *Frequency of occurrence of the letters in English*

Letter	Probability	Letter	Probability
A	.082	N	.067
B	.015	O	.075
C	.028	P	.019
D	.043	Q	.001
E	.127	R	.060
F	.022	S	.063
G	.020	T	.091
H	.061	U	.028
I	.070	V	.010
J	.002	W	.023
K	.008	X	.001
L	.040	Y	.020
M	.024	Z	.001

of occurrence of the letters of a memoryless model of the English language based on inspection of many samples of English text is shown in Table 5.1. The entropy of this memoryless source is 4.19 bits.

A better model of English is a source with memory. One way to capture memory is to regard the source alphabet as dependent pairs of letters but with the pairs generated independently. There are 27^2 pairs of letters, called *digrams*, if the alphabet $\mathcal{A}$ consists of the 26 letters augmented by a symbol for a space. By examining many samples of text, one can compile a list of the empirical probability of occurrence of these 27^2 digrams. The thirty most frequent digrams not containing a space in a decreasing order of frequency of occurrence are TH, HE, IN, ER, AN, RE, ED, ON, ES, ST, EN, AT, TO, NT, HA, ND, OU, EA, NG, AS, OR, TI, IS, ET, IT, AR, TE, SE, HI, and OF. A digram model of English consists of a probability for each of these pairs and all of the other pairs, a total of 27^2 digrams. An alternative to a digram model is a Markov model with the probability of a letter depending on the proceeding letter.

A more complicated model of English regards the source alphabet as triples of letters, called *trigrams*, with the succession of trigrams generated independently. If a space is included, there are 27^3 trigrams. The twelve most frequent *trigrams* in English text not containing a space and listed in decreasing order of frequency are THE, ING, AND, HER, ERE, ENT, THA, NTH, WAS, ETH, FOR, and DTH. One can compile a list of the empirical probability of all trigrams or only the most frequent, together with a default probability for the infrequent trigrams. A more ambitious set of trigrams would include spaces, and even other punctuation symbols.

Other models of English text can be defined. The symbols of the language can be regarded as words rather than letters, and a message is a sequence of n words from a fixed "alphabet," say of 4096 words. A memoryless model at the level of words regards successive words to be independent.

Once such a model has been constructed, the entropy of English text can be estimated. In this way one obtains estimates of the entropy ranging from 2.3 to 4.03 bits per letter.

5.7 Entropy expansion

If the source of the plaintext has low entropy, the unicity distance will be small. Then the cryptography system is vulnerable to a direct ciphertext attack. If only one key can account for a given ciphertext, then the plaintext can be recovered by a direct search. Possibly this direct search can be replaced by a hierarchical structure to reduce the work. In any case, a ciphertext-only attack can succeed. There are two possible ways to increase the unicity distance in order to reduce this vulnerability. The unicity distance, given by

$$n_u = \frac{H(\mathcal{K})}{\log_2 \|\mathcal{A}\| - H(\boldsymbol{p})},$$

can be increased by increasing the numerator. This means increasing the size of the keyspace or at least ensuring that keys are used equiprobably. Alternatively, the unicity distance can be increased by reducing the denominator, either by reducing $\log_2 \|\mathcal{A}\|$ or by increasing $H(\boldsymbol{p})$.

There are two ways to increase $H(\boldsymbol{p})$ in order to increase the unicity distance and so reduce the vulnerability. One way to increase the unicity distance is to increase the source entropy by artificially increasing the entropy of each symbol, which is discussed in this section. A second way is to reduce the length of the message by eliminating redundancy. The method of data compaction, which is discussed in the next section, changes the probability distribution on the alphabet so that it is more nearly uniform.

Thus, by using either of these methods to increase pre-symbol entropy, the number of typical sequences is reduced so that most remaining sequences are typical sequences. The contrast in these two approaches is illustrated in Figure 5.2. One can either reduce the size of the message space so that the number of atypical sequences is reduced, or one can increase the number of typical sequences.

Suppose that a source alphabet $\mathcal{A} = \{A, B, C\}$ has three symbols with the probability distribution $\boldsymbol{p} = \left(\frac{5}{8}, \frac{1}{4}, \frac{1}{8}\right)$. If $H(\boldsymbol{p}) = \log_2 \|\mathcal{A}\|$, the unicity distance is infinite, and a ciphertext-only attack is sure to fail even with infinite computational resources. To increase the entropy, the symbols can be split to form a new alphabet $\{A_1, A_2, A_3, A_4, A_5, B_1, B_2, C\}$, now with eight symbols. When symbol A occurs,

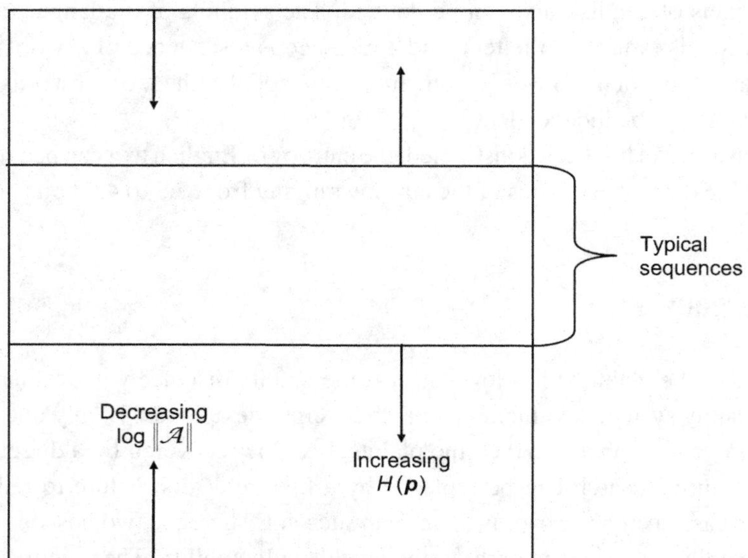

Figure 5.2 How to increase unicity distance

it is randomly and equiprobably replaced by A_1, A_2, A_3, A_4, or A_5. Similarly when symbol B occurs, it is randomly and equiprobably replaced by B_1 or B_2. Now, the new artificial source has entropy three, which is equal to $\log_2 \|\mathcal{A}\|$. Consequently, the unicity distance is infinite and a direct ciphertext-only attack must fail. The decryptor, however, can recover the original message simply by merging the artificial symbols in the obvious way to recover the actual symbols.

5.8 Data compaction

If $H(X) < \log J$, then the source output symbols (or output blocks) are not equiprobable. An encryption system that maps a block of source output symbols into a block of symbols of the same length, and in the same alphabet, must have a finite unicity distance, and so it is vulnerable to a known-ciphertext attack. This is because for each key, many ciphertexts are unlikely to occur. If the unicity distance is finite and the blocklength is larger, then only one plaintext message will account for that ciphertext. To eliminate redundancy in the block, one can use a *data compaction code*. A data compaction code encodes each source output sequence into a sequence of code symbols. A data compaction code can be a *fixed-length block code*, a *variable-length block code*, or a *tree code*. A block code compacts n message symbols into k compacted message symbols using $H(X^n) = H(Y^k)$. Because the block has been reduced from n symbols to k symbols, the entropy per symbol has been increased in the ratio of n to k.

We will describe only the class of variable-length block codes, known as *Huffman codes*. Consider a source whose alphabet $\{A, B, C, D, E, F, G, H\}$ has size eight. Suppose the probabilities are as follows:

$$\begin{aligned} p_A = p_B = p_C = p_D &= \tfrac{1}{32} \\ p_E = p_F &= \tfrac{1}{16} \\ p_G &= \tfrac{1}{4} \\ p_H &= \tfrac{1}{2}. \end{aligned}$$

The entropy $H(\boldsymbol{p}) = -\sum_j p_j \log_2 p_j$ of this source is $2\frac{1}{8}$ bits. The unicity distance is

$$\begin{aligned} n_u &= \frac{H(\mathcal{K})}{\log_2(\#\mathcal{A}) - H(\mathcal{M})} \\ &= \frac{H(\mathcal{K})}{3 - 2.125} \\ &= \frac{8}{7} H(\mathcal{K}). \end{aligned}$$

If $H(\mathcal{K})$ is 128 bits, then in principle, a known ciphertext consisting of about 147 ciphertext characters will be long enough to decrypt by using a direct ciphertext-only attack.

To reduce this vulnerability, a data compaction code can be used to increase the unicity distance. One such code is a Huffman code. An example of a Huffman code for the source is the map

$$\begin{aligned} A &\leftrightarrow 00000 \\ B &\leftrightarrow 00001 \\ C &\leftrightarrow 00010 \\ D &\leftrightarrow 00011 \\ E &\leftrightarrow 0010 \\ F &\leftrightarrow 0011 \\ G &\leftrightarrow 01 \\ H &\leftrightarrow 1. \end{aligned}$$

This code is a variable-length code, meaning that the codewords are not all the same length. However, because of the properties of the code, a string of codewords can always be uniquely parsed into the correct codewords. The average codeword length is $\overline{\ell} = \sum_j p_j \ell_j = 2\,\frac{1}{8}$ bits, which is exactly equal to the entropy $H(\boldsymbol{p})$. This equality occurs because, in this example, the probabilities are all negative powers of two. In this example, the unicity distance of the compacted data is infinite. Although 147 bits of ciphertext could be decoded in the absence of compaction, the ciphertext in the presence of compaction is secure from a direct attack.

In general, for memoryless source, a binary a Huffman code is constructed by a simple iterative procedure of combining the two least-probable symbols into a new artificial symbol, then treating the smaller code by the same method. The two combined symbols are distinguished in the compacted codeword by a zero or a one, respectively, at the end of that codeword.

The unicity distance for the Huffman code is larger than the uncompacted source data but, in general, is not infinite. It can be made yet larger by treating the source as a string of digrams.

5.9 The wiretap channel

A cryptosystem cannot have perfect secrecy unless the keyspace is at least as large as the message space, as is asserted by Theorem 5.3.1. That theorem, however, does not consider the possibility of noise contaminating the ciphertext. It may be that the adversary cannot observe the ciphertext as such, but can only observe a degraded version of the ciphertext. Practical communication channels are subject to noise and other impairments. For the usual purposes of data transmission, elaborate correction methods, introduced in Section 15.6, are employed to make the channel reliable and essentially error-free in the presence of these various impairments. However, when transmitting a secret key over a public channel, one may want to do more than simply make the channel error-free. One may also want to use the noise that is seen by the wiretapper to hide the transmitted message so that it cannot be intercepted. In order for this to be effective, even an adversary fully informed about the methods of error correction must be unable to remove the effects of channel noise. In theory, one can design a transmitted message so that the intended receiver can fully recover the message in the presence of noise, but with wiretap noise severe enough, the cryptanalyst cannot recover anything about the message or key. Such a channel is called a *wiretap channel.* The information-theoretic study of the wiretap channel proves that asymptotically good methods of communication exist that prevent eavesdropping on that channel. However, practical methods of communication that achieve such performance are evidently not yet in place.

The *secrecy capacity* will be defined below as the maximum rate at which the transmitter can reliably send information to the receiver provided that the information cannot be recovered by the wiretapper. The problem formulation requires that all participants are fully informed about all aspects of the channel including the channel noise statistics. Only the transmitted codeword is unknown. The information represented by the codeword may be a message or it may be a key to be used by an associated bulk encryption system.

A simple instance of a wiretap channel is a *degraded* wiretap channel. A degraded wiretap channel is a channel for which the wiretapper noise consists of both the noise of the main channel and additional noise, as is shown in Figure 5.3. These two

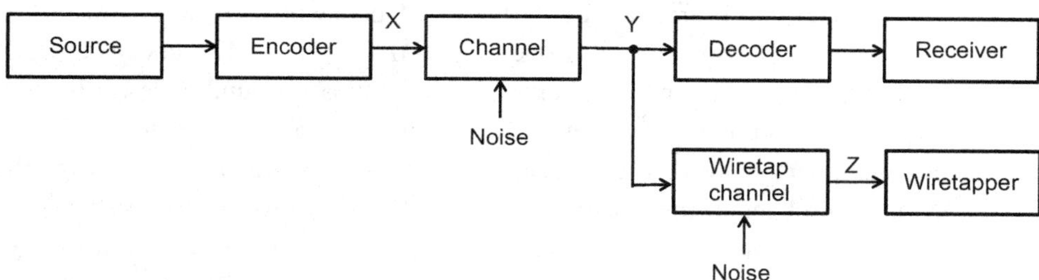

Figure 5.3 The degraded wiretap channel

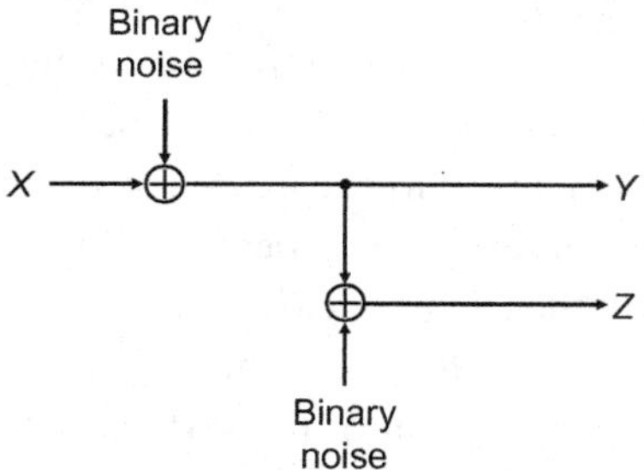

Figure 5.4 The degraded binary wiretap channel

noise terms are introduced consecutively. Only the additional noise in the wiretapper's received signal protects the encoded message from observation by the wiretapper. The encoder generates a coded message that must be completely reconstructed by the intended receiver even though it is contaminated by noise, yet the stronger noise on the codeword when it reaches the wiretapper must make the message completely illegible to the wiretapper. That stronger noise in the wiretapper's received message may be regarded as playing a role similar to the role of a one-time pad, although perfect secrecy is not required.

The secrecy capacity can be motivated by examining the simplest nontrivial example of a wiretap channel, which is the degraded binary symmetric wiretap channel shown in Figure 5.4. In this example, each of the three alphabets is a binary alphabet. The added binary noise in the direct channel is a one with probability ϵ' and is otherwise a zero. The additional binary noise added in the wiretap channel is a one with probability ϵ'' and is otherwise a zero. This means that the wiretapper sees a bit in error with probability $\epsilon = (1-\epsilon')\epsilon'' + \epsilon'(1-\epsilon'')$.

The task of the communication system on the binary wiretap channel can be motivated in terms of Hamming spheres in the binary vector space $\boldsymbol{F}_2^n$. A Hamming sphere of radius t about any element $\boldsymbol{c}$ of $\boldsymbol{F}_2^n$ is given by

$$\mathcal{S}_d = \{\boldsymbol{v} \in \boldsymbol{F}_2^n \mid d_H(\boldsymbol{c}, \boldsymbol{v}) \le t\}.$$

The encoder can be regarded as a list of M codewords $\{\boldsymbol{c}_m \in \boldsymbol{F}_2^n \mid m = 1, \ldots, M\}$. Each codeword corresponds to one of the M messages. The codewords are chosen so that the set of Hamming spheres of some radius t around codewords are pairwise disjoint. The number of codewords M is limited by the choice of t. The law of large numbers states that, for large n, when $\boldsymbol{c} \in \boldsymbol{F}_2^n$ is transmitted through the binary symmetric channel, the decoder will almost always receive a noisy vector $\boldsymbol{v}$ that is approximately at Hamming distance $n\epsilon'$ from $\boldsymbol{c}$. Accordingly, one wants t to be a little larger than $n\epsilon'$. Then a sphere of radius t centered on $\boldsymbol{v}$ will contain only the correct codeword.

The wiretapper, however, will receive a noisy vector $\boldsymbol{v}'$ that is approximately at Hamming distance $n\epsilon$ from $\boldsymbol{c}_m$, which is larger than $n\epsilon'$. Thus, one wants to choose the set of codewords $\{\boldsymbol{c}_m\}$ such that each $\boldsymbol{v} \in \boldsymbol{F}_2^n$ at distance $n\epsilon$ from a codeword $\boldsymbol{c}_m$ is also at distance $n\epsilon$ from an exponentially large number of other codewords.

Methods of packing points into $\boldsymbol{F}_2^n$ to achieve this purpose are not well developed. One might hope that some sort of regular structure for the codewords would suffice, but perhaps this is not so. Every codeword is at the center of two concentric Hamming spheres, the smaller sphere of radius t_1 and the larger sphere of radius t_2. Every smaller sphere must not contain any other codeword. Every point of the surface of the larger sphere must be the center of a sphere that contains an exponentially large number of other codewords.

The binary case is representative of the general case. In the general case, a discrete memoryless wiretap channel has three alphabets: an input alphabet $\mathcal{A}$, an output alphabet $\mathcal{B}$, and a wiretap alphabet $\mathcal{C}$. A channel input $a_j \in \mathcal{A}$ produces two outputs $b_k \in \mathcal{B}$ and $\boldsymbol{c}_i \in \mathcal{C}$ with probability $P(b_k, c_i|a_j)$ abbreviated $P_{ki|j}$. This can be put in the form of an array: $\boldsymbol{P} = \{P_{ki|j}\}$. The channel is *memoryless* if a block input $\boldsymbol{a} = (a_{j_1}, a_{j_2}, \ldots, a_{j_n})$ results in the two block outputs $\boldsymbol{b} = \{b_{k_1}, b_{k_2}, \ldots, b_{k_n}\}$ and $\boldsymbol{c} = \{c_{i_1}, c_{i_2}, \ldots, c_{i_n}\}$ with the product probability

$$P(\boldsymbol{b}, \boldsymbol{c}|\boldsymbol{a}) = \prod_{\ell=1}^{n} P(b_{h_\ell}, c_{i_\ell}|a_{j_\ell}),$$

which is abbreviated $\Pi_{\ell=1}^{n} P_{k_\ell i_\ell | j_\ell}$. A discrete memoryless wiretap channel with a conditional transition probability that factors as

$$P_{ki|j} = P'_{k|j} P''_{i|j}$$

is a degraded discrete memoryless wiretap channel.

The maximum-likelihood decoder, when given the channel output $\boldsymbol{v}$, decides that codeword $\boldsymbol{c}_m$ was transmitted if $P(\boldsymbol{v} \mid \boldsymbol{c}_m) \geq P(\boldsymbol{v} \mid \boldsymbol{c}_{m'})$ for all $m' \neq m$. A tie for the largest probability can be broken in any convenient way, but one presumes that this is always wrong when calculating the probability of error. Given any $\epsilon > 0$, the needs of the primary channel are satisfied by choosing M and n such that the probability of decoding error is less than ϵ. To satisfy the need for security, M is further constrained by the additional condition that the expected number of m' such that

$$P(\boldsymbol{v}' \mid \boldsymbol{c}_{m'}) \geq P(\boldsymbol{v}' \mid \boldsymbol{c}_m)$$

is exponentially large. This means that the wiretapper will find exponentially many codewords that, from its data, are more likely than the correct codeword.

The *secrecy capacity* is an asymptotic statement that requires two limits. The probability of decoding error and the probability of correct wiretap decoding must each be smaller than a specific p_e that goes to zero asymptotically as blocklength n goes to infinity. Accordingly, the secrecy capacity is defined operationally as

$$C_s = \lim_{p_e \to 0} \left[\lim_{n \to \infty} \frac{1}{n} \log_2 M(n, p_e) \right]$$

where $M(n, p_e)$ is the cardinality of the largest set of codewords of blocklength n for which the probability of decoding error is not larger than p_e and the number of false codewords at the wiretapper is at least $p_e M$.

The formal mathematical methods of information theory show that the secrecy capacity can be expressed as

$$C_s = \max_{X-Y-Z} I(X;Y) - I(X;Z),$$

where the *mutual informations* are $I(X;Y) = H(X) - H(X|Y)$ and $I(X;Z) = H(X) - H(X|Z)$, and where the maximum is taken over the random variables Y and Z that are in a Markov relationship with the channel input X, which Markov chain is denoted $X - Y - Z$. We will not derive this information-theoretic statement.

The replacement of the operational definition of secrecy capacity by the information-theoretic definition of secrecy capacity does not complete the analysis. It is still necessary to evaluate the maximum stated above, which is a nontrivial task.

The special case of the gaussian wiretap channel has a closed-form solution for the secrecy capacity. This instance of a wiretap channel consists of gaussian noise of power N on the direct channel and gaussian noise of power N_a on the wiretapper's channel. A single input to the channel is a real number s, and a block input is a vector of real numbers $\boldsymbol{s} = \{s_1, s_2, \ldots, s_n\}$ with average power S. This we can define either as $S = \frac{1}{n} \sum s_i^2$, or as an expectation. The secrecy capacity of the gaussian wiretap channel is

$$C_s = \frac{1}{2} \log \left(1 + \frac{S}{N}\right) - \frac{1}{2} \log \left(1 + \frac{S}{N_a}\right).$$

There are two versions of the gaussian wiretap channel shown in Figure 5.5. In the first case, the decryptor and the eavesdropper receive the same signal, but each contaminated by independent gaussian noise, and the eavesdropper's noise has the larger power. In the

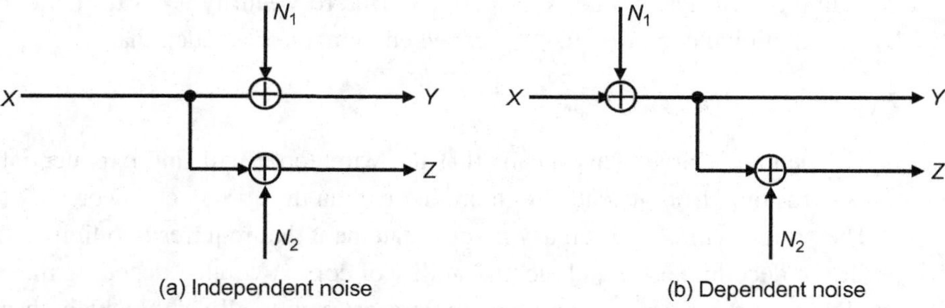

Figure 5.5 The gaussian wiretap channel

second case, referred to as the degraded gaussian broadcast channel, the eavesdropper's noise is dependent on the noise in the intended receiver's signal. These two cases are the two extremes of the general case in which the gaussian noise has the covariance matrix

$$\Sigma = \begin{vmatrix} \sigma_1^2 & \rho\sigma_1\sigma_2 \\ \rho\sigma_1\sigma_2 & \sigma_2^2 \end{vmatrix}$$

with σ_2 larger than σ_1.

Problems for Chapter 5

5.1 Prove the elementary inequality $\log_e x \leq x - 1$.

5.2 **a** Give a detailed procedure for constructing a Huffman code.

b Construct a Huffman code for an example with six symbols and probability distribution

$$\boldsymbol{p} = (0.3, 0.2, 0.2, 0.1, 0.1, 0.1).$$

5.3 A simple computer executes four commands called ADD, SUB, MPY, and STO, which are designated by the symbols A, S, M, and T. A computer program consists, in part, of a string of command symbols, and we consider only this part. Based on an examination of a number of computer programs, it is determined that the instruction types are used independently with probabilities $\frac{3}{4}$, $\frac{1}{16}$, $\frac{1}{16}$, and $\frac{1}{8}$.

a This sequence is encrypted with a block substitution cipher on blocks of length 3. What is the unicity distance?

b Now, suppose it is determined that the instruction sequence is a first-order Markov source with a transition matrix and equilibrium probability distribution

$$P = \begin{bmatrix} \frac{13}{16} & 0 & \frac{1}{4} & 1 \\ \frac{1}{16} & 0 & \frac{1}{4} & 0 \\ \frac{1}{16} & 0 & \frac{1}{4} & 0 \\ \frac{1}{16} & 1 & \frac{1}{4} & 0 \end{bmatrix} \qquad p = \begin{bmatrix} \frac{3}{4} \\ \frac{1}{16} \\ \frac{1}{16} \\ \frac{1}{8} \end{bmatrix}.$$

(Note that $p_j = \sum_i P_{j\,|\,i} p_i$, as is required for the equilibrium distribution.) Using the model, repeat part ***a***.

c Construct a Huffman code for datablocks of length two.

d What is the unicity distance now?

5.4 Prove that a cryptosystem with $\#\mathcal{P} = \#\mathcal{K}$ can have perfect secrecy only if it uses every possible ciphertext with equal probability.

5.5 Prove that a cryptosystem achieves perfect secrecy for all plaintext probability distributions if it achieves perfect secrecy for one plaintext probability distribution.

5.6 **a** Prove that

$$H(X, Y) = H(Y) + H(X \mid Y).$$

b Prove that

$$H(X, Y) \leq H(X) + H(Y)$$

with equality if, and only if, X and Y are independent.

c Conclude from parts (a) and (b) that

$$H(X \mid Y) \leq H(X)$$

with equality if, and only if, X and Y are independent.

5.7 Suppose the plaintext alphabet consists of the 26 letters of the English alphabet used independently with probability distribution $\boldsymbol{p}$. The key consists of a maximally random sequence of letters of the English alphabet of the same length as the message. The encryption rule is

$$y = x + k \pmod{26}$$

applied componentwise with letters represented by their numerical equivalent. Prove that this system has perfect secrecy.

5.8 **(Asymptotic approximation of the multinominal coefficient)** *Stirling's approximation* is

$$\sqrt{2n\pi}\left(\frac{n}{e}\right)^n < n! < \sqrt{2n\pi}\left(\frac{n}{e}\right)^n\left(1 + \frac{1}{12n-1}\right).$$

Prove that the multinominal coefficient can be approximated as

$$\frac{n!}{\prod_k (n_k!)} \approx e^{nH(\mathbf{p})},$$

where $\sum_k n_k = n$ and $p_k = n_k/n$.

5.9 The method of entropy expansion increases the entropy of the message space by inserting additional randomness into the messages. This randomness is easily removed from the decrypted message by remerging artificially distinguished symbols. Can the cryptanalyst use knowledge of the symbols to be remerged to counter the effectiveness of entropy expansion? Why?

5.10 The degraded gaussian wiretap channel consists of a real number with average power S as the channel input, to which gaussian noise of variance σ_1^2 is added on the primary channel, and to which additional gaussian noise of variance σ_2^2 is added on the wiretap channel. Accordingly, the wiretapper sees noise of variance $\sigma_1^2 + \sigma_2^2$ of which, a portion is correlated with the noise at the primary receiver.

a How would the secrecy capacity change if the wiretap noise were independent of the noise seen by the primary receiver, but still with the same variance: $\sigma_1^2 + \sigma_2^2$?

b Would you expect the answer to be different if the primary receiver had a feedback channel to the transmitter? Why?

Notes for Chapter 5

Claude Elwood Shannon (1916–2001) introduced the methods of information theory into cryptography. His work (1949) is often regarded as the foundation of modern cryptography. Shannon, early on, saw that the theory of cryptography is closely related to the theory of communication. The theory of communication studies how to convey information, and the theory of cryptography studies how to deny information. Shannon's formulation of information theory has applications to both topics. He introduced the notion of perfect secrecy and also recognized the relationship between the entropy function and cryptography. Shannon's 1949 paper on cryptography was a major impetus initiating the development of a formal mathematical theory of cryptography. Shannon proved that the one-time pad, introduced much earlier by Miller (1882), Vernam (1926), and others, is the only method of achieving perfect secrecy. Shannon also introduced the terms "data diffusion" and "data confusion."

Data compaction is now a fully developed subject and many data compaction codes are now available, the Huffman code being an early example. The primary purpose usually given for data compaction is to reduce the size of a data record, either for storage or transmission. However, by removing redundancy, data compaction also plays an important role in security.

The wiretap channel was introduced by Wyner (1975). The gaussian instance of the wiretap channel was studied by Leung-Yan-Cheong and Hellman (1978).

6 Block ciphers

A block cipher is a symmetric-key cipher that breaks a plaintext message into segments of fixed length. Each segment, called a *plaintext block*, consists of a fixed number of plaintext symbols. Using a secret key k known to both encryptor and decryptor, each plaintext block is encrypted independently into a *ciphertext block* of blocklength n. In many common block ciphers, but not all, the length n of the ciphertext block is equal to the length of the plaintext block. The ciphertext blocks are concatenated to form the ciphertext message, which is then sent to the decryptor. The decryptor receives the ciphertext message, then breaks it into the sequence of ciphertext blocks of length n and, using a corresponding key k, decrypts each ciphertext block of length n independently into the corresponding plaintext block. The plaintext blocks leaving the decryptor are concatenated to reform the plaintext message, which is then sent to the user.

The encryption function and the key k can remain the same from block to block, but the encryption of the ℓth block does not depend on the data within other blocks. Each block is encrypted independently. A symmetric-key block cipher requires that, for the same block, the encryptor and the decryptor use the same key. This means that a secure method of distributing this key to both the encryptor and decryptor, or of exchanging this key between them, is required. Any method of key distribution, including public-key cryptography, can be used.

6.1 Block substitution

A block cipher is described as a mapping $\boldsymbol{y} = e_k(\boldsymbol{x})$ from a plaintext block $\boldsymbol{x}$ of blocklength r to a ciphertext block $\boldsymbol{y}$ of blocklength n. Although r need not equal n, for most block ciphers of this chapter, we will set r equal to n. More precisely, a block cipher is a set of such maps $\{e_k\}$, indexed by a key k from a set of keys $\mathcal{K}$. We will usually think of $\boldsymbol{x}$ and $\boldsymbol{y}$ as binary blocks, perhaps of blocklength n equal to 64 or 128 bits. If the blocklength is 64 bits, then there are 2^{64} possible plaintexts and 2^{64} possible ciphertexts. The encryption $e_k(\boldsymbol{x})$ maps the set of 2^{64} binary blocks of length n onto this same set of 2^{64} binary blocks. This means that $e_k(\boldsymbol{x})$ is a substitution cipher on the set of blocks. Each of 2^{64} binary plaintext blocks is assigned a 64-bit binary ciphertext

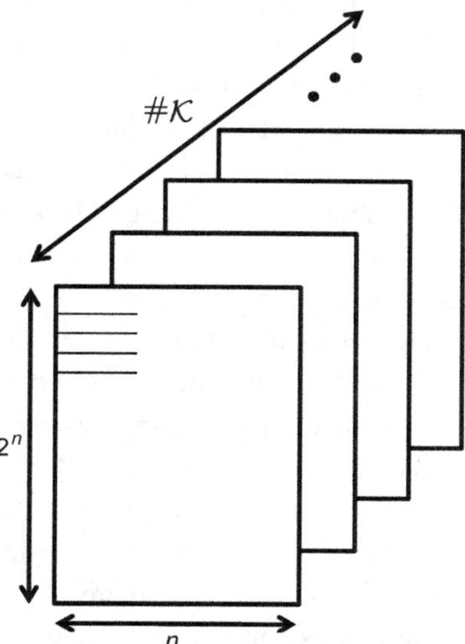

Figure 6.1 Conceptualizing a block cipher

block to represent it. Thus from an abstract point of view, the function $e_k(\boldsymbol{x})$ for each k is simply a fixed permutation on the space of all plaintext messages. A different key k gives a different permutation of the plaintext blocks. If there are $\#\mathcal{K}$ keys, then a given block cipher consists of $\#\mathcal{K}$ permutations of message blocks, and the key k specifies one of these permutations. If the keys are 64 bits long, for example, then there are (at most) 2^{64} keys. While this is a large number, the total number of permutations on the 2^{64} plaintext messages is $(2^{64})!$. This is an immense number, much larger then 2^{64}. The design of a block cipher is the task of choosing a set of these permutations that provides both security and simplicity. These two goals pull in opposite directions, and the challenge is to find a balance between them.

This abstract structure of a cryptosystem is illustrated in Figure 6.1. This figure gives a conceptualization of a block cipher as a collection of $\#\mathcal{K}$ encryption tables, one such table for each value of the key k. The key specifies the encryption table that is to be used. The specified table is then addressed by an n-bit plaintext block, and at each address of this table is stored an n-bit ciphertext block. Because the encryption must be inverted, each n-bit block occurs once as a ciphertext in each table. Therefore each encryption table, specified by k, is a permutation of the set of n-bit blocks. Of course, each such table shown in the figure would be immense, and so the figure is only conceptual. In practice, the individual entries in the tables cannot be stored and must be computed only when needed, so a computational structure is necessary, yet the structure of the tables should somehow prevent any significant inference about the message or the chosen key by an adversary who observes one or more ciphertexts.

The goal of simplicity means that there should be a simple procedure for combining the bits of the key with the bits of the plaintext so as to form the ciphertext. This requirement excludes many block permutations, all those that do not correspond to a simple computational procedure. For example, a random permutation of the blocks, stored as such, would require a table with 2^{64} entries each of length 64, and, evidently, could not, in general, be expressed by a simple algorithm. Such a table is not feasible. Only block ciphers in which the permutation is expressible by a tractable algorithm are feasible, and yet that permutation should be as inscrutable as a random permutation.

The goal of security requires that the algorithm not have any weakness by which a cryptanalyst can deduce the key or the plaintext. This objective is generally thought to impose a number of broad requirements. Every ciphertext bit should be a function both of every bit of the plaintext and every bit of the key. Changing any plaintext bit or any key bit should change any particular ciphertext bit about half of the time. Similarly, a change in any ciphertext bit should change the deciphered plaintext in an unpredictable way. In particular, about half of the plaintext bits should change, on average, when a ciphertext bit is reversed. More generally, there should be no evident statistical relationship between the ciphertext and the plaintext. The word "evident" is important in this statement because, of course, for each key, the relationship between key and ciphertext is ultimately deterministic. But this deterministic relationship must only be visible within the global view of the entire system, not in limited data sets.

6.2 The Feistel network

An accepted principle of many of the digital block ciphers in common use is to simultaneously achieve both security and computational simplicity by iterating an elementary core computation. This principle is embodied in a popular structure with many attractive features called a *Feistel network*. A Feistel network is shown in Figure 6.2. The plaintext is a binary dataword with an even number of bits that is broken into two halves. The core computation of a Feistel network interchanges the two halves of the dataword, performing a nonlinear keyed function on one-half of the dataword during this interchange and then adding it to the other half. The Feistel network achieves security by using multiple passes through this core computation. Each pass is called a *Feistel round*. Each round is keyed, either by the full key k or by a subkey $k^{(i)}$ formed from a portion of k. Because each round is identical, and computationally simple, the encryption is simple as well as secure. It is essential that the keyed function be nonlinear because linear operations can be interchanged and merged. If the operation were linear, then it may be that all of the multiple rounds could be merged into a single equivalent round, which would be much easier to attack.

The structure of the Feistel network has the interesting and useful feature that decryption is identical to encryption, differing only in that the subkeys are used in the

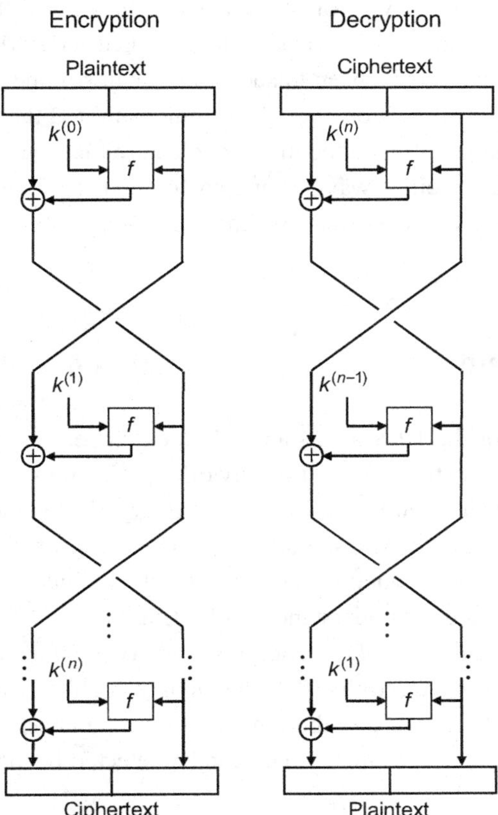

Figure 6.2 The Feistel network

reverse order. The choice of the keyed function within the core computation is not specified at the level of the network, and does not affect the structure and reversibility of the network. This keyed function is any fixed nonlinear function of the key and data, but is not otherwise specified by the structure of the Feistel iteration.

Although the simple structure of the Feistel network does resemble the structure of a cascade cipher, there is a considerable difference because of the intervening movement of the data between the two half-words. This movement is described by the suggestive term of "data diffusion." Likewise, the function f, which folds the key bits into the data bits, is described by the suggestive term of "data confusion." Thus the goal of the Feistel network is to provide both "data confusion and data diffusion."

The Feistel network has been popular and successful as a structure in practical block cryptosystems. Among those popular applications, only a portion of the full key is used in each round. In part, this is to avoid any unintentional self-canceling that might conceivably occur inadvertently if the same full key were to be used in every round.

The Feistel network is subject to several types of attack. Attacks known as differential cryptoanalysis attacks and linear cryptoanalysis attacks will be defined and studied later in the chapter. Such attacks, evidently, have been unsuccessful against popular block ciphers that use a form of Feistel network. Moreover, the fragmentation of the key into subkeys also creates a potential vulnerability. One might conjecture that rounds could be attacked individually in some unspecified way because each round is only protected by the shorter subkey, but no such attacks appear to be known.

6.3 The Data Encryption Standard

The *Data Encryption Standard*, or the DES, is a standard block cipher based on the Feistel network that has seen very extensive use for network applications. It is now largely outdated and has been superceded by the Advanced Encryption Standard, or the AES. Although DES is no longer in widespread use, a discussion of it remains pedagogically important, and provides a good introduction to block ciphers. A good understanding of AES begins with a good understanding of DES.

Using a binary key of blocklength 56 bits, DES encrypts a binary plaintext block of blocklength 64 bits into a binary ciphertext block of blocklength 64 bits. The 56-bit key is lengthened to 64 bits for storage or distribution by inserting a check bit after every 7 key bits so that a single bit error within the key can be detected in each set of 8 bits.

To introduce the DES design requirements, suppose that the binary plaintext source has an entropy of 0.5 bits per bit. Then the unicity distance of a binary block code with a 56-bit key is 112 bits. This is slightly less than two ciphertext blocks, so we know that it is virtually impossible for two typical binary sequences each of two 64-bit blocks encoded with different keys to be the same ciphertext. Thus for such a source, it would usually be possible, in principle, to decrypt by simply trying all possible keys on a ciphertext message two blocks long in order to recover the only two-block plaintext that is typical of the source. Because there are 2^{56} keys, this search was once considered computationally intractable unless the computation could be structured, deducing only a few key bits or plaintext bits at a time, and thereby solved by an algorithm much faster than direct search. Thus the challenge is to choose the encryption function so that a structured search does not exist. Despite intentions, this attack is now computationally feasible, but only with immense computational resources.

The space of 64-bit words has cardinality 2^{64}. The number of ways of mapping this space onto itself is $2^{64}!$. Each such map is a permutation on the set of 2^{64} blocks, and each key selects one of these permutations. Only 2^{56} permutations can be addressed by a 56-bit key. One can say that the design requirement of DES was to select 2^{56} permutations from a set of $2^{64}!$ permutations in a way that is practical to implement yet does not present any weakness for the cryptanalyst to exploit.

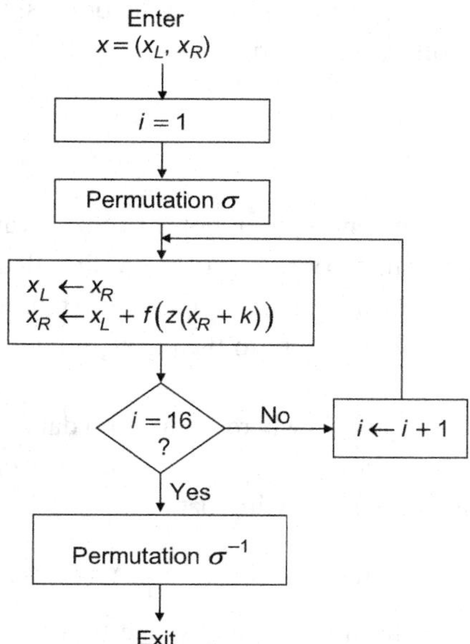

Figure 6.3 The Data Encryption Standard

The DES algorithm achieves its practicality by using multiple iterations of a basic core computation as represented by a Feistel network. The purpose of the iterations is to create a kind of mixing that is designed to combine the basic steps in a computationally simple, but mathematically complex way. The goal of the algorithm is a fast and simple implementation that gives high security.

We will describe the DES as an iterative algorithm consisting of sixteen passes through a core encryption computation, as shown in Figure 6.3. Each iteration is called a *round*. The encryption core consists of a function $f(z, k')$, described below, where z is a 48-bit subblock of data bits rearranged in a specified pattern, and k' is a 48-bit subkey formed by a subset of bits taken from the 56-bit key k and arranged in a specific pattern.

The full encryption consists of sixteen rounds of the iteration

$$x_L^{(i)} = x_R^{(i-1)}$$
$$x_R^{(i)} = x_L^{(i-1)} + f(z(x_R^{(i-1)}), k^{(i)}),$$

where $x_L^{(i)}$ and $x_R^{(i)}$ are each 32 bits, which we think of as the left half and the right half of the 64-bit block $x^{(i)}$, writing $x^{(i)} = (x_L^{(i)} \mid x_R^{(i)})$. The first round is preceded by a fixed permutation of the plaintext message. This operation will be written as $x^{(0)} = \sigma(x)$ where σ denotes the fixed permutation. The last round is followed by the inverse of the initial permutation $y = \sigma^{-1}(x^{(16)})$. It is clear then that the decryption must begin with the permutation σ and must end with the permutation σ^{-1}. Because this permutation σ is publicly known, the permutation σ has no cryptographic significance and no evident purpose.

It is clear that, for any function $f(z, k)$, the ith round can be undone simply by repeating the same steps according to the following pattern:

$$x_L^{(i-1)} = x_R^{(i)} + f(z(x_L^i), k^{(i)})$$
$$x_R^{(i-1)} = x_L^{(i)}.$$

Hence, the decryption has the same form as the encryption, but the indices run backwards. This means that the last encryption round corresponds to the first decryption round. The only other difference in the equations is the interchange of right and left. Indeed, the decryption is otherwise identical in structure to the encryption except for the subkey sequence.

The inner encryption consists of the function $f(z, k)$ where z is a 48-bit data subblock and k is a 48-bit subkey. The 48-bit word z is computed from the 32-bit word x by simply repeating some of the bits according to the following pattern:

$$z = (x_{32}, x_1, x_2, x_3, x_4, x_5, x_4, x_5, x_6, x_7, x_8, x_9, x_8, x_9, x_{10}, x_{11}, \ldots, x_{31}, x_{32}, x_1).$$

The pattern is created by first cyclically shifting the 32 bits by one bit of the 32-bit word to place x_{32} in the first position, then after every sixth bit, repeating the previous two bits in the manner shown.

Each 48-bit subkey consists of a subset of the 56 bits of the full key arranged in a permuted order. The most direct way to describe the subkeys in terms of the full key is by explicitly listing the bits of each of the sixteen subkeys. Let k_ℓ for $\ell = 1, \ldots, 64$ denote the 64 bits of the full key k. The sixteen subkeys are

$$k^{(1)} = (k_{10}, k_{51}, k_{34}, k_{60}, k_{49}, \ldots, k_{62}, k_{55}, k_{31})$$
$$k^{(2)} = (k_{02}, k_{43}, k_{26}, k_{52}, k_{41}, \ldots, k_{54}, k_{47}, k_{23})$$
$$k^{(3)} = (k_{51}, k_{27}, k_{10}, k_{36}, k_{25}, \ldots, k_{38}, k_{31}, k_{07})$$
$$\vdots$$
$$k^{(16)} = (k_{18}, k_{59}, k_{42}, k_{04}, k_{57}, \ldots, k_{06}, k_{63}, k_{39}).$$

A more concise way to specify the subkeys is by a mathematical expression. Thus the sixteen subkeys are defined by the expression $k_{ij} = k_{18-8j+41i}$, for $i = 0, \ldots, 47$, and $j = 1, \ldots, 16$, where the subscript is reduced modulo 64 and k_{ij} on the left side denotes the ith bit of subkey $k^{(j)}$. Each of these sixteen subkeys is a 48-bit subkey. The set of subkeys needed for the sixteen rounds of decryption is this same set of sixteen keys, but used in the opposite order starting with $k^{(16)}$.

Next, for the ith round, using subkeys $k^{(i)} = k^{(1)}, \ldots, k^{(16)}$, write the binary sum $z + k^{(i)}$. This is then partitioned into eight parts as

$$z + k^{(i)} = (B_1, B_2, \ldots, B_8),$$

where B_ℓ is a 6-bit word, and $(B_1, B_2, \ldots, B_8)$ is the 48-bit word formed by the concatenation of these eight 6-bit words. The final step to define the confusion function is to define the 32-bit vector $f(z, k^{(i)})$ as

$$f(z, k^{(i)}) = (S_1(B_1), S_2(B_2), \ldots, S_8(B_8)),$$

where each function $S_\ell(B_\ell)$ maps six bits into four bits. This means that each $S_\ell(B_\ell)$ can be regarded as a table of 64 4-bit entries. The 64 entries of each $S_\ell(B_\ell)$ are conventionally represented as a 4-by-16 array of hexadecimal characters, which is called a *substitution box*, or for short, an *S-box*. There are eight such S-boxes. The 4-bit entries of each S-box expressed in hexadecimal notation are:

$$S_1 = \begin{bmatrix} E & 4 & D & 1 & 2 & F & B & 8 & 3 & A & 6 & C & 5 & 9 & 0 & 7 \\ 0 & F & 7 & 4 & E & 2 & D & 1 & A & 6 & C & B & 9 & 5 & 3 & 8 \\ 4 & 1 & E & 8 & D & 6 & 2 & B & F & C & 9 & 7 & 3 & A & 5 & 0 \\ F & C & 8 & 2 & 4 & 9 & 1 & 7 & 5 & B & 3 & E & A & 0 & 6 & D \end{bmatrix}$$

$$S_2 = \begin{bmatrix} F & 1 & 8 & E & 6 & B & 3 & 4 & 9 & 7 & 2 & D & C & 0 & 5 & A \\ 3 & D & 4 & 7 & F & 2 & 8 & E & C & 0 & 1 & A & 6 & 9 & B & 5 \\ 0 & E & 7 & B & A & 4 & D & 1 & 5 & 8 & C & 6 & 9 & 3 & 2 & F \\ D & 8 & A & 1 & 3 & F & 4 & 2 & B & 6 & 7 & C & 0 & 5 & E & 9 \end{bmatrix}$$

$$S_3 = \begin{bmatrix} A & 0 & 9 & E & 6 & 3 & F & 5 & 1 & D & C & 7 & B & 4 & 2 & 8 \\ D & 7 & 0 & 9 & 3 & 4 & 6 & A & 2 & 8 & 5 & E & C & B & F & 1 \\ D & 6 & 4 & 9 & 8 & F & 3 & 0 & B & 1 & 2 & C & 5 & A & E & 7 \\ 1 & A & D & 0 & 6 & 9 & 8 & 7 & 4 & F & E & 3 & B & 5 & 2 & C \end{bmatrix}$$

$$S_4 = \begin{bmatrix} 7 & D & E & 3 & 0 & 6 & 9 & A & 1 & 2 & 8 & 5 & B & C & 4 & F \\ D & 8 & B & 5 & 6 & F & 0 & 3 & 4 & 7 & 2 & C & 1 & A & E & 9 \\ A & 6 & 9 & 0 & C & B & 7 & D & F & 1 & 3 & E & 5 & 2 & 8 & 4 \\ 3 & F & 0 & 6 & A & 1 & D & 8 & 9 & 4 & 5 & B & C & 7 & 2 & E \end{bmatrix}$$

$$S_5 = \begin{bmatrix} 2 & C & 4 & 1 & 7 & A & B & 6 & 8 & 5 & 3 & F & D & 0 & E & 9 \\ E & B & 2 & C & 4 & 7 & D & 1 & 5 & 0 & F & A & 3 & 9 & 8 & 6 \\ 4 & 2 & 1 & B & A & D & 7 & 8 & F & 9 & C & 5 & 6 & 3 & 0 & E \\ B & 8 & C & 7 & 1 & E & 2 & D & 6 & F & 0 & 9 & A & 4 & 5 & 3 \end{bmatrix}$$

$$S_6 = \begin{bmatrix} C & 1 & A & F & 9 & 2 & 6 & 8 & 0 & D & 3 & 4 & E & 7 & 5 & B \\ A & F & 4 & 2 & 7 & C & 9 & 5 & 6 & 1 & D & E & 0 & B & 3 & 8 \\ 9 & E & F & 5 & 2 & 8 & C & 3 & 7 & 0 & 4 & A & 1 & D & B & 6 \\ 4 & 3 & 2 & C & 9 & 5 & F & A & B & E & 1 & 7 & 6 & 0 & 8 & D \end{bmatrix}$$

$$S_7 = \begin{bmatrix} 4 & B & 2 & E & F & 0 & 8 & D & 3 & C & 9 & 7 & 5 & A & 6 & 1 \\ D & 0 & B & 7 & 4 & 9 & 1 & A & E & 3 & 5 & C & 2 & F & 8 & 6 \\ 1 & 4 & B & D & C & 3 & 7 & E & A & F & 6 & 8 & 0 & 5 & 9 & 2 \\ 6 & B & D & 8 & 1 & 4 & A & 7 & 9 & 5 & 0 & F & E & 2 & 3 & C \end{bmatrix}$$

$$S_8 = \begin{bmatrix} D & 2 & 8 & 4 & 6 & F & B & 1 & A & 9 & 3 & E & 5 & 0 & C & 7 \\ 1 & F & D & 8 & A & 3 & 7 & 4 & C & 5 & 6 & B & 0 & E & 9 & 2 \\ 7 & B & 4 & 1 & 9 & C & E & 2 & 0 & 6 & A & D & F & 3 & 5 & 8 \\ 2 & 1 & E & 7 & 4 & A & 8 & D & F & C & 9 & 0 & 3 & 5 & 6 & B \end{bmatrix}.$$

For each S-box, the 6-bit input specifies one element of that S-box. The specified element of the S-box specifies the 4-bit output expressed in hexadecimal notation.

Each row of each S-box consists of a permutation of the sixteen hexadecimal characters. Otherwise, the entries in the S-boxes are supposedly arbitrary, as might be generated by random selection. While a structured pattern – either intentional or accidental – could exist in the choice of entries in the S-boxes, no evidence that this is so has ever been made known. Some have conjectured that there may be a trapdoor hidden in the pattern of these integers that allows the designers of DES to readily decrypt a message without knowing the key. No evidence of such a trapdoor has ever been presented. It is important, however, to the study of cryptography to emphasize that this covert trapdoor is a possibility. If a trapdoor did exist, it would be a more subtle trapdoor than most other trapdoors that we encounter.

The set of S-boxes can be expressed as a set of eight look-up tables, each consisting of 64 words of 4 bits. The eight look-up tables, of 64 words each, together comprise 512 words and are addressed by 9 bits consisting of the 6 bits of B_ℓ and the 3 bits specifying ℓ. The 9 bits can be regarded as a single address at which a 4-bit number is stored.

6.4 Using the Data Encryption Standard

Although it has been described as a pure block cipher, and one might presume that this use is the most common, the Data Encryption Standard also can be used in a variety of other ways. To use it as a pure block cipher with a single key, the key is inserted once, and a longer message is broken into 64-bit blocks. The blocks are fed into the encryptor one by one and encrypted one by one with the same key. The successive block encryptions do not interact in any way other than that they use the same key. The individual 64-bit ciphertext blocks are then concatenated to form the entire ciphertext message. To decrypt, the process is reversed.

A minor objection to the use of DES in this way as a pure block cipher is that whenever a plaintext block is repeated, it produces a repeated ciphertext block. This might happen, for example, if an uncompacted document has many blank pages, or if two documents have an identical set of pages.

There are several alternatives to the elementary way of operating DES that are chosen at the discretion of the user. These alternatives do not necessarily provide any more security, but are preferred by some users. Some of these options, including common block encryption, are shown in Figure 6.4. These methods can be used with any block cipher.

One alternative way of using DES is to componentwise-add each 64-bit ciphertext block to the next 64-bit plaintext block before that plaintext block is encrypted. This

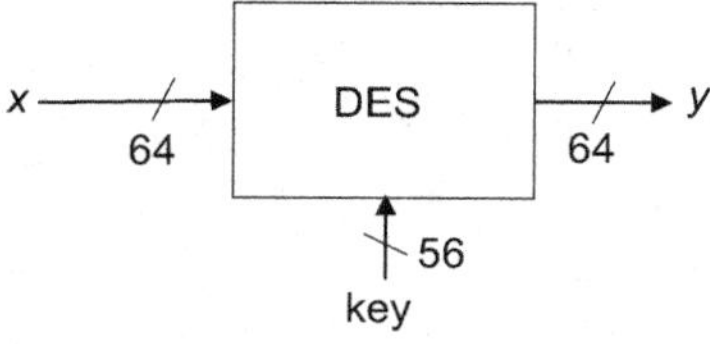

(a) Common block encryption

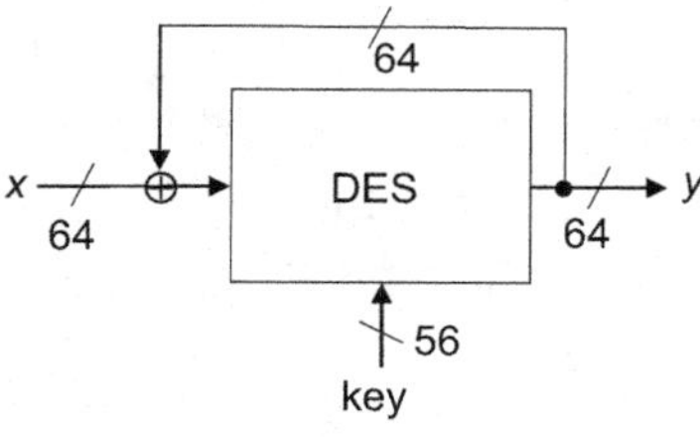

(b) Modified block encryption

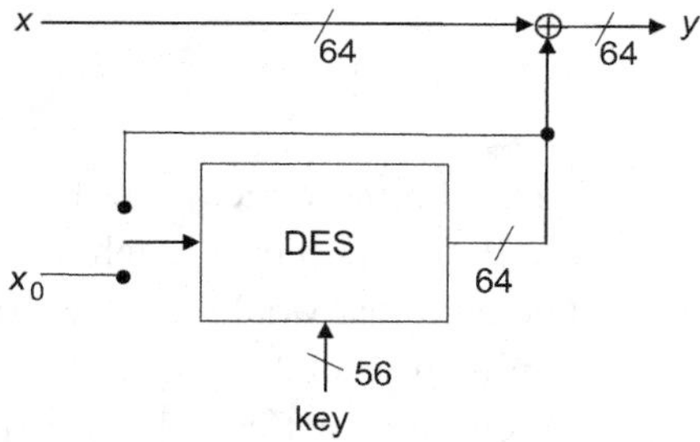

(c) Keystream encryption

Figure 6.4 Alternative standardized schemes for DES

procedure is trivial to invert in the decryptor because the plaintext blocks are recovered one by one in sequence. A variation of this is to add 56 bits of the ciphertext to the key for encoding the next block. This procedure is also trivial to invert in the decryptor, requiring only that the blocks are decrypted in the same order in which they are encrypted.

Another alternative is to use DES to generate a semi-infinite binary keystream to be used as an additive stream cipher, as discussed in Chapter 7. Simply initialize the encryptor with any 64-bit vector x_0 and the assigned key k to produce the first 64 bits of the keystream by encrypting the initialization vector. Then at each step, encrypt the previous 64-bit output of the encryptor using the same key k. Each output of the encryptor, in turn, is used as the next 64 bits of the keystream, and also as the next input to the encryptor. In this way an unending sequence of keystream bits is generated and is added, bit by bit, to the plaintext. Decryption inverts this encryption process by forming the same keystream and subtracting it, bit by bit, from the ciphertext.

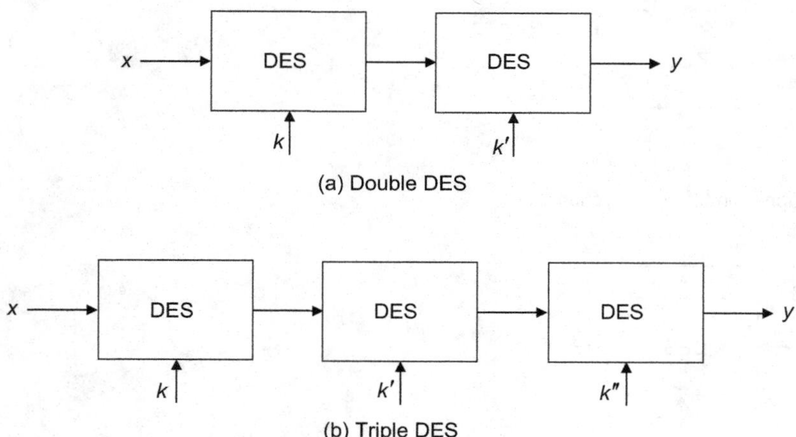

Figure 6.5 Cascaded encryption using DES

6.5 Double and triple DES encryption

The best attack presently known against DES may be the direct brute-force attack of simply trying all 2^{56} keys. This is about 64×10^{15} keys. For a rough calculation, suppose that a special-purpose computer could be built that would try, say, 10 billion keys per second. This would then require only 6 million seconds to try all keys, which is about 60 days. This hypothetical attack may seem to be extreme, but it is possible and can be seen as a future vulnerability. Calculations such as this one suggest that the size of the key is unsatisfactory for some or many applications, so the key must be lengthened.

Any block cipher can be cascaded, so the discussion of multiple encryption in this section applies to any block cipher. We refer specifically to DES because the 56-bit key size of DES leads some to use double and triple encryption in order to lengthen the key.

Double DES encryption consists of the application of the same encryption process twice with two different keys in the form $y = e_{k'}(e_k(x))$, as shown in Figure 6.5. Triple DES encryption consists of the application of the same encryption process three times with either two or three different keys. Encryption with three keys has the form $y = e_{k''}(e_{k'}(e_k(x)))$. However, one may choose to set $k'' = k$ so that then $y = e_k(e_{k'}(e_k(x)))$. In this case, triple encryption uses only two keys, thereby imitating triple encryption while only doubling the total key length. This is the same key length as is used with double encryption, but the triply encrypted ciphertext is judged to be more secure. Two-key triple encryption is not as secure as three-key triple encryption, at least against a brute-force attack. Apparently, nothing more damaging than a brute-force attack is known against three-key encryption.

Double encryption is less popular because it is vulnerable to a known-plaintext attack called a *meet-in-the-middle attack*. Given the known pair x and y, satisfying $y = e_{k'}(e_k(x))$, one must find k and k'. This expression can be rewritten as $d_{k'}(y) = e_k(x)$. The meet-in-the-middle attack is to decrypt y for all 2^{56} decryption functions $d_{k'}(y)$, and to encrypt x for all 2^{56} encryption functions $e_k(x)$. From these, prepare two lists, each of length 2^{56}, then find the common entry on the two lists. In this way, by finding the single common entry on the two lists, the two keys k and k' are found. Although this attack appears to be intractable because it requires an enormous amount of work and requires an enormous memory, the amount of work is actually only a small multiple of the amount of work required to break a single-key encryption by a known-plaintext attack. We conclude that double encryption is more secure than single encryption, but it does not have the additional security that one should expect from a keylength of 112 bits. For this reason, it is deemed not worth the additional effort and additional keylength. Triple DES encryption is not threatened by a meet-in-the-middle attack and so is preferred.

Triple DES is often implemented with the cute idea of using DES decryption in place of the middle encryption so that $y = e_k(d_{k'}(e_k(x)))$. With this implementation, single-DES encryption is identical to triple DES encryption if $k' = k$. This has no real effect on triple-DES security. It might be attractive in some applications because it allows a triple-DES encryptor to send messages to a single-DES decryptor without any change other than how the keys are specified. It is only necessary to set k' equal to k in the triple-DES encryption. This is one way to provide backwards compatible technology in which a newer cryptographic system that uses triple DES can be used transparently in conjunction with an older cryptographic system that uses single DES.

6.6 The Advanced Encryption Standard

The *Advanced Encryption Standard*, or AES, is a block cipher that, in its preferred form, uses datalengths of 128 bits and keylengths of 128, 192, or 256 bits. The encryption procedure consists of the repetition of a core computation. Each repetition of this core computation is called a *round*. The number of rounds, which equals ten, twelve, or fourteen, depends on the keylength. We will refer primarily to the case in which the keylength is 128 bits, for which ten rounds are used.

The structure of the Advanced Encryption Standard is inspired by the structure of the Feistel network. However, AES goes well beyond breaking datawords into two half-words. It is not a Feistel network, but can be regarded as a generalization of a Feistel network.

Let k be a key of length 128 bits and regard the 128 bits of the key as consisting of 16 bytes, each byte consisting of 8 bits. Let s be a plaintext block of length 128 bits

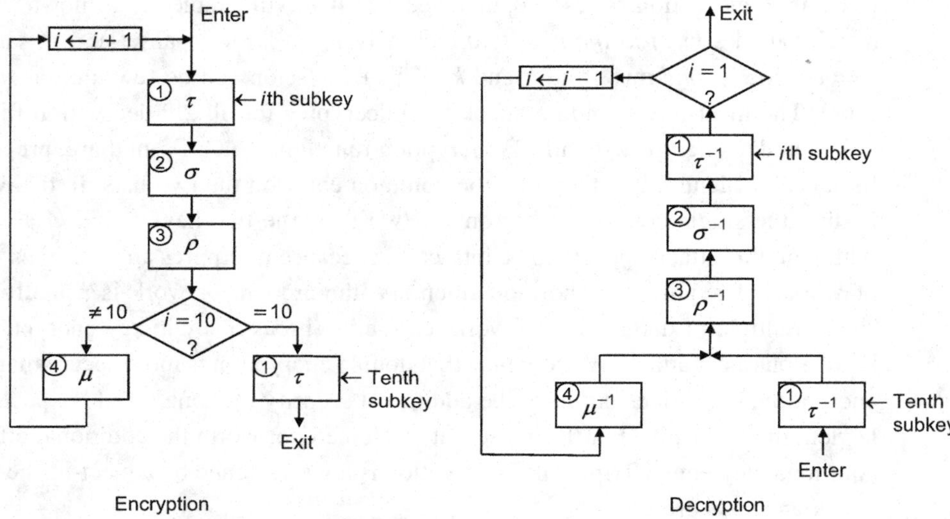

Figure 6.6 The Advanced Encryption Standard

and regard the 128 bits of the plaintext block as consisting of 16 bytes, each byte consisting of 8 bits. To facilitate description of the computation, the 16 plaintext bytes are regarded as arranged in a four-by-four array as follows:

$$s = \begin{bmatrix} s_{00} & s_{01} & s_{02} & s_{03} \\ s_{10} & s_{11} & s_{12} & s_{13} \\ s_{20} & s_{21} & s_{22} & s_{23} \\ s_{30} & s_{31} & s_{32} & s_{33} \end{bmatrix}.$$

Because each element of the array s represents 8 bits, the complete array consists of the 128 plaintext bits at the start of the encryption process. Each round replaces the matrix s by a new matrix s. At the end of the ten rounds of the encryption process, the array s will contain the ciphertext block. Part of the core computation involves computations combining individual elements of the array s with the key k. This can be regarded as the task of "data confusion." Part of the core computation involves rearrangement of the elements of the array. This can be regarded as the task of "data diffusion."

Each round begins with a partially encrypted block produced by the previous round, as shown in Figure 6.6. A round begins by combining a 128-bit subkey with the most recent partially encrypted block s. The round then computes a new partially encrypted block of 128 bits from the current partially encrypted block of 128 bits. That new block of 128 bits is again regarded in the same form as a four-by-four matrix of bytes, which continues to be called s.

Each round consists of four steps, as shown in Figure 6.6.

Step 1 This step is the encryption step. It is the only step that makes use of the key. The 128-bit subkey designated for that round and the 128-bit iterate are regarded as two 128-bit vectors over $\boldsymbol{F}_2$ and are added componentwise. Step 1 is clearly invertible simply by adding the 128-bit subkey to the 128-bit cipherword. This is because $s + k + k = s$.

Step 2 This step is a nonlinear step in which each entry of the array is replaced by a function of itself. Thus each byte s_{ij} of the four-by-four array s is replaced by

$$s_{ij} \leftarrow \sigma(s_{ij}),$$

where σ is an invertible nonlinear function that maps an 8-bit number into an 8-bit number. This function will be described later. For the moment, the function $\sigma(s)$ can be described simply as a table look-up as given in Figure 6.7, where the pair of hexadecimal symbols (x, y) represents the byte s. It can be checked that $\sigma(s)$ is invertible by verifying that each 8-bit number appears in the table exactly once. Thus $\sigma(s)$ is a permutation of bytes.

Step 3 This step consists of a byte-level cyclic shift on each of the four rows of $\boldsymbol{s}$. Denote a row as $s(x) = s_3x^3 + s_2x^2 + s_1x + s_0$ with the polynomial coefficients given by the elements of the row. Then the function $\rho(s)$ is represented by the polynomial $xs(x) \pmod{x^4 - 1}$. It is obvious that this step can be inverted.

Step 4 This step consists of an operation on each column of matrix $\boldsymbol{s}$. Regard each element of $\boldsymbol{s}$ as an element of $\boldsymbol{F}_{256}$ represented as a polynomial in z with multiplication in $\boldsymbol{F}_{256}$ based on the irreducible polynomial $z^8 + z^4 + z^3 + z + 1$. Identify the four elements of a column with a polynomial in x denoted $s(x) = s_3x^3 + s_2x^2 + s_1x + s_0$, where the coefficients of this polynomial are the elements of the column. Let $c(x)$ and $d(x)$ be fixed polynomials of $\boldsymbol{F}_{256}[x]$ given by

$$c(x) = (z + 1)x^3 + x^2 + x + z$$
$$d(x) = (z^3 + z + 1)x^3 + (z^3 + z^2 + 1)x^2 + (z^3 + 1)x + (z^3 + z^2 + z),$$

where the coefficients written as polynomials in z denote elements of $\boldsymbol{F}_{256}$. An evaluation of the polynomial product $c(x)d(x)$ by elementary polynomial multiplication, with addition and multiplication of coefficients as operations in $\boldsymbol{F}_{256}$, gives $c(x)d(x) = 1 \pmod{x^4 - 1}$. Accordingly, the multiplication of $s(x)$ by $c(x)$ is inverted by multiplying the product $c(x)s(x)$ by $d(x) \pmod{x^4 - 1}$.

The encryption executes ten rounds, each with the four steps as described. At the end of ten rounds, an eleventh subkey is added to the iterate to form the final ciphertext. This completes the description of the four steps of the encryption function. To complete the description of AES, we must now describe the computations of the eleven subkeys and the permutation function σ.

The sequence of eleven 128-bit subkeys, $k^{(0)}, k^{(1)}, \ldots, k^{(10)}$, is computed as a sequence of forty-four 32-bit quarter keys, given by $\boldsymbol{w} = (w_0, w_1, \ldots, w_{43})$, where the 128-bit master key is partitioned as $k = (w_0, w_1, w_2, w_3)$. If $j \neq 0 \pmod 4$, then the quarter key w_j is defined by

$$w_j = w_{j-1} + w_{j-4}.$$

If $j = 0 \pmod 4$, then the quarter key w_j is defined by

$$w_j = \xi(w_{j-1}) + w_{j-4} + (x^{(j-4)/4}, 0, 0, 0),$$

where the function ξ is given by

$$\xi(a_0, a_1, a_2, a_3) = (\sigma(a_1), \sigma(a_2), \sigma(a_3), \sigma(a_0)).$$

Then $k^{(i)}$ consists of four quarter keys, given by $(w_{4i}, \ldots, w_{4i+3})$.

The function σ is defined in two parts: the first part regards s as an element of the field $\boldsymbol{F}_{256}$ and sets $s \leftarrow f(s)$, where

$$f(s) = \begin{cases} s^{-1} & \text{if } s \neq 0 \\ 0 & \text{if } s = 0. \end{cases}$$

The second part then regards s as a polynomial, denoted $s(x)$, in the ring $\boldsymbol{F}_2[x]/\langle x^8 - 1\rangle$. Then $s(x)$ is mapped to $r(x)$ by

$$r(z) = (z^4 + z^3 + z^2 + z + 1)s(z) + (z^6 + z^5 + z + 1) \pmod{z^8 - 1},$$

and $r(z)$ replaces $s(z)$ in the array s. This computation can be written explicitly as

$$\begin{bmatrix} r_0 \\ r_1 \\ r_2 \\ r_3 \\ r_4 \\ r_5 \\ r_6 \\ r_7 \end{bmatrix} = \begin{bmatrix} 1 & 0 & 0 & 0 & 1 & 1 & 1 & 1 \\ 1 & 1 & 0 & 0 & 0 & 1 & 1 & 1 \\ 1 & 1 & 1 & 0 & 0 & 0 & 1 & 1 \\ 1 & 1 & 1 & 1 & 0 & 0 & 0 & 1 \\ 1 & 1 & 1 & 1 & 1 & 0 & 0 & 0 \\ 0 & 1 & 1 & 1 & 1 & 1 & 0 & 0 \\ 0 & 0 & 1 & 1 & 1 & 1 & 1 & 0 \\ 0 & 0 & 0 & 1 & 1 & 1 & 1 & 1 \end{bmatrix} \begin{bmatrix} s_0 \\ s_1 \\ s_2 \\ s_3 \\ s_4 \\ s_5 \\ s_6 \\ s_7 \end{bmatrix} + \begin{bmatrix} 1 \\ 1 \\ 0 \\ 0 \\ 0 \\ 1 \\ 1 \\ 0 \end{bmatrix}.$$

By setting $s(x)$, in turn, to each of its 256 possible values, the entries of Figure 6.7 can now be computed.

For decryption it is necessary to invert this nonlinear step. The relationship

$$s(z) = (z^6 + z^3 + z)r(z) + z^2 + 1 \pmod{z^8 - 1}$$

provides the inverse relationship to the second part because

$$(z^4 + z^3 + z^2 + z + 1)(z^6 + z^3 + z) = 1 \pmod{z^8 - 1},$$

$x\backslash y$	0	1	2	3	4	5	6	7	8	9	A	B	C	D	E	F
0	63	7C	77	7B	F2	6B	6F	C5	30	01	67	2B	FE	D7	AB	76
1	CA	82	C9	7D	FA	59	47	F0	AD	D4	A2	AF	9C	A4	72	C0
2	B7	FD	93	26	36	3F	F7	CC	34	A5	E5	F1	71	D8	31	15
3	04	C7	23	C3	18	96	05	9A	07	12	80	E2	EB	27	B2	75
4	09	83	2C	1A	1B	6E	5A	A0	52	3B	D6	B3	29	E3	2F	84
5	53	D1	00	ED	20	FC	B1	5B	6A	CB	BE	39	4A	4C	58	CF
6	D0	EF	AA	FB	43	4D	33	85	45	F9	02	7F	50	3C	9F	A8
7	51	A3	40	8F	92	9D	38	F5	BC	B6	DA	21	10	FF	F3	D2
8	CD	0C	13	EC	5F	97	44	17	C4	A7	7E	3D	64	5D	19	73
9	60	81	4F	DC	22	2A	90	88	46	EE	B8	14	DE	5E	0B	DB
A	E0	32	3A	0A	49	06	24	5C	C2	D3	AC	62	91	95	E4	79
B	E7	C8	37	6D	8D	D5	4E	A9	6C	56	F4	EA	65	7A	AE	08
C	BA	78	25	2E	1C	A6	B4	C6	E8	DD	74	1F	4B	BD	8B	8A
D	70	3E	B5	66	48	03	F6	0E	61	35	57	B9	86	C1	1D	9E
E	E1	F8	98	11	69	D9	8E	94	9B	1E	87	E9	CE	55	28	DF
F	8C	A1	89	0D	BF	E6	42	68	41	99	2D	0F	B0	54	BB	16

Figure 6.7 The AES *S*-box

and

$$(z^6 + z^3 + z)(z^6 + z^5 + z + 1) + z^2 + 1 = 0 \pmod{z^8 - 1}.$$

The two parts of Step 1 taken together may be regarded simply as a permutation of the set of 256 8-bit bytes, thereby suppressing any mention of the field $\boldsymbol{F}_{256}$. As such, the function σ may be precomputed and stored as a table. The conventional way of storing this table, shown in Figure 6.7, is called an S-box in the literature of AES. In this representation of an S-box, each 8-bit byte is expressed as a pair of hexadecimal symbols. These are symbols from the hexadecimal alphabet $\{0, 1, 2, \ldots, A, B, C, D, E, F\}$ corresponding to the sixteen binary numbers $\{0000, 0001, \ldots, 1111\}$. The input to the S-box is the byte x, y, where x consists of 4 bits and specifies the row of the array, and y consists of 4 bits and specifies the column. The output of the S-box, denoted $b(x, y)$ is stored at that (x, y) position of the S-box array.

There is widespread trust in the computational security of AES, although there is no formal mathematical proof to that effect. There is no mathematical certainty that there is not a trapdoor, either intentional or accidental, in the AES encryption procedure. However, there is no evidence whatsoever to suggest that such a trapdoor does exist.

6.7 Differential cryptanalysis

Differential cryptanalysis is an attack on DES that can be used to attack any Feistel network. It has been unsuccessful against DES. It succeeds if the number of Feistel rounds is not too large. Indeed, it appears that it is necessary that DES uses at least sixteen rounds for encryption in order to ensure that differential cryptanalysis has no possibility of an advantage over a brute-force attack.

Differential cryptanalysis is a selected plaintext attack in which two plaintexts are chosen in such a way that there will be a certain amount of cancellation if the two ciphertext blocks are subtracted bitwise. The hope is that this differential ciphertext can be used to make some inference, however slight, about the key. By repeating this process many times, perhaps a massive number of times, one hopes to gradually learn the key.

In order to explain differential cryptanalysis, we will describe its use to attack a modified DES that uses only three rounds. Then the encryption of the plaintext $(x_L^{(0)}, x_R^{(0)})$ can be written out in detail, first as

$$\begin{aligned} x_R^{(3)} &= x_L^{(2)} + f(x_R^{(2)}, k^{(3)}) \\ &= x_R^{(1)} + f(x_R^{(2)}, k^{(3)}), \end{aligned}$$

and finally

$$x_R^{(3)} = x_L^{(0)} + f(x_R^{(0)}, k^{(1)}) + f(x_R^{(2)}, k^{(3)}).$$

A similar expansion holds for the encryption of a second plaintext $(x_L^{*(0)}, x_R^{*(0)})$. This is

$$x_R^{*(3)} = x_L^{*(0)} + f(x_R^{*(0)}, k^{(1)}) + f(x_R^{*(2)}, k^{(3)}).$$

Let $\Delta x_R = x_R - x_R^*$, and $\Delta x_L = x_L - x_L^*$. Then

$$\begin{aligned}\Delta x_R^{(3)} = \Delta x_L^{(0)} &+ f(x_R^{(0)}, k^{(1)}) - f(x_R^{*(0)}, k^{(1)}) \\ &+ f(x_R^{(2)}, k^{(3)}) - f(x_R^{*(2)}, k^{(3)}).\end{aligned}$$

Now, choose the plaintexts with $x_R^{(0)} = x_R^{*(0)}$ so that the two middle terms cancel. Then the equation becomes

$$\Delta x_R^{(3)} - \Delta x_L^{(0)} = f(x_R^{(2)}, k^{(3)}) + f(x_R^{*(2)}, k^{(3)}).$$

But $\Delta x_R^{(2)} = \Delta x_L^{(3)} = \Delta y_L$ and $\Delta x_L^{(0)} = \Delta x_L$, so we have

$$\Delta y_R - \Delta x_L = f(y_L, k^{(3)}) - f(y_L^*, k^{(3)}).$$

Because this is a chosen plaintext attack, the plaintext is known and so Δx_L is known. In addition, both ciphertexts are known. Now, everything in the equation is known except the key $k^{(3)}$. Brute-force, trial-and-error testing of all keys now eventually gives $k^{(3)}$ either in its entirety or in part. In this way, the effective size of the keyspace has been reduced to the 48 bits of the subkey. A brute-force attack with this smaller keyspace is more tractable than a brute-force attack on the original keyspace, though still onerous.

A differential cryptanalysis attack against the sixteen rounds of DES is far more difficult, and very tedious, but it can be done. It has been claimed that a differential cryptanalysis attack using 2^{43} plaintexts will permit the DES key to be deduced. While this complexity may be considered marginally feasible, it is certainly unreasonable for typical applications. If there were fewer than sixteen rounds, a differential cryptanalysis attack would be less complicated and the security of DES would then be in question.

6.8 Linear cryptanalysis

The cascade of two linear functions can always be replaced by a single linear function. Therefore linearity in a cryptosystem with multiple rounds is a potential weakness that a cryptanalyst can attack. Modern block ciphers such as DES employ complicated nonlinear operations in order to counter possible linearity attacks.

Another reason that linearity is a vulnerability is that the order of execution of linear operations can be interchanged in order to bring like operations together. Perhaps the like operations then can be combined to effectively reduce complexity. Because the cryptanalyst recognizes that a linear system is vulnerable to attack in this way, one proposed attack against a nonlinear system is to partially suppress the nonlinearity by a strategy of approximating the nonlinear function by a linear function. In the case of a

Feistel network, such as DES, the goal is to reduce the main attack into a simpler attack against a linear or partially linear cryptosystem, but combined with some additional techniques to counter the residual nonlinearity.

In DES, for example, a linear cryptanalysis attack would model the function f, or some part of the function f, in a suitable arithmetic system, then fit a linear function to that function. In this way, the original nonlinear function would be modeled as a linear function combined with a residual nonlinear function. Attacks of this kind have been attempted against DES but have (apparently) failed. Although linear cryptanalysis has not succeeded as a method of attack against DES, it appears to be potentially better than a brute-force attack, but waiting to be augmented by new ideas, though none are expected.

Problems for Chapter 6

6.1 The Advanced Encryption Standard (AES) encrypts datablocks of 128 bits with a keylength of (up to) 256 bits. How many permutations of the set of 128-bit words are there? How many keys of length 256 bits are there? Estimate the fraction of the set of possible permutations that is actually used by AES.

6.2 Explain why triple DES is not vulnerable to a meet-in-the-middle attack.

6.3 The DES uses sixteen rounds of a core iteration and a 56-bit key. Suppose that this is regarded as a double encryption consisting of eight rounds forming the first encryption, followed by eight more rounds forming the second encryption. Taking this point of view, can a meet-in-the-middle attack give an attractive way of breaking DES? Give the reason for or against. How does this relate to a meet-in-the-middle attack against double DES?

6.4 Describe a modification of the DES that uses a key k of length 128 bits and otherwise, the encryptor is unchanged. Is your modification vulnerable to a meet-in-the-middle attack? In what way, if any, is your modification less secure than the Advanced Encryption Standard?

6.5 Suppose that the DES key and the plaintext are both (bitwise) complemented. Does this mean that the DES ciphertext is (bitwise) complemented as well? Is the AES ciphertext complemented under the same conditions?

6.6 Verify that the AES decryption does invert the AES encryption.

6.7 Combine a sequence of DES encryptions into a chain by the statement $y^{(i)} = e_k(x^{(i)} + y^{(i-1)})$, where $x^{(i)}$ is the ith plaintext block, and $y^{(i)}$ is the ith ciphertext block. Explain why this does not reduce security. Does it increase security? Show how to decrypt the chained ciphertext.

6.8 Design and sketch a variation of a Feistel network that breaks the 64-bit plaintext into four quarter-words instead of two half-words. Be sure that the encryption is invertible. Do you see any implementation advantage of your design? Do you see any security weakness?

Notes for Chapter 6

The Data Encryption Standard is a block cryptosystem that evolved from early work started at IBM Research by Feistel, and DES became an industry standard in 1977. A precursor version of DES known as Lucifer, described in several variations as by Feistel (1974) and Sorkin (1984), was an early commercial block cipher. These methods, based on the Feistel network, were published widely in and after 1971. It has apparently never been revealed how the constants in the DES substitution boxes were chosen. A history of DES has been given by Smid and Branstad (1992). Other block ciphers in common use for niche applications, such as IDEA and SAFER, have evolved from the notion of the Feistel network though with modifications such as breaking a 64-bit plaintext into 8 bytes instead of two halfblocks.

The method of differential cryptanalysis, though it may have been known to some at the time of the development of DES, did not appear in the open literature until it was published two decades later by Biham and Shamir (1993), and later advanced by Lai, Massey, and Murphy (1991). Linear cryptanalysis was introduced by Matsui (1994) in expectation that it would lead to a successful ciphertext-only attack. Although linear cryptanalysis has had some limited success, and has further weakened the confidence in DES, it has not led to practical attacks. The methods of differential cryptanalysis and linear cryptanalysis were combined by Langford and Hellman (1994) into a single attack, although still not a successful attack.

Despite the limited success of the differential cryptanalysis and linear cryptanalysis when applied to the DES, the real weakness of DES is that the keyspace is too small. Hellman (1980) showed that it is feasible to build a special-purpose computer that could open 56-bit-key DES cipherblocks by an essentially brute-force attack. This then led to the use of double encryption and triple encryption. The meet-in-the-middle attack was introduced by Merkle and Hellman (1981) to debunk the merits of any double encryption. A meet-in-the-middle attack is not effective against triple encryption.

The Data Encryption Standard has now been superceded by the Advanced Encryption Standard, which was designed after an open call for proposals. This standard is based on a proposal of Daemen and Rijmen (2000, 2001), originally going by the name “Rijndael” from a contrived combination of their two names. It became a standard under its current name only after modifications were incorporated into the original proposal. Because of its increased key length, neither differential cryptanalysis, linear cryptanalysis, nor a brute-force attack appear to have any hope of breaking AES. As for the DES, the sceptic can always suspect that there is a cleverly hidden trapdoor in the mathematics of the AES through which the message can be read by a cryptanalyst who does not have the key, but does know the trapdoor. There is no evidence to sustain such a suspicion, but neither is there a proof that such a trapdoor does not exist. The standard treatment of AES has been given in this chapter. There are also treatments of AES that give algebraic descriptions, one by Lenstra (2002), and one by Rosenthal (2003).

7 Stream ciphers

In contrast to a block cipher, which regards messages as segmented into datablocks of a fixed length, a *stream cipher* has a fixed beginning, but it has no fixed end. A stream cipher encrypts a plaintext datastream of indefinite length, such as a long bitstream, which we can regard as an infinite datastream. An infinite stream of plaintext symbols is fed into an encryptor, using an encryption key of length m bits, and the infinite stream of ciphertext symbols leaves the encryptor. Later, the infinite stream of ciphertext symbols enters the decryptor, which uses the same key of length m bits, and the infinite stream of plaintext symbols leaves the decryptor.

Stream ciphers have a long history, perhaps inspired by the one-time pad, and are in widespread use in many applications. A common form of stream cipher, called an *additive stream cipher*, converts the m-bit key into a semi-infinite binary stream, called a *keystream*. The encryptor adds the binary keystream modulo two to the binary datastream, bit by bit, to form the codestream. The decryptor adds the same keystream, modulo two, to the codestream, again bit by bit, and in this way recovers the datastream.

Recall that a binary one-time pad consists of a semi-infinite random sequence of equiprobable bits that is known to both the encryptor and the decryptor. There is no constraint on the random binary sequence used as a one-time pad other than that it be maximally random. This means that the bits of the sequence are independent and equiprobable.

An additive binary keystream, in practice, is specified by a master key consisting of a block of m bits. This means that there can be only 2^m keystreams in the cipher, and the block key specifies which of these 2^m keystreams is to be used. The goal when designing a stream cipher is to choose these 2^m keystreams so that they appear to be arbitrary and mimicking randomness, with no evident accessible structure, and yet are simple to construct from the m-bit key. Although the bits of the keystream are dependent, this dependence must not be easy to extract.

A binary keystream, itself generated by a key of length m, may also be used in a different way as a method of sequentially replacing the keys used by an adjacent block cipher. Simply segment the semi-infinite binary keystream into a sequence of m-bit blocks, each block now regarded as a new m-bit key for the adjacent block cipher system. In this way, each block of the data can be block-encrypted with a different key, or a key that is frequently changed. Although the sequence of such computed keys is generated by a single initialization of the keystream generator by a master key, one presumes that the block cipher is further protected because it continually changes its key.

7.1 State-dependent encryption

To be practical, a stream cipher must be realizable, so the encryptor must have a finite memory. Therefore, in general, the encryptor is a finite-state machine controlled by a key. We suppose that the finite memory can hold L symbols taking values in some finite alphabet $\mathcal{A}$, usually the binary alphabet. The set of the current values of the L symbols in the memory is called the *state* of the memory. The state takes its values in a block alphabet of size $(\#\mathcal{A})^L$. For a binary alphabet, the state can take on 2^L values.

For example, the keyed function

$$y_i = e_k(x_i, x_{i-1}, x_{i-2}, \ldots, x_{i-L})$$

forms a stream cipher with a state memory of length L and an encryption key k. In this example, the state is equal to the L most recent symbols of the plaintext. The state memory is simply used to store these L most recent symbols of the plaintext. The ith symbol of the ciphertext is computed directly from these L stored symbols of the plaintext. This kind of encryptor is called a *feedforward encryptor*, the name referring to the fact that the window on the past, from i to $i - L$, moves forward as i increases, and past plaintext symbols are eventually discarded by the encryptor. The encryption of the current symbol is not affected by the plaintext symbols that are older than the L most recent symbols.

More generally, let the memory at time i hold the *state vector*, defined as $(s_1^{(i)}, s_2^{(i)}, \ldots, s_L^{(i)}) = \mathbf{s}^{(i)}$. At each i, the new state is computed from the current state history and the new input symbol. Then the finite-state machine is described as

$$y_i = e_k(x_i, \mathbf{s}^{(i-1)})$$
$$\mathbf{s}^{(i)} = f(x_i, \mathbf{s}^{(i-1)}).$$

Even more generally, the function f that updates the state may also depend on the key k. Then the encryptor, as a finite-state machine, is described as follows:

$$y_i = e_k(x_i, \mathbf{s}^{(i-1)})$$
$$\mathbf{s}^{(i)} = f_k(x_i, \mathbf{s}^{(i-1)}).$$

The general form of an encryptor for a steam cipher is shown in Figure 7.1.

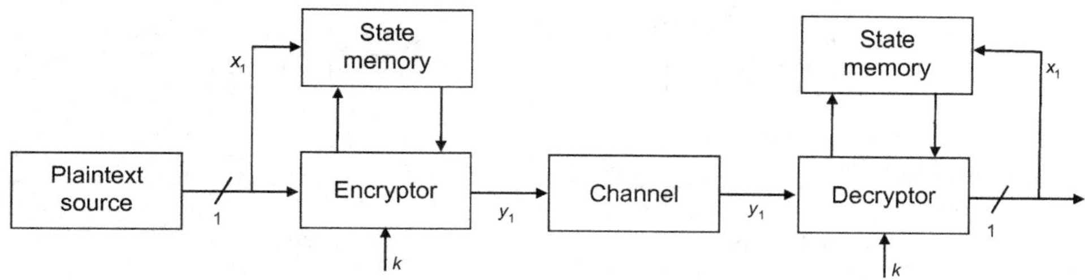

Figure 7.1 State-dependent encryption

The decryptor has a similar structure. It is described as a finite-state machine as follows:

$$\begin{aligned} x_i &= d_k(y_i, s^{(i-1)}) \\ s^{(i)} &= f_k(x_i, s^{(i-1)}). \end{aligned}$$

Notice that in the decryptor the function defining the state $s^{(i)}$ is the same as in the encryptor. Moreover, the plaintext symbol x_i is recovered without reference to the new state $s^{(i)}$. All of the values needed by the decryptor to compute the new state $s^{(i)}$ are then available after the plaintext symbol x_i is computed.

One method of forming a state-dependent encryptor for a stream cipher is to use a block encryptor as the core of the encryptor, augmented by simple external operations that turn the block encryptor into a stream encryptor. For example, as was discussed in Section 6.4, let $\boldsymbol{y} = e_k(\boldsymbol{x})$ denote a block encryption function on blocks of length n. Let $\boldsymbol{x}_\ell$, for $\ell = 1, \ldots,$ denote a series of data blocks, each of length n. Then define a stream encryption as the stream of blocks $\boldsymbol{y}_\ell = e_k(\boldsymbol{x}_\ell + \boldsymbol{y}_{\ell-1})$, starting with any specified initial block $\boldsymbol{y}_0$. Thus each ciphertext block is added to the next plaintext block prior to block encryption. The ciphertext blocks are concatenated to form the ciphertext message. The stream decryption is $\boldsymbol{x}_\ell = d_k(\boldsymbol{y}_\ell) + \boldsymbol{y}_{\ell-1}$, which removes the previous ciphertext block after each block is decrypted.

7.2 Additive stream ciphers

A simple and widely used stream cipher is the *additive stream cipher*, which is a special case of a stream cipher. An additive stream cipher is sometimes said to be inspired by Shannon's one-time pad. A one-time pad requires a semi-infinite, truly random binary sequence. In contrast, an additive binary stream cipher uses a periodic binary sequence of very large period that imitates a one-time pad and superficially has the appearance of a random binary sequence.

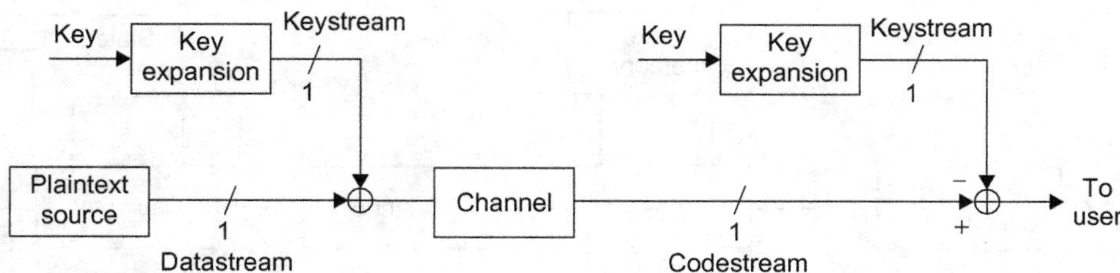

Figure 7.2 Additive keystream cipher

An additive stream cryptography system over the abelian group G encrypts by adding componentwise an infinite sequence of symbols of G to the plaintext. Thus

$$y_i = x_i + z_i^{(k)},$$

where the sequence $\{z_i^{(k)}\}$, called the *keystream*, is an infinitely long sequence with symbols from the group G. The binary field, $\boldsymbol{F}_2$, which uses modulo-two addition, usually suffices for G. The keystream to be used is specified by a master block key k from a set of keys $\mathcal{K}$. There is one such keystream for each master key $k \in \mathcal{K}$. If the keys are m-bit binary numbers, then there are 2^m keystreams. The task of key expansion is to stretch a block key k into a semi-infinite keystream. Although the keystreams are generated from the master keys deterministically by a finite-state machine, one attempts to design this process so that the keystreams appear to be maximally random, and so that the future values of the keystream cannot be predicted from a segment of the past values of the keystream. Accordingly, such keystreams are often called *pseudorandom keystreams*.

An additive keystream cryptographic system is shown in Figure 7.2. To form the keystream, denoted $z^{(k)}$, we compute the ith bit by the rule

$$z_i^{(k)} = e_k(\boldsymbol{s}^{(i-1)})$$
$$\boldsymbol{s}^{(i)} = f_k(\boldsymbol{s}^{(i-1)}),$$

where k is the master key, i is the time index, and $\boldsymbol{s}^{(i)}$ is the state vector $(s_1^{(i)}, s_2^{(i)}, \ldots, s_L^{(i)})$. The ciphertext is the sequence of bits given by

$$y_i = x_i + z_i^{(k)} = x_i + e_k(s^{(i-1)}) \pmod 2.$$

The additive keystream cipher is a special case of a stream cipher in that f_k and e_k do not have the databit x_i as an argument, and so the keystream does not depend on the data. Because there are only $\#\mathcal{K} = 2^m$ keys, there are only 2^m keystreams, each of which is an infinitely long sequence. A practical keystream generator must have finite complexity, so it uses a finite-state machine to form the infinite keystream. Because a finite-state machine has only a finite number of states, it must eventually repeat a state. Once it repeats a state, it must repeat the same history of states following that state. Thus every keystream generated by a finite-state machine is eventually periodic.

An additive keystream cipher is *linear* if the sum of the keystream and any cyclic translate of that keystream is equal to yet some other cyclic translate of the original keystream. That is, for any ℓ, there exists an ℓ' such that

$$z_i + z_{((i+\ell))} = z_{((i+\ell'))}$$

for all i, where the double parentheses mean modulo-n indexing and n is the period of the cyclic sequence. Because linear sequences can be generated by a shift register, they are called linear shift-register sequences. The theory of linear additive keystreams begins with the study of linear shift-register sequences.

An additive keystream must be synchronized at the encryptor and decryptor. This means that the decryption must subtract the keystream symbol z_i from the ciphertext symbol y_i. If it subtracts the keystream symbol $z_{i'}$, for $i' \neq i$, then we have the computation

$$y_i - z_{i'} = x_i + z_i - z_{i'},$$

which would be unintelligible to the user. This requirement is not unique to additive keystreams. Any state-dependent stream cipher must be given an initial state, and this must be synchronized with the first symbol of the keystream. Indeed, synchronization is always a requirement in any system that communicates digital bitstreams. Even a block cipher must be properly segmented into blocks for decryption.

While these points may be obvious, what may not be obvious is that the method of keystream synchronization may possibly provide a vulnerability that could be exploited by the cryptanalyst, although we have no evidence for this concern. One must always be wary of secondary functions that support the encryption.

7.3 Linear shift-register sequences

A periodic sequence over $\boldsymbol{F}_q$ is an infinite sequence $\boldsymbol{v} = (v_0, v_1, v_2, \ldots)$ of field elements that satisfies $v_{i+n} = v_i$ for some integer n, called the *period* of the sequence. A *shift-register sequence* $\boldsymbol{v} = (v_0, v_1, \ldots, v_0, v_1, \ldots)$ is a periodic sequence of elements that can be generated by a feedback shift-register circuit. By this broad definition, however, every periodic sequence would be a shift-register sequence. The term is usually used to refer only to a periodic sequence that can be generated by a *nontrivial* feedback shift-register circuit, whose period is $q^m - 1$ or a divisor of $q^m - 1$ for some value of m. A feedback shift-register circuit is another name for a kind of recursion.

A *shift register* is an ordered set of q-ary memory cells, with each cell used to hold one element of the field. Each cell is updated at each clock cycle by taking its new value from its leftmost neighbor. The cell at the left end has no neighbor on the left. That cell takes its new value as a linear combination of the contents of the other cells. This linear combination is called the *feedback*. To generate a periodic sequence of period

$q^m - 1$ requires a finite-state machine with at least m memory cells and at most $q^m - 1$ memory cells. A shift-register circuit can be represented as a recursion

$$v_i = f(v_{i-1}, v_{i-2}, \ldots, v_{i-L}),$$

where, for a sequence of period $q^m - 1$, L is at least m and at most $q^m - 1$.

A *linear* sequence over the field $\boldsymbol{F}_q$ is a sequence formed by a shift register whose feedback is a linear function of the contents of the cells. A *maximal* linear shift-register sequence over the field $\boldsymbol{F}_q$ is a linear sequence of period $q^m - 1$ formed by a shift register with only m memory cells. A maximal shift-register sequence is also called an *m-sequence*.

We consider only shift-register sequences over the binary field $\boldsymbol{F}_2$. A maximal sequence over $\boldsymbol{F}_2$ is closely related to an irreducible polynomial over $\boldsymbol{F}_2$, and therefore related to the field $\boldsymbol{F}_{2^m}$. A periodic sequence over $\boldsymbol{F}_2$ of period $q^m - 1$ can be constructed by using an irreducible polynomial $p(x)$ of degree m over $\boldsymbol{F}_2$. The sequence will have period $2^m - 1$ if the irreducible polynomial $p(x)$ is required to have the property that the element x has order $2^m - 1$ under modulo $p(x)$ arithmetic. This means that $p(x)$ is a primitive polynomial and the element x is a primitive element of $\boldsymbol{F}_{2^m}$. Primitive polynomials and primitive elements are discussed in Chapter 9.

Each nonzero element of $\boldsymbol{F}_{2^m}$ can be represented as an m-bit binary word or as a power of a primitive element α. This means that there is a unique m-bit binary word corresponding to α^ℓ for each ℓ from zero to $2^m - 2$. For each ℓ, take the high-order bit of the binary word corresponding to α^ℓ to form a sequence of $2^m - 1$ bits. This periodic sequence – or any of its cyclic shifts – is an m-sequence. Such an m-sequence can be constructed from any primitive polynomial. Despite the rigid structure of such sequences, we will see that they display many properties that one might like to see in a random sequence.

Let $p(x)$ be a primitive polynomial with the elements of $\boldsymbol{F}_{2^m}$ represented as polynomials in x. Then $x = \alpha$ is a primitive element, and

$$x^m + p_{m-1}x^{m-1} + p_{m-2}x^{m-2} + \cdots + p_1 x + p_0 = 0$$

where the coefficients are elements of $\boldsymbol{F}_2$. Therefore

$$x^\ell = p_{m-1}x^{\ell-1} + p_{m-2}x^{\ell-2} + \cdots + p_1^{\ell-m+1}x + p_0^{\ell-m}$$

for $\ell = m, m+1, \ldots$. This describes exactly the operation of a linear-feedback shift register of length m with taps given by $p(x)$. The m-sequence is the sequence of feedback bits generated by this circuit.

All of the structure of an m-sequence follows easily from the fact that the primitive element x has order $2^m - 1$. This requires that, at each of the $2^m - 1$ steps, the linear-feedback shift register must hold a different m-bit binary word; otherwise the pattern would repeat earlier. Each of the $2^m - 1$ nonzero m-bit binary numbers must appear exactly once in the shift register. This means that the number of zeros and ones is nearly

Table 7.1 *The cycle of α*

α^0	=	0	0	0	1
α^1	=	0	0	1	0
α^2	=	0	1	0	0
α^3	=	1	0	0	0
α^4	=	0	0	1	1
α^5	=	0	1	1	0
α^6	=	1	1	0	0
α^7	=	1	0	1	1
α^8	=	0	1	0	1
α^9	=	1	0	1	0
α^{10}	=	0	1	1	1
α^{11}	=	1	1	1	0
α^{12}	=	1	1	1	1
α^{13}	=	1	1	0	1
α^{14}	=	1	0	0	1

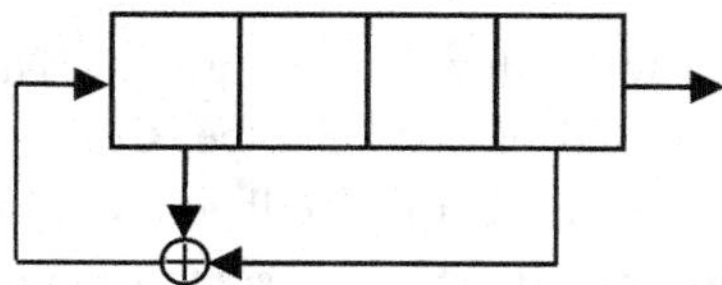

Figure 7.3 A simple linear-feedback shift register

balanced. Similar reasoning about the high-order r-bit subsequence in the shift register leads to similar conclusions about all r-bit subsequences for r at most m.

The m-sequence is a linear sequence, meaning that the sum of two cyclic translates of the sequence is equal to another cyclic translate of the sequence. That is, because the low-order bit of α^{i+a} and α^{i+b}, satisfy

$$\alpha^{i+a} + \alpha^{i+b} = \alpha^{i+c}$$

for some c. The linearity property follows from the field structure because for any a and b in $\boldsymbol{F}_q$, there is a c in $\boldsymbol{F}_q$ such that

$$\alpha^a + \alpha^b = \alpha^c.$$

For example, let α be a zero of the primitive polynomial $p(x) = x^4 + x + 1$. Then α is a primitive element of $\boldsymbol{F}_{16}$. The shift-register circuit that multiplies by the element α is shown in Figure 7.3. Table 7.1 shows the orbit of α in $\boldsymbol{F}_{16}$ with the elements expressed in the polynomial basis. The leftmost column of Table 7.1 gives the periodic sequence 000100110101111. Every other column is a cyclic translate of this column. This periodic sequence is a linear shift-register sequence.

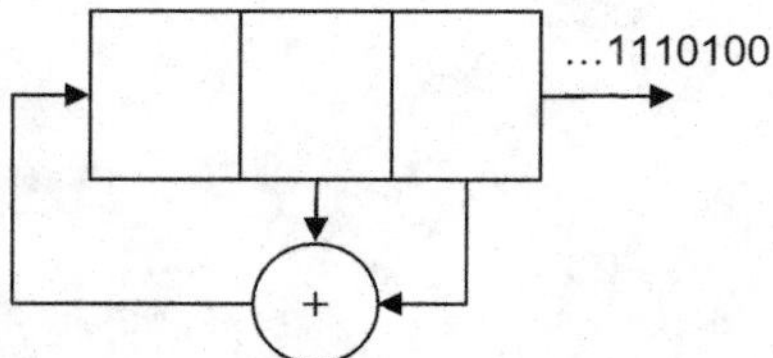

Figure 7.4 Generation of an m-sequence of period 7

A maximal sequence emulates many properties of a random sequence. Approximately half of the bits of a random binary sequence will be ones and approximately half of the bits will be zeros.[1] Similarly, the doublets 00, 01, 10, and 11 each will occur about one-fourth of the time. Likewise, each 3-bit pattern will occur about one-eighth of the time, and a similar remark holds for longer multituples not longer than m. The m-sequence mentioned above satisfies these conditions. Indeed, one might say that it satisfies the conditions too well. It has seven zeros and eight ones, and when viewed cyclically, three instances of 00, and four each of 01, 10, and 11; one instance of 000, and two each of 001, 010, 011, 100, 101, 110, and 111. This is as close to balanced as is possible for a sequence of period 15.

Another example of linear-feedback shift register is shown in Figure 7.4. The polynomial $p(x) = x^3 + x + 1$ is a primitive polynomial of degree three. Consequently, an m-sequence of length seven can be generated using the feedback shift register of Figure 7.4. When initializing the shift register with x^0, represented from left to right as 100, the sequence of generated bits at the output, now from left to right, is

$$0\,0\,1\,0\,1\,1\,1\;\ldots,$$

which repeats periodically. This is an m-sequence of period seven. Again, the number of zeros and ones in one period differs by one, which is as nearly equal as is possible for a sequence of odd blocklength. With the sequence viewed cyclically, the overlapping 2-bit subsequences are 00, 01, 10, 01, 11, 11, 10. Except for the subsequence 00, there are two sequences of each kind, which is as nearly equal as possible for a sequence of odd blocklength. The 3-bit subsequences, viewed cyclically, are also balanced; each appears once except for the subsequence 000.

This behavior holds for any m-sequence. For example, choose a primitive polynomial $p(x)$ of degree thirty. The m-sequence will have a period of length $2^{30} - 1$, or about 10^9 bits. For each r smaller than thirty, all subsequences of length r except the all-zero subsequence of length r will occur with equal frequency.

The set of all maximal linear shift-register sequences is the set of sequences that can be obtained in this way by the choice of the primitive polynomial $p(x)$. The degree m

[1] Of course, a sequence is not itself random. This phrasing refers to the selection from an ensemble of sequences. The sequence is the outcome of this selection.

of $p(x)$ determines both the extension field $\boldsymbol{F}_{2^m}$ and the period $2^m - 1$ of the maximal sequence. The number of primitive polynomials $p(x)$ of degree m is limited, as will be counted in Theorem 9.13.5 of Section 9.13, so the number of maximal sequences of length $2^m - 1$ is limited as well.

7.4 The linear-complexity attack

The length of the shortest linear recursion that produces the periodic sequence $\boldsymbol{v} = (v_0, v_1, \ldots, v_{n-1})$ is called the *linear complexity* of the periodic sequence. If the linear recursion is known, then the entire periodic sequence is known except for the starting phase. A linear-complexity attack is a partially-known-plaintext attack against an additive keystream. From any known segment of the keystream, a linear-complexity attack attempts to compute a linear recursion

$$v_i = -\sum_{k=1}^{L} \Lambda_k v_{i-k}$$

that will produce all of the keystream from that known segment.

Because a known plaintext is easily removed from an additive stream cipher to reveal the keystream, the linear-complexity attack can be used whenever a sufficiently long segment of the plaintext is known. The attack first finds the shortest linear recursion that produces the known segment of the keystream, and then uses that linear recursion to produce a continuation of the known keystream in the hope that the continuation will be the entire keystream. This attack will fail if the known segment of the keystream is too short.

The linear-complexity attack is an attack on the keystream itself, not on the procedure that generated the keystream. The attack does not inquire how the keystream was actually generated. The keystream need not have been generated by a linear recursion, but if it is a periodic sequence, it can always be regarded as having been so generated by a sufficiently long linear recursion. This is because every periodic sequence over the field $\boldsymbol{F}$ of period n can be produced by the trivial linear recursion $x_i = x_{i-n}$ for $i = n+1, n+2, \ldots$. Thus the linear complexity of every periodic sequence in $\boldsymbol{F}$ of period n is not larger than n, and often is much smaller.

Neither the linear recursion nor its initial state nor its length L is known prior to the attack. The unknowns Λ_i for $i = 1, \ldots, L$, specify the unknown recursion and the unknowns v_i, for $i = 0, \ldots, L-1$, determine the unknown initial state. Both must be determined by the linear-complexity attack when given only a subsequence of $2L$ consecutive symbols of the keystream. If the linear complexity of the keystream is indeed at most L, then the recursion gives a system of L linear equations in the L unknown Λ_j. This system of linear equations can be written in matrix form as

$$\begin{bmatrix} v_{L+1} \\ v_{L+2} \\ \vdots \\ v_{2L} \end{bmatrix} = \begin{bmatrix} v_L & v_{L-1} & \cdots & v_1 \\ v_{L+1} & v_L & \cdots & v_2 \\ \vdots & & \vdots & \\ v_{2L-1} & & \cdots & v_{L-1} \end{bmatrix} \begin{bmatrix} \Lambda_1 \\ \Lambda_2 \\ \vdots \\ \Lambda_L \end{bmatrix},$$

where $v_1, v_2, \ldots, v_{2L}$ is the observed segment of the keystream, a total of $2L$ consecutive symbols. All v_i needed by the stated equations are known, and the Λ_i are the unknowns. The recursion is found by solving this matrix equation for the vector of tap weights Λ_i. The recursion can then be executed to produce the entire periodic sequence.

The linear-complexity attack will succeed whenever the length of the observed segment of the keystream is at least twice the actual linear complexity of the keystream (defined in the next section) because then all elements of the above matrix equation except the coefficients of $\Lambda(x)$ are known. It is not necessary that the linear complexity L be known prior to the attack. The attack will succeed if the length of the observed segment of the keystream satisfies the necessary condition that it is at least twice as large as the linear complexity; otherwise, it will fail. For this reason, a linear-complexity attack is a threat only if the linear complexity is significantly smaller than one-half of the period of the keystream. If the linear complexity is nearly half of the period or length of the keystream, then the linear-complexity attack requires that nearly all of the keystream be known in order to produce the rest. In this case, a linear-complexity attack is not a suitable attack.

A linear-complexity attack is both tractable and meaningful because of the properties of the matrix equation describing the recursion. A linear-complexity attack is tractable because there are efficient computational algorithms that will solve this form of matrix equation. The matrix itself is known as a *Toeplitz matrix*. There are fast algorithms to invert a matrix of this form. Moreover, because of the special form of the system of linear equations, even faster algorithms are available to solve this system of equations without inverting a matrix. A linear-complexity attack is meaningful because any $2L$ consecutive known components of the sequence will do to reconstruct the entire sequence if the linear complexity is indeed L. Even some patterns of nonconsecutive blocks of components will sometimes suffice to set up an invertible matrix. Such patterns can be analyzed in terms of the heft and rank of the matrix describing the recursion. These terms are defined in Section 9.12 of Chapter 9.

7.5 Analysis of linear complexity

We have stated that an additive keystream may be vulnerable to a linear-complexity attack, which is an attack that produces an entire keystream from a small known segment of the keystream by setting up a linear recursion. If this attack were to succeed, then the keystream is only as secure as is that keystream segment. A linear-complexity attack requires that some segment of the keystream be received in the clear, or can be otherwise deduced by some means from an observed segment of the ciphertext. This amounts to a (partially) known-plaintext attack.

A *linear recursion* of length L in the field $\boldsymbol{F}$ – or the ring R – is an expression of the form

$$v_i = -\sum_{k=1}^{L} \Lambda_k v_{i-k} \qquad i = L, L+1, \ldots .$$

A linear recursion of length L *produces* the entire sequence $\boldsymbol{v}$ from its first L components. For example, the elementary linear recursion $v_i = v_{i-1} + v_{i-2}$ in the ring $\boldsymbol{Z}$ produces the well-known *Fibonacci sequence* $(1, 1, 2, 3, 5, 8, 13, \ldots)$.

The *linear complexity* of any sequence $\boldsymbol{v}$ is defined to be the length of the shortest linear recursion that will produce that sequence. For example, the linear complexity of the Fibonacci sequence is two because the linear recursion $v_i = v_{i-1} + v_{i-2}$ will produce the Fibonacci sequence, but no shorter linear recursion will produce the Fibonacci sequence.

A *linear-complexity attack* is an attempt to find a linear recursion that produces a sequence, when given a segment of that sequence. To protect against a linear-complexity attack, one wants sequences of large linear complexity, but to be practical, the structure of these sequences must not be too complicated. Otherwise, the sequences would be too difficult to generate.

The linear-complexity attack is based on the observation that if a sequence $\boldsymbol{v}$ satisfies the recursion

$$v_i = -\sum_{k=1}^{L} \Lambda_k v_{i-k} \qquad i = L, L+1, \ldots,$$

and v_i is known for $2L$ consecutive values of the index i, then the recursion gives a system of L linear equations in the L unknown coefficients Λ_i.

There are several theorems that are useful for studying linear complexity. We begin with a theorem that states a condition under which two linear recursions will continue to agree if they agree up to a certain point. This theorem is used in the proofs of the two theorems that follow.

Theorem 7.5.1 (Agreement theorem) *If two linear recursions $\Lambda(x)$ and $\Lambda'(x)$ of length L and L', respectively, both produce the sequence $v_0, v_1, \ldots, v_{r-1}$, and if $r \geq L + L'$, then both linear recursions produce the same sequence thereafter.*

Proof It is enough to show that both recursions produce the same rth term under the conditions of the theorem. We must show that

$$-\sum_{k=1}^{L} \Lambda_k v_{r-k} = -\sum_{k=1}^{L'} \Lambda'_k v_{r-k},$$

under the assumption that

$$v_i = -\sum_{k=1}^{L} \Lambda_k v_{i-k} \qquad i = L, \ldots, r-1$$

and

$$v_i = -\sum_{k=1}^{L'} \Lambda'_k v_{i-k} \qquad i = L', \ldots, r-1.$$

Because $r \geq L + L'$, we can set $i = r - j$ in each of these two equations. Then we can write

$$v_{r-j} = -\sum_{k=1}^{L} \Lambda_k v_{r-j-k} \qquad j = 1, \ldots, L'$$

and

$$v_{r-j} = -\sum_{k=1}^{L'} \Lambda'_k v_{r-j-k} \qquad j = 1, \ldots, L,$$

with all indicated terms from the given sequence $v_0, v_1, \ldots, v_{r-1}$. Finally, we have

$$\begin{aligned} -\sum_{k=1}^{L} \Lambda_k v_{r-k} &= \sum_{k=1}^{L} \Lambda_k \sum_{j=1}^{L'} \Lambda'_j v_{r-k-j} \\ &= \sum_{j=1}^{L'} \Lambda'_j \sum_{k=1}^{L} \Lambda_k v_{r-k-j} \\ &= -\sum_{j=1}^{L'} \Lambda'_j v_{r-j}, \end{aligned}$$

as was to be proved. □

Theorem 7.5.2 (Massey's theorem) *If the shortest linear recursion that produces the sequence $(v_0, v_1, \ldots, v_{r-2})$ has length L and does not produce the sequence $\mathbf{v} = (v_0, v_1, \ldots, v_{r-2}, v_{r-1})$, then every linear recursion that produces $\mathbf{v}$ has a length of at least $r - L$.*

Proof Suppose a linear recursion of length L' satisfying $L' < r - L$ exists that produces the sequence $(v_0, \ldots, v_{r-1})$. Then we have two recursions that produce $(v_0, \ldots, v_{r-2})$. But $L + L' \leq r - 1$, so by the agreement theorem, both must produce the same value at the next iteration, contrary to the premise of the theorem. □

For example, the recursion $v_i = v_{i-1} + v_{i-2}$ produces the Fibonacci sequence. In particular, it produces the finite-length sequence (1, 1, 2, 3, 5, 8, 13, 21, 34), but it does not produce the longer sequence $(1, 1, 2, 3, 5, 8, 13, 21, 34, A)$ unless $A = 55$. For other values of A, Massey's theorem states that every linear recursion that produces the sequence has length at least 8. Likewise, the smallest linear recursion over $\boldsymbol{F}_2$ that produces the finite-length sequence $1, 1, 0, 1, 1, 0, 1, 1, 0, 0$ has a length at least 8.

Let ω be an element of order n in $\boldsymbol{F}$ or possibly in an extension field of $\boldsymbol{F}$. Such an ω always exists if $\boldsymbol{F}$ has characteristic zero, and such an ω always exists if the characteristic p of the finite field $\boldsymbol{F}_p$ is coprime with n. Furthermore, such an ω of order n exists in $\boldsymbol{F}_{p^m}$ if, and only if, n divides $p^m - 1$. Then any vector $\boldsymbol{v}$ of length n has a Fourier transform, $\boldsymbol{V}$, given by

$$V_j = \sum_{i=0}^{n-1} \omega^{ij} v_i \qquad j = 0, \ldots, n-1.$$

The properties of the Fourier transform are summarized in Section 9.12.

The next theorem characterizes the Hamming weight of a Fourier transform, where the Hamming weight of a vector is defined as the number of nonzero components of that vector.

Theorem 7.5.3 (Blahut's theorem) *The linear complexity of a periodic sequence over* $\boldsymbol{F}$ *of period n is equal to the Hamming weight of its Fourier transform, provided a Fourier transform of blocklength n exists.*

Proof Let $\boldsymbol{v}$ be the periodic sequence of blocklength n, and let $\boldsymbol{V}$ be the Fourier transform of $\boldsymbol{v}$ given in terms of its components as

$$V_j = \sum_{i=1}^{n-1} v_i \omega^{ij}$$

for any ω of order n in $\boldsymbol{F}$ or in an extension of $\boldsymbol{F}$, provided such an ω of order n exists. Consider any recursion of the form

$$v_i = -\sum_{k=1}^{L} \Lambda_k v_{((i-k))}$$

where the double parentheses denote modulo n. Let $\Lambda(x) = 1 + \sum_{i=1}^{L} \Lambda_i x^i$ and $v(x) = \sum_{i=0}^{n-1} v_i x^i$. The recursion can be rewritten as a polynomial product

$$\Lambda(x)v(x) = 0 \pmod{n},$$

from which we can write

$$\Lambda(\omega^{-i})v(\omega^{-i}) = 0$$

where ω is any element of order n, possibly in an extension field. Therefore, if $\lambda_i = \Lambda(\omega^{-i})$ and $v_i = V(\omega^{-i})$, we see that $\lambda_i = 0$ whenever $v_i \neq 0$. Because $\Lambda(x)$ cannot have more zeros than its degree L, we conclude that $L \geq \text{wt}(\boldsymbol{v})$. Moreover, we can always construct a $\Lambda(x)$ of degree $\text{wt}(\boldsymbol{v})$ with zeros at the nonzeros of $\boldsymbol{v}_i$. This $\Lambda(x)$ corresponds to a recursion of minimum length, and the length of this recursion must be $L = \text{wt}(\boldsymbol{v})$. □

7.6 Keystreams from nonlinear feedback

A linear shift-register sequence is an elementary example of a keystream. It is perhaps the most commonly used mathematical structure out of which additive keystream ciphers are constructed, but it is normally augmented with embellishments that are included to increase the security. The embellishments are needed because a linear shift-register sequence, although rather elegant with a clean mathematical structure, is vulnerable to a linear-complexity attack. There are many such embellishments of linear shift-register sequences that are intended to counter the linear-complexity attack. We will consider constructions that are based on nonlinear shift-register feedback in this section, constructions that combine several linear shift-register sequences in Section 7.7, and constructions that are based on nonlinear operations on the output of a linear-feedback shift register in Section 7.8.

A recursion of the form $v_i = -f(v_{i-1}, v_{i-2}, \ldots, v_{i-m})$ is called a *nonlinear recursion* if the function f is not linear. A binary recursion is one in which all the variables and operations are in $\boldsymbol{F}_2$. To specify an arbitrary nonlinear binary recursion, observe that for each of 2^m possible states of an m-bit memory, the value of f can be either zero or one. Thus the function f can be described as a binary number of length 2^m. There are 2^{2^m} such binary numbers, and so there are 2^{2^m} such binary functions. Of these, 2^m are linear functions. Every recursion is eventually periodic because the state corresponding to the argument of the function f can only take a finite number of values and so must repeat. Some such recursions can have a starting transient (which depends on the initialization) prior to entering the periodic behavior. These recursions are called *singular recursions*. Recursions that are periodic for any initialization are called *nonsingular recursions*.

As we have seen, a linear-feedback shift register of length m cannot produce a sequence with period larger than $2^m - 1$, even though a memory of m bits could hold 2^m values. This is because, to produce a periodic sequence of period 2^m using m bits of state memory, the linear-feedback shift register must never contain all zeros. If ever it did contain all zeros, it would continue to hold all zeros thereafter. A nonlinear periodic sequence, however, can have period 2^m and can be produced by a finite-state machine with m bits of memory. It cannot produce a sequence with a period that is larger than 2^m.

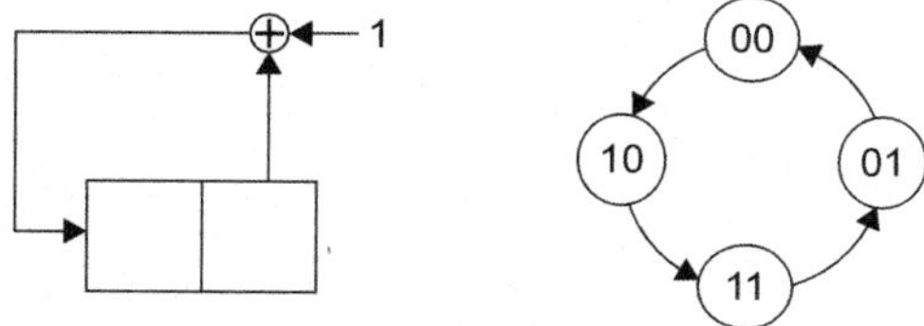

Figure 7.5 A shift register for a deBruijn sequence

Definition 7.6.1 *A deBruijn sequence is any periodic sequence over a q-ary alphabet of period q^m in which every pattern of m symbols begins once in each segment of length q^m.*

The definition recognizes the fact that q^m m-tuples begin in any segment of length m, and some will extend beyond that segment. Indeed, those q^m m-tuples will be in an interval of length $q^m + m - 1$. Because we restrict the continuing discussion to binary sequences, there will be 2^m m-tuples beginning in any period.

Every deBruijn sequence can be formed by a nonlinear boolean feedback function of the form

$$\begin{aligned} v_i &= f(v_{i-1}, v_{i-2}, \ldots, v_{i-m}) \\ &= v_{i-m} + g(v_{i-1}, \ldots, v_{i-m+1}). \end{aligned}$$

Only those functions g for which the period is 2^m give deBruijn sequences. There are $2^{(2^{m-1}-m)}$ binary deBruijn sequences of period 2^m, up to translation.

There is only one deBruijn sequence of length four, two of length eight, sixteen of length sixteen, and 2048 of length thirty-two. Some examples are

$m = 2$ 0011

$m = 3$ 00010111

$m = 4$ 0000111101100101

$m = 5$ 00000100011001010011101011011111.

Because a deBruijn sequence has period 2^m and m bits of memory, every possible state of the memory must recur during every period. Inspection of the above sequences, regarded cyclically, will reveal that the sequence of length 2^m displays every m-bit pattern as a subsequence. There can be no starting transient in a deBruijn sequence.

An example of a shift-register circuit that produces the only deBruijn sequence of length four is shown in Figure 7.5. One period of the deBruijn sequence is 0011, which then repeats. The sequence of states of the 2-bit memory is also shown in Figure 7.5. This periodic sequence is clearly not linear because $0011 + 0110 = 0101$ which is not a cyclic translate of 0011. Nevertheless, this nonlinear sequence with $m = 2$ can be produced by a linear-feedback shift register of length three, as is shown in Figure 7.6.

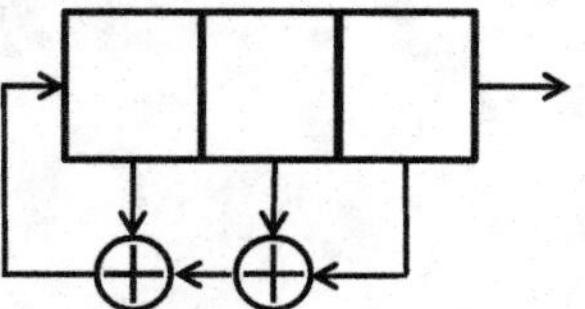

Figure 7.6 A linear-feedback shift register for a deBruijn sequence

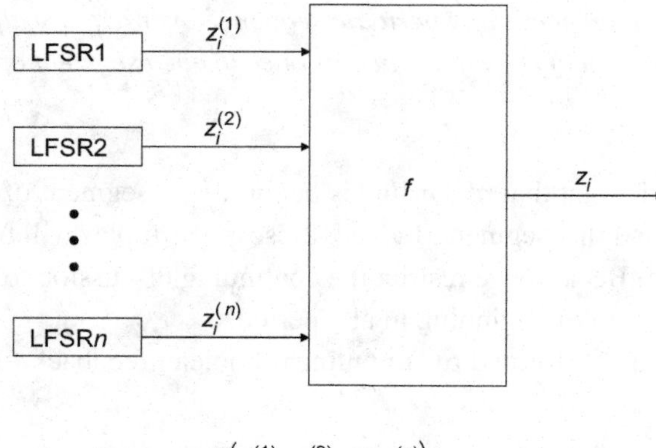

Figure 7.7 Nonlinear combining of shift-register sequences

It cannot be produced by a linear-feedback shift register of length two, so the linear complexity of this sequence is three. No simple expression for the linear complexity of a general deBruijn sequence is known.

The linear-feedback shift register shown in Figure 7.6 produces a nonlinear sequence and the feedback is described by the polynomial equation $z^3 = z^2 + z + 1$ corresponding to $v_i = v_{i-1} + v_{i-2} + v_{i-3}$. We recall that an irreducible polynomial must produce a linear sequence and this is not a linear sequence. But $z^3 + z^2 + z + 1$ is not an irreducible polynomial, so the discussion of Section 7.3 does not assert that this recursion produces a linear sequence, and indeed it does not.

7.7 Keystreams from nonlinear combining

A deBruijn sequence introduces nonlinearity in order to produce a periodic sequence with the largest possible period that can be produced with m bits of state memory. However, maximizing the sequence period is usually not the only important consideration. It is much more important to balance between maximizing security and minimizing complexity. This balance might be obtained by some more elaborate nonlinearity, as by combining several linear-feedback shift-register sequences. Figure 7.7 shows a

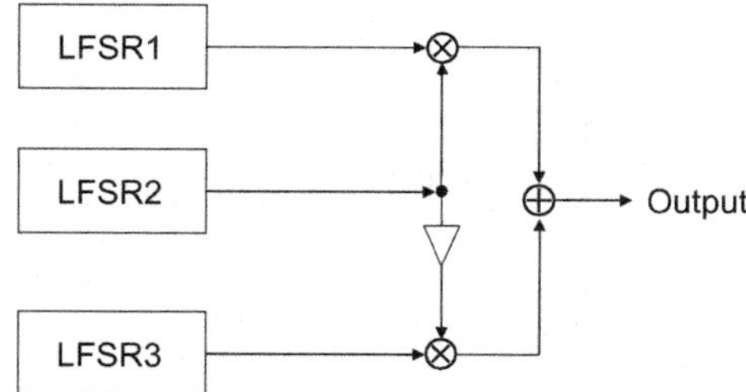

Figure 7.8 Geffe keystream generator

general form that uses nonlinear operations on the outputs of the linear-feedback shift registers. Some of these elementary methods of combining multiple recursions have names.

A *Geffe keystream* is the combination of two shift-register sequences $\{u_i\}$ and $\{v_i\}$, with periods n_1 and n_2, under the control of a third shift-register sequence with period n_3, as shown in Figure 7.8. A Geffe keystream is formed by the nonlinear, but memoryless, combining of multiple linear-feedback shift registers. The Geffe keystream is defined in $\boldsymbol{F}_2$ as $z_i = w_i u_i + \overline{w}_i v_i$ where $\overline{w}_i = 1 + w_i$. The period of the Geffe keystream is $n = \text{LCM}[n_1, n_2, n_3]$. The linear complexity of the Geffe keystream is $(n_1 + 1)n_2 + n_1 n_3$, which is a direct consequence of the forthcoming Theorem 7.8.3. If n_1, n_2, and n_3 are pairwise coprime and approximately equal, then $n \approx n_1^3$. Therefore the linear complexity is about $2n^{2/3}$, which is comparable to n. Accordingly, the Geffe keystream is not especially vulnerable to a plaintext linear-complexity attack. However, it is vulnerable to a correlation attack, as is described in Section 7.9.

A *Beth–Piper keystream* uses two shift-register sequences $\{u_i\}$ and $\{v_i\}$. It repeats the bits of $\{u_i\}$ while v_i is zero. Thus for each i:

$$\begin{aligned} &\text{If } v_i = 0, \quad \text{then } z_i = u_\ell \text{ and } \ell \leftarrow \ell \\ &\text{If } v_i = 1, \quad \text{then } z_i = u_\ell \text{ and } \ell \leftarrow \ell + 1. \end{aligned}$$

Each time a bit of $\boldsymbol{v}$ is a zero, it causes the most recent symbol of $\boldsymbol{z}$ to be repeated. Each time a bit of $\boldsymbol{v}$ is a one, it selects the next unused symbol of $\boldsymbol{u}$ for the next symbol of $\boldsymbol{z}$. The linear complexity of the Beth–Piper keystream is apparently unknown, but evidently this keystream falls to a correlation attack.

An *alternating keystream* uses a generator with three shift registers as shown in Figure 7.9. Each bit of one sequence is used to enable the clock of one of the other two shift registers, but not both. The two output sequences are then added.

A *shrinking keystream* uses two linear shift-register sequences $\{u_i\}$ and $\{v_i\}$. It *deletes* the bit u_i whenever v_i is zero. Thus for each i:

$$\begin{aligned} &\text{If } v_i = 0, \quad \text{then } i \leftarrow i + 1 \\ &\text{If } v_i = 1, \quad \text{then } z_i = u_i, i \rightarrow i + 1, \ell \rightarrow \ell + 1. \end{aligned}$$

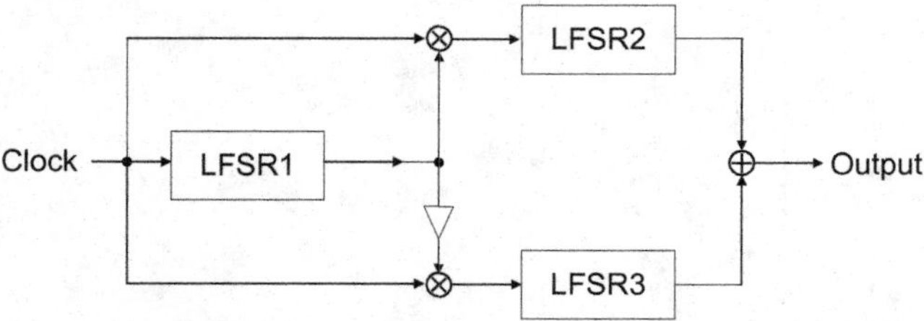

Figure 7.9 The alternating-step generator

Each time a bit of $\boldsymbol{v}$ is a zero, it causes the most recent symbol of $\boldsymbol{u}$ to be skipped. Each time a bit of $\boldsymbol{v}$ is a one, it causes the next symbol of $\boldsymbol{u}$ to be the next symbol of z. The linear complexity of the shrinking keystream appears to be unknown. A shrinking keystream is believed to be immune to a correlation attack.

For examples of a Beth–Piper keystream and a shrinking keystream, let $\boldsymbol{v}$ and $\boldsymbol{u}$ be the periodic sequences

$$\boldsymbol{v} = 010011010111100\ldots$$
$$\boldsymbol{u} = 0001011111000101111\ldots$$

of period 15 and of period 7, respectively. Then

$$z = 00000110111\ldots$$

is the Beth–Piper keystream and

$$z = 00110001\ldots$$

is the shrinking keystream.

A *majority-clocking keystream* is based on an alternative method that combines the outputs of three binary linear-feedback shift registers of different lengths by modulo-two addition. Each shift register also provides a designated control bit taken from one interior stage of the shift register. Accordingly, this control bit is a bit of that sequence delayed from the current output bit. The three control bits are combined in some fashion to form three clock-enabling bits, one for each of the three shift registers. A typical rule is that the current output bit of any shift register is repeated without change for as long as the control bit of that sequence differs from both of the control bits of the other two sequences. This is done by disabling the clock of that shift register as is shown in Figure 7.10. The three control bits are designated a, b, and c in the figure. From the three control bits, a, b, and c, simple logic computes the three clocking bits, A, B, and C, given by

$$A = a \cdot b + \overline{a} \cdot \overline{b} + a \cdot c + \overline{a} \cdot \overline{c}$$
$$B = b \cdot c + \overline{b} \cdot \overline{c} + b \cdot a + \overline{b} \cdot \overline{a}$$
$$C = c \cdot a + \overline{c} \cdot \overline{a} + c \cdot b + \overline{c} \cdot \overline{b}.$$

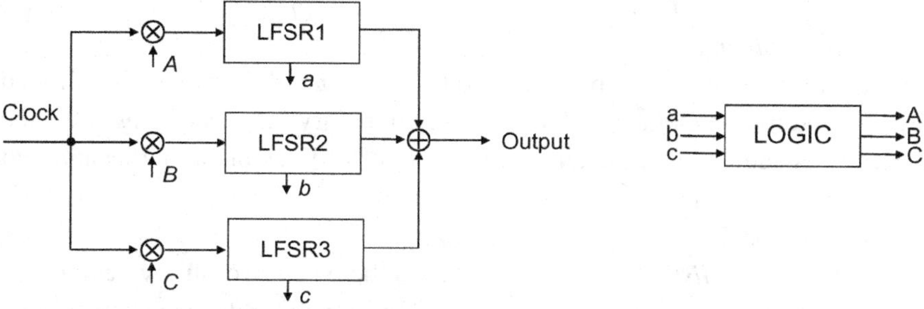

Figure 7.10 Majority-clocking keystream

By the logical multiplication of each clock-enabling bit with the clocking bit, the clocking bit is enabled to reach the appropriate shift register. This means that each shift register is clocked only about half of the time and always in an apparently irregular pattern. One intends this irregularity to counter a linear-complexity attack.

7.8 Keystreams from nonlinear functions

A keystream can also be formed from a linear-feedback shift register by a nonlinear function operating on the state of that linear-feedback shift register. We are interested in binary sequences formed by nonlinear binary shift registers of length n, which means that we are interested in binary functions from $\boldsymbol{F}_2^n$ to $\boldsymbol{F}_2$.

Any function $f(x_1, \ldots, x_n)$ from $\boldsymbol{F}_2^n$ to $\boldsymbol{F}_2$ is called a *boolean function.* Every boolean function can be expressed in a standard form, known as the *algebraic normal form*, or more simply, as the *normal form*, which is given as the following sum of products:

$$\begin{aligned} f(x_1, \ldots, x_n) & \\ &= [a_0] + [a_1x_1 + \cdots + a_nx_n] \\ &\quad + [a_{1,2}x_1x_2 + a_{1,3}x_1x_3 + \cdots + a_{n-1,n}x_{n-1}x_n] + \cdots + [a_{1,2,\ldots,n}x_1x_2 \cdots x_n]. \end{aligned}$$

The *order* of any term in the sum is defined to be the number of variables in that term. The terms are grouped by brackets according to order. The first bracket, $[a_0]$, contains the only zero-order term. The second bracket, $[a_1x_1 + \cdots + a_nx_n]$ contains all first-order terms. The $(i + 1)$th bracket contains all ith-order terms. The largest order of any term in the standard form with a nonzero coefficient is called the *nonlinear order* λ of the boolean function $f(x_1, \ldots, x_n)$. An *affine boolean function* is a boolean function of nonlinear order one. A *linear boolean function* is an affine boolean function with coefficient a_0 equal to zero. A linear boolean function satisfies $f(x_1, x_2, \ldots, x_n) +$

$f(x'_1, x'_2, \ldots, x'_n) = f(x_1 + x'_1, x_2 + x'_2, \ldots, x_n + x'_n)$. All other boolean functions are *nonlinear boolean functions.*

Any binary function can be described by a *truth table*. The entries of this truth table are indexed by the elements of $\boldsymbol{F}_2^n$. These are binary vectors of length 2^n and can be regarded as addresses of the table. The contents of the table at the location addressed by $(x_1, x_2, \ldots, x_n)$ is the value of the binary function f at $(x_1, x_2, \ldots, x_n)$. This value is a zero or a one.

The *Hamming distance*, denoted $d_H(\boldsymbol{v}, \boldsymbol{u})$, between two binary vectors $\boldsymbol{v}$ and $\boldsymbol{u}$ is the number of places in which they differ. The Hamming distance between $\boldsymbol{v}$ and $\boldsymbol{u}$ is related to the Hamming weight by

$$d_H(\boldsymbol{v}, \boldsymbol{u}) = w_H(\boldsymbol{v} - \boldsymbol{u}).$$

The *Hamming distance,* denoted $d_H(f, g)$, between two boolean functions f and h from $\boldsymbol{F}_2^n$ to $\boldsymbol{F}_2$ is equal to the number of values of the vector argument $(x_1, \ldots, x_n)$ for which $f(x_1, \ldots, x_n)$ differs from $h(x_1, \ldots, x_n)$. The Hamming distance between two such boolean functions can be as large as 2^n. The *degree of nonlinearity* of the boolean function f is the Hamming distance from f to the set of all affine boolean functions, defined as

$$n(f) = \min_h d_H(f, h),$$

where the minimum is over the set consisting of all affine boolean functions on n variables. The degree of nonlinearity measures how closely the nonlinear function f can be approximated by a linear function.

An example of a boolean function is the function from $\boldsymbol{F}_2^6$ to $\boldsymbol{F}_2$ given by

$$f(x_1, x_2, x_3, x_4, x_5, x_6) = 1 + x_3 + x_1x_2 + x_5x_6.$$

This is a boolean function with nonlinear order $\lambda = 2$. This nonlinear function can be used to alter the output of a linear-feedback shift register. A linear-feedback shift register with feedback polynomial $x^6 + x + 1$ and the above nonlinear function used to produce a nonlinear output is shown in Figure 7.11, with an initialization 100000.

A second example is the function from $\boldsymbol{F}_2^4$ to $\boldsymbol{F}_2$ given by

$$f(x_1, x_2, x_3, x_4) = 1 + x_2 + x_4 + x_1x_2x_3$$

which has nonlinear order $\lambda = 3$.

Definition 7.8.1 *A nonlinear-output linear-feedback shift register sequence is a sequence over* $\boldsymbol{F}_2$ *given by the nonlinear binary function* $z_i = f(x_1^{(i)}, \ldots, x_m^{(i)})$, *where* $(x_1^{(i)}, \ldots, x_m^{(i)})$ *is the binary representation of the orbit of an element* α *of the field* $\boldsymbol{F}_{2^m}$, *as given by* $(\alpha^1, \alpha^2, \alpha^3, \ldots, \alpha^i, \ldots)$.

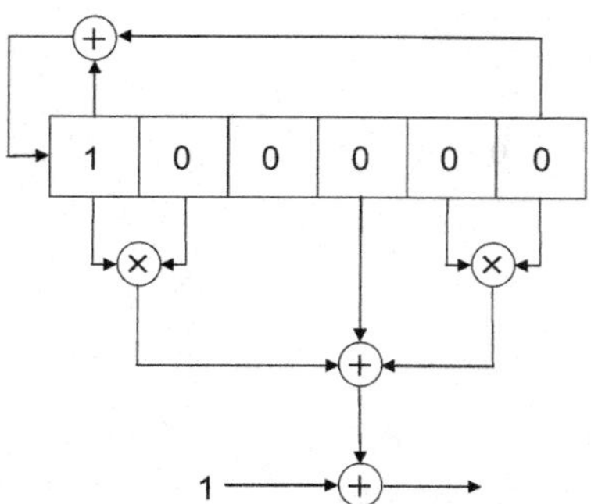

Figure 7.11 A nonlinear-output linear-feedback shift register

Theorem 7.8.2 (Key's theorem) *The linear complexity of a nonlinear-output linear-feedback shift register sequence produced by a shift register of length L and nonlinear order λ is not larger than* $\sum_{i=1}^{\lambda} \binom{L}{i}$.

Proof The proof is based on Theorem 7.5.3 which says that the linear complexity of a periodic sequence $\boldsymbol{v}$ of period n is equal to the Hamming weight of its Fourier transform $\boldsymbol{V}$. If any linear-feedback shift register that produces $\boldsymbol{v}$ has length L, then the linear complexity of the sequence $\boldsymbol{v}$ is at most L. Moreover, by superposition of the feedback connections, the linear complexity of the sum of two sequences is not larger than the larger of their two individual linear complexities. This means that it is enough to prove the theorem for a nonlinear function, $f(v_1, \ldots, v_n)$, that has only a single nonlinear term.

We begin with a proof for the special case of a nonlinear function that has only a single nonlinear term of order two. As a function of i, the single term of order two produces the cyclic sequence $w_i = v_{i-a}v_{i-b}$ abbreviated $w_i = v_i'v_i''$ where $v_i' = v_{i-a}$ and $v_i'' = v_{i-b}$. The cyclic sequence $\boldsymbol{w}$ is a componentwise product, so the convolution theorem says that

$$\boldsymbol{W} = \boldsymbol{V}' * \boldsymbol{V}''$$

in the Fourier transform domain, where $\boldsymbol{V}'$ and $\boldsymbol{V}''$ are the Fourier transforms of $\boldsymbol{v}'$ and $\boldsymbol{v}''$, respectively, which are translates of each other. Let $V'(x) = \sum_{j=0}^{n-1} V_j'x^j$ and $V''(x) = \sum_{j=0}^{n-1} V_j''x^j$. The polynomials $V'(x)$ and $V''(x)$ are sparse, each having weight at most L, and, by the translation property of the Fourier transform, both are nonzero in exactly the same components. By Theorem 7.5.3, the linear complexity of the sequence $\boldsymbol{w}$ with components $w_i = v_{i-a}v_{i-b}$ is equal to the weight of polynomial $W(x)$ where $W(x) = V'(x)V''(x)$. There are L^2 terms in the raw polynomial product $V'(x)V''(x)$, but many terms will have the same power of x and so will combine into one term, or even cancel. The linear complexity of the nonlinear term of order two is not larger than the number of nonzero coefficients of $W(x)$.

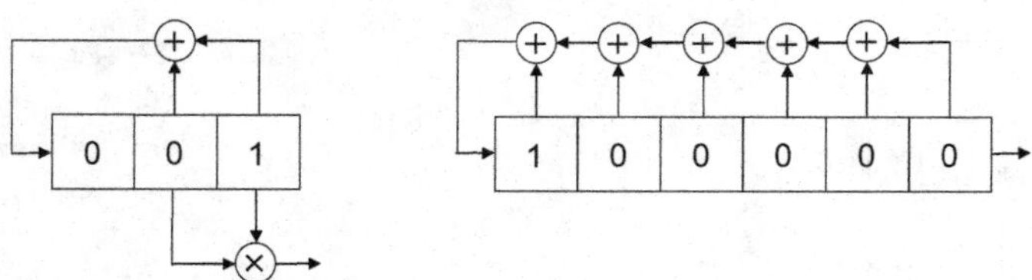

Figure 7.12 Illustrating Key's theorem

To determine the number of nonzero coefficients of $W(x)$, let ℓ index only the nonzero coefficients of $V(x)$ so that the product can be written as

$$W(x) = \sum_{\ell'=1}^{L} \sum_{\ell''=1}^{L} V'_{j_{\ell'}} V''_{j_{\ell''}} x^{j_{\ell'}+j_{\ell''}}.$$

The product terms $V'_{j_{\ell'}} V''_{j_{\ell''}}$ and $V'_{j_{i'}} V''_{j_{i''}}$ add whenever $j_{\ell'} + j_{\ell''} = j_{i'} + j_{i''}$. The L^2 terms under the sum may be written in matrix form. The diagonal terms, with ℓ' equal to ℓ'', have a power of x that appears once. The off-diagonal terms, those with ℓ' not equal to ℓ'', have a power of x that appears twice, once on each side of the diagonal. This means that the number of such terms is

$$L_W = \binom{L}{1} + \binom{L}{2},$$

because there are $\binom{L}{1}$ terms on the diagonal and $\binom{L}{2}$ terms above the diagonal. Possibly, because of cancellation, some of the terms may be zero. The linear complexity of an output sequence is not larger than L_W and, because some of the terms may cancel, it may be smaller. This completes the proof for nonlinear functions of order two.

A similar analysis for a term of the form $v'_i v''_i v'''_i$ would show that

$$L_W = \binom{L}{1} + \binom{L}{2} + \binom{L}{3}.$$

This analysis is easily extended to a higher-order term.

Finally, because the linear complexity of a sum of terms is not larger than the largest linear complexity of any one term, the theorem follows. □

An example of Key's theorem is shown in Figure 7.12. On the left is a nonlinear-output linear-feedback shift register of length three and nonlinear order two. The period of the shift-register sequence is seven, and so the period of the output sequence is seven. The linear shift-register sequence itself is

$$\boldsymbol{x} = (1101001),$$

and from this, by multiplying two terms, the nonlinear output sequence is

$$\boldsymbol{z} = (1100000),$$

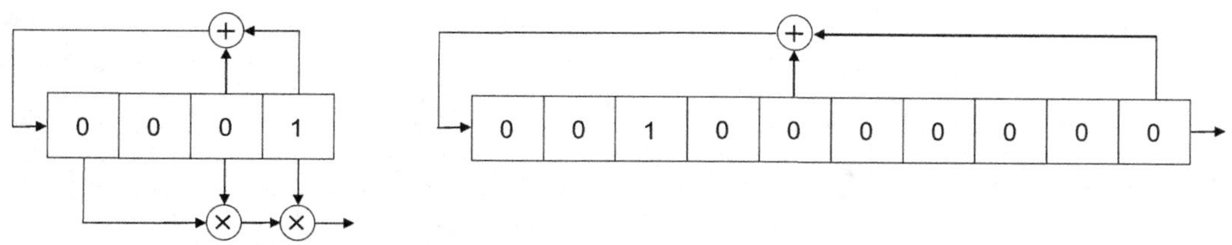

Figure 7.13 Illustrating nontightness of Key's theorem

each sequence read from right to left. Key's theorem says that the linear complexity of the output sequence is at most six. On the right of Figure 7.12 is a linear-feedback shift register of length six that gives the same output when initialized as shown. No shorter linear-feedback shift register produces this sequence. Thus the linear complexity of the output sequence is actually six, which meets Key's theorem with equality.

A second example of Key's theorem is shown in Figure 7.13. On the left is a nonlinear-output linear-feedback shift register of length four and nonlinear order three. The linear shift register itself produces the sequence

$$\boldsymbol{x} = (111010110010001),$$

and from this, by multiplying three terms, is computed the nonlinear periodic output sequence

$$\boldsymbol{z} = (001000010000000)$$

from left to right, respectively. On the right side of Figure 7.13 is a linear-feedback shift register that produces this same periodic sequence. In this case, Key's theorem says that the linear complexity is not larger than fourteen. In fact, the linear complexity is ten because we have exhibited a linear-feedback shift register of length ten that produces it, and there is no smaller linear-feedback shift register that does so. This example shows that Key's theorem is not tight. By referring to the proof of Key's theorem, this can be understood as a consequence of various components canceling in the Fourier transform domain.

Another approach is to use multiple linear-feedback shift registers with componentwise multiplication of the outputs to generate a sequence. Using nonlinear combining of only one of the outputs of each shift register is not as restrictive as it might appear because the statement allows the use of multiple copies of the same shift register with different phases. The left side of Figure 7.14 shows two linear-feedback shift registers whose outputs are multiplied to form a keystream. The right side of Figure 7.14 shows a single linear-feedback shift register that produces the same sequence. No shorter such shift register exists. Thus the linear complexity of the nonlinear keystream is six. This is equal to the product of the lengths of the two original shift registers, which is consistent with the forthcoming theorem.

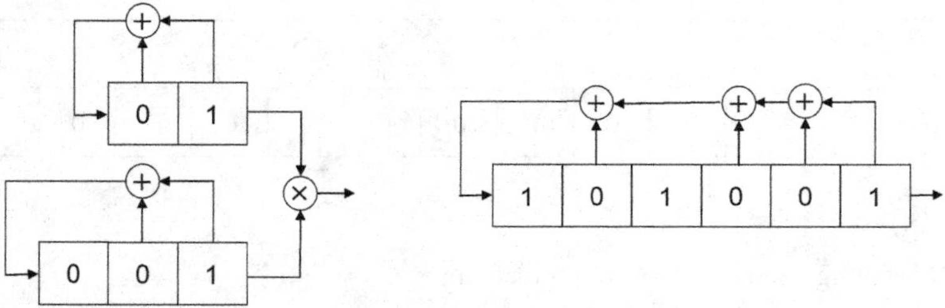

Figure 7.14 A nonlinear combination generator and its equivalent

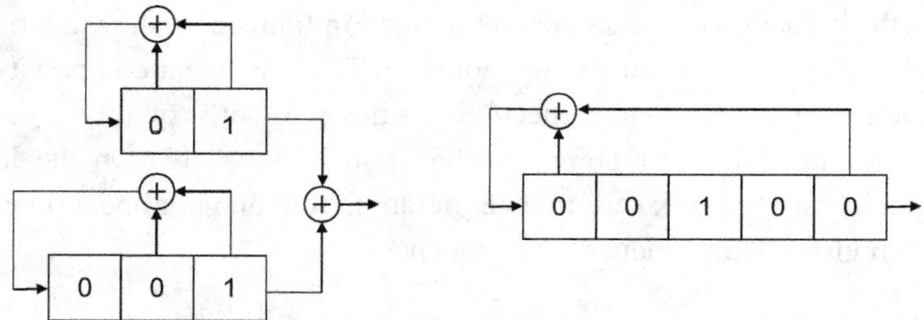

Figure 7.15 A linear combination generator and its equivalent

A similar conclusion holds if the outputs of the shift registers are added to form a keystream, as shown in Figure 7.15. Again, the right side of the figure shows a single linear-feedback shift register of length five that produces the same sequence. No shorter such shift register exists. Thus the linear complexity of the linear keystream is five. This is equal to the sum of the lengths of the original two sequences.

We are now motivated for the following theorem.

Theorem 7.8.3 *The linear complexity of a nonlinear combination of n shift-register sequences with linear complexity $L_1, L_2, \ldots, L_n$, respectively, using a boolean combining function*

$$z_i = f(z_i^{(1)}, z_i^{(2)}, \ldots, z_i^{(n)})$$

in normal form has linear complexity not larger than $f(L_1, L_2, \ldots, L_n)$, where f is now interpreted as a function on the integers.

Proof Let $L = f(L_1, L_2, \ldots, L_n)$. We must show that the collection of linear-feedback shift registers of lengths $L_1, L_2, \ldots, L_n$ can be replaced by one linear-feedback shift register of length not larger than L. Because the boolean combining function can be executed as a sequence of binary additions and binary multiplications, it is only necessary to prove the theorem for a simple componentwise binary addition $z_i = x_i + y_i$ and for a simple componentwise binary multiplication $z_i = x_i y_i$. In each case, the method of proof is to transform the problem into the Fourier domain and to then refer to Theorem 7.5.3, which says that the Fourier transforms of $\boldsymbol{x}$ and $\boldsymbol{y}$ have weights L_1 and L_2, respectively.

Let the periodic sequences $\boldsymbol{x}$ and $\boldsymbol{y}$ have coprime periods n_1 and n_2, respectively. Now, regard the two periodic sequences of length n_1 and n_2 to each be a periodic sequence of period $n = n_1 n_2$ as follows. Repeat the first sequence n_2 times to form a periodic sequence of period n. By Theorem 9.12.3, the Fourier transform of the new sequence consists of the components of the original Fourier transform with $n_2 - 1$ zeros inserted after every component of the original sequence. Similarly, repeat the second sequence n_1 times to form a periodic sequence of length n. Again, by Theorem 9.13.3, the Fourier transform of the longer sequence of period n consists of the components of the original sequence with $n_1 - 1$ zeros inserted after every component of the original sequence.

First consider the case of multiplication with $z_i = f(x_i, y_i) = x_i y_i$, where as described above, $\boldsymbol{x}$ and $\boldsymbol{y}$ are periodic sequences with coprime periods n_1 and n_2. Both sequences are regarded as lengthened to the common period $n = n_1 n_2$. The two new sequences of period n are multiplied componentwise, so their Fourier transforms are convolved. Because the two Fourier transforms have L_1 and L_2 nonzero components, respectively, the convolution has at most $L_1 L_2$ nonzero components. By Theorem 7.5.3, the inverse Fourier transform then has linear complexity at most $L_1 L_2$. Indeed, if n_1 and n_2 are coprime, there will be exactly $L_1 L_2$ nonzero components in the convolution. Therefore if n_1 and n_2 are coprime the linear complexity will be exactly $L_1 L_2$.

Next, consider the case of addition with $z_i = f(x_i, y_i) = x_i + y_i$, where as described above, $\boldsymbol{x}$ and $\boldsymbol{y}$ are periodic sequences with common period $n = n_1 n_2$. The two sequences have L_1 and L_2 nonzero components, respectively, so the sum has at most $L_1 + L_2$ nonzero components. Because n_1 and n_2 are coprime, the nonzero components in the Fourier transforms of the two sequences do not occur in the same components. This means that there can be no cancellation, so there are exactly $L_1 + L_2$ nonzero components. Again, by Theorem 7.5.3, the inverse Fourier transform has linear complexity at most $L_1 + L_2$. This holds with equality if n_1 and n_2 are coprime. □

The next theorem states that Theorem 7.8.3 holds with equality under certain satisfying conditions.

Theorem 7.8.4 *The inequality of the previous theorem holds with equality if the component linear-feedback shift-register sequences are maximal-length sequences.*

Proof The proof merely consists of inspection of the equality conditions in the proof of Theorem 7.8.3, and will not be discussed further. □

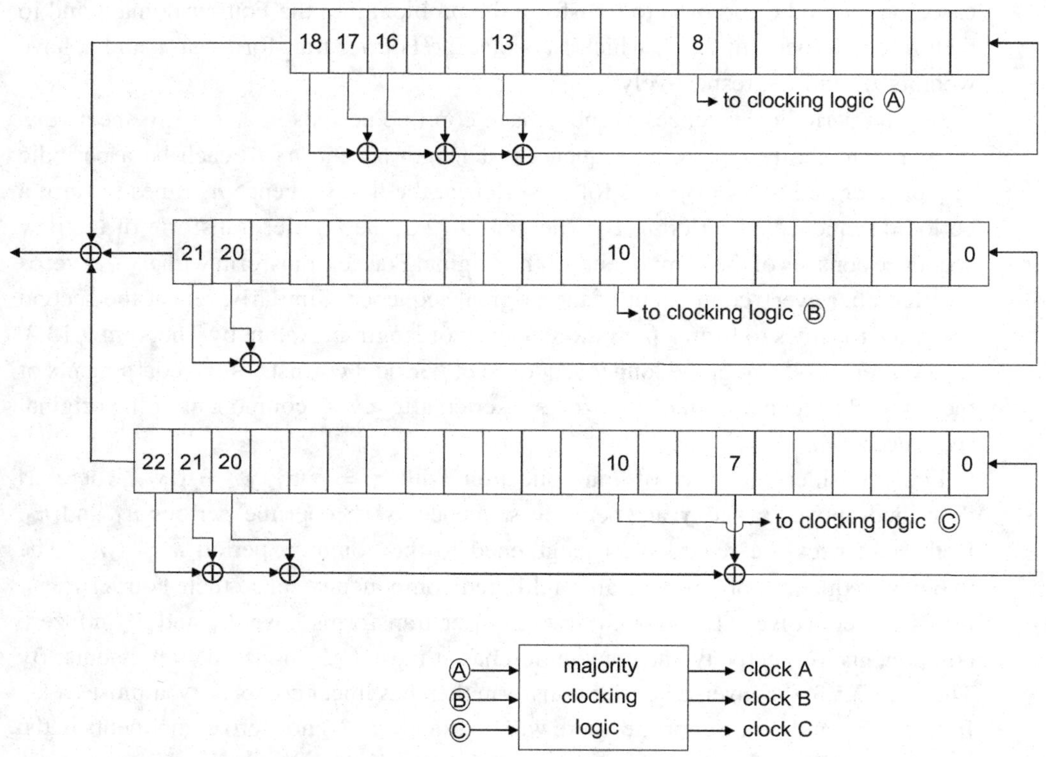

Figure 7.16 The GSM stream cipher

As an example of the theorem, the Geffe cipher has a nonlinear combining function in normal form $f(z_1, z_2, z_3) = z^{(3)} + z^{(1)}z^{(2)} + z^{(2)}z^{(3)}$ and period $\text{LCM}(L_1, L_2, L_3)$. By Theorem 7.8.3, the linear complexity is at most $L_3 + L_1L_2 + L_2L_3$, which, by Theorem 7.8.4, holds with equality if all shift registers are of maximum length. Thus with $L_1 = 30$, $L_2 = 31$, $L_3 = 29$, the linear complexity of this linear combination is 1858. A linear-complexity attack will succeed if at least 3716 consecutive bits of the Geffe keystream are observed. The period of this Geffe keystream is 26,970, so the linear-complexity attack requires that more than 10% of one period of the sequence be observed.

A related construction intended to suppress any structure that may be useable to the cryptanalyst is to combine the outputs of multiple shift registers that are individually clocked in a stop-and-go manner with some irregularity in the pattern of clocking. In order to have a keystream with a period at least in the billions, well over 30 bits of memory will be required. This keystream may be obtained by nonlinear or irregular combining of three linear-feedback shift registers. An example is the *GSM stream cipher*, shown in Figure 7.16, consisting of the combination of three shift-register circuits. This cipher uses three linear-feedback shift registers of lengths 19, 22, and

23, respectively. It requires a total of 64 bits to initialize these three shift registers. The three connection polynomials used by the GSM cipher are

$$p_1(x) = x^{18} + x^{17} + x^{16} + x^{13} + 1$$
$$p_2(x) = x^{21} + x^{20} + 1$$
$$p_3(x) = x^{22} + x^{21} + x^{20} + x^7 + 1.$$

The irregular nonlinear behavior of this stream cipher is provided by a self-clocking scheme that is suggestive of the Beth–Piper keystream.

To initialize the GSM stream cipher generator in a simple way, the 64-bit key for the GSM cipher is shifted into all three shift registers by adding the key to the feedback term at the right side of each shift register. The insertion of the 64-bit key fills the three registers with the 64 initialization bits. Because of the feedback, the initialization bits residing in the registers after the key is shifted in are not the same as the key, but do depend on the key.

The three outputs of the three linear-feedback shift-register circuits are added together to produce the keystream. To protect the keystream sequence from a linear-complexity attack, the three shift-register circuits are not all shifted at every clock time. Instead, the three clocking commands are generated by the three shift registers working in concert. Each shift register has a special clocking bit, as shown in the figure. Each shift register is clocked only if its special clocking bit agrees with at least one of the two special clocking bits of the other two shift registers. Thus the shift register whose clocking bit is in the minority, if there is one, will not be shifted. It is easy to see that each shift register will be clocked three times out of four, and no shift register can be permanently stagnant.

7.9 The correlation attack

One kind of known-plaintext attack for breaking certain forms of nonlinear stream ciphers is known as a *correlation attack.* We have already discussed how one might choose to use a binary keystream that is based on the nonlinear combining of several shift registers in order to counter a linear-complexity attack. A Geffe cipher is a simple example. However, this keystream might be vulnerable to attack in a different way. It might be vulnerable to a correlation attack. Nonlinear combining, as used in the Geffe cipher, can be attacked using statistical-estimation methods to penetrate the nonlinearities, at least in part, so that something about the underlying linear-feedback shift registers can be revealed. In response to this threat of a correlation attack, the cryptographer will avoid some forms of nonlinear combining functions. Correlation immunity should be built into the keystream generator to thwart a correlation attack.

A correlation attack requires that a significant segment of the plaintext be known so that it can be used to find the key. This again has the form of an inverse problem. Given partial knowledge of the keystream e_k, find k. Here the key k corresponds to the unknown parameters that determine the outputs of the multiple underlying linear-feedback shift registers. In the standard version of this attack, the linear-feedback shift registers and the nonlinear functions that combine their outputs are fully known. The key then consists only of the initialization of the shift registers, and the goal of the attack is to find the initialization. A correlation attack requires a significant correlation between the keystream and the specific parameters of the underlying linear-feedback shift registers.

We will describe a correlation attack for a Geffe keystream. This Geffe keystream consists of the merging of two subkeystreams, one subkeystream from each of the two linear-feedback shift registers. We will suppose that the structures of the two linear-feedback shift registers are known to the cryptanalyst. The two linear-feedback shift registers have lengths m_1 and m_2, respectively, so the key consists of the $m_1 + m_2$ bits that were used to initialize these two shift registers. This means that there are $2^{m_1+m_2}$ possible keys, several of which, such as an all-zero initialization, are trivial and cannot be used. The correlation attack will be directed against one of the two shift registers, presumably the shorter one, which we can take to have length m_1. There are only $2^{m_1} - 1$ possible nonzero initializations of the linear-feedback shift register of length m_1. That shift register can be initialized with any nonzero value and will eventually cycle through all possible nonzero values. The output of that shift register is a periodic sequence of period $2^{m_1} - 1$. If the parameters of the shift register are known, then that periodic sequence is known to the cryptanalyst, and only its phase as determined by the initialization is not known.

Each bit of the Geffe keystream consists of the bit value corresponding to one of the two subkeystreams, the choice between them being controlled by a third shift-register sequence. For our analysis, we will regard the control sequence as a random sequence. Thus the Geffe keystream is randomly chosen from the first subkeystream for about half of the bits, and randomly chosen from the second subkeystream for the other half. When the second subkeystream is chosen, about half of the bits, on average, will agree with the first subkeystream. Thus the Geffe keystream agrees with the first of its subkeystreams about 75% of the time. Similarly, it agrees with the second of its subkeystreams about 75% of the time.

A correlation attack requires that a subsequence of the Geffe keystream be known. It correlates the known segment of the keystream with a reference subkeystream consisting of one period of the output of the first linear-feedback shift register with an arbitrary initialization. The reference cyclic subkeystream will contain a segment that is equal to the actual m_1-bit initialization used in that shift register, but that segment is (cyclically) delayed by some integer value ℓ_0 with respect to the reference keystream. At the value of delay ℓ equal to ℓ_0, the content of the shift register is equal to the initialization key, and the reference subkeystream is fully correlated with the actual subkeystream underlying the known segment. When ℓ is equal to ℓ_0, about 75% of the bit positions in the two sequences will agree. When ℓ is not equal to ℓ_0, about 50% of the bit positions will agree. Then the expected value of the normalized correlation is 0.75. Thus, looking for a correlation with respect to the reference keystream peak determines the phase of the shorter Geffe subkeystream.

Let z_i for $i = 0, \ldots, t-1$ denote the t bits of the observed keystream that has been uncovered from the known plaintext expressed in the bipolar alphabet $\{-1, +1\}$. Let w_i be the sequence of keystream bits that are produced by the first linear-feedback shift register also expressed in the bipolar alphabet $\{-1, +1\}$. The correlation is defined as

$$\phi(\ell) = \sum_{i=0}^{t-1} z_i w_{i+\ell}.$$

When $\ell = \ell_0$, about 75% of the terms will be $+1$ and the remainder will be -1. Thus

$$\mathrm{E}\left[\phi(\ell)\right] = \begin{cases} \frac{1}{2}t & \text{if} \quad \ell = \ell_0 \\ 0 & \text{if} \quad \ell \neq \ell_0. \end{cases}$$

The variance in $\phi(\ell)$ is given by

$$\mathrm{E}\left[\phi(\ell)^2\right] = n.$$

The estimated sequence phase is given by

$$\widehat{\ell_0} = \operatorname{argmax}_k \phi(\ell).$$

The probability of error is the probability that some ℓ other than ℓ_0 achieves the maximum. To simplify the calculation of the probability of error by using an approximate calculation, consider instead n gaussian random variables of variance $\sigma^2 = n$, of which one random variable has mean $A = n/2$, and the remainder have mean zero. This approximation can be motivated by appealing to the central limit theorem.

The probability p_e that a gaussian random variable with mean A and variance σ^2 is not larger than each of $n-1$ other independent gaussian random variables, each with zero mean and variance σ^2, is

$$p_e = 1 - \int_{-\infty}^{\infty} \frac{1}{\sqrt{2\pi\sigma^2}} e^{-(z-A)^2/2\sigma^2} \left[\int_{-\infty}^{z} \frac{1}{\sqrt{2\pi\sigma^2}} e^{-x^2/2\sigma^2} dx \right]^{n-1} dz,$$

which is written as one minus the probability that the random variable with mean A is larger than every other random variable. This expression can be integrated numerically.

This expression is hard to interpret as it stands. To understand what it says, use the standard inequality

$$\int_z^{\infty} \frac{1}{\sqrt{2\pi}} e^{-x^2/2} dx < \frac{1}{2} e^{-z^2/2} \qquad \text{for } z > 0.$$

Then

$$1 - p_e = \int_{-\infty}^{\infty} \frac{1}{\sqrt{2\pi\sigma^2}} e^{-(z-A)^2/2\sigma^2} \left[1 - \int_{z}^{\infty} \frac{1}{\sqrt{2\pi\sigma^2}} e^{-x^2/2\sigma^2} dx\right]^{n-1} dz$$
$$> \int_{-\infty}^{\infty} \frac{1}{\sqrt{2\pi\sigma^2}} e^{-(z-A)^2/2\sigma^2} \left[1 - \frac{1}{2} e^{-z^2/2\sigma^2}\right]^{n-1} dz$$
$$> \int_{-\infty}^{\infty} \frac{1}{\sqrt{2\pi\sigma^2}} e^{-(z-A)^2/2\sigma^2} \left[1 - \frac{(n-1)}{2} e^{-z^2/2\sigma^2}\right] dz.$$

This can now be rewritten as

$$p_e \le \frac{n-1}{2} \int_{-\infty}^{\infty} \frac{1}{\sqrt{2\pi\sigma^2}} e^{-(2z^2-2Az+A^2)/2\sigma^2} dz.$$

Complete the square in the exponent to write

$$p_e < \frac{n-1}{2} \int_{-\infty}^{\infty} e^{-2(z-A/2)^2/2\sigma^2} e^{-A^2/2\sigma^2} dz$$
$$= \frac{n-1}{2} \sqrt{2\pi\sigma^2} e^{-A^2/2\sigma^2}.$$

Because $A = n/2$ and $\sigma^2 = n$, this gives

$$p_e < (n-1)\sqrt{\pi n/2}\, e^{-n/16}.$$

Ignoring constants, the dominant behavior can be coarsely expressed as $e^{-(n-\log n)}$. This coarse expression shows that the probability of failure of a correlation attack eventually goes to zero exponentially fast in the length of the available plaintext n.

7.10 Pseudorandom sequences

The task of additive keystream encryption requires the availability of a long and seemingly random sequence of numbers. We will deal only with binary sequences of this kind. The binary sequence should be many millions or even billions of bits in length, and the same sequence must be known and synchronized at both the encryptor and the decryptor. Such a sequence is called a *one-time pad* if it is maximally random. There are severe practical problems in requiring such a long sequence to be maximally random because a true maximally random sequence cannot be compacted; it must be stored at a rate of one stored bit per sequence bit. To generate, distribute, and store millions or billions of random bits securely may be impractical. The problem of generating and communicating these bits to both the encryptor and decryptor, which, presumably, are not normally at the same place is an important consideration. The transfer of the random sequence must be through a secure channel, such as a courier, that uses no encryption because if encryption were used, that encryption itself requires a random sequence. The key must also be protected while waiting to be used. Consequently, in most practical applications, a shorter binary sequence is repeatedly modified and reused in both the transmitter and the receiver according to a fixed procedure in order to generate a long sequence, albeit one that is not maximally random. The long sequence so generated is then called a *pseudorandom sequence*, and the short sequence from which it is generated is called the *key sequence*. This chapter has considered such sequences in considerable depth in terms of their generation. This raises some questions about the generality of those sequences. The purpose of this section is to briefly consider pseudorandom sequences as such, and not as the results of the methods of generation.

Loosely, a pseudorandom binary sequence is a periodic sequence of ones and zeros with period n that over small intervals has the appearance of a random sequence. Roughly, the contiguous segments of r bits from the sequence will take on each of its 2^r possible patterns almost equally often. A pseudorandom sequence is generated from a key sequence by a specific rule. Therefore if the key is known, pseudorandom sequences are quite deterministic and not very complex; nevertheless, they exhibit a superficial appearance of randomness. One tries to design the function that maps the key sequence into the pseudorandom sequence so that the superficial randomness is as unfathomable as would be true randomness.

How should one define a pseudorandom sequence? The theory of the generation of pseudorandom sequences is quite extensive and full of subtlety. We have presented an extensive introduction in this chapter. Apparently, to mimic randomness, we should require that about one half of the bits are ones and about one half are zeros; that about one fourth of all pairs of bits are of each possibility (00, 01, 10, and 11); and in general, we should require that about 2^{-r} of all r-tuples of bits are of each possibility. However, we cannot insist on this requirement for an arbitrarily large r if the sequence is generated by a finite-state machine, because such a pseudorandom sequence must eventually be periodic.

Moreover, if we are too insistent on making every sequence appear to be fully random, we eliminate many sequences from consideration. It is a philosophical curiosity that the more constraints we put on a sequence in order to call it pseudorandom, the less freedom there is in the sequence and the more deterministic it becomes. If we impose strong enough constraints on the frequency of occurrence of subsequences, then we reduce the possible choices to those sequences known as *deBruijn sequences*. These are the periodic sequences of period 2^m for which every possible subsequence of length m occurs. By deleting one symbol from the all-zero subsequence, one obtains a periodic sequence of period $2^m - 1$ in which, except for the all zero subsequence, every subsequence of length m occurs once. Among these shortened deBruijn sequences are the maximal-length sequences which are those produced by a cycle of an element of a finite field. In general, there are many such shortened deBruijn sequences of period 2^m, many of which have linear complexity much larger than m, which means that they cannot be generated by a linear-feedback shift register of length

m or less. Moreover, we lack a general theory for generating the full family of long deBruijn sequences.

We can define a pseudorandom sequence in terms of the periodic correlation function. Let c_k for $k = 0, \ldots, n-1$ be a binary sequence taking the values in the bipolar alphabet $\{-1, +1\}$. The periodic autocorrelation function is

$$\phi_i = \sum_{k=0}^{n-1} c_k c_{((k+i))},$$

where the double parentheses denote modulo n. A pseudorandom sequence of length n is defined loosely as a binary sequence of length n taking values in the bipolar alphabet such that for $i \neq 0$, $|\phi_i|$ is small compared to ϕ_0. This definition avoids firm constraints on subsequences, yet it does impose some control over them. The imprecision in the definition arises because we have not specified what it means to be small compared to n. The rule of thumb is that small means on the order of $\sqrt{n}$.

The definition can be generalized from one sequence to a set of sequences. A pseudorandom set of sequences of length n is defined imprecisely as a set of pseudorandom sequences of length n such that the cross-correlation between any two distinct sequences is small compared to n.

7.11 Nonlinear sets of sequences

To complete the study of sequences in this chapter, we will briefly discuss sets of sequences. Although sets of sequences do not play a significant role in cryptography, they are a visible part of the literature of binary sequences, and so provide additional context for the study of individual sequences. They are also a bridge into other related topics.

A set of binary sequences consists of multiple sequences of the same length n defined by a common rule. For example, the set of cyclic translations of a maximal sequence of blocklength n, studied in previous sections, forms a set of sequences of blocklength n. This set is an example of a set of binary sequences constructed in $\boldsymbol{F}_2$. It is a linear set of $\boldsymbol{F}_2$ sequences because the sum of two such $\boldsymbol{F}_2$ sequences is another such $\boldsymbol{F}_2$ sequence.

A maximal sequence can be used to construct other sets of sequences. We will construct nonlinear sets of $\boldsymbol{F}_2$ sequences, meaning that the sum of two sequences in such a set is not, in general, an element of that set of sequences. In this section, we describe two nonlinear sets of sequences, namely the *Kasami sequences* and the *Gold sequences*, which are sets of sequences that have good correlation properties when represented as sequences in the bipolar alphabet $\{-1, +1\}$. Two bipolar sequences from the same set of Kasami sequences or the same set of Gold sequences are unlikely to be confused even when observed in the presence of severe noise and interference. The Kasami sequences and the Gold sequences are constructed in $\boldsymbol{F}_2$ from certain combinations of the maximal sequences.

A designated set of Kasami sequences is also called a *Kasami code*. A standard set of Gold sequences is also called a *Gold code*. First, we will describe the Kasami sequences. Let $n = 2^m - 1$ for m even, and choose an m-sequence $\boldsymbol{a}$ of blocklength $n = 2^m - 1$. Because m is even, n factors as

$$n = (2^{m/2} - 1)(2^{m/2} + 1).$$

Consequently, start with any bit position of $\boldsymbol{a}$, and take every $(2^{m/2} + 1)$th bit of $\boldsymbol{a}$ cyclically repeated to obtain n bits. This gives a sequence $\boldsymbol{b}$ of blocklength n that has a period of length $2^{m/2} - 1$ with $2^{m/2} + 1$ repetitions of this period. In other words, sequence $\boldsymbol{b}$ is obtained by cyclically decimating $\boldsymbol{a}$ by $2^{m/2} + 1$. There are $2^{m/2} - 1$ distinct cyclic translations of $\boldsymbol{b}$. The Kasami code is the set of these cyclic translations together with $\boldsymbol{a}$ itself. Thus

$$\mathcal{C} = \{\boldsymbol{a}, \boldsymbol{a} + T^{\ell}\boldsymbol{b} : \quad \ell = 0, \ldots, 2^{m/2} - 2\},$$

where the operator T denotes a cyclic shift by one bit position. There are $2^{m/2}$ Kasami sequences in a Kasami code, each sequence having blocklength n. We can evaluate the Kasami code using the cross-correlation function. Given two sequences $\boldsymbol{c}$ and $\boldsymbol{c}'$ of length N, the periodic cross-correlation function is

$$\phi_i(\boldsymbol{c}, \boldsymbol{c}') = \sum_{k=0}^{N-1} c_k c'_{((k+i))}.$$

The cross-correlation functions and the autocorrelation functions (except for the main peak) of sequences of a Kasami code (in the bipolar alphabet) take values only in the set $\{-1, 2^{m/2} - 1, -2^{m/2} - 1\}$. Hence, a pair of sequences from a Kasami code has a cross-correlation function whose magnitude is never larger than $2^{m/2} + 1$.

For example, a Kasami code with $m = 10$ has 32 sequences of blocklength 1023. Any pair of Kasami sequences in this Kasami code has a cross-correlation function with a magnitude not larger than 33, whereas the central value of each autocorrelation function has magnitude 1023.

A *Gold code* is a different set of sequences that are also constructed from maximal sequences, but by a different rule. These are the Gold sequences. For a given m, let $\boldsymbol{a}$ and $\boldsymbol{b}$ be a pair of distinct maximal sequences that will be specified later. The Gold code is the set

$$\mathcal{C} = \{\boldsymbol{a}, \boldsymbol{b}, \boldsymbol{a} + T^{\ell}\boldsymbol{b} : \quad \ell = 0, \ldots, 2^m - 2\},$$

where the operator T denotes a cyclic translation by one bit position. The Gold code $\mathcal{C}$ contains $2^m + 1$ sequences of blocklength $2^m - 1$.

To complete the description, we need to specify $\boldsymbol{a}$ and $\boldsymbol{b}$. These are two maximal sequences of blocklength $2^m - 1$, chosen so that their cross-correlation function has a maximum value $2^{\lfloor (m+2)/2 \rfloor} + 1$. Such a pair of maximal sequences will always exist. The cross-correlation functions and – except for the main peak – the autocorrelation functions of elements of a Gold code in the bipolar alphabet only take values in the set $\{-1, -2^{\lfloor (m+2)/2 \rfloor} - 1, 2^{\lfloor (m+2)/2 \rfloor} - 1\}$. Hence, the largest magnitude of any cross-correlation function of any pair of sequences from $\mathcal{C}$ is $2^{\lfloor (m+2)/2 \rfloor} + 1$.

For example, a Gold code with $m = 10$ consists of 1025 sequences each of blocklength 1023. Any two of the Gold sequences in this code have a cross-correlation function whose magnitude is not larger than 65. Each autocorrelation function of a Gold sequence has a central value equal to 1023, and elsewhere is not larger than 65.

Problems for Chapter 7

7.1 By showing that it cannot be divided by any first-degree or second-degree polynomial, prove that $x^5 + x^2 + 1$ is an irreducible polynomial over $\boldsymbol{F}_2$. Why is this a proof? Is it a primitive polynomial? Sketch a linear-feedback shift register based on this polynomial. What is the length of the maximal-length sequence generated by this shift register? Plot the periodic autocorrelation function for the sequence expressed in the bipolar alphabet.

7.2 Two periodic binary sequences with periods n_1 and n_2 are componentwise added modulo two. What is the period of the resulting sequence?

7.3 Let $p(x)$ be a primitive binary polynomial of degree m. Let c be an m-sequence of length $2^m - 1$ generated by $p(x)$.

a Prove that every cyclic shift of c is also an m-sequence generated by $p(x)$.

b Prove that every linear combination of two such m-sequences is also such an m-sequence.

c Prove that each of these m-sequences has 2^{m-1} ones and $2^{m-1} - 1$ zeros.

d Prove that any two of these sequences agree in exactly $2^{m-1} - 1$ places.

7.4 Prove that a maximal-length binary shift-register sequence of length $2^m - 1$ displays every r-tuple 2^{m-r} times, except for the all-zero r-tuple, which it displays $2^{m-r} - 1$ times.

7.5 Let $f(v_1, v_2, \ldots, v_m)$ be any function from $\boldsymbol{F}_2^m$ to $\boldsymbol{F}_2$. Define the sequence $\boldsymbol{v} = (v_1, v_2, \ldots)$, by

$$v_{m+i} = f(v_i, v_{i+1}, \ldots, v_{i+m-1})$$

for $i = 1, 2, 3, \ldots$.

a Prove that the sequence is eventually periodic.

b Prove that there is a minimum value of L (called the *linear complexity* of the sequence) for which a linear recursion

$$v_i = -\sum_{k=1}^{L} \Lambda_k v_{i-k}$$

exists.

c Conclude that every finite-state sequence can be regarded as a shift-register sequence. What can be said about the relationship between m and L?

7.6 **a** Does a periodic binary deBruijn sequence of length sixteen contain every binary 4-tuple as a segment?

b Can a periodic binary deBruijn sequence of length sixteen be generated by a linear-feedback shift register?

7.7 Show that it is easy to convert any maximal linear-feedback shift-register sequence into a deBruijn sequence by adding $\prod_{\ell=1}^{m-1} \overline{v}_{i-\ell}$ to the feedback, where $\overline{\boldsymbol{v}}$ is the complement of $\boldsymbol{v}$.

7.8 **a** Is the binary recursion with boolean feedback function

$$f(v_{j-1}, v_{j-2}, v_{j-3}) = 1 + v_{j-2} + v_{j-3} + v_{j-1}v_{j-2}$$

a deBruijn sequence? Write out the first sixteen outputs.

b Show that the deBruijn sequence 0011 with period four can be produced by a linear-feedback shift register of length three, but cannot be produced by a linear-feedback shift register of length two.

7.9 How many periodic binary sequences of period four are there? How many deBruijn sequences of length four are there? How many maximal-length sequences of length four are there? Can every m-sequence be converted to a deBruijn sequence by inserting an additional zero? Can every deBruijn sequence be obtained in this way?

7.10 Prove that a binary nonlinear recursion $v_j = -f(v_{j-1}, v_{j-2}, \ldots, v_{j-L})$ with boolean function f is nonsingular if, and only if,

$$f = v_{j-L} + g(v_{j-1}, \ldots, v_{j-L+1})$$

for some boolean function $g(v_{j-1}, \ldots, v_{j-L+1})$.

7.11 Let $p(x)$ be an irreducible polynomial over $\boldsymbol{F}_2$ of degree n, and let β be a zero of $p(x)$ in a suitable extension field. What can be said about the order of β? Express the other zeros of $p(x)$ in terms of β.

7.12 How many cyclic binary maximal-length sequences of period n are there? How many zeros are there in one period? How many ones? How many times does each subsequence of length r bits appear (regarding the sequence cyclically)?

7.13 Prove that the GSM stream cipher clocking strategy will always eventually clock each of the three shift registers, and so cannot stagnate.

7.14 Prove Key's theorem for an arbitrary output order λ.

7.15 A *self-shrinking keystream* consists of a linear-feedback shift-register sequence that has been postprocessed two bits at a time by the following rule:

$$01 \to 0$$
$$11 \to 1$$
$$00 \to \text{skip}$$
$$10 \to \text{skip}.$$

Starting with a maximal sequence of length fifteen, determine the corresponding self-shrinking keystream. What is its period? Are there any advantages in using overlapping or noncontiguous bit pairs? How would you attack a self-shrinking keystream?

7.16 Can you construct a periodic binary sequence of length fifteen that contains all fifteen nonzero subsequences but is not a maximal shift-register sequence?

7.17 Is it meaningful to construct a self-shrinking keystream from a binary deBruijn sequence?

7.18 Prove that

$$\int_z^\infty \frac{1}{\sqrt{2\pi}} e^{-x^2/2} dx < \frac{1}{2} e^{-z^2/2}$$

for $z > 0$.

Notes for Chapter 7

A comprehensive early study of linear-feedback shift-register sequences can be found in the books of Golomb (1964, 1967), who early on advocated the role of shift-register sequences in cryptography. Shift-register sequences, also known as linear-recurring sequences, have been studied by Carmichael (1920), Ward (1933), and Hall (1938). An important special case consists of those sequences called maximal-length sequences or, more simply, m-sequences. Maximum-period nonlinear sequences were introduced by deBruijn (1946) and Flye Sainte-Marie (1894), who gave the formula for the number of such sequences. The linear complexity of deBruijn sequences was studied by Etzion (1999). A general treatment of stream ciphers can be found in the book by Rueppel (1986).

Many good algorithms, such as the Berlekamp–Massey algorithm, are known for computing the shortest linear recursion that produces a given sequence. Massey gave a statement relating the linear complexity of a sequence to the linear complexity of a truncated subsequence. Blahut (1979) gave a statement relating the linear complexity of a sequence to the Hamming weight of its Fourier transform. Key (1976) gave a bound on the linear complexity of a nonlinear recursion. We have emphasized the role of the Fourier transform in the proof of Key's theorem. Nonlinear recursions are also studied in the work of Chan, Goresky, and Klapper (1990). Klapper (1994) shows that the linear complexity of a binary sequence can be large when viewed as a binary sequence, yet can be small when the sequence is regarded as a sequence of symbols in an extension field.

The nonlinear combining of multiple shift-register sequences, or of multiple translates of a single shift-register sequence, has been widely studied, but this is a vast topic with many aspects that have not yet been explored. A bound on the linear complexity of a binary sequence obtained by nonlinear combining of shift-register sequences was obtained by Selmer (1966) and Herlestam (1986). This bound was shown to be tight under certain nonrestrictive conditions by Rueppel and Slaffelbach (1987). The Geffe (1973) cipher, the Beth–Piper (1984) cipher, and the shrinking cipher that was proposed by Coppersmith, Krawczyk, and Mansour (1993) are elementary nonlinear ciphers. The correlation attack on nonlinear sequences was introduced by Siegenthaler (1984), with related work by Meier and Staffelbach (1988, 1989). Brynielsson (1985) argues that binary sequences must have a conflict between good linear complexity and strong correlation immunity because there are so few binary functions. Sequence sets with good correlation properties are discussed in the books by Golomb and Gong (2005) and by Goresky and Klapper (2012).

The GSM stream cipher is widely used in cellular telephony. That cipher, and its two variations designated A5/1 and A5/2, were developed in 1987. All details of the GSM cipher were kept secret, eventually becoming public only indirectly and over time. The GSM cipher has received widespread and intense scrutiny and has been successfully attacked, although it appears that the known attacks still require considerable effort to recover a plaintext message and are far beyond the casual user.

8 Authentication and ownership protection

Authentication studies methods for the verification of the signatory of a message. These methods are based on cryptography, but involve additional considerations. An authentic message, then, is one that is *signed* in a cryptographically secure way; otherwise, it is not acceptable. An authentic message need not be secret, and a secret message need not be authentic. This is sometimes referred to as the *separation principle* of secrecy and authentication. Of course, a message can be both secret and authentic, but each must be assured individually. A message can be encrypted after it is signed, or it can be signed after it is encrypted, or both, depending on the needs of the application. A signed message may even contain one or more embedded subsections that are themselves independently signed or encrypted by a third party, or are to be decrypted by yet another party.

Authentication is not the same as identification. Identification is a topic that studies methods for verifying or determining the identity of the transmitter of a message. While authentication verifies that the message did indeed come from the indicated sender, identification verifies that the sender does have a recognized position in the appropriate community. Authentication is studied in this chapter, and identification is studied in Chapter 14.

Ownership protection is another topic based on, and related to, the topic of cryptography. Ownership protection studies methods for the control of who owns a message or document, and of how the owner can control the use of the document. Ownership protection includes the two distinct but related topics of *watermarking* and *fingerprinting*, which we do not discuss.

Authentication, identification, and ownership protection all deal with verification, but they are actually quite different. The difference between authentication and ownership protection lies in which party requires the verification. Authentication is the study of methods by which the *receiver* can verify the *source* of a message so as to ensure that the message is indeed a message from the correct source. Ownership protection is the study of methods by which the *transmitter* can subsequently verify to others that it is the true source of a message so as to establish its ownership of the message. Ownership protection also includes methods to protect against – or to detect – unauthorized use or unauthorized copying. Finally, identification is the study of methods by which the *receiver* can verify the *identity* of the transmitter of a message as recognized by the relevant community.

8.1 Authentication

An authentication protocol is a formal procedure by which the originator or the transmitter of a message establishes the validity of the message by attaching a signature. The message is deemed to be valid if it displays a signature that is known to belong to the appropriate source. The purpose of authentication is to prevent an imposter from attaching the signature of that source to a fraudulent document. Authentication and secrecy are separate functions. Secrecy can be provided without authentication, and authentication can be provided without secrecy. The signature at the heart of an authentication protocol is usually based on a mathematical problem that is believed to be intractable. However, the introduction of an intractable mathematical problem is not sufficient to ensure authentication security. It may be that the protocol by which it is used is not secure, even though the underlying mathematical problem is intractable. An authentication protocol must also be secure against many other kinds of attacks directed against the way in which the signature is applied.

A long message, such as a legal document, should be signed in its entirety. This is because an unsigned segment of a document could be changed at a later time without invalidating the signature on the remainder of the document. Moreover, if different sections of one document are signed separately, it might be possible that a section of the document could be deleted, thereby creating a false appearance or deception.

For a long message, it will be computationally expensive to sign the entire message as such. Instead, a *message digest* is usually computed from the entire message and the signature is applied to the message digest. The signed message digest is then appended to the original message. The reason for signing a message digest instead of the message itself is to make the computational burden acceptable and to facilitate standard signature procedures. This also prevents some kinds of tampering that could be used to disrupt a long message that is broken into individual segments for signing. Separately signed blocks can be deleted or rearranged. The remaining message segments will still be properly signed and possibly could be assembled into an apparently valid message, but individually signed with a different or diluted meaning.

8.2 Identification

There is an unavoidable difficulty underlying authentication that requires comment and careful thought. This is the issue of identity. An authentication protocol can only verify that a given message has come from an already recognized identity, but it cannot establish the identity of a remote, isolated, uncertified, and unauthorized stranger encountered by chance, as on a network. Identification is an additional requirement separate from authentication. Authentication validates the message as coming from a recognized source. Identification validates the source of the message as a recognized member of a relevant community.

The identification protocol must prevent with certainty the transfer of identity from one individual to another, as by theft. To prevent theft, a modern identification protocol requires the source to prove that it possesses appropriate certification without revealing that certification. This process is referred to as *zero-knowledge identification*, and is discussed in Chapter 14.

The notion of the identity of an individual is a subtle notion, and can only be defined by the collective attributes of the individual and the available certification. Only certification by one or more trusted certification authorities, or other trusted third parties, can establish an identity by binding an appropriate certification to a public identity code of that individual. Thus an identification protocol, in itself, is used only to ensure that a message came from a properly certified or previously recognized source, meaning one that possesses the necessary certificates. These certificates bind the individual to the larger community. In effect, the certificates become the identity of that individual.

8.3 Authentication signatures

A digital message x is to be signed in order to authenticate the source of that message. This requires a secure signature protocol. The signature protocols described in this section allow the message recipient to confirm that the message did indeed originate from a designated source. They do not in themselves provide identification of the source. Accordingly, these signature protocols are called authentication signatures. Identification signatures are described in Section 14.5.

A digital message may be a text document, an image, a video, or any other digital file. The message to be signed may be plaintext or ciphertext. Because the message may be very long, the plaintext or the ciphertext may be condensed prior to signing, and usually is condensed, as we shall discuss later under the topic of *hashing*. The process of hashing replaces the actual message x with a *message digest*, denoted $\text{hash}(x)$. The message digest is then signed and the signed message digest, denoted $\text{sign}_k(\text{hash}(x))$, is attached to the message x. A signed message then is the pair $y = (x, \text{sign}_k(\text{hash}(x)))$ where k is a key of some sort. The nature of the key k depends on the method of signature.

If desired, a signed message y can be encrypted in its entirety, including the signature. This can be done even if the unsigned message is already encrypted. Indeed, the two levels of ciphertext might require two independent decryptors. Moreover, verification of a signature only asserts that the message came from the owner of that signature. It does not, itself, identify the owner.

We will discuss the topic of hashing in detail in later sections, after first discussing the topic of signatures. In this section, we will study only the methods of signing either the message itself or the digest of the message.

Two methods of forming a secure digital authentication signature are the *RSA signature scheme* and the *Elgamal signature scheme*. The RSA signature scheme uses the RSA cryptosystem, modified, as a mechanism to sign a document. The Elgamal signature scheme uses the Elgamal cryptosystem, modified, as a mechanism to sign a document.

For either signature scheme, the signature is usually applied to a message digest, denoted hash(x), rather than to the message x itself. This limitation is a consequence of the fact that the message x has a variable and unbounded length. For this reason, the message x is condensed to a message digest, denoted hash(x), by using a suitable public hash function. The signature is then applied to hash(x) rather than to x.

RSA signature

We first describe the RSA signature scheme, which incorporates the same structure as the RSA encryption scheme, but the process is reversed. As before, $n = pq$, where p and q are two primes held in secret by the signer, and a and b are two integers such that $ab = 1 \pmod{\phi(n)}$. The integers n and b are the public signature key, while a, p, and q are the private key, known only to the signer. To sign a message x, define the *digital signature* as

$$y = \text{sign}_a(x) \;=\; x^a \pmod{n}.$$

The signed message is then given by $(x, y) = (x, \text{sign}_a(x))$. To verify that the signature is valid, compute $\text{ver}_b(y) = y^b \pmod{n}$ and compare it to the message x. If $y^b = x$, then the message is authentic because the signature could only have been made by one who knows the integer a. One can conclude that the owner of the integer a, perhaps a stranger to the verifier and with no other identity other than the possession of a, is the source of the signed message.

In practice, the signature is not applied to x, but to the message digest hash(x), so $y = (\text{hash}(x))^a$, where a is secret, and verification is provided by testing whether $\text{hash}(x) = y^b$. This test can be performed in the open, and by anyone, because both b and the hash function are publicly known.

There is a minor issue in the RSA signature that should be mentioned. Because the verification function ver(y) is public and it inverts the signature function, an adversary can create a false signed message taking the form of $(x, \text{sign}_a(x))$ by choosing a random y as a false signature and forming $(\text{ver}_b(y), y)$. Thus y is a signature for the "message" ver(y). Of course, the false message is a meaningless string of symbols but it does successfully pass the signature verification and must be recognized to be nonsense by other layers of the protocol. Thus an RSA signature, by itself, does not protect against a well-designed flooding attack.

Elgamal signature

The Elgamal signature scheme uses the intractability of the discrete-log problem to form a secure signature scheme. For any prime p, the set $\boldsymbol{Z}_p^*$ is a cyclic group of order $p-1$, and can be generated by some element α of $\boldsymbol{Z}_p$. The prime p is chosen sufficiently large so that the discrete-log problem in $\boldsymbol{Z}_p^*$ is intractable. Both the prime p and the generator α are public parameters of any system standard embodying the Elgamal signature protocol.

To generate an Elgamal signature, the individual signer uses two randomly chosen keys. The first key is a permanent secret identity key i. The signer chooses this integer i once and computes $I = \alpha^i$. The integer i is kept secret. The integer I is made public and becomes the permanent public authentication code of the signer. More specifically, the individual becomes recognized as the keeper of $\log_p I$. Presumably the integer $\log_p I$ is only known to the one who computed I because to compute the logarithm of I in $\boldsymbol{F}_p$ is intractable. To sign a message $x \in \boldsymbol{Z}_p$, the signer chooses a document key, which is unique to that document and never used again, consisting of a random integer $k \in \boldsymbol{Z}_p^*$. This document key is required to satisfy the condition GCD(k, $p-1) = 1$, which means that several values of the document key k may need to be tried to find one that is satisfactory, though this will be very rare. Then define $K = \alpha^k$.

To sign the message x, the signer computes

$$\Delta = k^{-1}(x - iK) \pmod{(p-1)}$$

and tests that Δ is nonzero. In the very rare event that Δ is zero, a new document signature key, k is chosen and a new value of Δ is computed. This process must be repeated, as necessary, until Δ is nonzero. The signature $\text{sign}_k(x)$ of the message x is the pair (K, Δ). The signed message is

$$(x, \text{sign}_k(x)) = (x, (K, \Delta)).$$

Verification, upon receipt of the signed message $(x, \text{sign}_k(x))$, is done by checking that $\alpha^x = K^\Delta I^K$. The proof of the verification procedure follows from the definition of Δ by writing

$$x = \Delta k + iK \pmod{(p-1)}.$$

Therefore

$$\begin{aligned} \alpha^x &= \alpha^{\Delta k}\alpha^{iK} \\ &= K^\Delta I^K. \end{aligned}$$

All quantities on the right side are known, so the verification is well defined.

Because the message is hashed, the actual signed message is $(x, (K, \Delta))$ where $\Delta = k^{-1}(\text{hash}(x) - iK)\pmod{(p-1)}$. The verification then is $\alpha^{\text{hash}(x)} = K^{\Delta}I^{k}$. Because the hash function hash(x) and the signature verification function $\text{ver}_I(z, y)$ are public functions, anyone can verify the signature by computing $z = \text{hash}(x)$ and then computing the required test.

A third party can forge signatures in the obvious way by learning the signer's secret key. The key might be learned by a direct attack on the underlying cryptographic technique or by an indirect backdoor attack. To obstruct certain kinds of attempts to do this, the signature key includes a portion that is used only once.

8.4 Hash functions

A *hash function*, denoted hash(x) or $h(x)$, is any well-defined deterministic procedure that reduces a long message of arbitrary length – perhaps even millions of bits or symbols – to a short *message digest* of a fixed length. Hash functions have many applications. We are interested here only in those *one-way hash functions* that are used in cryptography. These are called *cryptographic hash functions*, the terminology only referring to the application to cryptography including authentication, and not necessarily implying any secret key. We may have either keyless hash functions or keyed hash functions. For current use in cryptography, a popular minimum value of hash length for a message digest has been 160 bits. In this case, the hash function maps any arbitrary binary message of arbitrary length into a 160-bit message digest. Longer message digests are now becoming preferred. Cryptographic hash functions are called one-way hash functions to imply that it is intractable to compute the inverse in order to find one or more messages corresponding to a given message digest.

Let $\mathcal{A}^*$ denote the set of all sequences of arbitrary length of elements from the alphabet $\mathcal{A}$. A message $\boldsymbol{x}$ of arbitrary length over the alphabet $\mathcal{A}$ is an element of $\mathcal{A}^*$. A hash function, hash(x), is a function that maps $\mathcal{A}^*$ onto $\mathcal{A}^n$, where n is the length of the message digest. We usually regard messages to be binary strings, in which case a hash function can be described as a map from $\{0, 1\}^*$ onto $\{0, 1\}^n$. A signature of the message x is a function $y = \text{sign}(\text{hash}(x))$. A signed message is the pair (x, y).

A third party may attempt to forge signatures by transferring a signature $\text{sign}_k(\text{hash}(x))$ from its true document to a false document. To do this, the false document must be designed to have the same hash function as the true document. This is attempted by making innocuous changes in the false document such as extra word spaces or paragraph indentations. Each modified version of the fraudulent message can be hashed by the adversary because the hash function is public. If ever a modified fraudulent message is found whose hash function is equal to hash(x), then $\text{sign}_k(\text{hash}(x))$ can be transferred from the true message to the fraudulent message. The occurrence of equality between the hash of the true message and the hash of a fraudulent message is called a *collision* of hash functions. To counter this, the hash function must be resistant to collisions.

A good hash function for cryptographic applications should be easy to compute for any message. It should not be feasible to invert the hash function in order to find the message. It should not be feasible to modify a message without changing its hash function, and it should not be feasible to find two messages with the same hash function. In particular, it should not be feasible for an adversary to successfully transfer a signed hash function to another document.

To meet these goals, there are a number of broad properties that a cryptographic hash function should have. These properties may appear to be conservative and more stringent than is required in any application. However, they are considered to be sufficiently general to allow confidence that the given hash function will resist unspecified attacks.

Definition 8.4.1 *A hash function hash(x) is preimage resistant if, given a hash value h, it is not computationally feasible to find a message x such that hash(x) $= h$. A preimage-resistant hash function is a one-way function.*

Preimage resistance is an important property of a hash function. The definition of preimage resistance requires that it not be possible to find any x satisfying the hash function, even a meaningless x. This is a conservative requirement. The notion of preimage resistance involves inverting the hash function, and this is not easy to analyze. In contrast, there are other notions of hash performance involving only the direct hash function that may be easier to analyze. Because these other notions are more stringent than preimage resistance, they provide conservative bounds on performance.

Definition 8.4.2 *A hash function* hash(x) *is strongly collision-free if it is not computationally feasible to find two different messages x and x' for which* $\text{hash}(x) = \text{hash}(x')$.

A *strongly collision-free hash function* is one for which the definition refers to any pair of different messages, x and x', even if they are not legitimate messages. The definition does not concern itself with how such a collision can be used by an adversary. In contrast, if the statement refers only to messages x and x' for which one or both are "typical" in some sense, then the hash function is sometimes called a *weakly collision-free hash function.* A more precise definition follows.

Definition 8.4.3 *A hash function is weakly collision-free if, when given a legitimate message x, it is not computationally feasible to find another message x' such that hash(x) $=$ hash(x').*

As it turns out, a strongly collision-free hash function is always a one-way hash function.

Definition 8.4.4 *A hash function is resistant to length extension if it is not computationally feasible when given hash(x) to find an x' such that the concatenation (x, x') satisfies* $\text{hash}(x, x') = $ *hash*(x).

A *random hash function* consists of a random assignment of elements of $\mathcal{A}^n$ to each element of $\mathcal{A}^*$. A random hash function must be known to all parties, so it must be stored in a (massively) large table. A random hash function is an abstract notion of a good hash function – although only in a conceptual way – because a random hash function is exponentially complicated in the maximum message length. For messages of length N, a random hash function requires a table of 2^N entries, where N may be on the order of many thousands, or even larger. The reason for introducing a random hash function is that it establishes a notional goal that a practical hash function should strive to meet. A practical hash function should look like a random hash function to a casual observer, and to the cryptanalyst.

A random binary message digest of length k bits for messages of length N consists of 2^N binary words of length n with the n bits in each word selected randomly and independently. We can regard these 2^N words as stored in a table of size 2^N. To hash an N-bit message, use the N-bit message as an address to look up the value of the hash function. Of course, even if N is as short as 4096 bits, the memory has 2^{4096} entries, so the random hash function is far from realistic. It is only a conceptual device that helps to clarify the design goals of a practical hash function. A practical hash function should aspire to be a deterministic algorithm that is indistinguishable from a random hash function by any tractable inspection.

Because the space of messages is always much larger than the space of message digests, it is clear that collisions must exist. Indeed each n-bit message digest will have, on average, 2^{N-n} messages of length N that hash into that message digest. The goal in defining a hash function is to make it computationally intractable to find a collision.

8.5 The birthday attack

A *collision attack* on a signature scheme is a probabilistic attack that tries to fraudulently sign a message when given a properly signed message $(x, \text{sign}_k(\text{hash}(x)))$. This attack uses the publicly known functions $\text{hash}(x)$ and $\text{ver}_k(y)$. Given a false message x' that is to be fraudulently signed by transferring the signature $\text{sign}_k(\text{hash}(x))$ from its proper document, the forger computes both $\text{hash}(x) = \text{ver}_k(\text{sign}_k(\text{hash}(x)))$ and $\text{hash}(x')$. If $\text{hash}(x) = \text{hash}(x')$, then $(x', \text{sign}_k(\text{hash}(x))$ is a successful forgery of a signature on message x'. Otherwise, the forger makes inconsequential changes in the false message x' and tries again. If x' is a long text file, for example, it will be easy to make inconsequential changes, as by inserting or deleting extraneous words, spaces, or punctuation. There may even be a part of the file that is dummy text set aside for this purpose. If the probability of a collision is high enough, this attack will eventually succeed. To counter this collision attack, the probability of a collision must be made very small. This leads to a requirement for a key size of more than 128 bits. A commonly accepted minimum signature standard uses 160 bits to prevent this kind of attack. Newer standards now use 256, 384, or 512 bits for the signature length.

A variation of the collision attack allows the forger access to both the true message x and the false message x'. Inconsequential changes are made in both. This gives a list of candidate hash(x) and a list of hash(x'). If there is a common entry on the two lists, the forger presents x to the signer to be signed, then transfers the signature to x'. This attack presumes that the forger has the opportunity to tamper with the valid message before it is signed.

In an artificial variation, the forger simply generates a large random collection of messages in hopes that two have the same hash function. One is the message that is submitted to be signed, and the signature is then transferred to the other message. If this contrived collision of meaningless messages is statistically rare, then more realistic collisions are also rare.

This collision attack is called a birthday attack because its analysis uses the same arguments as are used to derive the well-known *birthday surprise* of probability theory. This is the surprising statement that in any set of 23 randomly chosen people, the probability that there is at least one common birthday (out of 365 days) is greater than one half. To show this, let $n = 365$ and let p_c be the probability of a collision. The probability that two randomly chosen people do not have the same birthday is $(n-1)/n$. The probability that three randomly chosen people do not have the same birthday is $((n-1)/n)((n-2)/n)$. The probability that there is no birthday collision in a set of k people is

$$
\begin{aligned}
1 - p_c &= \left(1 - \frac{1}{n}\right)\left(1 - \frac{2}{n}\right)\cdots\left(1 - \frac{k-1}{n}\right) \\
&= \prod_{i=1}^{k-1}\left(1 - \frac{i}{n}\right).
\end{aligned}
$$

This can be evaluated numerically for small values of k. For large values of k, it is more convenient to use an approximation. For this purpose, recall that for small x, $1 - x \approx e^{-x}$, which leads to

$$
\begin{aligned}
1 - p_c &= \prod_{i=1}^{k-1}\left(1 - \frac{i}{n}\right) \\
&\approx \prod_{i=1}^{k-1} e^{-i/n} = e^{-\sum_{i=1}^{k-1} i/n}.
\end{aligned}
$$

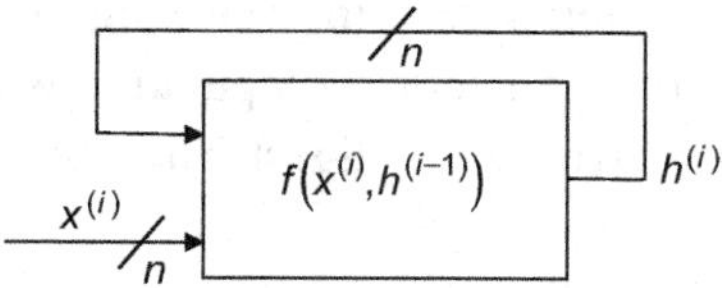

Figure 8.1 Iterative computation of message digest

Therefore

$$\begin{aligned} p_c &\approx 1 - e^{-k(k-1)/n} \\ &\approx 1 - e^{-k^2/n}. \end{aligned}$$

If $p_c = 0.5$ and $n = 365$, then $k = 22.3$, which is the birthday surprise. If $n = 2^{40}$, as for a 40-bit message digest, and $p_c = 0.5$, then k is about 10^6. Thus a collision attack would need about 1 million trials to have a probability of 0.5 of successfully finding two messages with a hash collision for a 40-bit message digest. If $n = 2^{160}$ as for a 160-bit message digest, a collision attack would need about 10^{24} random trials to obtain a collision of two random messages, presumably an intractable amount of computation. It would be even more difficult to find a collision of two meaningful messages. Thus an authentication scheme using a 160-bit hash function may be regarded as secure against a birthday attack.

8.6 Iterated hash constructions

Because messages are of variable length and very long, it is natural and convenient to break a message into small blocks of fixed length and to fold these blocks into the hash function one by one. Modern cryptographic hash functions are usually constructed by means of such an iterated procedure, referred to as the *Merkle–Damgaard hash construction.* The Merkle–Damgaard hash construction simply refers to the use of multiple rounds of any core computation to compute a message digest, and is validated by the statement that overall security is guaranteed if the core computation is secure. A Merkle–Damgaard hash function is any hash function that is computed by this structure of a block-organized iteration, as shown in Figure 8.1. If $f(y)$ is a function that is computationally intractable to invert, then $f(f(x))$ is a function that is also computationally intractable to invert.

The Merkle–Damgaard construction breaks a message of arbitrary length into blocks of length n bits, padding the full message, if necessary, with extra bits so that the message length is a multiple of the blocklength n. In each iteration, it updates a temporary message digest. In each iteration, a new block is folded into the core computation, which computes a new temporary message digest, usually of blocklength n as well, by combining the next block with the previous temporary message digest. Each new block, in its turn, is paired with the most recent temporary message digest, and the two are passed again through the core computation. The core computation is a secure one-way function that computes the ith temporary message digest from the $(i-1)$th temporary message digest and the ith block of the message.

The core computation including the method of folding in new blocks of data can take many forms. Any block encryption function $e_k(x)$ can be used as the core computation. Because a block encryption function $e_k(x)$ has two inputs, the key k and the plaintext x, these two inputs can be assigned a variety of alternative roles to comprise the core computation and so to define a Merkle–Damgaard iteration.

For example, let $x^{(i)}$ denote the ith block of the message x being hashed and let $h^{(i)}$ denote the message digest iterate at the ith iteration. The iterate $h^{(i)}$ will become the message digest when the iterations are complete. The equation

$$h^{(i)} = e_{x^{(i)}}(h^{(i-1)} + x^{(i-1)}) \qquad i = 1, \ldots, I$$

defines a Merkle–Damgaard iteration that uses any block encryption function $e_k(x)$ with key k. An alternative method of hashing, also with the block encryption function $e_k(x)$, is given by the equation

$$h^{(i)} = e_{h^{(i-1)}}(x^{(i)} + h^{(i-1)}) \qquad i = 1, \ldots, I.$$

This equation also defines a Merkle–Damgaard iteration, but this Merkle–Damgaard iteration is different from the iteration described previously. Both of these examples inherit their collision-resistant security from the security of the underlying block encryption function $e_k(x)$. The Merkle–Damgaard principle is that the iterative use of $e_k(x)$ does not dilute the collision resistance. This principle is certainly in accord with intuition. This is validated by the remark that any encryption function that is secure to begin with cannot be made less secure by encrypting a second time with an arbitrary key.

8.7 Formal hash functions

Hash functions that are formally secure under appropriate stated conditions can be defined in terms of formal mathematical structures. The stated conditions can be based on the presumption that some well-studied problem such as the discrete-log problem or the biprime factoring problem is indeed intractable. However, these formally secure algebraic hash functions generally involve computations that are deemed to be too burdensome for practical applications of hashing. This is because, in practice, rapid hashing of very large files, perhaps millions of bits in length, is often required. We will study the formally secure hash functions, not for their practicability, but as examples of desirable hash functions and to round out the theory.

The *Gibson hash function* is based on the intractability of integer factoring. In this sense, it is formally secure. The Gibson hash function uses a biprime $n = pq$ where p and q are distinct large primes of the form, $p = 2p' + 1$ and $q = 2q' + 1$, and both p' and q' are primes. Then $\boldsymbol{Z}_n^*$ has order $4p'q'$ and has a cyclic subgroup of order $2p'q'$ generated by an element α. The biprime n is formed by randomly choosing large integers as candidates for p' and q' and testing them for primality, then testing $2p' + 1$ and $2q' + 1$ for primality. If all tests are successful, one has a suitable n. Otherwise, start over for either or both factors of n.

Define the Gibson hash of the message x by the exponentiation

$$\text{hash}(x) = \alpha^x \pmod{n}.$$

To this purpose, the message x, which may be presented in the form of a long binary sequence, is regarded as an integer. If the task of biprime factoring is intractable, as is generally believed, then it is intractable to invert the Gibson hash function.

The *Chaum, van Heijst, Pfitzmann hash function* is based on the intractability of the discrete-log problem. Let p be a large prime for which $q = (p - 1)/2$ is also a prime. Let α and β be two primitive elements of $\boldsymbol{F}_p$, where it is required that $\log_\alpha \beta$ is not known and is intractable to compute. The message x is partitioned as $x = (x_1, x_2)$, which can be regarded as two nonnegative integers, each not larger than $q - 1$. These two integers can be expressed as two binary words each of length $\lfloor \log_2 q - 1 \rfloor$ bits. The hash function $\{0, \ldots, q - 1\} \times \{0, \ldots, q - 1\}$ is defined by

$$h(x_1, x_2) = \alpha^{x_1} \beta^{x_2} \pmod{p},$$

which is a nonnegative integer smaller than p.

The proof of the security of this hash function rests on the belief that it is computationally intractable to compute $\log_\alpha \beta$. This follows because any method to find a hash collision can be easily converted into a method to compute $\log_\alpha \beta$ as follows. If one can find distinct pairs (x_1, x_2) and (x_1', x_2') such that $\alpha^{x_1} \beta^{x_2} = \alpha^{x_1'} \beta^{x_2'} \pmod{p}$ then

$$\alpha^{x_1} (\alpha^{\log_\alpha \beta})^{x_2} = \alpha^{x_1'} (\alpha^{\log_\alpha \beta})^{x_2'} \pmod{p}$$

so that

$$x_1 + x_2 \log_\alpha \beta = x_1' + x_2' \log_\alpha \beta \pmod{p - 1}$$

which becomes

$$\log_\alpha \beta = \frac{x_1' - x_1}{x_2 - x_2'} \pmod{p - 1}.$$

Thus, if this hash function can be broken, then the discrete-log problem in $\boldsymbol{F}_p^*$ can be solved.

Neither of these two formal secure hash functions of this section are suitable for hashing large files quickly. In the next section, we examine practical hash functions.

8.8 Practical hash functions

A hash function is judged both by its simplicity and by its security. The simplicity of a hash function is judged in large measure by the speed at which it can hash large files. The security of a hash function is judged in large measure by two criteria: inversion resistance and collision resistance. Because the hash function in practical applications is routinely used to hash large files, possibly consisting of megabytes, the computations must be simple. Because the hash function must resist attack, it must be secure. Thus there is always a tension between the conflicting goals of simplicity and security. To achieve simplicity, practical hash functions employ a massive number of simple operations, primarily componentwise operations and additions, and they do not rest on apparently intractable mathematical problems. For this reason, proofs of security are lacking for practical hash functions in common use. Verification of security is anecdotal, not formal.

This section introduces some of the practical hash functions in wide use. The purpose of this section is to provide only an overview and general understanding of practical hash functions in common use. The discussion is not intended to provide the full details required for an implementation.

An early keyless hash function that was simple enough to be practical, and of the same vintage as the DES, is the MD4 hash function, taking its name from the phrase "message digest." This hash function was designed for ease of computation rather than for provable security. It was soon judged to be insecure because of published vulnerabilities. An improved version is known as MD5, but this version is now also regarded as insecure. These two cryptographic hash functions – MD4 and MD5 – are designed to be extremely fast, not to be provably secure. Although the MD4 hash function has been compromised, it is still in use in many legacy applications, and it can be broken only with effort. The underlying ideas have continued to evolve leading to a more modern and presumably secure hash function. The more recent class of hash functions is formally called SHA (for secure hash algorithm), which appears in a series of improvements.

Although the MD4 and MD5 cryptographic hash functions are now regarded to be potentially insecure, they are of interest for pedagogical purposes because many cryptographic hash functions now in common use have evolved from their concepts. Thus the description of the MD4 hash function is given here because it provides an appropriate introduction to this class of hash functions. The MD5 is similar and is not discussed here.

The MD4 hash function produces a 128-bit message digest, a hash length that is now regarded as insufficient. The MD4 hash function consists of a sequence of major iterations at the end of which is produced the 128-bit message digest. Each major iteration folds a frame, consisting of 512 bits organized as sixteen blocks each of length 32 bits, into the running computation of the 128-bit message digest. The iterations continue until the entire message has been digested.

Let x be a variable-length binary message of length m bits that is to be hashed by the MD4 hash function. The message x is first lengthened by the rule

$$x \leftarrow (x, 1, 0, \ldots, 0, m'),$$

where m' is a 64-bit word specifying the length m of the message in bits, modulo 2^{64}, and there are enough zeros inserted between the single one and the 64-bit representation of m so that the new x has a length divisible by 512. This extension, comprised of the single one, the 64-bit word, and the variable number of zeros, ensures that when the message is broken into 32-bit blocks, the number of such blocks will be divisible by sixteen. This extended message is the input to the hash computation. It will be processed one frame at a time, each frame consisting of sixteen 32-bit blocks.

The output of the hash computation is a 128-bit message digest. The 128-bit message digest is formed as the concatenation of four 32-bit registers, denoted A, B, C, and D, which together hold the running message digest as it is computed. Each major iteration takes one 512-bit frame of the message, computes a frame digest consisting of four 32-bit words, and adds this frame digest to the 128-bit running message digest stored in the four registers A, B, C, and D. The four registers are initialized by the hexadecimal constants

$$\begin{aligned} A &\leftarrow 67452301 \\ B &\leftarrow efcdab89 \\ C &\leftarrow 98badcfe \\ D &\leftarrow 10325476. \end{aligned}$$

The frame digest consists of 128 bits regarded as four blocks, each block of length 32 bits. The blocks are denoted r_A, r_B, r_C, and r_D, respectively, and initialized to zero at the start of each frame. The sixteen 32-bit blocks of the message frame are folded into the working variables three times, each time called a round, and each round consisting of sixteen minor iterations, thereby comprising a total of forty-eight minor iterations. During round ℓ for $\ell = 1, 2, 3$, as shown in Figure 8.2, the four registers are updated during each of sixteen iterations by stepping through the following four statements four times for $j = 1, 2, 3$

$$\begin{aligned} r_A &\leftarrow r_A + f_\ell(r_B, r_C, r_D) + x_{4j} && \pmod{2^{32}} \quad \pmod{x^{k_{\ell 1}} - 1} \\ r_B &\leftarrow r_B + f_\ell(r_C, r_D, r_A) + x_{4j+1} && \pmod{2^{32}} \quad \pmod{x^{k_{\ell 2}} - 1} \\ r_C &\leftarrow r_C + f_\ell(r_D, r_A, r_B) + x_{4j+2} && \pmod{2^{32}} \quad \pmod{x^{k_{\ell 3}} - 1} \\ r_D &\leftarrow r_D + f_\ell(r_A, r_B, r_C) + x_{4j+3} && \pmod{2^{32}} \quad \pmod{x^{k_{\ell 4}} - 1} \end{aligned}$$

with the functions $f_\ell(x, y, z)$ defined below. The additions here are integer additions modulo 2^{32}. After the additions, the resulting binary number is cyclically shifted by $k_{\ell i}$ bits, where the $k_{\ell i}$ indicated by the modulo $x^k - 1$ terms are specified by a table.

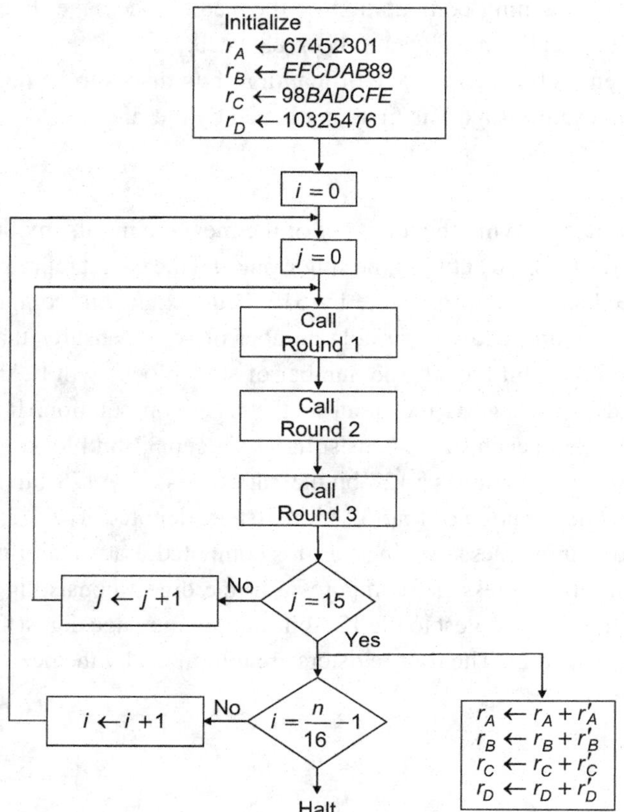

Figure 8.2 The MD4 hash function

During each round, the set of four equations is stepped through four times to fold in the sixteen 32-bit new data blocks of that frame. This is repeated for three rounds with $\ell = 1, 2, 3$. The functions $f_\ell(x, y, z)$ are prescribed functions of the four registers given by the three componentwise boolean functions

$$f_1(x, y, z) = (x \wedge y) \vee (\overline{x} \wedge z)$$
$$f_2(x, y, z) = (x \wedge y) \vee (x \wedge z) \vee (y \wedge z)$$
$$f_3(x, y, z) = x \oplus y \oplus z,$$

where $\wedge$ and $\vee$ denote componentwise "and" and componentwise "or," respectively, $\oplus$ denotes modulo-two addition, and the overbar represents binary complementation.

The SHA hash function is a more recent and more secure hash function that was developed to remedy the deficiencies of the MD4 and MD5 hash function. The cryptographic hash algorithm known as SHA-1 retains the spirit of the MD4 hash function, but with improvements. The SHA hash function produces a 160-bit message digest. This length is now regarded as a minimal acceptable hash length for a message digest.

The SHA hash function has continued to evolve and now exists in many variations, some with longer hash lengths, such as SHA-224, SHA-256, SHA-384, and SHA-512.

As with the MD4 hash function, the SHA hash function lengthens the message as necessary so that it consists of a variable number of 512-bit frames. The SHA-1 hash function consists of a sequence of major iterations at the end of which it produces a 160-bit message digest. As with MD4, each major iteration folds a frame consisting of 512 bits organized as sixteen blocks, each of length 32 bits, into the running computation of the 160-bit message digest. The iterations continue until the entire message has been digested. This much is similar to MD4. An important difference, however, is that during each iteration, that frame is combined with the four previous frames during the processing. In this way, at each iteration, the next 512-bit frame is folded into the running message digest.

The 160-bit running message digest is contained in five 32-bit registers denoted A, B, C, D, and E. The initial values of these five registers in hexadecimal notation are

$$\begin{aligned}
A &\leftarrow 67452301\\
B &\leftarrow efcdab89\\
C &\leftarrow 98badcfe\\
D &\leftarrow 10325476\\
E &\leftarrow c3d2e1f0.
\end{aligned}$$

In addition to the running message digest, the algorithm employs a temporary list of eighty 32-bit words, denoted w_i for $i = 0, \ldots, 79$. At the start of each major iteration the list is moved down sixteen places, discarding the oldest sixteen 32-bit words, and inserting the sixteen words of the next frame at the top of the list. Then the other w_i on the list are replaced one by one by the rule

$$w_i = \text{cyclicshift}(w_{i-3} \oplus w_{i-8} \oplus w_{i-14} \oplus w_{i-16}) \qquad i = 16, \ldots, 79,$$

where, as indicated, w_i is the one-bit cyclic shift of the 32-bit word in parentheses, and where $\oplus$ denotes componentwise modulo-two addition.

Next, define the four componentwise boolean functions

$$\begin{aligned}
f_1(x, y, z) &= (x \wedge y) \vee (\overline{x} \wedge z)\\
f_2(x, y, z) &= x \oplus y \oplus z\\
f_3(x, y, z) &= (x \wedge y) \vee (x \wedge z) \vee (y \wedge z)\\
f_4(x, y, z) &= x \oplus y \oplus z.
\end{aligned}$$

Each of these four functions is used for twenty minor iterations, thereby giving eighty iterations in total. Thus $f_1(x, y, z)$ is used for the first twenty iterations, $f_2(x, y, z)$ is used for the second twenty iterations, $f_3(x, y, z)$ is used for the third twenty iterations, and $f_4(x, y, z)$ is used for the fourth twenty iterations. These eighty minor iterations in each frame of SHA-1 replace the three rounds of sixteen minor iterations used in each frame of MD4. The SHA-1 hash function also uses a displacement by a constant during the computations of each frame. The displacement values, in hexadecimal notation, are the constants

$$
\begin{aligned}
K_1 &= \text{5a827999}\\
K_2 &= \text{6ed9eba1}\\
K_3 &= \text{8f1bbcdc}\\
K_4 &= \text{ca62c1d6},
\end{aligned}
$$

each of which is used for twenty iterations.

To complete the definition of the SHA-1 hash function, define

$$Z = \text{cyclicshift}(A) + f_i(B, C, D) + E + W_i + K_i,$$

in which the contents of A are cyclically shifted by five places. Next store a copy of A, B, C, D, E as A', B', C', D', E'. Then make the word-level shift of the five registers in the pattern

$$Z \to A \to B \to C \to D \to E$$

so that A, B, C, D, E is replaced by (Z, A, B, C, D), then followed by a cyclic shift of the new C by 30 bits. Finally, perform the additions

$$
\begin{aligned}
A &\leftarrow A + A'\\
B &\leftarrow B + B'\\
C &\leftarrow C + C'\\
D &\leftarrow D + D'\\
E &\leftarrow D + E'.
\end{aligned}
$$

This completes the summary of SHA-1. There is no formal mathematical proof that this complicated hash function is secure. Security is assured by the continued scrutiny of the community and by the presumption that a long repetitive sequence of simple, but apparently structure-free computations cannot be inverted. Attacks, all of which apparently have been unsuccessful, attempt to undo the computations step by step.

Problems for Chapter 8

8.1 Let $h(x)$ be defined by

$$h(x) = x^2 + ax + b \pmod{2^m},$$

where the constants a and b are elements of $\mathbf{Z}_{2^m}$, and $x \in \mathbf{Z}_{2^m}$ is a message to be hashed. Given any x, find another message $x' \neq x$ such that $h(x') = h(x)$. Is $h(x)$ a satisfactory hash function?

8.2 Given the hash function $y = h(x)$ on the elements x of the set $\mathcal{X}$, let

$$N_y = \#\{x | h(x) = y\}$$

denote the number of x that hash into $y \in \mathcal{Y}$. Let

$$N = \#\{(x_1, x_2) | x_1 \neq x_2, h(x_1) = h(x_2)\}$$

denote the total number of collisions (with double counting).

a Show that

$$\sum_{y \in Y} (N_y - \overline{N}_y)^2 = 2N + \#\mathcal{X} - \frac{(\#\mathcal{X})^2}{\#\mathcal{Y}},$$

where

$$\overline{N}_y = \frac{1}{\#N_y} \sum_y N_y.$$

b Next, prove that

$$N \geq \frac{(\#\mathcal{X})(\#\mathcal{X} - \#\mathcal{Y})}{2(\#\mathcal{Y})}$$

with equality if, and only if,

$$N_y = \frac{\#\mathcal{X}}{\#\mathcal{Y}}$$

for all $y \in \mathcal{Y}$.

c Suppose that messages x_1 and x_2 are chosen randomly and equiprobably. State and prove an inequality relating the probability of a collision p_e to $\#\mathcal{Y}$. When is this inequality satisfied with equality?

8.3 Prepare a list of your thirty closest relatives (or friends, or classmates, or presidents) and their birthdays. Is there a birthday collision? Discuss. Should this collision be a surprise?

8.4 Using the birthday surprise approximation

$$p_c = 1 - e^{-k^2/n},$$

show that the probability that the same object will be randomly drawn twice (with replacement) from a set of n objects will be one-half if the number of draws is equal to $1.177\sqrt{n}$.

8.5 Let $p = 2p' + 1$ and $q = 2q' + 1$ be distinct secret primes such that p' and q' are also both prime. Let $n = pq$, and let α be an element of $2p'q'$ in $\mathbf{Z}_n^*$. Let h be a hash function given by the rule $h(x) = \alpha^x \pmod{n}$. Given the three-way collision $h(1294755) = h(52738737) = h(80115359)$, find p and q.

8.6 Prove that any algorithm for constructing a collision in the Gibson hash function can be converted into an algorithm for factoring biprimes. Does this proof assume a special form for the prime factors? Does this proof provide assurance that the Gibson hash function is secure?

8.7 Describe how to use any cryptographic hash function together with a Feistel network to construct a block cryptosystem comparable to DES. Explain why your encryption is invertible even though the hash function is a one-way function.

8.8 To provide additional security, it is common practice to store a hash of a password in the host system rather than store the password itself. Explain this practice. Why might the system extend the password with additional public information prior to hashing?

Notes for Chapter 8

The Elgamal signature scheme was described by Elgamal (1985a) and is a starting point for the development of later enhancements and for the eventual industry standards. The Elgamal signature scheme is rarely used in its original form, but it is widely used in modified form. In particular, a modified form of the Elgamal signature scheme is widely used as an authorized signature standard under the name Digital Signature Standard (DSS), and also as the Digital Standard Algorithm (DSA). The DSS uses a subgroup of $\boldsymbol{Z}_p^*$ of order q, where q is about 160 bits and the field size p is 512 to 1024 bits. Nyberg and Rueppel (1996) have described another variant. There is also a version of the Elgamal signature scheme based on elliptic curves, known as Elliptic Curve DSA (ECDSA).

The Merkle–Damgaard notion of iterative hash construction was published independently by Merkle (1990) and Damgaard (1990). Simon (1998) formally studied the use of one-way functions to construct hash functions. The secure cryptographic hash function based on the integer factoring problem was published by Gibson (1991). The secure cryptographic hash function based on the discrete-log problem was published by Chaum, van Heijst, and Pfitzmann (1992). Hash functions for use in security applications form a special class within the general study of hash functions because of the demands for security as well as simplicity. The MD4 hash function was designed for simplicity by Rivest (1991), and quickly followed by the improved MD5 and SHA. A modification by Rivest (1991) influenced further developments. The hash function known as the Secure Hash Algorithm (SHA) was made a standard by the National Bureau of Standards in 1993, based on the results of an open and public competition. It has this name because it is anecdotally secure, not because it is theoretically secure. Another open and public competition has led to a new and improved hash function.

The use of block ciphers for hash functions was studied by Preneel, Govaerts, and Vandewalle (1993). There have been many attacks on hash functions, but these are not often published in the refereed literature. Published samples of attacks are given in the work by Wang, Lai, Feng, Chen, and Yu (2005), and by Wang and Yu (2005). The herding attack is due to Kelsey and Kohno (2006).

9 Groups, rings, and fields

Several different algebraic systems will often share common structural properties. Accordingly, it is efficient and productive to gather together all algebraic systems with a similar structure, and to study them collectively as a single category. Any properties that follow directly from the common structure will hold for all algebraic systems with that same structure, and these properties can be developed and studied for a given category, and then applied to all algebraic systems in that same category.

The three most important algebraic categories are the category of *groups*, the category of *rings*, and the category of *fields*. We will discuss each of these in turn. Each of these algebraic categories plays a different but powerful role in the development of the subject of cryptography.

The most familiar example of a group is the set of integers $\boldsymbol{Z}$ under the usual operation of addition. The set of integers is a group with an infinite number of elements. An example of a group with a finite number of elements is the group of nonnegative integers smaller than n with addition modulo n, and denoted $\boldsymbol{Z}_n$.

The most familiar example of a ring is again the set of integers $\boldsymbol{Z}$, but now with two operations. These are the two usual operations of addition and multiplication of integers. Many other rings are also important. Much of the mathematical structure of the ring of integers is mimicked by a similar mathematical structure within the ring of univariate polynomials over a field $\boldsymbol{F}$. To enlarge our collection of cryptographic techniques, we will turn to the structure of polynomial rings, especially the ring of bivariate polynomials, under polynomial addition and polynomial multiplication.

The most familiar examples of a field are the set of real numbers and the set of rational numbers under the usual operations of addition and multiplication, now with the operation of division. The set of complex numbers is also a familiar example of a field. Much of the mathematical structure of the set of reals, the set of rationals, and the set of complex numbers is mimicked by the structure and behavior of other mathematical systems that we will study. All mathematical systems with this common algebraic structure are known as fields.

9.1 Groups

An algebraic system consisting of a set G and a single operation on pairs of elements of G is called a group whenever the operation has certain standard and required properties. The notion of a group has already been introduced in Section 2.1. The set of integers under addition is a familiar example of a group. Many properties of the set of integers under addition are properties that hold in any group G. These properties are best studied in an abstract setting using the language of groups. Then it will be clear that the consequences that follow directly from the defining properties of a group are valid in every group.

The definition is as follows. A *group* G is any set having an operation $*$ defined on pairs of elements of the set that satisfies four standard properties. These properties are the following. A group is closed under the operation $*$. Thus for any two elements a and b of G, possibly the same element, $a * b$ is an element of G. A group G has an element e, called the *identity element* and satisfying $a * e = e * a$ for every element a of the group G. Every element a of the group G has an *inverse*, denoted a^{-1}, with the property that $a * a^{-1} = a^{-1} * a = e$. Finally, the group operation is associative, meaning that $(a * b) * c = a * (b * c)$.

It is easy to show that, in every group, the identity element is unique. Also, the inverse of each element of the group is unique, and the inverse of the inverse of an element is the element itself. Thus $(a^{-1})^{-1} = a$.

Whenever the group operation is used to combine the same element with itself two or more times, an exponential notation can be used. Thus $a^2 = a * a$, and

$$a^{\ell} = a * a * \cdots * a,$$

where there are ℓ copies of a on the right. This can be written $a^{\ell} = a * a^{\ell-1}$.

A group may have a finite number of elements, in which case the number of elements is called the *order* of the group and denoted $\#G$. A group may also have an infinite number of elements.

In general, any subset of G that is a group under the same operation is called a *subgroup* of G. The smallest integer k, if there is one, such that $g^k = 1$ is called the *order* of the element g of the group G. For any element g of finite order, the *cycle* $\{g^1, g^2, \ldots, g^k\}$, where $g^k = 1$, is itself a group contained in G, sharing the same operation. It is called a *cyclic subgroup* of G of order k, and g is a *generator* of this cyclic subgroup. Any nonzero element g of a finite group G is the generator of a cyclic subgroup of G, so if G is a finite group, then every element of G has finite order. Thus the order of the cyclic subgroup is equal to the order of any element g that generates the cyclic subgroup. If the cycle of any element of G is equal to G, then G itself is called a *cyclic group*.

There are also general terms in use for algebraic systems that have some of the properties of a group, but not necessarily all of them. These terms are "semigroup" and "monoid." A *monoid* is an algebraic system with an operation $*$ that satisfies all properties of a group except for the property that every element has an inverse. A *semigroup* is an algebraic system with an operation $*$ that is closed under $*$ and satisfies the associative law, but a semigroup is not required to have an identity element and so need not have an inverse for every element.

The commutative property $a * b = b * a$ is not required to hold in a group. Therefore, it may be that $a * b \neq b * a$ for some or most elements of the group G. If, however, $a * b = b * a$ for every a and b in G, then the group G is called an *abelian group*. A cyclic subgroup is always abelian. It is often convenient to use the symbol $+$ in place of $*$ when the group is an abelian group. Then we may write

$$\ell a = a + a + \cdots + a,$$

where there are ℓ copies of a on the right. Moreover, we write

$$(\ell_1 + \ell_2)a = \ell_1 a + \ell_2 a.$$

For an example of a nonabelian group, the set of six matrices

$$\begin{bmatrix} 1 & 0 & 0 \\ 0 & 1 & 0 \\ 0 & 0 & 1 \end{bmatrix}, \begin{bmatrix} 1 & 0 & 0 \\ 0 & 0 & 1 \\ 0 & 1 & 0 \end{bmatrix}, \begin{bmatrix} 0 & 1 & 0 \\ 1 & 0 & 0 \\ 0 & 0 & 1 \end{bmatrix}, \begin{bmatrix} 0 & 1 & 0 \\ 0 & 0 & 1 \\ 1 & 0 & 0 \end{bmatrix}, \begin{bmatrix} 0 & 0 & 1 \\ 1 & 0 & 0 \\ 0 & 1 & 0 \end{bmatrix}, \begin{bmatrix} 0 & 0 & 1 \\ 0 & 1 & 0 \\ 1 & 0 & 0 \end{bmatrix},$$

forms a group under matrix multiplication, which is easy to verify. The group has order six, and by Lagrange's theorem, the order of each element divides six. It is a nonabelian group because, for example,

$$\begin{bmatrix} 0 & 1 & 0 \\ 0 & 0 & 1 \\ 1 & 0 & 0 \end{bmatrix} \begin{bmatrix} 0 & 0 & 1 \\ 0 & 1 & 0 \\ 1 & 0 & 0 \end{bmatrix} \neq \begin{bmatrix} 0 & 0 & 1 \\ 0 & 1 & 0 \\ 1 & 0 & 0 \end{bmatrix} \begin{bmatrix} 0 & 1 & 0 \\ 0 & 0 & 1 \\ 1 & 0 & 0 \end{bmatrix}.$$

This nonabelian group is known as the *permutation group* on three objects.

For an example of an abelian group, we can take $\mathbf{Z}_n$. This is the set $\{0, 1, \ldots, n-1\}$ under the operation of modulo n addition. The group operation in $\mathbf{Z}_n$ is conventionally denoted by the symbol $+$. This group is a cyclic group of order n under addition, and is generated by the element 1. It is clearly an abelian group because every cyclic group is abelian. Moreover, this group has a cyclic subgroup for each prime factor of n. In fact, every subgroup of $\mathbf{Z}_n$ is a cyclic subgroup whose order divides n.

A fundamental property of a finite group is stated by Lagrange's theorem, which was given earlier in Chapter 2, and is restated here.

Theorem 9.1.1 (Lagrange) *The order of any element β of a finite group divides the order of the group.*

Proof This theorem was proved in Chapter 2 as Theorem 2.1.1. □

The Lagrange theorem is closely related to the notion of a coset. Let H be a subgroup of G, and let h_i denote the elements of H indexed by i. The *coset* of any element g of the abelian group G is the set $\{g + h_i | h_i \in H\}$. The set of all cosets is denoted G/H and is referred to as a *quotient group*. This term is a reference to the fact that a subgroup H of G divides G into cosets. If G is not abelian, then the cosets of G are called *left cosets* or *right cosets* according to whether they are defined as $g * h_i$ or as $h_i * g$.

The *direct sum*, denoted $G \oplus G'$ or $G + G'$, of two abelian groups G and G' is the set of pairs $\{(a, b)\}$ with $a \in G$ and $b \in G'$, together with the group operation

$$(a, a') * (b, b') = (a * b, a' * b'),$$

where $*$ has three different meanings in this expression. On the left, $*$ denotes the group operation in $G \oplus G'$; while on the right, the first $*$ denotes the group operation in G and the second $*$ denotes the group operation in G'. The order of the direct sum $G \oplus G'$ is the product of the orders of the two summands. That is,

$$\#(G \oplus G') = (\#G)(\#G').$$

The fundamental theorem of finite abelian groups states that every finite abelian group can be expressed as the direct sum[1] of cyclic subgroups, each with an order equal to a prime power. A cyclic group of order n is isomorphic[2] to $\mathbf{Z}_n$. A cyclic group of order mn is isomorphic to the direct sum $\mathbf{Z}_n \oplus \mathbf{Z}_m$ if, and only if, m and n are coprime. Moreover, every finite abelian group satisfies the isomorphism

$$G \simeq \mathbf{Z}_{k_1} \oplus \mathbf{Z}_{k_2} \oplus \cdots \oplus \mathbf{Z}_{k_n}$$

where $k_1, \ldots, k_n$ are primes or powers of primes, not necessarily distinct. If m and n are not coprime, then $\mathbf{Z}_m \oplus \mathbf{Z}_n$ is not cyclic, so it is not isomorphic to $\mathbf{Z}_{mn}$.

For example, every abelian group of order eight is isomorphic to either $\mathbf{Z}_2 \oplus \mathbf{Z}_2 \oplus \mathbf{Z}_2$, or to $\mathbf{Z}_2 \oplus \mathbf{Z}_4$, or to $\mathbf{Z}_8$. Of these, only $\mathbf{Z}_8$ is cyclic, because neither $\mathbf{Z}_2 \oplus \mathbf{Z}_2 \oplus \mathbf{Z}_2$ nor $\mathbf{Z}_2 \oplus \mathbf{Z}_4$ has an element of order 8. On the other hand, every abelian group of order fifteen is isomorphic to $\mathbf{Z}_3 \oplus \mathbf{Z}_5$ and is cyclic because three and five are coprime.

Definition 9.1.2 *A torsion group is a group in which every element has a finite order.*

A torsion group need not be a finite group, but a finite group is always a torsion group. An element of G whose order divides m is called an m-torsion point of G. The set of all m-torsion points of G is denoted $G[m]$. The set $G[m]$ itself is a group, but it

[1] For arbitrary groups, not necessarily abelian, this concept is called a *direct product* and is denoted $G = G_1 \times G_2$. The notation $\oplus$ is often preferred to $\times$ for abelian groups.

[2] Two groups are *isomorphic* if they have identical structure but are represented differently. An invertible map from the first group onto the second group that respects the group structure is called an *isomorphism*.

need not be a cyclic group. For example, if $G = \mathbf{Z}_3 \oplus \mathbf{Z}_3 \oplus \mathbf{Z}_5$, then $G[3] = \mathbf{Z}_3 \oplus \mathbf{Z}_3$, which is not cyclic.

9.2 Rings

A ring R is an algebraic system with two operations $+$ and $\times$, called addition and multiplication, that satisfy certain required properties. These are the properties that are so familiar in the exemplar ring, the ring of integers $\mathbf{Z}$. The first property required of a ring is that the elements of R together with the operation that is denoted $+$ must form an abelian group. The operation $+$ is called *addition* in the ring. The identity element under addition is called *zero*, and is denoted by 0. The second property required of a ring is that R must be closed under the operation $\times$. The operation $\times$ is called *multiplication* in the ring. The multiplication symbol is occasionally omitted, and indicated by simple juxtaposition of two elements. The next property that a ring must satisfy is *associativity* of multiplication, meaning that $(ab)c = a(bc)$. The final two properties that a ring must satisfy are the two *distributivity* properties $(a + b)c = ac + bc$ and $c(a + b) = ca + cb$. The distributivity properties create an interaction between the operation of addition and the operation of multiplication, which leads to many important consequences.

The addition operation in a ring is required to be commutative. That is, $a + b = b + a$ for all a and b in R because the ring is an abelian group under addition. However, the multiplication operation is not required to be commutative. That is, there may be elements a and b in R such that $ab \neq ba$. A ring in which the multiplication operation satisfies $ab = ba$ for all a and b is called a *commutative ring*.

Every ring must have an identity element under addition because the ring is a group under addition. However, a ring need not have an identity element under multiplication. If the ring does have an identity element under multiplication, it is called a *ring with identity*, a *ring with unit*, or a *unital ring*. The identity element may be called *one*, but in some rings it is referred to only as the identity element.

If a ring has an identity under multiplication, then the notion of the *inverse* of an element can be defined. A nonzero element a of the unital ring R may have an inverse from the right under multiplication. This *right inverse* is an element b of the group such that $ab = 1$. A nonzero element a of the unital ring R may have an inverse from the left under multiplication. This *left inverse* is an element b of the group such that $ba = 1$. If an element a has both a left inverse and a right inverse, they must be the same element. This is proved by first writing $ab = 1$ and $ca = 1$ where b and c are the right inverse and the left inverse, respectively, of a. Then $b = cab = c$, so $b = c$. This element is then called, simply, the *inverse* of a, and is denoted a^{-1}. A ring need not have an inverse for every element. In a commutative ring, the left inverse of an element must also be a right inverse. A nonzero element with an inverse is called a *unit* of the ring. Not every nonzero element need be a unit.

In any ring R, the zero element satisfies $0a = a0 = 0$ for all $a \in R$. This is a consequence of the distributive law and the other axioms of a ring.

Some important examples of rings are the ring of integers $\boldsymbol{Z}$, and the ring of integers modulo n $\boldsymbol{Z}_n$. Other examples that are important to us are $\boldsymbol{F}[x]$ and $\boldsymbol{F}(x)$, the ring of polynomials over the field $\boldsymbol{F}$ and the ring of rational functions over the field $\boldsymbol{F}$. The ring $\boldsymbol{F}[x]$ consists of all polynomials with coefficients in $\boldsymbol{F}$, with the usual notions of addition and multiplication of polynomials. The ring $\boldsymbol{F}(x)$ consists of all rational functions over $\boldsymbol{F}$ with the usual notions of addition and multiplication of rational functions. A *rational function* (or univariate rational function) is an expression of the form $f(x)/g(x)$ where $f(x)$ and $g(x)$ are polynomials with coefficients in $\boldsymbol{F}$. Division of a rational function by $f(x)/g(x)$ is defined as multiplication by $g(x)/f(x)$ provided $f(x)$ is nonzero. A monic rational function is a rational function in which the leading coefficients of $f(x)$ and $g(x)$ are both equal to one. We will also make use of $\boldsymbol{F}[x, y]$ and $\boldsymbol{F}(x, y)$, the rings of bivariate polynomials and bivariate rational functions, respectively.

Any ring R is a set of elements and so has subsets, some of which have a special form. A *subring* of R is defined in the obvious way. Other special subsets, more general than a subring, play an important role. An *ideal* I is any subset of the ring R that satisfies two properties. The first property is that the ideal I is closed under addition. The second property is that any product of an element of I with an element of R is an element of I. A *proper ideal* of R is one that is not equal to the ring itself. A *prime ideal* of any commutative ring R is a proper ideal I of R for which, whenever the product ab of two elements of R is in I, then a and b are each in I as well. A prime ideal in $\boldsymbol{Z}$ is the set $\{\ell p\}$ of all integer multiples of a prime p. Thus a prime ideal is a generalization of a prime of $\boldsymbol{Z}$ to a general commutative ring.

For example, in the ring of integers $\boldsymbol{Z}$, the set of all even integers is a proper ideal because the sum of two even integers is always even, and the product of an even integer with any other integer is always an even integer. It is a prime ideal. Likewise, the set of all multiples of three is also a proper ideal, and indeed, a prime ideal. The set of multiples of six is a proper ideal, but not a prime ideal.

9.3 Fields

In any commutative ring, as in the ring of integers, we can add, subtract, and multiply, and these operations are always required to behave similarly. In a ring with identity, it is possible to divide by any element that is a unit. This is because an element a is called a unit whenever there is an element b for which $ab = 1$. To divide by a, simply multiply by that b. If every nonzero element is a unit, then it is possible to divide by any nonzero element. A ring with this property belongs to a special category. The general term denoting members of this category is *field*.

The set of rational numbers $\boldsymbol{Q}$, the set of real numbers $\boldsymbol{R}$, and the set of complex numbers $\boldsymbol{C}$ are each familiar examples of a field. Each is a number system in which one can add, multiply, subtract, and divide in a way for which all the usual properties of algebra apply. This means that the properties of commutativity, associativity, and distributivity all hold in the rational field $\boldsymbol{Q}$, in the real field $\boldsymbol{R}$, and in the complex field $\boldsymbol{C}$. There are many other algebraic systems that share these properties. For efficiency of understanding, we collect all of these algebraic systems together so that we can study their common properties simultaneously.

Definition 9.3.1 *A field $\boldsymbol{F}$ is a set containing at least two elements that has two operations on pairs of elements, called addition and multiplication, such that the following properties hold:*

(1) The set $\boldsymbol{F}$ is an abelian group under addition.
(2) The set $\boldsymbol{F}$ is closed under multiplication, and the set of nonzero field elements is an abelian group under multiplication.
(3) The distributive law

$$(a + b)c = ac + bc$$

holds for all a, b, and c in the set $\boldsymbol{F}$.

In particular, most of the familiar rules of linear algebra hold in any field. A subset of a field that itself is a field under the inherited operations of addition and multiplication is called a *subfield*. The real field is a subfield of the complex field.

We will be especially interested in fields with a finite number of elements. These fields are called *finite fields* or *Galois fields* and denoted $\boldsymbol{F}_q$ or $GF(q)$, where q is the number of elements of the finite field. We will be particularly interested in finite fields with p elements where p is an odd prime, and in finite fields with 2^m elements. These are called *prime fields* and *binary fields*, respectively. We will also be interested in finite fields with p^m elements where p is a prime. A finite field with p^m elements is called a field of *characteristic* p. No finite field exists with a number of elements q that is neither a prime nor a prime power.

The set of nonzero elements of $\boldsymbol{F}_q$ is denoted $\boldsymbol{F}_q^*$. Every element of $\boldsymbol{F}_q^*$ has an inverse. Thus $\boldsymbol{F}_q^*$ is a finite group under the operation of multiplication. Moreover, as will be proved in Section 9.8, $\boldsymbol{F}_q^*$ is a cyclic group for every q. This means that $\boldsymbol{F}_q^*$ can be generated by a single element.

Definition 9.3.2 *A primitive element of the finite field $\boldsymbol{F}_q$ is an element such that every nonzero field element can be expressed as a power of that element.*

The proof that every finite field has a primitive element is the topic of Section 9.8. For an example of a primitive element, note that the powers of 3 in the field $\boldsymbol{F}_7$ are: $3^1 = 3$, $3^2 = 2$, $3^3 = 6$, $3^4 = 4$, $3^5 = 5$, and $3^6 = 1$. Because all six nonzero elements of $\boldsymbol{F}_7$ appear, the element 3 is a primitive element. The element 2, however, is not primitive.

The following theorem is useful in a finite field with p^m elements.

Theorem 9.3.3 *In a finite field of characteristic p, the expression*

$$(a+b)^{p^m} = a^{p^m} + b^{p^m}$$

always holds.

Proof It is enough to first prove

$$(a+b)^p = a^p + b^p.$$

Recall that the binomial theorem gives

$$(a+b)^p = a^p + \sum_{\ell=1}^{p-1} \binom{p}{\ell} a^{p-\ell} b^\ell + b^p \pmod{p},$$

noting that addition is modulo p. But

$$\binom{p}{\ell} = \frac{p!}{(p-\ell)!\ell!}$$

which, because p is a prime, is a multiple of p unless $\ell = p$ or $\ell = 1$. Therefore $\binom{p}{\ell} \pmod{p} = 0$ for $\ell = 1, \ldots, p-1$, so

$$(a+b)^p = a^p + b^p$$

in $\boldsymbol{F}_{p^m}$.

This conclusion can be applied twice to give

$$(a+b)^{p^2} = ((a+b)^p)^p = (a^p + b^p)^p = a^{p^2} + b^{p^2}.$$

In turn,

$$(a+b)^{p^m} = (a^{p^{m-1}} + a^{p^{m-1}})^p = a^{p^m} + b^{p^m}$$

as was to be proved. □

9.4 Prime fields

An important example of a finite field is a field with p elements where p is a prime. First, recall that the ring of integers modulo n, denoted $\boldsymbol{Z}_n$, is a commutative ring with identity, and satisfies all of the properties required of a field except, possibly, the requirement that every nonzero element has an inverse under multiplication. Thus, to show that $\boldsymbol{Z}_n$ is a field if n is a prime p, it is enough to show that every nonzero element of $\boldsymbol{Z}_n$ has an inverse under multiplication if n is a prime p.

Theorem 9.4.1 *The ring of integers modulo n is a field if, and only if, n is a prime.*

Proof The proof follows from the extended euclidean algorithm. If n is the prime p, then every element $a \in \boldsymbol{Z}_p$ is coprime with p. Therefore for the coprime pair a and p, by a corollary to the extended euclidean algorithm, given as Corollary 2.2.3, there exist integers A and P such that

$$Aa + Pp = 1.$$

Therefore $Aa = 1 \pmod{p}$, so $a^{-1} = A \pmod{p}$, which is an element of $\boldsymbol{Z}_p$. This means that a does have an inverse a^{-1} under multiplication in the ring $\boldsymbol{Z}_p$. The statement is true for every nonzero element of $\boldsymbol{Z}_p$. Thus the ring $\boldsymbol{Z}_p$ is a field.

On the other hand, if n is composite, say $n = ab$, then $ab = 0$ modulo n even though a and b are both nonzero and smaller than n. If a has an inverse a^{-1}, then $b = 0$, which is a contradiction. Consequently, $\boldsymbol{Z}_n$ is not a field if n is composite. □

Because, for any prime p, $\boldsymbol{Z}_p$ is a field, it will usually be referred to henceforth by the label $\boldsymbol{F}_p$, although we may occasionally continue to use the label $\boldsymbol{Z}_p$ when only the addition properties are under discussion. The field $\boldsymbol{F}_p$ is the simplest example of a field of *characteristic* p, which is the term used to refer to any field in which the element one has order p under addition.

In a prime field $\boldsymbol{F}_p$, Fermat's little theorem becomes the statement that the polynomial $x^{p-1} - 1$ has $p - 1$ zeros. This implies that the factorization

$$x^p - x = x \prod_{\ell=1}^{p-1} (x - \ell)$$

holds in the field $\boldsymbol{F}_p$. But this is getting ahead of the story.

9.5 Binary fields and ternary fields

Every field has an identity under multiplication, denoted 1 and called one. Any field $\boldsymbol{F}$ for which $1 + 1 = 0$ is called a *binary field*,[3] or a field of *characteristic two*. Any field for which $1 + 1 + 1 = 0$ is called a *ternary field*, or a field of *characteristic three*. The

[3] With little risk of confusion, we call $\boldsymbol{F}_2$ *the* binary field and $\boldsymbol{F}_{2^m}$ *a* binary field.

Table 9.1 *Arithmetic tables for F_2 and F_3*

	+	0	1		×	0	1	
F_2 :	0	0	1		0	0	0	
	1	1	0		1	0	1	
	+	0	1	2	×	0	1	2
	0	0	1	2	0	0	0	0
F_3 :	1	1	2	0	1	0	1	2
	2	2	0	1	2	0	2	1

simplest binary field is $\boldsymbol{F}_2$. The simplest ternary field is $\boldsymbol{F}_3$. These fields have addition and multiplication tables, as given in Table 9.1. Because 2 and 3 are primes, $\boldsymbol{Z}_2$ and $\boldsymbol{Z}_3$ are the fields $\boldsymbol{F}_2$ and $\boldsymbol{F}_3$, as was discussed in Section 9.4. However, the rings $\boldsymbol{Z}_{2^m}$ and $\boldsymbol{Z}_{3^m}$ are not fields if m is larger than one. For example, in the ring $\boldsymbol{Z}_4$, $2 \cdot 2 = 0$. This violates the axioms of a field.

The field $\boldsymbol{F}_{2^m}$ and $\boldsymbol{F}_{3^m}$ do exist for any value of the integer m, and each is unique but for notation. These fields must be defined in a different way, as by using the general method of construction of extension fields to be given in Section 9.7. Anticipating this method, we construct $\boldsymbol{F}_{2^m}$ as follows. First select an irreducible polynomial over $\boldsymbol{F}_2$, denoted $p(x)$. Then define $\boldsymbol{F}_{2^m}$ as the set of all polynomials over $\boldsymbol{F}_2$ whose degree is at most $m - 1$. There are 2^m such polynomials, so $\boldsymbol{F}_{2^m}$ has 2^m elements. Define addition in $\boldsymbol{F}_{2^m}$ as addition of polynomials. Define multiplication in $\boldsymbol{F}_{2^m}$ as multiplication of polynomials modulo the irreducible polynomial $p(x)$.

Similarly, to construct $\boldsymbol{F}_{3^m}$, let $p(x)$ be an irreducible polynomial over $\boldsymbol{F}_3$ of degree m. Then define $\boldsymbol{F}_{3^m}$ as the set of all polynomials over $\boldsymbol{F}_3$ whose degree is at most $m - 1$. There are 3^m such polynomials in $\boldsymbol{F}_{3^m}$, so $\boldsymbol{F}_{3^m}$ has 3^m elements. Define addition in $\boldsymbol{F}_{3^m}$ as addition of polynomials. Define multiplication in $\boldsymbol{F}_{3^m}$ as multiplication of polynomials modulo the irreducible polynomial $p(x)$.

9.6 Univariate polynomials

A *univariate polynomial*, or *polynomial*, over the field $\boldsymbol{F}$ is a mathematical expression of the form $p(x) = p_n x^n + p_{n-1} x^{n-1} + \cdots + p_1 x + p_0$, where $p_n, p_{n-1}, \ldots, p_1, p_0$ are elements of the field $\boldsymbol{F}$, called *coefficients* of the polynomial and *scalars* of the field, and x is a formal symbol called an *indeterminate*. The subscript on a coefficient is called the *index* of the coefficient. The expressions x^j are called *univariate monomials* of the polynomial and the expressions $p_j x^j$ are called *terms*. The *zero polynomial* is the polynomial with all coefficients equal to zero. The *leading index* of the polynomial $p(x)$ is the largest value of j for which the coefficient p_j is nonzero. The *leading coefficient* of the polynomial $p(x)$ is that coefficient p_j for which j is the leading index. It must be finite. The *degree* of the polynomial $p(x)$, denoted deg $p(x)$, is the integer value of the leading index. By convention, the degree of the zero polynomial is $-\infty$. A *monic polynomial* is a polynomial whose leading coefficient is the field element one. The *formal derivative* of the polynomial $p(x) = p_n x^n + p_{n-1}x^{n-1} + \cdots + p_1 x + p_0$ is the polynomial $p'(x) = np_n x^{n-1} + (n-1)p_{n-1}x^{n-2} + \cdots + p_1$, where ip_i means the sum of i copies of p_i.

Polynomials over the field $\boldsymbol{F}$ can be added or multiplied in the usual way. Two polynomials over $\boldsymbol{F}$ can be added in the usual way of polynomial addition by adding the coefficients with the same index. Thus

$$p(x) + q(x) = \sum_i (p_i + q_i)x^i.$$

This operation of polynomial addition satisfies all of the axioms of a group. The polynomial with all coefficients equal to zero is the identity element of the set of polynomials regarded as a group. Likewise, two polynomials over the field $\boldsymbol{F}$ can be multiplied in the usual way of polynomial multiplication, given by

$$p(x)q(x) = \sum_i \sum_j p_i q_{j-i} x^j.$$

The set of all polynomials over $\boldsymbol{F}$, together with the operations of polynomial addition and polynomial multiplication, is a commutative ring with identity, called the *polynomial ring*, or the ring of univariate polynomials. The polynomial ring over $\boldsymbol{F}$ is denoted $\boldsymbol{F}[x]$.

Indeed, at the level of the formal algebraic manipulations, as determined by the properties of a commutative ring with identity, the ring $\boldsymbol{F}[x]$ and the ring $\boldsymbol{Z}$ behave similarly in many ways. We can, and will, develop many of the same notions and theorems within the ring of polynomials $\boldsymbol{F}[x]$ that were developed earlier in the ring of integers $\boldsymbol{Z}$.

The operation of polynomial multiplication, in general, does not have an inverse. In some instances, however, as when $c(x) = a(x)b(x)$, multiplication does have an inverse, called *division*, and written as $a(x) = c(x)/b(x)$ or $b(x) = c(x)/a(x)$. The polynomials $a(x)$ and $b(x)$ are called *factors* of $c(x)$. A *composite polynomial* $c(x)$ over $\boldsymbol{F}$ is a polynomial that can be written as $c(x) = a(x)b(x)$ where $a(x)$ and $b(x)$ are in $\boldsymbol{F}[x]$, but neither is an element of $\boldsymbol{F}$. An *irreducible polynomial* is one that is not composite. A *prime polynomial* is a polynomial of degree at least one that is both an irreducible polynomial and a monic polynomial. A field in which only polynomials of degree one are irreducible is called an *algebraically closed field*. The complex field $\boldsymbol{C}$ is algebraically closed. The real field $\boldsymbol{R}$ is not. The rational field $\boldsymbol{Q}$ is not. Every field is contained in an algebraically closed field.

Just as the ring of integers $\boldsymbol{Z}$ has a division algorithm, so too, each ring of polynomials $\boldsymbol{F}[x]$ has a division algorithm. Given any two polynomials $a(x)$ and $b(x)$ of the ring

$F[x]$, there exist two polynomials $Q(x)$ and $r(x)$, called the *quotient polynomial* and the *remainder polynomial*, such that

$$a(x) = Q(x)b(x) + r(x),$$

and $\deg r(x) < \deg b(x)$. Elementary algorithms for computing the quotient polynomial and the remainder polynomial are well known. These hold in any field.

Because $F[x]$ is a ring, it has ideals, just as does any commonplace ring.[4] Recall that an ideal I must be closed under addition and the ideal I must contain $a(x)b(x)$ whenever $b(x) \in F[x]$ and $a(x) \in I$. Let $g(x)$ be a nonzero monic polynomial of least degree in the ideal I. We can take $g(x)$ to be a monic polynomial because, whenever $g(x)$ is contained in I, all scalar multiples of $g(x)$ are contained in I. Then for any polynomial $a(x)$ contained in I, the division algorithm states that

$$a(x) = Q(x)g(x) + r(x),$$

where $\deg r(x) < \deg g(x)$. But $r(x)$ is in I and has a smaller degree than $g(x)$. Therefore $r(x)$ is the zero polynomial and $a(x)$ is a multiple of $g(x)$. Because $a(x)$ is arbitrary, we conclude that every polynomial of the ideal I of $F[x]$ is a multiple of $g(x)$. For this reason, $g(x)$ is called the *generator* of the ideal I.

In general, an ideal of $F[x]$ might be defined as a linear combination of several polynomials. Thus let $g_1(x)$ and $g_2(x)$ be any two fixed polynomials, and define

$$I = \{a_1(x)g_1(x) + a_2(x)g_2(x) \mid a_1(x), a_2(x) \in F[x]\}.$$

It is clear that the set I is an ideal because the sum of two elements of I is an element of I and the multiple of any element of I by any $a(x)$ is an element of I. But we have already observed that every element of I is a multiple of a generator $g(x)$. This means that there is a more succinct way of describing this ideal as

$$I = \{a(x)g(x) | a(x) \in F[x]\}.$$

Because $g_1(x)$ and $g_2(x)$ are themselves in I, they also must be multiples of $g(x)$, and so can be written $g_1(x) = b_1(x)g(x)$ and $g_2(x) = b_2(x)g(x)$. We can conclude that if $g_1(x)$ and $g_2(x)$ are coprime, then the only $g(x)$ that divides both of them is $g(x) = 1$. Because I then contains every polynomial multiple of $g(x)$, we conclude that $I = F[x]$ if $g_1(x)$ and $g_2(x)$ are coprime. If $g_1(x)$ and $g_2(x)$ are not coprime, then the greatest common divisor of $g_1(x)$ and $g_2(x)$ is the generator $g(x)$ of the ideal.

Given the ring $F[x]$ and a fixed polynomial $p(x)$ of $F[x]$, one can apply the division algorithm to every element of $F[x]$ to obtain the set of all remainders under division by $p(x)$. All elements of $F[x]$ with the same remainder belong to a subset of $F[x]$ called an *equivalence class* because these elements are equal modulo $p(x)$. The set of all such equivalence classes is called a *quotient ring* and denoted $F[x]/\langle p(x)\rangle$. The

[4] A field has no proper ideals.

quotient ring $F[x]/\langle p(x)\rangle$ is itself a ring under the natural definitions of addition and multiplication. The elements of $F[x]/\langle p(x)\rangle$, as subsets of $F[x]$, are sets of polynomials, the polynomials of each set having the same remainder $r(x)$. The remainder itself is a polynomial in that equivalence class. The canonical representative of an equivalence class is the remainder polynomial associated with that equivalence class. For computations, it is appropriate to regard the elements of the quotient ring simply to be those remainder polynomials. Then the quotient ring, more simply, is the set of remainders. Sometimes this view is emphasized by writing $F[x]_{p(x)}$ to denote the quotient ring regarded as a set of canonical representations.

If $a(x) = Q(x)b(x)$, then the remainder $r(x)$ under division by $b(x)$ is zero, and $b(x)$ is said to *divide* $a(x)$ or to be a *factor* of $a(x)$. The *greatest common divisor*, denoted $\text{GCD}[r(x), s(x)]$, of the two polynomials $r(x)$ and $s(x)$ is the monic polynomial of largest degree that divides both of them. Two polynomials $r(x)$ and $s(x)$ over the field F are called *coprime* if their greatest common divisor equals one. The *least common multiple*, denoted $\text{LCM}[r(x), s(x)]$, of the two polynomials $r(x)$ and $s(x)$ is the monic polynomial of smallest degree that both of them divide. A prime polynomial is a polynomial that is both monic and irreducible over the field F. Just as any natural integer can be factored uniquely into a product of primes and prime powers, so, too, any monic univariate polynomial over a field F can be factored uniquely into a product of prime polynomials and powers of prime polynomials as asserted by the following theorem.

Theorem 9.6.1 (Unique factorization for polynomials) *Every monic univariate polynomial $s(x)$ over the field F can be written uniquely (up to the order of the factors) as*

$$s(x) = p_1(x)^{e_1} p_2(x)^{e_2} \cdots p_k(x)^{e_k}$$

where the $p_i(x)$ are prime polynomials and the e_i are positive integers.

Proof Suppose that the theorem is false. Let $p(x)$ be a monic polynomial of the least degree for which the theorem fails. Then $p(x)$ has two different factorizations:

$$p(x) = a_1(x)a_2(x) \cdots a_k(x) = b_1(x)b_2(x) \cdots b_j(x),$$

where the $a_k(x)$ and $b_j(x)$ are prime polynomials, possibly with some repeated.

All of the $a_k(x)$ must be different from all of the $b_j(x)$ because, otherwise, the common terms could be canceled to give a polynomial of smaller degree that can be factored in two different ways.

Without loss of generality, suppose that the degree of $b_1(x)$ is not larger than the degree of $a_1(x)$. Then

$$a_1(x) = b_1(x)Q(x) + s(x),$$

where $\deg s(x) < \deg b_1(x) \leq \deg a_1(x)$. Then

$$s(x)a_2(x)a_3(x)\cdots a_k(x) = b_1(x)[b_2(x)\cdots b_j(x) - Q(x)a_2(x)\cdots a_k(x)].$$

Factor both $s(x)$ and the bracketed term on the right into their prime polynomial factors and, if necessary, divide by a field element to make all factors monic. Because $b_1(x)$ does not appear on the left side, we have two different factorizations of a monic polynomial whose degree is smaller than the degree of $p(x)$, contrary to the choice of $p(x)$. The contradiction proves the theorem. □

The unique factorization theorem makes it clear that for any polynomials $s(x)$ and $t(x)$, both GCD$[s(x), t(x)]$ and LCM$[s(x), t(x)]$ are unique because the greatest common divisor is the product of all prime factors common to both $s(x)$ and $t(x)$, each factor raised to the smallest power with which it appears in either $s(x)$ or $t(x)$, and because the least common multiple is the product of all prime factors that appear in either $s(x)$ or $t(x)$, each factor raised to the largest power that appears in either $s(x)$ or $t(x)$.

A polynomial over the field $\boldsymbol{F}$ can be *evaluated* at an element β of $\boldsymbol{F}$. This is done by substituting β for the indeterminate x in the statement of the polynomial and executing the indicated operations. Thus

$$p(\beta) = p_n\beta^n + p_{n-1}\beta^{n-1} + \cdots + p_1\beta + p_0.$$

If $p(\beta) = 0$, then β is called a *zero* of the univariate polynomial $p(x)$, or a *root* of the equation $p(x) = 0$. A polynomial over $\boldsymbol{F}$ does not necessarily have a zero in $\boldsymbol{F}$. The polynomial $x^2 + x + 1$ over $\boldsymbol{F}_2$ has no zeros in $\boldsymbol{F}_2$.

Theorem 9.6.2 *The polynomial $p(x)$ has the field element β as a zero if, and only if, $x - \beta$ is a factor of $p(x)$.*

Proof Suppose that β is a zero of $p(x)$. The division algorithm states that

$$p(x) = (x - \beta)Q(x) + s(x),$$

where the degree of $s(x)$ is less than the degree of $x - \beta$. This means that $s(x)$ is a field element s_0, so

$$p(x) = (x - \beta)Q(x) + s_0.$$

But then

$$p(\beta) = (\beta - \beta)Q(\beta) + s_0 = 0,$$

so $s_0 = 0$. This means that

$$p(x) = (x - \beta)Q(x)$$

as was to be proved. □

Theorem 9.6.3 (Fundamental theorem of algebra) *A polynomial $p(x)$ of degree n over $\boldsymbol{F}$ has, at most, n zeros in $\boldsymbol{F}$ or any extension of $\boldsymbol{F}$.*

Proof Factor $p(x)$ into its irreducible factors which factorization is unique by the unique factorization theorem. The degree of $p(x)$ equals the sum of the degrees of its irreducible factors and there is an irreducible factor of degree one for each zero. □

The polynomial ring $\boldsymbol{F}[x]$ has a euclidean algorithm[5] that is similar to the euclidean algorithm for integers. Its structure rests on the division algorithm.

Theorem 9.6.4 (Euclidean algorithm for polynomials) *The greatest common divisor of two distinct polynomials $r(x)$ and $s(x)$ over the field $\boldsymbol{F}$ is a scalar multiple of the final nonzero remainder $r_n(x)$ computed by the iteration*

$$\begin{aligned}
s(x) &= Q_1(x)r(x) + r_1(x)\\
r(x) &= Q_2(x)r_1(x) + r_2(x)\\
r_1(x) &= Q_3(x)r_2(x) + r_3(x)\\
&\vdots\\
r_{n-2}(x) &= Q_n(x)r_{n-1}(x) + r_n(x)\\
r_{n-1}(x) &= Q_{n+1}(x)r_n(x).
\end{aligned}$$

Proof We will show that $r_n(x)$ divides GCD$[r(x), s(x)]$ and that GCD$[r(x), s(x)]$ divides $r_n(x)$. This means that $r_n(x)$ must be a scalar multiple of GCD$[r(x), s(x)]$. In particular, reading the list of equations from the bottom to the top, the last equation shows that $r_n(x)$ divides $r_{n-1}(x)$. Then the next to last equation shows that $r_n(x)$ must divide $r_{n-2}(x)$ because it divides both $r_{n-1}(x)$ and $r_n(x)$. Continuing in this way to proceed up the list of equations shows that $r_n(x)$ divides both $r(x)$ and $s(x)$. Therefore $r_n(x)$ divides GCD$[r(x), s(x)]$.

Next, reading the list of equations from top to bottom, because GCD$[r(x), s(x)]$ divides both $r(x)$ and $s(x)$, the first equation implies that GCD$[r(x), s(x)]$ also divides $r_1(x)$. But now the second equation implies that GCD$[r(x), s(x)]$ divides $r_2(x)$. Continue in this way until the second to last equation implies that GCD$[r(x), s(x)]$ divides $r_n(x)$.

Because $r_n(x)$ both divides GCD$[r(x), s(x)]$ and GCD$[r(x), s(x)]$ is divisible by $r_n(x)$, the two must be the same, but for a scalar multiple. Thus the theorem is proved. □

[5] The euclidean algorithm can be used in any euclidean domain. A *euclidean domain* is an algebraic category in which a norm is defined, which in turn means that a division algorithm can be defined. This leads to the notion of a factor and the subsequent notions of "smallness" or "smoothness."

The euclidean algorithm for polynomials has the following powerful consequence. It closely follows the similar statement accompanying the euclidean algorithm for integers.

Corollary 9.6.5 (Extended euclidean algorithm for polynomials) *For any two distinct polynomials $r(x)$ and $s(x)$ over the field $\boldsymbol{F}$, there exist two polynomials $R(x)$ and $S(x)$ over $\boldsymbol{F}$ such that*

$$R(x)r(x) + S(x)s(x) = \text{GCD}[r(x), s(x)].$$

Proof The last remainder polynomial $r_n(x)$ in the statement of the theorem satisfies the expression

$$r_n(x) = r_{n-2}(x) - Q_n(x)r_{n-1}(x),$$

which gives $r_n(x)$ in terms of $r_{n-1}(x)$ and $r_{n-2}(x)$. In the same way, $r_{n-1}(x)$ can be expressed in terms of $r_{n-2}(x)$ and $r_{n-3}(x)$. Then, by back-substitution, all remainders can be eliminated except $r_n(x)$. This gives $r_n(x)$ as a linear combination of $r(x)$ and $s(x)$ with polynomial coefficients. Because $r_n(x)$ is a scalar multiple of GCD$[r(x), s(x)]$, the proof is complete. □

The equation in this corollary is called the *Bézout identity* and the polynomials $R(x)$ and $S(x)$ are called *Bézout polynomials.*

Corollary 9.6.6 *For any two coprime polynomials $r(x)$ and $s(x)$ over the field $\boldsymbol{F}$, there exist two polynomials $R(x)$ and $S(x)$ over $\boldsymbol{F}$ such that*

$$R(x)r(x) + S(x)s(x) = 1.$$

Proof This follows by setting GCD$[r(x), s(x)] = 1$ in Corollary 9.6.5. □

Just as it is in the integer ring $\boldsymbol{Z}$, there is a chinese remainder theorem in the polynomial ring $\boldsymbol{F}[x]$, and it is developed in the same way.

Theorem 9.6.7 (Chinese remainder theorem) *Given a set of pairwise coprime polynomials $m^{(0)}(x), m^{(1)}(x), \ldots, m^{(k)}(x)$, the system of congruences*

$$c^{(i)}(x) = c(x) \pmod{m^{(i)}(x)}$$

has exactly one solution satisfying

$$\deg c(x) < \sum_{i=0}^{k} \deg m^{(i)}(x).$$

Proof Suppose that $c(x)$ and $c'(x)$ both satisfy all the stated congruences. Then

$$c(x) = Q^{(i)}(x)m^{(i)}(x) + c^{(i)}(x)$$
$$c'(x) = Q'^{(i)}(x)m^{(i)}(x) + c^{(i)}(x).$$

Therefore, $c(x) - c'(x)$ is a multiple of $m^{(i)}(x)$ for each i. Because the $m^{(i)}(x)$ are pairwise coprime, this means that $c(x) - c'(x)$ is a multiple of the product of the $m^{(i)}(x)$. But the degree of $c(x) - c'(x)$ is less than the sum of the degrees of the $m^{(i)}(x)$. Therefore, $c(x) - c'(x) = 0$. We conclude that $c(x) = c'(x)$. □

This system of congruences can be solved in a way that is similar to the case of the integer ring. The method is based on the fact given in Corollary 9.6.6, that in any ring of polynomials over a field, for any given $a(x)$ and $b(x)$, there exist polynomials $A(x)$ and $B(x)$ such that

$$A(x)a(x) + B(x)b(x) = \text{GCD}[a(x), b(x)]$$

as used in the following theorem.

Theorem 9.6.8 *Let $M(x) = \Pi_{r=0}^{k} m^{(r)}(x)$ be a product of pairwise coprime polynomials and let $M^{(i)}(x) = M(x)/m^{(i)}(x)$. Then the system of congruences*

$$c_i(x) = c(x) \pmod{m^{(i)}(x)} \qquad i = 0, \ldots, k$$

is uniquely solved by

$$c(x) = \sum_{i=0}^{k} c^{(i)}(x)N^{(i)}(x)M^{(i)}(x) \pmod{M(x)},$$

where $N^{(i)}(x)$ satisfies

$$N^{(i)}(x)M^{(i)}(x) + n^{(i)}(x)m^{(i)}(x) = 1.$$

Proof We must show that the stated $c(x)$ satisfies every congruence in the given system of congruences. But

$$c(x) = c^{(i)}(x)N^{(i)}(x)M^{(i)}(x) \pmod{m^{(i)}(x)}$$

because $M^{(r)}(x)$ has $m^{(i)}(x)$ as a factor if $r \neq i$. Then, as asserted by Corollary 9.6.6,

$$N^{(i)}(x)M^{(i)}(x) + n^{(i)}(x)m^{(i)}(x) = 1,$$

so we have

$$N^{(i)}(x)M^{(i)}(x) = 1 \pmod{m^{(i)}(x)}.$$

Therefore for each i,

$$c(x) = c^{(i)}(x) \pmod{m^{(i)}(x)}$$

which concludes the proof of the theorem. □

9.7 Extension fields

There is a simple construction that, in general, will *extend*, or enlarge, a field to a larger field. For example, the binary field $\boldsymbol{F}_2$ can be extended to the field $\boldsymbol{F}_{2^m}$, which accordingly is called an *extension field* of $\boldsymbol{F}_2$. This same simple construction will extend the ternary field $\boldsymbol{F}_3$ to the field $\boldsymbol{F}_{3^m}$. Indeed, the construction will extend the field $\boldsymbol{F}_q$ to the field $\boldsymbol{F}_{q^m}$, where q is a prime or a prime power. The construction is similar to the familiar construction that extends the real field $\boldsymbol{R}$ to the complex field $\boldsymbol{C}$.

The construction can be introduced by first describing the formal construction of the complex field $\boldsymbol{C}$ from the real field $\boldsymbol{R}$. It is well known that not every polynomial over the real field $\boldsymbol{R}$ can be factored into a product of first-degree polynomials over $\boldsymbol{R}$. More specifically, any polynomial of the ring $\boldsymbol{R}[x]$ can be factored into a product of first-degree and second-degree factors. Moreover, some polynomials of $\boldsymbol{R}[x]$ do not have only first-degree factors over $\boldsymbol{R}$. The polynomial $x^2 + 1$, for example, is an irreducible polynomial of degree two over the real field $\boldsymbol{R}$. It is an element of the polynomial ring $\boldsymbol{R}[x]$, but it is not the product of two first-degree polynomials from the ring $\boldsymbol{R}[x]$. Accordingly, we define the extension field, called the *complex field* $\boldsymbol{C}$, as the quotient ring $\boldsymbol{R}[x]/\langle x^2 + 1\rangle$. Although this definition of the complex field may appear unfamiliar, it is actually the usual definition phrased in a different and more formal language. The elements of the field are polynomials in x of degree less than two, and polynomial multiplication is reduced modulo $x^2 + 1$. This means that $x^2 = -1$, and $(ax + b)(cx + d) = (ad + bc)x + (bd - ac)$. This is the familiar definition of multiplication of complex numbers if x is replaced by $i = \sqrt{-1}$. In this way, the complex field is defined as $\boldsymbol{C} = \boldsymbol{R}[x]/\langle x^2 + 1\rangle$.

The construction is very general. Given any field $\boldsymbol{F}$, for this purpose called a *ground field*, and an irreducible polynomial $p(x)$ over $\boldsymbol{F}$ of degree m, an extension field of $\boldsymbol{F}$ can be constructed as the quotient ring $\boldsymbol{F}[x]/\langle p(x)\rangle$. This quotient ring formally is a set of equivalence classes formed by the equivalence relationship $p(x) \equiv 0$. However, we will choose a canonical representative from each equivalence class and regard the extension field to be the set of these canonical representatives. To emphasize this interpretation, we might refer to the extension field as $\boldsymbol{F}[x]_{p(x)}$ when it is viewed as the set of canonical representatives. The elements of $\boldsymbol{F}[x]_{p(x)}$ are the polynomials in x of degree smaller than the degree of $p(x)$. Addition is addition of polynomials. Multiplication is multiplication of polynomials modulo $p(x)$. Because $p(x)$ is an irreducible polynomial, every polynomial $a(x)$ of degree smaller than the degree of $p(x)$ is coprime with $p(x)$. Hence the Bézout relationship

$$A(x)a(x) + P(x)p(x) = 1$$

gives $A(x)a(x) = 1\,(\mathrm{mod}\ p(x))$. Therefore every $a(x)$ has an inverse modulo $p(x)$, given by $a(x)^{-1} = A(x)\,(\mathrm{mod}\ p(x))$, so the quotient ring is a field. The *degree* of this extension field is equal to the degree of $p(x)$.

The complex field $\boldsymbol{C}$ is an example of a field that cannot be so extended because there are no irreducible polynomials of degree two or greater over $\boldsymbol{C}$. In contrast, the rational field $\boldsymbol{Q}$ does have extension fields of every degree, denoted $\boldsymbol{Q}[x]/\langle p(x)\rangle$ where $p(x)$ is an irreducible polynomial. The *degree* of the extension field is equal to the degree of $p(x)$. An extension of the rational field of finite degree is called a *number field*. A number field constructed using an irreducible polynomial $p(x)$ with a zero at ξ is also denoted $\boldsymbol{Q}(\xi)$ as an alternative notation to $\boldsymbol{Q}[x]/\langle p(x)\rangle$. Again, the *degree* of the number field is equal to the degree of $p(x)$. A number field constructed using a cyclotomic polynomial as the irreducible polynomial is called a *cyclotomic field*. The real field $\boldsymbol{R}$ is an extension of $\boldsymbol{Q}$, but it is not an extension field of finite degree, so it is not a number field. Every number field is a subfield of the complex field.

A simple example of a number field is the field of gaussian rationals $\boldsymbol{Q}(i)$, given by $\boldsymbol{Q}(i) = \{a + ib \mid a, b \in \boldsymbol{Q}\}$, where addition and multiplication are as complex numbers using $i^2 = -1$. The field $\boldsymbol{Q}(i)$ is formally obtained from $\boldsymbol{Q}$ by defining the quotient ring

$$\boldsymbol{Q}(i) = \boldsymbol{Q}[x]/\langle x^2 + 1\rangle,$$

recalling the elementary fact that $x^2 + 1$ is irreducible over the rationals as well as over the reals. Thus $x^2 = -1$ in this quotient ring. Once again, to make the structure more familiar, the symbol x can be replaced by the symbol i, and $i^2 = -1$. Then $\boldsymbol{Q}(i) = \{a + bi\}$ where a and b are both rationals.

Some elements of a number field have only integer coefficients in their polynomial representation. The set of all elements of the number field $\boldsymbol{Q}(\xi)$ with integer coefficients is called a *number ring* and denoted $\mathbf{Z}[\xi]$. The elements of $\mathbf{Z}[\xi]$ are called the *integers* of $\boldsymbol{Q}(\xi)$. Moreover, an element of $\boldsymbol{Q}(\xi)$ that is a zero of a monic polynomial of $\mathbf{Z}[x]$ is called an *algebraic integer* of $\boldsymbol{Q}(\xi)$. It is possible that an algebraic integer of $\boldsymbol{Q}(\xi)$ is not in $\mathbf{Z}[\xi]$ and so is not an integer of $\boldsymbol{Q}(\xi)$, as is discussed in Section 10.17. The subset of $\boldsymbol{Q}(i)$ with integer coefficients is called the *ring of gaussian integers*.

For a second example of a number field, notice that $x^4 + 1$ is irreducible over the rational field $\boldsymbol{Q}$ because its zeros want to be at the four points $(\pm 1 \pm i)/\sqrt{2}$, none of which is rational. Therefore $\boldsymbol{Q}[x]/\langle x^4 + 1\rangle$ is a number field, and called $\boldsymbol{Q}((1 + i)/\sqrt{2})$. The elements of this number field are polynomials of degree three or less over the rationals. Addition and multiplication are as polynomials with reduction modulo $x^4 + 1$. This means that x^4 becomes -1 in the reduction process. To emphasize that this is the same method that was used to construct $\boldsymbol{Q}(i)$, we write this as

$$\boldsymbol{Q}[x]/\langle x^4 + 1\rangle = \{a + bi + ci^2 + di^3\},$$

and multiplication of elements of this extension field uses the reduction that $i^4 = -1$.

For a third example of a number field, let φ be the *golden ratio*, given by

$$\varphi = \frac{1 + \sqrt{5}}{2},$$

which is a zero of the polynomial $x^2 - x - 1$. Because the polynomial $x^2 - x - 1$ is irreducible over $\boldsymbol{Q}$ and has φ as a zero, $\boldsymbol{Q}[x]/\langle x^2 - x - 1\rangle$ is a number field, also called $\boldsymbol{Q}(\varphi)$. In this number field, $x^2 = x + 1$. Again, replacing x by i, the field $\boldsymbol{Q}[x]/\langle x^2 - x - 1\rangle$ is the set $\{a + bi\}$, where $a, b \in \boldsymbol{Q}$, but now with i used in place of x the reduction is $i^2 = i + 1$.

Any extension of a field $\boldsymbol{F}$ based on a polynomial of degree two is called a *quadratic extension field*. A quadratic extension of $\boldsymbol{Q}$ is called a *quadratic number field*. For an example of a quadratic number field, observe that $x^2 + D$ is an irreducible polynomial over the rationals if D is a square-free integer, as is $x^2 - D$. The number field based on the polynomial $x^2 + D$ is denoted $\boldsymbol{Q}(\sqrt{-D})$. It consists of elements of the form $a + ib\sqrt{D}$ where a and b are any elements of $\boldsymbol{Q}$ and $i = \sqrt{-1}$. In this case, the field is called an *imaginary quadratic number field*, although the elements are actually complex. Similarly, $x^2 - D$ is also an irreducible polynomial over $\boldsymbol{Q}$ if D is a square-free integer. The field elements of $\boldsymbol{Q}[x]/\langle x^2 - D\rangle$ are real numbers of the form $a + b\sqrt{D}$. This field is called a *real quadratic number field* and is denoted $\boldsymbol{Q}(\sqrt{D})$.

A finite field $\boldsymbol{F}_q$ also can be extended to a larger finite field, denoted $\boldsymbol{F}_{q^m}$, whenever an irreducible polynomial of degree m exists over $\boldsymbol{F}_q$. We will see in Section 9.13 that an irreducible polynomial of every degree does exist over every finite field $\boldsymbol{F}_q$. To illustrate, we will construct the quadratic extension field $\boldsymbol{F}_{p^2}$ from the finite field $\boldsymbol{F}_p$. If p is a prime of the form $4k + 3$, then $p - 1$ is a nonsquare modulo p. This is easy to see. There are $4k + 2$ nonzero elements in $\boldsymbol{Z}_p$, so $1 = \alpha^{4k+2}$ for some primitive element α. Therefore $\alpha^{2k+1} = -1$, so the square root of -1 would be $\alpha^{(2k+1)/2}$, which is not an element of $\boldsymbol{Z}_p$. This means that if $p = 4k + 3$, then $x^2 + 1$ is an irreducible polynomial in $\boldsymbol{Z}_p$, so the quadratic extension field can be defined as polynomial arithmetic modulo $x^2 + 1$. Multiplication in $\boldsymbol{F}_{p^2}$ for any prime of the form $4k + 3$ is given by $(a + bx)(c + dx) = (ad + bc) + (bd - ac)x$, which again, by using i in place of x, emulates multiplication in the complex field $\boldsymbol{C}$.

If p is of the form $4k + 1$, then $x^2 + 1$ is not irreducible. This polynomial cannot be used to construct an extension of a field of the form $\boldsymbol{F}_{4k+1}$. For such fields, another choice of polynomial such as $x^2 + x + 1$, $x^2 - x + 1$, or $x^2 + x - 1$ can be used, provided it is irreducible in that field. For example, $x^2 + x + 1$ is irreducible in $\boldsymbol{F}_{17}$ but not in $\boldsymbol{F}_{13}$.

To summarize, the field $\boldsymbol{F}_{p^2}$ consists of the set $\{a + ib \mid a, b \in \boldsymbol{F}_p\}$, with addition defined by

$$(a + ib) + (c + id) = (a + c) + i(b + d) \pmod p$$

and if p is of the form $4k + 3$, multiplication is defined by

$$(a + ib)(c + id) = (ac - bd) + i(ad + bc) \pmod p$$

whereas, if p is a prime of the form $4k + 1$, multiplication is defined by

$$(a + ib)(c + id) = (ac - bd) + i(ad + bc - bd),$$

provided $x^2 + x + 1$ is an irreducible polynomial in $\boldsymbol{F}_p$ (which is so if $p = 2 \pmod 3$).

The unique factorization theorem says that $x^{q^m} - x$ factors uniquely over $\boldsymbol{F}_q$ into monic irreducible polynomials. Every element of $\boldsymbol{F}_{q^m}$ is a zero of $x^{q^m} - x$, so every element of $\boldsymbol{F}_{q^m}$ is a zero of an irreducible factor of $x^{q^m} - x$.

Definition 9.7.1 *The minimal polynomial over the ground field $\boldsymbol{F}_q$ of an element γ in the extension field $\boldsymbol{F}_{q^m}$ is the monic polynomial over $\boldsymbol{F}_q$ of smallest degree with γ as a zero.*

The minimal polynomial of γ over $\boldsymbol{F}_q$ is unique. This is because two minimal polynomials of the same degree, both monic, can be subtracted to form a nonzero polynomial of smaller degree that has γ as a zero, contradicting the definition of a minimal polynomial. The minimal polynomial of a primitive element is called a *primitive polynomial.*

Theorem 9.7.2 *The minimal polynomial over $\boldsymbol{F}_q$ of γ, an element of the extension field $\boldsymbol{F}_{q^m}$, is also the minimal polynomial of γ^q.*

Proof Because q is a power of the field characteristic p, we can use Theorem 9.3.3 to write

$$[m(x)]^q = \left[\sum_{i=0}^{\deg m(x)} m_i x^i\right]^q = \sum_{i=0}^{\deg m(x)} m_i^q x^{qi}.$$

But the coefficient m_i is any element of $\boldsymbol{F}_q$, so $m_i^q = m_i$. Thus we can conclude that

$$[m(x)]^q = m(x^q).$$

Because γ is a zero on the left, γ^q is a zero on the right. □

Definition 9.7.3 *Two elements of the extension field F_{q^m} that share the same minimal polynomial over F_q are called conjugates (with respect to F_q).*

Two elements that are both conjugate to the same element are themselves conjugates to each other. This means that the notion of conjugacy separates the elements of F_{q^m} into subsets, called *conjugacy classes*. Thus, the elements of a conjugacy class are conjugates.

Definition 9.7.4 *The q-ary trace of an element β of F_{q^m} is the sum*

$$\text{trace}(\beta) = \sum_{i=0}^{m-1} \beta^{q^i}.$$

It is clear that the qth power of the q-ary trace of any element of F_{q^m} is equal to the q-ary trace of that element. It follows from the definition that

$$\text{trace}(\beta + \gamma) = \text{trace}(\beta) + \text{trace}(\gamma).$$

Theorem 9.7.5 *The q-ary trace of the elements of F_{q^m} takes on each value of F_q equally often, that is, q^{m-1} times.*

Proof Suppose that $\beta \in F_{q^m}$ has q-ary trace equal to $\gamma \in F_q$. Then β is a zero of the polynomial

$$x^{q^{m-1}} + x^{q^{m-2}} + \cdots + x^q + x - \gamma.$$

This polynomial has degree of q^{m-1} and so it has at most q^{m-1} zeros. But there are only q such polynomials and every element of F^{q^m} is a zero of one of them. The theorem follows. □

Theorem 9.7.6 *The quadratic equation*

$$x^2 + x + a = 0,$$

where a is an element of F_{2^m}, has a root in F_{2^m} if, and only if, the binary trace of a equals zero.

Proof Let β be a root of the quadratic equation so $\beta^2 + \beta + a = 0$. Then the binary trace of both sides gives $\text{trace}(\beta^2 + \beta + a) = \text{trace}(0) = 0$. Hence $\text{trace}(a) = 0$. □

Let us next construct the binary extension field F_{2^4} from the field F_2. The polynomial $x^4 + x + 1$ is irreducible over F_2. Let $\alpha = x$ and list the powers of α in F_{2^4}. These are polynomials in z reduced modulo $z^4 + z + 1$. They are

$$\begin{aligned}
\alpha^1 &= z\\
\alpha^2 &= z^2\\
\alpha^3 &= z^3\\
\alpha^4 &= z + 1\\
\alpha^5 &= z^2 + z\\
\alpha^6 &= z^3 + z^2\\
\alpha^7 &= z^3 + z + 1\\
\alpha^8 &= z^2 + 1\\
\alpha^9 &= z^3 + z\\
\alpha^{10} &= z^2 + z + 1\\
\alpha^{11} &= z^3 + z^2 + z\\
\alpha^{12} &= z^3 + z^2 + z + 1\\
\alpha^{13} &= z^3 + z^2 + 1\\
\alpha^{14} &= z^3 + 1\\
\alpha^{15} &= 1.
\end{aligned}$$

These are the fifteen nonzero elements of the field $\boldsymbol{F}_{16}$. Of these, α, α^2, α^4, and α^8 share the same minimal polynomial, namely $z^4 + z + 1$. Therefore α, α^2, α^4, and α^8 are conjugates. Also α^3, α^6, α^9, and α^{12} share the same minimal polynomial, namely $z^4 + z^3 + z^2 + z + 1$. Therefore α^3, α^6, α^9, and α^{12} are conjugates. Also, α^{14}, α^{13}, α^{11}, and α^7 share the same minimal polynomial, namely $x^4 + x^3 + 1$. Also α^5 and α^{10} share the same minimal polynomial, namely $x^2 + x + 1$. The minimal polynomials are

$$\begin{aligned}
x^4 + x + 1 &= (x - \alpha)(x - \alpha^2)(x - \alpha^4)(x - \alpha^8)\\
x^4 + x^3 + x^2 + x + 1 &= (x - \alpha^3)(x - \alpha^6)(x - \alpha^{12})(x - \alpha^9)\\
x^4 + x^3 + 1 &= (x - \alpha^7)(x - \alpha^{11})(x - \alpha^{13})(x - \alpha^{14})\\
x^2 + x + 1 &= (x - \alpha^5)(x - \alpha^{10})\\
x + 1 &= (x - \alpha^0).
\end{aligned}$$

Definition 9.7.7 *The algebraic closure of the field $\boldsymbol{F}$, denoted $\overline{\boldsymbol{F}}$, is the union of all extension fields of $\boldsymbol{F}$.*

The algebraic closure is an algebraically closed field. Every polynomial of degree n in an algebraically closed field factors into a product of n polynomials of degree one.

The following theorem is a consequence of the fact that an extension of an extension of $\boldsymbol{F}$ is itself an extension of $\boldsymbol{F}$. An equivalent statement is that every polynomial of degree n in $\overline{\boldsymbol{F}}$ factors into n polynomials of degree one, some possibly repeated.

Theorem 9.7.8 *The algebraic closure of any field $\boldsymbol{F}$ has no proper algebraic extension fields.*

Proof Let $p(x)$ be any polynomial over $\overline{\boldsymbol{F}}$. Then $p(x)$ has a finite number of coefficients, and each coefficient must lie in an extension field of $\boldsymbol{F}$. Thus there is an extension field of $\boldsymbol{F}$ of finite degree containing all of these extension fields, and so containing all coefficients of $p(x)$. This means that $p(x)$ is a polynomial of finite degree in an extension field of $\boldsymbol{F}$, and so $p(x)$ factors in an extension of that extension field. Thus $p(x)$ factors completely in some extension of $\boldsymbol{F}$. This means that $\overline{\boldsymbol{F}}$ has no nontrivial irreducible polynomials, so it cannot be algebraically extended. □

9.8 The multiplication cycle in a finite field

The set of nonzero elements of the finite field $\boldsymbol{F}_q$ is a group, denoted $\boldsymbol{F}_q^*$, under the operation of multiplication. This set of nonzero field elements is closed under multiplication and contains the inverse of every nonzero element of $\boldsymbol{F}_q$. Indeed, as we shall see in this section, the nonzero elements of a finite field always form a cyclic group under the operation of multiplication. Accordingly, except for zero, the field $\boldsymbol{F}_q$ can be generated by a single element, denoted α, and called a primitive element. Any element of $\boldsymbol{F}_q^*$ can be written as a power of α. This is an important property of a finite field, which will be useful in many ways. The work of this section is to prove this assertion.

Theorem 9.8.1 *The group $\boldsymbol{F}_q^*$ consisting of the nonzero elements of the finite field $\boldsymbol{F}_q$ under multiplication is a cyclic group.*

Proof There are $q-1$ nonzero elements in $\boldsymbol{F}_q$. The proof is trivial if $q-1$ is a prime because then every element except zero and one has order $q-1$, and thus every such element generates a cyclic group.

Otherwise, consider the prime factorization of $q-1$, which can be written as

$$q-1=\prod_{i=1}^{s} p_i^{e_i}.$$

A polynomial over a field $\boldsymbol{F}$ cannot have more zeros than its degree. Therefore there must be at least one element of $\boldsymbol{F}_q$ that is not a zero of the polynomial $x^{(q-1)/p_i}-1$. This means that for each i, there exists a nonzero element a_i of $\boldsymbol{F}_q$ for which $a_i^{(q-1)/p_i}$ is not equal to one. Let $b_i=a_i^{(q-1)/p_i^{e_i}}$ and let $b=\prod_{i=1}^{s} b_i$. We will prove that b has order $q-1$, which means that $\boldsymbol{F}_q^*$ is cyclic and b is a primitive element.

Step 1 The element b_i has order $p_i^{e_i}$. *Proof* Clearly $(a_i^{(q-1)/p_i^{e_i}})^{p_i^{e_i}}$ equals one, so the order of $a_i^{(q-1)/p_i^{e_i}}$ divides $p_i^{e_i}$ and must be of the form $p_i^{\mu_i}$. But this order cannot be smaller than $p_i^{e_i}$ because $a_i^{(q-1)/p_i}$ is not equal to one, and $(a_i^{(q-1)/p_i^{e_i}})^{p_i^{e_i-1}} = a_i^{(q-1)/p_i} \neq 1$.

Step 2 The element b has order $q-1$. *Proof* Let n be the order of b. We must show that $n = q-1$. For each i, we know that

$$\left(b^{\prod_{k\neq i} p_k^{e_k}}\right)^n = 1.$$

But $b = \prod_{i=1}^{s} b_i$ and $b_j^{p_j^{\nu_j}} = 1$, so we conclude that

$$b_i^{n \prod_{j\neq i} p_j^{\nu_j}} = 1.$$

This means that $n \prod_{j\neq k} p_j^{e_j}$ is a multiple of $p_k^{e_k}$. Because the p_j are distinct primes, it follows that n is a multiple of $p_k^{e_k}$ for each k. Hence n is a multiple of $q-1 = \prod_{k=1}^{s} q_k^{e_k}$. Because the order of every element of $\boldsymbol{F}_p^*$ is not larger than $q-1$, it follows that b has order $q-1$. The proof is now complete. □

Corollary 9.8.2 *Every Galois field has a primitive element.*

Proof The set $\boldsymbol{F}_q^*$ is a cyclic group under multiplication, and every cyclic group has a generator α. Therefore the element α is a primitive element. □

Corollary 9.8.3 *A Galois field with q elements has $\phi(q-1)$ primitive elements.*

Proof A primitive element α of $\boldsymbol{F}_q$ has order $q-1$. Every element of $\boldsymbol{F}_q$ is a power of the primitive element α. If j is a proper factor of $q-1$, then α^j does not have order $q-1$. If j is coprime with $q-1$, then α^j must also have order $q-1$, and so α^j is also a primitive element. This means that $\boldsymbol{F}_q$ has $\phi(q-1)$ primitive elements, where $\phi(x)$ is the totient function. □

Theorem 9.8.4 *Over the field $\boldsymbol{F}_q$, we have the factorization*

$$x^{q-1} - 1 = \prod_{i=0}^{q-2}(x - \alpha^i),$$

where α is any primitive element of $\boldsymbol{F}_q$.

Proof Every element γ of $\boldsymbol{F}_q$ satisfies $\gamma^{q-1} = 1$, and so every element γ is a zero of the polynomial $x^{q-1} - 1$. Because this polynomial has degree $q-1$, it can have only $q-1$ zeros, and we have found all of them. □

One can conclude from Theorem 9.8.4 that every nonzero element of $\boldsymbol{F}_q$ is a zero of the polynomial $x^{q-1} - 1$, so every element β of $\boldsymbol{F}_q$ satisfies $\beta^q = \beta$. This motivates the following definition.

Definition 9.8.5 *The Frobenius map over $\overline{\boldsymbol{F}}_q$ is the function $\pi_q(x) = x^q$.*

The Frobenius map takes the elements of $\overline{\boldsymbol{F}}_q$ into $\overline{\boldsymbol{F}}_q$. Clearly, for $\beta \in \boldsymbol{F}_q$, $\pi_q(\beta) = \beta$, and for $\beta \notin \boldsymbol{F}_{q^m}$, $\pi_q(\beta) \neq \beta$. Thus an element β of $\overline{\boldsymbol{F}}_q$ is in the ground field $\boldsymbol{F}_q$ if, and only if, $\pi_q(\beta) = \beta$. The Frobenius map can be composed with itself as $\pi_q(\pi_q(\beta))$, which will be written as $\pi_q \circ \pi_q$ or as π_q^2. Then $\pi_q \circ \pi_q(x) = (x^q)^q = x^{q^2}$. The m-fold composition of π_q with itself, denoted $\pi_q \circ \pi_q \circ \cdots \circ \pi_q$, will be abbreviated π_q^m.

Theorem 9.8.6 *The Frobenius map over $\overline{\boldsymbol{F}}_q$ satisfies*

$$\pi_q(x + y) = \pi_q(x) + \pi_q(y)$$
$$\pi_q(xy) = \pi_q(x)\pi_q(y),$$

and $\pi^m(\beta) = \beta$ for every $\beta \in \boldsymbol{F}_{q^m}$.

Proof The first statement uses the fact that, for p a prime, $\binom{p}{i}$ is a multiple of p, so $\binom{p}{i} = 0 \pmod{p}$ unless $i = p$ or 0. Thus $\pi_q(x + y) = (x + y)^q = x^q + \Sigma_i \binom{p}{i} x^{q-i} y^i + y^q$. The second statement is trivial because

$$\pi_q(xy) = (xy)^q = x^q y^q = \pi_q(x)\pi_q(y).$$

The third statement follows because $\pi_q \circ \pi_q \circ \cdots \circ \pi_q(\beta) = \beta^{q^m} = \beta$ for $\beta \in \boldsymbol{F}_{q^m}$. □

9.9 Cyclotomic polynomials

The polynomial $x^n - 1$ is important in any field $\boldsymbol{F}$ because the n zeros of $x^n - 1$ are the nth roots of unity either in the field $\boldsymbol{F}$ itself or in an extension of the field $\boldsymbol{F}$. This polynomial can be expressed in terms of its prime factors as

$$x^n - 1 = p_1(x)p_2(x) \cdots p_k(x)$$

in any field $\boldsymbol{F}$. The prime factors depend on the field $\boldsymbol{F}$. If any one of these prime factors has degree larger than two, then it can be used to extend the field $\boldsymbol{F}$. That polynomial then has its zeros in that extension field.

In this section, we will be interested in the prime factors of $x^n - 1$ over the rational field $\boldsymbol{Q}$. When $x^n - 1$ is regarded as a polynomial over $\boldsymbol{Q}$, its n zeros all lie in the complex field $\boldsymbol{C}$, which is the algebraic closure of $\boldsymbol{Q}$. They can be written as ω_n^ℓ for $\ell = 0, \ldots, n-1$, where

$$\omega_n = e^{-2\pi i/n}$$

and $i = \sqrt{-1}$. The prime factorization over $\boldsymbol{C}$ is then given by

$$x^n - 1 = \prod_{\ell=0}^{n-1} (x - \omega_n^\ell).$$

However, each of these prime factors (except for $x - 1$ and, if n is even, $x + 1$) has a complex coefficient. We want to find the prime polynomial factors of $x^n - 1$ over $\boldsymbol{Q}$. These are the irreducible monic polynomial factors with only rational coefficients. The prime factors of $x^n - 1$ over $\boldsymbol{Q}$ are known as *cyclotomic polynomials*. Each such prime factor must be a product of those factors $x - \omega_n^\ell$ over the complex field that need to be multiplied together to get a polynomial with rational coefficients. Accordingly, the definition of a cyclotomic polynomial is

$$\Phi_n(x) = \prod_{\text{GCD}(i,n)=1} (x - \omega_n^i),$$

which has only rational coefficients. We will show this by showing the stronger fact that these polynomials only have integer coefficients. To see that the coefficients of $\Phi_n(x)$ are integers, note that

$$x^n - 1 = \Phi_n(x) \prod_{d|n} \Phi_d(x).$$

Writing this in the form

$$\Phi_n(x) = \frac{x^n - 1}{\prod_{d|n} \Phi_d(x)}$$

shows that $\Phi_n(x)$ has only integer coefficients if $\Phi_d(x)$ has only integer coefficients for all d that divide n. This is because division of a monic polynomial with only integer coefficients by another monic polynomial with only integer coefficients yields a monic polynomial with only integer coefficients. Because $\Phi_1(x) = x - 1$, we can conclude that all $\Phi_n(x)$ have integer coefficients. Hence the $\Phi_n(x)$ are factors of $x^n - 1$ over $\boldsymbol{Z}$, and so over $\boldsymbol{Q}$. They cannot be factored further in $\boldsymbol{Q}$.

It follows immediately from the definition that the degree of the nth cyclotomic polynomial satisfies

$$\deg \Phi_n(x) = \phi_n,$$

where ϕ_n is the totient function.

The cyclotomic polynomials also can be computed directly by factoring $x^n - 1$ over the integers. The first twelve cyclotomic polynomials are

$$\begin{aligned}
\Phi_1(x) &= x - 1\\
\Phi_2(x) &= x + 1\\
\Phi_3(x) &= x^2 + x + 1\\
\Phi_4(x) &= x^2 + 1\\
\Phi_5(x) &= x^4 + x^3 + x^2 + x + 1\\
\Phi_6(x) &= x^2 - x + 1\\
\Phi_7(x) &= x^6 + x^5 + x^4 + x^3 + x^2 + x + 1\\
\Phi_8(x) &= x^4 + 1\\
\Phi_9(x) &= x^6 + x^3 + 1\\
\Phi_{10}(x) &= x^4 - x^3 + x^2 - x + 1\\
\Phi_{11}(x) &= x^{10} + x^9 + x^8 + x^7 + x^6 + x^5 + x^4 + x^3 + x^2 + x + 1\\
\Phi_{12}(x) &= x^4 - x^2 + 1.
\end{aligned}$$

These polynomials have all of their coefficients equal either to 0 or to ± 1. Indeed, all cyclotomic polynomials up to $\Phi_{104}(x)$ have all of their coefficients equal to 0 or ± 1. (The cyclotomic polynomial $\Phi_{105}(x)$, however, has two coefficients equal to 2.) As we concluded earlier, every cyclotomic polynomial has only integer coefficients.

Although cyclotomic polynomials were introduced as irreducible polynomials over $\boldsymbol{Q}$, and are described in terms of their zeros in the extension field $\boldsymbol{C}$, they have coefficients only in $\boldsymbol{Z}$. Thus it may appear that there is no compelling reason to introduce $\boldsymbol{Q}$ or $\boldsymbol{C}$ other than to invoke the convenience of working in a field. Therefore, one can introduce the prime field $\boldsymbol{F}_p$ instead, provided this field has an element of order n. This only requires that n divides $p - 1$. Now the cyclotomic polynomials are viewed as polynomials over the field $\boldsymbol{F}_p$ with zeros in the field $\boldsymbol{F}_p$ or in the extension field $\boldsymbol{F}_{p^k}$.

For this purpose, write

$$x^n - 1 = \prod_{\ell=0}^{n-1}(x - \omega^\ell),$$

where now ω is an element of order n in the field $\boldsymbol{F}_p$ or in the extension field $\boldsymbol{F}_{p^k}$. Such an element exists in $\boldsymbol{F}_{p^k}$, for instance, if $p^k = an + 1$ for some a. The Legendre prime-number theorem assures us that for a chosen n and with $k = 1$, such a p always exists. Even more, we are also free to use larger k for which this form produces a prime power for the chosen value of n.

For an example of the formation of cyclotomic polynomials that divide $x^n - 1$ with $n = 6$, using the prime field $\boldsymbol{F}_7$, choose $a = 1$, so that $an + 1$ is equal to the prime 7.

Let $\omega = 5$, which has order six in $\boldsymbol{F}_7$. Then

$$\begin{aligned} x^6 - 1 &= (x-1)(x-5)(x-4)(x-6)(x-2)(x-3) \qquad \pmod 7 \\ &= [(x-5)(x-3)]\,[(x-4)(x-2)]\,(x-6)(x-1) \pmod 7 \\ &= \Phi_6(x)\Phi_3(x)\Phi_2(x)\Phi_1(x). \end{aligned}$$

We conclude that in $\boldsymbol{F}_7$ with $\omega = 5$, the cyclotomic polynomials are

$$\begin{aligned} \Phi_1(x) &= (x-1) \\ \Phi_2(x) &= (x-6) = x+1 \\ \Phi_3(x) &= (x-2)(x-4) = x^2+x+1 \\ \Phi_6(x) &= (x-3)(x-5) = x^2-x+1, \end{aligned}$$

which are the usual cyclotomic polynomials as were computed in $\boldsymbol{C}$.

An example with odd n chooses $a = 2$, $n = 3$, and $\omega = 2$, so that $an + 1$ is equal to the prime 7. Then

$$\begin{aligned} x^3 - 1 &= (x-1)(x-2)(x-4) \pmod 7 \\ &= (x-1)(x^2+x+1) \quad \pmod 7 \\ &= \Phi_1(x)\Phi_3(x). \end{aligned}$$

We conclude that in $\boldsymbol{F}_7$ with $\omega = 2$, the cyclotomic polynomials are

$$\begin{aligned} \Phi_1(x) &= (x-1) \\ \Phi_3(x) &= (x-2)(x-4) = x^2+x+1, \end{aligned}$$

which are the usual cyclotomic polynomials as were computed in $\boldsymbol{C}$.

Another example chooses $a = 2$, $n = 8$, and $\omega = 9$ so that $an + 1$ is equal to the prime 17. Then in $\boldsymbol{F}_{17}$, the factors of $x^8 - 1$ are $(x - 9^\ell)$ for $\ell = 1, \ldots, 8$. Then

$$\begin{aligned} x^8 - 1 &= (x-1)(x-9)(x-13)(x-15)(x-16)(x-8)(x-4)(x-2) \quad \pmod{17} \\ &= \Phi_8(x)\Phi_4(x)\Phi_2(x)\Phi_1(x). \end{aligned}$$

We conclude that in $\boldsymbol{F}_{17}$, with $\omega = 9$, the cyclotomic polynomials are

$$\begin{aligned} \Phi_1(x) &= (x-1) \\ \Phi_2(x) &= (x-16) = x+1 \\ \Phi_4(x) &= (x-13)(x-4) = x^2+1 \\ \Phi_8(x) &= (x-9)(x-15)(x-8)(x-2) = x^4+1, \end{aligned}$$

which again agree with the usual cyclotomic polynomials.

Every cyclotomic polynomial can be written in this way because, by Theorem 2.7.1, for every n, there is a prime p of the form $an + 1$. More generally, one can also form cyclotomic polynomials using fields of the form $\boldsymbol{F}_{p^m}$. For example, $\boldsymbol{F}_9$ contains an element ω of order eight of the form $\omega = 1 + i$ where $i^2 = -1$. Then one again finds that

$$x^8 - 1 = (x-\omega)(x-\omega^2)(x-\omega^3)(x-\omega^4)(x-\omega^5)(x-\omega^6)(x-\omega^7)(x-\omega^8)$$
$$= \Phi_8(x)\Phi_4(x)\Phi_2(x)\Phi_1(x).$$

Then in $\boldsymbol{F}_9$ with $\omega = 1+i$, and $i^2 = -1$, the cyclotomic polynomials are

$$\Phi_1(x) = (x-\omega^8) = x-1$$
$$\Phi_2(x) = (x-\omega^4) = x+1$$
$$\Phi_4(x) = (x-\omega^2)(x-\omega^6) = (x-2i)(x+2i) = x^2+1$$
$$\Phi_8(x) = (x-\omega^1)(x-\omega^3)(x-\omega^5)(x-\omega^7)$$
$$= (x-1-i)(x+2-2i)(x+1+i)(x-2+2i)$$
$$= x^4+1.$$

by a simple calculation in $\boldsymbol{F}_9$. Again, these agree with the usual cyclotomic polynomials.

9.10 Vector spaces

An n-tuple of field elements $(v_0, v_1, \ldots, v_{n-1})$, denoted $\boldsymbol{v}$, over a field $\boldsymbol{F}$ is called a *vector* of length n over the field $\boldsymbol{F}$ with components v_i. The set of all vectors of blocklength n over the field $\boldsymbol{F}$, together with two operations known as *vector addition* and *scalar multiplication*, to be defined below, is called a *vector space* over the field $\boldsymbol{F}$. In discussions of vector spaces, the elements of the underlying field $\boldsymbol{F}$ are called *scalars*.

Scalar multiplication is an algebraic operation that multiplies a vector $\boldsymbol{v}$ by a scalar c. Scalar multiplication for n-tuples is defined componentwise by

$$c(v_0, v_1, \ldots, v_{n-1}) = (cv_0, cv_1, \ldots, cv_{n-1}).$$

Vector addition is an operation that combines two vectors $\boldsymbol{v}'$ and $\boldsymbol{v}''$ to produce a third vector $\boldsymbol{v} = \boldsymbol{v}' + \boldsymbol{v}''$. Vector addition of n-tuples is defined componentwise by

$$(v_0', v_1', \ldots, v_{n-1}') + (v_0'', v_1'', \ldots, v_{n-1}'') = (v_0' + v_0'', v_1' + v_1'', \ldots, v_{n-1}' + v_{n-1}'').$$

The set of n-tuples over a field $\boldsymbol{F}$ is a concrete and common form of a vector space. A vector space of n-tuples is only one instance of a vector space, but it is the instance that is perhaps the most useful.

A vector space is defined abstractly as any set V together with the two operations, vector addition and scalar multiplication, such that the following properties are satisfied:

(1) The set V is an abelian group under vector addition.
(2) (*Distributivity*) For any two elements $\boldsymbol{v}_1$ and $\boldsymbol{v}_2$ of V and any scalar c,

$$c(\boldsymbol{v}_1 + \boldsymbol{v}_2) = c\boldsymbol{v}_1 + c\boldsymbol{v}_2.$$

(3) (*Distributivity*) For any vector $\boldsymbol{v}$ of V and any two scalars c_1 and c_2, $1\boldsymbol{v} = \boldsymbol{v}$ and

$$(c_1 + c_2)\boldsymbol{v} = c_1\boldsymbol{v} + c_2\boldsymbol{v}.$$

(4) (*Associativity*) For any vector $\boldsymbol{v}$ of V and any two scalars c_1 and c_2,

$$(c_1c_2)\boldsymbol{v} = c_1(c_2\boldsymbol{v}).$$

The zero element of V under vector addition is called the *origin* of the vector space and is denoted $\boldsymbol{0}$. Notice that there are two different uses for the symbol $+$: vector addition and addition within the field. Furthermore, the symbol $\boldsymbol{0}$ is used for the origin of the vector space and the symbol 0 is used for the zero of the field. In practice, these ambiguities cause no confusion.

A subset of a vector space is called a *vector subspace* if it is also a vector space under the original vector addition and scalar multiplication. Under the operation of vector addition, a vector space is a group, and a vector subspace is a subgroup. In order to check whether a nonempty subset of a vector space is a vector subspace, it is only necessary to check for closure under vector addition and closure under scalar multiplication. All other required properties are then inherited from the original space. Closure under scalar multiplication ensures that the zero vector is in the subset.

The set of all n-tuples over the field $\boldsymbol{F}$ is an n-dimensional vector space over $\boldsymbol{F}$, and is denoted by $\boldsymbol{F}^n$. We have already noted that, in the n-tuple space $\boldsymbol{F}^n$, vector addition and scalar multiplication are defined componentwise. In $\boldsymbol{F}^n$, there is another operation called the *componentwise product* of two n-tuples. If $\boldsymbol{u} = (a_0, \ldots, a_{n-1})$ and $\boldsymbol{v} = (b_0, \ldots, b_{n-1})$, the componentwise product is the vector defined as $\boldsymbol{uv} = (a_0b_0, a_1b_1, \ldots, a_{n-1}b_{n-1})$. It is a vector obtained by the componentwise product of the components of $\boldsymbol{u}$ and $\boldsymbol{v}$.

The *inner product* of two n-tuples of $\boldsymbol{F}^n$ is a scalar defined as

$$\begin{aligned}\boldsymbol{u} \cdot \boldsymbol{v} &= (a_0, \ldots, a_{n-1}) \cdot (b_0, \ldots, b_{n-1}) \\ &= a_0b_0 + \cdots + a_{n-1}b_{n-1}.\end{aligned}$$

It is immediately verified that $\boldsymbol{u} \cdot \boldsymbol{v} = \boldsymbol{v} \cdot \boldsymbol{u}$, that $(c\boldsymbol{u}) \cdot \boldsymbol{v} = c(\boldsymbol{u} \cdot \boldsymbol{v})$, and also that $\boldsymbol{w} \cdot (\boldsymbol{u} + \boldsymbol{v}) = (\boldsymbol{w} \cdot \boldsymbol{u}) + (\boldsymbol{w} \cdot \boldsymbol{v})$. If the inner product of two vectors is zero, the vectors are said to be *orthogonal*. There are some fields over which it is possible for a nonzero vector to be orthogonal to itself, but this cannot happen in a vector space over the real field. A vector orthogonal to every vector in a vector subspace is said to be orthogonal to the vector subspace.

The *norm* or *euclidean weight*, denoted $\|\boldsymbol{v}\|^2$, of a vector $\boldsymbol{v}$ in the real vector space $\boldsymbol{R}^n$ is defined by its square $\|\boldsymbol{v}\|^2 = \boldsymbol{v} \cdot \boldsymbol{v}$. The *norm* or *euclidean weight* of a vector $\boldsymbol{v}$ in the complex vector space $\boldsymbol{C}^n$ is defined by its square $\|\boldsymbol{v}\|^2 = \boldsymbol{v} \cdot \boldsymbol{v}^*$, where $\boldsymbol{v}^*$ denotes the componentwise complex conjugate of $\boldsymbol{v}$. The *euclidean distance* $d_e(\boldsymbol{v}, \boldsymbol{u})$ between two vectors $\boldsymbol{v}$ and $\boldsymbol{u}$ in $\boldsymbol{C}^n$ is the euclidean weight of the componentwise difference. Thus $d_e(\boldsymbol{v}, \boldsymbol{u}) = \|\boldsymbol{v} - \boldsymbol{u}\|$.

The notions of euclidean weight and euclidean distance do not carry over to $\boldsymbol{F}_q$. Instead, the *Hamming weight* $w_H(\boldsymbol{v})$ of a vector $\boldsymbol{v}$ in any n-tuple vector space $\boldsymbol{F}^n$ is defined as the number of nonzero components in $\boldsymbol{v}$. The *Hamming distance* between any two vectors $\boldsymbol{v}$ and $\boldsymbol{u}$ of $\boldsymbol{F}^n$ is defined as $d_H(\boldsymbol{v}, \boldsymbol{u}) = w_H(\boldsymbol{v} - \boldsymbol{u})$.

In a vector space V, a sum of the form $\boldsymbol{v} = a_1\boldsymbol{v}_1 + a_2\boldsymbol{v}_2 + \cdots + a_n\boldsymbol{v}_n$, where the a_i are scalars, is called a *linear combination* of the vectors $\boldsymbol{v}_1, \ldots, \boldsymbol{v}_n$. A set of vectors $\{\boldsymbol{v}_1, \ldots, \boldsymbol{v}_n\}$ is called a *linearly dependent* set if there are scalars $a_1, \ldots, a_n$, not all zero, such that $a_1\boldsymbol{v}_1 + a_2\boldsymbol{v}_2 + \cdots + a_n\boldsymbol{v}_n = \boldsymbol{0}$. A set of vectors that is not linearly dependent is called a *linearly independent* set.

A set of vectors of V is said to *span* the vector space V if every vector of V can be expressed as a linear combination of elements of the set in at least one way. If V is spanned by a finite set of vectors, then it is called a *finite-dimensional vector space*. A vector space that is spanned by n vectors cannot contain a linearly independent vector set of vectors containing more than n vectors. Two linearly independent sets of vectors that span the same finite-dimensional vector space have the same number of vectors.

9.11 Linear algebra

Linear algebra is concerned with systems of first-degree equations in multiple variables over a field $\boldsymbol{F}$. Most of the operations of linear algebra that are familiar in the real field or the complex field are valid in an arbitrary field. Indeed, these methods may sometimes apply, in part, to systems of first-degree equations over a commutative ring.

The methods of linear algebra are concisely expressed in the language of matrices. An *n-by-m matrix* over a field $\boldsymbol{F}$ consists of nm elements from the field $\boldsymbol{F}$ arranged in a rectangular array of n rows and m columns. Accordingly,

$$\boldsymbol{A} = \begin{bmatrix} a_{11} & a_{12} & \ldots & a_{1m} \\ \vdots & & \vdots & \\ a_{n1} & a_{n2} & & a_{nm} \end{bmatrix} = [a_{ij}].$$

The *transpose* of an n-by-m matrix $\boldsymbol{A}$ is an m-by-n matrix, denoted $\boldsymbol{A}^T$, with elements $a_{ij}^T = a_{ji}$. The set of all n-by-m matrices over $\boldsymbol{F}$ is a group under matrix addition. The additive identity, denoted $\boldsymbol{0}$, is the n-by-m matrix with zeros for all elements. The addition of two n-by-m matrices, denoted $\boldsymbol{C} = \boldsymbol{A} + \boldsymbol{B}$, is defined elementwise by $c_{ij} = a_{ij} + b_{ij}$.

The multiplication of an n-by-m matrix and an m-by-ℓ matrix, denoted $\boldsymbol{C} = \boldsymbol{AB}$, is defined elementwise by

$$c_{ij} = \sum_{k=1}^{m} a_{ik} b_{kj}.$$

A *square matrix* is a matrix with the same number of rows and columns. The set of all n by n square matrices over $\boldsymbol{F}$ is a ring under matrix addition and matrix multiplication. It is a noncommutative ring with identity. The identity under multiplication, denoted $\boldsymbol{I}$, is the n-by-n matrix with ones in all diagonal positions and zeros elsewhere.

A matrix can be multiplied by a scalar α by multiplying every element of the matrix by α. A nonnegative scalar function on the set of n-by-n square matrices $\boldsymbol{A}$, denoted $\|\boldsymbol{A}\|$, that satisfies the property $\|\alpha \boldsymbol{A}\| = |\alpha| \, \|\boldsymbol{A}\|$ is called a *norm*.

A square matrix over the field $\boldsymbol{F}$ is associated with two important scalars called the *trace* and the *determinant*, which are defined as follows.

Definition 9.11.1 *The trace of a square n-by-n matrix $\boldsymbol{A}$ with elements a_{ij} in the field $\boldsymbol{F}$ is given by*

$$\text{trace } \boldsymbol{A} = \sum_{i=1}^{n} a_{ii}.$$

Definition 9.11.2 *The determinant of a square n-by-n matrix $\boldsymbol{A}$ with elements a_{ij} in the field $\boldsymbol{F}$ is given by*

$$\det \boldsymbol{A} = \sum \xi_{i_1,\ldots,i_n} a_{1i_1} a_{2i_2} a_{3i_3} \cdots a_{ni_n},$$

where $i_1, i_2, \ldots, i_n$ is a permutation of the integers $1, 2, \ldots, n$, the sum is over all possible permutations, and $\xi_{i_1,\ldots,i_n}$ is ± 1 according to whether the permutation is even or odd.

An odd permutation is one that can be obtained as a sequence of an odd number of transpositions. An even permutation is one that cannot be obtained by an odd number of transpositions. A transposition is the interchange of two terms.

If $\boldsymbol{A}$ is a square matrix with a nonzero determinant, then there exists another matrix, denoted $\boldsymbol{A}^{-1}$ and called the *inverse* of $\boldsymbol{A}$, satisfying the property $\boldsymbol{A}^{-1}\boldsymbol{A} = \boldsymbol{A}\boldsymbol{A}^{-1} = \boldsymbol{I}$. An n-by-n matrix with a zero determinant does not have an inverse.

For example, the determinant of the matrix

$$\boldsymbol{M} = \begin{bmatrix} 1 & 0 & 2 \\ 2 & 1 & 0 \\ 0 & 2 & 1 \end{bmatrix}$$

is zero in a field of characteristic three. Therefore this matrix $\boldsymbol{M}$ does not have an inverse in any field of characteristic three. In every other field, the determinant is nonzero, and the matrix $\boldsymbol{M}$ does have an inverse in those fields.

Definition 9.11.3 *For any matrix* $\boldsymbol{A}$ *with at least as many columns as rows.*

The rank of $\boldsymbol{A}$ *is the largest value of* r *such that some set of* r *columns of* $\boldsymbol{A}$ *is linearly independent.*

The heft of $\boldsymbol{A}$ *is the largest value of* r *such that every set of* r *columns of* $\boldsymbol{A}$ *is linearly independent.*

Clearly, for any matrix $\boldsymbol{A}$, the inequality

$$\text{heft}\,\boldsymbol{A} \leq \text{rank}\,\boldsymbol{A}$$

holds. The rank can also be described as the smallest value of r such that *every* set of $r+1$ columns is linearly dependent, and the heft can also be described as the smallest value of r such that *some* set of $r+1$ columns is linearly dependent.

Every square matrix $\boldsymbol{A}$ is associated with a certain polynomial, called the *characteristic polynomial* of $\boldsymbol{A}$, and defined by

$$p_{\boldsymbol{A}}(\lambda) = \det[\boldsymbol{A} - \lambda \boldsymbol{I}].$$

Two of the coefficients of the characteristic polynomial of $\boldsymbol{A}$ are the determinant of $\boldsymbol{A}$ and the trace of $\boldsymbol{A}$. The zeros of the characteristic polynomial are known as the *eigenvalues* of $\boldsymbol{A}$.

We will give a very explicit proof of the next theorem only for the case of a 2-by-2 matrix. We omit the general proof for an n-by-n matrix because it becomes rather involved, and we only have application for the case of a 2-by-2 matrix.

Theorem 9.11.4 (Cayley–Hamilton) *Any square matrix* $\boldsymbol{A}$ *satisfies its own characteristic polynomial.*

Proof The proof will be given only for 2-by-2 matrices. Let

$$\boldsymbol{A} = \begin{bmatrix} a & b \\ c & d \end{bmatrix}$$

be a 2-by-2 matrix over the field F. The characteristic equation of $\boldsymbol{A}$, defined as $\det[\boldsymbol{A} - \lambda \boldsymbol{I}] = 0$, is

$$\lambda^2 - (a+d)\lambda + (ad - bc) = 0.$$

The statement of the theorem is that the matrix $\boldsymbol{A}$ can be inserted in place of λ and the resulting matrix polynomial is equal to the zero matrix. That is,

$$\boldsymbol{A}^2 - (a+d)\boldsymbol{A} + (ad - bc)\boldsymbol{I} = \boldsymbol{0},$$

where $\boldsymbol{I}$ is the identity matrix and $\boldsymbol{0}$ is the zero matrix. But

$$\boldsymbol{A}^2 = \begin{bmatrix} a^2+bc & ab+bd \\ ca+cd & cb+a^2 \end{bmatrix}.$$

Direct substitution of $\boldsymbol{A}^2$ and $\boldsymbol{A}$ into the characteristic polynomial gives

$$\begin{bmatrix} a^2+bc & ab+bd \\ ca+cd & cb+a^2 \end{bmatrix} - \begin{bmatrix} a(a+d) & b(a+d) \\ c(a+d) & d(a+d) \end{bmatrix} + \begin{bmatrix} ad-bc & 0 \\ 0 & ad-bc \end{bmatrix} = \begin{bmatrix} 0 & 0 \\ 0 & 0 \end{bmatrix},$$

as was to be proved. □

9.12 The Fourier transform

Whenever an element ω of order n exists in the field $\boldsymbol{F}$, every vector $\boldsymbol{v}$ of blocklength n over $\boldsymbol{F}$ is associated with another vector of blocklength n over $\boldsymbol{F}$ called the *Fourier transform* of $\boldsymbol{v}$ and denoted $\boldsymbol{V}$. The Fourier transform is defined componentwise by

$$V_j = \sum_{i=0}^{n-1} v_i \omega^{ij} \qquad j = 0, \ldots, n-1.$$

Both the mapping itself from $\boldsymbol{v}$ to $\boldsymbol{V}$ and the image $\boldsymbol{V}$ of this mapping are referred to as the Fourier transform.

For example, in $\boldsymbol{F}_5$ the element 3 has order four, so

$$\begin{bmatrix} V_0 \\ V_1 \\ V_2 \\ V_3 \end{bmatrix} = \begin{bmatrix} 1 & 1 & 1 & 1 \\ 1 & 3 & 4 & 2 \\ 1 & 4 & 1 & 4 \\ 1 & 2 & 4 & 3 \end{bmatrix} \begin{bmatrix} v_0 \\ v_1 \\ v_2 \\ v_3 \end{bmatrix}$$

is a Fourier transform of blocklength 4 in $\boldsymbol{F}_5$ expressed as a matrix equation. Similarly, in $\boldsymbol{F}_7$ the element 3 has order six, so

$$\begin{bmatrix} V_0 \\ V_1 \\ V_2 \\ V_3 \\ V_4 \\ V_5 \end{bmatrix} = \begin{bmatrix} 1 & 1 & 1 & 1 & 1 & 1 \\ 1 & 3 & 2 & 6 & 4 & 5 \\ 1 & 2 & 4 & 1 & 2 & 4 \\ 1 & 6 & 1 & 6 & 1 & 6 \\ 1 & 4 & 2 & 1 & 4 & 2 \\ 1 & 5 & 4 & 6 & 2 & 3 \end{bmatrix} \begin{bmatrix} v_0 \\ v_1 \\ v_2 \\ v_3 \\ v_4 \\ v_5 \end{bmatrix}$$

is a Fourier transform of blocklength 6 in $\boldsymbol{F}_7$ expressed as a matrix equation. Finally, the element α^3 has order five in $\boldsymbol{F}_{16}$ if α is a primitive element, so

$$\begin{bmatrix} V_0 \\ V_1 \\ V_2 \\ V_3 \\ V_4 \end{bmatrix} = \begin{bmatrix} 1 & 1 & 1 & 1 & 1 \\ 1 & \alpha^3 & \alpha^6 & \alpha^9 & \alpha^{12} \\ 1 & \alpha^6 & \alpha^{12} & \alpha^3 & \alpha^9 \\ 1 & \alpha^9 & \alpha^3 & \alpha^{12} & \alpha^6 \\ 1 & \alpha^{12} & \alpha^9 & \alpha^6 & \alpha^3 \end{bmatrix} \begin{bmatrix} v_0 \\ v_1 \\ v_2 \\ v_3 \\ v_4 \end{bmatrix}$$

is a Fourier transform of blocklength 5 in $\boldsymbol{F}_{16}$.

The finite field $\boldsymbol{F}_q$ always can be generated by a primitive element of order $q-1$, so the finite field $\boldsymbol{F}_q$ always has a Fourier transform of blocklength $q-1$. Moreover, the field $\boldsymbol{F}_q$ has a Fourier transform of blocklength n for every n that divides $q-1$. The field $\boldsymbol{F}_q$ does not have a Fourier transform of blocklength n for any n that does not divide $q-1$, although a Fourier transform of this blocklength will exist in an extension field $\boldsymbol{F}_{q^m}$ if n is coprime with q.

The Fourier transform has a rich set of properties. The vector $\boldsymbol{v}$ can be recovered from its Fourier transform $\boldsymbol{V}$ by the *inverse Fourier transform*

$$v_i = \frac{1}{n}\sum_{j=0}^{n-1} \omega^{-ij} V_j \quad i = 0, \ldots, n-1.$$

This expression recovers the vector $\boldsymbol{v}$ because, for any field $\boldsymbol{F}$,

$$x^n - 1 = (x-1)(x^{n-1} + x^{n-2} + \cdots + x + 1),$$

from which we can write $\sum_{j=0}^{n-1} \omega^{rj} = 0$ for $r \neq 0 \pmod{n}$ because ω is an element of order n, and $\sum_{j=0}^{n-1} \omega^{rj} = n$ if $r = 0$. Therefore

$$\begin{aligned} \frac{1}{n}\sum_{j=0}^{n-1} \omega^{-ij} V_j &= \frac{1}{n}\sum_{k=0}^{n-1} v_k \sum_{j=0}^{n-1} \omega^{(k-i)j} \\ &= v_i \end{aligned}$$

for $i = 0, \ldots, n-1$.

The *cyclic convolution* of two vectors $\boldsymbol{f}$ and $\boldsymbol{g}$ of blocklength n is defined componentwise as

$$e_i = \sum_{k=0}^{n-1} f_{((i-k))} g_k \qquad i = 0, \ldots, n-1,$$

where the double parentheses denote modulo n on the indices. The cyclic convolution of $\boldsymbol{f}$ and $\boldsymbol{g}$ is denoted $\boldsymbol{f} * \boldsymbol{g}$.

Theorem 9.12.1 (Convolution theorem) *The vector $\boldsymbol{e}$ is given by the cyclic convolution of the vectors $\boldsymbol{f}$ and $\boldsymbol{g}$ if, and only if, the components of the Fourier transforms $\boldsymbol{F}$*

and $\boldsymbol{G}$ *satisfy*

$$E_j = F_j G_j \qquad j = 0, \ldots, n-1.$$

Proof This holds because

$$\begin{aligned} e_i &= \sum_{k=0}^{n-1} f_{((i-k))} \left(\frac{1}{n} \sum_{j=0}^{n-1} \omega^{-jk} G_j \right) \\ &= \frac{1}{n} \sum_{j=0}^{n-1} \omega^{-ij} G_j \left(\sum_{k=0}^{n-1} \omega^{(i-k)j} f_{((i-k))} \right) = \frac{1}{n} \sum_{j=0}^{n-1} \omega^{-ij} G_j F_j. \end{aligned}$$

We now recognize the right side of this equation as an inverse Fourier transform, which means that $E_j = F_j G_j$, as was to be proved. □

Theorem 9.12.2 (Modulation and translation) *If* $\boldsymbol{v} \leftrightarrow \boldsymbol{V}$ *is a Fourier transform pair, then the following, specified componentwise, are Fourier transform pairs*

$$\begin{aligned} \omega^k v_i &\leftrightarrow V_{((j+k))} \\ v_{((i-k))} &\leftrightarrow \omega^k V_j. \end{aligned}$$

Proof The theorem is proved by immediate substitutions. □

Theorem 9.12.3 (Repetition) *If* $\boldsymbol{v}' \leftrightarrow \boldsymbol{V}'$ *is a Fourier transform pair over the field* $\boldsymbol{F}$ *of blocklength* n' *and the vector* $\boldsymbol{v}'$ *is repeated* n'' *times to form a vector* $\boldsymbol{v}$ *of blocklength* $n = n'n''$*, then the Fourier transform of* $\boldsymbol{v}$ *is a vector* $\boldsymbol{V}$ *of blocklength* n *with components* V_j *given by* $V_{nj'} = V'_j$ *and* $V_j = 0$ *for all other* j.

Proof Let ω be an element of order n in $\overline{\boldsymbol{F}}$. Then $\beta = \omega^{n''}$ is an element of order n' in $\overline{\boldsymbol{F}}$, and $\gamma = \omega^{n'}$ is an element of order n'' in $\overline{\boldsymbol{F}}$. Let

$$\begin{aligned} i &= i' + n'i'' \qquad i' = 0, \ldots, n'-1; \quad i'' = 0, \ldots, n''-1 \\ j &= n''j' + j'' \qquad j' = 0, \ldots, n'-1; \quad j'' = 0, \ldots, n''-1. \end{aligned}$$

With this alternative indexing, the equation

$$V_j = \sum_{i=0}^{n-1} \omega^{ij} v_i$$

becomes

$$V_{n''j'+j''} = \sum_{i'=0}^{n'-1} \sum_{i''=0}^{n''-1} \beta^{i'j'} \gamma^{i''j''} \omega^{i'j''} v_{i'+n'i''}.$$

Now regard $\boldsymbol{v}$ as a two-dimensional n'-by-n'' array indexed by i' and i''. By the premise of the theorem, all rows of the array are identical and equal to $\boldsymbol{v}'$. Thus $v'_{i'+n''i''} = v'_{i'}$ independent of i'', so that

$$V_{n''j'+j''} = \sum_{i'=0}^{n'-1} \beta^{i'j'} \left[\omega^{i'j} \sum_{i'=0}^{i'-1} \gamma^{i''j''} v'_{i'} \right].$$

Each columnwise Fourier transform corresponding to the inner sum is the Fourier transform of a constant, and so is equal to v_i in the top row, and is equal to zero in all other rows. This means that all rowwise Fourier transforms, except for the first row, are all zero. The Fourier transform of the first row is $\boldsymbol{V}'$. □

The following theorem is a companion to Theorem 9.12.3. The summation in the equation of the theorem is referred to as *aliasing*. The theorem then can be summarized as the statement that decimation of $\boldsymbol{v}$ corresponds to aliasing of $\boldsymbol{V}$.

Theorem 9.12.4 (Decimation) *If $\mathbf{v} \leftrightarrow \mathbf{V}$ is a Fourier transform pair over the field $\boldsymbol{F}$ of blocklength $n = n'n''$ and $\mathbf{v}$ is decimated by selecting every n''th component of $\mathbf{v}$ starting with v_0 to form the decimated vector $\mathbf{v}'$ of blocklength n' with components $v'_{i'} = v_{n''i'}$ for $i' = 0, \ldots, n' - 1$, then the Fourier transform $\mathbf{V}'$ of $\mathbf{v}'$ is given by*

$$V'_j = \frac{1}{n''} \sum_{j''=0}^{n''-1} V_{((j'+n'j''))}.$$

Proof Write the spectral index j in terms of a vernier index j' and a course index j''

$$j = j' + n'j''; j' = 0, \ldots, n' - 1 \quad \text{and} \quad j'' = 0, \ldots, n'' - 1.$$

Then, because $n'n'' = 1$ and ω has order n, we can write

$$\begin{aligned} v'_i = v_{n''i'} &= \frac{1}{n} \sum_{j'=0}^{n'-1} \sum_{j''=0}^{n''-1} \omega^{-n''i'j'} \omega^{-n''n'i'j''} V_{((j'+n'j''))} \\ &= \frac{1}{n'} \sum_{j'=0}^{n'-1} \beta^{-i'j'} \left[\frac{1}{n''} \sum_{j''=0}^{n''-1} V_{((j'+n'j''))} \right], \end{aligned}$$

where $\beta = \omega^{n''}$ is an element of order n'. The inverse Fourier transform completes the proof. □

9.13 Existence of finite fields

We have asserted that the field $\boldsymbol{F}_q$ exists if q is a prime or a power of a prime, and that a field $\boldsymbol{F}_q$ does not exist if q is not a prime or a power of a prime. Now we turn to the proof of these two assertions.

First, we deal with the positive assertion regarding existence. We have given a construction to obtain the extension field $\boldsymbol{F}_{p^m}$ from the ground field $\boldsymbol{F}_p$ using an irreducible univariate polynomial of degree m over $\boldsymbol{F}_p$. Thus, a proof of the existence of $\boldsymbol{F}_{p^m}$ now comes down to a proof that for every positive integer m, there does exist an irreducible polynomial of degree m over $\boldsymbol{F}_p$. Indeed, we shall see that over every $\boldsymbol{F}_p$, there are many irreducible polynomials of each degree.

To count the irreducible polynomials, we make use of the *Möbius function* $\mu(n)$, which is defined on $\boldsymbol{Z}$ in terms of the factorization of the integer n as

$$\mu(n) = \begin{cases} 0 & \text{if } n \text{ has a repeated prime factor} \\ 1 & \text{if } n \text{ has an even number of distinct prime factors} \\ -1 & \text{if } n \text{ has an odd number of distinct prime factors.} \end{cases}$$

To avoid any misinterpretations, we remark specifically that $\mu(1)$ is equal to one, which does follow from the definition by regarding one to have no prime factors and zero to be an even number. Thus for $n = 1, 2, 3, 4, 5, 6, 7, 8, \ldots$, we find that $\mu(n) = 1, -1, -1, 0, -1, 1, -1, 0, \ldots$.

Theorem 9.13.1 *For each value of n, the Möbius function satisfies*

$$\sum_{d|n} \mu(d) = \begin{cases} 1 & \textit{if} \quad n = 1 \\ 0 & \textit{if} \quad n > 1, \end{cases}$$

where the expression $d|n$ ranges over all positive integers d that divide n including 1 *and n.*

Proof Every nonempty finite set has as many subsets with an even number of elements as it has subsets with an odd number of elements, where the empty set is regarded as having an even number of elements. This is a manifestation of the fact that there are as many binary sequences of length n with an even number of ones as with an odd number of ones.

The theorem is trivial with $n = 1$, so we may consider only $n > 1$. Let $\mathcal{S}_n$ be the set of prime factors of n, and suppose that there are no repeated prime factors. Every d that divides n itself has a set of prime factors that is a subset of the set of prime factors of n. The number of subsets of $\mathcal{S}_n$ with an even number of primes is equal to the number of subsets with an odd number of primes, so the sum given in the theorem equals zero.

Now suppose that n has at least one repeated prime factor. Then $\mu(d)$ is zero for terms involving a repeated prime. Therefore the sum can be restricted to only those d with no repeated prime factors. This amounts to considering only subsets of the set of *distinct* prime factors, which is now equivalent to the case already considered. □

For every integer d that divides n there is another integer n/d that also divides n. Therefore, replacing d by n/d, the summation of Theorem 9.13.1 can be rewritten as

$$\sum_{(n/d)|n} \mu\left(\frac{n}{d}\right) = \begin{cases} 1 & \text{if} \quad n = 1 \\ 0 & \text{if} \quad n > 1 \,. \end{cases}$$

This seems like a needlessly indirect way of writing the theorem, but it will prove to be useful. It leads easily to the *Möbius inversion formula* in the next theorem.

Given any function $g(n)$ on the positive integers, the *Möbius transform* of $g(n)$ is defined as

$$G(n) = \sum_{d|n} g(d).$$

To facilitate the proof of the next theorem, we define two sets based on the factors of n, given by $\mathcal{S}(n) = \{(e, m) : m|e|n\}$ and $\mathcal{T}(n) = \{(m, r) : (m|n, r|(n/m)\}$. These two sets are equivalent under the direct map $(e, m) \mapsto (m, e/m)$ and the inverse map $(m, r) \mapsto (mr, m)$. To sum over all integer pairs in $\mathcal{S}(n)$ is equivalent to summing over all integer pairs in $\mathcal{T}(n)$.

Theorem 9.13.2 (Möbius inversion formula) *Any function $g(n)$ on the integers can be recovered from its Möbius transform $G(n) = \sum_{d|n} g(d)$ by the inverse Möbius transform*

$$g(n) = \sum_{e|n} \mu\left(\frac{n}{e}\right) G(e).$$

Proof

$$\begin{aligned} \sum_{e|n} \mu\left(\frac{n}{e}\right) G(e) &= \sum_{e|n} \mu\left(\frac{n}{e}\right) \sum_{m|e} g(m) \\ &= \sum_{e|n} \sum_{m|e} g(m) \mu\left(\frac{n}{e}\right). \end{aligned}$$

This sum is a sum over the elements of $\mathcal{S}(n)$. This is equivalent to a sum over the elements of $\mathcal{T}(n)$. Therefore

$$\begin{aligned} \sum_{e|n} \mu\left(\frac{n}{e}\right) G(e) &= \sum_{m|n} \sum_{r|(n/m)} g(m) \mu\left(\frac{n}{e}\right) \\ &= \sum_{m|n} g(m) \sum_{r|(n/m)} \mu\left(\frac{n/m}{r}\right) \end{aligned}$$

because $e = mr$. But, by Theorem 9.13.1, the second summation is zero unless $n/m = 1$. Therefore $\sum_{e|n} \mu\left(\frac{n}{e}\right) G(e) = g(n)$, as was to be proved. □

Theorem 9.13.3 *The number of irreducible polynomials of degree n over $\boldsymbol{F}_q$, denoted $L_q(n)$, is given by*

$$L_q(n) = \frac{1}{n} \sum_{d|n} \mu\left(\frac{n}{d}\right) q^d.$$

Proof For any $\boldsymbol{F}_q$, the polynomial $x^{q^m} - x$ is the product of all monic polynomials that are irreducible over $\boldsymbol{F}_q$ and whose degree d divides m. There are $L_q(d)$ polynomials of degree d and their product has degree $dL_q(d)$. Therefore, equating the degree of the polynomial $x^{q^m} - x$ and the degree of the product of its polynomial factors gives

$$q^m = \sum_{d|m} dL_q(d).$$

The Möbius inversion formula then states that

$$dL_q(d) = \sum_{r|d} \mu\left(\frac{d}{r}\right) q^r,$$

which provides the completion of the proof. □

For example, inspection shows there are three irreducible polynomials of degree four over $\boldsymbol{F}_2$. These are $x^4 + x^3 + 1$, $x^4 + x + 1$, and $x^4 + x^3 + x^2 + x + 1$. There are no other irreducible polynomials of degree four over $\boldsymbol{F}_2$. To reach this same conclusion from Theorem 9.13.3, write

$$\begin{aligned} L_2(4) &= \frac{1}{4} \sum_{d|4} \mu\left(\frac{4}{d}\right) 2^d \\ &= \frac{1}{4}[\mu(4)2^1 + \mu(2)2^2 + \mu(1)2^4] \\ &= \frac{1}{4}[0 - 2^2 + 2^4] = 3, \end{aligned}$$

so the theorem states that there are three irreducible polynomials of degree four over $\boldsymbol{F}_2$.

For a second example, let $n = 5$ to obtain

$$\begin{aligned} L_2(5) &= \frac{1}{5}[\mu(5)2^1 + \mu(1)2^5] \\ &= \frac{1}{5}[-2 + 32] = 6, \end{aligned}$$

so there are six irreducible polynomials of degree five over $\boldsymbol{F}_2$.

For a final example, let $n = 6$ to obtain

$$\begin{aligned} L_2(6) &= \frac{1}{6}[\mu(6)2^1 + \mu(3)2^2 + \mu(2)2^3 + \mu(1)2^6] \\ &= \frac{1}{6}[2 - 4 - 8 + 64] \\ &= 9, \end{aligned}$$

so there are nine irreducible polynomials of degree six over $\boldsymbol{F}_2$.

To show that irreducible polynomials do exist for every value of n, we now only need to show that $L_q(n)$, as given in Theorem 9.13.3, is not equal to zero for any value of n.

Corollary 9.13.4 *Irreducible polynomials of every degree exist over $\boldsymbol{F}_q$.*

Proof Theorem 9.13.3 states that the number of irreducible polynomials is

$$L_q(n) = \frac{1}{n}\sum_{d:d|n} \mu\left(\frac{n}{d}\right)q^d.$$

Because $\mu(1) = 1$, this leads to

$$\begin{aligned} L_q(n) &= \frac{1}{n}\left(q^n + \sum_{\substack{d:d|n \\ d\neq n}} \mu\left(\frac{n}{d}\right)q^d\right) \\ &\geq \frac{1}{n}\left[q^n - \sum_{\substack{d:d|n \\ d\neq n}} q^d\right] \geq \frac{1}{n}\left[q^n - \sum_{i=0}^{n-1} q^i\right] \\ &= \frac{1}{n}\left[q^n - \frac{1-q^n}{1-q}\right] = \frac{1}{n}\left[\frac{q^{n+1} - 2q^n + 1}{q^n - 1}\right]. \end{aligned}$$

The numerator of the final term is equal to one if $q = 2$, and is clearly positive if q is larger than 2. Therefore, $L_q(n) > 0$ for all n. This means that there is at least one irreducible polynomial of degree n over $\boldsymbol{F}_q$, as was to be proved. □

Some of the monic irreducible polynomials have the special property that their zeros in the extension field are primitive elements of the extension field. These are called *primitive polynomials.* Whereas irreducible polynomials are counted by using the Möbius function $\mu(n)$, primitive polynomials are counted by using the totient function $\phi(n)$.

Theorem 9.13.5 *For every positive integer m and for every finite field $\boldsymbol{F}_q$, there exist exactly $\phi(q^m - 1)/m$ primitive polynomials of degree m over $\boldsymbol{F}_q$.*

Proof Corollary 9.13.4 shows that the finite field $\boldsymbol{F}_{q^m}$ does exist. Because $\boldsymbol{F}^*_{q^m}$ is a cyclic group under multiplication, there is a primitive element α with order $q^m - 1$, and this element has a minimal polynomial $m_\alpha(x)$ of degree m. Moreover the element α^i also has order $q^m - 1$ for every i that is coprime with $q^m - 1$. Every such α^i is also a primitive element, and so every α^i also has a minimal polynomial $m_{\alpha^i}(x)$ of degree m. Thus, there are $\phi(q^m - 1)$ primitive elements, and each primitive element has a minimal polynomial of degree m, and so that polynomial must be the minimal polynomial for the m conjugates of α^i, all of which themselves are primitive elements. Therefore, there are $\phi(q^m - 1)/m$ such minimal polynomials. These minimal polynomials are the indicated primitive polynomials. □

The theorem allows one to conclude that $\phi(q^m - 1)/m$ must be an integer whenever q is a prime power, as is indeed the case.

For an example of the theorem, $\phi(2^4 - 1)/4 = 2$, so there are only two primitive polynomials of degree four over $\boldsymbol{F}_2$. These are $x^4 + x + 1$ and $x^4 + x^3 + 1$. As primitive polynomials, these must both be irreducible polynomials. The third irreducible polynomial of degree four is $x^4 + x^3 + x^3 + x + 1$, which is not a primitive polynomial.

Furthermore, $\phi(2^5 - 1)/5 = 6$, so there are six primitive polynomials of degree five over $\boldsymbol{F}_2$. We saw earlier that there are six irreducible polynomials of degree five over $\boldsymbol{F}_2$, which means that every irreducible polynomial of degree five over $\boldsymbol{F}_2$ is a primitive polynomial. Finally, $\phi(2^6 - 1)/6 = 6$, so there are six primitive polynomials of degree six over $\boldsymbol{F}_2$. The other three irreducible polynomials of degree six over $\boldsymbol{F}_2$ are not primitive.

The following theorem states that the number of primitive polynomials is nonzero for all m.

Theorem 9.13.6 *For every prime power q, primitive polynomials of every degree exist over $\boldsymbol{F}_q$.*

Proof Theorem 9.13.5 states that there are $\phi(q^m - 1)/m$ primitive polynomials of degree m over $\boldsymbol{F}_q$, so it is enough to observe that the totient function $\phi(n)$ is never zero. The totient function $\phi(n)$ must be at least one because the integer one is always a positive integer that divides n. □

To complete the discussion of the existence of finite fields, we will show that the finite field $\boldsymbol{F}_q$ is unique up to *isomorphism*. This is a formal way of saying the finite field $\boldsymbol{F}_q$ is unique but for notation. This means that if two finite fields have the same number of elements, then they have the same internal structure although this equivalence may be obscured by the presentation. Thus they are the same finite field, possibly expressed differently and with different notation. From a computational point of view, although this statement is true, it may be misleading. For practical computational purposes, two different notational representations may be regarded as very different fields. In Chapter 12, we will look closely at the relationship between the notational representation of a field and the complexity of implementation.

Finally, we turn to the statement regarding the nonexistence of a field with n elements if n is not a power of a prime. We express this as the two positive statements that every finite field contains a prime field $\boldsymbol{F}_p$, and that a finite field with q elements exists only if q is a prime or a power of a prime.

Theorem 9.13.7 *Every finite field has a unique smallest subfield and this subfield has a prime number of elements.*

Proof The subfield must contain the elements zero and one. The element one generates a cyclic group under addition because any element of the finite field generates a cyclic group under addition. If n is the order of the group, this is the group $\boldsymbol{Z}_n$. Let α and β be two elements in the group generated under addition by the element one. Then by the distributive law,

$$\begin{aligned}\alpha\beta &= (1 + 1 + \cdots + 1)\beta \\ &= \beta + \beta + \cdots + \beta,\end{aligned}$$

where there are α copies of β in the sum, and the addition is modulo n. Hence, multiplication is also modulo n. In turn, we see that the multiplicative sequence β, β^2, β^3, is a cyclic subgroup of S. Thus S contains the identify element one under multiplication, and so β has an inverse under multiplication. This means that $\boldsymbol{Z}_n$ is a field, so by Theorem 9.4.1, n is a prime p. □

We next show that, because a finite field must contain a subfield $\boldsymbol{F}_p$ with p a prime, it must consist of p^m elements for some m.

Theorem 9.13.8 *The number of elements in any finite field $\boldsymbol{F}_q$ is a power of a prime p.*

Proof Theorem 9.13.7 tells us that the field $\boldsymbol{F}_q$ must contain a prime field $\boldsymbol{F}_p$. Let α be a primitive element of $\boldsymbol{F}_q$. Let m be the degree of the minimal polynomial of α over $\boldsymbol{F}_p$. Then every element of $\boldsymbol{F}_q$ can be represented uniquely as a polynomial of degree at most $m - 1$, with coefficients in $\boldsymbol{F}_p$. There are m coefficients and so there are p^m such polynomials. □

9.14 Bivariate polynomials

A *bivariate polynomial* over the field $\boldsymbol{F}$ is a mathematical object given by

$$p(x, y) = \sum_i \sum_j p_{ij} x^i y^j,$$

where the p_{ij} are elements of the field $\boldsymbol{F}$, called *coefficients*, and x and y are formal symbols, called *indeterminates*. The expressions $x^i y^j$ are called the *bivariate monomials* of the polynomial, and the expressions $p_{ij}x^i y^j$ are called the *terms* of the polynomial. The subscript on a coefficient is called the *bivariate index* of the coefficient. The *zero bivariate polynomial* is the bivariate polynomial with all coefficients equal to zero. The set of all bivariate polynomials over the field $\boldsymbol{F}$ is denoted $\boldsymbol{F}[x, y]$. A *bivariate rational function* is a formal expression of the form $p(x, y)/q(x, y)$ where $p(x, y)$ and $q(x, y)$ are bivariate polynomials. The set of all bivariate rational functions over $\boldsymbol{F}$ is denoted $\boldsymbol{F}(x, y)$.

Two bivariate polynomials, $p(x, y)$ and $q(x, y)$, can be added by the rule

$$p(x, y) + q(x, y) = \sum_i \sum_j (p_{ij} + q_{ij})x^i y^j.$$

With this definition of polynomial addition, the set $\boldsymbol{F}[x, y]$ of all bivariate polynomials over the field $\boldsymbol{F}$ forms a group. Two bivariate polynomials $p(x, y)$ and $q(x, y)$ can be multiplied by the rule

$$p(x, y)q(x, y) = \sum_i \sum_j \sum_{i'} \sum_{j'} p_{i'j'}q_{(i-i')(j-j')}x^i y^j.$$

With this definition of polynomial multiplication, the set $\boldsymbol{F}[x, y]$ of bivariate polynomials over the field $\boldsymbol{F}$ is a commutative ring with identity, also called a *ring with unit* or a *unital ring*. To this limited extent, $\boldsymbol{F}[x, y]$ can be regarded as a generalization of $\boldsymbol{F}[x]$. However, many familiar properties of $\boldsymbol{F}[x]$ do not hold in $\boldsymbol{F}[x, y]$. Indeed, $\boldsymbol{F}[x, y]$ has considerable complexity, which is sometimes an impediment and is sometimes quite useful.

The operation of multiplication of bivariate polynomials, in general, does not have an inverse. In some instances however, as when $c(x, y) = a(x, y)b(x, y)$ multiplication does have an inverse operation, called *division*, and written $a(x, y) = c(x, y)/b(x, y)$. The polynomials $a(x, y)$ and $b(x, y)$ are called *factors* of $c(x, y)$. A polynomial with no nontrivial bivariate factors is called an *irreducible bivariate polynomial*. In contrast to polynomials in $\boldsymbol{F}[x]$, which always factor in a sufficiently large extension field, polynomials in $\boldsymbol{F}[x, y]$ usually do not factor in any extension field.

Whereas the notion of degree for polynomials in a single variable is straightforward, the notion of degree for polynomials in two variables takes several forms. The x-degree of $p(x, y)$ is the largest value of i for which p_{ij} is nonzero for some value of j. The y-degree of $p(x, y)$ is the largest value of j for which p_{ij} is nonzero for some value of i. The *total degree*, or simply *degree*, of a bivariate polynomial $p(x, y)$ is the largest value of $i + j$ of any term of $p(x, y)$ with a nonzero coefficient. The total degree defines only a partial order because if $i + j = i' + j'$, then $x^i y^j$ and $x^{i'} y^{j'}$ are not ordered by the total degree. To make it into a total order, monomials of the same total degree are then ordered by the value i. This means that (i, j) is greater than (i', j') if $i + j$ is greater than $i' + j'$ or if $i + j = i' + j'$ and i is greater than i'. This total order on the elements of $\boldsymbol{F}[x, y]$ is called the *graded order*. The *leading monomial* of $p(x, y)$ is the monomial with nonzero coefficient and largest graded order. A bivariate polynomial is called a *monic bivariate polynomial* if the coefficient of the leading monomial is a one.

A bivariate polynomial can be *evaluated* at any point (β, γ) of the affine plane $\boldsymbol{F}^2$ by replacing the indeterminates x and y by the field elements β and γ. Then the field element

$$p(\beta, \gamma) = \sum_i \sum_j p_{ij} \beta^i \gamma^j$$

is called the *evaluation* of $p(x, y)$ at the point (β, γ).

If $p(\beta, \gamma) = 0$, then the point (β, γ) of $\boldsymbol{F}^2$ is called a *zero* of the bivariate polynomial $p(x, y)$. In contrast to a univariate polynomial, which can have only a finite number of zeros, a bivariate polynomial over $\boldsymbol{F}$ can have an infinite number of zeros. The set of zeros of $p(x, y)$ in the plane $\boldsymbol{F}^2$ is called a *plane curve*. In an algebraically closed field, to see only a finite number of zeros, one must look for the simultaneous zeros of two bivariate polynomials. A simultaneous zero of $p(x, y)$ and $q(x, y)$ is called an *intersection* of the plane curves defined by $p(x, y)$ and $q(x, y)$.

The two bivariate polynomials $p(x, y)$ and $q(x, y)$ of degree m and n, respectively, may both have a zero at a common point (β, γ). Indeed, they may both have a zero at each of multiple common points, but the number of possible common zeros is limited by the degrees of those two polynomials unless they have a common polynomial factor. The familiar statement that a univariate polynomial cannot have more zeros than its degree generalizes to the statement that two bivariate polynomials with no common polynomial factor cannot have more common zeros than the product of their degrees. This statement is known as *Bézout's theorem*. Moreover, Bézout's theorem says that if multiple zeros (which we do not formally define) are counted as such, then there are exactly mn common zeros of the two polynomials in the algebraically closed projective plane, as discussed in Section 10.10. Some of these zeros may be at points at infinity. Some of these zeros may be multiple zeros, and so counted with their multiplicity.

For example, Bézout's theorem says that the two polynomials $x^2 + y^2 - 1$ and $ax^2 + by^2 - c^2$ over $\boldsymbol{R}$, if distinct, have not more than four common zeros in $\boldsymbol{R}^2$, and exactly four zeros in the algebraically closed projective plane over $\boldsymbol{R}$. Because none of these common zeros are in the affine plane $\boldsymbol{C}^2$ if, for example, $a = b = 1$ and $c = 2$, they must then be zeros at infinity, possibly multiple zeros.

To visualize the meaning of a multiple zero, consider the situation in which the second bivariate polynomial $g(x, y)$ is the trivial polynomial y. Then we want to find the solutions to

$$\begin{aligned} p(x, y) &= 0 \\ y &= 0, \end{aligned}$$

which becomes $p(x, 0) = 0$. This situation reduces to the task of finding the zeros of the univariate polynomial $p(x, 0)$. There is a zero of multiplicity m at point $P = (x, 0)$ if $p(x, 0)$ and its first $m - 1$ derivatives all vanish at P. More generally, to visualize two curves with a multiple zero at point P, the two curves can be visualized as tangent to each other at the point P, one artificially straightened and the coordinate system rotated to make this case resemble the trivial case with $y = 0$. To do this rigorously, especially for finite fields, becomes rather technical. This is not important to us because, except for the points at infinity, we deal only with zeros of multiplicity one.

The ring of bivariate polynomials $\boldsymbol{F}[x, y]$ has ideals, just as does the ring of univariate polynomials $\boldsymbol{F}[x]$. In general, an ideal of $\boldsymbol{F}[x, y]$ is generated by a fixed set of bivariate polynomials $\{g_1(x, y), \ldots, g_m(x, y)\}$, as is given by

$$I = \{a_1(x, y)g_1(x, y) + \cdots + a_m(x, y)g_m(x, y) \mid a_j(x, y) \in \boldsymbol{F}[x, y]\}.$$

The polynomials $a_j(x, y)$ are any elements of the ring $\boldsymbol{F}[x, y]$. In contrast to an ideal of $\boldsymbol{F}[x]$, an ideal of $\boldsymbol{F}[x, y]$, in general, is not generated by a single polynomial, although some ideals of $\boldsymbol{F}[x, y]$ are so generated. As a result of this fact, the ring of bivariate polynomials $\boldsymbol{F}[x, y]$ has a much richer structure than the ring $\boldsymbol{F}[x]$.

The generators of a given ideal I are not unique. For example, we may replace $g_1(x, y)$ by $g_1(x, y) - g_2(x, y)$ without changing the set I of all polynomial combinations. There are, in general, many sets of generator polynomials for a given ideal of bivariate polynomials. Is there a best set, or at least a standard set? This question requires us to consider a total order on the set of bivariate polynomials based on the degree, such as the graded order.

Because the graded order is a total order on the monomials of a bivariate polynomial, every polynomial of the ideal I has a leading monomial. Some monomials, however, are not the leading monomial of any polynomial of the ideal. The set of all monomials that are not the leading monomial of any polynomial of the ideal will be called the *footprint* of the ideal I, and denoted $\Delta(I)$. Thus

$$\Delta(I) = \{(i, j) | x^i y^j \text{ is not the leading monomial of any element of } I\}.$$

Every element of $\Delta(I)^c$ corresponds to a monomial, $x^i y^j$, that is the leading monomial of an element of the ideal I. An *exterior corner* of $\Delta(I)$ is a point of $\Delta(I)^c$ such that both $(i - 1, j)$ and $(i, j - 1)$ are not points of $\Delta(I)^c$. A *minimal basis* for the ideal I is a set of polynomials, $\{g_\ell(x, y)\} \subset I$, that consists of one monic polynomial corresponding to each exterior corner of the footprint $\Delta(I)$ of I. The Buchberger algorithm, which we will not describe, computes a minimal basis of the ideal I from a given basis for I. The Buchberger algorithm for bivariate polynomials can be regarded as playing the same role in $\boldsymbol{F}[x, y]$ that the euclidean algorithm for univariate polynomials plays in $\boldsymbol{F}[x]$.

A *standard basis*, or *Groebner basis*, of an ideal is a minimal basis for which no leading monomial of a basis polynomial appears in any other basis polynomial. It is straightforward to compute a standard basis from a minimal basis. There can be only one monic polynomial in a standard basis with a leading element corresponding to a given exterior corner, because, if there were two, their difference would be a polynomial in the ideal I whose leading monomial was not in $\Delta(I)^c$.

9.15 Modular reduction and quotient groups

Many rings or groups can be used to construct a smaller ring or group by a generalization of the method of *modular reduction*, a method that we have seen several times. This is a powerful technique that can construct a new algebraic structure with desirable properties from an algebraic structure that is already defined. This procedure is based on the notion of an equivalence relation. Given any set S, an *equivalence relation* on S is a partition of S into subsets. Elements a and b of S are then said to be equivalent if they are in the same subset, and the subset is now called an *equivalence class*. In the special case in which the set S is a group, the partition can be defined to respect the group operation. An equivalence class is then called a *coset*. For example, if $S = \mathbf{Z}$, then an equivalence class, modulo n, is defined as the *modular equivalence* $\{i\} = \{i + \ell n\}$, where i is fixed and ℓ varies over $\mathbf{Z}$.

An algebraic structure consisting of the set of all cosets of a group (or a ring) with respect to a subgroup is called a *quotient group* (or a *quotient ring*). The most familiar example is the construction of the quotient group $\{\{i\} | i = 0, \ldots, n-1\}$ from $\mathbf{Z}$. This quotient group is denoted $\mathbf{Z}/\langle n \rangle$ or $\mathbf{Z}_n$ which are commonly regarded as equivalent labels. We make a slight distinction. When convenient, we will use $\mathbf{Z}/\langle n \rangle$ to refer to the formal notion of a set of *equivalence classes*, or the set of *cosets*, $\mathbf{Z}/\langle n \rangle = \{\{0\}, \{1\}, \{2\}, \ldots, \{n-1\}\}$ where $\{i\} = \{i + \ell p\}$. We will use the notation $\mathbf{Z}_n$ to refer to the computational structure $\mathbf{Z}_n = \{0, 1, 2, 3, \ldots, n-1\}$ with modulo p addition. The elements of $\mathbf{Z}_n$ are the *canonical representatives* of the elements of $\mathbf{Z}/\langle n \rangle$.

Modular reduction in the ring of polynomials $\boldsymbol{F}[x]$ with respect to a polynomial $p(x)$ is defined in a similar way. The equivalence class of $a(x)$ is given by $\{a(x)\} = \{a(x) + \ell(x)p(x) : \ell(x) \in \boldsymbol{F}[x]\}$. Because this equivalence class respects the group operation, it is a coset. The set of equivalence classes will be denoted by $\boldsymbol{F}[x]/\langle p(x) \rangle$. The notation $\boldsymbol{F}[x]_{p(x)}$ will sometimes be used to denote the set of canonical representatives, given by

$$\boldsymbol{F}[x]_{p(x)} = \{a(x) \mid \deg a(x) < \deg p(x)\},$$

with addition and multiplication as polynomials reduced modulo $p(x)$. Each equivalence class of $\boldsymbol{F}[x]/\langle p(x) \rangle$ contains one and only one canonical representative and each canonical representative appears in only one unique equivalence class. Because

Table 9.2 *Some instances of modular reduction*

Host ring	Ring of equivalence classes	Ring of representatives
Integers $\mathbf{Z}$	$\mathbf{Z}/\langle n \rangle$	$\mathbf{Z}_n$
Polynomials $F[x]$	$F[x]/\langle p(x) \rangle$	$F[x]_{p(x)}$
	Coordinate ring	
Bivariate polynomials $F[x, y]$	$F[x, y]/\langle p(x, y) \rangle$	$F[x, y]_{p(x,y)}$
	Jacobian	
Divisors	D/D°	jac($\boldsymbol{X}$)
Bivariate polynomials $F[x, y]$	$F[x, y]/I$ $I = \langle p_1(x, y), \ldots, p_n(x, y) \rangle$	$F[x, y]_I$ $I = \langle p_1(x, y), \ldots, p_n(x, y) \rangle$

$\boldsymbol{F}[x]/\langle p(x) \rangle$ and $\boldsymbol{F}[x]_{p(x)}$, as defined here, are trivially isomorphic, we often refer to them interchangeably. As we have seen, the quotient ring $\boldsymbol{F}[x]_{p(x)}$ is a field if, and only if, $p(x)$ is irreducible. In this case, the appropriate name of the field will be preferred to the label $\boldsymbol{F}[x]_{p(x)}$.

Given any irreducible bivariate polynomial $p(x, y) \in \boldsymbol{F}[x, y]$, one can define the quotient ring $\boldsymbol{F}[x, y]/\langle p(x, y) \rangle$. Formally, this is the set of equivalence classes of bivariate polynomials in which two bivariate polynomials are equivalent if their difference is a multiple of $p(x, y)$. This quotient ring is called a *coordinate ring*. While the corresponding construction in the ring of univariate polynomials does form a field, this construction in the case of bivariate polynomials does not form a field. While this might seem to be a disappointment at first, the coordinate ring does have a different and important destiny.

The canonical representatives of a coordinate ring are easy to specify for a line, $y = ax + b$, or a parabola $y = ax^2 + bx + c$. This is because y appears in a trivial way in these expressions, so it can be used to eliminate y from every bivariate polynomial. Therefore, for these examples, the elements of the coordinate ring are polynomials in x. Thus we have the isomorphism $\boldsymbol{F}[x, y]_{y-ax-b} \simeq \boldsymbol{F}[x]$. This simple structure is not the general case. For example, the canonical representatives for the coordinate ring corresponding to the polynomial $x^2 + y^2 - 1$ consist of polynomials of degree one in y and arbitrary degree in x because $y^2 = 1 - x^2$ can be used to eliminate y^2 and all higher powers of y.

The powerful method of modular reduction will continue to recur frequently in later chapters. Table 9.2 summarizes some instances of this construction that will arise. In each instance, because the underlying set is a group, the construction makes use of the notion of a coset. The quotient group is the set of cosets, or equivalence classes induced by a specified subgroup. For example, in Section 11.7, the set of divisors of degree zero of a hyperelliptic curve will be partitioned in this way into equivalence classes called *divisor classes*. The set of these classes is called the *divisor class group* or the *jacobian*. In Section 10.17, the set of fractional ideals of a quadratic number field will be partitioned in this way into equivalence classes called *ideal classes*. The set of these ideal classes is called the *ideal class group*.

9.16 Factoring of univariate polynomials

The task of polynomial factoring in the field $\boldsymbol{F}$ is to compute all the prime factors of a given univariate monic polynomial in the field $\boldsymbol{F}$. These factors are the monic irreducible polynomial factors of the given polynomial. They are prime elements of $\boldsymbol{F}[x]$. For polynomials over the complex field (or any algebraically closed field), all prime polynomial factors are of degree one, and the task of factoring polynomial $s(x)$ amounts to the task of finding the zeros of $s(x)$. Finding the zeros of polynomials in the complex field or the real field usually employs numerical methods that do not resemble the algorithms for the factoring of integers. When the field is the real field, none of the prime factors will have degree larger than two. When the field is the rational field, the prime factors can have any degree. For example, the factors of $x^n - 1$ over the rationals are the cyclotomic polynomials studied in Section 9.9. There is a cyclotomic polynomial of every degree. Likewise, when the field is the finite field $\boldsymbol{F}_q$, there can be prime factors of every degree. In the cases of a finite field or the rational field, the factoring of polynomials begins to more closely resemble the factoring of integers because the algebra is discrete.

Given the monic univariate polynomial $f(x)$, the task of factoring is to compute the polynomial factorization

$$f(x) = p_1(x)^{e_1} p_2(x)^{e_2} \cdots p_k(x)^{e_k},$$

where the $p_i(x)$ are prime polynomials and the exponents e_i are integers.

It is enough to consider the case in which all exponents e_i are equal to one, because if any e_i is larger than one, then a preprocessing step that computes GCD$[f(x), f'(x)]$ will remove that factor. This is because the formal derivative $f'(x)$ is divisible by $p_i(x)^{e_i-1}$ whenever $f(x)$ is divisible by $p_i(x)^{e_i}$. Consequently, it is enough to consider factoring only polynomials of the "square-free" form

$$f(x) = p_1(x)p_2(x) \cdots p_k(x),$$

where all irreducible factors are distinct. The task of factoring polynomials of this form over the rational field or over a finite field may be somewhat suggestive of the task of factoring integers, but the methods in use are different.

Several algorithms for factoring polynomials in the finite field $\boldsymbol{F}_q$ are available. The *Berlekamp polynomial factoring algorithm* and the *Cantor–Zassenhaus polynomial factoring algorithm* are two appropriate algorithms. Of these, we will describe only the Berlekamp polynomial factoring algorithm.

An important application of any polynomial factoring algorithm will be found in the index-calculus method of computing discrete logarithms over the finite field $\boldsymbol{F}_{2^m}$, where m is larger than one or over $\boldsymbol{F}_{p^m}$ where p is an odd prime. The index-calculus method is the fastest method known for computing discrete logarithms over a finite field. It was described in Section 4.9 for computing discrete logarithms in a prime field, but it can be reformulated for computing discrete logarithms in any finite field. For a prime field, this method involves the factorization of integers into primes of $\boldsymbol{Z}$. Similarly, the elements of $\boldsymbol{F}_{p^m}$ regarded as polynomials over the base field $\boldsymbol{F}_{p^m}$ of degree at most m, must be factored into the prime polynomials of $\boldsymbol{F}[x]$. This factorization can be provided by the Berlekamp factoring algorithm.

The input to the Berlekamp factoring algorithm is a square-free monic polynomial $f(x)$ of degree n with coefficients in the finite field $\boldsymbol{F}_q$. The output of the Berlekamp factoring algorithm, unless $f(x)$ itself is a prime polynomial, is a polynomial $g(x)$ that properly divides $f(x)$ and has coefficients in the same field as $f(x)$. This gives the partial factorization $f(x) = g(x)(f(x)/g(x))$. The algorithm may then be applied, in turn, both to $g(x)$ and to $f(x)/g(x)$, and to all of their subsequent factors until reaching the decomposition of $f(x)$ into its irreducible polynomial factors.

To develop the Berlekamp factoring algorithm, recall that the quotient ring $\boldsymbol{F}_q[x]/\langle f(x)\rangle$ can be regarded as the set of all polynomial remainders under division by $f(x)$. This quotient ring also can be regarded as an n-dimensional vector space over $\boldsymbol{F}_q$ under polynomial addition, and as such will be denoted $\boldsymbol{V}$. Observe that all prime polynomial factors of $f(x)$, other than $f(x)$ itself, are contained within the quotient ring $\boldsymbol{F}_q[x]/\langle f(x)\rangle$. Some of the polynomials $g(x)$ of the quotient ring $\boldsymbol{F}_q[x]/\langle f(x)\rangle$ will satisfy the congruence,

$$g(x)^q = g(x) \pmod{f(x)},$$

others will not. Because of Theorem 9.3.3, we see that the set of polynomials solving this congruence is a vector subspace of $\boldsymbol{V}$. It will be denoted $\boldsymbol{A}$. This vector subspace $\boldsymbol{A}$ is not empty – it clearly contains both $g(x) = 0$ and $g(x) = 1$. It may also contain other polynomials. Indeed, some of these other $g(x)$ of $\boldsymbol{V}$ may also divide $f(x)$. We will look into this subspace in hopes of finding such a factor of $f(x)$. To this end, we make use of the following theorem.

Theorem 9.16.1 *If* $g(x) \in \boldsymbol{F}_q[x]/\langle f(x)\rangle$ *satisfies*

$$g(x)^q = g(x) \pmod{f(x)},$$

then

$$f(x) = \prod_{\beta \in \boldsymbol{F}_q} \text{GCD}(f(x), g(x) - \beta).$$

Proof Write the congruence defining the subspace in the form

$$g(x)^q - g(x) = 0 \pmod{f(x)}$$

so every $g(x)$ in $\boldsymbol{A}$ can be written in $\boldsymbol{F}_q[x]$ as

$$g(x)^q - g(x) = a(x)f(x)$$

for some polynomial $a(x)$. Next, recall that every element of $\boldsymbol{F}_p$ is a zero of $z^p - z$, so

$$z^p - z = z(z-1)(z-2)(z-3)\cdots(z-p+1).$$

Therefore, by similar reasoning,

$$\begin{aligned} g(x)^p - g(x) &= g(x)(g(x)-1)(g(x)-2)(g(x)-3)\cdots(g(x)-p+1) \\ &= \prod_{\beta \in \boldsymbol{F}_q} (g(x) - \beta). \end{aligned}$$

We can now write

$$a(x)f(x) = \prod_{\beta \in \boldsymbol{F}_q} (g(x) - \beta).$$

Therefore, each polynomial $g(x)$ that is contained in the vector subspace $\boldsymbol{A}$ must satisfy

$$f(x) = \prod_{\beta \in \boldsymbol{F}_q} \text{GCD}(f(x), g(x) - \beta).$$

as was to be proved. □

Every term of the product over β on the right side of the equation in the theorem must be a factor of $f(x)$. Possibly the only such nontrivial proper factor is $f(x)$ itself, in which case the algorithm fails for this $g(x)$. For other $g(x)$, there may be a nontrivial factor. Thus by computing $\text{GCD}(f(x), g(x) - \beta)$ for each element $g(x)$ of $\boldsymbol{A}$ and each field element β, one hopes to find a proper factor of $f(x)$.

The task now becomes the task of finding the appropriate elements of the vector space $\boldsymbol{A}$ so that the greatest common divisor in Theorem 9.16.1 can be computed. The Berlekamp algorithm finds such polynomials $g(x)$ by computing a basis for $\boldsymbol{A}$. This is achieved by the observation that the vector space $\boldsymbol{A}$ is the null space of a certain n-by-n matrix over $\boldsymbol{F}_q$, denoted $\boldsymbol{M}$. The matrix element m_{ij} is the coefficient of the jth-power term in the reduction of x^{iq} modulo $f(x)$. That is, each x^{iq} is written

$$x^{iq} \equiv m_{i,n}x^n + m_{i,n-1}x^{n-1} + \cdots + m_{i,1}x + m_{i,0} \pmod{f(x)}.$$

With each polynomial $g(x) = g_n x^n + g_{n-1}x^{n-1} + \cdots + g_1 x + g_0$, we associate the row vector $\boldsymbol{g} = (g_0, g_1, \ldots, g_n)$. In the same way, the row vector $\boldsymbol{g}$ corresponds to the reduction of $g(x)^q$ modulo $f(x)$. Consequently, a polynomial $g(x) \in R$ is in the set A if, and only if, $g(\boldsymbol{M} - \boldsymbol{I}) = \boldsymbol{0}$. This is so if, and only if, $\boldsymbol{g}$ is in the null space of $\boldsymbol{M} - \boldsymbol{I}$. To find a basis for the vector subspace A and thereby construct polynomials $g(x)$ in it, first compute the matrix $\boldsymbol{M} - \boldsymbol{I}$ and transform it to the reduced row–echelon form. Finally, from the matrix in this form, simply read off a basis for the null space. This gives the candidate $g(x)$. Then the greatest common divisors of the form described in Theorem 9.16.1 are computed in succession until a nontrivial factor of $f(x)$ is found.

Problems for Chapter 9

9.1 **a** Prove that in any group, the identity element is unique.
b Prove that in any group, the inverse of any element is unique.
c Prove that in any group, if b is the inverse of a, then a is the inverse of b.
d Prove that in any group, $(a^{-1})^{-1} = a$.

9.2 Show that if the ring R has an identity under multiplication from the left, that element is also an identity under multiplication from the right. That is, if 1 is the identity, and

$$1 \cdot a = a$$

for all a, then

$$a \cdot 1 = a$$

for all a.

9.3 **a** Demonstrate that $\mathbf{Z}_3 \oplus \mathbf{Z}_3$ has four nontrivial cyclic subgroups. Do they have the same order? What is their common order?
b How many cyclic subgroups does $\mathbf{Z}_p \oplus \mathbf{Z}_p$ have? Describe them. Do they have a nonempty intersection?

9.4 Prove that the identity element of a monoid is unique.

9.5 Suppose that H is a subgroup of the finite group G. Show that for any g_1 and g_2 in G, either $g_1 * H = g_2 * H$ or $(g_1 * H) \bigcap (g_2 * H)$ is empty. By the notation $g * H$ is meant the "left coset" of H, meaning the set $\{g * h : h \in H\}$.

9.6 Prove that the zero element in any ring R satisfies $a0 = 0a = 0$ for all $a \in R$.

9.7 Prove that in $\boldsymbol{F}[x]$, the ring of polynomials over the field $\boldsymbol{F}$, the only units are the polynomials of degree zero.

9.8 The ring of polynomials $\boldsymbol{C}[x]$ over the complex field includes polynomials with only real coefficients. Is the subset of polynomials with real coefficients a subring? Is it an ideal?

9.9 **a** Is the polynomial $x^4 + x^3 + x^2 + x + 1$ a prime polynomial over $\boldsymbol{F}_2$?
b Use this polynomial to construct a multiplication table for $\boldsymbol{F}_{16}$.
c With this representation of $\boldsymbol{F}_{16}$, does the element represented by the polynomial x have order fifteen?
d With this representation of $\boldsymbol{F}_{16}$, find a generator for the cyclic group $\boldsymbol{F}_{16}^*$ under multiplication.

9.10 **a** Prove that $x^2 + 1$ is irreducible over $\boldsymbol{F}_p$ if $p = 3 \pmod 4$.
b Prove that $x^2 + x + 1$ is irreducible over $\boldsymbol{F}_p$ if $p = 2 \pmod 3$.

9.11 Show that the quotient ring $\boldsymbol{F}[x]/\langle p(x)\rangle$ is indeed a ring under the natural definition of ring addition and ring multiplication. What are these definitions of ring addition and ring multiplication?

9.12 For the formal derivative of polynomials over a field $\boldsymbol{F}$, prove that

$$[r(x)s(x)]' = r'(x)s(x) + r(x)s'(x),$$

and that if $a(x)^2$ divides $r(x)$, then $a(x)$ divides $r'(x)$.

9.13 Show that if p is a prime of the form $4k + 3$, then $p - 1$ is a quadratic nonresidue. Show that for such p, $x^2 + 1$ is an irreducible polynomial, and that with $x^2 + 1$ used to construct $\boldsymbol{F}_{p^2}$, multiplication in $\boldsymbol{F}_{p^2}$ looks like multiplication in $\boldsymbol{C}$.

9.14 Prove that if $\text{GCD}[g_1(x), \ldots, g_n(x)] = 1$, then there exist polynomials $b_1(x), \ldots, b_n(x)$ satisfying $b_1(x)g_1(x) + \cdots + b_n(x)g_n(x) = 1$.

9.15 Write out the first twenty values of the function $L_q(n)$ given in Theorem 9.13.3. Write out the first twenty values of the totient function $\phi(n)$. Prove that $\phi(n) \leq L_q(n)$ for all n. Why is this to be expected?

9.16 Prove that for any prime p, a field $\boldsymbol{F}_{p^m}$ can always be constructed as an extension of $\boldsymbol{F}_p$, provided that an irreducible polynomial of degree m over $\boldsymbol{F}_p$ exists.

9.17 Prove that any polynomial $g(x)$ over $\boldsymbol{F}_q$ satisfies

$$g(x)^q - g(x) = \prod_{\beta \in \boldsymbol{F}_q} (g(x) - \beta).$$

9.18 Prove the Möbius inversion formula.

9.19 Let ω be an element of order n in the field $\boldsymbol{F}$, and let $\boldsymbol{v}$ be a vector of length n over the field $\boldsymbol{F}$. Define the *Fourier transform* of $\boldsymbol{v}$ as the vector $\boldsymbol{V}$, given by

$$V_j = \sum_{i=0}^{n-1} \omega^{ij} v_i \qquad j = 0, \ldots, n-1.$$

a Define and prove an inverse Fourier transform that computes $\boldsymbol{v}$ from $\boldsymbol{V}$.
b State and prove a convolution theorem.

9.20 Calculate the cyclotomic polynomial $\Phi_5(x)$ by working in the field $\boldsymbol{F}_{11}$.

9.21 Prove the following:
(i) Under ring multiplication, the set of units of a ring forms a group.
(ii) If $c = ab$ and c is a unit, then a has a right inverse and b has a left inverse.

(iii) If $c = ab$ and a has no right inverse or b has no left inverse, then c is not a unit.

9.22 Let G be a cyclic subgroup of $\boldsymbol{F}^*_{q^k}$ with prime order and with k larger than one that is not a subgroup of $\boldsymbol{F}^*_q$. Prove that G contains no element of $\boldsymbol{F}_q$ except the identity element.

9.23 Is it true that

$$\overline{\boldsymbol{F}}_{p^n} = \overline{\boldsymbol{F}}_p$$

for any value of the integer n?

9.24 Let $\boldsymbol{A}$ and $\boldsymbol{B}$ be two compatible square matrices over the field $\boldsymbol{F}$. Prove that the characteristic polynomials of $\boldsymbol{AB}$ and $\boldsymbol{BA}$ are equal.

9.25 Is factorization unique in a number ring? Hint: Factor 6 in $\boldsymbol{Z}[\sqrt{-5}]$.

9.26 An *algebraic integer* is a complex number that is a zero of a monic element of $\boldsymbol{Z}[x]$. A monic element of $\boldsymbol{Z}[x]$ is a monic univariate polynomial with all coefficients in $\boldsymbol{Z}$. An *integer* of the number field $\boldsymbol{Q}(\xi)$ is an element of $\boldsymbol{Q}(\xi)$ that is also an algebraic integer. The set of all integers of $\boldsymbol{Q}(\xi)$ is a ring. The ring of integers of $\boldsymbol{Q}(\xi)$ contains $\boldsymbol{Z}[\xi]$, the set of polynomials in ξ with integer coefficients, but it may be larger.

a Prove that the set of all algebraic integers is closed under addition and multiplication, and therefore is a subring of the set of complex numbers regarded as a ring.

b Prove that the set of integers of $\boldsymbol{Q}(\xi)$ is a ring, and contains $\boldsymbol{Z}[\xi]$.

c Suppose that $p(x) = x^2 - D$, where D is a square-free integer. Show that $p(x)$ is irreducible over $\boldsymbol{Q}$. Show that the ring of integers of $\boldsymbol{Q}(\sqrt{D})$ is given by

$$\boldsymbol{Z}[\sqrt{D}] = \{m + \alpha n \mid m, n \in \boldsymbol{Z}\},$$

where α is given by $\alpha = \sqrt{D}$ if $D \neq 1 \pmod 4$ and $\alpha = (1 + \sqrt{D})/2$ if $D = 1 \pmod 4$.

d Why does the definition of an integer of a number field require that the polynomial be monic?

Notes for Chapter 9

The formal subject of abstract algebra is a fundamental pillar of modern mathematics. Much of the basic structure was discovered and first described by Evariste Galois (1811–32) and by Niels Henrik Abel (1802–29). The abelian groups and Galois fields are so-named in their honor. Our understanding of much of the structure of the ring of univariate polynomials, and especially of the ring of bivariate polynomials, we owe to the work of Etienne Bézout (1730–83).

A great many textbooks are available that treat the material of this chapter in full detail. Most of these books develop this material from the point of view of formal mathematics, rather than from the point of view of an engineering application, or emphasizing the computational algebra. A time-honored classic is the two-volume work by van der Waerden (1950). The standard reference for the subject of finite fields is the book by Lidl and Niederreiter (1983). The books by Cox, Little, and O'Shea (1992) and by Blahut (2008) treat some of these topics from a computational or engineering point of view.

The task of factoring polynomials over finite fields mimics the task of factoring integers, but the methods are different. The Berlekamp algorithm (1967, 1970), and the Cantor–Zassenhaus algorithm (1981) are two practical and widely used methods of factoring polynomials over finite fields. The latter and more recent of the two algorithms is now usually preferred.

10 Cryptography based on elliptic curves

A *plane curve* $\mathcal{X}$ over the field $\boldsymbol{F}$ is the set of points (x, y) in the plane $\boldsymbol{F}^2$ that are the zeros of some fixed irreducible bivariate polynomial $p(x, y)$ over $\boldsymbol{F}$. If one can define a pairwise operation $(x, y) + (x', y')$ taking any two points (x, y) and (x', y') of the curve into a third point of the curve so as to form an abelian group, then one can use this group operation to define a public-key cryptography system in various ways. Of course, one then requires assurance that such a cryptosystem is secure. These topics comprise the subjects of elliptic-curve cryptography and elliptic-curve cryptanalysis. Together they form the subject of elliptic-curve cryptology.

Elliptic curves on finite fields are a very attractive class of plane curves that allow one to define a well-behaved operation on any two points of the curve. This operation, called *point addition*, forms a finite abelian group whose cyclic subgroups are used to form public-key cryptosystems, called *elliptic-curve cryptosystems*. Elliptic-curve cryptography is attractive because, in part, index calculus methods of attack have not been found for elliptic curves and are not expected because the notion of a smooth integer does not have a parallel for the points of an elliptic curve.[1] In fact, no satisfactory subexponential algorithm is known for solving the discrete-log problem on an elliptic curve. All the usual cryptography protocols that are based on the discrete-log problem, such as the Diffie–Hellman key exchange, the Elgamal cryptosystem, and the digital signature algorithm, can be used in any cyclic subgroup of the group of the elliptic curve. In each case, security is assured by the apparent intractability of the discrete-log problem on the group of an elliptic curve.

[1] From the point of view of formal mathematics, any two cyclic groups of the same prime order p are equivalent in that both are equivalent to $\boldsymbol{Z}_p$ under addition. However, for our purposes, $\boldsymbol{Z}_p$ is an unsatisfactory cyclic group of order p because the additional multiplicative structure of $\boldsymbol{Z}_p$ somehow leaks through into the additive group structure in a way that can be exploited when studying the cyclic structure of the additive group. This multiplicative structure is not used by the cryptosystem itself, but it does give the cryptanalyst an additional weapon with which to attack the cryptographic system. We prefer a cyclic group of order p, with p a large prime, in which no additional algebraic structure is useful to a cryptanalyst. Although the cyclic group to be developed will be isomorphic to $\boldsymbol{Z}_p$, it must be fundamentally different. What is needed to complete such a discussion is a statement that the isomorphism between the two cyclic groups is computationally intractable. What we will say instead is the weaker statement that no tractable method is known for solving the discrete-log problem on point addition over an elliptic curve.

10.1 Elliptic curves

Let $\boldsymbol{F}$ be a field. We will be briefly interested in the fields $\boldsymbol{R}$ and $\boldsymbol{C}$ in this section, and we will be primarily interested in the finite fields $\boldsymbol{F}_q$ here and in later sections. A *plane curve* $\mathcal{X}$ is the set of zeros in the plane $\boldsymbol{F}^2$ of a bivariate polynomial, $p(x, y)$. A plane curve is written as

$$\mathcal{X} = \{(x, y) \in \boldsymbol{F}^2 \mid p(x, y) = 0\}.$$

There are a great many bivariate polynomials over $\boldsymbol{F}$, so there are a great many plane curves over $\boldsymbol{F}$. In this chapter, we will restrict our discussion to the class of plane curves that is useful to us.

A plane curve can be defined to include extra points outside of the plane that are appended to the plane by a formal procedure which will be described later. These special points are called the *points at infinity*. When there is only one such point of the curve at infinity, as will be the case in this chapter, it is denoted ∞. Now, the plane curve is written:

$$\mathcal{X} = \{(x, y) \in \boldsymbol{F}^2 \ : \ p(x, y) = 0\} \cup \{\infty\}.$$

The point at infinity will be a very useful adjunct to the curves of this chapter.

Some curves have troublesome points called *singular points* that lead to annoying exceptions to many otherwise general mathematical statements that might be made. We will avoid these exceptions the easy way by considering only those curves that have no singular points. To define the notion of a nonsingular plane curve, define a singular point of the bivariate polynomial $p(x, y)$ as a point $P = (x, y)$ at which

$$\frac{\partial p(x, y)}{\partial x} = \frac{\partial p(x, y)}{\partial y} = p(x, y) = 0.$$

A curve $\mathcal{X}$ is called a *smooth curve* and the polynomial $p(x, y)$ is called a *nonsingular polynomial* if $\mathcal{X}$ has no singular points (including the point at infinity) in $\boldsymbol{F}$ or in any finite extension of $\boldsymbol{F}$. Although we must ensure that no point at infinity is a singular point, we are not yet ready to discuss this. A polynomial with one or more singular points is called a *singular polynomial*.

Every plane curve is associated with a certain scalar, called the *genus* of that curve. The genus is a useful parameter of a plane curve. It is complicated to define the genus in the general case of a plane curve, but for a nonsingular plane curve, an elementary definition of genus can be given. The genus of a nonsingular plane curve where $p(x, y)$ has degree d is given by the *Plücker formula* $g = \binom{d-1}{2}$. We will not define the genus for singular curves, as we have no need to do so. Because we only deal with nonsingular plane curves, we can take the Plücker formula as the definition of genus. An elliptic curve, defined below, has genus one.

An *elliptic curve* over the field $\boldsymbol{F}$ is the plane curve

$$\mathcal{X}(\boldsymbol{F}) = \{(x, y) \in \boldsymbol{F}^2 : y^2 + a_1xy + a_3y = x^3 + a_2x^2 + a_4x + a_6\} \cup \{\infty\},$$

provided the defining polynomial $p(x, y)$ is a nonsingular polynomial. The formal definition of an elliptic curve follows.

Definition 10.1.1 *An elliptic curve over the field $\boldsymbol{F}$ consists of the set of zeros of a nonsingular bivariate polynomial of the form*

$$p(x, y) = y^2 + a_1xy + a_3y - x^3 - a_2x^2 - a_4x - a_6,$$

together with the symbol ∞, which is called the point at infinity, and where $a_1, a_2, a_3, a_4,$ and a_6 are elements of the field $\boldsymbol{F}$.

The points of the curve $\mathcal{X}$ not at infinity are called the *affine points* of $\mathcal{X}$. All zeros of an elliptic curve over the field $\boldsymbol{F}$ are affine zeros except for the single zero at infinity. Accordingly, the plane $\boldsymbol{F}^2$ is called the *affine plane*. When the points at infinity are appended to the affine plane, the augmented plane is called the *projective plane*. If the field $\boldsymbol{F}$ is contained in a larger field $\boldsymbol{K}$, such as the extension field $\boldsymbol{F}_{q^m}$, or the algebraic closure $\overline{\boldsymbol{F}}$ of the field $\boldsymbol{F}$, then the points of $\mathcal{X}(\boldsymbol{K})$ that are also in $\mathcal{X}(\boldsymbol{F})$ are called the *rational points* of $\mathcal{X}(\boldsymbol{K})$.

A polynomial with the specific standard form of the coefficient indices given in Definition 10.1.1 is called the *Weierstrass form* of the elliptic polynomial.

Because an elliptic curve is required to be nonsingular, the two expressions formed from the partial derivatives, given by

$$a_1y = 3x^2 + 2a_2x + a_4$$

and

$$2y + a_1x + a_3 = 0,$$

must not be simultaneously satisfied at any point of the curve

$$\mathcal{X} : y^2 + a_1xy + a_3y = x^3 + a_2x^2 + a_4x + a_6.$$

Otherwise, the corresponding curve is not an elliptic curve, and so is not of interest to us. We must also be concerned that the point at infinity must not be a singular point, but we shall see in Section 10.10 that this point is never singular.

The zeros of the polynomial $p(x, y)$ are the points of the curve. A point may be designated as P or Q. Later, we will define an operation called point addition on the points of an elliptic curve so that $\chi(\boldsymbol{F})$ forms an abelian group.[2] Then the point at

[2] A theorem of Riemann and Roch implies that any smooth curve with a group structure given by rational maps must be an elliptic curve. Elliptic curves are the only curves for which point addition defines a group.

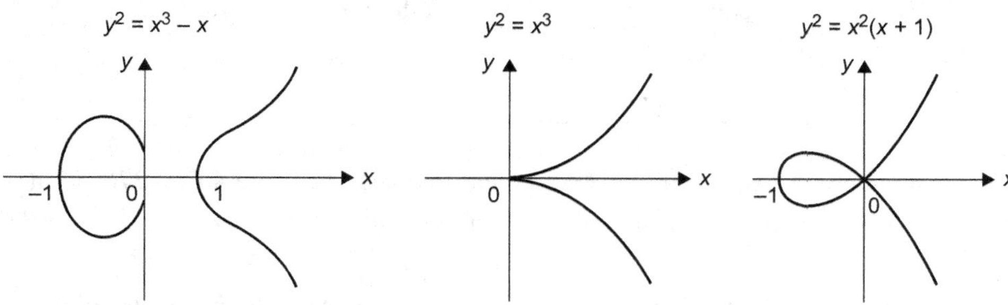

Figure 10.1 Three plane curves

infinity will play the role of the group identity. This is why we are careful to adjoin this extra point to the curve.

Examples of three curves over $\boldsymbol{R}$ based on polynomials in the form

$$\mathcal{X} : y^2 = x^3 + ax + b$$

are shown in Figure 10.1. For these examples, the coefficients of $p(x, y)$ are real numbers and the points of the curve lie in $\boldsymbol{R}^2$. Two of the examples in Figure 10.1, those with the polynomials $y^2 - x^3$ and $y^2 - x^2(x - 1)$, have singular points and so are not smooth curves. These singular points are the points of the curve at which both partial derivatives are equal to zero. Because these curves are not smooth, they are not elliptic curves, and they are of no interest to us.

If the characteristic of the field is neither two nor three, then by a suitable change in variables, we can always write the elliptic curve $\mathcal{X}$ as

$$\mathcal{X} : y^2 = x^3 + ax + b.$$

This is called the *short Weierstrass form* of the elliptic polynomial. An elliptic curve over $\boldsymbol{R}$ or $\boldsymbol{C}$ can always be written in the short Weierstrass form. Any elliptic curve over $\boldsymbol{F}_p$ can be written in the short Weierstrass form if p is larger than three.

When the polynomial is written in the short Weierstrass form, the condition for a nonsingular polynomial is easy to state in terms of the *discriminant* of the polynomial, or of the elliptic curve, defined as $4a^3 + 27b^2$, as follows.

Theorem 10.1.2 *The polynomial $y^2 - x^3 - ax - b$ is nonsingular over any field $\boldsymbol{F}$ of characteristic not two or three if, and only if, $4a^3 + 27b^2$ is nonzero in the field $\boldsymbol{F}$.*

Proof It is easy to see that the polynomial is nonsingular if, and only if, the cubic term on the right, $x^3 + ax + b$, has no repeated zeros. There will be a repeated zero if, and only if, we can write

$$\begin{aligned} x^3 + ax + b &= (x - x_1)^2(x - x_2) \\ &= x^3 - (2x_1 + x_2)x^2 + (x_1^2 + 2x_1x_2)x - x_1^2x_2. \end{aligned}$$

Therefore by equating coefficients on both sides, we obtain $2x_1 + x_2 = 0$, so this equation reduces to

$$x^3 + ax + b = x^3 - 3x_1^2 x + 2x_1^3$$

and neither two nor three is congruent to zero. Therefore $4a^3 + 27b^2 = 4(-3x_1^2)^3 + 27(2x_1^3)^2 = 0$ in $\boldsymbol{F}$ if, and only if, the polynomial is singular. □

Because the coefficients of a polynomial over the field $\boldsymbol{F}$ are also elements of any extension field of $\boldsymbol{F}$, a polynomial over $\boldsymbol{F}$ can also be regarded as a polynomial over any extension of $\boldsymbol{F}$. This means that a polynomial not only defines a curve in the field of the polynomial, but it also defines a curve in any extension of the field of the polynomial. When it is desired to emphasize that only the points in the field of the polynomial coefficients are under discussion, these points of the curve are the rational points of the curve, the terminology coming from the fact that a polynomial such as $x^2 + y^2 = 1$, regarded as a polynomial over $\boldsymbol{Q}$, has only rational points such as $\left(\pm\frac{3}{5}, \pm\frac{4}{5}\right)$ as its zeros. In particular, an elliptic curve over the rational field $\boldsymbol{Q}$, denoted $\mathcal{X}(\boldsymbol{Q})$, is a subset of the elliptic curve over the real field $\boldsymbol{R}$, denoted $\mathcal{X}(\boldsymbol{R})$, which, in turn, is a subset of the elliptic curve over the complex field $\boldsymbol{C}$, denoted $\mathcal{X}(\boldsymbol{C})$, based on the same polynomial. The elliptic curve over $\boldsymbol{C}$, however, cannot be easily visualized because the two variables x and y are each a complex number, each having both a real part and an imaginary part.

Another standard form of a polynomial for an elliptic curve, known as the *Legendre form*, is

$$y^2 = x(x-1)(x-\lambda)$$

where $\lambda \in \boldsymbol{F}$, a field whose characteristic is not two. Every elliptic curve over a field whose characteristic is not two can be put in the Legendre form by a suitable change of variables, but the parameter λ may be in an extension field of the field $\boldsymbol{F}$, and so the Legendre form may be undesirable.

An invertible coordinate transformation of the form

$$x = u^2 x' + r$$
$$y = u^3 y' + s u^2 x' + t,$$

where r, s, t, and u are arbitrary elements of $\boldsymbol{F}$ and u is nonzero, will give a different form of the elliptic curve, now in the variables x' and y'. In this case, we regard the new elliptic curve as equivalent to the original elliptic curve because they differ only by choice of the coordinate axes. The two curves are said to be *isomorphic* to each other. To identify that two elliptic curves $\mathcal{X}(\boldsymbol{F}_q)$ and $\mathcal{X}'(\boldsymbol{F}_q)$ are isomorphic, a constant called the *j-invariant* of the elliptic curve, with the field characteristic larger than three, is defined as

$$j(\mathcal{X}) = 1728\frac{4a^3}{4a^3 + 27b^2}$$

with the curve expressed in the short Weierstrass form. We will see that two elliptic curves in the same field with the same j-invariant are isomorphic. The denominator in the expression for the j-invariant is always nonzero if the polynomial is nonsingular, so $j(\mathcal{X})$ is finite for all elliptic curves. Of course, the j-invariant can be stated for an elliptic curve with coefficients expressed as in the long Weierstrass form, but then the expression is unwieldy and uninformative, and will not be presented here.

Theorem 10.1.3 *If two elliptic curves over the field $\boldsymbol{F}$ defined by polynomials with coefficients (a, b) and (a', b') in the short Weierstrass form have the same j-invariant, then there exists a nonzero u in $\overline{\boldsymbol{F}}$ such that*

$$x' = u^2 x \qquad y' = u^3 y$$

and

$$a' = u^4 a \qquad b' = u^6 b,$$

and so the two curves are isomorphic in the algebraic closure $\overline{\boldsymbol{F}}$.

Proof If a is zero, then the j-invariant is zero, so a' is zero as well. This means that b and b' are nonzero. In this case, b' is simply a multiple of b. If a is nonzero, let $u^4 = a/a'$, which means that $u^2 = \pm\sqrt{a/a'}$. Then, because the j-invariant is the same,

$$\frac{4a^3}{4a^3 + 27b^2} = \frac{4a'^3}{4a'^3 + 27b'^2} = \frac{4u^{-12}a^3}{4u^{-12}a^3 + 27b'^2} = \frac{4a^3}{4a^3 + 27u^{12}b'^2}.$$

This means that

$$b'^2 = (u^6 b)^2,$$

so that $b' = \pm u^6 b$. Accordingly, choose the sign of u^2 so that $b' = (a/a')^{3/2} b$, which completes the proof. □

If two elliptic curves have the same j-invariant, then they are isomorphic,[3] and can be changed one to the other by a coordinate transformation of the stated form. If u is an element of $\boldsymbol{F}$, then the isomorphism is visible in the ground field $\boldsymbol{F}$. If the parameter u is in an extension field, then even though the new curve remains in the ground field $\boldsymbol{F}$, the isomorphism is hidden from view in the ground field. The isomorphism can be seen, however, by stepping up into the extension field.

[3] Each value of the j-invariant corresponds to one isomorphism class in the algebraic closure $\overline{\boldsymbol{F}}_q$. When viewed only in $\boldsymbol{F}_q$, there are two isomorphism classes, which are referred to as twists of each other.

10.2 Elliptic curves over finite fields

An elliptic curve can be defined over any field $\boldsymbol{F}$. In particular, an elliptic curve can be defined over a finite field $\boldsymbol{F}_q$. If $\boldsymbol{F}$ is any field whose characteristic is not two, then by a suitable change of variables, the coefficients a_1 and a_3 become zero, so the Weierstrass form can be reduced to

$$\mathcal{X} : y^2 = x^3 + a_2x^2 + a_4x + a_6.$$

If $\boldsymbol{F}$ is any field whose characteristic is not three, then by a suitable change of variables, the coefficient a_2 becomes zero, so the Weierstrass form can be reduced to

$$\mathcal{X} : y^2 + a_1xy + a_3y = x^3 + a_2x^2 + a_6.$$

If $\boldsymbol{F}$ is any field whose characteristic is neither two nor three, then by a suitable change of variables, the Weierstrass form can be reduced to the short Weierstrass form

$$y^2 = x^3 + ax + b$$

as defined earlier.

The reason that the short Weierstrass form cannot be realized by a change in variables for a field whose characteristic is two or three is because a change of variables involving the form $(y + A)^2$ (mod 2) or of the form $(x + B)^3$ (mod 3) would not produce the cross terms needed to effect the desired cancellation. Of course, polynomials in the short Weierstrass form do exist in fields of characteristic two or three. The point here is that, in such fields, other polynomials, possibly singular, also exist that cannot be reduced to this form by a change in variables.

Several examples of elliptic curves in finite fields are shown in Figure 10.2. On the left side of the figure is a curve based on the polynomial

$$y^2 = x^3 + x + 1$$

in the field $\boldsymbol{F}_{11}$. The curve $\mathcal{X}(\boldsymbol{F}_{11}) : y^2 = x^3 + x + 1$ then has twelve points $(0, \pm 1)$, $(1, \pm 5)$, $(2, 0)$, $(3, \pm 3)$, $(4, \pm 5)$, $(6, \pm 5)$, and ∞, as can be found by direct search. These twelve points are shown in the figure. On the right side of Figure 10.2, shown symbolically, is an elliptic curve over a large prime field $\boldsymbol{F}_p$. The curve in $\boldsymbol{F}_p$ is shown only symbolically because, if p is a very large prime, there are so many points that it is impossible to explicitly display the curve. It is not in $\boldsymbol{R}^2$, and does not have the geometry of $\boldsymbol{R}^2$.

The bivariate polynomial $y^2 = x^3 + x + 1$ over $\boldsymbol{F}_{11}$ is also a bivariate polynomial in any extension field of $\boldsymbol{F}_{11}$. This means that the elliptic curve in $\boldsymbol{F}_{11}$ can be regarded as an elliptic curve in any extension field of $\boldsymbol{F}_{11}$. To extend the field $\boldsymbol{F}_{11}$, notice that, because $\sqrt{-1}$ does not exist in $\boldsymbol{F}_{11}$, the polynomial $x^2 + 1$ is irreducible. The extension field $\boldsymbol{F}_{11^2}$ can be written as the set $\{a + ib\}$ where $a, b \in \boldsymbol{F}_{11}$ and $i^2 = -1$.

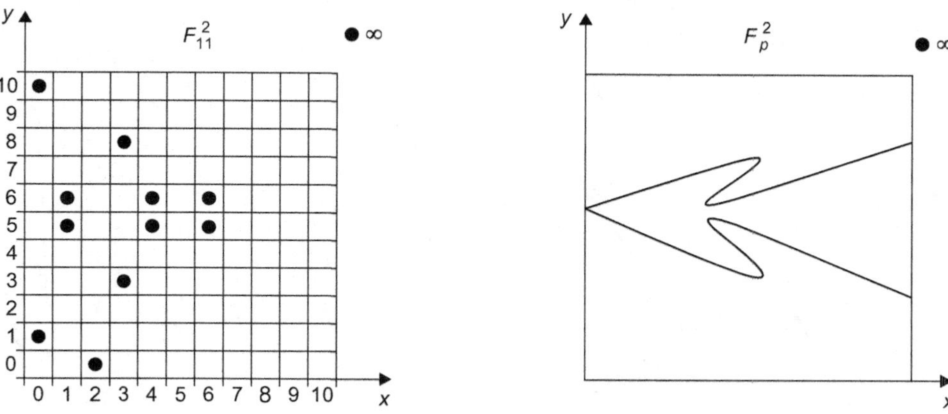

Figure 10.2 Some elliptic curves in the fields $\boldsymbol{F}_{11}$ and $\boldsymbol{F}_p$

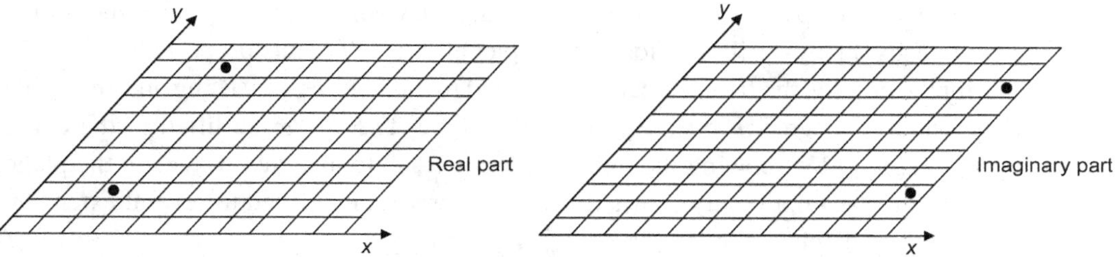

Figure 10.3 Some points of the curve $\mathcal{X}(\boldsymbol{F}_{11^2})$

There are 121 points in the field $\boldsymbol{F}_{11^2}$ and 121^2 points in the affine plane $\boldsymbol{F}_{11^2}^2$ plus a point at infinity, for a total of 14,642 points in the projective plane. The curve of Figure 10.3, when lifted to $\boldsymbol{F}_{11^2}$ is a set of points in this "complex plane." In the larger field, the elliptic curve is denoted $\mathcal{X}(\boldsymbol{F}_{11^2}) : y^2 = x^3 + x + 1$. With patience, one can count that there are 120 points on this curve in the plane $\boldsymbol{F}_{11^2}^2$. Of these 120 points, the rational points are the twelve points with both x and y in the ground field $\boldsymbol{F}_{11}$.

Even for this small example, one may conclude that there are now too many points on the curve to conveniently list all of them or to find all of them by direct search. A few examples of points on the curve $\mathcal{X}(\boldsymbol{F}_{11^2})$ are $(-1, \pm i)$, $(5, \pm i)$, and $(1 - i, \pm 2(1 + i))$, which can be verified to be on the curve by substitution into the defining polynomial. The latter two points are shown in Figure 10.3. The plane at the left shows the "real parts" $(x_R, y_R) = (1, \pm 2)$ of the two points, and the plane on the right shows the "imaginary parts" $(x_I, y_I) = (-i, \pm 2i)$ of the two points. It should be emphasized that a point $(x, y) = (x_R + ix_I, y_R + iy_I)$ consists of both its two real parts (x_R, y_R) and its two imaginary parts (x_I, y_I), but Figure 10.3 fails to indicate which of the two points in the real part of the plane on the left goes with which of the two points

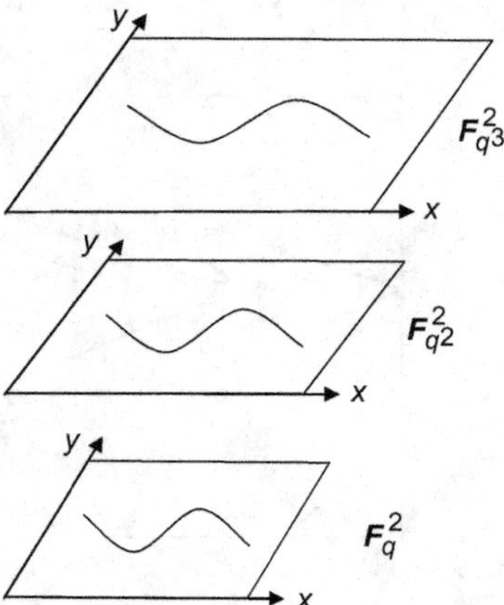

Figure 10.4 An elliptic curve lifted to some extension fields

in the imaginary part of the plane on the right. It would require a four-dimensional 11-by-11-by-11-by-11 illustration to so depict all the 120 points properly.

More generally, the field $\boldsymbol{F}_q$ can be extended to the field $\boldsymbol{F}_{q^m}$ by using an irreducible polynomial of degree m over $\boldsymbol{F}_q$. Then the curve $\mathcal{X}(\boldsymbol{F}_q)$ can be lifted to the curve $\mathcal{X}(\boldsymbol{F}_{q^m})$ simply by allowing the indeterminates (x, y) to also take values in the plane $\boldsymbol{F}^2_{q^m}$. Figure 10.4 notionally shows the elliptic curve $\chi(\boldsymbol{F}_q)$ lifted to several extension fields, and so displayed in the planes $\boldsymbol{F}^2_q$, $\boldsymbol{F}^2_{q^2}$ and $\boldsymbol{F}^2_{q^3}$.

The process of field extension can be continued indefinitely. Thus in this same way, the field $\boldsymbol{F}_{11}$ can be extended to $\boldsymbol{F}_{11^m}$ by using an irreducible polynomial $p(z)$ of degree m over $\boldsymbol{F}_{11}$. The elements of the field $\boldsymbol{F}_{11^m}$ then have m parts, each part being an element of $\boldsymbol{F}_{11}$. These field elements are conveniently represented as polynomials of degree $m - 1$ in the variable z. A point of the affine plane $\boldsymbol{F}_{11^m}$ has two coordinates, and $(x(z), y(z))$ is a point of the curve if

$$y(z)^2 = x(z)^3 + x(z) + 1 \pmod{p(z)},$$

where all coefficients are elements of $\boldsymbol{F}_{11}$ and are reduced modulo 11, and polynomials in z are reduced modulo $p(z)$. In this way one can test a point $(x(z), y(z))$ to see if it is on the curve. It may sometimes be difficult to find even one such nonrational point on the curve in this way if m is large.

In the ground field $\boldsymbol{F}_q$, it is trivial that if $(x, y) \in \mathcal{X}(\boldsymbol{F}_q)$, then $(x^q, y^q) \in \mathcal{X}(\boldsymbol{F}_q)$ because every element β of $\boldsymbol{F}_q$ satisfies $\beta^q = \beta$, so $(x^q, y^q) = (x, y)$ for all points (x, y) of the plane. It is also true, though less obvious, that if $(x, y) \in \mathcal{X}(\boldsymbol{F}_{q^m})$, then $(x^q, y^q) \in \mathcal{X}(\boldsymbol{F}_{q^m})$. This is because, for any β and γ of $\boldsymbol{F}_{q^m}$, $(\beta + \gamma)^q = \beta^q + \gamma^q$. Accordingly, consider any $(x, y) \in \mathcal{X}(\boldsymbol{F}^2_{q^m})$. This means that $y^2 = x^3 + ax + b$, where $(a, b) \in \boldsymbol{F}_q$. Therefore

$$(y^2)^q = (x^3 + ax + b)^q,$$

from which we can write

$$(y^q)^2 = (x^q)^3 + ax^q + b,$$

so $(x^q, y^q) \in \mathcal{X}(\boldsymbol{F}_{q^m})$. This function on the point $P = (x, y)$ is called the *Frobenius function* and denoted $\pi_q(P) = \pi_q(x, y) = (x^q, y^q)$.

To illustrate this fact by means of the earlier example, recall that $(5, \pm i) \in \mathcal{X}(\boldsymbol{F}_{11^2})$. Consequently, $(5^{11}, (\pm i)^{11}) \in \mathcal{X}(\boldsymbol{F}_{11^2})$, which is the point $(5, \mp i)$, as can be verified by direct computation. This is one way to find more points of the curve after one point is found, but only a small number of new points can be found in this way because the orbit under the Frobenius map is small.

10.3 The operation of point addition

We now define the operation known as *point addition* on the elliptic curve $\mathcal{X}$. Two points of the curve are "added" to get a third point of the curve. The word "add" is used merely as a convenient term for the new operation. Point addition is distinct from the operation of addition in the field.

For brevity, in this section we treat only the polynomial in the form

$$y^2 = x^3 + ax + b.$$

It is straightforward to expand the discussion to the more general form of the polynomial. The operation of point addition, denoted $+$, is such that if P and Q are points of the elliptic curve $\mathcal{X}$, then $P + Q$ is also a point of the curve $\mathcal{X}$. This operation of point addition will be defined so as to form an abelian group. Before defining point addition, we will specify the group identity and the group inverse. The point ∞ of the elliptic curve will be the group identity, meaning that $P + \infty = P$ for all $P \in \mathcal{X}$. To emphasize its role as the group identity, the point ∞ will also be denoted $\mathcal{O}$ in this chapter. The origin of the plane, denoted $(0, 0)$, is *not* the identity element of the group, and need not even be a point of the elliptic curve.

The inverse of point P, denoted $-P$, will be defined so that $P + (-P) = \mathcal{O}$. We first specify $-P$ explicitly for the case in which the curve is symmetric in y, as it is in the shortened form of $p(x, y)$, given by $y^2 = x^3 + ax + b$. Then if $P = (x, y)$, the inverse of P is $-P = (x, -y)$, which is also a point of the curve. We regard the line passing through both (x, y) and $(x, -y)$ as a vertical line. It intersects the curve for the third time at the identity element $\mathcal{O}$. Thus, point addition will be defined so that $(x, y) + (x, -y) = \mathcal{O}$. For the general case in which the polynomial cannot be written in the form $y^2 = x^3 + ax + b$, the expression for $-P$ is a little different, as will be seen later.

The definition of point addition will refer to the intersection of a "line" in the projective plane over $\boldsymbol{F}$ with the curve $\mathcal{X}$ in that same plane. The notion of a line, however, must be seen as a purely algebraic notion, defined as the set of zeros of a polynomial of degree one. There need not be a corresponding geometric notion of a line in a general field. An exception to this statement is the real field, in which case one can combine the algebraic notion of a line with the geometric notion of a line in order to sharpen one's insights. This geometrical interpretation is not available for most fields. Nevertheless, some of the geometrical terms like "line" and "intersect" carry over into discussions of the general case, even though these terms may have no geometrical interpretation.

The operation of point addition will be based on the fact that a straight line must intersect the elliptic curve $\mathcal{X}$ in three points whenever it intersects the curve in at least two points. To make the statement precise, a point of tangency must be regarded as a double intersection, and the line through a point and its negative is regarded as also intersecting the curve at the point at infinity. Then every straight line that intersects $\mathcal{X}$ at least twice intersects $\mathcal{X}$ exactly three times; one point of intersection may be the point at infinity, and two of the points may be the same point intersected twice.

We have already defined the negative of P by the statement $P + (-P) = \mathcal{O}$. We will now define point addition by the condition that whenever P, Q, and R are any three colinear points of $\mathcal{X}$, their addition satisfies

$$P + Q + R = \mathcal{O}.$$

These two conditions mean that $P + Q = -R$. Thus the sum of $P + Q$ is the negative of R, where R is the third point on the common line.

The use here of the symbol $+$ should be carefully distinguished from the use of the symbol $+$ to denote addition of elements of $\boldsymbol{F}_q$. Thus we will have (at least) three different operations under discussion that are denoted $+$: the field addition, the addition of points, and the addition of polynomials. We also have the operation of field multiplication, denoted $\times$.

To describe the operation of point addition graphically for a symmetric elliptic curve in $\boldsymbol{R}^2$, draw a straight line through the two points P and Q, and find the third point at which that line intersects the curve. Then, for a symmetric elliptic curve, the sum $P + Q$ is defined as the mirror image with respect to the x axis of the third point of intersection of the straight line with the curve. In the special case in which P and Q have the same x coordinate, then the straight line through both P and Q is a vertical line and the third intersection with the curve is the point at infinity. Finally, to add P to itself, draw a straight line tangent to the curve at P. The point $P + P$ will be defined as the negative of the intersection of that tangent line with the curve, and will be referred to as $2P$ or as $[2]P$.

Figure 10.5 shows how the two points P_1 and P_2 are "added" in the affine plane $\boldsymbol{R}^2$ to form the point P_3. The points P_2 and P_3 can then be added to form point P_4. This

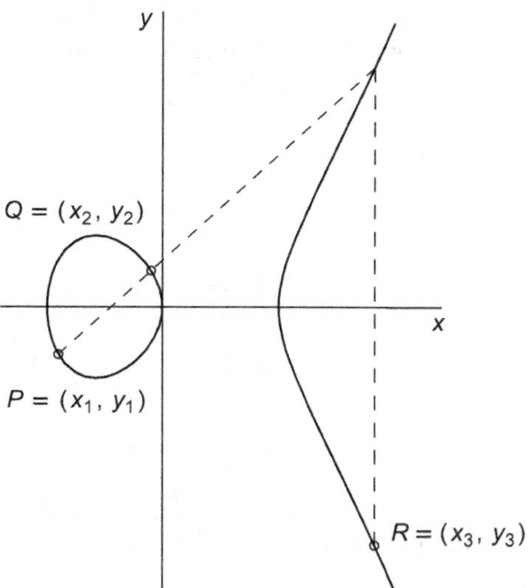

Figure 10.5 Addition of points of an elliptic curve

process can be continued indefinitely, adding the points $P_{\ell-1}$ and $P_{\ell-2}$ to compute the point P_ℓ, thereby computing an unending sequence of points of the curve. Possibly some elements of the sequence will be repeated. This must always happen if the curve has only a finite number of points.

Consider the plane curve $\mathcal{X}(\boldsymbol{R}) : y^2 = x^3 + ax + b$ over the familiar real field $\boldsymbol{R}$, and its intersection with any line in the real plane $\boldsymbol{R}^2$. The equation of the line is $y = mx + c$ or, if the line is vertical, $x = c$. In the latter case, the curve $\mathcal{X}$ reduces to y^2 equal to a constant, so the line intersects the curve in two affine points or has no intersection with the curve at an affine point. The point at infinity is the third intersection of the vertical line with the curve. Therefore $(x, y) + (x, -y) = \mathcal{O}$, so $-(x, y) = (x, -y)$ with $P = (x, y)$ and $-P = (x, -y)$. If the line is not vertical, the values of x at which the line $y = mx + c$ intersects the elliptic curve $\mathcal{X}$ are given by the solutions of

$$(mx + c)^2 = x^3 + ax + b.$$

This is a polynomial in x of degree three with coefficients in $\boldsymbol{R}$, so it must have three solutions in the complex field $\boldsymbol{C}$. If β is a complex solution, then the complex conjugate β^* is also a solution. Therefore, either two of the solutions are complex or all three solutions are real. If the line is chosen to intersect $\mathcal{X}(\boldsymbol{R})$ in two points of the plane $\boldsymbol{R}^2$, then two of the solutions for x must be real, so the third solution must also be real. This means that the line through any two specified points of the curve $\mathcal{X}(\boldsymbol{R})$, say $P = (x_1, y_1)$ and $Q = (x_2, y_2)$ with $x_1 \neq x_2$, must always intersect the plane curve

$\mathcal{X}(\boldsymbol{R})$ in exactly one other point of the real plane. This point is denoted $(x_3, -y_3)$ because it is equal to $-(P + Q)$. This third point can be found by factoring the cubic equation

$$\prod_{\ell=1}^{3}(x - x_\ell) = x^3 + ax + b - (mx + c)^2.$$

Equating the coefficients of x^2 on the left and on the right gives $x_1 + x_2 + x_3 = m^2$. We conclude that

$$x_3 = m^2 - x_1 - x_2$$
$$y_3 = m(x_1 - x_3) - y_1.$$

From these equations and the value of the slope m, the point $P_3 = (x_3, y_3)$ is easily computed. If the straight line intersects the curve $\mathcal{X}$ in two distinct points, say $P = (x_1, y_1)$ and $Q = (x_2, y_2)$, then the slope of the line is

$$m = \frac{y_2 - y_1}{x_2 - x_1}.$$

If instead, the point P_1 is to be added to itself, then the straight line will be the line tangent to the elliptic curve at the point P_1. This notion is defined, even in a finite field, by two properties: the point P_1 lies on both the line and the curve, and the line and the curve have the same slope at the point P_1. The slope is defined as the value of the derivative as given by

$$m = \left.\frac{dy}{dx}\right|_{P_1} = \frac{3x_1^2 + a}{2y_1}.$$

The line of slope m tangent to $\mathcal{X}$ at P_1 always intersects the curve $\mathcal{X}$ in exactly one other point, which is denoted $(x_3, -y_3)$. Therefore

$$x_3 = m^2 - 2x_1$$
$$y_3 = m(x_1 - x_3) - y_1.$$

If (x_1, y_1) and (x_2, y_2) are points of the curve, then (x_3, y_3) is also a point of the curve.

Figure 10.6 shows how the point P is added to itself. The point $2P$ is defined as the sum of P with itself, so that $2P = P + P$. The computation of $2P$ from P is called *point doubling*.

Although we referred to the real field while deriving the equation of point addition, the derivation is general and, for an elliptic curve of the form

$$\mathcal{X} : y^2 = x^3 + ax + b$$

is valid for any field. Simply replace $\boldsymbol{R}$ by the field $\boldsymbol{F}$, replace $\boldsymbol{C}$ by the algebraic closure $\overline{\boldsymbol{F}}$, and replace the derivative by the formal derivative. Unless the field characteristic is

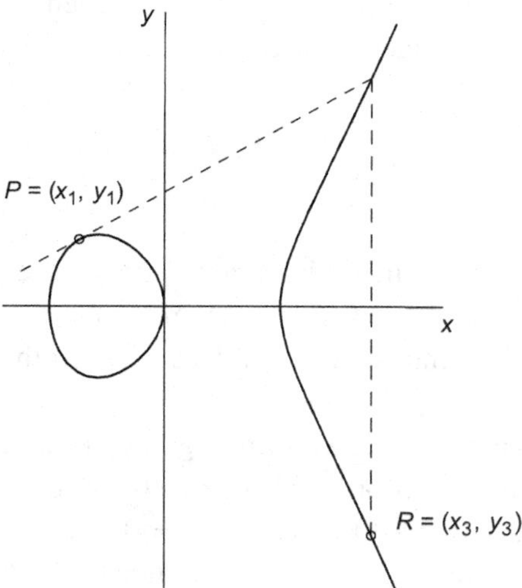

Figure 10.6 Doubling a point of an elliptic curve

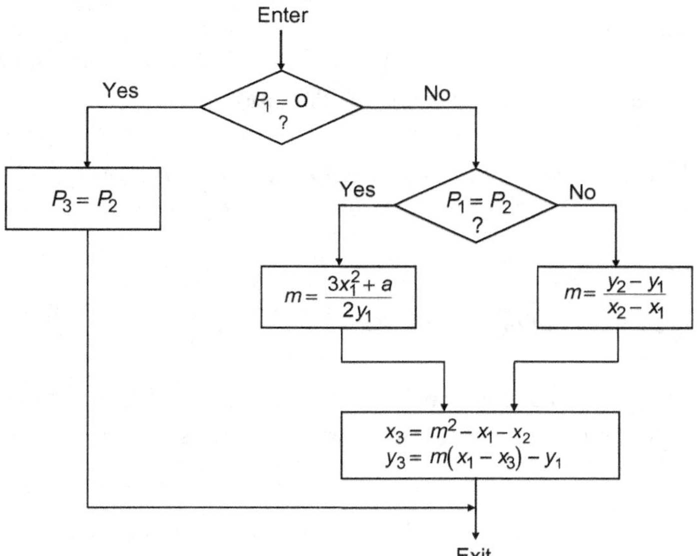

Figure 10.7 Partial flow chart for point addition on some elliptic curves

two or three, every elliptic curve can be written in this form by a change of variables, so this is a rather general description of point addition.

A partial flow chart for point addition on an elliptic curve for polynomials of this form is shown in Figure 10.7. To add two distinct points of $\mathcal{X}$ requires three multiplications and one inversion in the field $\boldsymbol{F}$. Doubling a point requires four multiplications and one inversion in $\boldsymbol{F}$. Some details dealing with special cases, such as the requirement that P_3 must equal $\mathcal{O}$ when $P_1 = P_2$ have been omitted from the simplified flow chart. These special cases must be included in a complete flow chart.

10.4 The order of an elliptic curve

Suppose that $\mathcal{X}$ is an elliptic curve over $\boldsymbol{F}_q$, a field of characteristic p. There are q^2 points in the affine plane over $\boldsymbol{F}_q$. We want to know how many of these points belong to the curve $\mathcal{X}$, and we want to know this integer exactly. This leads to the task of determining the order of an elliptic curve.

Under point addition, the points of an elliptic curve form a group. The number of points on the curve $\mathcal{X}$, denoted by $\#\mathcal{X}(\boldsymbol{F}_q)$, is its order. Heuristically, we may observe that $p(x, y)$ has a degree of two in y, so for a fixed x, say $x = \beta$, the polynomial reduces to a quadratic in only one indeterminate y. There are q choices for β, and so there are q such quadratic polynomials in y. We expect that for about half of the β, the quadratic polynomial will have two solutions in $\boldsymbol{F}_q$, and for the other β, it will have no solutions. Thus, we may expect that there are approximately q affine points on $\mathcal{X}$ plus one point at infinity.

Based on this heuristic discussion, the difference between the two integers $\#\mathcal{X}(\boldsymbol{F}_q)$ and $q + 1$ is of interest, leading to the following definition:

Definition 10.4.1 *The Frobenius trace of the elliptic curve $\mathcal{X}(\boldsymbol{F}_q)$ over the finite field $\boldsymbol{F}_q$ is the integer*

$$t = q + 1 - \#\mathcal{X}(\boldsymbol{F}_q),$$

where $\#\mathcal{X}$ denotes the order of the elliptic curve $\mathcal{X}$.

The Frobenius trace may be either positive or negative. To determine $\#\mathcal{X}(\boldsymbol{F}_q)$, it is enough to determine the integer t. Thus the number of points on the curve $\mathcal{X}$ can be written

$$\#\mathcal{X}(\boldsymbol{F}_q) = q + 1 - t.$$

It can be difficult to determine the exact value of t, or of $\#\mathcal{X}(\boldsymbol{F}_q)$.

A theorem of Hasse[4] says that the integer t satisfies $|t| \leq 2\sqrt{q}$. This is called the *Hasse bound.* We will state a version of this theorem that is strengthened to apply to any plane curve, not just an elliptic curve. It is a remarkable theorem of algebraic geometry, which is given here without proof.

[4] This can be written in the attractive form

$$(\sqrt{q} - 1)^2 \leq \#\mathcal{X}(\boldsymbol{F}_q) \leq (\sqrt{q} + 1)^2.$$

Theorem 10.4.2 (Serre improvement[5] to the Hasse–Weil bound) *The number of rational points of a plane curve $\mathcal{X}$ of genus g over $\boldsymbol{F}_q$ satisfies*

$$q + 1 - g\lfloor 2\sqrt{q}\rfloor \leq \#\mathcal{X}(\boldsymbol{F}_q) \leq q + 1 + g\lfloor 2\sqrt{q}\rfloor.$$

Because the genus of an elliptic curve is one, the theorem says that the number of points on an elliptic curve over the field $\boldsymbol{F}_q$ differs from $q + 1$ by at most $\pm\lfloor 2\sqrt{q}\rfloor$.

The Hasse–Weil bound cannot be improved as such, because it is possible to find a curve $\mathcal{X}$ with a value of t at either end of the $\pm g\lfloor 2\sqrt{q}\,\rfloor$ interval, as well as many values of t within the stated interval. Indeed, an elliptic curve always exists for a given t allowed by the theorem provided that $\text{GCD}(p, t) = 1$, and an elliptic curve exists for all t allowed by the theorem if q is a power of two. For odd primes p, the values of t for which a curve does not exist are technical but modest in number. For most values of t allowed by the Hasse–Weil bound, an elliptic curve does exist if p is large.

A simple way to change the curve in order to change the sign of the Frobenius trace is to "twist" the elliptic curve into a different elliptic curve. Given the elliptic curve

$$\mathcal{X}(\boldsymbol{F}_q) : y^2 = x^3 + ax + b,$$

define its twist as

$$\mathcal{X}'(\boldsymbol{F}_q) : y^2 = x^3 + d^2ax + d^3b,$$

where d is any nonsquare of $\boldsymbol{F}_q$. The j-invariant of the twisted curve $\mathcal{X}'(\boldsymbol{F}_q)$ is given by

$$j(\mathcal{X}') = \frac{4(d^2a)^3}{4(d^2a)^3 + 27(d^3b)^2} = 1728\frac{4a^3}{4a^3 + 27b^2},$$

which is equal to $j(\mathcal{X})$. Therefore $\mathcal{X}(\boldsymbol{F}_q)$ and $\mathcal{X}'(\boldsymbol{F}_q)$ are isomorphic in the algebraic closure $\overline{\boldsymbol{F}}_q$, yet the next theorem says that they do not have the same number of zeros in the ground field $\boldsymbol{F}_q$, and so are different curves when viewed in the ground field.

We shall see that the Frobenius trace t' for the curve $\mathcal{X}'(\boldsymbol{F}_q)$ satisfies $t' = -t$. This fact can be useful when counting the number of points on a large curve because it gives the number of points on two curves at the cost of counting the points on one curve. The following theorem establishes this property.

Theorem 10.4.3 *The total number of rational points on an elliptic curve $\mathcal{X} : y^2 = x^3 + ax + b$ over $\boldsymbol{F}_q$ and its twist $\mathcal{X}'$ satisfy*

$$\#\mathcal{X} + \#\mathcal{X}' = 2q + 2.$$

Proof The polynomial $y^2 = x^3 + ax + b$ is quadratic in y. Therefore the elliptic curve has two points for every value of x for which $x^3 + ax + b$ is a nonzero square.

[5] The floor function.

Let $z = xd$ where d is a nonzero element of $\boldsymbol{F}_q$. Because x can be recovered by $x = z/d$, this is a permutation of the nonzero points of $\boldsymbol{F}_q$. Thus, as x steps through all the nonzero points of $\boldsymbol{F}_q$, z also steps through all the nonzero points of $\boldsymbol{F}_q$, but in a different order. Suppose that $x^3 + ax + b$ is a square in $\boldsymbol{F}_q$ and d is a nonsquare. Then $d^3(x^3 + ax + b)$ is a nonsquare, and with $dx = z$, this means that $z^3 + ad^2z + bd^3$ is a nonsquare. Similarly, if $x^3 + ax + b$ is a nonsquare, then $z^3 + ad^2z + bd^3$ is a square. In this way, as x steps through the nonzero points of $\mathcal{X}(\boldsymbol{F}_q)$, z steps through the nonzero points of $\mathcal{X}'(\boldsymbol{F}_q)$ as well, and one polynomial or the other, but not both, is a square; and each square leads to two points on one curve or the other, but not both. This accounts for $2q$ points on the union of the two curves. Additionally, each curve has one point at infinity, which accounts for two additional points, or $2(q + 1)$ points in total. □

For example, over the field $\boldsymbol{F}_{131}$, the elliptic curve

$$\mathcal{X}(\boldsymbol{F}_{131}) : y^2 = x^3 - 3x + 8$$

has 110 points, as can be verified by straightforward enumeration. Therefore, the Frobenius trace can be evaluated as $t = q + 1 - \#\mathcal{X}(\boldsymbol{F}_{131}) = 22$. Then by recalling that -1 is a nonsquare in $\boldsymbol{F}_{131}$ because $131 = 4k + 3$, we immediately know that the elliptic curve $\mathcal{X}'(\boldsymbol{F}_{131}) : y^2 = x^3 - 3x - 8$, which is the twist of $\mathcal{X}(\boldsymbol{F}_{131})$, has $q + 1 + 22 = 154$ points on it. To go further with the field $\boldsymbol{F}_{131}$, we may study other elliptic curves on this same field. Up to isomorphism, it is enough to consider only curves defined by polynomials in the short Weierstrass form $y^2 = x^3 + ax + b$. Because there are only two free parameters a and b in this polynomial, and b is nonzero, there are only $130 \cdot 131$ possibilities. However, the j-invariant is an element of $\boldsymbol{F}_{131}$, and can only take on 131 values, so there are not more than 130 isomorphism classes. An elliptic curve in the finite field $\boldsymbol{F}_q$, with characteristic not two or three, has a j-invariant

$$j(\mathcal{X}) = 1728 \frac{4a^2}{4a^3 + 27b^2},$$

which is an element of $\boldsymbol{F}_q$, and so takes only q values. Curves with nonzero $j(\mathcal{X})$ are often, but not always, ordinary elliptic curves as defined in Section 10.6. There are $q - 1$ nonzero values of $j(\mathcal{X})$, and so there are $q - 1$ isomorphic classes of such elliptic curves over $\boldsymbol{F}_q$.

10.5 The group of an elliptic curve

We have asserted that, under the operation of point addition, the points of an elliptic curve form an abelian group. Clearly, the set of points of the curve is closed under point addition, there is an inverse of point addition, and there is an identity element. The only property required of a group that is not immediately evident is the associativity property. We must show that

$$(P_1 + P_2) + P_3 = P_1 + (P_2 + P_3).$$

The associativity property can be verified by elementary algebra. However, verification in this way is tedious because the special cases in which $P_2 = P_1$, $P_2 = -P_1$, or $P_1 = \mathcal{O}$ must be checked separately. We leave this verification as an exercise. A more sophisticated, and more delicate method is to first prove the associativity property in the complex field by a geometric argument, and then to argue that if the associativity property is true for elliptic curves in one field, it must be true for elliptic curves in all fields because the algebra is the same. We will return to this method of abstract reasoning in Section 10.16. The associativity property will be revisited then.

An example of an elliptic curve over the finite field $\boldsymbol{F}_{23}$ is $\mathcal{X}(\boldsymbol{F}_{23}) : y^2 = x^3 + x + 1$. The points of $\mathcal{X}(\boldsymbol{F}_{23})$ can be found directly by testing each point of the plane $\boldsymbol{F}_{23}^2$ to see if it is a zero of $p(x, y)$. A more informative procedure is to find the points of the curve by using the operation of point addition. Clearly, the point (0, 1) is an element of $\mathcal{X}(\boldsymbol{F}_{23})$. Using the formulas for point doubling and point addition, we can easily compute

$$\begin{aligned}
2P &= P + P &&= (0, 1) + (0, 1) &&= (6, 19)\\
3P &= 2P + P &&= (6, 19) + (0, 1) &&= (3, 13)\\
4P &= 2P + 2P &&= (6, 19) + (6, 19) &&= (13, 16)\\
5P &= 4P + P &&= (13, 16) + (0, 1) &&= (18, 3)\\
6P &= 3P + 3P &&= (3, 13) + (3, 13) &&= (7, 11)\\
&\ \vdots\\
27P &= 26P + P &&= (6, 4) + (0, 1) &&= (0, -1)\\
28P &= 27P + P &&= (0, -1) + (0, 1) &&= \mathcal{O}.
\end{aligned}$$

Thus the point (0, 1) has order 28, as shown by direct enumeration of its multiples. As we have asserted that $\mathcal{X}(\boldsymbol{F}_{23})$ is a group, Lagrange's theorem states that 28 must divide $\#\mathcal{X}(\boldsymbol{F}_{23})$. But x takes on only 23 values in $\boldsymbol{F}_{23}$, and y can have at most two values for each such x. There is also one point at infinity, so there cannot be more than 47 points on the curve. Because the only positive integer not larger than 47 that is divisible by 28 is 28 itself, we conclude that $\#\mathcal{X}(\boldsymbol{F}_{23}) = 28$. Thus the orbit of the point (0, 1) under the operation of point addition includes every point of the curve. We conclude that $\mathcal{X}(\boldsymbol{F}_{23})$ is a cyclic group under point addition, and the point (0, 1) is a generator for this group. Finally, we note that because the point P has order 28, the point $4P$ has order seven, which is a prime. Hence the point $4P = (13, 16)$ generates a cyclic subgroup whose order is a prime.

Because $\mathcal{X}(\boldsymbol{F}_{23})$ is a cyclic group of order 28, it is isomorphic to $\boldsymbol{Z}_{28}$, which means that $\mathcal{X}(\boldsymbol{F}_{23})$ and $\boldsymbol{Z}_{28}$ have identical abstract structures as groups. However, this isomorphism, when given as $kP \in \mathcal{X}(\boldsymbol{F}_{23})$ corresponding to $k \in \boldsymbol{Z}_{28}$, is very explicit, and we have not stated any rule that is more concise than the explicit statement. We do not give an inverse statement relating an arbitrary point $Q \in \mathcal{X}(\boldsymbol{F}_{23})$ to a point of $\boldsymbol{Z}_{28}$. Indeed, such a statement would be a solution to the discrete-log problem on an elliptic curve, a problem that we examine later in more detail and assert to be an intractable problem for elliptic curves of large order.

In general, an elliptic curve $\mathcal{X}(\boldsymbol{F}_q)$ is an abelian group under the operation of point addition, and so it may have cyclic subgroups. Given a cyclic subgroup G of large prime order, and an element $Q = aP$ for some known integer P, the elliptic-curve discrete-log problem is to compute a when given Q and P. For large group orders, it is computationally tractable to compute aP from a and P, but is computationally intractable to compute a from aP and P, or so we believe.

10.6 Supersingular elliptic curves

We now introduce the class of supersingular elliptic curves. These curves have many interesting properties, and as such have received special scrutiny. Their special structure leads to a slight simplification in the structure of the curve and the equations for point addition.

Definition 10.6.1 *An elliptic curve $\mathcal{X}$ over the finite field $\boldsymbol{F}_q$, where q is a power of the prime p, is called a supersingular elliptic curve if the Frobenius trace t is a multiple of p. The elliptic curve $\mathcal{X}$ is called an anomalous elliptic curve if the Frobenius trace t is equal to one. The elliptic curve $\mathcal{X}$ is called an ordinary elliptic curve if it is not a supersingular or anomalous elliptic curve.*

The term "supersingular" is perhaps an unfortunate choice of terminology in this instance, because it has nothing to do with the curve being singular. Every elliptic curve must be nonsingular by definition.

An elliptic curve over $\boldsymbol{F}_q$ is also an elliptic curve over any extension of $\boldsymbol{F}_q$. One would expect that a curve that is supersingular in a ground field is also supersingular in any extension field. Corollary 10.11.3 of Section 10.11 implies that this is true for $\mathcal{X}(\boldsymbol{F}_{q^2})$ and Theorem 10.11.6 then implies that it is true for $\mathcal{X}(\boldsymbol{F}_{q^m})$ for all m.

In general, the number of points on an elliptic curve over $\boldsymbol{F}_{p^m}$ satisfies

$$\#\mathcal{X}(\boldsymbol{F}_{p^m}) = p^m + 1 - t.$$

For a supersingular elliptic curve, t is a multiple of p, so the number of points of the curve in the field $\boldsymbol{F}_{p^m}$ is always of the form $ap + 1$ for some integer a. The curve always has an odd order if the polynomial order has an even number of affine points because there is only one more point at infinity. If p^m is a power of two, then every curve with an odd order has an even t and so is supersingular.

Over the binary field $\boldsymbol{F}_{2^m}$, there are precisely three classes of supersingular elliptic curves. These classes under isomorphism are represented by

$$\mathcal{X}_1 : y^2 + y = x^3,$$
$$\mathcal{X}_2 : y^2 + y = x^3 + x,$$
$$\mathcal{X}_3 : y^2 + y = x^3 + x + 1.$$

Every supersingular curve over $\boldsymbol{F}_{2^m}$ can be transformed into one of these three curves by a suitable change of variables. In contrast, the curve

$$\mathcal{X}_4 : y^2 + xy = x^3 + x^2 + 1$$

is an ordinary elliptic curve. The curve

$$\mathcal{X}_5 : y^2 = x^3 + x + 1$$

is singular. It is not an elliptic curve.

More generally, over the binary extension field $\boldsymbol{F}_{2^m}$, the curve

$$\mathcal{X}(\boldsymbol{F}_{2^m}) : y^2 + a_3 y = x^3 + a_4 x + a_6,$$

and with both a_3 and a_6 nonzero coefficients of the field $\boldsymbol{F}_{2^m}$, is a supersingular curve. Similarly, over the ternary extension field $\boldsymbol{F}_{3^m}$, the curve

$$\mathcal{X}(\boldsymbol{F}_{3^m}) : y^2 = x^3 + a_4 x + a_6$$

with a_4 and a_6 nonzero elements of $\boldsymbol{F}_{3^m}$, is always supersingular, and any supersingular curve over $\boldsymbol{F}_{3^m}$ can be put in this form by a suitable change of variables.

Theorem 10.6.2 *A supersingular elliptic curve over the prime field $\boldsymbol{F}_p$, where p is not equal to two or three, is an abelian group with $p + 1$ points.*

Proof Every elliptic curve is an abelian group under point addition. By the Hasse–Weil bound, $p - 1 - \#\mathcal{X}(\boldsymbol{F}_p) = t$ where $t \leq \lfloor 2\sqrt{p} \rfloor \leq 2\sqrt{p}$. By the definition of supersingular curve, t is divisible by p. Therefore t^2 is an integer not larger than $4p$ and is divisible by p^2. This is not possible if p is larger than three unless $t = 0$, so we can conclude that $t = 0$ and $\#\mathcal{X}(\boldsymbol{F}_p) = p + 1$. □

Theorem 10.6.3 *The elliptic curve $\mathcal{X}(\boldsymbol{F}_p) : y^2 = x^3 + b$ with $p = 2 \pmod 3$ and b nonzero is a supersingular elliptic curve.*

Proof The statement requires that $p - 2$ is a multiple of three, so $p - 1$ is not a multiple of three. Because the order of every element must divide $p - 1$, no element of $\boldsymbol{F}_p$ has order three. This means that β^3 and γ^3 cannot be equal if β and γ are distinct nonzero elements, because otherwise $(\beta/\gamma)^3 = 1$, and this cannot be. Thus the $p - 1$

nonzero elements have $p-1$ distinct cubes, so every element of $\boldsymbol{F}_p$ has a unique cube root in $\boldsymbol{F}_p$. This means that for each value of y, there is exactly one x satisfying $x^3 = y^2 - b$, namely the unique cube root of $y^2 - b$. Because there are p values of y, there are p affine points on the curve and one point at infinity. Therefore there are $p+1$ points on the elliptic curve, so the curve is supersingular. □

Theorem 10.6.4 *The elliptic curve $\mathcal{X}(\boldsymbol{F}_p): y^2 = x^3 + ax$ where p is a prime of the form $p = 3 \pmod 4$ and a is a nonzero element of $\boldsymbol{F}_p$ is a supersingular elliptic curve.*

Proof The elliptic curve over the field $\boldsymbol{F}_p$ for which $p = 3 \pmod 4$ has a polynomial of the form $y^2 = x^3 + ax$ with no constant term. The nonzero coefficient a is either a square or a nonsquare. Because -1 is a nonsquare when p has the form $p = 4\ell + 3$, one sees that either $+a$ is a square or $-a$ is a square. In either case, a change in variables can be made to reduce the polynomial to the form

$$y^2 = x^3 \pm x,$$

with the sign depending on whether or not a is a square. Write $\boldsymbol{F}_p$ as $\{0, \pm 1, \pm 2, \ldots, \pm(p-1)/2\}$. Then because -1 is not a square, $\beta^3 \pm \beta$ is a square if, and only if, $(-\beta)^3 \pm (-\beta) = (-1)(\beta^3 \pm \beta)$ is not a square. This means that of the $p-1$ nonzero elements of $\boldsymbol{F}_p$, exactly $(p-1)/2$ yield a square on the right side of $y^2 = x^3 \pm 1$. Each such square leads to two points of the curve, a total of $p-1$ points. Because $(0, 0)$ and ∞ are two more points of the curve, one can conclude that

$$\#\mathcal{X}(\boldsymbol{F}_p) = p + 1.$$

This means that $t = 0$. We conclude that any such curve is a supersingular elliptic curve. □

Theorem 10.6.5 *The elliptic curve $\mathcal{X}(\boldsymbol{F}_p): y^2 = x^3 + 1$, with p an odd prime, is a supersingular curve if, and only if, $p = 2 \pmod 3$.*

Proof This proof is not given. □

We have seen earlier that a supersingular elliptic curve over the prime field $\boldsymbol{F}_p$ has no rational points of order p. The following theorem says that a supersingular curve $\mathcal{X}(\boldsymbol{F}_q)$ has no points of order p in any extension field.

Theorem 10.6.6 *If q is a power of the prime p, then the elliptic curve $\mathcal{X}(\boldsymbol{F}_q)$ is a supersingular elliptic curve if, and only if, there are no points of order p in the algebraic closure of $\boldsymbol{F}_q$.*

Proof A supersingular elliptic curve over a field of characteristic p is an elliptic curve for which t is a multiple of p. Therefore for some integer a,

$$\begin{aligned}\#\mathcal{X}(\boldsymbol{F}_q) &= q + 1 - t \\ &= ap + 1.\end{aligned}$$

Therefore, p does not divide $\#\mathcal{X}(\boldsymbol{F}_q)$, so $\mathcal{X}(\boldsymbol{F}_q)$ has no subgroup with p elements. Because p does not divide $\#\mathcal{X}(\boldsymbol{F}_q)$, it follows from Corollary 10.11.3 and Theorem 10.11.6 that p does not divide the order of the curve in any extension field. □

10.7 Elliptic curves over binary fields

An elliptic curve is defined in any field. In this section, we discuss elliptic curves in fields of characteristic two.[6] The Weierstrass canonical form of an elliptic curve is

$$y^2 + a_1xy + a_3y = x^3 + a_2x^2 + a_4x + a_6.$$

By a suitable change of variables, any ordinary elliptic curve over the binary field $\boldsymbol{F}_{2^m}$ can be expressed in the simpler form

$$y^2 + xy = x^3 + a_2x^2 + a_6,$$

with a_6 nonzero. The coefficients a_2 and a_6 are elements of $\boldsymbol{F}_{2^m}$, with a_6 nonzero. Let $q = 2^m$. There are $q(q-1)$ such choices for the pair (a_2, a_6), but many of these choices are curves that are isomorphic to each other. The j-invariant, which for fields of characteristic two has the simple form $j(\mathcal{X}) = a_6^{-1}$, can only take on $q - 1$ nonzero values, so there are only $q - 1$ isomorphism classes of ordinary elliptic curves over $\overline{\boldsymbol{F}}_q$, each consisting of two isomorphism classes over $\boldsymbol{F}_q$.

There are precisely $2(q-1)$ isomorphic classes of ordinary elliptic curves over $\boldsymbol{F}_q$ with $q = 2^m$. These are labeled by the $2^m - 1$ nonzero j-invariants in $\boldsymbol{F}_{2^m}$ with two twist variants for each j-invariant. The curves in one class are given by

$$y^2 + xy = x^3 + a_2x^2 + a_6,$$

where $a_6 \in \boldsymbol{F}_{2^m}^*$, and $a_2 \in \{0, \gamma\}$ where γ is any element of $\boldsymbol{F}_{2^m}$ with trace one.

The elliptic curve denoted $\mathcal{X}$ is the set of points (x, y) that satisfy an elliptic polynomial. In the general case, one may consider points (x, y) in any extension field of $\boldsymbol{F}_{2^m}$. However, we are usually interested in the rational points, which are the points of the curve with components in $\boldsymbol{F}_{2^m}$ itself. The order of $\mathcal{X}$ must satisfy the Hasse–Weil bound

$$|\#\mathcal{X} - q - 1| \le \lfloor 2\sqrt{q} \rfloor.$$

[6] We refer to a field of characteristic two as a binary field and the field $\boldsymbol{F}_2$ as the binary field.

Thus the number of points on the curve, approximately equal to 2^m, satisfies

$$|\#\mathcal{X} - 2^m| \approx 2\sqrt{2^m}.$$

This means that for large m, we have a very good estimate of the order of any elliptic curve over $\boldsymbol{F}_{2^m}$. However, the Hasse–Weil bound does not give the order of the curve exactly. For applications to cryptography, we require the exact number because we need to know the largest prime factor of the order of the curve and therefore the order of the largest cyclic subgroup. We will address this task in Section 10.11 and Section 10.14. Section 10.11 gives a method to compute $\#\mathcal{X}(\boldsymbol{F}_{q^m})$ from $\#\mathcal{X}(\boldsymbol{F}_q)$ if the order of the smaller group is known. Section 10.14 gives a method to compute $\#\mathcal{X}(\boldsymbol{F}_q)$ directly.

We will now examine the operation of point addition on an elliptic curve over a binary field. For an elliptic curve defined by the Weierstrass canonical form, the negative of point $P_1 = (x_1, y_1)$, denoted $P_2 = -P_1$ is the other point at which a vertical line through P_1 intersects the curve. This means that, with x_1 held constant, y_1 and y_2 satisfy

$$(y - y_1)(y - y_2) = y^2 + a_1x_1y + a_3y - x_1^3 - a_2x_1^2 - a_4x_1 - a_6.$$

Then

$$y^2 - (y_1 + y_2)y + y_1y_2 = y^2 + (a_1x_1 + a_3)y - (x_1^3 + a_2x_1^2 + a_4x_1 + a_6).$$

Therefore

$$y_1 + y_2 = -a_1x_1 - a_3,$$

from which it follows that

$$-P_1 = (x_1, -y_1 - a_1x_1 - a_3).$$

For any pair of points P_1 and P_2, we use point addition to obtain a third point P_3, and we denote the operation of point addition as

$$P_3 = P_1 + P_2$$

or as

$$P_1 + P_2 - P_3 = \mathcal{O}.$$

It is a matter of elementary algebra to write the equation of the line through P_1 and P_2 (or the line tangent to the curve at P_1) and to find the third point at which this line intersects the curve $\mathcal{X}(\boldsymbol{F}_p)$. If the points are written $P_1 = (x_1, y_1)$, $P_2 = (x_2, y_2)$, and $P_3 = (x_3, y_3)$, then for a field of characteristic two and a curve in the short Weierstrass form $y^2 + xy = x^3 + a_2x^2 + a_6$, the point P_3 is found to satisfy

$$x_3 = m^2 + m + a_2 + x_1 + x_2$$

$$y_3 = m(x_1 + x_3) + x_3 + y_1,$$

where, if $P_1 \neq P_2$,

$$m = \frac{y_1 + y_2}{x_1 + x_2}$$

is the slope, and if $P_1 = P_2$,

$$m = \frac{x_1^2 + y_1}{x_1}$$

is the slope. These are the equations for point addition in a field of characteristic two.

The elliptic curve defined over the finite field $\boldsymbol{F}_{2^m}$ has a finite number of points in the plane $\boldsymbol{F}^2_{2^m}$. The curve $\mathcal{X}$, together with the operation of point addition $+$, forms a finite abelian group, and so, as does any finite abelian group, has the standard decomposition into cyclic subgroups. To describe this finite group, it is necessary to know exactly how many points are on the curve $\mathcal{X}$. If the order of $\mathcal{X}$ contains a large prime factor p, then the group contains a cyclic subgroup of order p. The following theorem shows that the full group does not have prime order.

Theorem 10.7.1 *Over fields of characteristic two, if $\mathcal{X}$ is an ordinary elliptic curve, then the order of $\mathcal{X}$ is divisible by two.*

Proof An ordinary elliptic curve over a field of characteristic two can be given in the form

$$\mathcal{X}(\boldsymbol{F}_{2^m}): y^2 + xy = x^3 + a_2x^2 + a_6.$$

The point $(0, \sqrt{a_6})$ is clearly an element of the curve $\mathcal{X}(\boldsymbol{F}_{2^m})$. This point is easily verified to have order two under point addition. □

This theorem can be combined with the Hasse–Weil bound to enumerate the possible values that the order of $\mathcal{X}$ might take. For example, consider $\boldsymbol{F}_{16}$. Then the Hasse–Weil bound states that

$$16 + 1 - 2\sqrt{16} \leq \#\mathcal{X}(\boldsymbol{F}_{16}) \leq 16 + 1 + 2\sqrt{16}.$$

Because $\#\mathcal{X}(\boldsymbol{F}_{16})$ is known to be a multiple of 2 for an ordinary elliptic curve, we conclude that

$$\#\mathcal{X}(\boldsymbol{F}_{16}) \in \{10, 12, 14, 16, 18, 20, 22, 24\}$$

for such curves. This statement cannot be made stronger. Every one of these cases will occur for some pair of values of a_2 and a_6. This situation is similar for elliptic curves over $\boldsymbol{F}_{2^m}$ for other values of m.

Theorem 10.7.2 *For any value of* $a_6 \in \boldsymbol{F}_{2^{2\ell+1}}$*, the two elliptic curves over* $\boldsymbol{F}_{2^{2\ell+1}}$*, defined by the two equations*

$$y^2 + xy = x^3 + a_6$$
$$y^2 + xy = x^3 + x^2 + a_6,$$

together have a total of $2 \cdot 2^{2\ell+1} + 4$ *rational points, each point counted according to its multiplicity.*

Proof Each curve has one point at infinity. Thus we must show that together the two curves have $2 \cdot 2^{2\ell+1} + 2$ affine points.

First, consider $x \neq 0$. By substituting $y = xz$, the two polynomials can be written as

$$z^2 + z = x + \frac{a_6}{x^2}$$
$$z^2 + z = x + 1 + \frac{a_6}{x^2}.$$

Theorem 9.7.6 states that $z^2 + z = \beta$ has two solutions, counting multiplicity if, and only if, the binary trace of β equals zero. Otherwise the equation has no solutions. Hence the first equation has a solution if, and only if,

$$\text{trace}\left(x + \frac{a_6}{x^2}\right) = 0.$$

The second equation has a solution if, and only if,

$$\text{trace}\left(x + 1 + \frac{a_6}{x^2}\right) = 0,$$

which can be rewritten as

$$\text{trace}\left(x + \frac{a_6}{x^2}\right) = \sum_{i=0}^{2\ell} 1^{2^i} = 1,$$

because the addition is modulo two. For each value of $x \in \boldsymbol{F}_{2^{2\ell+1}}$, the trace is equal to either zero or one. Therefore, for each nonzero value of x, one of these two curves gives two points and the other gives none. This accounts for $2(2^{2\ell+1} - 1)$ points.

Finally, if $x = 0$, both curves reduce to

$$y^2 = a_6,$$

which for each curve gives the additional point $\left(0, \sqrt{a_6}\right)$ with multiplicity two. This accounts for four additional points. Therefore we have a total of $2 \cdot 2^{2\ell+1} + 2$ points in the affine plane. Together with the two points at infinity, this establishes the statement of the theorem. □

Definition 10.7.3 *Elliptic curves of the form*

$$\mathcal{X}(\boldsymbol{F}_{2^m}) : y^2 + xy = x^3 + 1$$

or

$$\mathcal{X}(\boldsymbol{F}_{2^m}) : y^2 + xy = x^3 + x + 1$$

are called Koblitz elliptic curves.

A Koblitz curve is an ordinary elliptic curve over a binary field, and so has even order. Therefore the order cannot be a prime. The order can, however, be a small multiple of a prime.

For example

$$\#\mathcal{X}(\boldsymbol{F}_{2^{101}}) : y^2 = x^3 + x + 1 = 2 \cdot 1267650600228230886142808508011,$$

which is twice a prime, and

$$\#\mathcal{X}(\boldsymbol{F}_{2^{103}}) : y^2 = x^3 + 1 = 4 \cdot 2535301200456459535862530067069,$$

which is four times a prime. Koblitz curves are attractive because the equations for point addition simplify, and point addition can be implemented in a simple way.

10.8 Computation of point multiples

We have asserted that under the operation of point addition, the elements of an elliptic curve form an abelian group. Clearly, the set of points is closed under point addition. Moreover, there is an identity element under point addition, which is the point at infinity, denoted $\mathcal{O}$. Furthermore, every point $P = (x, y)$ of the curve has a negative $-P = (x, -y - a_1x - a_3)$ with the property that $P + (-P) = \mathcal{O}$. This leads to the notion of subtraction of points, given by $P - Q = P + (-Q)$. If the polynomial $p(x, y)$ has a_1 and a_3 that are both equal to zero, then the negative of $P = (x, y)$ is simply $-P = (x, -y)$.

The point multiple kP is defined as the sum of k copies of P. Thus

$$kP = P + P + \cdots + P,$$

where there are k terms in the sum, all equal to P. Any point P can be multiplied by the integer k by using the recursion $kP = (k-1)P + P$. The recursion, however, is not a suitable method to calculate kP if k is very large. Instead, any point P can be "doubled" simply by writing $2P = P + P$, and "redoubled" by writing $4P = 2P + 2P$. If k is large, the simplest way to compute kP is by successive doubling and adding. First compute $2P, 4P, 8P, \ldots, 2^mP$. Then use the binary expansion of k to select and add an appropriate subset of these points. Thus for example, $29P = 16P + 8P + 4P + P$.

In general, write the binary expansion of k as

$$k = k_0 + k_1 2 + k_2 2^2 + \cdots + k_{m-2} 2^{m-2} + k_{m-1} 2^{m-1} ,$$

where $k_i \in \{0, 1\}$. Then because $k_i \in \{0, 1\}$, we have the obvious expression

$$\begin{aligned} kP &= \left(\sum_{i=0}^{m-1} k_i 2^i \right) P \\ &= \sum_{i=0}^{m-1} k_i (2^i P) \\ &= \sum_{i:k_i=1} 2^i P. \end{aligned}$$

To compute kP using the last expression will take $m - 1$ point-doublings to compute the set $\{2^i P : i = 0, \ldots, m-1\}$ and at most $m - 1$ additions to execute the indicated sum. Thus to compute kP from k and P in this way has a complexity proportional to $\log_2 n$ where n is the order of the group on $\mathcal{X}$.

Despite its simplicity, this procedure has immense consequences in cryptography because it is feasible to compute kP by this method, even if k is a 100-bit binary number, whereas it would not be feasible to compute kP by the recursion $kP = (k-1)P + P$.

On the other hand, no comparable tractable procedure is known for inverting this calculation. To compute the value of k satisfying $kP = Q$, when given the points P and Q, is apparently hard; having a complexity proportional to $\sqrt{n}$, where again n is the order of the group $\mathcal{X}(\boldsymbol{F}_q)$. This is the discrete-log problem on elliptic curves. One can always solve the equation $kP = Q$ by computing iP for $i = 1, 2, 3, \ldots$, until Q is obtained, but this is intractable if k is large. One can do a little better than this clumsy method, but as far as we know, not very much better.

Thus the difference in computational complexity between computing kP from k and computing k from kP is immense. Suppose that the order of P is a 100-bit prime. Then to compute kP from k has a complexity proportional to 100, while to compute k from kP evidently has a complexity proportional to 2^{50}. This asymmetry is the basis of elliptic-curve cryptography.

10.9 Elliptic curve cryptography

An elliptic curve on a finite field is an abelian group under point addition, and so can be used to form a cryptography system using any cryptographic technique based on a finite abelian group. Usually, one requires a large cyclic group, preferably of prime order, so one chooses an elliptic curve that is a large cyclic group of prime order under point addition, or that contains a large cyclic subgroup of prime order under point addition.

Supersingular curves may be considered attractive for elliptic-curve cryptography because doubling a point requires only one inversion and four multiplications. That is, if $P = (x_1, y_1)$, then $2P = (x_3, y_3)$ where

$$x_3 = \left(\frac{x_1^2 + a_4}{a_3}\right)^2,$$

$$y_3 = \left(\frac{x_1^2 + a_4}{a_3}\right)(x_1 + x_3) + y_1 + a_3$$

and, by choosing $a_3 = 1$, we can even avoid the need for a division. This simplification is the motivation for using supersingular curves. However, this simplification is now viewed as too slight an advantage to justify using a form of elliptic curve with a perceived potential weakness. Supersingular elliptic curves are sometimes avoided in the design of cryptography systems because of evidence which suggests that a cryptosystem using a supersingular a curve may be less secure. Specifically, as will be discussed in Chapter 12, the group of point additions on any elliptic curve can be mapped into a discrete-log problem in the multiplicative group of a sufficiently large extension field of $\boldsymbol{F}_q$. An ordinary elliptic curve can be chosen so that this extension field is large and the discrete-log problem cannot be attacked in this way. For a supersingular curve, the appropriate extension field is not as large, and so the discrete-log problem in that extension field may be insufficiently hard. This conclusion might be overly cautious because the massive size of the extension field of $\boldsymbol{F}_q$ will partially compensate for any weakness of the discrete-log problem, but it is considered by some to be an adequate reason to avoid the use of supersingular elliptic curves. To some, the advantage of the simpler computations is not compelling.

For example, a supersingular elliptic curve over $\boldsymbol{F}_{2^{173}}$ is given by the curve $\mathcal{X}$: $y^2 + y = x^3 + x$. We will see later, in Section 12.8, that an attack known as the MOV attack can map the discrete-log problem on the elliptic curve $\mathcal{X}(\boldsymbol{F}_{2^{173}})$ to a discrete-log problem in $\boldsymbol{F}^*_{2^{692}}$, the multiplication group of the finite field $\boldsymbol{F}_{2^{692}}$. The elements of $\boldsymbol{F}_{2^{692}}$ are 692-bit binary numbers. The discrete logarithm in this field can be found with about 10^{18} field operations in the 692-bit arithmetic of $\boldsymbol{F}_{2^{692}}$. A computation of this complexity is generally regarded as impractical, but perhaps not intractable.

An elliptic curve cryptosystem is commonly used to create a symmetric key for a bulk encryption system using the Diffie–Hellman key exchange. It also can be used directly to encrypt a message using an analog of biprime cryptography, though this is less common. In this case, the message must be represented as a point P on the curve $\mathcal{X}(\boldsymbol{F}_q)$. A natural way to do this is to use the message to choose the x coordinate of the point P. However, not every value of x corresponds to a point of the curve. Only those x for which $x^3 + ax + b$ is a square can be a point of the curve. To protect against the message x not being a square, one can reserve, say, an 8-bit segment of the message coordinate x. These bits are not part of the message, but are chosen to assure that the message corresponds to a valid point of P. Each choice of these 8 bits yields a point of the curve with probability $1/2$ and there are 256 choices, so one can estimate that the probability that no choice of these 8 bits yields a point of the curve is 2^{-256}.

It is more common to use the elliptic curve for some other method of key exchange which we now describe.

Diffie–Hellman cryptosystem

Any large cyclic subgroup with prime order can be used to construct a Diffie–Hellman public-key cryptosystem by any of several general methods. A public-key cryptosystem using an elliptic curve can be constructed by first using the *Diffie–Hellman key exchange* to create the key, and then using that key with any satisfactory block cipher or stream cipher for the bulk encryption. The Diffie–Hellman key exchange on an elliptic curve is an immediate transfer of the Diffie–Hellman procedure from the cyclic group $\boldsymbol{Z}_p^*$ to any cyclic subgroup of $\mathcal{X}(\boldsymbol{F}_q)$. The curve $\mathcal{X}(\boldsymbol{F}_q)$ and a point $P \in \mathcal{X}(\boldsymbol{F}_q)$ form a standard, which is known to all users of the cryptosystem and is presumably also known to the cryptanalyst. The point P generates a cyclic subgroup of $\mathcal{X}(\boldsymbol{F}_q)$ and a key is an element of this subgroup. Two users, called user A and user B, mutually create the secret key, known only to the two of them, by exchanging messages over a public channel. User A randomly chooses a large integer a and computes aP, which is publicly transmitted while keeping the integer a secret. User B randomly chooses a large integer b and computes bP, which is publicly transmitted while keeping the integer b secret. Only user A knows a, and only user B knows b. Upon receiving aP, user B computes $b(aP) = (ab)P$. Upon receiving bP, user A computes $a(bP) = (ab)P$. Now both parties have computed the same point of the curve. This is the point $(ab)P$. Some aspect of the point, such as the x coordinate of P (or the y coordinate) provides the common secret key.

The direct attack is to compute a from P and aP. This requires the solution of the discrete-log problem on an elliptic curve. The only information that is available to all third parties, including the cryptanalyst, is P, aP, and bP. The security of the system depends on the presumption that it is intractable to compute $(ab)P$ from P, aP, and bP. While this presumption is unproven, it appears that all attempts to find such a method of computing $(ab)P$ from P, aP, and bP have been unsuccessful.

Elgamal cryptosystem

One can also define a public-key cryptosystem by using the elliptic curve $\mathcal{X}(\boldsymbol{F}_q)$ to form an Elgamal cryptosystem as follows. A generator P of a cyclic subgroup of $\mathcal{X}(\boldsymbol{F}_q)$ is made public, possibly a public standard. A message such as a key is represented as a point M of the curve. To receive a message, the decryptor randomly chooses an integer

a, then computes the point $A = aP$, which is then sent to the encryptor over a public channel. The point $A = aP$ is the public encryption key for any message to be sent to this decryptor, and could be published as such in a public directory. To encode message M, an encryptor chooses a random integer b, computes the two points $Q_1 = bP$ and $Q_2 = M + bA$, and sends the pair (Q_1, Q_2) as the encrypted message. The decryptor then computes $Q_2 - aQ_1 = (M + b(aP)) - b(aP) = M$. All operations are either point additions, point subtractions, or point multiplies in $\mathcal{X}(\boldsymbol{F}_p)$. All operations are straightforward. The cryptanalyst, however, does not know a, and it is intractable to compute a, so the task of cryptanalysis is evidently intractable.

The Elgamal cryptosystem on an elliptic curve relies on the assumption that the discrete-log problem is hard in the group of the elliptic curve. This means that it does not provide an alternative to the Diffie–Hellman key exchange that protects against the discovery of a solution to the discrete-log problem on the elliptic curve. Both methods of encryption will fail if the elliptic-curve discrete-log problem is broken.

10.10 The projective plane

An elliptic curve $\mathcal{X}$ has been defined with one extra point at infinity. Because of the way that it was introduced, this point at infinity may appear to be an artificial appendage to the curve, not really a natural part of the curve. Moreover, the point at infinity is not visible as a zero of the polynomial $p(x, y)$ that defines the curve $\mathcal{X}$. It is possible, however, to enlarge the plane over $\boldsymbol{F}$ in a natural way to a larger object called the *projective plane* over $\boldsymbol{F}$. To contrast the original plane with the projective plane, the original plane $\boldsymbol{F}^2$ is called the *affine plane*. Whereas the affine plane consists of all pairs (x, y), the projective plane consists of all triples (x, y, z) with the condition that the rightmost nonzero element of the triple is a one. Thus the projective plane is defined as $\{(x, y, 1) \mid x, y \in \boldsymbol{F}\} \bigcup \{(x, 1, 0) \mid x \in \boldsymbol{F}\} \bigcup \{(1, 0, 0)\}$. The first set in this union can be regarded as a copy of the affine plane that lies within the projective plane. The union of the second two sets is called the *line at infinity*. The single element in the third set is the point at infinity that is attached to the line at infinity.

To define the elliptic curve in the projective plane over $\boldsymbol{F}$, the *bivariate polynomial* $p(x, y)$ must be re-expressed as a homogeneous *trivariate polynomial* $p(x, y, z)$. A *homogeneous polynomial* is defined as a polynomial for which every monomial has the same total degree. To rewrite the bivariate polynomial $p(x, y) = \sum_{ij} p_{ij} x^i y^j$ of degree n as the trivariate polynomial $p(x, y, z)$ of degree n, simply insert a power of z into each monomial so that each monomial has the same degree. Then

$$\begin{aligned} p(x, y, z) &= \sum_{ij} p_{ij} x^i y^j z^{n-i-j} \\ &= y^2 z + a_1 xyz + a_3 yz^2 - x^3 - a_2 x^2 z - a_4 xz^2 - a_6 z^3. \end{aligned}$$

The affine zeros of $p(x, y)$ are the zeros of $p(x, y, z)$ with $z = 1$. There may be additional zeros of $p(x, y, z)$ on the line at infinity. The projective zeros of $p(x, y)$ are the zeros of $p(x, y, z)$ with $z = 0$.

For an elliptic curve, there is only one zero on the line at infinity and it is at $(0, 1, 0)$. To see this, recall that, for an elliptic curve,

$$p(x, y) = y^2 + a_1xy + a_3y - x^3 - a_2x^2 - a_4x - a_6.$$

The trivariate form of this polynomial is

$$p(x, y, z) = y^2z + a_1xyz + a_3yz^2 - x^3 - a_2x^2z - a_4xz^2 - a_6z^3.$$

Each monomial now has degree three, which is the degree of the bivariate polynomial $p(x, y)$. The trivariate polynomial $p(x, y, z)$ retains all the affine zeros of $p(x, y)$, occurring where $z = 1$. To see the zeros on the line at infinity, let $z = 0$. Then $p(x, 1, 0) = x^3$, which is zero only if $x = 0$. Finally, notice that $(1, 0, 0)$ is not a zero. Thus the elliptic curve always has a zero on only one point of the line at infinity. This is the point $(0, 1, 0)$.

The z derivative of $p(x, y, z)$ is

$$y^2 + a_1xy + 2a_3yz - a_2x^2 - 2a_4x - 3a_6z^2,$$

which at the point at infinity $\infty = (0, 1, 0)$ reduces to simply $y = 1$, which is not zero. Hence, the only point at infinity of an elliptic curve is never a singular point.

It may be helpful to discuss the zeros of $p(x, y, z)$ in the three-dimensional affine space $\boldsymbol{F}^3$. Because $p(x, y, z)$ is a homogeneous polynomial, if (α, β, γ) is a zero of $p(x, y, z)$, then $(\lambda\alpha, \lambda\beta, \lambda\gamma)$ is a zero of $p(x, y, z)$ as well. This means that all points of the line passing through the origin and any zero of $p(x, y, z)$ are also zeros of $p(x, y, z)$. Thus the set of zeros in $\boldsymbol{F}^3$ consists entirely of "rays" through the origin. It is enough to state only one point on each ray because one point determines the entire ray. These representative points are the projective zeros. If z is nonzero the representative point is taken to be $(x, y, 1)$. This can be regarded as a point of the affine plane. If z is zero, then the representative point is $(x, 1, 0)$ unless y is zero. Then it is $(1, 0, 0)$.

Thus, when the elliptic curve $\mathcal{X}$ is viewed in the projective plane, the point at infinity takes on a more natural appearance. If the points of the elliptic curve are regarded as rays from the origin in three-dimensional space, then the points at infinity are simply the rays with $z = 0$. These are the rays lying in the (x, y) plane. Moreover, under an invertible coordinate transformation the zeros, including the zero at infinity, can be relocated to new positions in the projective plane.

For example, let $\boldsymbol{F}$ be any field whose characteristic is not two or three. The bivariate polynomial $p(x, y) = y^2 - x^3 - ax - b$ becomes the homogeneous trivariate polynomial $p(x, y, z) = y^2z - x^3 - axz^2 - bz^3$. The zeros of this polynomial form an elliptic curve. The only zero on the line at infinity is at $(0, 1, 0)$. Under any invertible coordinate transformation, this homogeneous polynomial changes into another homogeneous polynomial whose zeros also form an elliptic curve. The coordinate transformation

$$\begin{bmatrix} x \\ y \\ z \end{bmatrix} = \begin{bmatrix} 0 & 1 & 1 \\ 1 & 0 & 1 \\ 1 & 1 & 1 \end{bmatrix} \begin{bmatrix} u \\ v \\ w \end{bmatrix},$$

is invertible because the determinant of the matrix is equal to one for any field characteristic. With this coordinate transformation, the trivariate polynomial in (x, y, z) becomes another homogeneous trivariate polynomial in (u, v, w), which can be reduced to a bivariate polynomial by setting w equal to one. This polynomial in u and v does describe an elliptic curve, but no longer in Weierstrass form. The original point at infinity $(x, y, z) = (0, 1, 0)$ has been moved to $(u, v, w) = (0, -1, 1)$.

For another example, let $\boldsymbol{F}$ be a field of characteristic two. The bivariate polynomial $x^3 + x^2 + xy + y^2 + 1$ becomes the homogeneous trivariate polynomial $x^3 + x^2z + xyz + y^2z + z^3$. Again, an invertible coordinate transformation will relocate the zeros of this polynomial to new positions in the projective plane.

10.11 Point counting in an extension field

The polynomial $p(x, y)$ that defines the elliptic curve $\mathcal{X}(\boldsymbol{F}_q)$ can also be regarded as a polynomial in the extension field $\boldsymbol{F}_{q^m}$, and so $p(x, y)$ also defines the elliptic curve $\mathcal{X}(\boldsymbol{F}_{q^m})$ in the larger field. Figure 10.4 on p. 302 shows this notionally. For each extension field of $\boldsymbol{F}_q$, an enlarged version of the elliptic curve appears. Of course, the ground field is contained in each extension field, so the points of the curve in the ground field are also part of the curve in the extension field. Our purpose is to count the number of points of the curve in the extension field not by actual counting, but by computation from the number of points in the ground field.

To count the number of rational points on an elliptic curve $\mathcal{X}(\boldsymbol{F}_q)$, it is enough to compute the Frobenius trace, which is defined as the difference between $q + 1$ and $\#\mathcal{X}(\boldsymbol{F}_q)$. Thus

$$t = q + 1 - \#\mathcal{X}(\boldsymbol{F}_q).$$

This statement, of course, also holds for the same curve viewed in the extension field $\boldsymbol{F}_{q^m}$. Thus

$$t_m = q^m + 1 - \#\mathcal{X}(\boldsymbol{F}_{q^m}),$$

where t_m is the Frobenius trace of $\mathcal{X}(\boldsymbol{F}_{q^m})$. Our task in this section is to compute t_m from t.

The second-degree polynomial $z^2 - tz + q$, in which the integer coefficients t and q are taken from an elliptic curve $\mathcal{X}(\boldsymbol{F}_q)$ with t equal to the Frobenius trace, can be factored in the complex field as

$$z^2 - tz + q = (z - \alpha)(z - \beta).$$

Because we know by the Hasse–Weil bound that $t^2 - 4q$ is never positive, we know that the zeros α and β are complex, and so they are conjugates. Furthermore, α and β satisfy $\alpha + \beta = t$ and $\alpha\beta = q$. The two complex numbers, α and β, will play an important role in the counting of the number of points on an elliptic curve in an extension field. Indeed, the forthcoming Theorem 10.11.2 will show that

$$z^2 - t_m z + q^m = (z - \alpha^m)(z - \beta^m).$$

Let $\#\mathcal{X}(\boldsymbol{F}_{q^m})$ denote the cardinality of the elliptic curve over $\boldsymbol{F}_{q^m}$. Define the Frobenius trace in the extension field as $t_m = q^m + 1 - \#\mathcal{X}(\boldsymbol{F}_{q^m})$ and, as was done in the ground field, define the polynomial $z^2 - t_m z + q^m$. This polynomial can be factored over the complex field as

$$z^2 - t_m z + q^m = (z - \alpha_m)(z - \beta_m),$$

for some complex conjugates α_m and β_m.

The following lemma uses s_m to denote $\alpha^m + \beta^m$, but it will be shown later that $s_m = t_m$.

Lemma 10.11.1 *Let $z^2 - tz + q = (z - \alpha)(z - \beta)$ in the complex field where $t = \alpha + \beta$ is the Frobenius trace and $q = \alpha\beta$. Let $s_m = \alpha^m + \beta^m$. Then s_m satisfies the integer recursion*

$$s_{m+1} = ts_m - qs_{m-1},$$

where $s_1 = \alpha + \beta$ and $s_0 = \alpha^0 + \beta^0 = 2$, and so s_m is an integer for all m.

Proof The quadratic factorization $z^2 - tz + q = (z - \alpha)(z - \beta)$ means that the parameter α satisfies $\alpha^2 - t\alpha + q = 0$. Multiply this by α^{m-1} so that

$$\alpha^{m+1} - t\alpha^m + q\alpha^{m-1} = 0.$$

In the same way, we have

$$\beta^{m+1} - t\beta^m + q\beta^{m-1} = 0.$$

Add these two equations to obtain

$$\alpha^{m+1} + \beta^{m+1} - t(\alpha^m + \beta^m) + q(\alpha^{m-1} + \beta^{m-1}) = 0.$$

With $s_m = \alpha^m + \beta^m$, the equation becomes $s_{m+1} - ts_m + qs_{m-1} = 0$. This gives the stated recursion. Because s_0 and s_1 are integers, all s_m are integers and so $\alpha^m + \beta^m$ is an integer. The proof is complete. □

The following theorem relates this factorization of s_m for arbitrary m to the factorization for $m = 1$. This theorem provides a powerful relationship because it says that it is enough to count zeros in the ground plane $\boldsymbol{F}_q^2$. The number of zeros in any extension plane $\boldsymbol{F}_{q^m}^2$ then can be computed using a simple formula.

Theorem 10.11.2 *An elliptic curve $\mathcal{X}(\boldsymbol{F}_q)$ over the field $\boldsymbol{F}_q$, when lifted to the extension field $\boldsymbol{F}_{q^m}$, has*

$$\#\mathcal{X}(\boldsymbol{F}_{q^m}) = q^m + 1 - (\alpha^m + \beta^m)$$

points in the projective plane, where α and β are the zeros of $z^2 - tz + q$.

Proof The proof consists of showing that

$$z^{2m} - t_m z^m + q^m = z^{2m} - s_m z^m + q^m$$

where in the polynomial on the right z^m is used as the polynomial indeterminate in place of z. First observe that $(z - \alpha)$ and $(z - \beta)$ divide $(z^m - \alpha^m)$ and $(z^m - \beta^m)$, respectively. Moreover, $(z - \alpha)(z - \beta) = z^2 - tz + q$ and $(z^m - \alpha^m)(z^m - \beta^m) = z^{2m} - (\alpha^m + \beta^m) + q^m$. We can conclude that $z^{2m} - s_m z^m + q^m = Q(z)(z^2 - tz + q)$ for some polynomial $Q(z)$, where all coefficients are integers.

Although the polynomial $z^{2m} - (\alpha^m + \beta^m)z^m + q^m$ resembles the polynomial $z^{2m} - (\alpha_m + \beta_m)z^m + q^2$, no reason has been given to prove that the middle terms are equal. There can be no such proof unless the first polynomial is given an operational significance. For this purpose we reference ahead to Theorem 10.13.3. By invoking this theorem in $\boldsymbol{F}_{q^m}$ and noting that the Frobenius map satisfies $\pi_{q^m} = \pi_q^m$, write the two polynomials as

$$\pi_{q^m}^2(P) - [t_m]\pi_{q^m}(P) - [q^m]P = \mathcal{O}$$

and

$$\begin{aligned}\pi_{q^m}^2(P) - [s_m]\pi_{q^m}(P) - [q^m]P &= Q(\pi_q)(\pi_q^2(P) - [t]\pi_q(P) - [q]P)\\ &= \mathcal{O}.\end{aligned}$$

Subtracting these two expressions gives

$$[s_m - t_m]\pi_{q^m}(P) = \mathcal{O}.$$

This can hold only if $s_m = t_m$ modulo the order. Therefore $\alpha^m + \beta^m$ is the Frobenius trace t_m of the curve in the extension field $\boldsymbol{F}_{q^m}$. The proof is complete. □

To use the theorem to compute $\#\mathcal{X}(\boldsymbol{F}_{q^m})$, we must first know $\#\mathcal{X}(\boldsymbol{F}_q)$. If q is small, then $\#\mathcal{X}(\boldsymbol{F}_q)$ can be obtained by evaluating the defining elliptic polynomial at every point of the plane and counting zeros. Then, because $\#\mathcal{X}(\boldsymbol{F}_q) = q + 1 - t$, we know t.

From t we can compute t_m from α and β which are obtained by factoring the polynomial $z^2 - tz + q$. Then $\#\mathcal{X}(\boldsymbol{F}_{q^m}) = q^m + 1 - t_m$.

For example, an elliptic curve over $\boldsymbol{F}_2$ is given by

$$\mathcal{X}(\boldsymbol{F}_2) : y^2 + y = x^3 + x$$

and

$$\#\mathcal{X}(\boldsymbol{F}_2) = q + 1 - t.$$

If $x = 0$, then $y = 0$ or 1. If $x = 1$, then again $y = 0$ or 1. Thus the curve $\mathcal{X}(\boldsymbol{F}_2)$ consists of five points $(0, 0)$, $(0, 1)$, $(1, 0)$, $(1, 1)$, and ∞. Therefore $t = 2 + 1 - 5 = -2$, which leads to the polynomial $x^2 - tx + q = x^2 + 2x + 2 = 0$. In turn, the two zeros of this polynomial are $\alpha, \beta = -1 \pm \sqrt{-1}$. From this we can write, for example,

$$\#\mathcal{X}(\boldsymbol{F}_{2^{173}}) = 2^{173} + 1 - (\alpha^{173} + \beta^{173}).$$

This gives a tractable way to compute the number of points on $\mathcal{X}(\boldsymbol{F}_{2^{173}})$. However, this expression involves complex numbers, which, in general, can lead to complications due to numerical imprecision. An alternative that uses only integer arithmetic may be preferred, as will be developed next.

First consider the case with $m = 2$, which is known as a quadratic field extension. In this special case, the statement of the theorem takes a simple form, as described by the following corollary.

Corollary 10.11.3 *If the elliptic curve $\mathcal{X}(\boldsymbol{F}_q)$ has $\#\mathcal{X}(\boldsymbol{F}_q) = q + 1 - t$ points in the ground field $\boldsymbol{F}_q$, then in the quadratic field extension $\boldsymbol{F}_{q^2}$, the elliptic curve $\mathcal{X}(\boldsymbol{F}_{q^2})$ has*

$$\#\mathcal{X}(\boldsymbol{F}_{q^2}) = (q + 1 - t)(q + 1 + t)$$

points.

Proof Theorem 10.11.2 states that

$$\#\mathcal{X}(\boldsymbol{F}_{q^2}) = q^2 + 1 - (\alpha^2 + \beta^2),$$

where $\alpha + \beta = t$ and $\alpha\beta = q$. Therefore $\alpha^2 + \beta^2 = (\alpha + \beta)^2 - 2q = t^2 - 2q$. From this, we have

$$\begin{aligned}\#\mathcal{X}(\boldsymbol{F}_{q^2}) &= q^2 + 1 - t^2 + 2q \\ &= (q + 1)^2 - t^2 \\ &= (q + 1 - t)(q + 1 + t),\end{aligned}$$

as was to be proved. □

For example, the elliptic curve $\mathcal{X}(\boldsymbol{F}_{131})$ with polynomial $y^2 = x^3 - 3x + 8$, considered in Section 10.4, has order 110 and $t = 22$. Therefore

$$\begin{aligned}\#\mathcal{X}(\boldsymbol{F}_{131^2}) &= (132 - 22)(132 + 22)\\ &= 16940.\end{aligned}$$

Because -1 is not a square in the field $\boldsymbol{F}_{131}$, we can write the elements of the extension field $\boldsymbol{F}_{131^2}$ as $a + ib$ where $i = \sqrt{-1}$ in the extension field $\boldsymbol{F}_{131^2}$. Then a point P of the curve can be written as $(a + ib, c + id)$ where $x = a + ib$ and $y = c + id$ are the two "complex" coordinates of a point in $\boldsymbol{F}_{131^2}$.

Corollary 10.11.3 can be used a second time to find $\#\mathcal{X}(\boldsymbol{F}_{q^4})$ in terms of $\#\mathcal{X}(\boldsymbol{F}_{q^2})$. First, write

$$\begin{aligned}\#\mathcal{X}(\boldsymbol{F}_{q^2}) &= (q + 1)^2 - t^2\\ &= q^2 + 1 - (t^2 - 2q).\end{aligned}$$

This has the same form as $\#\mathcal{X}(\boldsymbol{F}_q)$, but with q replaced by q^2 and t replaced by $t^2 - 2q$. Now refer to Corollary 10.11.3 with $\boldsymbol{F}_q$ replaced by $\boldsymbol{F}_{q^2}$, with q replaced by q^2, and with t replaced by $t^2 - 2q$. Then

$$\begin{aligned}\#\mathcal{X}(\boldsymbol{F}_{q^4}) &= (q^2 + 1 - t^2 + 2q)(q^2 + 1 + t^2 - 2q)\\ &= (q + 1 - t)(q + 1 + t)(q^2 - 2q + 1 + t^2)\end{aligned}$$

is the number of points that the curve $\mathcal{X}$ has in $\boldsymbol{F}_{q^4}$.

To continue the example, because $q = 131$ and $t = 22$, we know that $q^2 = 17161$ and $q^2 + 1 - \#\mathcal{X}(\boldsymbol{F}_{131^2}) = 222$, so we can compute that

$$\#\mathcal{X}(\boldsymbol{F}_{131^4}) = (q^2 + 1)^2 - (t^2 - 2q)^2 = 291,484,960.$$

The elements of this extension field $\boldsymbol{F}_{131^4}$ have the form $a + ib + i^2c + i^3d$, with multiplication of polynomials in i reduced modulo an irreducible polynomial $p(i)$ of degree four and integer coefficients reduced modulo 131. A point of the curve $\mathcal{X}(\boldsymbol{F}_{q^4})$ then has the form $(a + ib + i^2c + i^3d, e + if + i^2g + i^3h)$.

In general, for any even m, Corollary 10.11.3 allows us to write

$$\begin{aligned}\#\mathcal{X}(\boldsymbol{F}_{q^m}) &= (q^{m/2} + 1 - t_{m/2})(q^{m/2} + 1 + t_{m/2})\\ &= \#\mathcal{X}(\boldsymbol{F}_{q^{m/2}})[\#\mathcal{X}(\boldsymbol{F}_{q^{m/2}}) - q^{m/2} - 1].\end{aligned}$$

If $m/2$ is again even, this statement can be repeated once again, finally expressing $\#\mathcal{X}(\boldsymbol{F}_{q^m})$ in terms of $\#\mathcal{X}(\boldsymbol{F}_{q^{m/2^\ell}})$ for the largest possible such ℓ.

If m is composite but not a power of two, then an alternative formula must be used, which is not developed until later in this section. Anticipating this formula, given in Theorem 10.11.6, so that the example for the elliptic curve $\#\mathcal{X}(\boldsymbol{F}_{131})$ with $t = 22$ can

Table 10.1 *Some point counts for a curve on $\mathbf{F}_{131^m}$*

m	$\#\mathcal{X}(\mathbf{F}_{q^m})$
1	110
2	16, 940
3	2, 246, 090
4	291, 484, 960

be continued, we have

$$\#\mathcal{X}(\mathbf{F}_{q^3}) = (q+1)^3 - t^3 - \binom{3}{1} q[\#\mathcal{X}(\mathbf{F}_q)].$$

Setting $q = 131$, $t = 22$, and $\#\mathcal{X}(\mathbf{F}_q) = 110$, this becomes

$$\begin{aligned}\#\mathcal{X}(\mathbf{F}_{131^3}) &= 132^3 - 22^3 - \binom{3}{1} \cdot 131 \cdot 110 \\ &= 2,246,090.\end{aligned}$$

A summary of some counts of the number of points on the curve $\mathcal{X}(\mathbf{F}_{131^m})$ is shown in Table 10.1. In each case, except $m = 1$, the count includes points already counted in smaller subfields. This means that the points of $\mathcal{X}(\mathbf{F}_{q^3})$ include the points of $\mathcal{X}(\mathbf{F}_q)$, but not the points of $\mathcal{X}(\mathbf{F}_{q^2})$ because $\mathbf{F}_{q^2}$ is not a subfield of $\mathbf{F}_{q^3}$. The points of $\mathcal{X}(\mathbf{F}_{q^4})$ include the points of $\mathcal{X}(\mathbf{F}_{q^2})$, because $\mathbf{F}_{q^2}$ is a subfield of $\mathbf{F}_{q^4}$, but not the points in $\mathbf{F}_{q^3}$. Notice also that in each case $\#\mathcal{X}(\mathbf{F}_{q^m})$ is divisible by $\#\mathcal{X}(\mathbf{F}_q)$, as is required by Lagrange's theorem. Finally, notice that the number of points on $\mathcal{X}(\mathbf{F}_{q^m})$ is growing roughly as q^m, in keeping with the Hasse–Weil bound.

A powerful tool in the study of an elliptic curve is the *zeta function*[7] of the elliptic curve. The zeta function is defined in terms of the number of points that $\mathcal{X}$ has in the extension field $\mathbf{F}_{q^r}$, which is denoted $N_r = \#\mathcal{X}(\mathbf{F}_{q^r})$.

Definition 10.11.4 *The zeta function of the elliptic curve $\mathcal{X}$ over the field $\mathbf{F}_q$ is given by*

$$Z(x) = e^{\sum_{k=1}^{\infty} N_r x^r / r}$$

where N_r is the number of points of $\mathcal{X}(\mathbf{F}_{q^r})$.

As defined, the zeta function of the elliptic curve involves an infinite sum, so it does not appear to be useful, at least not in this form. The following theorem provides a more attractive and more useful expression for the zeta function.

[7] The term "zeta function" is a standard term for any generator function of this form. The term is used in this way in other contexts.

Theorem 10.11.5 (Hasse) *The zeta function of an elliptic curve over $\boldsymbol{F}_q$ can be expressed in the simple form of a rational function of x, given by*

$$Z(x) = \frac{1 - tx + qx^2}{(1-x)(1-qx)},$$

where $t = t_1$ is the Frobenius trace.

Proof We begin with the factorization

$$x^2 - tx + q = (x - \alpha)(x - \beta).$$

Then by Theorem 10.11.2,

$$N_n = q^n + 1 - (\alpha^n + \beta^n).$$

We now have

$$\begin{aligned}\log Z(x) &= \sum_n N_n x^n/n \\ &= \sum_n (q^n + 1 - \alpha^n - \beta^n)x^n/n \\ &= \sum_n \frac{1}{n} q^n x^n + \sum_n \frac{1}{n} x^n - \sum_n \frac{1}{n}\alpha^n x^n - \sum_n \frac{1}{n}\beta^n x^n .\end{aligned}$$

Therefore using the series expansion

$$-\log(1-x) = \sum_{n=1}^{\infty} \frac{1}{n} x^n,$$

we have

$$\begin{aligned}\log Z(x) &= -\log(1-qx) - \log(1-x) + \log(1-\alpha x) + \log(1-\beta x) \\ &= \log \frac{(1-\alpha x)(1-\beta x)}{(1-x)(1-qx)}\end{aligned}$$

from which the theorem follows. □

The expression for the zeta function in Theorem 10.11.5 depends only on the single unknown parameter t. To evaluate the expression, t can be determined by counting the rational points of $\mathcal{X}$. The value of t is then given by $t = q + 1 - N_1$, where $N_1 = \#\mathcal{X}(\boldsymbol{F}_q)$. Then the zeta function gives the number of points of $\mathcal{X}(\boldsymbol{F}_{q^r})$ for all values of r.

The following theorem expresses the number of points on an elliptic curve in an extension field in terms of the number of points in smaller fields.

Theorem 10.11.6 *If* $\#\mathcal{X}(\boldsymbol{F}_q) = q + 1 - t$, *then for odd* m,

$$\#\mathcal{X}(\boldsymbol{F}_{q^m}) = (q+1)^m - t^m - \sum_{i=1}^{(m-1)/2} \binom{m}{i} q^i \left[\#\mathcal{X}(\boldsymbol{F}_{q^{m-2i}})\right].$$

Proof Recalling that $\alpha + \beta = t$ and $\alpha\beta = q$, the expression for $\#\mathcal{X}(\boldsymbol{F}_{q^m})$ can be manipulated as follows: Make the change of variables to write

$$\begin{aligned} q^m + 1 - \alpha^m - \beta^m &= (q+1)^m - (\alpha+\beta)^m + q^m + 1 - (q+1)^m + (\alpha+\beta)^m - \alpha^m - \beta^m \\ &= (q+1)^m - t^m - \sum_{i=1}^{m-1} \binom{m}{i} q^i + \sum_{i=1}^{m-1} \binom{m}{i} \alpha^i \beta^{m-i} \\ &= (q+1)^m - t^m - \sum_{i=1}^{m-1} A_i, \end{aligned}$$

where

$$A_i = \binom{m}{i} (q^i - \alpha^i \beta^{m-i}).$$

Recall that m is odd and break the sum on A_i into two sums, one sum running from 1 to $(m-1)/2$, and one sum running from $(m+1)/2$ to $m-1$:

$$\sum_{i=1}^{m-1} A_i = \sum_{i=1}^{(m-1)/2} A_i + \sum_{i=(m+1)/2}^{m-1} A_i.$$

Next make a change in variables in the second sum by setting $i = m - j$ so that j runs from $(m-1)/2$ to 1. Then

$$\sum_{i=1}^{m-1} A_i = \sum_{i=1}^{(m-1)/2} A_i + \sum_{j=1}^{(m-1)/2} A_{m-j} = \sum_{i=0}^{(m-1)/2} (A_i + A_{m-i}).$$

Accordingly,

$$\begin{aligned} \sum_{i=1}^{m-1} A_i &= \sum_{i=1}^{(m-1)/2} \binom{m}{i} (q^i + q^{m-i} + \alpha^i \beta^{m-i} + \alpha^{m-i} \beta^i) \\ &= \sum_{i=1}^{(m-1)/2} \binom{m}{i} q^i (q^{m-2i} + 1 - \alpha^{m-2i} - \beta^{m-2i}). \end{aligned}$$

Therefore,

$$q^m + 1 - \alpha^m - \beta^m = (q+1)^m - t^m - \sum_{i=1}^{(m-1)/2} \binom{m}{i} q^i \left[q^{m-2i} + 1 - \alpha^{m-2i} - \beta^{m-2i}\right].$$

The ith term in brackets under the sum at the right can be recognized as $\#\mathcal{X}(\boldsymbol{F}_{q^{m-2i}})$, which completes the proof of the theorem. □

Finally, notice that $q+1-t$ clearly divides $(q+1)^m - t^m$, and by an induction hypothesis, $q+1-t$ divides every term within the square brackets on the right. Therefore $q+1-t$ divides the whole of the right side. This is consistent with Corollary 10.11.3, which states that $\#\mathcal{X}(\boldsymbol{F}_{q^2})$ is a multiple of $q+1-t$, and which shows that $\#\mathcal{X}(\boldsymbol{F}_{q^{2m}})$ is a multiple of $q+1-t$ whenever $\#\mathcal{X}(\boldsymbol{F}_{q^m})$ is a multiple of $q+1-t$. All of this is simply a manifestation of the fact that $\mathcal{X}(\boldsymbol{F}_q)$ is a subgroup of the group $\mathcal{X}(\boldsymbol{F}_{q^m})$, so Lagrange's theorem asserts that $\#\mathcal{X}(\boldsymbol{F}_q)$ divides $\#\mathcal{X}(\boldsymbol{F}_{q^m})$.

10.12 Morphisms of elliptic curves over the rationals

Any elliptic curve $\mathcal{X}$ over the field $\boldsymbol{F}$ can be mapped into itself or into another elliptic curve over the same field $\boldsymbol{F}$ by mapping each point of the curve to a point of that same curve or to a point of another elliptic curve. A mapping is called a *homomorphism*, or simply a *morphism*, if it preserves relevant algebraic structure, in our case the algebraic structure of the elliptic curve. Specifically, an elliptic-curve homomorphism is a homomorphism that preserves point addition. The mapping $\phi(P)$ preserves point addition if $\phi(P_1)+\phi(P_2)=\phi(P_3)$ whenever $P_1+P_2=P_3$. A mapping that preserves point addition has special properties that we will want to study. The mapping may also have other structural requirements that will be indicated by more specific terminology.

One way to define a morphism on an elliptic curve is in terms of two rational functions $R_1(x,y)$ and $R_2(x,y)$, mapping the point $P=(x,y)$ of one elliptic curve to a point of another elliptic curve by the expression $\phi(P)=(R_1(x,y),R_2(x,y))$. This case is the subject of the following definition.

Definition 10.12.1 *An endomorphism of an elliptic curve $\mathcal{X}$ is a homomorphism from $\mathcal{X}$ to itself that preserves point addition and is described by rational functions.*

Because an endomorphism must preserve point addition, the only constant endomorphism is the zero endomorphism for which $\phi(P)=\mathcal{O}$ for all P. An endomorphism that maps an elliptic curve onto itself is a special instance of the morphism that maps one elliptic curve $\mathcal{X}_1$ onto another elliptic curve $\mathcal{X}_2$, not necessarily the same curve.

Definition 10.12.2 *An isogeny of elliptic curves is a morphism described by rational functions from an elliptic curve $\mathcal{X}_1$ to a second elliptic curve $\mathcal{X}_2$ that preserves point addition.*

Two curves $\mathcal{X}_1$ and $\mathcal{X}_2$ are *isogenous* if there exists an isogeny ϕ from $\mathcal{X}_1$ to $\mathcal{X}_2$ and also an isogeny ϕ' from $\mathcal{X}_2$ to $\mathcal{X}_1$. In this case, the composition $\phi' \circ \phi$ is an endomorphism on $\mathcal{X}$. A special case of an invertible isogeny is as follows.

Definition 10.12.3 *An isomorphism of two elliptic curves over* $\boldsymbol{F}$ *is an invertible morphism from an elliptic curve* $\mathcal{X}_1$ *to an elliptic curve* $\mathcal{X}_2$ *that preserves point addition and is given by an affine coordinate conversion. An affine coordinate conversion is a coordinate conversion of the form*

$$x = u^2x' + r$$
$$y = u^3y' + su^2x' + t,$$

where r, s, t, *and* u *are elements of the field* $\boldsymbol{F}$ *and* u *is nonzero.*

Two elliptic curves that are related by an isomorphism are called *isomorphic*. Two elliptic curves that are each isomorphic to a third elliptic curve are isomorphic to each other. This means that the notion of isomorphism partitions the set of elliptic curves over $\boldsymbol{F}$ into equivalence classes. Each of these equivalence classes is designated by the single field element called the j-invariant, which was defined in Section 10.1. Two elliptic curves in an algebraically closed field with the same j-invariant are in the same equivalence class, and so are isomorphic.

The set of all endomorphisms on $\mathcal{X}$ is denoted end($\mathcal{X}$). The set end($\mathcal{X}$) clearly is not empty because it always contains the identity endomorphism, defined by $\phi(P) = P$. Another obvious instance of an endomorphism for curves defined by a polynomial of the form $y^2 = x^3 + ax + b$ is $\phi(P) = -P$. In addition to these two endomorphisms, the set end($\mathcal{X}$) also contains many other endomorphisms.

An endomorphism that is easy to describe is defined by the multiplication of each point of $\mathcal{X}$ by a positive integer n. This mapping takes the point $P \in \mathcal{X}$ to the point $[n]P$, where $[n]P$ is the sum of n copies of P, the sum being point addition, as defined in Section 10.3. The point $[n]P$ is clearly an element of $\mathcal{X}$. Moreover, $[n](P + P') = [n]P + [n]P'$, so this map preserves point addition. Therefore this map is an endomorphism. In this way, each element of $\boldsymbol{Z}$ corresponds to an element of end($\mathcal{X}$). We say that end($\mathcal{X}$) contains an image of $\boldsymbol{Z}$, meaning that every element of $\boldsymbol{Z}$ corresponds to an element of end($\mathcal{X}$), though not necessarily a unique element of end($\mathcal{X}$).

One can define addition and multiplication within the set of endomorphisms. If ϕ_1 and ϕ_2 are endomorphisms, then define the sum of endomorphisms $\phi_1 + \phi_2$ by $(\phi_1 + \phi_2)(P) = \phi_1(P) + \phi_2(P)$. Define the product of endomorphisms $\phi_1\phi_2$ by $(\phi_1\phi_2)(P) = \phi_1(\phi_2(P))$. Therefore with these definitions, the set of endomorphisms of an elliptic curve is a ring. Because end($\mathcal{X}$) is a ring, we can conclude that for any $\phi \in$ end($\mathcal{X}$), the ring end($\mathcal{X}$) also contains the endomorphism $p(\phi)$ where $p(z)$ is any polynomial with integer coefficients.

For example, if ϕ is the endomorphism that multiplies point P by the integer n, then ϕ^2 multiplies the point P by the integer n^2 and $p(\phi)$ multiplies point P by the integer $p(n)$. Because $p(n)$ is just another integer, this process of combining endomorphisms corresponding to integer multiplication in this way does not create a new instance of an endomorphism.

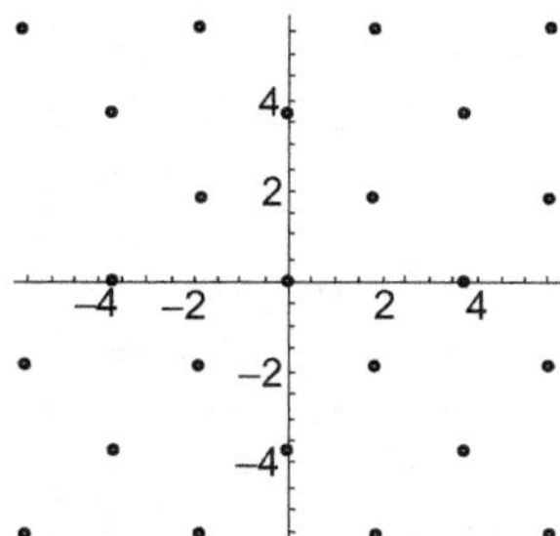

Figure 10.8 A lattice with 90° rotational symmetry

One may ask whether end($\mathcal{X}$) contains other endomorphisms that cannot be described as multiplication by an integer. We will see that for some elliptic curves over a quadratic extension of $\boldsymbol{Q}$, but not for every elliptic curve over a quadratic extension of $\boldsymbol{Q}$, the ring end($\mathcal{X}$) does contain other endomorphisms. For many elliptic curves, however, the ring end($\mathcal{X}(\boldsymbol{Q})$) does not contain any endomorphism other than multiplication by an integer.

For an example of an endomorphism that does not correspond to integer multiplication, consider the elliptic curve over the complex rational field $\boldsymbol{Q}(i)$:

$$\mathcal{X}(\boldsymbol{Q}(i)) : y^2 = x^3 + x,$$

with the map $\phi(P) = \phi(x, y) = (-x, iy)$, where $i^2 = -1$. If $P = (x, y)$ is a point of $\mathcal{X}$, then $(-x, iy)$ is a point of $\mathcal{X}$ as well, as can easily be checked. Thus the map $\phi(x, y)$ is an element of end($\mathcal{X}(\boldsymbol{Q}(i))$). Moreover, $\phi^2 = -1$, which is abbreviated symbolically as $\phi = \sqrt{-1}$, or as $\phi = i$. This means that the ring end($\mathcal{X}(\boldsymbol{Q}(i))$) contains an isomorphic copy, as rings, of the ring of gaussian integers $\boldsymbol{Z}[i]$. We then say that this elliptic curve has complex multiplication by $i = \sqrt{-1}$ in its endomorphism ring. The endomorphism labeled i means to map the point $P = (x, y)$ of the elliptic curve $\mathcal{X}$ to the point $(-x, iy)$. This we write as $[i]P = (-x, iy)$, which is also a point of the curve.

In contrast to the earlier example of an endomorphism consisting of multiplication by an element of $\boldsymbol{Z}$, this complex endomorphism depends on the specific polynomial defining the elliptic curve. We will see that some elliptic curves over $\boldsymbol{Q}$ have extra endomorphisms of this kind, and some elliptic curves do not have these extra endomorphisms. In the latter case, end($\mathcal{X}$) is isomorphic to $\boldsymbol{Z}$, as rings. In the former case, end($\mathcal{X}$) contains an image of $\boldsymbol{Z}$, but it also contains other endomorphisms.

At first sight, it may appear peculiar that some elliptic curves over $\boldsymbol{Q}$ have complex multiplication, and others do not. To make this circumstance less surprising, we refer to the more familiar situation of a lattice, as shown in Figure 10.8. The lattice of that figure is invariant under appropriate translations, as is true for any lattice. In addition, this lattice is invariant under a 90° rotation. Not every lattice has such a rotational invariance. Most do not, but some do.

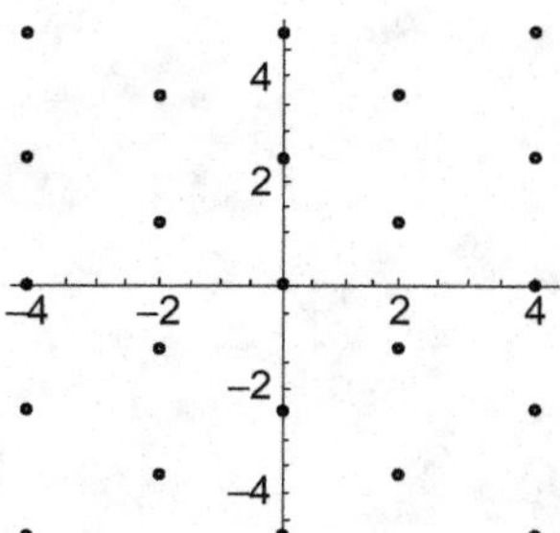

Figure 10.9 A lattice with 60° rotational symmetry

Just as some lattices have a rotational symmetry, but most do not, so too some elliptic curves on the rational field have a complex multiplication, but most do not.

This parallel is more than just an analogy. In Section 10.16, we will describe an isomorphism between complex elliptic curves and lattices in $\boldsymbol{C}^2$. Then, because each elliptic curve over $\boldsymbol{C}$ is closely associated with a lattice in $\boldsymbol{C}^2$ and every elliptic curve over $\boldsymbol{Q}$ can be embedded into $\boldsymbol{C}$, we can expect that the existence of complex multiplication on $\mathcal{X}(\boldsymbol{Q})$ is somehow related to some property of a complex lattice. This is, indeed, the case. An elliptic curve has complex multiplication, if and only if, the corresponding complex lattice has a suitable symmetry other than translation.

A second example of a lattice with a rotational symmetry is the lattice shown in Figure 10.9. This lattice is invariant under a 60° rotation. Figure 10.9 shows the lattice corresponding to the elliptic curve $\mathcal{X}(\boldsymbol{Q}) : y^2 + y = x^3$. In both this and the earlier example, even though the lattice actually lies in $\boldsymbol{C}^2$, it can be depicted in the real plane because the x component of every lattice point is purely real and the y component of every lattice point is purely imaginary. Most other lattices in $\boldsymbol{C}^2$ have both a nonzero real part and a nonzero imaginary part in both the x component and the y component and cannot be depicted so easily in $\boldsymbol{R}^2$.

Some examples of elliptic curves with complex multiplication are the following.

(1) The curve $\mathcal{X}$ defined by the polynomial $y^2 = x^3 + x$ has complex multiplication by $\sqrt{-1}$. The endomorphism of order four that corresponds to $i = \sqrt{-1}$ is given by $(x, y) \longmapsto (-x, iy)$. The cycle is

$$(x, y) \longmapsto (-x, iy) \longmapsto (x, -y) \longmapsto (-x, -iy) \longmapsto (x, y).$$

(2) The curve $\mathcal{X}$ defined by the polynomial $y^2 = x^3 + 4x^2 + 2x$ has complex multiplication by $\sqrt{-2}$. The endomorphism that corresponds to $\sqrt{-2}$ is given by

$$(x, y) \longmapsto \left(-\frac{y^2}{2x^2}, -\frac{y(x^2 - 2)\sqrt{-2}}{4x^2}\right).$$

Table 10.2 *Some curves with complex multiplication*

Number field	Elliptic polynomial
$Q(\sqrt{-1})$	$y^2 = x^3 + x$
$Q(\sqrt{-2})$	$y^2 = x^3 + 4x^2 + 2x$
$Q(\sqrt{-3})$	$y^2 + y = x^3$
$Q(\sqrt{-7})$	$y^2 + xy = x^3 - x^2 - 2x - 1$
$Q(\sqrt{-11})$	$y^2 + y = x^3 - x^2 - 7x + 10$
$Q(\sqrt{-19})$	$y^2 + y = x^3 - 38x + 90$
$Q(\sqrt{-43})$	$y^2 + y = x^3 - 860x + 9707$
$Q(\sqrt{-67})$	$y^2 + y = x^3 - 7370x + 243528$
$Q(\sqrt{-163})$	$y^2 + y = x^3 - 2174420x + 123413669.$

(3) The curve $\mathcal{X}$ defined by the polynomial $y^2 = x^3 + 1$ has complex multiplication by $\sqrt{-3}$. The cycle of points in $Q(\sqrt{-3})$ is

$$(x, y) \longmapsto (e^{i2\pi/3}x, y) \longmapsto (e^{i\pi/3}x, y) \longmapsto (x, y).$$

Because $e^{i\pi/3} = \cos\frac{\pi}{3} + i\sin\frac{\pi}{3} = \frac{1}{2} + i\frac{\sqrt{3}}{2}$, $Q(e^{i\pi/3})$ can be expressed as $Q(\sqrt{-3})$.

(4) The curve $\mathcal{X}$ defined by the polynomial $y^2 + 5xy = x^3 - x^2 + 7x$ has complex multiplication by $\sqrt{-7}$.

These and several other polynomials defining curves with complex multiplication are given in Table 10.2. In each case listed, there is an endomorphism from the curve to itself that behaves the same as multiplication of the points of the curve by the indicated complex number.

10.13 Morphisms of elliptic curves over finite fields

The endomorphism ring $\mathrm{end}(\mathcal{X}(\boldsymbol{F}_q))$ of the elliptic curve $\mathcal{X}(\boldsymbol{F}_q)$ on the finite field $\boldsymbol{F}_q$ is similar in some ways to the endomorphism ring $\mathrm{end}(\mathcal{X}(\boldsymbol{Q}))$ of an elliptic curve over the rational field $\boldsymbol{Q}$, but is different in other ways. On the one hand, $\mathrm{end}(\mathcal{X}(\boldsymbol{Q}))$ sometimes only contains an image of $\boldsymbol{Z}$, and sometimes other endormorphisms as well as an image of $\boldsymbol{Z}$. On the other hand, to see an isomorphic copy of $\boldsymbol{Z}$ in the endomorphism ring over a finite field, one must look to the algebraic closure $\overline{\boldsymbol{F}}_q$, and to the ring $\mathrm{end}(\mathcal{X}(\overline{\boldsymbol{F}}_q))$. The ring $\mathrm{end}(\mathcal{X}(\overline{\boldsymbol{F}}_q))$ always contains an isomorphic copy of $\boldsymbol{Z}$, and it always contains other endomorphisms as well.

Recall that the Frobenius map on the field $\boldsymbol{F}_q$, or on any extension of $\boldsymbol{F}_q$, is given by $\pi_q(x) = x^q$. Recall also that every point β of $\boldsymbol{F}_q$ is a zero of the polynomial $x^q - x$, so every β of $\boldsymbol{F}_q$ satisfies $\beta^q = \beta$, and so $\pi_q(\beta) = \beta$. The Frobenius map can be applied to both the x and y coordinates of any point of the curve $\mathcal{X}$ in the finite field $\boldsymbol{F}_q$, or in any extension of $\boldsymbol{F}_q$. Thus

$$\pi_q(x, y) = (\pi_q(x), \pi_q(y)) = (x^q, y^q),$$

where $(x, y) \in \mathcal{X}(\boldsymbol{F}_q)$, and by convention, the point at infinity goes to itself. Because $(x^q, y^q) = (x, y)$, the Frobenius map takes every point of the curve $\mathcal{X}(\boldsymbol{F}_q)$ to itself. This is the identity map, and so the Frobenius map is trivial if one considers only points in $\mathcal{X}(\boldsymbol{F}_q)$. However, the Frobenius map also applies to points of the curve $\mathcal{X}(\overline{\boldsymbol{F}}_q)$ on the algebraic closure of $\boldsymbol{F}_q$, in general mapping a point of $\mathcal{X}(\overline{\boldsymbol{F}}_q)$ to another point of $\mathcal{X}(\overline{\boldsymbol{F}}_q)$. It is easy to see that π_q maps a point P of $\mathcal{X}(\overline{\boldsymbol{F}}_q)$ to P itself if, and only if, P is an element of $\mathcal{X}(\boldsymbol{F}_q)$. Indeed, this map can be used as a way of recognizing the points of $\mathcal{X}(\boldsymbol{F}_q)$. For these and for other points of $\mathcal{X}(\overline{\boldsymbol{F}}_q)$, the following theorem applies.

Theorem 10.13.1 *The Frobenius map $\pi_q(P)$ on $\mathcal{X}(\overline{\boldsymbol{F}}_q)$ is an endomorphism. Moreover, for any $P \in \mathcal{X}(\overline{\boldsymbol{F}}_q)$, the Frobenius map $\pi_{q^k}(P) = P$ if, and only if, $P \in \mathcal{X}(\boldsymbol{F}_{q^k})$.*

Proof To prove the first statement of the theorem, it is necessary to show that for every $P \in \mathcal{X}(\overline{\boldsymbol{F}}_q)$, $\pi_q(P)$ is a point of the curve $\mathcal{X}(\overline{\boldsymbol{F}}_q)$, and that $\pi_q(P)$ preserves point addition: $\pi_q(P + P') = \pi_q(P) + \pi_q(P')$. For brevity, we give the proof of the theorem only for a curve whose defining polynomial is of the short Weierstrass form

$$y^2 = x^3 + ax + b.$$

The proof for the more general Weierstrass form of an elliptic polynomial is essentially the same.

Suppose that $(x, y) \in \mathcal{X}(\overline{\boldsymbol{F}}_q)$. Raising the polynomial to the qth power gives

$$(y^2)^q = (x^3 + ax + b)^q.$$

Because a and b are elements of the field $\boldsymbol{F}_q$, Theorem 9.3.3 states that this becomes

$$(y^2)^q = (x^q)^3 + ax^q + b,$$

which means that $(x^q, y^q) \in \mathcal{X}(\overline{\boldsymbol{F}}_q)$, which is the first property to be proved.

Next, to see by a direct argument that point addition is preserved under the Frobenius map, one can explicitly work through the formulas for point addition under the Frobenius map. For a more geometric argument, observe that if the points P_1, P_2, and P_3 are on a line, namely the line $y = ax + b$, then the points $\pi_q(P_1)$, $\pi_q(P_2)$, and $\pi_q(P_3)$ are also on a line, namely the line $y = a^q x + b^q$, which is the second property to be proved.

The second statement of the theorem is immediate because $\pi_{q^k}(x, y) = (\pi_{q^k}(x), \pi_{q^k}(y)) = (x, y)$ holds if, and only if, $(x, y) \in \boldsymbol{F}^2_{q^k}$. □

The theorem can be expressed succinctly as $\pi_q(\mathcal{X}(\overline{\boldsymbol{F}}_q)) = \mathcal{X}(\overline{\boldsymbol{F}}_q)$. This is an equality rather than a containment because every point has finite order so for every P, there will be an m such that $\pi_{q^m}(P) = P$. In view of this theorem, the Frobenius map when applied to the points of an elliptic curve over a finite field is an endomorphism. However, a different, more sophisticated endomorphism, described next, is the one called the Frobenius endomorphism.

The Frobenius map $\pi_q(P)$ takes every point of $\mathcal{X}(\overline{\boldsymbol{F}}_q)$ to another point of $\mathcal{X}(\overline{\boldsymbol{F}}_q)$ and is closed under addition. Moreover $\pi_q^2(P)$ is defined as $\pi_q(\pi_q(P))$. Therefore, any polynomial $p(x)$ with integer coefficients gives another map $p(\pi_q)$. The following definition gives an important instance of this fact based on the polynomial $p(x) = x^2 - tx + q$.

Definition 10.13.2 *The characteristic endomorphism of Frobenius is a function on the points of the plane* $\overline{\boldsymbol{F}}_q^2$*, given by*

$$\psi_q(P) = \pi_q^2(P) - [t]\pi_q(P) + [q]P,$$

where $\pi_q^2(P) = \pi_q(\pi_q(P))$*, and* t *is the Frobenius trace of the curve.*

It is easy to see that every point P of the elliptic curve $\mathcal{X}(\boldsymbol{F}_p)$ in the ground field $\boldsymbol{F}_q$ satisfies the *characteristic equation of Frobenius*, given by

$$\pi_q^2(P) - [t]\pi_q(P) + [q]P = \mathcal{O}.$$

As we now show, this is just a manifestation of the fact that in any finite abelian group, the group order times any group element gives the identity element. The order of point P is the smallest integer r such that $[r]P = \mathcal{O}$. The order of the group $\mathcal{X}(\boldsymbol{F}_q)$ is $\#\mathcal{X}(\boldsymbol{F}_q) = q + 1 - t$, and so the order of any element P divides $q + 1 - t$. Thus $[q + 1 - t]P = \mathcal{O}$ for every point P of $\mathcal{X}(\boldsymbol{F}_q)$. Then using the standard properties of addition in an abelian group, the equation can be rewritten as

$$P - [t]P + [q]P = \mathcal{O}.$$

For $P \in \mathcal{X}(\boldsymbol{F}_q)$, $(x^{q^2}, y^{q^2}) = (x^q, y^q) = (x, y)$, so this equation can be rewritten as

$$(x^{q^2}, y^{q^2}) - [t](x^q, y^q) + [q](x, y) = \mathcal{O},$$

for all $(x, y) \in \mathcal{X}(\boldsymbol{F}_q)$. Briefly, this becomes $\psi_q(P) = \mathcal{O}$ for all $P \in \mathcal{X}(\boldsymbol{F}_q)$. Thus the characteristic polynomial maps every point of $\mathcal{X}(\boldsymbol{F}_q)$ to the group identity element $\mathcal{O}$.

The following theorem extends this conclusion to all points of $\mathcal{X}(\overline{\boldsymbol{F}}_q)$. It says that $\psi_q[\mathcal{X}(\overline{\boldsymbol{F}}_q)] = \{\mathcal{O}\}$. The proof is based on the notion of a torsion point. As defined in Section 12.7, an m-torsion point is any point P of $\mathcal{X}(\overline{\boldsymbol{F}}_q)$ for which $[m]P = \mathcal{O}$ provided m is coprime to q. The set of all m-torsion points is a subgroup of $\mathcal{X}(\boldsymbol{F}_q)$ and is denoted $\mathcal{X}(\boldsymbol{F}_q)[m]$.

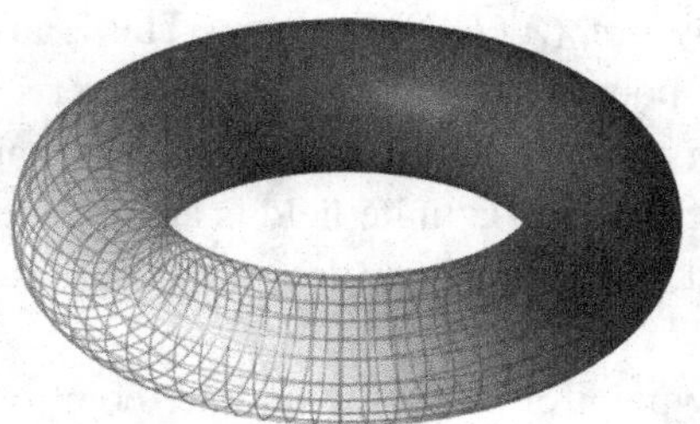

Figure 10.10 A discrete torus

The subgroup consisting of all m-torsion points always has order m^2 and satisfies the isomorphism

$$\mathcal{X}(\overline{\boldsymbol{F}}_q)[m] \simeq \mathbf{Z}_m \times \mathbf{Z}_m.$$

This is a powerful statement that will be fully explained in Section 12.8. We will use this fact here, and forward-reference to Theorem 12.8.2 for its justification.

Because each copy of the set $\mathbf{Z}_m$ on the right of the above isomorphism is integer addition modulo m, each copy of $\mathbf{Z}_m$ can be visualized as m discrete points arranged in a circle. Then $\mathbf{Z}_m \times \mathbf{Z}_m$ can be visualized as a discrete torus. Simply take an m-by-m section of the $\mathbf{Z}^2$ lattice and wrap it into a torus by attaching the left edge to the right edge, then attaching the top edge to the bottom edge. This forms the torus $\mathbf{Z}_m \times \mathbf{Z}_m$. Figure 10.10 depicts a torus inscribed with a discrete grid portraying $\mathbf{Z}_m \times \mathbf{Z}_m$. The proof of the next theorem will refer to this discrete torus.

Theorem 10.13.3 *The characteristic endomorphism of Frobenius $\psi_q(P)$ satisfies*

$$\pi_q^2(P) - [t]\pi_q(P) + [q]P = \mathcal{O}$$

for all points P of the curve $\mathcal{X}(\overline{\boldsymbol{F}}_q)$.

Proof We have already observed that this statement is straightforward for all $P \in \mathcal{X}(\boldsymbol{F}_q)$ because it is an immediate manifestation of the statement that every element of a finite group is taken to the identity element by multiplying it by its order. Accordingly, the task of the proof is to lift this conclusion to the curve $\mathcal{X}(\overline{\boldsymbol{F}}_q)$ over the algebraic closure $\overline{\boldsymbol{F}}_q$. Because every point of $\mathcal{X}(\overline{\boldsymbol{F}}_q)$ is an m-torsion point for some prime integer m, it is enough to fix the prime integer m and to prove the theorem for all m-torsion points.

We will work with the isomorphism $\mathcal{X}(\overline{\boldsymbol{F}}_q)[m] \simeq \mathbf{Z}_m \times \mathbf{Z}_m$, as given in Theorem 12.8.2. This isomorphism uniquely pairs one point P of $\mathcal{X}(\overline{\boldsymbol{F}}_q)[m]$ with one element $\Omega = (z_1, z_2)$ of $\mathbf{Z}_m \times \mathbf{Z}_m$. But if P is an m-torsion point of $\mathcal{X}(\overline{\boldsymbol{F}}_q)$, then $\pi_q(P)$ is a point of order m of $\mathcal{X}(\overline{\boldsymbol{F}}_q)$. If P corresponds to Ω, then the image of $\pi_q(P)$ in $\mathbf{Z}_m \times \mathbf{Z}_m$ will be denoted $\widehat{\pi}_q(\Omega)$. Then $\pi_q(P)$ corresponds to $\widehat{\pi}_q(\Omega)$ in the manner illustrated by the following diagram:

$$\begin{array}{ccc} \mathcal{X}(\overline{\boldsymbol{F}}_q)[m] & \leftrightarrow & \boldsymbol{Z}_m \times \boldsymbol{Z}_m \\ \pi_q(\cdot)\downarrow & & \downarrow \widehat{\pi}_q(\cdot) \\ \mathcal{X}(\overline{\boldsymbol{F}}_q)[m] & \leftrightarrow & \boldsymbol{Z}_m \times \boldsymbol{Z}_m , \end{array}$$

which is given pointwise by

$$\begin{array}{ccc} P & \leftrightarrow & \Omega \\ \downarrow & & \downarrow \\ \pi_q(P) & \leftrightarrow & \widehat{\pi}_q(\Omega) . \end{array}$$

With this definition, the map from Ω to $\widehat{\pi}_q(\Omega)$ can be regarded as the composition of the other three arrows of the diagram. We will determine the behavior of $\pi_q(P)$ by observing the behavior of its counterpart $\widehat{\pi}_q(\Omega)$ in $\boldsymbol{Z}_m \times \boldsymbol{Z}_m$.

Let (β_1, β_2) be a basis for $\boldsymbol{Z}_m \times \boldsymbol{Z}_m$, and let $\widehat{\pi}_q$ be the map from $\boldsymbol{Z}_m \times \boldsymbol{Z}_m$ to $\boldsymbol{Z}_m \times \boldsymbol{Z}_m$ induced by π_q. The proof now consists of an application of the Cayley–Hamilton theorem of linear algebra to the function $\widehat{\pi}_q$. To this end, observe that the Frobenius map $\pi_q(x, y)$ on $\mathcal{X}(\overline{\boldsymbol{F}}_q)[m]$ satisfies the linearity condition

$$[\alpha]\pi_q(P_1) + [\beta]\pi_q(P_2) = \pi_q([\alpha]P_1 + [\beta]P_2),$$

so $\pi_q(P)$ is a linear function on $\mathcal{X}(\overline{\boldsymbol{F}}_q)[m]$. This means that its image, $\widehat{\pi}_q(\Omega)$, is a linear function on $\boldsymbol{Z}_m \times \boldsymbol{Z}_m$. It must have the form

$$\begin{aligned} \widehat{\pi}_q(\beta_1, \beta_2) &= \begin{bmatrix} a & b \\ c & d \end{bmatrix} \begin{bmatrix} \beta_1 \\ \beta_2 \end{bmatrix} \pmod{m} \\ &= \boldsymbol{A} \begin{bmatrix} \beta_1 \\ \beta_2 \end{bmatrix} \pmod{m}, \end{aligned}$$

for some two-by-two matrix of integers $\boldsymbol{A}$. The Cayley–Hamilton theorem applies to matrices in the real field, so it applies to $\boldsymbol{A}$ regarded as a matrix over $\boldsymbol{Z}$. The *characteristic polynomial* of the matrix $\boldsymbol{A}$ is the polynomial $p(\lambda) = \det(\boldsymbol{A} - \lambda \boldsymbol{I})$. This is the polynomial

$$p(\lambda) = \lambda^2 - (a + d)\lambda + (ad - bc).$$

Therefore $\boldsymbol{A}$ satisfies

$$\boldsymbol{A}^2 - (a + d)\boldsymbol{A} + (ad - bc)\boldsymbol{I} = \boldsymbol{O},$$

as asserted by the Cayley–Hamilton theorem.

Thus the preimage $\pi_q(P)$ must satisfy the polynomial equation

$$\pi_q^2(P) - [a + d]\pi_q(P) + [ad - bc]P = \mathcal{O}.$$

It only remains to show that $a + d = t$ and $ad - bc = q$. It is enough to show that $ad - bc = q$ because then it is easy to see that this implies that $a + d = t$. But we know that $\pi_q^k(P) = P$ for some k, which means that

$$\begin{bmatrix} a & b \\ c & d \end{bmatrix}^k = \boldsymbol{I}.$$

Then $[\det \boldsymbol{A}]^k = 1 \pmod{q^k - 1}$ and $[\det \boldsymbol{A}]^j \neq 1 \pmod{q^j - 1}$ for j smaller than k, which implies that $\det \boldsymbol{A} = q$.

We conclude that, for every value of the prime integer m, the theorem applies to all m-torsion points of $\mathcal{X}(\overline{\boldsymbol{F}}_q)$. Because every point of $\mathcal{X}(\overline{\boldsymbol{F}}_q)$ is an m-torsion point for some value of m, the theorem applies to every point of $\mathcal{X}(\overline{\boldsymbol{F}}_q)$. □

A few comments may be helpful to illustrate the theorem. A polynomial over $\boldsymbol{F}_q$ is also a polynomial over $\boldsymbol{F}_{q^2}$, which we may choose instead to regard as the base field. Recall from Corollary 10.11.3 that

$$\#\mathcal{X}(\boldsymbol{F}_{q^2}) = (q + 1 - t)(q + 1 + t).$$

Therefore the order of the group of $\mathcal{X}(\boldsymbol{F}_{q^2})$ is $(q + 1 - t)(q + 1 + t)$, and

$$[(q + 1 - t)(q + 1 + t)]P = \mathcal{O}.$$

The characteristic polynomial for $\mathcal{X}(\boldsymbol{F}_{q^2})$ can be factored as

$$(\pi_q^2 + [t]\pi_q + [q])(\pi_q^2 - [t]\pi_q + [q])P = \mathcal{O}.$$

Let $P' = (\pi_q^2 - [t]\pi_q + [q])P$. Then either $P' = \mathcal{O}$, or P has been mapped into a point P' for which $(\pi_q^2 + [t]\pi_q + [q])P' = \mathcal{O}$. The theorem states that the first option is always the right one.

More generally, one can regard $\boldsymbol{F}_{q^m}$ as the base field and use Corollary 10.11.3 to write

$$Q(\pi_q)(\pi_q^2 - [t]\pi_q + [q])P = \mathcal{O}.$$

As before, let

$$P' = (\pi_q^2 - [t]\pi_q + [q])P.$$

Then either $P' = \mathcal{O}$, or $Q(\pi_q)P' = \mathcal{O}$. The theorem states that the first option is always the right one.

10.14 Point counting in a ground field

In Section 10.11, we described a powerful method of counting the number of points on an elliptic curve in the large extension field $\boldsymbol{F}_{q^m}$, provided the defining polynomial has all of its coefficients in the small ground field $\boldsymbol{F}_q$. That method is initialized with the number of rational points of the curve. These are the points of the curve both of whose coordinates are in the ground field $\boldsymbol{F}_q$. If the ground field is small, then the number of rational points on the curve can be found by elementary methods such as a direct search. However, a direct search is not a suitable method for counting the points of an elliptic curve whose defining polynomial has coefficients in a large field because the number of rational points is then extremely large, approximately equal to the field size. When the defining field of the curve is the field $\boldsymbol{F}_p$ for a large prime p, a point-counting algorithm is needed on $\boldsymbol{F}_p$. For such a field, point counting can be based on a method known as the *Schoof algorithm*, which we will describe here only in general terms, and only as an introductory overview.

The Schoof algorithm computes $\#\mathcal{X}(\boldsymbol{F}_q)$, which is an extremely large integer, but what we really want to know is the largest prime factor of $\#\mathcal{X}(\boldsymbol{F}_q)$. Usually, one wants the largest prime factor to be comparable to q, but methods of factoring large integers are not known. The need for factoring $\#\mathcal{X}(\boldsymbol{F}_q)$ is bypassed in the following way. First, by using trial division, remove as many small prime factors from $\#\mathcal{X}(\boldsymbol{F}_q)$ as is reasonable. Then the resulting integer is tested for primality. If the primality test fails, the curve is discarded and a new curve is examined. The search process continues to examine curves in this way until a curve with a large prime factor is found. With this procedure, an attempt at a complete factorization of $\#\mathcal{X}(\boldsymbol{F}_q)$ never really takes place.

The Schoof algorithm finds the Frobenius trace of the elliptic curve by breaking the large problem into many small subproblems that are combined by using the chinese remainder theorem. Recall that the Frobenius trace t is defined as $t = q + 1 - \#\mathcal{X}(\boldsymbol{F}_q)$. It is the difference between the size of the field plus one, and the order of the elliptic curve. The Hasse–Weil bound tells us that $|t| \leq 2\sqrt{q}$. Because t is much smaller than q, the Hasse–Weil bound assures us that the number of points on an elliptic curve over a large finite field is approximately equal to the size of the field, the difference being proportional to the square root of the field size. The Hasse–Weil bound, however, only gives a square-root bound on t. It does not give the exact number of points on the curve, and it is necessary to know the number of points exactly.

With this observation, our task is reduced to the task of computing t, which is a much smaller number than $\#\mathcal{X}(\boldsymbol{F}_q)$, but still very large. The Schoof algorithm circumvents the difficulty of computing t directly by computing $t_i = t \pmod{\ell_i}$ for many small primes ℓ_i such that $\Pi_i \ell_i$ is larger than $4\sqrt{q}$, which is the size of the interval in which t must lie. The chinese remainder theorem is then used to compute t from the set of the residues t_i. From t, it is easy to compute $\#\mathcal{X}(\boldsymbol{F}_q)$.

The characteristic equation of Frobenius can be written

$$(x^{q^2}, y^{q^2}) + [q](x, y) = [t](x^q, y^q),$$

where $[q](x, y)$ and $[t](x^q, y^q)$ are defined as the qth multiple of the point $(x, y) \in \mathcal{X}(\overline{\boldsymbol{F}}_q)$ and the tth multiple of the point $(x^q, y^q) \in \mathcal{X}(\overline{\boldsymbol{F}}_q)$ under point addition, and where $+$ denotes point addition. The characteristic equation of Frobenius appears to give a way to compute t. Pick any point $P = (x, y)$, and compute both (x^{q^2}, y^{q^2}) and (x^q, y^q). Then find the value of t that satisfies $[t](x^q, y^q) = (x^{q^2}, y^{q^2}) + [q](x, y)$. However, this equation has the form $[*]P = Q$. To solve this equation, one must compute the discrete logarithm in the group of the curve, which is an intractable computation. The Schoof algorithm instead selects only points $P = (x, y)$ of order ℓ_i for various small primes ℓ_i, and carries out this computation restricted to only those points. For these small problems, the computation is no longer intractable. In this way, the Schoof algorithm breaks a large intractable problem into small pieces that are tractable, and later combines the small pieces using the chinese remainder theorem.

The Frobenius trace lies in an interval of length $4\sqrt{q}$ because it satisfies

$$-2\sqrt{q} \leq t \leq 2\sqrt{q}$$

for an elliptic curve. Therefore by the chinese remainder theorem, any set of pairwise coprime integers ℓ_i satisfying $\Pi_i \ell_i > 4\sqrt{q}$ can be used as a set of moduli such that t can be recovered from the corresponding residues. To this end, one should choose the ℓ_i as primes. It follows from the prime number theorem that on the order of $\log q$ primes are needed for this purpose.

Theorem 10.14.1 *Let P be an element of $\mathcal{X}(\overline{\boldsymbol{F}}_q)$ of prime order ℓ. Then P satisfies*

$$\pi_q^2(P) - [t_\ell]\pi_q(P) + [q_\ell]P = \mathcal{O},$$

where $t_\ell = t \pmod{\ell}$ and $q_\ell = q \pmod{\ell}$.

Proof Theorem 10.13.3 states that $\pi_q^2(P) - [t]\pi_q(P) + [q]P = \mathcal{O}$ for all elements of $\mathcal{X}(\overline{\boldsymbol{F}}_q)$. If P is a point of order ℓ, where ℓ is a prime, then $[q]P = [q_\ell]P$. Moreover, if P is a point of order ℓ, then because $[t]\pi_q(P) = \pi_q([t]P)$, we also have $[t]\pi_q(P) = \pi_q([t_\ell]P) = [t_\ell]\pi_q(P)$. The theorem follows. □

The main idea of the Schoof algorithm is now established. However, it is not yet in a sufficiently mature form to begin computations. The remaining task is to find a point P of order ℓ_i for each prime ℓ_i. Because such a point P could be in an extremely large extension field, it may be difficult, or effectively impossible, to find and explicitly state such a point. Fortunately, the points can be treated implicitly knowing only that they exist and using only their properties needed to apply Theorem 10.14.1. This is done by using a certain quotient ring, denoted R_ℓ, as a proxy for any and all points in the torsion

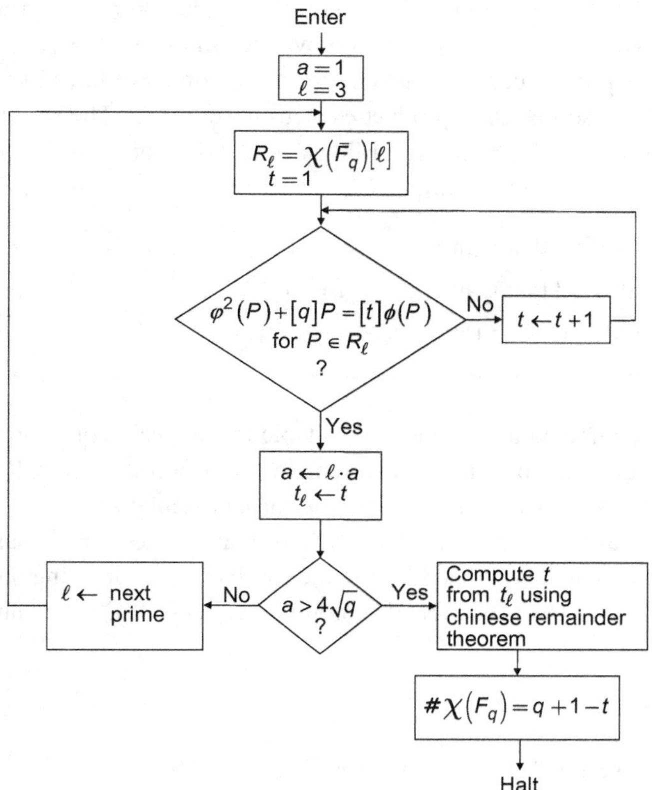

Figure 10.11 The structure of the Schoof algorithm

group $\mathcal{X}(\overline{\boldsymbol{F}}_q)[\ell]$, which is the group consisting of all $P \in \mathcal{X}(\overline{\boldsymbol{F}}_q)$ such that $[\ell]P = \mathcal{O}$. To carry out this procedure, substantial extensions of the theory are necessary, but are not provided here.

The basic structure of the Schoof algorithm is outlined in Figure 10.11. The ℓ-torsion group of $\mathcal{X}(\overline{\boldsymbol{F}}_q)$ is represented by the set R_ℓ. The core step of the algorithm tests points of R_ℓ against the characteristic polynomial of Frobenius without actually computing these points, and which we do not describe.

As an example, we will briefly discuss the use of the Schoof algorithm to compute the number of points on the elliptic curve

$$\mathcal{X} : y^2 + xy = x^3 + 1$$

over the field $\boldsymbol{F}_{64}$. Because 64 is a power of two and the base field is $\boldsymbol{F}_2$, we could calculate the number of points on the curve easily by using the recurrence based on the methods of Section 10.11 and would find that $\#\mathcal{X}(\boldsymbol{F}_{64}) = 56$ and $t = 9$. Instead, we will treat this curve as if it were too large to employ these simple methods.

To calculate the order of this curve using Schoof's algorithm, first choose a sufficient number of small primes so that their product exceeds $4\sqrt{64} = 32$. The set of primes 2, 3, 5, and 7 satisfies this condition. Accordingly, one must compute $t \pmod{\ell}$ for each ℓ. These are the following residues:

Let $\ell = 2$ We would find that $t \pmod 2 = t_2 = 1$

Let $\ell = 3$ We would find that $t \pmod 3 = t_3 = 0$

Let $\ell = 5$ We would find that $t \pmod 7 = t_5 = 2$

Let $\ell = 7$ We would find that $t \pmod 7 = t_7 = 2$.

Each residue can be computed by any method suitable for that value of ℓ. From these residues, one then computes by the chinese remainder theorem that $t = 9$, and so $\#\mathcal{X}(\boldsymbol{F}_{64}) = q + 1 - t = 56$, which agrees with the earlier calculation.

To compute the required residues, different methods may be used for various ℓ_i. For $\ell = 2$, an immediate conclusion is that, because the curve is not a supersingular curve, the trace is odd, so $t(\text{mod } 2) = t_2 = 1$. Next, to compute t_3, let $\ell = 3$, so we must solve the equation

$$[t_3]\pi_q(P) = \pi_q^2(P) + P,$$

where the point P is any point of $\mathcal{X}(\overline{\boldsymbol{F}}_q)$ of order three, and so $[64]P = P$. We might suppose that a point of order three could lie in a massively large extension field. If so, we would not want to see this point. Furthermore, because we do not yet know the order of the curve $\mathcal{X}(\boldsymbol{F}_q)$, we cannot know the size of the extension field. Nor can we know how to find a point of order three. We must extend the theory to refer to points of order three without computing them. Similarly, we must extend the theory to points of prime order ℓ without computing them. We do not describe how to do this, but only make a few comments.

The Schoof algorithm requires the computation of $t \pmod{\ell}$ for each chosen prime ℓ. To this end, the following procedure is helpful. Determine whether the polynomial $x^2 - tx + q$ factors modulo the prime ℓ. If it does factor, determine $t \pmod{\ell}$ by a method called the *Elkies procedure*. Otherwise, if the polynomial $x^2 - tx + q$ does not factor over ℓ, determine $t \pmod{\ell}$ by a method called the *Atkins procedure*. The terminology may improperly suggest that this distinction between the two procedures is a property of the particular choice of ℓ, but it depends as well on t and q. Thus a given prime ℓ may lead to the Elkies procedure for some elliptic curves and the Atkins procedure for others, depending on the t and q. Of course, we do not know t, so again we must use an indirect method to determine whether the polynomial $x^2 - tx + q$ factors.

The key to the Elkies and Atkins procedures is that an ℓ-torsion point is a zero not only of the polynomial defining the elliptic curve, but also of other polynomials. If any polynomial $f(x, y)$ can be found for which an ℓ-torsion point P is a zero, then that polynomial can be used to reduce the complexity by eliminating some of the terms. This means that for any chosen ℓ, we want to find a polynomial for which all ℓ-torsion points are zeros. Methods for doing this are known, but would take us too far afield to describe.

10.15 The method of xedni calculus

Xedni calculus is an unsuccessful attack on the discrete-log problem on the group of a finite elliptic curve on a prime field. We discuss it here because the failure of this attack is part of the lore establishing confidence in the security of elliptic-curve cryptography. The xedni-calculus attack is motivated by the index-calculus attack on the discrete logarithm in a finite field. However, whereas the index-calculus attack does threaten the discrete-log problem in $\boldsymbol{F}_p^*$, the xedni-calculus attack is believed to be too weak to threaten elliptic-curve cryptography. Because it is neither larommi nor lagelli, we will discuss xedni calculus here. We will treat only the case where the field size q is a prime p. The conclusion of the cryptology community that xedni calculus does not threaten elliptic-curve cryptography provides an anecdotal validation of elliptic-curve cryptography.

Let $\mathcal{X}$ be an elliptic curve over $\boldsymbol{F}_p$ and let P and Q be points of $\mathcal{X}(\boldsymbol{F}_p)$. The elliptic-curve discrete-log problem is to find the integer a such that $[a]P = Q$, where $[a]P$ denotes the point addition of a copies of P. This is similar to the discrete-log problem in $\boldsymbol{F}_p^*$, which is vulnerable to an index-calculus attack, but the index-calculus attack is based on the existence of a large number of small primes and on the notion of a smooth integer. In contrast, elliptic curves do not provide a corresponding notion of a large number of small points because there is no notion of a small point. Xedni calculus attempts to get around this, but fails.

The xedni-calculus attack uses the fact that an elliptic curve $\mathcal{X}(\boldsymbol{F}_q)$ over a finite field can be lifted to an elliptic curve $\mathcal{X}(\boldsymbol{Q})$ over the rational field, as stated below in Theorem 10.15.1. Indeed, there may be many such elliptic curves over the rationals, many with very large coefficients, to which $\mathcal{X}(\boldsymbol{F}_q)$ can be lifted. The *lift* of a curve $\mathcal{X}(\boldsymbol{F}_p)$ is a curve $\mathcal{X}(\boldsymbol{Q})$ whose points can be reduced to $\mathcal{X}(\boldsymbol{F}_p)$ by a modulo-p reduction.

To mimic index calculus, one would first lift the curve $\mathcal{X}(\boldsymbol{F}_p)$ to a curve $\mathcal{X}(\boldsymbol{Q})$, then lift the two points P and Q to lie on $\mathcal{X}(\boldsymbol{Q})$. As stated, this task is judged to be probably harder than the discrete-log problem that it is asked to replace. This is because the lifts of P and Q, although in $\boldsymbol{Q}$, might be expressed as the ratio of two extremely large integers, even intractably large integers. Instead, one first lifts a small number of randomly chosen points of the elliptic curve $\mathcal{X}(\boldsymbol{F}_q)$ to the plane $\boldsymbol{Q}^2$, then fits an elliptic curve $\mathcal{X}(\boldsymbol{Q})$ through these points in hopes that the lifted points will be linearly dependent, which would allow the method of xedni calculus to be applied. It is at this point that the xedni calculus fails because it is so highly unlikely that the lifted points will be linearly dependent. It is now generally believed that there is no computationally tractable method to ensure that the lifted points are linearly dependent.

The validity of lifting is based on the following theorem.

Theorem 10.15.1 *Let P and Q be elements of the elliptic curve $\mathcal{X}(\boldsymbol{F}_p) : y^2 = x^3 + Ax + B$. Then over the rational field $\boldsymbol{Q}$ there exists an elliptic curve $\mathcal{X}(\boldsymbol{Q}) : y^2 = x^3 + \widehat{A}x + \widehat{B}$ with integers $\widehat{A}$ and $\widehat{B}$ and points $\widehat{P}$ and $\widehat{Q}$ such that, under modular p equivalence, $A \equiv \widehat{A}$, $B \equiv \widehat{B}$, $P \equiv \widehat{P}$, and $Q \equiv \widehat{Q}$.*

Proof Let $P = (x_1, y_1)$ and $Q = (x_2, y_2)$ be points of $\mathcal{X}(\boldsymbol{F}_p) : y^2 = x^3 + Ax + B$. To prove the theorem we must choose appropriate integers $\widehat{A}$ and $\widehat{B}$, as well as points $\widehat{P} = (\widehat{x}_1, \widehat{y}_1)$ and $\widehat{Q} = (\widehat{x}_2, \widehat{y}_2)$ with integers for all coordinates, such that $\widehat{P} = P \pmod p$, $\widehat{Q} = Q \pmod p$, and such that $\widehat{P}$ and $\widehat{Q}$ are on the curve $\mathcal{X}(\boldsymbol{Q}) : y^2 = x^3 + \widehat{A}x + \widehat{B}$. The two cases $x_1 = x_2$ and $x_1 \neq x_2$ must be considered separately.

Suppose that the x coordinates of P and Q are different: $x_1 \neq x_2$. Choose integers $\widehat{x}_1$ and $\widehat{x}_2$ such that $\widehat{x}_1 = x_1 \pmod p$ and $\widehat{x}_2 = x_2 \pmod p$. Choose $\widehat{y}_1$ such that $\widehat{y}_1 = y_1 \pmod p$. Choose $\widehat{y}_2$ such that $\widehat{y}_2 = y_2 \pmod p$ and $y_2^2 - y_1^2 = 0 \pmod{x_2 - x_1}$. It is possible to satisfy the constraint by the chinese remainder theorem because $|x_2 - x_1| < p$ so $\mathrm{GCD}(p, x_2 - x_1) = 1$.

Next we must choose $\widehat{A}$ and $\widehat{B}$ to satisfy

$$\widehat{y}_1^2 = \widehat{x}_1^3 + \widehat{A}\widehat{x}_1 + \widehat{B}$$
$$\widehat{y}_2^2 = \widehat{x}_2^3 + \widehat{A}\widehat{x}_2 + \widehat{B}.$$

We can solve these equations for $\widehat{A}$ and $\widehat{B}$, which gives

$$\widehat{A} = \frac{\widehat{y}_2^2 - \widehat{y}_1^2}{\widehat{x}_2 - \widehat{x}_1} - \frac{\widehat{x}_2^3 - \widehat{x}_1^3}{\widehat{x}_2 - \widehat{x}_1},$$
$$\widehat{B} = \widehat{y}_1^2 - \widehat{x}_1^3 - \widehat{A}\widehat{x}_1.$$

But $\widehat{y}_2^2 - \widehat{y}_1^2$ and $\widehat{x}_2^3 - \widehat{x}_1^3$ are both divisible by $\widehat{x}_2 - \widehat{x}_1$, which means that all quantities on the right sides of these equations are integers. Therefore $\widehat{A}$ and $\widehat{B}$ are both integers. We conclude that the points $\widehat{P}$ and $\widehat{Q}$ do lie on the curve

$$\mathcal{X}(\boldsymbol{Q}) : y^2 = x^3 + \widehat{A}x + \widehat{B},$$

as required.

Now, suppose that the x coordinates for P and Q are equal: $x_1 = x_2$. In this case $P = \pm Q$. Choose integers $\widehat{x}_1$, $\widehat{y}_1$, and $\widehat{A}$ such that $x_1 = x_2 = \widehat{x}_1 \pmod p$, $y_1 = \widehat{y}_1 \pmod p$, and $A = \widehat{A} \pmod{p}$. Next, let $\widehat{B} = \widehat{y}_1^2 - \widehat{x}_1^3 - \widehat{A}\widehat{x}_1$. Then $\widehat{P} = (\widehat{x}_1, \widehat{y}_1)$ and $\widehat{Q} = \pm\widehat{P}$ are points of $\widehat{\mathcal{X}}(\boldsymbol{Q}) : y^2 = x^3 + \widehat{A}x + \widehat{B}$.

Finally, we must verify that $\widehat{\mathcal{X}}(\boldsymbol{Q})$ has no singular points. But

$$4\widehat{A}^3 + 27\widehat{B}^2 = 4A^3 + 27B^2 \pmod{p},$$

which is nonzero, so $\widehat{\mathcal{X}}(\boldsymbol{Q})$ has no nonsingular points. It is an elliptic curve over $\boldsymbol{Q}$ that meets the required conditions. □

Notice that if the prime p underlying $\boldsymbol{F}_p$ has 100 decimal digits, then the coefficients $\widehat{A}$ and $\widehat{B}$ may have 300 decimal digits. This is the reason that an alternative approach has been proposed.

The xedni-calculus attack on the discrete-log problem $[a]P = Q$ is developed as follows. An arbitrary bivariate cubic polynomial $ax^3 + bx^2y + cxy^2 + dy^3 + ex^2 + fxy + gy^2 + hx + iy + j$ has ten coefficients, but if the coefficient of y^3 is required to be zero, then only nine coefficients are arbitrary. This means that a cubic polynomial in the Weierstrass form can be fit through any nine points, but in general, not more than nine.

Choose r pairs of integers (s_i, t_i) for $i = 1, \ldots, r$, and where r is at most nine, but typically five or six. This means that the computation is underdetermined, allowing flexibility. Then, to solve $[a]P = Q$, form the r trial points

$$P_i = [s_i]P - [t_i]Q$$

of the curve $\mathcal{X}(\boldsymbol{F}_p)$. Lift each such point P_i to a point $\widehat{P_i}$ of the rational plane $\boldsymbol{Q}^2$ and form the curve $\mathcal{X}(\boldsymbol{Q})$ through these points. This is always possible because the Weierstrass form has nine free coefficients. By choice of the curve $\mathcal{X}(\boldsymbol{Q})$, all of the lifted points $\widehat{P_i}$ now have representatives on the curve $\mathcal{X}(\boldsymbol{Q})$.

If such exists, find a dependence relationship in the rational field $\boldsymbol{Q}$ among the lifted points $\widehat{P_i}$ of the curve $\mathcal{X}(\boldsymbol{Q})$.

$$[n_1]\widehat{P_1} + \cdots + [n_r]\widehat{P_r} = \mathcal{O}.$$

If no such relationship exists, which is almost certainly the case, the attack has failed. When the attack fails, one may either restart the attack with new trial points or quit.

The dependence relationship, if it exists, can be written in terms of point multiples written as

$$\sum_{i=1}^{r} n_i(s_i\widehat{P} - t_i\widehat{Q}) = \mathcal{O}.$$

This can be written as

$$\sum_{i=1}^{r} n_i(s_i\widehat{P} - t_i a\widehat{P}) = \mathcal{O}$$

or as

$$\sum_{i=1}^{r} n_i(s_i - t_i a)\widehat{P} = \mathcal{O}.$$

This means that $\sum_{i=1}^{r} n_i(s_i - t_i a)$ divides the order m of $\mathcal{X}(\boldsymbol{F}_p)$. Now the task is one of solving an integer problem. Given n_i, s_i, t_i, and m, and that $\sum_{i=1}^{r} n_i(s_i - t_i a)$ divides m, what is a? Thus the task of solving a presumably intractable elliptic curve discrete-log problem has been replaced by an integer problem which is far easier to solve. The need for the points of $\mathcal{X}(\boldsymbol{Q})$ to be linearly dependent, however, is where the attack fails.

For a simple example of this method, let $p = 11$ and consider the elliptic curve $\mathcal{X} : y^2 = x^3 + x + 6$ in the field $\boldsymbol{F}_{11}$. Let $P = (3, 6)$ and $Q = (7, 2)$. Our task is to solve the equation $[a]P = Q$ for a. For this small example, it is simple to verify that P and Q are both points of the curve $\mathcal{X}(\boldsymbol{F}_{11})$ and that $Q = [4]P$ is the solution to be found. Although we already know that the answer is 4, we will outline a computation of this value for a by using xedni calculus. For this calculation, we list the thirteen points of $\mathcal{X}(\boldsymbol{F}_{11})$ as follows:

$$\begin{aligned}
P &= (3, 6)\\
2P &= (8, 8)\\
3P &= (5, 2)\\
4P &= (7, 2)\\
5P &= (2, 4)\\
6P &= (10, 2)\\
7P &= (10, 9)\\
8P &= (2, 7)\\
9P &= (7, 9)\\
10P &= (5, 9)\\
11P &= (8, 3)\\
12P &= (3, 5)\\
&\mathcal{O}.
\end{aligned}$$

Of course, in the general case, this table would not be known. It is given here only for convenience.

Arbitrarily choose pairs of integers (s_i, t_i) for $i = 1, 2$ as $(2, 1)$ and $(3, 2)$. Then by reference to the table for $\mathcal{X}(\boldsymbol{F}_{11})$, we can compute $P_1 = [s_1]P - [t_1]Q = (8, 3)$ and $P_2 = [s_2]P - [t_2]Q = (2, 7)$. Of course, these two points must be computed directly, without reference to the table of $\mathcal{X}(\boldsymbol{F}_{11})$, which would not be known in the general case. Now, regard these points not in $\boldsymbol{F}_{11}$, but in $\boldsymbol{Q}$ and fit them with an elliptic curve. Because we have selected only two points, we can use a polynomial of the form $y^2 = x^3 + Ax + B$. Then solve for the coefficients by writing

$$\begin{bmatrix} A \\ B \end{bmatrix} = \begin{bmatrix} 8 & 1 \\ 2 & 1 \end{bmatrix}^{-1} \begin{bmatrix} -503 \\ 41 \end{bmatrix},$$

and the desired elliptic curve is

$$\mathcal{X}(Q) : y^2 = x^3 - \frac{272}{3}x + \frac{667}{3}.$$

The two lifted points lie on this curve. Now, find integers n_i, if they exist, such that

$$\Sigma_i [n_i] P_i = \mathcal{O}.$$

Even for this small example, this is too much computation to do by hand, and may not be possible. Instead, to continue to illustrate the method, we pretend that a solution is

$$\Sigma_i [n_i] P_i = [3]P_1 + [4]P_2 = \mathcal{O}.$$

Were this a solution, it would mean that $3(2 - a) + 4(3 - 2a)$ divides 13. That is, $18 - 11a$ divides 13. A solution is $a = 4$ because $18 - 11a = -26$ which is a multiple of 13. We would then conclude that $a = 4$, which could be checked by computing $[4]P$ to find that it would be Q.

10.16 Elliptic curves and the complex field

The subject of elliptic curves has richness and versatility, and a seemingly never-ending theory. Elliptic curves can be made to morph into other mathematical structures that might seem to have little to do with elliptic curves. Although our interest is in elliptic curves over finite fields, the complex field does remain prominent in the background, both mathematically and historically. Elliptic curves in the real field or the complex field can be understood geometrically, as well as algebraically. Geometric concepts in the real field or the complex field are an aide to developing intuition that is helpful for understanding elliptic curves in a general field.

We have asserted that the operation of point addition on an elliptic curve defines a group, and this assentation requires a proof, but only a partial proof was given earlier. Surprisingly, the proof of associativity is not easily forthcoming because the direct method of proof is tedious, and the elegant method of proof is difficult. The most direct way to verify that the group properties hold is by elementary algebraic manipulations of the underlying equations. It can be quite tedious, however, to verify the associative law $(P_1 + P_2) + P_3 = P_1 + (P_2 + P_3)$ by writing out the equations because, in part, there are special cases in which $P_1 = P_2$ or $P_1 = -P_2$ or $P_1 = \mathcal{O}$ that must be verified. Moreover, this direct method of proof by elementary algebraic manipulation, though

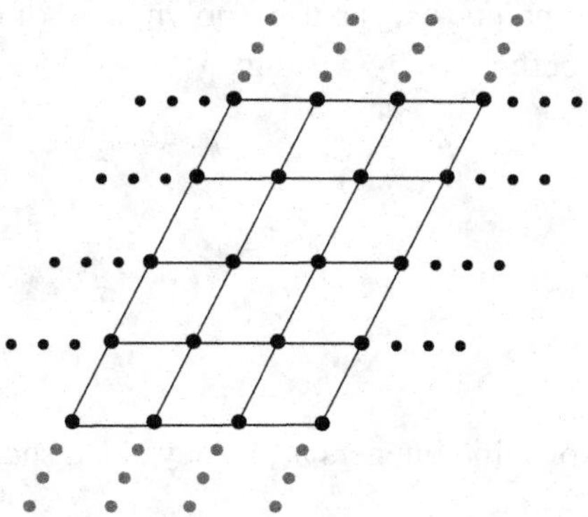

Figure 10.12 A lattice

perfectly rigorous, would not give any new insight. Instead, one can show indirectly that the group properties hold by reasoning in a much different setting.

A *torus* in the real number system is a well-understood and easily visualized geometric object as a two-dimensional surface embedded in three-dimensional euclidean space. A real torus is shown in Figure 10.10. A torus in the complex number system cannot be visualized in three-dimensional euclidean space. However, it is a torus in the complex number system that is important to us because, as we shall assert, it is isomorphic to an elliptic curve over the complex field. This means that a complex torus and a complex elliptic curve can be mapped one into the other in a way that respects point addition. Because the algebraic properties of the two objects mimic each other, reasoning about one object usually applies to the other.

In addition, we will make the rather sophisticated, but subtle, remark that point addition on an elliptic curve over $\boldsymbol{F}$ has the structure of a set of identities over $\boldsymbol{F}$ whose validity depends only on the formal axioms that every field satisfies. Properties unique to a particular field do not play a role in any proof of these identities. Therefore if the definition of point addition on an elliptic curve over any field forms a group, then the definition of point addition surely forms a group on an elliptic curve over every field. This means that it is enough to prove the group structure for elliptic curves on the complex field, and this can be done by reasoning about the complex torus.

A lattice in the plane $\boldsymbol{C}^2$ cannot be easily visualized as such, but a lattice, denoted $\boldsymbol{\Lambda}$, in the plane $\boldsymbol{R}^2$ can be easily visualized, as shown in Figure 10.12. Consider the quotient group $\boldsymbol{R}^2/\boldsymbol{\Lambda}$, which is the set of equivalence classes induced by the lattice $\boldsymbol{\Lambda}$ on the real plane $\boldsymbol{R}^2$. The points of any one cell (including two edges) of Figure 10.12 serve as the set of canonical representatives of $\boldsymbol{R}^2/\boldsymbol{\Lambda}$. This cell is periodically replicated to fill the real plane as shown in Figure 10.12. To form the quotient group $\boldsymbol{R}^2/\boldsymbol{\Lambda}$, points on the left edge of any cell are identified with points on the right edge of that cell, and points on the bottom edge are identified with points on the top edge, so the quotient group $\boldsymbol{R}^2/\boldsymbol{\Lambda}$ is actually a real torus.

A similar construction using a lattice in the plane $\boldsymbol{C}^2$ leads to a complex torus. Because every point of the complex torus consists of two complex numbers, this complex torus cannot be embedded into three-dimensional euclidean space, so it is much harder to visualize than a real torus. In the same way as for a lattice in $\boldsymbol{R}^2$, the representative cell of the complex lattice $\boldsymbol{\Lambda}$ is periodically replicated to fill the plane $\boldsymbol{C}^2$.

The main theorem of this section states that an elliptic curve over any field forms a group under the operation of point addition. This theorem can be proved directly by elementary algebraic manipulations. Instead, we will digress to introduce the topic of a complex lattice, and we invoke the fact that the structure of elliptic curves is related to the structure of lattices. While this may seem like an extraordinarily indirect method of proof, it does move the subject of elliptic curves to a higher level of abstraction, which eventually leads to sharper insights.

For this purpose, choose $\boldsymbol{F}$ as the complex field $\boldsymbol{C}$, and let $\boldsymbol{\Lambda}$ be the complex lattice in $\boldsymbol{C}^2$ generated by the two complex numbers ω_1 and ω_2. That is, define

$$\boldsymbol{\Lambda} = \{(m\omega_1 + n\omega_2) : m, n \in \boldsymbol{Z}\}.$$

This is expressed concisely as

$$\boldsymbol{\Lambda} = \boldsymbol{Z}\omega_1 + \boldsymbol{Z}\omega_2.$$

The lattice $\boldsymbol{\Lambda}$ is a lattice in $\boldsymbol{C}^2$ because ω_1 and ω_2 are complex. Consider the quotient group $\boldsymbol{C}^2/\boldsymbol{\Lambda}$, which is the set of equivalence classes induced by the complex lattice $\boldsymbol{\Lambda}$ on the complex plane $\boldsymbol{C}^2$. Because $\boldsymbol{C}$ is a group under addition and $\boldsymbol{\Lambda}$ is a subgroup, the set $\boldsymbol{C}^2/\boldsymbol{\Lambda}$ is also a group. We will give a map from the group $\boldsymbol{C}^2/\boldsymbol{\Lambda}$ to an elliptic curve that preserves addition, taking addition in $\boldsymbol{C}^2/\boldsymbol{\Lambda}$ into point addition on the curve. Therefore point addition on the elliptic curve over $\boldsymbol{C}^2$ forms a group because it inherits the group structure from $\boldsymbol{C}^2/\boldsymbol{\Lambda}$. This, in turn, implies that point addition forms a group on an elliptic curve in any field.

A complex function $f(z)$ is called *doubly periodic* with period lattice $\boldsymbol{\Lambda}$ if $f(z+\lambda) = f(z)$ for all $\lambda \in \boldsymbol{\Lambda}$. Such functions are also called *elliptic functions*.[8] An example of an elliptic function is the *Weierstrass function*, which for the lattice $\boldsymbol{\Lambda}$ is defined as

$$p_{\boldsymbol{\Lambda}}(z) = \frac{1}{z^2} + \sum_{\substack{\lambda\in\Lambda \\ \lambda\neq 0}} \left(\frac{1}{(z-\lambda)^2} - \frac{1}{\lambda^2} \right).$$

The complex function $p_{\boldsymbol{\Lambda}}(z)$ depends on the lattice $\boldsymbol{\Lambda}$. Because $p_{\boldsymbol{\Lambda}}(z)$ is doubly periodic, it gives a well-defined function $\boldsymbol{C}/\boldsymbol{\Lambda} \longrightarrow \boldsymbol{C} \cup \{\infty\}$. The sum on λ converges absolutely and uniformly for z in any compact subset of $\boldsymbol{C}$ not containing a point of the lattice $\boldsymbol{\Lambda}$.

[8] Not to be confused with the polynomials defining elliptic curves.

Next define the mapping

$$C/\mathbf{\Lambda} \to (C \times C) \cup \{\infty\}$$

by

$$z \mapsto \begin{Bmatrix} (p_{\mathbf{\Lambda}}(z), p'_{\mathbf{\Lambda}}(z)) & \text{if} \quad z \notin \mathbf{\Lambda} \\ \infty & \text{if} \quad z \in \mathbf{\Lambda}. \end{Bmatrix}$$

where $p'_{\mathbf{\Lambda}}(z)$ denotes the derivative of $p_{\mathbf{\Lambda}}(z)$. We will show below that the function $p_{\mathbf{\Lambda}}(z)$ satisfies the differential equation

$$p'_{\mathbf{\Lambda}}(z)^2 = 4p_{\mathbf{\Lambda}}(z)^3 - g_2 p_{\mathbf{\Lambda}}(z) - g_3,$$

where the complex constants g_2 and g_3 depend on the lattice $\mathbf{\Lambda}$. Thus the point $(p_{\mathbf{\Lambda}}(z), p'_{\mathbf{\Lambda}}(z))$ is a point of the elliptic curve $\mathcal{X}(C) : y^2 = 4x^3 - g_2 x - g_3$.

This is a complex analytic isomorphism, and the image is (isomorphic to) an elliptic curve. In fact, any elliptic curve over C arises in this way, so we say that $C/\mathbf{\Lambda}$ *is* an elliptic curve.[9] But $C/\mathbf{\Lambda}$ has a group structure, so the map transports the group structure from $C/\mathbf{\Lambda}$ to $\mathcal{X}(C)$. Thus the group properties of point addition on elliptic curves can be inferred from the group structure of $C/\mathbf{\Lambda}$. In particular, point addition on the elliptic curve satisfies the associative property.

It remains to show that $p_{\mathbf{\Lambda}}(z)$ satisfies the above differential equation. The main line of proof is as follows. The series expansion

$$\frac{1}{(1 - z/\lambda)^2} = 1 + 2\frac{z}{\lambda} + 3\frac{z^2}{\lambda^2} + 4\frac{z^3}{\lambda^3} + \cdots$$

leads to the Laurent series for the Weierstrass function

$$p_{\mathbf{\Lambda}}(z) = z^{-2} + 3G_4 z^2 + 5G_6 z^4 + 7G_8 z^6 + \cdots,$$

because λ and $-\lambda$ are both elements of $\mathbf{\Lambda}$. Then

$$p'_{\mathbf{\Lambda}}(z) = -2z^{-3} + 6G_4 z + 20G_6 z^3 + \cdots,$$

where the coefficients G_k, known as the *Eisenstein coefficients*, depend on the lattice $\mathbf{\Lambda}$. Cubing and squaring these two expressions gives

$$p_{\mathbf{\Lambda}}(z)^3 = z^{-6} + 9G_4 z^{-2} + 15G_6 + \cdots$$
$$p'_{\mathbf{\Lambda}}(z)^2 = 4z^{-6} - 24G_4 z^{-2} - 80G_6 + \cdots.$$

Now, define the power series

$$f(z) = p'_{\mathbf{\Lambda}}(z)^2 - 4p_{\mathbf{\Lambda}}(z)^3 + 60G_4 p_{\mathbf{\Lambda}}(z) + 140G_6.$$

[9] Which depends on the meaning of the word "is."

This is clearly a power series in z with constant term zero because all terms with nonpositive powers of z cancel. But, by inspection of $p_\Lambda(z)$, there can be no terms with positive powers of z, so $f(z) = 0$. Finally, replacing $p'_\Lambda(z)$ by y and $p_\Lambda(z)$ by x leads to an equation for the desired elliptic curve.

Theorem 10.16.1 *The points of an elliptic curve over the field **F** form a group under point addition.*

Proof The proof relies on the isomorphism between $\boldsymbol{C}^2/\boldsymbol{\Lambda}$ and $\mathcal{X}(\boldsymbol{C})$. Addition is associative in the two-dimensional plane $\boldsymbol{C}^2$. It remains associative under modular reduction by a lattice $\boldsymbol{\Lambda}$. Next, recall that every elliptic curve $\mathcal{X}(\boldsymbol{C})$ over $\boldsymbol{C}$ is isomorphic to a complex torus $\boldsymbol{C}^2/\boldsymbol{\Lambda}$ for some lattice $\boldsymbol{\Lambda}$. Because $\boldsymbol{C}^2/\boldsymbol{\Lambda}$ is a group under modular addition, $\mathcal{X}(\boldsymbol{C})$ is also a group under point addition, so the theorem is proved for $\boldsymbol{F} = \boldsymbol{C}$. But the group properties of point addition on an elliptic curve must be a consequence of the equations for point addition and these equations are the same for any field. Hence, the theorem must hold for any field $\boldsymbol{F}$. □

10.17 Curves constructed using complex multiplication

An elliptic curve in a finite field can be formed in the obvious way by first choosing a prime power q for the field size, then choosing the Weierstrass coefficients, so as to form a suitable polynomial over that field. In the absence of any guidelines, one picks these quantities arbitrarily, then counts the number of points on the resulting elliptic curve, and examines the curve in other ways to see whether the curve is suitable. Because one wants a curve whose order has a very large prime factor, it may be necessary to examine a great number of curves before finding one that is satisfactory. One would prefer to have a direct approach. More focused methods of curve construction are available. One method first specifies some desired attribute of the curve, such as a large prime order, then constructs a specific elliptic curve with that attribute. One such method is described in this section, and others in Chapter 12.

In this section, we will discuss a rather advanced method of curve construction based on the theory of complex multiplication on elliptic curves, but giving only an outline of the method. The method of curve construction using complex multiplication on elliptic curves first executes some integer computations so as to choose a suitable order for the desired curve, and only later constructs the elliptic curve corresponding to these parameters. It constructs an elliptic curve over a finite field with a prescribed number of points as a reduction of an elliptic curve over an imaginary quadratic number field that has complex multiplication. Such a curve must have an endomorphism ring larger than $\boldsymbol{Z}$.

To outline the main idea of the method, we first present the key steps without explanation. The explanation comes later. The procedure starts by choosing a square-free integer D called the *discriminant* of the elliptic curve. Associated with each value of D is another integer h_D called the *class number* of D, which is defined later. The procedure finds a prime p that satisfies the diophantine equation[10] $4p = x^2 + Dy^2$, where x and y are integers. (The factor of 4 is conventional, though unneeded.) A solution consists of the three integers x, y, and p. The solution then implies the existence of h_D pairs of elliptic curves, up to isomorphism, with orders $m = p + 1 \pm x$. The two curves in each pair are twists of each other.

For a simple example outlining the method, choose the square-free integer $D = 7$, which has a class number $h_D = 1$. Because $h_D = 1$, there is only one pair (up to isomorphism) of associated elliptic curves over $\boldsymbol{F}_p$ with orders $m = p + 1 \pm x$. The prime

$$p = 781221660082682887337352611537$$

is a solution to the equation $4p = x^2 + Dy^2$ with $x = 1676624638069870$. We conclude that there are two elliptic curves over $\boldsymbol{F}_p$, related by a twist, with orders $p + 1 \pm x$. One curve has order

$$m = p + 1 - x = 781221660082681210712714541668,$$

which is four times an odd prime. The twist of this curve has order

$$m' = p + 1 + x = 781221660082684563961990681408,$$

which is not as attractive as the first curve because m' does not have such a large prime factor. In this way, the integer m has been calculated and examined for primality before the curve itself is computed. Because m is satisfactory, the curve can be computed next, as will be described below. The two curves are

$$\begin{aligned} \mathcal{X}(\boldsymbol{F}_p) : y^2 = x^3 &+ 586337137088968521507562977329x \\ &+ 470612877688284093511930750213 \end{aligned}$$

and

$$\begin{aligned} \mathcal{X}(\boldsymbol{F}_p) : y^2 = x^3 &+ 384410658135923325515205253294x \\ &+ 777088212145737475235038576554 \end{aligned}$$

up to isomorphism. It is straightforward to determine that the first of these curves has order m and the second has order m'. The first curve, then, is the curve of interest.

Although a curve obtained using the method of complex multiplication is not an arbitrary curve nor a typical curve, there is no evidence that such a curve has any special vulnerability that can be exploited by a cryptanalyst. Nevertheless, there remains the possibility that knowledge of the method of constructing the curve can be used by the

[10] A *diophantine equation* is a polynomial equation for which the coefficients and the indeterminates take only integer values.

adversary in some undiscovered way to reduce the complexity of the elliptic-curve discrete-log problem.

It is appropriate next to discuss methods to solve the given diophantine equation: $4p = x^2 + Dy^2$. Cornacchia's algorithm is a method to solve the more general diophantine equation $x^2 + dy^2 = n$ for x and y where d and n are given and coprime. First find any r_0 such that $r_0^2 = -d \pmod{n}$. If no such r_0 exists, then the given equation has no solution. If there is a solution, use the euclidean algorithm as follows:

$$\begin{aligned} n &= Q_0 r_0 + r_1 \\ r_0 &= Q_1 r_1 + r_2 \\ &\vdots \\ r_{k-2} &= Q_{k-1} r_{k-1} + r_k, \end{aligned}$$

stopping when $r_k \leq \sqrt{n}$. Then $x = r_k$ and $y = \sqrt{(n - x^2)/d}$, provided that y is an integer. If y is not an integer, the diophantine equation has no solution.

To use this method of curve construction for a chosen D, one chooses various suitable primes p and tries them until a prime is found for which the equation $x^2 + Dy^2 = 4p$ has a solution, as can be found by Cornacchia's algorithm. The expected number of tries before finding a solution is $1/h_D$ where, for each D, h_D is the class number of D, as discussed below.

Our next task is to explain the origin of the stated diophantine equation. To this purpose, first recall that the order m of an elliptic curve over $\boldsymbol{F}_p$ is given by $m = \#\mathcal{X}(\boldsymbol{F}_p) = p + 1 \pm t$, where $t = \alpha + \beta$ and α and β are zeros of the second-degree equation

$$z^2 - tz + p = 0.$$

To construct an elliptic curve with an attractive order, we begin with this equation, noting that the complex numbers

$$\alpha = (x + i\sqrt{D}y)/2$$

and

$$\beta = (x - i\sqrt{D}y)/2$$

satisfy $\alpha + \beta = x$. Furthermore, $(z - \alpha)(z - \beta) = z^2 - xz + (x^2 + Dy^2)/4$. This polynomial has the required form if $x = t$ and $(x^2 + Dy^2)/4 = p$, which is the diophantine equation discussed above. If, for a given square-free integer D, we can find three integers x, y, and p that satisfy this equation, and $p + 1 \pm x$ has a large prime factor, then we can hope to find a corresponding elliptic curve. This observation does not show that such an elliptic curve must exist, only that it might exist. It takes more explanation to say that it does exist.

Having such a solution, our task now becomes to find an elliptic curve of the form

$$\mathcal{X}(\boldsymbol{F}_p) : y^2 = x^3 + ax + b$$

that has the Frobenius trace t which was chosen above, and this curve will be found by the method of complex multiplication on elliptic curves. Again, we first describe the method, and explain it later, but without fully developing the underlying theory.

As we have seen, the elliptic curve $\mathcal{X}(\boldsymbol{F}_p)$ is associated with the discriminant

$$\Delta = 4a^3 + 27b^2$$

and the j-invariant

$$j = 1728 \frac{4a^3}{4a^3 + 27b^2},$$

where Δ is nonzero. Elliptic curves with $j = 0$ or 1728 are atypical and require special consideration. We will exclude such curves from further discussion. Up to isomorphism, there are only two elliptic curves over the same finite field with the same j-invariant, and they are twists of each other. An elliptic curve with a specified j-invariant is easy to construct. In the field $\boldsymbol{F}_p$, let $k = j_0/(1728 - j_0)$. Then the curve $\mathcal{X}(\boldsymbol{F}_p) : y^2 = x^3 + 3kx + 2k$ is an elliptic curve whose j-invariant is j_0, as is easily checked. Moreover, the curve $\mathcal{X}(\boldsymbol{F}_p) : y^2 = x^3 + 3kc^2x + 2kc^3$ is also an elliptic curve with the same j-invariant, where c is any nonzero element of $\boldsymbol{F}_p$. For each j, this curve in $\boldsymbol{F}_p$ is either isomorphic to the same curve or to a twist of that curve, depending on whether c is a square or a nonsquare in $\boldsymbol{F}_p$. These are the curves that are found by the method of complex multiplication.

Thus it is only necessary to find the possible j-invariants corresponding to the curves with the values of p, m, and m' that were obtained by solving the diophantine equation $4p = x^2 + Dy^2$. These j-invariants are obtained as the zeros of yet another polynomial, $H_D(x)$, called the *Hilbert class polynomial*, a polynomial that has degree h_D. The Hilbert class polynomial $H_D(x)$ has h_D zeros, all in $\boldsymbol{Z}$, and so specifies h_D j-invariants by its zeros, and each j-invariant leads to a pair of (isomorphism classes of) elliptic curves. Of this pair, one must have order m and one must have order m'. It is easy to determine which curve has which order so as to resolve this ambiguity.

The Hilbert class polynomial $H_D(x)$ provides a magical doorway through which information about an elliptic curve over $\boldsymbol{Q}(\sqrt{-D})$ can be passed to become information about the corresponding elliptic curve over $\boldsymbol{F}_p$. This is because the Hilbert class polynomial is the same for both curves. Its zeros in $\boldsymbol{Q}(\sqrt{-D})$ are the j-invariants of the elliptic curves in $\boldsymbol{Q}(\sqrt{-D})$, and its zeros in $\boldsymbol{F}_p$ are the j-invariants of the elliptic curve in $\boldsymbol{F}_p$. Knowing the j-invariants of the curve in one field allows one to compute $H_D(x)$, then find the zeros of $H_D(x)$ in the other field in order to find the j-invariants in that field.

At this point, a potential computational burden of the method of curve construction can be recognized. To find one j-invariant, the procedure computes a polynomial $H_D(x)$ of degree h_D and finds any one of its zeros. Although we only need to know one of the j-invariants because we want only one of the elliptic curves over $\boldsymbol{F}_p$, we must first compute a polynomial that may have a very large degree, only to find one zero. This can be a computational burden if $H_D(x)$ has very large coefficients or if h_D is very large, as it may be for some D. If h_D is small, as it is for other D, the computational burden is not an issue.

An example of a Hilbert class polynomial is

$$H_{23}(x) = x^3 + 3491750x^2 - 5151296875x + 12771880859375.$$

This Hilbert class polynomial corresponds to discriminant $D = 23$. The degree of $H_{23}(x)$ is three, which means that the class number h_{23} is 3. The three zeros of the polynomial $H_{23}(x)$ are the three j-invariants. Finally, the three pairs of elliptic curves are given by

$$\mathcal{X}(\boldsymbol{F}_p): y^2 = x^3 + 3kx + 2k$$
$$\mathcal{X}'(\boldsymbol{F}_p): y^2 = x^3 + 3kc^2x + 2kc^3,$$

where $k = j/(1728 - j)$ in $\boldsymbol{F}_p$ for each of the three values of j that are zeros of $H_{23}(x)$, and where c is any nonsquare of $\boldsymbol{F}_p$. For each value of j, the corresponding curve $\mathcal{X}'(\boldsymbol{F}_p)$ is a twist of $\mathcal{X}(\boldsymbol{F}_p)$.

The remainder of this section is a summary of the mathematics underlying the method of complex multiplication. This summary is intended as a survey to provide a brief introduction to the main notions of this deep topic. We first refer to the rich structure of quadratic number fields. Because the imaginary number $\sqrt{-D}$ plays a role in the expression for the complex number α that is a zero of the polynomial $x^2 - tx + q$, it is natural to consider the imaginary quadratic number field $\boldsymbol{Q}(\sqrt{-D})$ in the search for an elliptic curve. Such fields were discussed in Section 9.7. To find the desired curve by this method requires understanding of the structure of the field $\boldsymbol{Q}(\sqrt{-D})$.

Recall that an imaginary quadratic number field is a field of the form

$$\boldsymbol{Q}(\sqrt{-D}) = \{a + ib\sqrt{D} \mid a, b \in \boldsymbol{Q}\},$$

where D is a square-free positive integer now called the *discriminant* of the quadratic field, a and b are rationals, and $i = \sqrt{-1}$. (If $D = 1$, this is the field composed of those complex rationals known as the *gaussian rationals*.) Some elements of the number field $\boldsymbol{Q}(\sqrt{-D})$ are called *algebraic integers*[11] (or simply *integers*). The ring of algebraic

[11] The algebraic integers of the complex field $\boldsymbol{C}$, in general, are the zeros in $\boldsymbol{C}$ of the monic polynomials in $\boldsymbol{Z}[x]$. These polynomials are the monic polynomials with integer coefficients.

integers of $\boldsymbol{Q}(\sqrt{-D})$, denoted $\mathcal{O}_D$ (or $\mathcal{O}_{\boldsymbol{Q}(\sqrt{-D})}$), consists of those algebraic integers of $\boldsymbol{C}$ that are also in $\boldsymbol{Q}(\sqrt{-D})$. Thus

$$\mathcal{O}_D = \{\beta \in \boldsymbol{Q}(\sqrt{-D}) \mid p(\beta) = 0 \quad \text{for some monic polynomial } p(x) \in \boldsymbol{Z}[x]\}.$$

The set $\mathcal{O}_D$ plays a role in $\boldsymbol{Q}(\sqrt{-D})$ similar to the role that $\boldsymbol{Z}$ plays in $\boldsymbol{Q}$.

The set $\mathcal{O}_D$ includes all elements of the set $\boldsymbol{Z}(\sqrt{-D}) = \{a + ib\sqrt{D} \mid a, b \in \boldsymbol{Z}\}$, and possibly other elements of $\boldsymbol{Q}(\sqrt{-D})$. This is because, for any integers a and b of $\boldsymbol{Z}$, the number $a + b\sqrt{-D})$ is always a zero of the polynomial $x^2 - 2ax + a^2 + b^2 D$. Thus $\boldsymbol{Z}[\sqrt{-D}]$ provides an infinite number of algebraic integers to the set $\boldsymbol{Q}(\sqrt{-D})$. However, not every element of $\mathcal{O}_D$ needs to be an element of $\boldsymbol{Z}(\sqrt{-D})$. There could be other elements, as noted in the following theorem.

Theorem 10.17.1 *The set $\mathcal{O}_D$ of algebraic integers of $\boldsymbol{Q}(\sqrt{-D})$ satisfies*

$$\mathcal{O}_D = \begin{cases} \boldsymbol{Z}[(1 + \sqrt{-D})/2] & \text{if} \quad D = 3 \pmod 4) \\ \boldsymbol{Z}[\sqrt{-D}] & \text{if} \quad D = 1, 2 \pmod 4). \end{cases}$$

Proof Any element s of $\mathcal{O}_D$ is an algebraic integer of $\boldsymbol{C}$ also lying in the set $\boldsymbol{Q}(\sqrt{-D})$, so $s = a + b\sqrt{-D}$ where a and b are rationals. If $b = 0$, then s is an integer. If b is nonzero, then s is a zero of the polynomial

$$\begin{aligned} f(x) &= (x - (a + b\sqrt{-D}))(x - (a - b\sqrt{-D})) \\ &= x^2 - 2ax + (a^2 + Db^2), \end{aligned}$$

where $2a$ and $a^2 + Db^2$ are required to be integers. If s is a zero of any other polynomial $g(x)$, then $f(x)$ divides $g(x)$ because s is a zero of $f(x)$. Thus we only need to consider the circumstances under which $2a$ and $a^2 + Db^2$ are both integers.

Because $2a$ is an integer, either a itself is an integer or a is a half integer. If a is an integer, then $a^2 + Db^2$ is an integer if, and only if, Db^2 is an integer, which means that the rational b itself is an integer because D is a square-free integer.

If a is a half integer, then $a = c + \frac{1}{2}$ where c is an integer. Then $a^2 = c^2 + c + \frac{1}{4}$, so Db^2 must be equal to $d + \frac{3}{4}$ where d is an integer. Let $D = 4k + \ell$, where $\ell = 1, 2$, or 3. Then $Db^2 = 4kb^2 + \ell b^2 = d + \frac{3}{4}$, which can be satisfied if, and only if, b is one-half of an odd integer and ℓ is equal to three.

Thus if $\ell = 1$ or 2, then a and b must both be integers, so $\mathcal{O}_D = \boldsymbol{Z}[\sqrt{-D}]$. If $\ell = 3$, an element of $\mathcal{O}_D$ has the form $a + b\sqrt{-D}$, where either a and b are each an arbitrary integer, or are each an arbitrary integer (with b odd) plus one-half. This means that an element of $\mathcal{O}_D$ has the form $A + B(1 + \sqrt{-D})/2$ where A and B are arbitrary integers. Thus $\mathcal{O}_D = \boldsymbol{Z}[(1 + \sqrt{-D})/2]$ as was to be proved. □

An imaginary quadratic number field $\boldsymbol{Q}(\sqrt{-D})$ has an infinite number of elements, so this field is an infinite field. However, a finite group called the *ideal class group* can be constructed by starting from an imaginary quadratic number field. The elements of the ideal class group are called *ideal classes*, which we will discuss first and explain later. The number of ideal classes in the ideal class group is given by the *class number* h_D, which was introduced earlier. Thus for each square-free integer D, the field $\boldsymbol{Q}(\sqrt{-D})$ is associated with a class number h_D. Moreover, if there are h_D ideal classes in the ideal class group corresponding to $\boldsymbol{Q}(\sqrt{-D})$, then the number of pairs of nonisomorphic elliptic curves $\mathcal{X}(\boldsymbol{F}_p)$ such that the endomorphism ring $\operatorname{end}(\mathcal{X}(\boldsymbol{F}_p)) = \mathcal{O}_D$ is equal to h_D.

The relationship between the discriminant D and the class number h_D is not a simple one. For some D, the class number h_D can be small, for others it can be quite large. The only values of the discriminant D for which $h_D = 1$ are the nine values $D = 1, 2, 3, 7, 11, 19, 43, 67$, and 163. There is no evident pattern that describes this sequence. In particular, $h_7 = 1$, which means that, up to isomorphism, there is only one pair of elliptic curves with endomorphism ring of the form $\operatorname{end}(\mathcal{X}(\boldsymbol{F}_p)) = \boldsymbol{Z} + \boldsymbol{Z}\omega$, where, by Theorem 10.17.1, $\omega = (1 + \sqrt{-7})/2$ because $7 = 3 \pmod 4$. In contrast, $h_5 = 2$, which means that up to isomorphism, there are exactly two pairs of elliptic curves with endomorphism ring $\operatorname{end}(\mathcal{X}(\boldsymbol{F}_p)) = \boldsymbol{Z} + \boldsymbol{Z}\omega$, where $\omega = \sqrt{-5}$ because $5 = 1 \pmod 4$. In further contrast, $h_{23} = 3$, which means that up to isomorphism, there are exactly three pairs of elliptic curves with endomorphism ring $\operatorname{end}(\mathcal{X}(\boldsymbol{F}_p)) = \boldsymbol{Z} + \boldsymbol{Z}\omega$, where $\omega = (1 + \sqrt{-23})/2$ because $23 = 3 \pmod 4$.

Theorem 10.17.1 tells us that

$$\boldsymbol{Q}[\sqrt{-D}] \supset \mathcal{O}_D \supseteq \boldsymbol{Z}[\sqrt{-D}]$$

and the containment on the right is strict if $D = 3 \pmod 4$. In any case, $\boldsymbol{Q}[\sqrt{-D}]$ contains an infinite number of algebraic integers. Our next task is to impose a structure on the set of algebraic integers so that the infinite set $\mathcal{O}_D$ can be broken into a finite number of sets. To do so, first enlarge $\mathcal{O}_D$ to its field of fractions $K = \{a/b \mid a, b \in \mathcal{O}_D, b \neq 0\}$. The field K has the same relationship to the ring $\mathcal{O}_D$ that the rational field $\boldsymbol{Q}$ has to the integer ring $\boldsymbol{Z}$. An element of K is defined so that division is possible as the ratio of two algebraic integers, provided the denominator is nonzero. Multiplication is defined as $(a/b)(c/d) = ac/bd$. Division by (c/d) is defined as multiplication by (d/c), provided c is nonzero.

To define the ideal class group, one first defines a fractional ideal. A *fractional ideal* of $\mathcal{O}_D$ is defined as a subset I of K that is closed under multiplication by elements of $\mathcal{O}_D$, and for which there exists a nonzero $r \in \mathcal{O}_D$, such that $(r)I \subseteq \mathcal{O}_D$, where $(r)I = \{ri \mid i \in I\}$. The element r can be thought of as clearing out the denominators in I. Define an equivalence relation on the set of all fractional ideals of $\mathcal{O}_D$ such that two ideals I and J are equivalent if there exist $a, b \in \mathcal{O}_D$ such that $(a)I = (b)J$. In this way the set of all fractional ideals is partitioned into equivalence classes, denoted (I). These equivalence classes are the ideal classes. Finally, notice that the ideal classes form a group because there is a group operation defining $(I)(I')$. Under this operation, the set of all ideal classes is called the *ideal class group*. The number of ideal classes is finite, so it is useful for computational purposes. This fact is one reason for giving the elaborate construction of the ideal class group.

For an example of a fractional ideal, the set

$$I = \left\{ \frac{2a_1 + a_2 - 5a_4 + (a_2 + 2a_3 + a_4)\sqrt{-5}}{3 + \sqrt{-5}} \mid a_i \in \boldsymbol{Z} \right\}$$

is a fractional ideal in $\boldsymbol{Q}(\sqrt{-5})$ because $(3 + \sqrt{-5})I = \langle 2, 1 + \sqrt{-5} \rangle$.

An *order*[12] in an imaginary quadratic number field is a subring R of $\mathcal{O}_D$ that contains $\boldsymbol{Z}$. Thus

$$\boldsymbol{Z} \subset R \subseteq \mathcal{O}_D$$

for any order R. An order always has the form

$$R = \boldsymbol{Z} + \boldsymbol{Z} f\delta,$$

where f is a positive integer and δ is given by

$$\delta = \begin{cases} (1 + \sqrt{-D})/2 & \text{if} \quad D = 3 \pmod 4 \\ \sqrt{-D} & \text{if} \quad D = 1, 2 \pmod 4). \end{cases}$$

By choosing an integer f larger than one, the order R is a proper subset of $\mathcal{O}_D$, but always properly containing $\boldsymbol{Z}$.

An elliptic curve $\mathcal{X}(\boldsymbol{F}_q)$ that is not supersingular has an endomorphism ring that is either equal to $\boldsymbol{Z}$ or is equal to an order in an imaginary quadratic number field $\boldsymbol{Q}(\sqrt{-D})$. If the endomorphism ring of the curve is equal to an order in an imaginary quadratic number field, then the curve is said to have complex multiplication. We can construct such a curve by first choosing the square-free integer D so as to determine the number field $\boldsymbol{Q}(\sqrt{-D})$, as described earlier.

Finally, the *Hilbert class polynomial* $H_D(x) \in \boldsymbol{Z}[x]$ is defined as the minimal polynomial of the j-invariant of the complex elliptic curve $\boldsymbol{C}/\mathcal{O}_D$, where $\mathcal{O}_D$ is the imaginary quadratic order with discriminant D. Thus

$$H_D(x) = \prod_{\alpha} (x - j(\alpha)).$$

Because a j-invariant is always an algebraic integer, $H_D(x) \in \boldsymbol{Z}[x]$. The polynomial $H_D(x)$ factors completely over $\boldsymbol{F}_p$, and its zeros are precisely the j-invariants of the elliptic curve $\mathcal{X}(\boldsymbol{F}_p)$ whose endomorphism ring is isomorphic to $\mathcal{O}_D$.

[12] An order in the theory of rings is a concrete noun and is not to be confused with the order of a group.

Problems for Chapter 10

10.1 Verify that the plane curves in $\boldsymbol{R}^2$ defined by the polynomials $y^2 = x^3$ and $y^2 = x^2(x+1)$ are singular. Verify that the plane curve in $\boldsymbol{R}^2$ defined by the polynomial $y^2 = x^3 - x$ is nonsingular.

10.2 Describe the group $\mathcal{X}(\boldsymbol{F}_5)$ for the following elliptic curves:

$$\mathcal{X}(\boldsymbol{F}_5)\colon y^2 = x^3 + 1$$
$$\mathcal{X}(\boldsymbol{F}_5)\colon y^2 = x^3 + x$$
$$\mathcal{X}(\boldsymbol{F}_5)\colon y^2 = x^3 + 2x.$$

Can you find an elliptic curve over $\boldsymbol{F}_5$ with more points than any of these?

10.3 **a** Let $\mathcal{X}$ be the curve $y^2 + y = x^3 + x + 1$ defined over $\boldsymbol{F}_2$. Show that $\mathcal{X}$ is a supersingular elliptic curve.

b Let $q = 2^m$. Let P and Q be two points of the curve $\mathcal{X}(\boldsymbol{F}_q)$. Give formulae with proofs for the coordinates of $P + Q$ (make sure that all possible cases are treated).

c Give a formula for the order of the group $\mathcal{X}(\boldsymbol{F}_q)$ in terms of m.

d Given that a discrete-log problem on $\mathcal{X}(\boldsymbol{F}_q)$ can be reduced to a discrete-log problem in the multiplicative group of $\boldsymbol{F}_{q^m}$, find a value of m that would yield a reasonably secure elliptic-curve cryptosystem based on $\mathcal{X}(\boldsymbol{F}_q)$ that is suitable for a smart card.

10.4 Let $\mathcal{X}$ be the elliptic curve over $\boldsymbol{F}_3$ given by

$$\mathcal{X} : y^2 + y = x^3 - x + 1.$$

How many rational points does $\mathcal{X}$ have? How many points are there on the curve in the extension field $\boldsymbol{F}_{3^m}$?

10.5 Show that the bivariate polynomial $y^2 = x^3 + x + 1$ is nonsingular in the field $\boldsymbol{F}_{11}$. How many zeros might this polynomial have as allowed by the Hasse–Weil bound? Find the zeros explicitly. How many points are there on the elliptic curve? Do they form a cyclic group? What is the order of the largest cycle?

10.6 Show explicitly that if the characteristic of the field is two or three, then it is not possible, in general, to reduce the Weierstrass form of an elliptic polynomial to the short Weierstrass form.

10.7 Let $\mathcal{X} : y^2 = x^3 + x + 6$ be an elliptic curve over the field $\boldsymbol{F}_{11}$. Find the points of $\mathcal{X}$. Do the points of $\mathcal{X}$ form a cyclic group under point addition? How many generators are there?

10.8 Show, for fields whose characteristic is not two or three, that the polynomial $x^3 + ax + b$ has no repeated zeros if, and only if, $4a^3 + 27b^2$ is nonzero. Why must fields of characteristic two or three be excluded?

10.9 Is it true that for every odd prime p, the elliptic curve

$$\mathcal{X} : y^2 = x^3 + x$$

satisfies

$$\#\mathcal{X}(\boldsymbol{F}_p) = 0 \pmod 4?$$

10.10 Consider the elliptic curve $\mathcal{X} : y^2 = x^3 + ax + b$ where $a, b \in \boldsymbol{F}_q$, a field of characteristic larger than three. Derive the algebraic formulas for point addition on this elliptic curve.

10.11 Verify the associative law

$$(P_1 + P_2) + P_3 = P_1 + (P_2 + P_3)$$

for point addition on elliptic curves by elementary manipulations of the explicit algebraic equations. This will require consideration of a number of special cases.

10.12 Demonstrate the associative law

$$(P_1 + P_2) + P_3 = P_1 + (P_2 + P_3)$$

for point addition on an elliptic curve in $\boldsymbol{R}^2$ by carefully sketching a graph of an elliptic curve in $\boldsymbol{R}^2$ with three points P_1, P_2, and P_3 labeled. Then draw appropriate lines to demonstrate both sides of the equation.

10.13 Starting with the Weierstrass form of the elliptic curve given in Definition 10.1.1, derive a set of equations for point addition on an elliptic curve that apply for any field characteristic.

10.14 Find the order of the point $P = (2, 3)$ on the elliptic curve $\mathcal{X} : y^2 = x^3 + 1$. Is it possible to solve this problem without specifying the field? For what fields does the problem make sense?

10.15 Given the elliptic curves $\mathcal{X}(\boldsymbol{F}_p) : y^2 = x^3 + ax + b$ and $\mathcal{X}'(\boldsymbol{F}_p) : y^2 = x^3 + d^2ax + d^3b$, show that $\mathcal{X}(\boldsymbol{F}_p)$ can be converted to $\mathcal{X}'(\boldsymbol{F}_p)$ by a simple change of variables if d is a nonzero square in $\boldsymbol{F}_q$.

10.16 **a** Let $\mathcal{X}(\boldsymbol{F}_q)$ be an elliptic curve over $\boldsymbol{F}_q$. Show that $\#\mathcal{X}(\boldsymbol{F}_{q^3})$ is an integer multiple of $\#\mathcal{X}(\boldsymbol{F}_q)$.

b More precisely, show that

$$\#\mathcal{X}(\boldsymbol{F}_{q^3}) = (q^2 - q + qt + 1 + t + t^2)\#\mathcal{X}(\boldsymbol{F}_q),$$

where t is the Frobenius trace.

10.17 A *Koblitz elliptic curve* is an elliptic curve over the binary field $\boldsymbol{F}_{2^p}$ based on the polynomial $p(x, y) = y^2 + xy + x^3 + a_2x^2 + 1$, with $a_2 = 0$ or 1. Find at least one example of a Koblitz curve over $\boldsymbol{F}_{2^p}$, p an odd prime, such that $\#\mathcal{X}(\boldsymbol{F}_{2^p})$ is equal to twice or four times a prime number according to whether a_2 is one or zero, respectively.

10.18 Show that the elliptic curve

$$\mathcal{X}(\boldsymbol{F}_{43}) : y^2 = x^3 + 39x^2 + x + 41$$

has 43 points in $\boldsymbol{F}_{43}$. Is this an anomalous elliptic curve?

10.19 Suppose that $\boldsymbol{F}_p$ contains an element β of order three, and let $\Phi(x, y) = (\beta x, y)$. Let P be an element of the elliptic curve $\mathcal{X}(\boldsymbol{F}_p) : y^2 = x^3 + 1$.

a Show that $\Phi(P)$ is an element of $\mathcal{X}(\boldsymbol{F}_p)$.

b Show that $\Phi(P_1 + P_2) = \Phi(P_1) + \Phi(P_2)$ for all P_1 and P_2 on $\mathcal{X}(\boldsymbol{F}_p)$.

10.20 Let $\pi_q(P)$ denote the Frobenius map on an elliptic curve $\overline{\boldsymbol{F}}_q$ as defined by $\pi_q(x, y) = (x^q, y^q)$. By treating the equations for point addition on an elliptic curve, explicitly verify that $\pi_q(P_1 + P_2) = \pi_q(P_1) + \pi_q(P_2)$. Does this imply that $\pi_q([a]P) = [a]\pi_q(P)$?

10.21 Let k be a 128-bit number, and let P be any point of the elliptic curve $\mathcal{X}(\boldsymbol{F}_q)$. The point $[k]P$ can be computed from P by the method of successive doubling and adding.

a Develop a method of computing $[k]P$ by successive doubling, adding, and subtracting. What are the benefits of this alternative method?

b Develop an algorithm that uses *nonsuccessive* doubling and addition, but no subtraction. For example, $28P = 2(2(4P + 2P + P))$. Does your algorithm have any merit for an arbitrary k with 128 bits?

10.22 Develop the equations for a key-exchange cryptography system that uses *point halving* on an elliptic curve. Point halving, which is an alternative to point doubling, requires the computation of $Q = \frac{1}{2}P$. That is, given P, one must find the point Q satisfying $2Q = P$. Compare the computations needed for point halving with those required for point doubling.

10.23 Find $\#\mathcal{X}(\boldsymbol{F}_{64}) : y^2 + xy = x^3 + 1$.

10.24 Suppose that p is a prime not smaller than 5. Prove that $\mathcal{X}(\boldsymbol{F}_p)$ is a supersingular elliptic curve if, and only if,

$$\#\mathcal{X}(\boldsymbol{F}_p) = p + 1.$$

By finding a counterexample, show that the statement cannot be extended to $p = 3$.

10.25 Prove that the only constant endomorphism on an elliptic curve is the zero endomorphism, which is given by $\phi(P) = \mathcal{O}$ for all P.

10.26 Let $\mathcal{X}(\boldsymbol{F}_q)$ be an elliptic curve in the field $\boldsymbol{F}_q$, and let $\mathcal{X}(\boldsymbol{F}_{q^r})$ be that same curve lifted to the extension field $\boldsymbol{F}_{q^r}$.

a Show that for any point $(x, y) \in \mathcal{X}(\boldsymbol{F}_{q^r})$, $(x^q, y^q) \in \mathcal{X}(\boldsymbol{F}_{q^r})$.

b For any $(x, y) \in \mathcal{X}(\boldsymbol{F}_{q^r})$, define the trace of the point (x, y) as

$$\text{trace}(x, y) = (x, y) + (x^q, y^q) + (x^{q^2}, y^{q^2}) + \cdots + (x^{q^r}, y^{q^r}).$$

Prove that the trace of (x, y) is an element of $\mathcal{X}(\boldsymbol{F}_q)$.

10.27 **a** Prove that if $\mathcal{X}(\boldsymbol{F}_{q^m})$ is a supersingular elliptic curve for $m = 1$, then $\mathcal{X}(\boldsymbol{F}_{q^m})$ is supersingular for every value of m.

b Prove that if $\mathcal{X}(\boldsymbol{F}_{q^m})$ is an anomalous elliptic curve for $m = 1$, then $\mathcal{X}(\boldsymbol{F}_{q^m})$ could be nonanomalous for some other values of m.

10.28 The zeta function of the elliptic curve $\mathcal{X}(\boldsymbol{F}_q)$ can be expressed as

$$Z(x) = \frac{1 - tx + qx^2}{(1-x)(1-qx)},$$

where t is the Frobenius trace. By differentiation of the logarithm, show that

$$\#\mathcal{X}(\boldsymbol{F}_{q^n}) = q^n + 1 - \alpha^n - \beta^n,$$

where α and β are the zeros of $1 - tx + qx^2$.

10.29 Suppose that q is not a power of two. The *Edwards representation* of an elliptic curve over $\boldsymbol{F}_q$ is defined by the Edwards polynomial

$$x^2 + y^2 = 1 + dx^2y^2,$$

where a is a nonzero, nonsquare element of $\boldsymbol{F}_q$. Show that this curve is (birationally) equivalent to the polynomial

$$y^2 = (x - d - 1)(x^2 - 4d)$$

which, when multiplied out, is in Weierstrass form for an elliptic curve.

10.30 A polynomial over $\boldsymbol{F}_3$ of the form $y^2 = x^3 - x \pm 1$ is a supersingular polynomial. Let m be coprime with 6 (and so be even). Show that

$$\#(\boldsymbol{F}_{3^m}) = 3^m \pm 3^{(m+1)/2} + 1,$$

where the sign agrees with the sign in the above polynomial. **Hint:** Evaluate $(3^m + 3^{(m+1)/2} + 1)(3^m - 3^{(m+1)/2} + 1)$.

10.31 Verify Cornacchia's algorithm.

10.32 Prove that the elliptic curve over $\boldsymbol{F}_p$ with $p = 3 \pmod 4$ and polynomial $y^2 = x^3 + ax$ is a supersingular elliptic curve for any nonzero value of a.

10.33 Prove that a supersingular elliptic curve over $\boldsymbol{F}_p$, for p not equal to two or three, has no points of order p in any extension field.

Notes for Chapter 10

Elliptic curves take their name from elliptic functions, which are used in the study of elliptic integrals and are themselves defined in terms of ellipses. This is the convoluted historic origin of the name. Elliptic curves in the real or complex number system were studied intensively early on by using a geometric framework. It is easy to depict an elliptic curve in the real-number system geometrically as a curve in the plane. Beyond this, there is a parallel structural relationship between an elliptic curve in the complex plane and a torus in the complex number system. All of the notions of elliptic curves have been reworked using purely algebraic arguments so that the results apply in any field. However, the geometric insight of the real-number elliptic curves provides guidance and intuition for the purely algebraic structure of elliptic curves in other fields.

The early studies of plane curves, including elliptic curves, considered only curves in the real field and the complex field. Later, André Weil (1906–98) insisted on discarding the geometry and deriving all properties of plane curves by using only algebraic methods. This policy helped to open the way to the study of plane curves such as elliptic curves in a finite field. Indeed, curves, surfaces, and other varieties can be studied in any field. Notwithstanding this modern viewpoint, the geometric intuition associated with the real field is, to most of us, a great aid to understanding and a guide to further developments.

The idea of using point addition on an elliptic curve to define a cryptographic system was proposed independently by Miller (1985) and Koblitz (1987), who were motivated by the observation that there is no obvious analogy of the index calculus for the elliptic curves. Evidently, the then unpublished work using elliptic curves for integer factoring by Lenstra (1987) had drawn their attention to elliptic curves.

The remarkable Hasse–Weil bound originated with Helmet Hasse (1898–1979). Hasse proved this bound for elliptic curves in 1936, and in 1949 Weil proved the generalization for curves of arbitrary genus. We have stated the bound in a form with a slight improvement due to Serre. The study of algebraic integers within number fields, which is known as *algebraic number theory*, was developed in part by Richard Dedekind (1831–1916).

Strong methods of computing the number of rational points on a general elliptic curve over a finite field are now available. The Schoof algorithm (1985) with enhancement by Elkies (1991) and Atkin (1988) made it possible to count points on an elliptic curve. Without such methods, modern large elliptic-curve cryptosystems would be very difficult, and perhaps infeasible, to design. The records on counting the rational points of elliptic curves of very large order continue to advance. For example, Lercier and Morain (1995), using the Schoof algorithm, have computed the number of points on some elliptic curves over $\boldsymbol{F}_q$ for $q = 10^{499} + 153$ and $q = 2^{1301}$. More recently, the Satoh algorithm (2002) has been introduced as an alternative, and now preferred, method of point counting based on the theory of p-adic fields.

The methods of xedni calculus were introduced and whimsically named by Silverman (2000), who was motivated by the methods of index calculus. Xedni calculus is an attempt to attack elliptic-curve cryptography by mimicking index calculus. It now appears, based on the work of Jacobson and others (2000), that elliptic-curve cryptography cannot be successfully attacked from this direction, though we still have neither a formal statement nor a proof of this anecdotal remark. This negative remark should be seen as a positive statement regarding the power of elliptic-curve cryptography.

Although it is technically meaningless to say that one countably infinite set is larger than another, from a practical point of view, the number of cryptosystems based on elliptic curves over finite fields is vastly larger than the number of cryptosystems based on factoring integers. This is because the practical options within the ring $\boldsymbol{Z}_p$ are fewer than are the options within the set of elliptic curves over the many finite fields. One may suppose that there is a virtually unlimited amount of work remaining to be done in order to sort out all of these possibilities and to identify any unique advantages or disadvantages for various choices of elliptic curves.

The use of complex multiplication to construct elliptic curves on a finite field originated with Atkin and Morain (1993). The use of point halving in place of point doubling was suggested independently by Knudsen (1999) and Schroeppel (2000). Point halving leads to a small decrease in the computational burden of finding point multiples. An extensive treatment of the subject of elliptic-curve cryptography can be found in the monograph by Blake, Seroussi, and Smart (1999), and in the books of Silverman (1986) and Washington (2008) as well as the book by Hankerson, Menezes, and Vanstone (2004). A history of elliptic-curve cryptography is given in the paper by Koblitz, Koblitz, and Menezes (2011). Standardized applications of elliptic-curve cryptography are given in the *Federal Information Processing Standards* (FIPS).

11 Cryptography based on hyperelliptic curves

The widespread success of cryptography based on elliptic curves motivates the investigation of other curves for possible cryptographic uses. However, elliptic curves are the only plane curves that admit a definition of point addition in such a way that the points of the curve form a group. This does not mean that other curves cannot be used. It only means that the points of those curves must be organized to form a group in some other way. It is more complicated to find group structures based on other curves. In general, the curve $\mathcal{X}$ must be embedded into a larger algebraic structure on which a suitable group operation can be defined. Hyperelliptic curves are a class of curves that lead to such a group structure. A hyperelliptic curve is associated in a natural way with an abelian group called the jacobian of the hyperelliptic curve. In contrast to the curve itself, the jacobian of a hyperelliptic curve does admit a suitable group structure. Based on the group structure of its jacobian, a hyperelliptic curve can be used to construct a cryptographic system. Most of the chapter is devoted to the task of defining the jacobian of a hyperelliptic curve and its relevant computational algorithms. The usual methods of cryptography constructed on a large finite group are then immediately applicable.

Because an elliptic curve is a special case of a hyperelliptic curve, this chapter also serves to extend our understanding of elliptic curves. In particular, by working with the poles and zeros of elliptic curves in Section 11.3, we will study the notions of divisors and rational functions defined on elliptic curves.

11.1 Hyperelliptic curves

A hyperelliptic curve over the field $\boldsymbol{F}$ is a generalization of an elliptic curve over the field $\boldsymbol{F}$. Hyperelliptic curves form a very large class of plane curves, and very few of them have been studied in detail for applications in cryptography. The definition is as follows.

Definition 11.1.1 *A hyperelliptic curve $\mathcal{X}$ of genus g over the field $\boldsymbol{F}$ is the set of rational projective zeros of a polynomial of the form* $p(x, y) = y^2 + h(x)y - f(x)$

that has no affine singularities in any extension of $\boldsymbol{F}$*, where the univariate polynomials* $h(x)$ *and* $f(x)$ *are elements of* $\boldsymbol{F}[x]$ *that satisfy* $\deg h(x) \leq g$ *and* $\deg f(x) = 2g + 1$.

More explicitly, a hyperelliptic curve over the finite field $\boldsymbol{F}_q$ is

$$\mathcal{X}(\boldsymbol{F}_q) = \{(x, y) \in \boldsymbol{F}_q^2 : y^2 + h(x)y = f(x)\} \cup \{\infty\},$$

where all zeros are affine zeros except for the single point at infinity, the polynomial is nonsingular at all affine points, and with $h(x)$ and $f(x)$ satisfying the required degree conditions[1] that $\deg h(x) \leq g$ and $\deg f(x) = 2g + 1$.

A hyperelliptic curve of genus one is an elliptic curve. Except for the case of an elliptic curve, a hyperelliptic curve always has a singularity at the point $(0, 1, 0)$, which is its only point at infinity. This singularity is an unavoidable consequence of the definition. Only in the affine plane is a hyperelliptic curve required to have no singular points. Accordingly, the polynomial $y^2 + h(x)y - f(x)$ will be called nonsingular if it has no singular affine points.

If the field $\boldsymbol{F}$ is contained in a larger field $\boldsymbol{K}$, such as the extension field $\boldsymbol{F}_{q^m}$ or the algebraic closure $\overline{\boldsymbol{F}}$ of the field $\boldsymbol{F}$, then the hyperelliptic curve is also defined in $\boldsymbol{K}$ as

$$\mathcal{X}(\boldsymbol{K}) = \{(x, y) \in \boldsymbol{K}^2 : y^2 + h(x)y = f(x)\} \cup \{\infty\},$$

because the same polynomial $y^2 + h(x)y - f(x)$ can also be regarded as a polynomial over any larger field $\boldsymbol{K}$ that contains $\boldsymbol{F}$. This notion is the same as the familiar notion of evaluating a polynomial with coefficients in the rational field $\boldsymbol{Q}$ to find its zeros in the real field $\boldsymbol{R}$ or in the complex field $\boldsymbol{C}$. The points of $\mathcal{X}(\boldsymbol{K})$ that are in $\mathcal{X}(\boldsymbol{F})$ are the rational points of $\mathcal{X}(\boldsymbol{K})$.

The polynomial $p(x, y)$ satisfies the requirements of the definition if it has no singular affine points in $\boldsymbol{F}$ or in any extension of $\boldsymbol{F}$. The polynomial is nonsingular in the affine plane if the formal partial derivatives are not both zero at any point of the affine plane at which $p(x, y)$ equals zero. This means that the two equations

$$\begin{aligned} 2y + h(x) &= 0 \\ h'(x)y &= f'(x) \end{aligned}$$

must not be simultaneously satisfied at any affine point at which $y^2 + h(x)y = f(x)$, where $h'(x)$ and $f'(x)$ denote the formal derivatives of $h(x)$ and $f(x)$. Otherwise, the corresponding curve is not a hyperelliptic curve, and so is of no interest to us.

Because the point at infinity of every hyperelliptic curve of genus larger than one is always a singular point, there may sometimes be technical difficulties related to that point. Rather than dealing with these difficulties, we will turn a blind eye in that direction with some hope that our trusting approach will have no unfortunate consequences.

[1] We do not consider hyperelliptic curves for which the degree of $f(x)$ is even.

Lemma 11.1.2 *A hyperelliptic polynomial is irreducible.*

Proof Suppose that the polynomial $p(x, y) = y^2 + h(x)y - f(x)$ is reducible. Then

$$y^2 + h(x)y - f(x) = (y - a(x))(y - b(x))$$

so that

$$\begin{aligned}\deg[a(x)b(x)] &= \deg a(x) + \deg b(x)\\ &= \deg f(x) = 2g + 1,\end{aligned}$$

which is odd, so $a(x)$ and $b(x)$ cannot both have the same degree and they cannot both have a degree that is larger than g. One of them has a degree that is not larger than g and the other has a degree that is larger than g. But

$$\deg[a(x) + b(x)] = \deg h(x) \leq g.$$

These two statements are not compatible. Therefore the polynomial is irreducible. □

Theorem 11.1.3 *If q is a power of two, then $h(x)$ must be nonzero if the curve is a hyperelliptic curve.*

Proof Suppose that the characteristic of $\boldsymbol{F}_q$ is two and $h(x) = 0$. Then the equation of the curve and its two partial derivatives reduce to the two equations

$$\begin{aligned} y^2 &= f(x)\\ f'(x) &= 0.\end{aligned}$$

Bézout's theorem states that these two equations must have a common zero in some extension field. More directly, if β is a zero of $f'(x)$, then $y = \sqrt{f(\beta)}$ exists in some extension field so there is a simultaneous solution of the two equations. Therefore the polynomial is singular if $h(x) = 0$. □

Suppose that the characteristic of $\boldsymbol{F}_q$ is an odd prime. Then, by a change of variables, $h(x)$ can be set to zero as follows. Let $y = z - \frac{1}{2}h(x)$. Then

$$\begin{aligned} y^2 + h(x)y - f(x) &= (z - \tfrac{1}{2}h(x))^2 + h(x)(z - \tfrac{1}{2}h(x)) - f(x)\\ &= z^2 - \tfrac{1}{4}h(x)^2 - f(x)\\ &= z^2 - g(x),\end{aligned}$$

where $g(x) = f(x) + \frac{1}{4}h(x)^2$. Because $\deg g(x) = 2q + 1$, the polynomial $z^2 - g(x)$ also gives a hyperelliptic curve, which is equivalent to the original curve. Thus the polynomial $p(x, y) = y^2 - f(x)$, if it is nonsingular, gives a fully general hyperelliptic curve in a field of odd characteristic. Because of this fact, if q is not a power of two, we can, and usually will, take $h(x) = 0$ by an appropriate change of variables. Then the hyperelliptic curve will be written

$$\mathcal{X} : y^2 = f(x).$$

In contrast, if q is a power of two, an appropriate change of variables will not be available. Then the hyperelliptic curve must be written in the original form

$$\mathcal{X} : y^2 + h(x)y = f(x),$$

where $2 \deg h(x) + 1 \leq \deg f(x)$.

Theorem 11.1.4 *If q is an odd prime or a power of an odd prime, then $\mathcal{X} : y^2 = f(x)$ is a hyperelliptic curve if, and only if, $f(x)$ has no repeated zeros.*

Proof Because we can set $h(x)$ equal to zero without loss of generality, the condition for a nonsingular curve reduces to the statement that the three equations

$$\begin{aligned} y^2 &= f(x) \\ 2y &= 0 \\ 0 &= f'(x) \end{aligned}$$

have no common solution. This reduces further to the statement that $f(x)$ and $f'(x)$ have no common zeros. But $f(x)$ and $f'(x)$ have a common zero if, and only if, $f(x)$ has a repeated zero. Therefore the bivariate polynomial $y^2 - f(x)$ is singular if, and only if, $f(x)$ has a repeated zero. □

An example of a hyperelliptic curve over the field $\boldsymbol{F}_7$ is the curve

$$\mathcal{X} : y^2 + xy = x^5 + 5x^4 + 6x^2 + x + 3.$$

Because $\deg f(x) = 2g + 1$, the genus of $\mathcal{X}$ is two. It is easy to check that the polynomial is nonsingular so the curve is, indeed, a hyperelliptic curve. The affine plane over $\boldsymbol{F}_7$ contains 49 affine points. By direct substitution, one can determine that the rational affine points $(1, 1)$, $(1, 5)$, $(2, 2)$, $(2, 3)$, $(5, 3)$, $(5, 6)$, $(6, 4)$, and no others, satisfy the equation defining $\mathcal{X}$ and the curve has one point at infinity. This leads to the enumeration of the rational points of the hyperelliptic curve over $\boldsymbol{F}_7$ as the set

$$\mathcal{X}(\boldsymbol{F}_7) = \{(1, 1), (1, 5), (2, 2), (2, 3), (5, 3), (5, 6), (6, 4), \infty\}$$

including the one point at infinity, denoted ∞. Thus the curve $\mathcal{X}(\boldsymbol{F}_7)$ consists of eight points of the projective plane. There is no group structure on these points, and so ∞ is not a group identity.

A hyperelliptic curve will usually intersect the line $y = mx + c$ in more than two points, but it intersects the vertical line $x - c = 0$ in either two points or not at all. This is because, with the substitution $x = c$, the polynomial defining the curve becomes a polynomial in y of degree two, which has either two rational solutions or none. Any other line, $y = mx + c$, intersects the hyperelliptic curve in at most $2g + 1$ rational points, which is seen by substituting $mx + c$ for y in the equation of the curve. The x coordinates of the points of intersection can now be found as the zeros of the univariate polynomial

$$p(x) = (mx + c)^2 + h(x)(mx + c) - f(x).$$

This is a univariate polynomial of degree $2g + 1$, so it has at most $2g + 1$ zeros. Indeed, the univariate polynomial has exactly $2g + 1$ zeros, counted with multiplicity, in a sufficiently large extension field. The y coordinate of each zero is then easily computed from the x coordinate as $y = mx + c$.

Definition 11.1.5 *The opposite of point $P = (x, y)$ on the hyperelliptic curve $\mathcal{X}$ is defined as the point $\widetilde{P} = (x, -y - h(x))$. The opposite of the point at infinity is itself, thus $\widetilde{\infty} = \infty$. A special point of $\mathcal{X}$ is defined as a point P for which $\widetilde{P} = P$. An ordinary point of $\mathcal{X}$ is a point of $\mathcal{X}$ that is not a special point.*

The opposite of a point of the curve is always a point of the curve, and is the point itself if that point is a special point. The opposite of the opposite of a point is the point itself. Thus the notion of an opposite ordinarily creates pairs of ordinary points, and it isolates special points.

Recall that our example of a hyperelliptic curve

$$\mathcal{X} : y^2 + xy = x^5 + 5x^4 + 6x^2 + x + 3$$

over the field $\boldsymbol{F}_7$ is given explicitly as

$$\mathcal{X}(\boldsymbol{F}_7) = \{(1, 1), (1, 5), (2, 2), (2, 3), (5, 3), (5, 6), (6, 4), \infty\}.$$

The curve $\mathcal{X}(\boldsymbol{F}_7)$ has only the affine point $(6, 4)$ and the point at infinity as special points and has six ordinary points coming in pairs, with the points in a pair having the same value of x. The left side of Figure 11.1 shows a detailed representation of this curve in the plane $\boldsymbol{F}_7^2$, including the point at infinity. The figure illustrates how the ordinary points come in pairs at certain values of x, while the special point $(6, 4)$ has no mate.

The right side of Figure 11.1 also shows, but now symbolically, a hyperelliptic curve over $\boldsymbol{F}_p$ where p is a large prime, say a 50-digit prime. In this case, the grid is so fine and the points of the curve $\boldsymbol{F}_p^2$ are so tiny that they cannot be resolved. Only a symbolic curve that represents the numerous points of the curve is shown. A better representation might be to show the plane as a shade of gray to represent the unresolved points of the curve $\mathcal{X}(\boldsymbol{F}_p)$. It would be a very light shade of gray because by the Hasse–Weil bound,

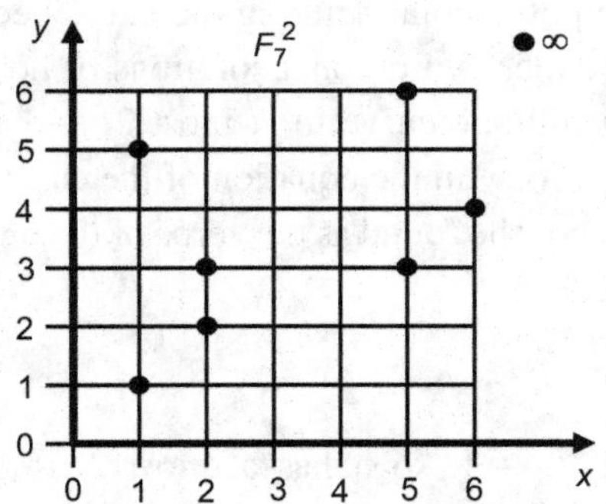

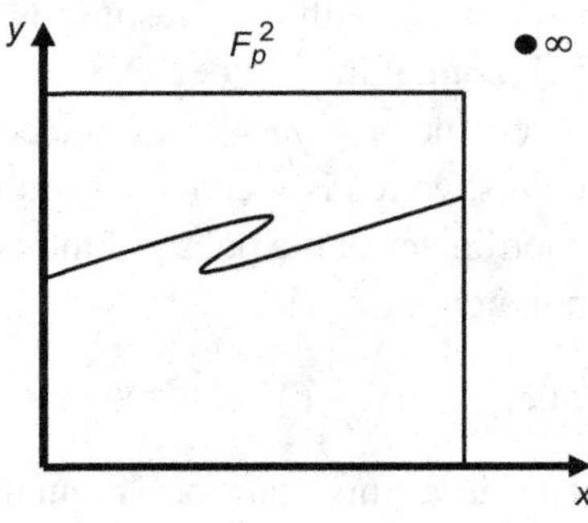

Figure 11.1 Hyperelliptic curves over $\boldsymbol{F}_7$ and $\boldsymbol{F}_p$

if p is a 50-digit prime, only about 10^{50} points are on the curve, while the plane $\boldsymbol{F}_p^2$ contains about $(10^{50})^2$ points.

11.2 Coordinate rings and function fields

The ring consisting of the set of all bivariate polynomials over the field $\boldsymbol{F}$, together with the operations of polynomial addition and polynomial multiplication, is denoted $\boldsymbol{F}[x, y]$ and called the ring of bivariate polynomials over $\boldsymbol{F}$. Associated with every plane curve $\mathcal{X}(\boldsymbol{F})$:$p(x, y) = 0$ is a modular reduction of $\boldsymbol{F}[x, y]$ called a *coordinate ring*. This modular reduction is a quotient ring denoted by $\boldsymbol{F}[x, y]/\langle p(x, y)\rangle$ (or by $\boldsymbol{F}[x, y]_{p(x,y)}$), where $p(x, y)$ is the polynomial defining the plane curve $\mathcal{X}(\boldsymbol{F})$. In the case of a hyperelliptic curve, $p(x, y) = y^2 + h(x)y - f(x)$, and the coordinate ring is denoted $\boldsymbol{F}[x, y]/\langle y^2 + h(x)y - f(x)\rangle$ (or perhaps by $\boldsymbol{F}[x, y]_{y^2+h(x)y-f(x)}$). It is also denoted, more concisely, by the notation $\boldsymbol{F}[\mathcal{X}]$.

The notion of a coordinate ring extends the notion of a function defined on the points of the line, denoted $f(x)$, to a function $f(P)$ defined on the points of the curve $\mathcal{X}$. Each element of the coordinate ring can be visualized as the set of all surfaces $f(x, y) = 0$ that have the same valuation $f(P)$ at all points P of the curve.

A coordinate ring formed from $\boldsymbol{F}[x, y]$ by $y^2 + h(x)y - f(x)$ is formally a set of equivalence classes of $\boldsymbol{F}[x, y]$. Two elements of $\boldsymbol{F}[x, y]$, say $q(x, y)$ and $r(x, y)$, are *equivalent* if $q(x, y) - r(x, y) = a(x, y)(y^2 + h(x)y - f(x))$ for some bivariate polynomial $a(x, y)$. Then $q(P) - r(P) = 0$ for $P \in \mathcal{X}(\boldsymbol{F})$. An equivalence class is the set of all polynomials that are equivalent, and the coordinate ring is the set of all equivalence classes. The process of forming a coordinate ring from $\boldsymbol{F}[x, y]$ using $p(x, y)$ as a modulus is analogous to forming $\boldsymbol{Z}/\langle p\rangle$ from $\boldsymbol{Z}$ using p as a modulus. Just as every element of $\boldsymbol{Z}/\langle p\rangle$ has a unique canonical representative consisting of a nonnegative integer smaller than p, every element of the coordinate ring $\boldsymbol{F}[x, y]/\langle y^2 + h(x)y - f(x)\rangle$ has a unique canonical representative consisting of a bivariate polynomial of degree at most one in y and of arbitrary degree in x. By repeatedly using $y^2 = -h(x)y + f(x)$ to eliminate powers of y of degree two or more, any element of the coordinate ring can be represented as a polynomial of the form $\beta(x, y) = a(x) + b(x)y$. We may regard the canonical representatives themselves to be the elements of the coordinate ring, and so we can suppress mention of the equivalence classes. Addition in the coordinate ring is addition of polynomials. Multiplication in the coordinate ring is multiplication of polynomials modulo $y^2 + h(x)y - f(x)$. This means that whenever the monomial y^2 is formed as a result of multiplying two polynomials, it is eliminated by setting $y^2 = -h(x)y + f(x)$. In summary,

$$F[x, y]_{y^2+h(x)y-f(x)} = \{a(x) + b(x)y \mid a(x), b(x) \in \boldsymbol{F}[x]\}$$

with multiplication of elements reduced by $y^2 = -h(x)y + f(x)$.

The definition of a coordinate ring $\boldsymbol{F}[x, y]/\langle p(x, y)\rangle$ applies for any bivariate polynomial $p(x, y)$. It resembles the method of constructing the monovariate quotient ring $\boldsymbol{F}[x]/\langle p(x)\rangle$, which we know is an extension field whenever $p(x)$ is an irreducible polynomial over $\boldsymbol{F}$. For bivariate polynomials, however, the quotient ring $\boldsymbol{F}[x, y]/\langle p(x, y)\rangle$ is not a field even if $p(x, y)$ is irreducible. This is because the necessary multiplicative inverses do not exist in $\boldsymbol{F}[x, y]/\langle p(x, y)\rangle$. To make this coordinate ring into a field, the multiplicative inverses must be explicitly included. Specifically, one starts with the larger ring, denoted $\boldsymbol{F}(x, y)$, consisting of all rational functions over $\boldsymbol{F}$, and reduces this larger ring modulo $p(x, y)$. This larger quotient ring is denoted $\boldsymbol{F}(x, y)/\langle p(x, y)\rangle$. The elements formally are equivalence classes of rational functions. Each equivalence class has a unique canonical representative, which is a rational function whose numerator is a polynomial over $\boldsymbol{F}$ with y degree at most one, and whose denominator is a nonzero polynomial over $\boldsymbol{F}$, also with y degree at most one. Then every nonzero rational function $f(x, y) = a(x, y)/b(x, y)$ has both $a(x, y)$ and $b(x, y)$ nonzero, and so has inverse $f(x, y)^{-1} = b(x, y)/a(x, y)$. We can conclude that $\boldsymbol{F}(x, y)/\langle p(x, y)\rangle$ is a field, because all the required properties of a field are satisfied. It is called a *function field*.

The coordinate ring $\boldsymbol{F}[x, y]/\langle p(x, y)\rangle$ is contained in the function field $\boldsymbol{F}(x, y)/\langle p(x, y)\rangle$. The function field has the cleaner and more elegant structure, but the coordinate ring has an interesting and useful structure of its own.

There is additional terminology used in conjunction with the coordinate ring of a hyperelliptic curve. Let $\boldsymbol{F}[x, y]_{p(x,y)} = \{a(x) + yb(x)\}$ denote the set of canonical representatives of the coordinate ring. As an element of $\boldsymbol{F}[x, y]/\langle p(x, y)\rangle$, the *degree* of $\beta(x, y) = a(x) + yb(x)$ is defined as

$$\deg[\beta(x, y)] = \max\{2\deg a(x), 2g + 1 + 2\deg b(x)\}.$$

The *conjugate* of the element $\beta(x, y) = a(x) + b(x)y$ is defined as

$$\beta^*(x, y) = a(x) - b(x)(y + h(x)).$$

If $h(x)$ is zero, the conjugate has the attractive form $\beta^*(x, y) = a(x) - b(x)y$. The *norm* of $\beta(x, y)$, viewed as an element of the coordinate ring $\boldsymbol{F}[x, y]/\langle y^2 + h(x)y - f(x)\rangle$, is defined as

$$\text{norm}(\beta(x, y)) = \beta(x, y)\beta^*(x, y).$$

The norm can be expressed as a univariate polynomial in x as follows:

$$\begin{aligned}\text{norm}(\beta(x, y)) &= (a(x) + b(x)y)(a(x) - b(x)(y + h(x))) \\ &= a(x)^2 - a(x)b(x)h(x) + b(x)^2(y^2 + h(x)y).\end{aligned}$$

Because $y^2 + h(x)y = f(x)$ for elements of the coordinate ring, this becomes

$$\text{norm}(\beta(x, y)) = a(x)^2 - a(x)b(x)h(x) - b(x)^2 f(x),$$

which does not depend on y. In this way, the norm is expressed as a polynomial in only one variable. The norm can be useful in studying some questions about polynomials on a plane curve by restating those questions as simpler questions about polynomials in one variable.

11.3 Poles and zeros

A rational function of one variable in the field $\boldsymbol{F}$ is given by

$$f(x) = \frac{a(x)}{b(x)},$$

where $a(x)$ and $b(x)$ are monic polynomials. This monovariate rational function can be described by its poles and zeros, possibly in an extension field. The zeros are the values of x at which $a(x)$ is equal to zero; the poles are the values of x at which $b(x)$ is equal to zero. The poles and zeros determine the function $f(x)$ up to a scalar multiplier, which is equal to one if the rational function is required to be monic.

The situation is different in two variables. A rational function of two variables in the field $\boldsymbol{F}$ is given by

$$f(x, y) = \frac{a(x, y)}{b(x, y)}.$$

This function cannot be described by a finite set of poles and zeros. The zeros of $a(x, y)$ form a plane curve, as do the zeros of $b(x, y)$. A plane curve has an infinite number of zeros, some of them in the ground field $\boldsymbol{F}$ and all of them in the algebraic closure $\overline{\boldsymbol{F}}$. The situation is different, however, in a coordinate ring or a function field.

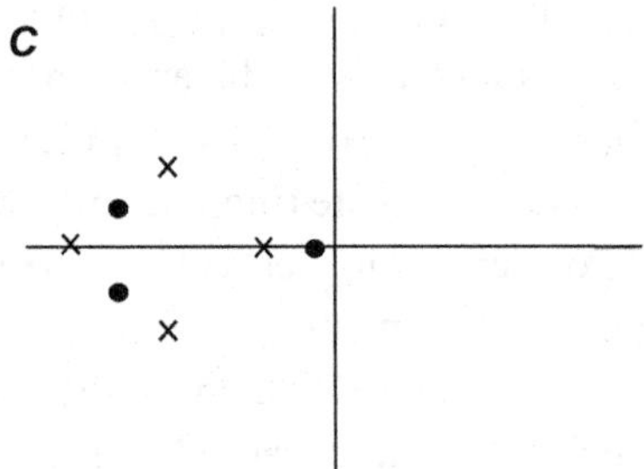

Figure 11.2 Poles and zeros

A familiar procedure that is used to define a univariate polynomial or a univariate rational function over the real field $\boldsymbol{R}$ is to design a pattern of poles and zeros for that rational function in the complex plane, as shown in Figure 11.2. Because the complex conjugate of each pole or zero is also a pole or a zero, respectively, the resulting polynomial has only real coefficients even though the poles and zeros are complex. Thus to get a rational function over the real field $\boldsymbol{R}$, we specify poles and zeros in the complex field $\boldsymbol{C}$, adhering to the appropriate conjugacy relationships. We will want to mimic this procedure in our new situation but it is different in two ways. We want the field to be a finite field $\boldsymbol{F}_q$ rather than the complex field $\boldsymbol{C}$, and we want to construct bivariate polynomials and bivariate rational functions rather than univariate polynomials and univariate rational functions.

The procedure to construct univariate polynomials over the finite field $\boldsymbol{F}_q$ is similar to the procedure over $\boldsymbol{R}$. It is similar, but several distinctions should be noted. One notable difference is that an element of the extension field $\boldsymbol{F}_{q^m}$ will normally have more than one conjugate. Indeed, if β is a zero in $\boldsymbol{F}_{q^m}$, then β will belong to a set with as many as m conjugates in $\boldsymbol{F}_{q^m}$, including β itself. Another distinction is that the algebraic notions do not lend themselves to an informative geometric illustration analogous to Figure 11.2. Perhaps for this reason, one usually describes the poles and zeros in the smallest extension field $\boldsymbol{F}_{q^m}$ that holds all of them, rather than in the algebraic closure $\overline{\boldsymbol{F}}_q$.

A procedure similar to the one above will be developed to specify a bivariate polynomial over the real numbers by specifying a set of two-dimensional complex zeros in the two-dimensional complex plane[2] $\boldsymbol{C}^2$. However, describing this procedure first requires more discussion. The zeros of a bivariate polynomial $a(x, y)$ do not, in general, form a discrete set of points, rather they form a curve. For this reason, one specifies only those zeros of the polynomial that are in common with a fixed polynomial. These common zeros can be thought of as the set of crossings of the curve defined by the polynomial $a(x, y)$ with a fixed reference curve. In our case, the fixed reference curve is the hyperelliptic curve under study. A set of points at which the curve defined

[2] The term "complex plane" traditionally refers to $\boldsymbol{C}$, so we refer to $\boldsymbol{C}^2$ as the two-dimensional complex plane.

by polynomial $a(x, y)$ crosses the hyperelliptic curve defined by polynomial $p(x, y)$ does specify an $a(x, y)$, but not a unique $a(x, y)$. Indeed, if the bivariate polynomial $a(x, y)$ has the required zeros, then $a(x, y) + b(x, y)p(x, y)$ does also. Thus, such a set of zeros on a curve corresponds to an element of the coordinate ring $\boldsymbol{F}[x, y]/\langle p(x, y)\rangle$.

The reverse need not be true. Not every set of zeros with respect to a fixed curve can be associated with a bivariate polynomial or an element of the coordinate ring $\boldsymbol{F}[x, y]/\langle p(x, y)\rangle$. For some sets of points on the fixed curve $\mathcal{X}$, there is no polynomial $a(x, y)$ with zeros on these and only these points of the curve. Consequently, not every pattern of poles and zeros in the two-dimensional complex plane $\boldsymbol{C}^2$ corresponds to a bivariate rational function. Some patterns of poles and zeros on a curve are not consistent with any bivariate rational function over $\boldsymbol{C}$. Similar remarks apply to bivariate polynomials and to bivariate rational functions over $\boldsymbol{F}_q$. A pole–zero pattern that does correspond to a function in the coordinate ring can be called a principal pole–zero pattern.

A conventional way of specifying an equivalence class of bivariate rational functions is by giving the poles and zeros that it has in common with a curve. This is expressed in a formal way by giving a tabulation of the poles and zeros. This tabulation is called the *divisor* of the rational function. The divisor, which is formally defined in the next section, is an integer-valued vector with components indexed by the points of the curve. A positive integer at point P of the curve indicates that $f(x, y)$ has a zero at point P of the curve with the value of the integer giving the order of the zero. A negative integer at point P indicates that $f(x, y)$ has a pole at point P of the curve with the value of the integer giving the order of the pole.

For a monovariate rational function in projective coordinates the divisor is simply the number of poles and zeros at each point. The polynomial

$$f(x) = \frac{a_i x^i + a_{i-1} x^{i-1} + \cdots + a_0}{b_j x^j + b_{j-1} x^{j-1} + \cdots + b_0}$$

can be written in *homogeneous form*, either as

$$f(x, z) = \frac{a_i x^i + a_{i-1} x^{i-1} z + \cdots + a_0 z^i}{b_j x^j z^{i-j} + b_{j-1} x^{j-1} z^{i-j+1} + \cdots + b_0 z^i}$$

if j is smaller than i, or as

$$f(x, z) = \frac{a_i x^i z^{j-i} + a_{i-1} x^{i-1} z^{j-i+1} + \cdots + a_0 z^j}{b_j x^j + b_{j-1} x^{j-1} z + \cdots + b_0 z^j}$$

if j is larger than i. Thus in the homogeneous form, every monomial in both numerator and denominator has the same degree. Affine points of the homogeneous function have the form $(x, 1)$ and the point at infinity is $(1, 0)$. Now, the meaning of a zero or a pole at infinity is clear. There can be a pole of order $i - j$ or a zero of order $j - i$, depending on which of these is nonnegative.

A similar description applies to bivariate functions. The rational function

$$f(x, y) = \frac{a_{ii'}x^i y^{i'} + a_{ii'-1}x^i y^{i'-1} + \cdots + a_{00}}{b_{jj'}x^j y^{j'} + b_{jj'-1}x^j y^{j'-1} + \cdots + b_{00}}$$

can be written in homogeneous form, either as

$$f(x, y, z) = \frac{a_{ii'}x^i y^{i'} + a_{i,i'-1}x^i y^{i'-1}z + \cdots + a_{00}z^{i+i'}}{b_{jj'}x^j y^{j'} z^{i+i'-j-j'} + b_{jj'-1}x^j y^{j'-1} z^{i+i'-j-j'+1} + \cdots + b_{00}z^{i+i'}}$$

if $j + j'$ is smaller than $i + i'$, or as

$$f(x, y, z) = \frac{a_{ii'}x^i y^{i'} z^{j+j'-i-i'} + a_{i,i'-1}x^i y^{i'-1} z^{j+j'-i-i'+1} + \cdots + a_{00}z^{j+j'}}{b_{jj'}x^j y^{j'} + b_{jj'-1}x^j y^{j'-1}z + \cdots + b_{00}z^{j+j'}}$$

if $j + j'$ is larger than $i + i'$. In the homogeneous form of the bivariate rational function, every monomial in the numerator and the denominator has the same degree.

Affine points of the homogeneous bivariate function $f(x, y, z)$ are points of the form $(x, y, 1)$. The points at infinity are those points of the form $(x, 1, 0)$ or $(1, 0, 0)$. Taken together, these points constitute the line at infinity. The point $(1, 0, 0)$ is the point at infinity of the line at infinity.

11.4 Divisors

In order to supply the background needed to construct a group on a hyperelliptic curve, we will require the development of the technical topic of divisors. Before we restate the definition of a divisor on a plane curve, we will recall the definition of a vector indexed by a set $\mathcal{X}$. Given any index set $\mathcal{X}$, a vector $\boldsymbol{v}$ over the field $\boldsymbol{F}$ can be written as a finite formal sum of points $\sum_i \alpha_i P_i$, where the P_i are elements of $\mathcal{X}$ used to index the components of the vector and the α_i are elements of $\boldsymbol{F}$, only a finite number of which are nonzero. The summation in the formal sum does not represent an executable operation. The sum of two such vectors $\boldsymbol{v} = \sum_i \alpha_i P_i$ and $\boldsymbol{v}' = \sum_i \alpha'_i P_i$ is defined as $\sum_i (\alpha_i + \alpha'_i) P_i$. The index set $\mathcal{X}$ could be any set such as the set of integers, or the set of rational points of an elliptic curve or a hyperelliptic curve. The requirement that only a finite number of coefficients are nonzero holds automatically for the rational points of a hyperelliptic curve over a finite field.

The following definition of a divisor should be compared to the definition of a vector. A divisor is like a vector, but the components are integers rather than field elements, and the index set always consists of the points of a curve, possibly only the rational points.

Definition 11.4.1 *A divisor (or Weil divisor) on the plane projective curve $\mathcal{X}$ is a finite formal sum of points of $\mathcal{X}$, denoted $D = \sum_i m_i P_i$, where the m_i are integers and the P_i are points of the curve.*

The divisor is formally defined as a *finite* sum,[3] $D = \sum_i m_i P_i = m_1 P_1 + m_2 P_2 + \cdots + m_n P_n$, where each m_i is an integer, possibly negative and possibly zero. A divisor is simply an assignment of an integer to each of a finite number of points of a curve. The summation sign appearing in a divisor is conventional, and does not represent an executable operation. The plus signs could just as well be replaced by commas. Because a divisor is a finite sum, only a finite number of m_i can be nonzero.

The addition of the two divisors $D = \sum_i m_i P_i$ and $D' = \sum_i m_i' P_i$ is defined as

$$D + D' = \sum_i (m_i + m_i')P_i.$$

The summation $m_i + m_i'$ in this equation is a summation of integers, but the summation $\sum_i$ is the formal sum corresponding to the notation defining a divisor. Two divisors are added by taking the union of the sets of points with nonzero integer coefficients, and adding the integer coefficients on any points that belong to both sets, possibly $m_i + m_i' = 0$ for some i. With this definition of addition, the set of divisors forms an abelian group over the points of the curve $\mathcal{X}$. (A group formed in this particular way on any set is called a *free abelian group* on that set.) From a computational point of view, a divisor is a finite set of points of $\mathcal{X}$, together with an integer associated with each of those points. The set of all divisors indexed by the points of $\mathcal{X}$ will be denoted $\mathcal{D}$.

We cannot define point addition on a hyperelliptic curve to form a finite group, but we will persist and instead try to define addition on the set of divisors on the hyperelliptic curve so as to form a finite group. Because two divisors can be added, the set of divisors resembles what we want to define, but not quite. Because the coefficients are arbitrary integers, the set of divisors on $\mathcal{X}$ is infinite, and we want a finite group. Our next step is to consolidate the set of divisors by forming equivalence classes by a kind of modular reduction analogous to the modular reduction used to form $\mathbf{Z}_p$ from

[3] There is an unfortunate ambiguity in notation here. Within a divisor, the symbol + is not an executable operation. The symbol + means "and" in the sense of "together with," while in point addition, the symbol + means "and" in the sense of "plus." The notation $m_1 P_1 + m_2 P_2$ may mean "point P_1 with multiplicity m_1 together with point P_2 with multiplicity m_2," or it may mean "multiply P_1 by the integer m_1, multiply P_2 by the integer m_2, and add." Both usages of + are well established, and the context resolves the ambiguity. Furthermore, a plus sign between divisors is executable $(m_1 P_1 + m_2 P_2) + (m_1' P_1 + m_2' P_2) = (m_1 + m_1')P_1 + (m_2 + m_2')P_2$. The sums inside the parentheses are executable, the one between is not. To resolve this ambiguity explicitly when the context is not sufficient, the notation $[m_1]P_1 + [m_2]P_2$ is used to express point addition, and $m_1(P_1) + m_2(P_2)$ is used to express a divisor. Thus, in computations, the expression $m_1(aP_1 + bP_2) + m_2(cP_1 + dP_2)$ should have a clear meaning as a divisor consisting of the two points, $[a]P_1 + [b]P_2$ and $[c]P_1 + [d]P_2$, but $m_1([a]P_1 + [b]P_2) + m_2([c]P_1 + [d]P_2)$ is more certain, though more burdensome. To this purpose, the central plus sign could be read as a comma. Moreover, the notation $[a + b]P$ means the point P multiplied by $a + b$, whereas the notation $(a + b)P$ means point P with multiplicity $(a + b)$.

$\mathbf{Z}$. This means that we need to define a method of reducing one divisor modulo another divisor. Before we give this definition, we will first refine and expand our terminology regarding divisors.

The *support* of divisor D, denoted sup D, is the set of points at which m_i is nonzero. An *effective divisor* is one for which all m_i are nonnegative. The *norm* of divisor D, denoted $|D|$, is defined as $|D| = \sum_{i \neq \infty} |m_i|$. The *degree* of divisor D, denoted deg D, is defined as $\deg D = \sum_i m_i$. The degree defines a map $\mathcal{D} \to \mathbf{Z}$, assigning an integer to each divisor.

The points of the curve at which a rational function $h(x, y)$ is to be evaluated can be designated by reference to a divisor D. If the divisor has coefficient m_i at point P_i, then the *evaluation* of the rational function $h(x, y)$ at the point $P_i = (\beta, \gamma)$ is defined as $h(P_i)^{m_i} = h(\beta, \gamma)^{m_i}$. More generally, the evaluation of the rational function $h(x, y)$ on the entire divisor $D = \sum_i m_i P_i$ is defined as $h(D) = \Pi_i h(P_i)^{m_i}$.

A divisor with degree zero is one for which $\sum_i m_i = 0$. This is often expressed as $m_\infty = -\sum_{i \neq \infty} m_i$, and the divisor with degree zero is written

$$D = \sum_i m_i P_i - (\sum_i m_i)\infty$$

where now the first sum is over only affine points.[4] When dealing only with divisors of degree zero, one sometimes omits explicit mention of m_∞ because then it is implied. One may also write a divisor of degree zero as $D = \sum_i m_i P_i - (*)\infty$, where $(*)$ is shorthand for $\sum_i m_i$, or, more briefly, as $D = \sum_i m_i P_i$, mentioning only the affine points and including the term $(*)\infty$ by implication. The set of all divisors with degree zero, denoted $\mathcal{D}^\circ$, is a subgroup of $\mathcal{D}$.

The *greatest common divisor*, denoted GCD(D, D'), of the two divisors $\sum_i m_i P_i$ and $\sum_i m'_i P_i$ is the zero-degree divisor

$$\text{GCD}(D, D') = \sum_i \min(m_i, m'_i) P_i - \left[\sum_i \min(m_i, m'_i)\right]\infty.$$

The greatest common divisor of two divisors is always an element of $\mathcal{D}^\circ$.

Each divisor of $\mathcal{D}^\circ$ can be identified with a finite set of points of the projective curve whose integer coefficients sum to zero. The set $\mathcal{D}^\circ$ is often described in a more formal way by considering the function that maps each divisor into its degree. Then $\mathcal{D}^\circ$ can be defined as the *kernel* of the degree map. The kernel of the degree map is the set of

[4] The notation here is slightly garbled. It might be clearer written as

$$D = \sum_i m_i P_i + \left(-\sum_i m_i\right)\infty.$$

The second summation sign denotes an integer addition. The first summation is a formal (nonexecutable) sum.

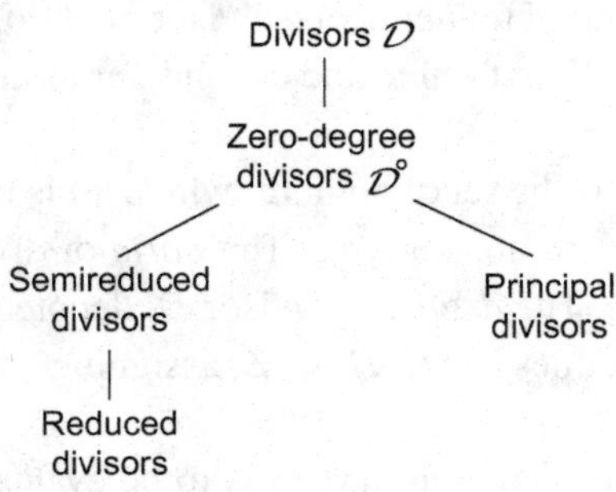

Figure 11.3 A taxonomy of divisors

divisors that map to degree zero under the degree map, so the alternative statement is just a more elegant way of saying the same thing.

The set of all divisors $\mathcal{D}$ is too large for our needs because it contains an infinite number of divisors. Even the smaller set $\mathcal{D}^\circ$ is too large for our needs. It, too, contains an infinite number of divisors. We must further reduce the set in order to get a finite set. In Sections 11.6 and 11.7, we will see that the reduced set that we will define becomes an abelian group under the appropriate definition of group operation. But first, in this section, we must define a reduced divisor, and a group operation on the set of reduced divisors. Accordingly, the set of reduced divisors will be called the *group of reduced divisors*, or the *jacobian*, and the group operation will be called *addition of reduced divisors*, or *jacobian addition*.

We will restrict the set of divisors in two steps as shown on the left side of Figure 11.3. First, we restrict the set of divisors of degree zero to what is called the set of semireduced divisors. These are defined in terms of the special points and the opposite of a point that were introduced in Definition 11.1.5. Then we further restrict the set of semireduced divisors to the set of reduced divisors. The set of reduced divisors, when given an appropriate group operation, is the desired group defined on a hyperelliptic curve.

Definition 11.4.2 *A semireduced divisor of a hyperelliptic curve is an element of $\mathcal{D}^\circ$ satisfying the first three of the following conditions: a reduced divisor of a hyperelliptic curve is an element of $\mathcal{D}^\circ$ satisfying all four of the conditions.*

(1) All m_i with $i \neq \infty$ are nonnegative.
(2) If P_i is a special point, then $m_i \leq 1$.
(3) A point and its opposite do not both have a nonzero coefficient.
(4) The sum of m_i over all affine points is not larger than the genus g.

Because all m_i are nonnegative, a reduced divisor can be defined as a semireduced divisor whose norm is at most equal to g. It is clear that the set of reduced divisors on $\mathcal{X}(\boldsymbol{F}_q)$ is a finite set because $\mathcal{X}(\boldsymbol{F}_q)$ consists of a finite set of points, and no integer coefficient of any point can be negative or larger than g. If the genus g is larger than one, then each reduced divisor can contain more than one point, but not more than g points.

For an elliptic curve, however, $g = 1$, so at most one m_i can be equal to one, and all other m_i are equal to zero. Thus for an elliptic curve, each nonempty reduced divisor is a set containing one point of the elliptic curve. A reduced divisor then is equivalent to a point of $\mathcal{X}$. The set of reduced divisors of an elliptic curve is equivalent to the elliptic curve itself. Thus the set of reduced divisors on a hyperelliptic curve can be seen as another generalization of an elliptic curve.

Reduced divisors can be added just as any divisors can be added, but the sum of two reduced divisors, in general, is not a reduced divisor. To obtain a group structure on the set of reduced divisors, we must now introduce a modular reduction. This is a method of reducing any semireduced divisor to an equivalent reduced divisor. First we must introduce the notion of a principal divisor.

11.5 Principal divisors

One way to obtain a divisor of degree zero on the points of a curve $\mathcal{X}$ over $\boldsymbol{F}$ is by using the poles and zeros of a rational function on the curve to specify the points of the divisor and to use the order of the poles and zeros to specify the integer coefficients of the divisor. A *rational function* on the field $\boldsymbol{F}$ is an expression of the form $H(x, y) = F(x, y)/G(x, y)$, where $F(x, y)$ and $G(x, y)$ are polynomials in $\boldsymbol{F}[x, y]$. We require that any common polynomial factor be cleared from the numerator and denominator and that the polynomials $F(x, y)$ and $G(x, y)$ are each not divisible by the polynomial defining the curve $\mathcal{X}$. If the numerator $F(x, y)$ equals zero at P, then P is called a *zero* of the rational function $H(x, y)$. If the denominator $G(x, y)$ equals zero at P, then P is called a *pole* of the rational function $H(x, y)$. By Bézout's theorem, $F(x, y)$ and $G(x, y)$ can each have only a finite number of zeros on $\mathcal{X}$. Therefore $H(x, y)$ can have only a finite number of poles and zeros on the curve $\mathcal{X}$.

It is possible for $F(x, y)$ and $G(x, y)$ to each have a zero at the same point P of the curve $\mathcal{X}$. Then $H(x, y)$ has both a zero and a pole at P. This cannot happen for a rational function $H(x) = F(x)/G(x)$ in one variable over any field $\boldsymbol{F}$ because a common zero implies a common factor of degree one, and we require that any common factor of $F(x)$ and $G(x)$ be canceled. However, the behavior is different for polynomials in two variables. A point of a curve can be both a zero and a pole of a rational function in two variables because $F(x, y)$ and $G(x, y)$ can have a common zero at P and yet not have a common polynomial factor in the field $\boldsymbol{F}$ or in any extension field of $\boldsymbol{F}$. In this case, the value of the rational function, as such, is not defined at the point P.

A *valuation*,[5] denoted v_P or $v_P(H)$, of $H(x, y)$ at a point P of the curve is defined as the number of zeros at P minus the number of poles at P. If $H(x, y)$ has only a zero of order n at P, then $v_P(H) = n$. If $H(x, y)$ has only a pole of order n at P, then $v_P(H) = -n$. If $H(x, y)$ has a zero of order n at P and a pole of order m at P, then $v_P(H) = n - m$.

The rational function $H(x, y)$ can be used to define a divisor on the hyperelliptic curve $\mathcal{X}$, denoted $\mathrm{div}H(x, y)$ or $\mathrm{div}\, H$, and given by

$$\mathrm{div}\,(H(x, y)) = \sum_P v_P(H)P - (*)\infty,$$

where now the indicated summation is only over the rational affine points and $(*)$ represents an integer that is given by the number of zeros at infinity minus the number of poles at infinity. It can be difficult to determine the number of zeros or poles at infinity, but if we accept the statement that every divisor constructed in this way from a rational function must have degree zero, then we can write $(*) = \sum_P v_P(H)$. This means that $\sum_P v_P(H) - (*) = 0$. For a divisor defined in this way, the integer coefficient, denoted $(*)$, does correspond to the number of poles minus the number of zeros of the point at infinity.[6] For our purposes, we simply define $(*)$ as the total number of affine poles minus the total number of affine zeros, suffering only a loss of eloquence, but not of rigor.

The order of a pole or a zero of a bivariate rational function is a generalization of the order of a pole or a zero of a monovariate rational function, but it is tricky to define the order in the bivariate case. It does correspond to the number of derivatives of the numerator or denominator of the rational function with respect to the curve that are equal to zero, but the notion of the derivative of one curve with respect to another curve crossing it or touching it is not straightforward to define. It is, however, an intuitively clear notion, and this intuition is probably enough for our needs.

One way to determine the order of a zero at the point P of the rational function $H(x, y)$ is to use the equation of the curve $p(x, y) = 0$ to algebraically manipulate $H(x, y)$ into the form

$$H(x, y) = A(x, y)^r B(x, y)$$

such that $B(P) \neq 0$ and $A(P) = 0$. Then r is the order of the zero at P.

For example, given the curve $\mathcal{X} : y^2 = x^3 - x$, does the rational function $f(x, y) = x/y$ have a zero at $(x, y) = (0, 0)$ and, if so, what is its order? The expression $f(x, y) = x/y$ is undefined as written at $(0, 0)$, but we can use the equation of the curve to rewrite it as

$$f(x, y) = \frac{x}{y} = \frac{y}{x^2 - 1},$$

[5] A valuation is not to be confused with an evaluation.

[6] This can be so simplified because a hyperelliptic curve has only one point at infinity.

from which we see that $f(0,0) = 0$. To find the order of this zero, rewrite $f(x, y)$ as

$$f(x, y) = y\frac{1}{x^2 - 1} = A(x, y)^r B(x, y).$$

The second term corresponding to $B(x, y)$ is nonzero at $(0, 0)$, and the first term y corresponding to $A(x, y)$ has exponent one, so $(0, 0)$ is a zero of order one.

For a second example, given the same curve, the rational function $f(x, y) = x$ has an obvious zero at $(x, y) = (0, 0)$. To see the order of the zero at $(0, 0)$ for this rational function, use the equation of the curve $y^2 = x^3 - x$ to write

$$f(x, y) = x = y^2\frac{1}{x^2 - 1} = A(x, y)^r B(x, y).$$

Again, the term corresponding to $B(x, y)$ is nonzero at $(0, 0)$, and the term corresponding to $A(x, y)$ has exponent two, so for this rational function, $(0, 0)$ is a zero of order two.

Any divisor that can be obtained from the poles and zeros of a rational function on a curve $\mathcal{X}$ in this way always has degree zero because the coefficient of ∞ ensures that the sum is zero. Moreover, it is given a special status among divisors of degree zero as follows.

Definition 11.5.1 *A divisor that corresponds to the poles and zeros, counted with multiplicity, of a rational function on the curve $\mathcal{X}$ is called a principal divisor of $\mathcal{X}$.*

There are several natural questions that arise. Is every divisor of degree zero a principal divisor? Can a principal divisor correspond to more than one (nonequivalent) monic rational function on $\mathcal{X}(\boldsymbol{F}_q)$? In the next section we will answer both of these questions negatively for elliptic curves, provided that all poles and zeros in the algebraic closure $\overline{\boldsymbol{F}}_q$ are considered. This is an important conclusion because it tells us that it is enough to specify all poles and zeros of a rational function in the algebraic closure of $\boldsymbol{F}$. The rational function on the curve is then fully determined except for a scalar multiplier. If there are no poles, then the set of zeros uniquely determine a polynomial $a(x, y)$ modulo $p(x, y)$. That is, the zeros determine a unique element of the coordinate ring $\boldsymbol{F}[x, y]_{p(x,y)}$.

11.6 Principal divisors on elliptic curves

Although this chapter, in general, deals with hyperelliptic curves, this section returns to the special case of hyperelliptic curves of genus one. These are the elliptic curves. The study of principal divisors for elliptic curves both illustrates the properties of divisors and gives deeper understanding of elliptic curves.

It is easy to give an example of a principal divisor on an elliptic curve because a straight line that intersects an elliptic curve $\mathcal{X}(\boldsymbol{F}_q)$ at least twice always intersects it at three points, counting multiple intersections as such. Thus the elementary rational function $\ell(x, y) = ax + by + c$ for which the denominator is equal to one, is equal to zero at three points of the elliptic curve $\mathcal{X}$, denoted P_1, P_2, and P_3. Therefore the principal divisor is $\text{div}(\ell(x, y)) = P_1 + P_2 + P_3 - 3\mathcal{O}$. The point at infinity must have a pole of order three because a principal divisor always has degree zero.

Similarly, the vertical line through $P_3 = (x_3, y_3)$ and $-P_3 = P_1 + P_2$ is given by the zeros of the trivial rational function $x - x_3$. The vertical line through the point P_3 is $v(x, y) = x - x_3$. The divisor of the vertical line $x - x_3$ is $\text{div}(x - x_3) = P_3 + (-P_3) + (-2)\mathcal{O}$ where the coefficient of minus two[7] is needed so that the degree of the divisor is zero. From these two divisors, we can conclude that

$$\begin{aligned} \text{div}\left(\frac{ax + by + c}{x - x_3}\right) &= \text{div}(ax + by + c) - \text{div}(x - x_3) \\ &= (P_1 + P_2 + P_3 - 3\mathcal{O}) - (P_3 + (P_1 + P_2) - 2\mathcal{O}) \\ &= P_1 + P_2 - (P_1 + P_2) - \mathcal{O}. \end{aligned}$$

This expression is the divisor of an elementary rational function so it is a principal divisor. We will see in time, that with an appropriate qualification, not only does the elementary function on the left determine the divisor on the right, but the divisor on the right also specifies the rational function on the left. The qualification is that the rational function is determined up to a scalar multiple.

We can also relate the divisors in this equation by

$$P_1 + P_2 = (P_1 + P_2) + \mathcal{O} + \text{div}\left(\frac{ax + by + c}{x - x_3}\right)$$

which is an equality relating divisors. The divisor on the left, however, is not a principal divisor. It does not correspond to any rational function, as asserted by Bézout's theorem. Therefore the right side does not describe a principal divisor.

Not every divisor is a principal divisor. For some divisors, there is no rational function whose pattern of poles and zeros on the elliptic curve $\mathcal{X}$ corresponds to that divisor. An example of a divisor that is not a principal divisor is an easy consequence of the following theorem. It is enough to prove this theorem for canonical rational functions. A *canonical rational function* on an elliptic curve is a rational function with monic polynomials in both the numerator and denominator, each with y degree at most one.

Theorem 11.6.1 *No rational function can have only one or two rational affine zeros or poles on the elliptic curve $\mathcal{X}(\boldsymbol{F})$ except for the two zeros or two poles lying on a vertical line.*

[7] To reconcile the single zero at ∞ with the -2 coefficient in the divisor at ∞, we regard ∞ to have three poles in addition to one zero, but omit the technical explanation.

Proof Let $y^2 + h(x)y = f(x)$ denote the equation of the elliptic curve, noting that if (β, γ) is a point of this curve, then $(\beta, -(\gamma + h(\beta)))$ is also a point of the curve. Let $f(x, y) = h(x, y)/g(x, y)$ be any rational function on the field $\boldsymbol{F}$. Then $h(x, y) = a(x) + yb(x)$. Let $P = (\beta, \gamma)$ be a zero of $h(x, y)$. Then $h(\beta, \gamma) = a(\beta) + \gamma b(\beta) = 0$. If the polynomial $b(x)$ is identically zero, then $h(\beta, \gamma) = a(\beta) = 0$ for all y including $y = \gamma$ and $y = -(\gamma + h(\beta))$ so $h(x, y)$ has more than one zero on the curve.[8] If the polynomial $b(x)$ is not identically zero, and $\boldsymbol{F}$ is not a binary field, then $h(x) = 0$ and $h(\beta, -\gamma) = 0$. If $\boldsymbol{F}$ is a binary field, then $h(x) \neq 0$ and $h(\beta, \gamma + h(\beta)) = 0$. The same conclusion follows, so $h(x, y)$ cannot have only one rational affine zero. By similar reasoning $f(x, y)$ cannot have only one rational affine pole.

Now, suppose that $f(x, y)$ has an affine divisor $D = P_1 + P_2$ with two rational points. Define $\widetilde{f}(x, y) = f(x, y)/(ax + by + c)$, where $ax + by + c$ is the equation of a line that passes through P_1 and P_2, so it passes through a third point P_3. Therefore $\widetilde{f}(x, y)$ is a rational function with the affine divisor $\widetilde{D} = -P_3$. But we have already shown that such a divisor does not exist. Therefore a principal divisor of the form $D = P_1 + P_2 + (-2)\infty$ with $P_2 \neq -P_1$ does not exist either unless $P_2 = -P_1$. Similar statements apply to $g(x, y)$, so we can conclude that the statement of the theorem holds. □

This line of thought can be repeated to find certain other divisors of larger degree that also cannot be principal divisors. However, instead of continuing in this way, which would be clumsy, we state and prove the general conclusion as a theorem.

Theorem 11.6.2 *A divisor $\sum_i a_i P_i$ of the elliptic curve $\mathcal{X}(\boldsymbol{F})$ is a principal divisor of $\mathcal{X}(\boldsymbol{F})$ if, and only if, the following two conditions are satisfied:*

(i) $\sum_i a_i = 0$ and,
(ii) as a sum of points, $\sum_i [a_i] P_i = \mathcal{O}$.

Proof The first condition, $\sum_i a_i = 0$, consists of an integer addition. By the definition of a principal divisor, it is a requirement for a principal divisor on any hyperelliptic curve.

The second condition refers to the operation of point addition, so it applies only to elliptic curves. Let $f(x, y)$ be a rational function with m zeros and m' poles, and with divisor D. We will change $f(x, y)$ to another rational function by removing any three zeros that lie on a line. To find three zeros on a line, we may need to create a pole–zero pair at some third point. Suppose that $f(x, y)$ is a rational function on $\mathcal{X}(\boldsymbol{F})$ with two zeros lying on the line $\ell(x, y)$ that intersects the elliptic curve at a third point. If $f(x, y)$

[8] Bézout's theorem states that there are $3\deg[a(x) + yb(x)]$ common zeros, but these need not be rational affine zeros.

has two zeros on the line $\ell(x, y)$ then the function $f'(x, y) = \frac{f(x,y)}{\ell(x,y)}$ has these two zeros removed, and has a new pole that $f(x, y)$ does not have, and has all the other zeros and poles of $f(x, y)$. Moveover, if $f(x, y)$ has only two zeros on the line $\ell(x, y)$, then

$$f'(x, y) = \frac{f(x, y)}{\ell(x, y)}$$

does have these two zeros $\ell(x, y)$ removed but has one new pole at the third zero of $\ell(x, y)$. The total number of zeros and poles is reduced by one. In either case $f'(x, y)$ has fewer total zeros and poles than does $f(x, y)$. Alternatively, if $f(x, y)$ has at least two poles, a similar reduction can be made, introducing one new zero. This process can be repeated multiple times to reduce the number of poles or zeros as long as the function has at least two zeros or two poles. The process must terminate with a function that has no zeros or poles, because by Theorem 11.6.1, it can neither terminate on a function with only one or two zeros nor on a function with only one or two poles, and it does not terminate on a function with more than two zeros or poles.

To finish the proof, notice that the value of the sum $\sum_i [a_i]P_i$ does not change at any step of the process, and that sum $\sum_i [a_i]P_i$ terminates with the value $\mathcal{O}$, so it must have been equal to $\mathcal{O}$ to start with. □

Our next task is to show that the canonical representative of a rational function on an elliptic curve is defined by its poles and zeros up to a constant multiplier. The statement that appears in the next theorem closely resembles the familiar statement that every rational function in one variable can be expressed in the form

$$f(x) = e\frac{\prod_\ell (a_\ell x + 1)}{\prod_{\ell'} (c_{\ell'} x + 1)},$$

where a_ℓ and c_ℓ are elements of the algebraically closed field $\overline{\boldsymbol{F}}$, and if $f(x)$ is monic, then $e = \prod_{\ell'} c_{\ell'} / \prod_\ell a_\ell$.

We have now arrived at the following statement.

Theorem 11.6.3 *Every rational function on an elliptic curve $\mathcal{X}(\overline{\boldsymbol{F}})$ on an algebraically closed field $\overline{\boldsymbol{F}}$ is equivalent to a function written as*

$$f(x, y) = e\frac{\prod_\ell (a_\ell x + b_\ell y + 1)}{\prod_{\ell'} (c_{\ell'} x + d_{\ell'} y + 1)},$$

where all constants are elements of $\overline{\boldsymbol{F}}$.

Proof Every rational function has a finite set of poles and zeros on $\mathcal{X}(\overline{\boldsymbol{F}})$, by Bézout's theorem, and corresponds to a unique principal divisor. Thus a rational function of the stated form will be found by reasoning from the principal divisor.

The total number of zeros of $f(x, y)$ can be reduced by at least two, possibly introducing one new pole, by defining the new function

$$g(x, y) = f(x, y)/(ax + by + 1)$$

for appropriate field elements a and b. Similarly, the total number of poles of $f(x, y)$ can be reduced by at least two, possibly introducing one new zero, by defining the new function

$$g(x, y) = f(x, y)(cx + dy + 1)$$

for appropriate field elements c and d.

This reduction can be repeated until there are no remaining zeros and no remaining poles. But any rational function with no zeros or poles is a constant. Thus the iterations terminate only when

$$f(x, y)\frac{\prod_{\ell'}(c_{\ell'}x + d_{\ell'}y + 1)}{\prod_{\ell}(a_{\ell}x + b_{\ell}y + 1)} = e,$$

which is equivalent to the statement to be proved. □

The representation given in the theorem need not be unique as written, but it will be unique after the numerator and denominator are multiplied out and reduced by the equation of the curve. Then the canonical representative of the original rational function is recovered. Stated more formally in the language of Section 11.2, the representation given in the theorem need not be unique as written, but it belongs to a unique element of the function field, and so corresponds to the unique canonical representative of that element of the function field.

This definition allows us to state the following corollary.

Corollary 11.6.4 *A principal divisor of the elliptic curve $\mathcal{X}(\boldsymbol{F}_q)$ corresponds uniquely to a canonical rational function on $\mathcal{X}(\boldsymbol{F}_q)$ having only rational poles and zeros.*

Proof Follows directly from the expression of the theorem because each line intersects the elliptic curve only at rational points. □

We close the section with the elegant Weil reciprocity theorem, which says that an appropriate product of the evaluations of $f(x, y)$ on the poles and zeros of $g(x, y)$ is equal to the corresponding product of the evaluations of $g(x, y)$ on the poles and zeros of $f(x, y)$. By limiting the statement to elliptic curves, the ideas of this section can be used in the proof.

Theorem 11.6.5 (Weil reciprocity) *Let $D = \operatorname{div} f(x, y)$ and $D' = \operatorname{div} g(x, y)$ be two disjoint divisors for the elliptic curve $\mathcal{X}(F)$ with no common points. Then $f(D') = g(D)$.*

Proof The proof is in two steps.

Step 1 The theorem is easy to prove if $f(x, y)$ and $g(x, y)$ are simply the equations of straight lines. First consider a vertical line and a horizontal line through the origin. Then the two polynomials are trivial and given by $f(x, y) = y$ and $g(x, y) = x$. The zeros that $f(x, y)$ has in common with $y^2 + xy = x^3 + ax^2 + bx + c$ are at the points where $y^2 = c$. Therefore the three points of intersection are $(0, \sqrt{c})$, $(0, -\sqrt{c})$, and $\mathcal{O}$. The zeros of $g(x, y)$ in common with $y^2 + xy = x^3 + ax^2 + bx + c$ are at the zeros of $x^3 + ax^2 + bx + c$. Therefore the three points of intersection are $(x_1, 0)$, $(x_2, 0)$, $(x_3, 0)$ where x_1, x_2, and x_3 are the zeros of the polynomial $x^3 + ax^2 + bx + c$. Moreover, any two lines can be converted to one of these two cases by a coordinate conversion, possibly nonorthogonal. Because an elliptic curve remains an elliptic curve under a coordinate conversion, the conclusion holds for $f(x, y) = ax + b$ and $g(x, y) = bx + c$. Then

$$\begin{aligned} f(D') &= f(x_1, y_1) f(x_2, y_2) f(x_3, y_3) \\ &= f(P_1') f(P_2') f(P_3') \\ &= x_1 x_2 x_3 = c \end{aligned}$$

and

$$\begin{aligned} g(D) &= g(P_1) g(P_2) g(P_3) \\ &= -\sqrt{c}\sqrt{c} = c. \end{aligned}$$

A similar proof holds for two lines parallel to the x axis. Moreover, any two lines can be converted to one of these two cases by a (possibly nonorthogonal) coordinate conversion. Because an elliptic curve remains an elliptic curve under a coordinate conversion, the conclusion holds for $f(x, y) = ax + b$ and $g(x, y) = bx + c$.

Step 2 The proof is now completed by referring to Theorem 11.6.3, which says that every rational function on an elliptic curve is equivalent to a product of lines, either in the numerator, in the denominator, or both. □

11.7 Jacobians as quotient groups

We now return to the study of hyperelliptic curves of arbitrary genus, emphasizing those hyperelliptic curves that are not elliptic curves and so have a genus larger than one. For a hyperelliptic curve of genus larger than one, we do not have a succinct theorem like Theorem 11.6.2 to help us recognize a principal divisor.

Let $\overline{D}$ denote a principal divisor on the hyperelliptic curve $\mathcal{X}$ and let $\overline{\mathcal{D}}$ denote the set of principal divisors on $\mathcal{X}$. Because for any two rational functions $H_1(x, y)$ and $H_2(x, y)$ the divisor of the product satisfies $\text{div}(H_1 H_2) = \text{div}(H_1) + \text{div}(H_2)$, we conclude that $\overline{\mathcal{D}}$ is a subgroup of $\mathcal{D}^\circ$, the set of divisors of degree zero. Moreover, it is a proper subgroup because not every divisor of degree zero is a principal divisor. This means that the quotient group $\mathcal{D}^\circ/\overline{\mathcal{D}}$ is defined and is nontrivial. In particular, a divisor of degree zero is equivalent to the all-zero divisor, if and only if it is a principal divisor.

We now have a formal definition of the jacobian as follows.

Definition 11.7.1 *The jacobian of the hyperelliptic curve $\mathcal{X}(\boldsymbol{F}_q)$, denoted* $\text{jac}(\mathcal{X})$ *or* $\text{jac}(\mathcal{X}(\boldsymbol{F}_q))$, *is the quotient group $\mathcal{D}^\circ/\overline{\mathcal{D}}$.*

Thus, the jacobian regards all principal divisors as comprising a single equivalence class that is equivalent to the all-zero divisor. All divisors that differ from each other by a principal divisor are regarded as equivalent, and so belong to the same equivalence class. These equivalence classes are also called by the more general term of *divisor classes*,[9] and together the divisor classes form a *divisor class group*, which for hyperelliptic curves is just another name for the jacobian, but it is the standard name for a more general group.

If $\mathcal{X}$ is an elliptic curve, then $\text{jac}(\mathcal{X})$ and the curve $\mathcal{X}$ are isomorphic because in the case of an elliptic curve, there is exactly one point P of $\mathcal{X}$ in each element $\{P\}$ of $\text{jac}(\mathcal{X})$. In this case, the jacobian is obviously a finite abelian group.

If the genus is larger than one, then the elements of $\text{jac}(\mathcal{X})$ are nontrivial equivalence classes of divisors. The elements of $\text{jac}(\mathcal{X})$ consist of finite sets of points of $\mathcal{X}$ with an integer weight m_i associated with each point. Each element of $\text{jac}(\mathcal{X})$ is a finite list of points of $\mathcal{X}$, with point P having a weight m and written $m(P)$. The sum of two elements D and D' of $\text{jac}(\mathcal{X})$ is defined as

$$\begin{aligned} D + D' &= \sum_i m_i P_i + \sum_i m_i' P_i \\ &= \sum_i (m_i + m_i') P_i \pmod{\overline{\mathcal{D}}}. \end{aligned}$$

The modulo $\overline{\mathcal{D}}$ operation can be executed simply by gathering any collection of terms of the sum $D + D'$ forming a principal divisor $\overline{D} \in \overline{\mathcal{D}}$, then setting that collection of terms equal to zero.

[9] A divisor class, which is a set of polynomials, should not be confused with a divisor, which is a set of poles and zeros.

Thus, although the points of $\mathcal{X}$ cannot be made into a group under any definition of a group operation, the elements of jac($\mathcal{X}$) can be made into a group by a suitable definition of a group operation, namely modular addition of divisors.

The jacobian jac($\mathcal{X}$) is a finite group, which is what is desired, so from a formal point of view, we are finished. However, we are not finished from a computational point of view because, to perform computations, we need a way to choose a principal divisor so that the modular reduction can be executed. This will require more definitions.

11.8 The group of a hyperelliptic curve

Let us introduce the jacobian in another and more practical way. A hyperelliptic curve with genus one is an elliptic curve and so it has a natural group structure on its points, as was studied in Chapter 10. In contrast, if the genus is larger than one, then the hyperelliptic curve has no natural group structure on its points. Point addition cannot be defined on the points of the curve to form a group. In particular, the method used for elliptic curves does not work here because a line can intersect a hyperelliptic curve at more than three points, and no other construction can be found to define addition of points, as such, to make this set into a group. With a bit of tenacity however, we enlarge the setting by embedding the hyperelliptic curve $\mathcal{X}$ into a larger set that does admit a group structure. We might attempt to do this by replacing each point of $\mathcal{X}(\boldsymbol{F}_q)$ by a set of points. Thus replace the point $P \in \mathcal{X}$ by the set $\{P\}$ that contains only the single point P. Then one might attempt to define point addition as a set union $\{P_1\} + \{P_2\} = \{P_1, P_2\}$. However, this definition would not work because $\{P_1\} \cup \{P_1, P_2\} = \{P_1, P_2\}$, which leads to the conclusion that this operation cannot be made into a group operation. Indeed, there would be no concept of the negative of $\{P\}$. A second attempt might define addition such that $\{P_1\} + \{P_1, P_2\} = \{2P_1, P_2\}$ and define negatives such that $-\{aP_1, bP_2\} = \{(-a)P_1, (-b)P_2\}$. This concept is almost what we need, but still it is not quite right. For one thing, the number of such sets grows to infinity, whereas we require a finite group that contains a large cyclic subgroup. The correct way to make this concept work is by means of the notion of a reduced divisor, as given in Definition 11.4.2.

Definition 11.8.1 *The jacobian of the hyperelliptic curve $\mathcal{X}(\boldsymbol{F}_q)$, denoted* jac($\mathcal{X}$) *or* jac($\mathcal{X}(\boldsymbol{F}_q)$)*, is the set of its reduced divisors.*

The definition of a jacobian, as given here, only defines a set, not a group. This is because a group operation has not yet been defined for this set of reduced divisors. It will be defined soon.

We have now given two definitions, Definition 11.7.1 and Definition 11.8.1, of different objects, and called both of them the jacobian. They are indeed different: one

Table 11.1 *The elements of a jacobian over* $\boldsymbol{F}_7$

0	$(6,4)+(1,1)-2\infty$
$(6,4)-\infty$	$(6,4)+(1,5)-2\infty$
$(1,1)-\infty$	$(6,4)+(2,2)-2\infty$
$(1,5)-\infty$	$(6,4)+(2,3)-2\infty$
$(2,2)-\infty$	$(6,4)+(5,3)-2\infty$
$(2,3)-\infty$	$(6,4)+(5,6)-2\infty$
$(5,3)-\infty$	$(1,1)+(2,2)-2\infty$
$(5,6)-\infty$	$(1,1)+(2,3)-2\infty$
$2(1,1)-2\infty$	$(1,1)+(5,3)-2\infty$
$2(1,5)-2\infty$	$(1,1)+(5,6)-2\infty$
$2(2,2)-2\infty$	$(1,5)+(2,2)-2\infty$
$2(2,3)-2\infty$	$(1,5)+(2,3)-2\infty$
$2(5,3)-2\infty$	$(1,5)+(5,3)-2\infty$
$2(5,6)-2\infty$	$(1,5)+(5,6)-2\infty$
	$(2,2)+(5,3)-2\infty$
	$(2,2)+(5,6)-2\infty$
	$(2,3)+(5,3)-2\infty$
	$(2,3)+(5,6)-2\infty$

defines a set of equivalence classes and one defines a set of reduced divisors. However, our next task is to show that every equivalence class contains exactly one reduced divisor and every reduced divisor is in an equivalence class, so the two definitions, although not the same, are equivalent. A reduced divisor is a canonical representative of an equivalence class.

To recapitulate Definition 11.8.1 by way of an example, let $\mathcal{X}$ be the hyperelliptic curve of genus two over $\boldsymbol{F}_7$ that was described earlier as the set of points

$$\mathcal{X}(\boldsymbol{F}_7) = \{(1,1),(1,5),(2,2),(2,3),(5,3),(5,6),(6,4),\infty\}.$$

The canonical representatives of the jacobian on this curve are easy to list. They are the reduced divisors of the form $\sum_i m_i P_i$. This means that every m_i must be nonnegative, and in each divisor, the m_i must sum to at most two because the genus of the curve is two. This means that zero, one, or two points of the curve $\mathcal{X}$ will appear in each reduced divisor, and so zero, one, or two points appear in each canonical representative of the jacobian. Moreover, a point and its opposite cannot both have a nonzero coefficient, and a special point cannot have a coefficient larger than one. These restrictions are satisfied by the elements given in Table 11.1 and by no other elements. These are all of the elements of the jacobian. Each element has degree zero. There are 32 entries in Table 11.1. These are all of the elements of the jacobian, and so this jacobian has order 32.

The reduced divisors can be added as divisors in the obvious way. For example, if $D = 2(1, 5) - 2\infty$ and $D' = (1, 1) + (5, 6) - 2\infty$, then

$$D + D' = (1, 1) + 2(1, 5) + (5, 6) - 4\infty.$$

The sum $D + D'$, however, is not a reduced divisor, so it is not an element of jac($\mathcal{X}$). It must be reduced in some way. The method of reduction will be defined so as to make the set of reduced divisors into a group. The operation of jacobian addition then is defined as divisor addition followed by modular reduction. The zero divisor plays the role of the group identity element, and so might also be denoted as $\mathcal{O}$ in the group of the jacobian. Notice that the identity element $\mathcal{O}$ of the jacobian is not (0, 0), which is the origin of the plane $\boldsymbol{F}_q^2$, nor is it ∞, which is the point at infinity of $\mathcal{X}(\boldsymbol{F}_q)$.

Our remaining task is to define (in the next section) a reduction operation on elements of jac($\mathcal{X}(\boldsymbol{F}_q)$) in order to form a group. The group operation then is the addition of the reduced divisors followed by this modular reduction. This group operation on the elements of the jacobian will be called jacobian addition. With the addition operation, jac($\mathcal{X}(\boldsymbol{F}_q)$) becomes a finite abelian group. For now, our task is to define this operation. Later, we will devise a convenient algorithm to execute the operation of jacobian addition.

11.9 Semireduced divisors and jacobians

Addition of points cannot be defined on a hyperelliptic curve of genus larger than one in such a way as to form a group. Instead, addition is defined on the elements of the jacobian of the hyperelliptic curve. Computation on a hyperelliptic curve consists of addition of elements of the jacobian.

To understand this computation, recall that for the hyperelliptic curve

$$\mathcal{X} : y^2 + yh(x) = f(x),$$

the jacobian is given by

$$\text{jac}(\mathcal{X}) = \{\text{divisors of degree zero}\}/\{\text{principal divisors}\}.$$

As formally defined, the elements of jac($\mathcal{X}(\boldsymbol{F}_{q^r})$) are equivalence classes, and addition within the jacobian is addition of equivalence classes. However, this interpretation as a set of equivalence classes is not suitable for practical computation. Instead, each equivalence class should be designated by a canonical representative. Just as the quotient group $\mathbf{Z}/\langle n\rangle$ is equivalent to $\mathbf{Z}_n$, the set of canonical representatives, so too the jacobian as a set of equivalence classes is equivalent to a set of canonical representatives. We have already mentioned this canonical representation several times, referring to the jacobian as the set of reduced divisors. We will see in this section that each equivalence class contains exactly one reduced divisor and each reduced divisor is in an equivalence class, so this canonical representation is indeed appropriate. The canonical representative of an equivalence class is the unique reduced divisor in that equivalence class. Accordingly, we may regard the jacobian not as a set of equivalence classes but as the set of reduced divisors. Our task now is to prove that this is true.

Recall that a reduced divisor is defined as a divisor of the form $\sum_i m_i P_i - (*)\infty$, all of whose coefficients m_i are nonnegative and satisfy certain other conditions. Recall that the *opposite* of the affine point $P = (a, b)$ is $\widetilde{P} = (a, -b - h(a))$. If $\widetilde{P}$ is the opposite of P, then P is the opposite of $\widetilde{P}$. The point P is a *special point* if $P = \widetilde{P}$, and otherwise is an *ordinary point*. The point ∞ is a special point with $\widetilde{\infty} = \infty$. If the vertical line $x - a = 0$ intersects the hyperelliptic curve $\mathcal{X}$ at the ordinary point P, then it also intersects $\mathcal{X}$ at its opposite $\widetilde{P}$. Recall also that if $H(x, y)$ is a rational function on the curve $\mathcal{X}$, then $\text{div}(H)$ denotes the divisor of $H(x, y)$. The rational function $x - a$, which defines a vertical line at $x = a$, either has two ordinary points or one special point on the curve. In the case that $P = (a, b)$ is an ordinary point, then $\text{div}(x - a) = P + \widetilde{P} - 2\infty$. In the case that $P = (a, b)$ is a special point, then $\text{div}(x - a) = 2P - 2\infty$.

A semireduced divisor, defined in Definition 11.4.2, is a divisor of the form $D = \sum_i m_i P_i - (\sum_i m_i)\infty$, each m_i is a nonnegative integer, where P_i and $\widetilde{P}_i$ do not both appear in D, and $m_i \leq 1$ if P_i is a special point. A reduced divisor, also defined in Definition 11.4.2, is a semireduced divisor for which the sum of m_i over all affine points is at most g.

The appropriate theorems, to be developed, result in a representation of $\text{jac}(\mathcal{X})$ as the set of reduced divisors. This is a concrete representation of a jacobian that makes computation convenient. First, in the next theorem, we show that every equivalence class has at least one semireduced divisor. Then, in a later section, we show that every equivalence class has exactly one reduced divisor. And, of course, every reduced divisor must be in an equivalence class because every divisor of zero degree is in an equivalence class. All this means that there is a one-to-one correspondence between equivalence classes and reduced divisors.

Theorem 11.9.1 *Every divisor D of degree zero and norm $|D|$ on a hyperelliptic curve is equivalent to at least one semireduced divisor D' with norm $|D'| \leq |D|$.*

Proof Let $D = \sum_P m_P P$ be a divisor of degree zero. Under the designated equivalence relationship, any principal divisor can be added to D or subtracted from D without changing the equivalence class. For the purpose of finding semireduced divisors equivalent to D, we will use only principal divisors corresponding to vertical lines. This will be enough. Partition the ordinary points of $\mathcal{X}$ into two sets, $\mathcal{X}_1$ and $\mathcal{X}_2$, such that distinct opposites P and $\widetilde{P}$ are in different sets, and P is in $\mathcal{X}_1$ if $m_P \geq m_{\widetilde{P}}$. This can always be done, not necessarily uniquely. Let $\mathcal{X}_0$ denote the set of special points.

Write the divisor D as

$$D = \sum_{P \in \mathcal{X}_0} m_P P + \sum_{P \in \mathcal{X}_1} m_P P + \sum_{\widetilde{P} \in \mathcal{X}_2} m_{\widetilde{P}} \widetilde{P} - (*)\infty$$

and recall that the jacobian is defined so that D is in the same equivalence class as $D - \operatorname{div} f(x, y)$ for any rational function $f(x, y)$ of the ring $\boldsymbol{F}(x, y)$. Accordingly, define a divisor D' as

$$D' = D - \sum_{P \in \mathcal{X}_0} \left\lfloor \frac{m_P}{2} \right\rfloor \operatorname{div}(x - a_P) - \sum_{P \in \mathcal{X}_2} m_P \operatorname{div}(x - a_P),$$

where, in each term of each sum, a_P is the first coordinate of $P = (a_P, b_P)$, and $x - a_P$ is the equation of a vertical line. Then D' is equivalent to D because the last two terms on the right side are principal divisors. Using the expressions for $\operatorname{div}(x - a)$, we now have

$$\begin{aligned} D' &= D - \sum_{P \in \mathcal{X}_0} \left\lfloor \frac{m_P}{2} \right\rfloor (2P - 2\infty) - \sum_{P \in \mathcal{X}_2} m_P (P + \widetilde{P} - 2\infty) \\ &= \sum_{P \in \mathcal{X}_0} m'_P P + \sum_{P \in \mathcal{X}_1} m'_P P, \end{aligned}$$

where clearly $0 \leq m'_P \leq 1$ for $P \in \mathcal{X}_0$. All coefficients m_P of points in $\mathcal{X}_2$ are transferred from those points in $\mathcal{X}_2$ to the opposite points in $\mathcal{X}_1$, but are otherwise unchanged. Then D' is a semireduced divisor equivalent to D and with $|D'| \leq |D|$, as was to be proved. □

Observe that the central idea of this proof is that for a pair of opposite points, P and $\widetilde{P}$, with coefficients m_P and $m_{\widetilde{P}}$, the lesser of the two coefficients can be seen to be zero in an equivalent divisor by adding (or subtracting) the same constant to both m_P and $m_{\widetilde{P}}$. This aspect of the proof will be used again in the proof of the forthcoming Theorem 11.11.1 to conclude that each divisor class contains a reduced divisor.

11.10 The Mumford transform

At this point it is helpful to introduce an alternative representation of a semireduced divisor. Just as a function of a real variable $s(t)$ can be usefully represented by its Fourier transform $S(f)$, so too a semireduced divisor $D = \sum_P m_P P - (*)\infty$ (abbreviated $\sum_P m_P P$) can be usefully represented by its *Mumford representation* or *Mumford transform*. The Mumford transform of a reduced divisor, denoted by $(a(x), b(x))$, is a polynomial representation that is suitable for certain kinds of computation. The jacobian can be redefined in the Mumford transform domain, although this definition obscures the defining property of a jacobian as a divisor class group consisting of divisors of degree zero modulo divisors of rational functions. To denote the usual representation of a divisor in terms of the Mumford transform, we will use the notation $D = \text{div}(a(x), b(x))$.

The divisor completely describes a rational function on a hyperelliptic curve up to a scalar multiplier. A divisor of a rational function may be considered as an alternative representation of the rational function on a hyperelliptic curve that is convenient for many kinds of computation. In turn, the Mumford transform is an alternative representation of a semireduced divisor that is convenient for some other kinds of computation such as adding the elements of a jacobian. The Mumford transform is not defined for divisors that are not semireduced divisors. For some purposes, it is easier to work with the pair of univariate polynomials $a(x)$ and $b(x)$ comprising the Mumford transform of a semireduced divisor than to work with a list of its poles and zeros.

Definition 11.10.1 *The Mumford transform of a semireduced divisor* $D = \Sigma_i m_i P_i - (*)\infty$ *with affine points denoted* $P_i = (x_i, y_i)$ *is a pair of polynomials* $(a(x), b(x))$, *where* $a(x) = \Pi_i (x - x_i)^{m_i}$, *and* $b(x)$ *is a polynomial of least degree such that* $y_i = b(x_i)$ *for each* i.

For a reduced divisor, the polynomial $a(x)$ must be a polynomial of degree at most g, and $b(x)$ is a polynomial of degree less than g. Otherwise, the Mumford transform corresponds to a semireduced divisor. We will describe a way to recover the semireduced divisor D from the Mumford transform $(a(x), b(x))$, and we will describe a way to obtain the Mumford transform $(a(x), b(x))$ from the semireduced divisor D. Thus the two representations of a semireduced divisor are equivalent. The zero divisor, which is the divisor with all m_i equal to zero, corresponds to the identity element of the jacobian. In turn, the zero divisor corresponds to the element $(1, 0)$ in the Mumford transform domain.

The Mumford transform is easy to compute from the expression $D = \sum_i m_i P_i$. Calculation of the polynomial $a(x)$ is explicit. If all nonzero m_i are equal to one, the polynomial $b(x)$ can be determined by Lagrange interpolation. Otherwise, it can be determined by the chinese remainder theorem as described below in Proposition 11.10.3. In turn, it clearly is also easy to recover the usual representation of D from the Mumford transform. To compute x_i and m_i, one factors $a(x)$. Then for each i, y_i is obtained by evaluating $b(x)$ at x_i.

Our next task is to show that $b(x)$ of the Mumford transform divides the polynomial $b(x)^2 + b(x)h(x) - f(x)$. This polynomial is the negative norm of $y - b(x)$ and is computed as follows. Because the polynomial $y - b(x)$ is an element of the coordinate ring $\boldsymbol{F}[x, y]_{p(x,y)}$, it has a norm given by

$$\begin{aligned}
\text{norm}(y - b(x)) &= (y - b(x)(y + h(x) + b(x))) \\
&= y^2 - b(x)^2 + yh(x) - b(x)h(x) \\
&= f(x) - b(x)^2 - b(x)h(x),
\end{aligned}$$

where the last line follows because $y^2 + yh(x) = f(x)$. It is conventional to carry forward the negative of this polynomial as $b(x)^2 + b(x)h(x) - f(x)$.

The precursor statement that leads to this important property of the Mumford transform is the following proposition regarding a single point of a curve. The proof of the proposition provides a straightforward prescription for computing the polynomial $b(x)$ for a one-point divisor. To extend this proposition to a general semireduced divisor, such $b(x)$ corresponding to one-point divisors are combined using the chinese remainder theorem.

Proposition 11.10.2 *Given any point $P = (x^*, y^*)$ of the hyperelliptic curve $\mathcal{X}(\boldsymbol{F})$: $y^2 + h(x)y = f(x)$ and any nonnegative integer m, not larger than one if P is a special point, there exists a unique polynomial $b(x)$ of degree less than m such that $b(x^*) = y^*$ and $b(x)^2 + b(x)h(x) - f(x)$ is divisible by $(x - x^*)^m$.*

Proof To verify the condition of the theorem for a special point, recall that if m is nonzero, it must be one. Then $b(x) = y^*$ is a polynomial of degree zero that satisfies the condition $b(x^*) = y^*$ and the second condition as shown by direct substitution into $b(x)^2 + b(x)h(x) - f(x)$ and using the equation of the curve.

To verify the condition of the theorem for an ordinary point, temporarily make the change of variables that replaces $x - x^*$ by x so that the required condition becomes

$$g(x) \triangleq b(x)^2 + b(x)h(x) - f(x) = 0 \pmod{x^m},$$

now with the translated point given by $P = (0, y^*)$. At the point $(0, y^*)$, the equation $y^2 + yh(x) = f(x)$ becomes $y^{*2} + y^*h_0 = f_0$. Because the curve is nonsingular, we know that $2y^* + h_0 \neq 0$.

To verify the condition that $g(x)$ is divisible by x^m requires showing that $g_\ell = 0$ for $\ell = 1, \ldots, m - 1$. To see that $g_0 = g(0) = 0$, note that $b_0 = y^*$ so

$$g(0) = g_0 = b_0^2 + b_0h_0 - f_0 = y^{*2} + y^*h_0 - f_0 = 0,$$

because $(0, y^*)$ is a point of the curve. To see that g_1 is zero, write

$$g(x) = (b_0 + b_1x + \ldots)^2 + (b_0 + b_1x + \ldots)(h_0 + h_1x + \ldots) - (f_0 + f_1x + \ldots),$$

from which we conclude that

$$g_1 = (2b_0 + h_0)b_1 - f_1.$$

This can be set equal to zero by choice of b_1 as $b_1 = (b_0h_1 - f_1)/(2b_0 + h_0)$, recalling that $2b_0 + h_0$ is not zero. In the same way, if $b_i = 0$ for $i = 1, \ldots, \ell - 1$, then for $\ell = 2, \ldots, m - 1$,

$$g(x) = (b_0 + b_\ell x^\ell + \ldots)^2 + (b_0 + b_\ell x^\ell + \ldots)(h_0 + h_1x + \ldots) - f(x),$$

from which we conclude that

$$g_\ell = (2b_0 + h_0)b_\ell + b_0 h_\ell - f_\ell$$

which can be set to zero by choice of b_ℓ because $2b_0 + h_0$ is not zero. Each g_ℓ is linear in b_ℓ and involves only those b_ℓ with $i \le \ell$. Thus each g_ℓ, in turn, can be set to zero by choice of b_ℓ because the coefficient of that b_ℓ is $2b_0 + h_0$ for every ℓ and so is nonzero. Thus $g_\ell = 0$ for $\ell = 1, \ldots, m-1$, and $\deg b(x) < m$. This process results in a unique polynomial $b(x)$ satisfying the requirements of the theorem.

Reversing the change of variables returns the point to (x^*, y^*). This completes the verification for an ordinary point that $(x - x^*)^m$ divides $b(x)^2 + b(x)h(x) - f(x)$. □

The next theorem uses the chinese remainder theorem to extend this statement to multiple points.

Proposition 11.10.3 *Let $D = \sum_i m_i P_i - (*)\infty$ be a semireduced divisor for hyperelliptic curve $\mathcal{X}(\boldsymbol{F}_p)$. There exists a unique univariate polynomial $b(x)$ of degree less than $\sum_i m_i$ such that $b(x_i) = y_i$ for all i and $\prod_i (x - x_i)^{m_i}$ divides $b(x)^2 + b(x)h(x) - f(x)$.*

Proof Proposition 11.10.2 states that for each affine point (x_i, y_i) of the divisor and for each m_i, there is a unique polynomial $b_i(x)$ of degree less than m_i such that $(x - x_i)^{m_i}$ divides $b_i(x)^2 + b_i(x)h(x) - f(x)$. For each point P_i of the divisor, let $b_i(x)$ be this polynomial. The chinese remainder theorem states that there is a polynomial $b(x)$ of degree less than $\sum_i m_i$ satisfying $b(x) = b_i(x) \pmod{(x - x_i)^{m_i}}$ for each i. We must show that $b(x)$ satisfies the required conditions. Let

$$b(x) = \sum_{j=0}^{n} b_j(x) N_j(x) \prod_{k \neq j} (x - x_k)^{m_k}$$

be the polynomial given by the chinese remainder theorem, where $N_j(x)$ is the polynomial satisfying

$$N_j(x) \prod_{k \neq j} (x - x_k)^{m_k} + n_j(x)(x - x_j)^{m_j} = 1$$

as asserted by Corollary 9.6.6.

Clearly, $b(x_i) = b_i(x_i) = y_i$. To verify that $b(x)^2 + b(x)h(x) - f(x) = 0 \pmod{\prod_i (x - x_i)^{m_i}}$ notice that because all cross terms are zero modulo $\prod_i (x - x_i)^{m_i}$, the square of $b(x)$ is

$$b(x)^2 = \sum_i b_i(x)^2 N_i(x)^2 \prod_{i' \neq i} (x - x_i)^{2m_i} \pmod{\prod_i (x - x_i)^{m_i}}.$$

But $N_i(x)\prod_{i'\neq i}(x-x_{i'})^{m_{i'}} = 1 - n_j(x)(x-x_i)^{m_i}$, so

$$\begin{aligned} b(x)^2 &= \sum_i b_i(x)^2 N_i(x)[1-n_i(x)(x-x_i)^{m_i}]\prod_{i'\neq i}(x-x_{i'})^{m_{i'}} \pmod{\prod_i (x-x_i)^{m_i}} \\ &= \sum_i b_i(x)^2 N_i(x)\prod_{i'\neq i}(x-x_{i'})^{m_{i'}} \pmod{\prod_i (x-x_i)^{m_i}}. \end{aligned}$$

Finally, let $f_i(x)$ be given by

$$f_i(x) = f(x) \mod (x-x_i)^{m_i}$$

so

$$f(x) = \sum_{i=0}^{n} f_i(x)N_i(x)\prod_{r\neq i}(x-x_r)^{m_r}.$$

This gives

$$b(x)^2 + b(x)h(x) - f(x) = \left[\sum_i \frac{b_i(x)^2 + b_i(x)h(x) - f_i(x)}{(x-x_i)^{m_i}}\right]\prod_i (x-x_i)^{m_i},$$

where, by Proposition 11.10.2, each term in the sum is a polynomial. This completes the proof of the theorem. □

The next theorem is a reformulation of Proposition 11.10.3.

Theorem 11.10.4 *The Mumford transform creates a one-to-one correspondence between semireduced divisors D on the hyperelliptic curve $\mathcal{X}(\boldsymbol{F}) : y^2 + h(x)y = f(x)$ over a field* **F** *and pairs of polynomials $(a(x), b(x))$ over* **F** *satisfying*

(1) $a(x)$ is a monic polynomial.
(2) $\deg b(x) < \deg a(x) = \deg D$.
(3) $b(x)^2 + h(x)b(x) - f(x)$ is a multiple of $a(x)$.

Moreover, those $(a(x), b(x))$ that also satisfy $\deg a(x) \leq g$ correspond to reduced divisors.

Proof Given a semireduced divisor $D = \sum_i m_i P_i - (*)\infty$ where $P_i = (x_i, y_i)$, define $a(x) = \Pi_i(x-x_i)^{m_i}$ and $b(x)$ of minimal degree such that $y_i = b(x_i)$. Then $a(x)$ is a monic polynomial and $\deg b(x)$ is less than $\deg a(x)$ because the degree of $b(x)$ is one less than the number of distinct points (x_i, y_i). The correspondence is one-to-one because the third condition ensures that $(x_i, b(x_i))$ is a point of the hyperelliptic curve. On the other hand, because $(x_i, b(x_i))$ is a point of the hyperelliptic curve, $b(x)$ must satisfy the third condition.

The final statement of the theorem is immediate because every semireduced divisor with $\deg a(x) \leq g$ is a reduced divisor. □

The theorem shows that $c(x) = (b^2(x) + h(x)b(x) - f(x))/a(x)$ is a polynomial. Accordingly, we may choose to express the Mumford transform by a redundant alternative form as

$$D = \operatorname{div}(a(x), b(x), c(x)).$$

This alternative statement contains the underlying hyperelliptic curve within itself because the polynomial defining the curve is $y^2 = b^2(x) - a(x)c(x)$.

The final theorem of this section states that an alternative way to recover the usual form of the divisor from the Mumford representation is by using the notion of the greatest common divisor of divisors.

Theorem 11.10.5 *The semireduced divisor* $D = \operatorname{div}(a(x), b(x))$ *satisfies*

$$\operatorname{div}(a(x), b(x)) = \operatorname{GCD}[\operatorname{div}(a(x)), \operatorname{div}(b(x) - y)].$$

Proof Let $\mathcal{X}_0$ be the set of special points in the support of D and let $\mathcal{X}_1$ be the set of ordinary points in the support of D. Let $\mathcal{X}_2$ be the set of all opposite points of the points in $\mathcal{X}_1$. The opposite points cannot be in the support by the definition of a semireduced divisor. Then

$$D = \sum_{P_i \in \mathcal{X}_0} P_i + \sum_{P_i \in \mathcal{X}_1} m_i P_i - (*)\infty.$$

Next observe that if $P = (x, y)$ is a zero of $a(x)$, then $-P = (x, -y - h(x))$ is a zero of $a(x)$ as well, so

$$\operatorname{div}(a(x)) = \operatorname{div} \prod_i (x - x_i)^{m_i} = \sum_{P_i \in \mathcal{X}_0} 2P_i + \sum_{P_i \in \mathcal{X}_1} m_i P_i + \sum_{P_i \in \mathcal{X}_2} m_i P_i - (*)\infty.$$

Also observe that

$$\operatorname{div}(b(x) - y) = \sum_{P_i \in \mathcal{X}_0} t_i P_i + \sum_{P_i \in \mathcal{X}_1} s_i P_i + \sum_{P_i \in \mathcal{X}_3} r_i P_i - (*)\infty,$$

where $\mathcal{X}_3$ is the set of all points of the curve that are not in $\mathcal{X}_0 \cup \mathcal{X}_1 \cup \mathcal{X}_2$. Hence

$$\operatorname{div}(a(x), b(x) - y) = \sum_{P_i \in \mathcal{X}_0} \min(2, t_i) P_i + \sum_{P_i \in \mathcal{X}_2} \min(m_i, s_i) P_i.$$

It remains only to show that $s_i \geq m_i$ which means that $\min(m_i, s_i) = m_i$, and that $t_i \leq 1$ which means that $\min(2, t_i) \leq 1$. To show that $s_i \geq m_i$, recall that the (negative) norm in the coordinate ring of the polynomial $b(x) - y$, is $b(x)^2 + h(x)b(x) - f(x)$. But $a(x) = (x - x_i)^{m_i}$ divides $b(x)^2 + h(x)b(x) - f(x)$, which means that $s_i \geq m_i$ and therefore $\min(m_i, s_i) = m_i$.

To see that the coordinate x_i of a special point x_i is a simple zero of $b(x)^2 + h(x)b(x) - f(x)$, observe that the derivative of this polynomial reduces as

$$\begin{aligned} 2b(x)b'(x) + b'(x)h(x) + b(x)h'(x) - f'(x) &= b'(x)(2y - h(x)) - (h'(x)y - f'(x)) \\ &= f'(x) - h'(x)y \neq 0. \end{aligned}$$

The second equality holds because $2y - h(x) = 0$ at a special point. The inequality in the last line holds because the curve is nonsingular. Therefore, for a special point, x_i is a simple zero, so $t_i = 1$ for $P_i \in \chi_0$. Therefore $\min(2, t_i) \leq 1$. □

11.11 The Cantor reduction algorithm

To show that every divisor class contains at least one reduced divisor, we will give an explicit computational algorithm to compute from a semireduced divisor a reduced divisor in that same divisor class. Every equivalence class is guaranteed to have a semireduced divisor by Theorem 11.9.1. We will use a procedure called the *Cantor reduction* for computing a reduced divisor from a semireduced divisor. We will then show that a divisor class cannot contain two reduced divisors, and so conclude that every divisor class contains exactly one reduced divisor.

The Cantor reduction, as discussed in this section, considers divisors in the Mumford transform domain only for those hyperelliptic curves of the form $\mathcal{X} : y^2 = f(x)$. The general case with nonzero $h(x)$ is essentially the same and is left for an exercise. In Section 11.13, the Cantor reduction will be used, possibly with nonzero $h(x)$, to define jacobian addition by reducing the combination of two reduced divisors to a single reduced divisor. The divisor formed in a jacobian addition is in the Mumford representation. In many applications, it can remain in the Mumford representation after the computation. In such applications, there is no need to reconvert the new divisor back to the pole–zero representation.

The Cantor reduction is as follows. Replace the semireduced divisor $D = \text{div}(a(x), b(x))$ by the equivalent divisor $D' = D - \text{div}(b(x) - y)$. This is an equivalent divisor because $\text{div}(b(x) - y)$ is a principal divisor. This divisor can be written in the more suggestive form $D' = -((b(x) - y) - D) = \text{div}(a'(x), b'(x))$. The first inequality is a trivial manipulation of signs. The second equality can be asserted because D' must have a Mumford transform, which can always be written as $(a'(x), b'(x))$ for some $a'(x)$ and $b'(x)$. Now we can recognize that

$$a'(x) = (f(x) - b^2(x))/a(x)$$

and

$$b'(x) = -b(x) \pmod{a'(x)}.$$

This is the Cantor reduction. We see that $\deg a'(x)$ is less than $\deg a(x)$ unless $(a'(x), b'(x))$ represents a reduced divisor, which means that the Cantor reduction reduces the degree of any semireduced divisor that is not a reduced divisor. This is the essence of the proof of the following theorem.

Theorem 11.11.1 *The Cantor reduction algorithm computes a reduced divisor that is equivalent to a given semireduced divisor.*

Proof There are two statements to be proved. In Step 1, we prove that the algorithm computes a reduced divisor. In Step 2, we prove that the reduced divisor is equivalent to the given semireduced divisor.

Step 1 Let $(a(x), b(x))$ denote a semireduced divisor D in the Mumford transform domain and let deg $a(x) = m$ and deg $b(x) = n < m$. Then because $a'(x) = (f(x) - b^2(x))/a(x)$ is a polynomial, as asserted in the previous section, we can write

$$\begin{aligned}\deg a'(x) &\leq \max(2g+1, 2n) - m \\ &= \max(2g+1-m, 2n-m).\end{aligned}$$

Suppose that $m \geq g+2$. Then $2g+1 \leq 2m-3$, and $2n-m \leq m-2$, so

$$\deg a'(x) \leq \max(m-3, m-2) = m-2$$

which is less than the degree of $a(x)$. Suppose that $m = g+1$. Then

$$\deg a'(x) \leq \max(g, g-1) = g.$$

We conclude that if the norm $|D|$ is larger than g, the reduction procedure will reduce $|D|$. The reduction procedure can be repeated as necessary until $|D|$ is not larger than g.

Step 2 To show that D' is equivalent to D, we must show that D' and D differ by a principal divisor. We will show that $D' = D - \text{div}(b(x) - y)$. The development will use the elementary equivalence

$$\sum_i m_i P_i - (*)\infty \sim -\sum_i m_i \widetilde{P}_i + (*)\infty.$$

This is the statement that the opposite, denoted $\widetilde{D}$, of the divisor D is equivalent to the negative of divisor D, meaning that both are in the same equivalence class. This equivalence follows from the divisor of a vertical line $\ell(x, y)$, which on a hyperelliptic curve always has the form

$$\text{div}\ell(x, y) = P + \widetilde{P} - (*)\infty.$$

The proof is broken into several parts. First $\operatorname{div}(a(x), b(x))$ is converted into expressions for $\operatorname{div}(a(x))$ and $\operatorname{div}(b(x))$. These are used to derive expressions for $\operatorname{div}(a'(x))$ and $\operatorname{div}(b'(x))$. Finally the equality $\operatorname{div}(a'(x), b'(x)) = \operatorname{GCD}(a'(x), b'(x))$ is used to characterize D'.

(i) Any semireduced divisor D can be written

$$D = \sum_{P_i \in \mathcal{X}_0} P_i + \sum_{P_i \in \mathcal{X}_1} m_i P_i - (*)\infty,$$

where $\mathcal{X}_0$ and $\mathcal{X}_1$ are the special and ordinary points of D, respectively. The Mumford representation of D is $(a(x), b(x))$, and the divisor of $a(x)$ satisfies

$$\begin{aligned}\operatorname{div}(a(x)) &= D + \widetilde{D} \\ &= \sum_{P_i \in \mathcal{X}_0} 2P_i + \sum_{P_i \in \mathcal{X}_1} m_i P_i + \sum_{\widetilde{P}_i \in \mathcal{X}_2} m_i \widetilde{P}_i - (*)\infty,\end{aligned}$$

where $\mathcal{X}_2$ is the set of points that are opposite of the points of $\mathcal{X}_1$.

(ii) Because $D = \operatorname{GCD}(a(x), b(x) - y)$, it must be true that $\operatorname{div}(b(x) - y)$ has the form

$$\operatorname{div}(b(x) - y) = \sum_{P_i \in \mathcal{X}_0} P_i + \sum_{P_i \in \mathcal{X}_1} n_i P_i + \sum_{P_i \in \mathcal{X}_2} 0\widetilde{P}_i + \sum_{P_i \in \mathcal{X}_3} s_i P_i - (*)\infty,$$

where $\mathcal{X}_3$ is a set of points disjoint from $\mathcal{X}_0, \mathcal{X}_1, \mathcal{X}_2$, and $\{\infty\}$, and where $n_i \geq m_i$ and $s_i = 1$ whenever P_i is a special point. But

$$\begin{aligned}\operatorname{norm}(b(x) - y) &= (b(x) - y)(b(x) + h(x) + y) \\ &= b^2(x) + b(x)h(x) - f(x),\end{aligned}$$

and the divisor of $b(x) + h(x) + y$ is the opposite of the divisor of $b(x) - y$. Therefore

$$\begin{aligned}\operatorname{div}(b^2(x) + b(x)h(x) - f(x)) = &\sum_{P_i \in \mathcal{X}_0} 2P_i + \sum_{P_i \in \mathcal{X}_1} n_i P_i + \sum_{\widetilde{P}_i \in \mathcal{X}_2} n_i \widetilde{P}_i + \sum_{P_i \in \mathcal{X}_3} s_i P_i \\ &+ \sum_{P_i \in \mathcal{X}_3} s_i \widetilde{P}_i - (*)\infty.\end{aligned}$$

(iii) We conclude from (i) and (ii) that $\operatorname{div}(a'(x))$ has the form

$$\begin{aligned}\operatorname{div}(a'(x)) &= \operatorname{div}(b^2(x) + b(x)h(x) - f(x)) - \operatorname{div}(a(x)) \\ &= \sum_{P_i \in \mathcal{X}_1} (n_i - m_i) P_i + \sum_{P_i \in \mathcal{X}_1} (n_i - m_i) \widetilde{P}_i + \sum_{P_i \in \mathcal{X}_3} s_i P_i + \sum_{P_i \in \mathcal{X}_3} s_i \widetilde{P}_i - (*)\infty.\end{aligned}$$

(iv) Recall that because $b'(x) = -h(x) - b(x) \pmod{a'(x)}$, we have $b'(x) = -h(x) - b(x) + s(x)a'(x)$ for some polynomial $s(x)$. Let $P_i = (x_i, y_i)$. Then

$b'(x_i) = -h(x_i) - b(x_i) + s(x_i)a'(x_i) = -h(x_i) - y_i$. It follows that

$$\operatorname{div}(b'(x) - y) = \sum_{P_i \in \mathcal{X}_i} 0P_i + \sum_{P_i \in \mathcal{X}_1} r_i \widetilde{P}_i + \sum_{P_i \in \mathcal{X}_3} 0P_i + \sum_{P_i \in \mathcal{X}_3} w_i \widetilde{P}_i + \sum_{P_i \in \mathcal{X}_4} z_i P_i - (*)\infty.$$

(v) We conclude that

$$\begin{aligned}\operatorname{div}(a'(x), b'(x)) &= \sum_{P_i \in \mathcal{X}_1} (n_i - m_i)\widetilde{P}_i + \sum_{P_i \in \mathcal{X}_1} s_i \widetilde{P}_i - (*)\infty \\ &\sim - \sum_{P_i \in \mathcal{X}_1} (n_i - m_i)P_i - \sum_{P_i \in \mathcal{X}_1} s_i P_i - (*)\infty \\ &= D - \operatorname{div}(b(x) - y).\end{aligned}$$

which completes the proof of the theorem. □

11.12 Reduced divisors and jacobians

We can now show that the jacobian is indeed represented as the set of reduced divisors. This fact has been mentioned several times, but not proved. It is an important statement because it replaces the abstract notion of a jacobian as a divisor class group with a concrete notion that is suitable for computation.

The converse statement, of course, is unnecessary to prove because a reduced divisor is a divisor of degree zero, and so is an element of an equivalence class.

Theorem 11.12.1 *Every divisor of degree zero is equivalent to a unique reduced divisor.*

Proof Because every divisor class contains at least one semireduced divisor, Theorem 11.11.1 shows that every divisor of degree zero is equivalent to at least one reduced divisor. Therefore we need only to prove that there cannot be two reduced divisors in the same equivalence class. These two statements then prove that a unique reduced divisor exists in every equivalence class.

Suppose that $D = \sum_P m_P P$ and $D' = \sum_P m'_P P$ are distinct reduced divisors that are in the same equivalence class. This means that $D - D'$ is a principal divisor, and so is equivalent to a semireduced divisor, which will be denoted D''. We will show that such a D'' cannot exist.

Step 1 We first show that D'' cannot be the zero divisor. Because D and D' are not equal, there must be a point Q such that $m_Q \neq m'_Q$. Because both of these integers are nonnegative we can, without loss of generality, suppose that $m_Q \geq 1$. We first show that the equivalent semireduced divisor D'' is not the zero divisor by examining the semireduced divisor D'' only at the point Q and its opposite $\widetilde{Q}$. To this purpose, recall that the proof of Theorem 11.9.1 forms a semireduced divisor from a zero-degree divisor using only the operation of adding (or subtracting) $m_{\widetilde{P}}$ from both m_P and $m_{\widetilde{P}}$ for each pair of opposite points $(P, \widetilde{P})$.

There are three cases to consider.

(i) If $m'_Q = m'_{\widetilde{Q}} = 0$, then $m''_Q = m_Q \geq 1$.
(ii) If $m'_Q \neq 0$ with $1 \leq m'_Q < m_Q$, then $m''_Q = (m_Q - m'_Q) \geq 1$.
(iii) If $m'_{\widetilde{Q}} \neq 0$ with $1 \leq m'_{\widetilde{Q}} \leq m_Q$, then $m''_Q = (m_Q + m'_{\widetilde{Q}}) \geq 1$.

(The third case does not occur if Q is a special point.) There are no other possibilities, so we conclude that $m''_Q \geq 1$. Therefore D'' is not the zero divisor.

Step 2 We next show that D'' cannot be a nonzero semireduced divisor. If D'' is a semireduced divisor and is nonzero, it must correspond to the poles and zeros of a nonzero rational function. Because D'' is a semireduced divisor, all of its coefficients are nonnegative, so this rational function has no poles. Therefore it is a polynomial, $f(x, y)$. As an element of the coordinate ring, the polynomial can be written as $f(x, y) = a(x) - yb(x)$, which has degree $2g + 1$ unless $b(x)$ is zero. But $\deg|D''| \leq 2g$, so we conclude that $b(x) = 0$ and $f(x, y) = a(x)$. This means that if the ordinary point $P = (x, y)$ is a point of the curve satisfying $f(P) = f(x, y) = 0$, then $\widetilde{P} = (x, -y)$ is a point of the curve satisfying $f(\widetilde{P}) = f(x, -y) = 0$, contradicting the fact that D'' is a semireduced divisor. Thus $D - D'$ is not equivalent to a nonzero semireduced divisor.

Because D'' can be neither the zero semireduced divisor nor a nonzero semireduced divisor, we conclude that D'' cannot be a semireduced divisor. The contradiction proves that the premise of the proof is false, which means that two reduced divisors cannot be in the same equivalence class. □

We now know that each equivalence class of the jacobian has exactly one reduced divisor. This means that we can identify each equivalence class with its reduced divisor and regard the jacobian to be the set of reduced divisors rather than the set of equivalence classes. From a computational point of view, it is enough to consider the jacobian to be the set of reduced divisors. To make the set into a group, addition is defined within the set of reduced divisors as addition modulo the set of principal divisors.

11.13 The Cantor–Koblitz algorithm

The elements of a jacobian of a hyperelliptic curve can be regarded either as *divisor classes*, meaning equivalence classes of divisors, or as reduced divisors, which are the canonical representatives of the divisor classes. The first interpretation puts the structure of the jacobian into the definition of the equivalence classes, which means that addition of equivalence classes is then trivial but abstract. In computational algebraic geometry, however, one cannot compute with equivalence classes as such. For computations, one requires something more concrete. The use of reduced divisors as canonical representatives is concrete and the structure of the jacobian as a set of reduced divisors is then simple. Now, however, the algebraic structure shows itself in the modular reduction that is needed in the addition operation. For the purpose of computation, this modular reduction must be described explicitly. To this end, the Mumford representation of a reduced divisor is convenient.

The *Cantor–Koblitz algorithm* is an efficient procedure that adds two elements of the jacobian to obtain a new element of the jacobian. It requires that each element of the jacobian be represented by the unique reduced divisor contained in that divisor class expressed in the Mumford representation. Whereas the formal definition of a divisor as a set of points is appropriate for developing the theory and the associated theorems, the Mumford representation is a convenient representation for the purpose of computation.

The Cantor–Koblitz algorithm will be stated for any hyperelliptic curve $\mathcal{X}(\boldsymbol{F}_q) : y^2 + h(x)y = f(x)$, but for nonbinary fields, we would set $h(x)$ equal to zero. When $h(x) = 0$, the algorithm is also called the *Cantor algorithm*. For a binary field, a nonzero $h(x)$ must be retained.

The Cantor–Koblitz algorithm makes heavy use of the extended euclidean algorithm for polynomials. Given the sum $D_1 + D_2$ of two reduced divisors $D_1 = \text{div}(a_1(x), b_1(x))$ and $D_2 = \text{div}(a_2(x), b_2(x))$, the equivalent reduced divisor $D_3 \sim D_1 + D_2$, denoted $\text{div}(a_3(x), b_3(x))$, is computed in two steps, a composition step and a reduction step. The reduction step was described earlier in Section 11.11, and Theorem 11.11.1 proved that the reduction step computes a reduced divisor that is equivalent to a given semireduced divisor. The composition step combines any two semireduced divisors, which could be any two reduced divisors, to form a semireduced divisor. First, the extended euclidean algorithm is used twice to compute the polynomial

$$d(x) = \text{GCD}[a_1(x), a_2(x), b_1(x) + b_2(x) + h(x)]$$

which is then expressed using the Bézout relation as

$$d(x) = s_1(x)a_1(x) + s_2(x)a_2(x) + s_3(x)(b_1(x) + b_2(x) + h(x)),$$

where $s_1(x)$, $s_2(x)$, and $s_3(x)$ are the Bézout polynomials computed by the extended euclidean algorithm for polynomials. A semireduced divisor, $\text{div}(a_3(x), b_3(x))$, is now computed by

$$a_3(x) = a_1(x)a_2(x)/d^2(x)$$
$$b_3(x) = \frac{s_1(x)a_1(x)b_2(x) + s_2(x)a_2(x)b_1(x) + s_3(x)(b_1(x)b_2(x) + f(x))}{d(x)} \pmod{a_3(x)},$$

where $a_3(x)$ and $b_3(x)$ are both polynomials, as will be shown.

In general, div$(a_3(x), b_3(x))$ will not be a reduced divisor, but we will show that it is always a semireduced divisor and so can be made into a reduced divisor by the Cantor reduction step that was given in Section 11.11. Accordingly, if $\deg a_3(x)$ is larger than g, reduce the degree by the reduction step

$$a_3'(x) = (f(x) - h(x)b_3(x) - b_3^2(x))/a_3(x),$$

followed by

$$b_3'(x) = -h(x) - b_3(x) \pmod{a_3'(x)}.$$

If $\deg a_3(x) > g$, repeat the reduction step, stopping when $\deg a_3(x) \leq g$. Then

$$D_3 = \text{div}(a_3(x), b_3(x))$$

is the unique reduced divisor satisfying $D_3 \sim D_1 + D_2$.

Theorem 11.13.1 *The Cantor–Koblitz algorithm results in a reduced divisor that is equivalent to* $D_1 + D_2$.

Proof It is necessary only to show that the composition step results in a semireduced divisor. This is because Theorem 11.11.1 shows that the reduction step produces a reduced divisor from a semireduced divisor. The proof of the composition step is lengthy and will be broken into several steps.

Step 1 The first step of the proof is to show that $a_3(x)$ and $b_3(x)$ are both polynomials. First notice that $a_1(x)$ and $a_2(x)$ are each divisible by $d(x)$ because $d(x)$ is so defined, which means that $a_1(x)a_2(x)$ is divisible by $d(x)^2$, and also the first two terms in the expression for $b_3(x)$ are each divisible by $d(x)$. The second factor in the third term in that expression can be written

$$b_1(x)b_2(x) + f(x) = b_1(x)(b_1(x) + b_2(x)) + (f(x) - b_1(x)^2).$$

The first term on the right side of this expression is divisible by $d(x)$ because $d(x)$ is so defined. By Proposition 11.10.3, the last term on the right side is a multiple of $a(x)$ and so is also divisible by $d(x)$ because $a(x)$ is divisible by $d(x)$. Therefore $a_3(x)$ and $b_3(x)$ are polynomials as was asserted, and $\deg b(x) < \deg a(x)$.

Step 2 The second step of the proof is to show that $D_3 = \text{div}(a_3(x), b_3(x))$ is a semireduced divisor. But, by Theorem 11.10.4, any $(a(x), b(x))$ is a semireduced divisor if $\deg b(x) < \deg a(x)$ and $a(x)$ divides $b(x)^2 + b(x)h(x) - f(x)$. (Notice that the line of thought that will be used in Step 3 can be formed into an alternative proof of this statement, but we prefer to use this more elegant proof.)

Step 3 The final step of the proof is to show that D_3 is equivalent to $D_1 + D_2$. This means that $\mathrm{div}(a_3(x), b_3(x))$ corresponds to the divisor $D_1 + D_2$ with suitable principal divisors removed. We will show that each pair of opposites P and $\widetilde{P}$ of D_3 has the same pair of weights m and $\widetilde{m}$ that one would obtain by adding divisors D_1 and D_2, then subtracting divisors of the form $P + \widetilde{P} - 2\mathcal{O}$ which is the principal divisor corresponding to a line through the two opposite points. We have already seen that the algorithm provides a solution satisfying $\deg b(x) < \deg a(x)$ and $a(x)$ divides $b(x)^2 + b(x)h(x) - f(x)$. Because any divisor satisfying these conditions is a semireduced divisor, the proof is complete. To show that D_3 is a semireduced divisor, we will examine each pair of opposite ordinary points, $(P, \widetilde{P})$, and each special point, P, to see that the required conditions are met for that point.

The ordinary points will be treated one pair of points at a time. For each pair of opposite ordinary points, $(P, \widetilde{P})$, there are four weights to be considered: two weights for the two opposite points for divisor D_1 and two weights for the two opposite points for divisor D_2. The four weights are here denoted m_1, $\widetilde{m}_1$, m_2, and $\widetilde{m}_2$, referring to the two opposite points P and $\widetilde{P}$ in the two divisors D_1 and D_2. Without loss of generality, we will require that points and divisors are labeled such that the first weight m_1 is not smaller than the other weights. Because D_1 and D_2 are each a semireduced divisor, each pair of opposite points $(P, \widetilde{P})$ can have only one nonzero weight in each of these two divisors. This means that the only quadruples of weights that we must consider for each pair of opposite points $(P, \widetilde{P})$ are either of the form $(m_1, 0, m_2, 0)$ or $(m_1, 0, 0, \widetilde{m}_2)$, and in both cases m_1 is at least as large as each of the other terms.

(i) Suppose that $m_1 = 0$. Then $\widetilde{m}_1 = m_2 = \widetilde{m}_2 = 0$ as well. This means that neither $a_1(x)$ nor $a_2(x)$ has the x coordinate of P as a zero, and so $a_1(x)a_2(x)$ does not have x as a zero. Thus neither P nor $\widetilde{P}$ appears in D_3.

(ii) Suppose that $\widetilde{m}_1 = \widetilde{m}_2 = 0$. Then $a_1(x)$ has $(x - x_P)$ but not $(x - x_{\widetilde{P}})$ as a factor, and $a_2(x)$ does not have $(x - x_P)$ as a factor, so $d(x)$ does not have $(x - x_P)$ or $(x - x_{\widetilde{P}})$ as a factor. Thus $a_1(x)a_2(x)/d(x)$ has $(x - x_P)^{m_1}$ as a factor, but not $(x - x_{\widetilde{P}})$. Thus P, but not $\widetilde{P}$, appears in D_3.

(iii) Suppose that m_1 and m_2 are both nonzero. Then $a_1(x)$ has $(x - x_P)^{m_1}$ as a factor, and $a_2(x)$ has $(x - x_P)^{m_2}$ as a factor, and so $a_1(x)a_2(x)$ has $(x - x_P)^{m_1+m_2}$ as a factor. But $b_1(x) + b_2(x)$ does not have $(x - x_P)$ as a factor because $b_1(x_P) = b_2(x_P) = y_P$, which is nonzero, so $b_1(x_P) + b_2(x_P) = 2y_P$. Thus $a_1(x)a_2(x)/d(x)$ has $(x - x_P)^{m_1+m_2}$ as a factor, but does not have $(x - x_{\widetilde{P}})$ as a factor. Thus P, but not $\widetilde{P}$, appears in D_3 and has weight $m_1 + m_2$.

(iv) Suppose that m_1 and $\widetilde{m}_2$ are nonzero. Then $a_1(x)$ has $(x - x_P)^{m_1}$ as a factor and $a_2(x)$ has $(x - x_{\widetilde{P}})^{\widetilde{m}_2}$ as a factor. Moreover $b_1(x_P) - b_2(x_P) = 0$, so $b_1(x_P) + b_2(x_P) = 0$. In fact, $b_1(x_P) + b_2(x_P)$ has $(x - x_P)^{\widetilde{m}_2}$ as a factor, so $d(x)$ has $(x - x_P)^{\widetilde{m}_2}$ as well, as is shown by taking appropriate derivatives. Then $a_1(x)a_2(x)/d(x)^2$ has weight $(x - x_P)^{m_1 - \widetilde{m}_2}$.

The special points must be considered as several cases that will be treated separately. In each case of a special point, $y = 0$, m_1 and m_2 are each zero or one. There is no distinct opposite point of a special point.

(i) Suppose that $m_1 = 1$ and $m_2 = 0$. Moreover $b_1(x_P) + b_2(x_P) = 2y_P$. In this case $d(x)$ is nonzero.
(ii) Suppose that $m_1 = m_2 = 1$. In this case the product $a_1(x)a_2(x)$ as well as $d(x)$ are each divisible by $(x - x_P)$.

In conclusion, the composition step does produce a semireduced divisor D_3 that is equivalent to $D_1 + D_2$. The reduction step, as shown in Theorem 11.11.1, does result in a reduced divisor that is equivalent to a semireduced divisor, so the proof is complete. □

For a simple example of the Cantor algorithm, we will compute the sum $D_3 = D_1 + D_2$ for two elements of the jacobian on the hyperelliptic curve $\mathcal{X}(\boldsymbol{F}_3) : y^2 = x^5 - 1$. This curve has four points: (1, 0), (2, 1), (2, 2), and ∞. Let $D_1 = \text{div}(x^2 + 2, x - 1)$ and $D_2 = \text{div}(x + 1, 1)$ be two divisors expressed using the Mumford representation. Then

$$\text{GCD}[x^2 + 2, x + 1, x] = 1$$

and the extended euclidean algorithm gives the Bézout relation

$$(x^2 + 2)(2) + (x - 2)(x) + (2)(x) = 1.$$

We then compute

$$a_3(x) = x^3 + x^3 + 2x + 2$$
$$b_3(x) = x + 2$$

and

$$a_3'(x) = x^2 + 2x + 1.$$

We conclude that the sum $D_1 + D_2$ is given by

$$(x - 1, 0) + (x^2 - x + 1, -x + 1) = (x^2 - x + 1, x - 1)$$

in the Mumford representation. Then $D_1 + D_2 = 2(2, 1)$.

For a second example, take the elliptic curve $\mathcal{X}(\boldsymbol{R}) : y^2 = x^5 - 4x^4 - 14x^3 + 36x^2 + 45x$ and the three points $P = (1, 8)$, $Q = (3, 0)$, and $R = (5, 0)$. Let $D_3 = D_1 + D_2$, where

$$\begin{aligned} D_1 &= P + Q - 2\mathcal{O} \\ &= \text{div}(x^2 - 4x + 3, -4x + 12) \\ D_2 &= P + R - 2\mathcal{O} \\ &= \text{div}(x^2 - 6x + 5, -2x + 10). \end{aligned}$$

The sum

$$D_1 + D_2 = 2P + Q + R - 4\mathcal{O}$$

is not a reduced divisor. It must be reduced using the Cantor algorithm, which is a bit tedious for a hand calculation.

11.14 Hyperelliptic-curve cryptography

Elliptic-curve cryptography is in widespread use and has been thoroughly scrutinized. It is believed to be secure. Hyperelliptic-curve cryptography is also believed to be secure, but it has not been scrutinized as thoroughly as elliptic-curve cryptography. Thus one may question whether there is a need for hyperelliptic-curve cryptography. The arguments in favor of its use are not compelling at this time, but the arguments in favor of its study are. It may be that some weakness or disadvantage of elliptic-curve cryptography may one day be found to which hyperelliptic-curve cryptography is not vulnerable. Moreover, study of hyperelliptic-curve cryptography may provide new insights into elliptic-curve cryptography. More practically, it may be that hyperelliptic-curve cryptography can achieve a desired level of secrecy with less complexity. This is because an element of the divisor consists of multiple points, as many points as the genus of the curve. A key of size b bits can be represented by the multiple points of a reduced divisor, each expressed using the elements of a field that is much smaller than b bits, and on the order of b/g bits. If computations involving g points of a curve over the small field $\boldsymbol{F}_{2^{b/g}}$ can be made simpler than computations involving a single point of the large field $\boldsymbol{F}_{2^b}$, then the complexity is reduced accordingly. However, this approach does have its dark side. These properties may also be of some advantage to the cryptanalyst. If the genus is too large, then the notion of smoothness of an element can be introduced, and presumably the method of index calculus becomes a viable attack. Of course, this sort of argument refers to asymptotic behavior, not necessarily practical security.

The jacobian of a hyperelliptic curve is a finite abelian group, and so it has cyclic subgroups. For the purpose of cryptography, we are interested in those cyclic subgroups with very large prime orders in which the discrete-log problem is intractable. Because the discrete-log problem in a large cyclic subgroup of the jacobian of a hyperelliptic curve is believed to be intractable, in general, any cryptographic method that is based on a finite group may be applied, provided that the hyperelliptic curve is carefully chosen. To this purpose, let the reduced divisor D be the public generator of G, a cyclic subgroup of $\text{jac}(\mathcal{X}(\boldsymbol{F}_q))$ of prime order r, under the operation of jacobian addition. The elements of the cyclic group are denoted ℓD for $\ell = 1, \ldots, r$.

Any cryptographic technique based on the abstract structure of a group can be used with the group of a jacobian. For example, a public-key cryptosystem using the jacobian of a hyperelliptic curve can be defined by using the Diffie–Hellman key-exchange protocol. This is a straightforward generalization of a public-key cryptosystem using an elliptic curve for the Diffie–Hellman key exchange that was described in Section 10.4.

To design a cryptosystem based on the Diffie–Hellman key exchange using the jacobian of a hyperelliptic curve, the hyperelliptic curve $\mathcal{X}$ and a designated element D of its jacobian are chosen as standard and public parameters. These parameters are publicly known to all users. The designated element D of the jacobian is given as a reduced divisor, and should have an order equal to a very large prime.

To exchange a key using the Diffie–Hellman procedure, party A randomly chooses a large integer a, then computes and transmits aD to party B where aD is a reduced divisor in the group G. Similarly, party B randomly chooses a large integer b, then computes and transmits the reduced divisor bD to party A. Party A computes the reduced divisor $a(bD)$, which is equal to $(ab)D$, and party B computes the reduced divisor $b(aD)$, which also is equal to $(ab)D$. Both parties now have the same key, which is an element of the jacobian, and so is a reduced divisor which consists of a set of points of the hyperelliptic curve. This reduced divisor is a secret known to both users. A binary sequence formed from the divisor by some standard convention then forms the key. For example, the key could be composed of the x coordinates of the points of $(ab)D$.

The cryptanalyst only has reduced divisors D, aD, and bD from which to compute $(ab)D$. Because it is believed that this task is as difficult as the discrete-log problem in the jacobian of an elliptic curve, the cryptanalyst will fail to find $(ab)D$.

Any generic attack on the discrete-log problem in a finite abelian group can be used to attack the discrete-log problem in the jacobian of a hyperelliptic curve. This includes the Pohlig–Hellman algorithm, the Shanks algorithm, and the Pollard algorithm for discrete logarithms. The Pohlig–Hellman algorithm only applies to groups whose order is composite. To avoid the Pohlig–Hellman attack, one cautiously chooses a hyperelliptic curve whose jacobian has an order that is a large prime or contains a large prime factor. However, the Pohlig–Hellman attack is not a threat if the order of the jacobian cannot be factored by the adversary by any known algorithm. Such a jacobian is probably not useful to the cryptographer either.

The threat that seems to cause the most concern is the Frey–Rück attack. This attack maps the jacobian into the cyclic group under multiplication of a finite field. The Frey–Rück attack appears to become a greater threat as the genus of the hyperelliptic curve increases. This is because of the increased granularity due to more and smaller points in the elements of the jacobian for a fixed key size. For this reason, hyperelliptic curves of genus two or three are commonly preferred.

11.15 Order of the hyperelliptic jacobians

In this section, we will discuss methods of counting the number of points on a hyperelliptic curve and counting the number of elements in the jacobian on that curve. The

Hasse–Weil bound tells us that the number of rational points on a hyperelliptic curve satisfies

$$q + 1 - g\lfloor 2\sqrt{q} \rfloor \leq \#\mathcal{X}(\boldsymbol{F}_q) \leq q + 1 + g\lfloor 2\sqrt{q} \rfloor.$$

A corresponding statement about the number of elements in the jacobian (the number of equivalence classes) is

$$(\sqrt{q} - 1)^{2g} \leq \#\mathrm{jac}(\mathcal{X}(\boldsymbol{F}_q)) \leq (\sqrt{q} + 1)^{2g}.$$

This latter statement is called the *Hasse–Weil interval* for the order of the jacobian. We do not prove this statement.

These statements are related to an unexpected and useful theorem that gives the number of points on a hyperelliptic curve over a finite field, and the number of elements in its jacobian in terms of a set of complex numbers. To reach this theorem, we first define the zeta function for a hyperelliptic curve. The zeta function for a hyperelliptic curve is defined in the same way as it is defined in the case of an elliptic curve.

Definition 11.15.1 *The zeta function of the hyperelliptic curve $\mathcal{X}$ over the field $\boldsymbol{F}_q$ is given by*

$$Z_{\mathcal{X}(\boldsymbol{F}_q)}(t) = e^{\sum_{r=1}^{\infty} N_r t^r / r},$$

where N_r is the number of points of $\mathcal{X}(\boldsymbol{F}_{q^r})$.

Theorem 11.15.2 (Weil) *The zeta function of a hyperelliptic curve is a rational function of t given by*

$$Z(t) = \frac{\psi(t)}{(1-t)(1-qt)},$$

where $\psi(t)$ is a polynomial of degree $2g$ with integer coefficients.

Proof This is a generalization of Theorem 10.11.5, which makes the same statement for the special case of elliptic curves. A proof of this more general statement will not be given. □

Theorem 11.15.3 *A hyperelliptic curve $\mathcal{X}(\boldsymbol{F}_q)$ of genus g over the field $\boldsymbol{F}_q$, when lifted to the extension field $\boldsymbol{F}_{q^m}$, has*

$$\#\mathcal{X}(\boldsymbol{F}_{q^m}) = q^m + 1 - (\alpha_1^m + \overline{\alpha}_1^m + \alpha_2^m + \overline{\alpha}_2^m + \cdots + \alpha_g^m + \overline{\alpha}_g^m)$$

points, and the cardinality of the jacobian is

$$\#\mathrm{jac}(\mathcal{X}(\boldsymbol{F}_{q^m})) = \prod_{i=1}^{g} |1 - \alpha_i^m|^2,$$

where each of the complex numbers $\alpha_1, \ldots, \alpha_g$ is of squared magnitude $|\alpha_i|^2 = q$.

Proof This is a generalization of Theorem 10.11.2, which makes the same statement restricted to elliptic curves. A proof of this statement for hyperelliptic curves will not be given. □

The reason that this theorem is so useful is that the g complex numbers $\alpha_1, \ldots, \alpha_g$ do not depend on m. The number of points on the curve $\mathcal{X}(\boldsymbol{F}_{q^m})$ can be found by enumeration for several small values of m. From these counts the complex numbers α_i can be determined by using the statement of the theorem. Then using the same theorem, the number of points of the jacobian can be computed for any larger value of m.

The theorem does suggest that, for a hyperelliptic curve, the Frobenius trace might be generalized to g parameters, denoted $t_1, t_2, \ldots, t_g$, with $t_i = \alpha_i + \overline{\alpha}_i$ for each i.

11.16 Some examples of the jacobian group

For our first example of a jacobian of a hyperelliptic curve, we will study the curve

$$\mathcal{X} : y^2 + y = x^5 + x^3 + x$$

over the field $\boldsymbol{F}_{2^m}$. We want to find the order of the jacobian, denoted $\#\text{jac}(\mathcal{X}(\boldsymbol{F}_{2^m}))$. In particular, we want to know whether $\text{jac}(\mathcal{X}(\boldsymbol{F}_{2^m}))$ has a subgroup whose order is a large prime. This will be so if $\#\text{jac}(\mathcal{X}(\boldsymbol{F}_{2^m}))$ has a large prime factor, or is itself a prime.

Because the genus g of this hyperelliptic curve equals two, we know from Theorem 11.15.3 that complex numbers ω_1 and ω_2 exist with conjugates $\overline{\omega}_1$ and $\overline{\omega}_2$ such that for all m

$$\#\mathcal{X}(\boldsymbol{F}_{2^m}) = q^m + 1 - \omega_1^m - \overline{\omega}_1^m - \omega_2^m - \overline{\omega}_2^m$$

and

$$\#\text{jac}(\mathcal{X}(\boldsymbol{F}_{2^m})) = |1 - \omega_1^m|^2 |1 - \omega_2^m|^2,$$

so we must find the complex numbers ω_1 and ω_2. This we can do explicitly from the cases with $m = 1$ and $m = 2$ because the hyperelliptic curve of our example is defined over the ground field $\boldsymbol{F}_2$, so we can count that $\#\mathcal{X}(\boldsymbol{F}_2) = 3$ and $\#\mathcal{X}(\boldsymbol{F}_4) = 9$ as follows.

In $\boldsymbol{F}_2$, $y \in \{0, 1\}$. For both values $y = 0$ and $y = 1$, we see that $y^2 + y = 0$, so we must solve $x^5 + x^3 + x = 0$ in $\boldsymbol{F}_2$. Clearly, $x = 1$ does not solve this equation in $\boldsymbol{F}_2$, so only $x = 0$ solves this equation in $\boldsymbol{F}_2$. The point at infinity, given by $\infty = (0, 1, 0)$, is also a zero of the homogeneous polynomial

$$p(x, y, z) = y^2z^3 + yz^4 + x^5 + x^3z^2 + zx^4,$$

so we have found that the curve has exactly three rational points over $\boldsymbol{F}_2$, namely $(0, 0)$, $(0, 1)$, and ∞.

In $\boldsymbol{F}_4$, $y \in \{0, 1, \alpha, \alpha + 1\}$ with $\alpha \in \boldsymbol{F}_4$ satisfying $\alpha^2 = \alpha + 1$, we easily see that $y^2 + y = 0$ if $y = 0$ or 1, and $y^2 + y = 1$ if $y = \alpha$ or $\alpha + 1$. We also easily see that $x^5 + x^3 + x = 0$ if $x = 0, \alpha$, or $\alpha + 1$ and that $x^5 + x^3 + x = 1$ if $x = 1$. Therefore the curve has nine points in $\boldsymbol{F}_4$, namely, $(0, 0)$, $(0, 1)$, $(1, \alpha)$, $(1, \alpha + 1)$, $(\alpha, 0)$, $(\alpha, 1)$, $(\alpha + 1, 0)$, $(\alpha + 1, 1)$, and ∞.

From the equation over the complex field $\boldsymbol{C}$,

$$\#\mathcal{X}(\boldsymbol{F}_{q^m}) = 2^m + 1 - \omega_1^m - \overline{\omega}_1^m - \omega_2^m - \overline{\omega}_2^m$$

with $m = 1$ or 2, we have that

$$\begin{aligned} 3 &= 2 + 1 - \omega_1 - \overline{\omega}_1 - \omega_2 - \overline{\omega}_2 \\ 9 &= 4 + 1 - \omega_1^2 - \overline{\omega}_1^2 - \omega_2^2 - \overline{\omega}_2^2 \end{aligned}$$

from which to solve for elements ω_1 and ω_2 of the complex field $\boldsymbol{C}$. These two equations can be rewritten

$$\begin{aligned} \omega_1 + \overline{\omega}_1 + \omega_2 + \overline{\omega}_2 &= 0 \\ (\omega_1 + \overline{\omega}_1)^2 - 2\omega_1\overline{\omega}_1 + (\omega_2 + \overline{\omega}_2)^2 - 2\omega_2\overline{\omega}_2 &= -4. \end{aligned}$$

Let $\gamma_1 = \omega_1 + \overline{\omega}_1$ and $\gamma_2 = \omega_2 + \overline{\omega}_2$, and recall that $\omega_1\overline{\omega}_1 = \omega_2\overline{\omega}_2 = q$ and $q = 2$ to obtain the equations

$$\begin{aligned} \gamma_1 + \gamma_2 &= 0 \\ \gamma_1^2 + \gamma_2^2 &= 4. \end{aligned}$$

Thus $\gamma_1 = \sqrt{2}$ and $\gamma_2 = -\sqrt{2}$, which means that $\omega_1 + \overline{\omega}_1 = \sqrt{2}$ and $\omega_1\overline{\omega}_2 = 2$. Therefore ω_1 is a zero of $x^2 - \sqrt{2}x + 2$. Hence

$$\begin{aligned} \omega_1, \overline{\omega}_1 &= \frac{\sqrt{2} \pm \sqrt{-6}}{2} \\ \omega_2, \overline{\omega}_2 &= \frac{-\sqrt{2} \pm \sqrt{-6}}{2}. \end{aligned}$$

Now it is straightforward to evaluate $\#\mathcal{X}(\boldsymbol{F}_{q^m})$ for any r. The first few cases are listed in Table 11.2. The values of $\#\text{jac}(\mathcal{X}(\boldsymbol{F}_{q^m}))$ are also listed.

The dependence of $\#\text{jac}(\mathcal{X}(\boldsymbol{F}_{2^m}))$ as a function of m can be studied by starting with the equation

$$\#\text{jac}(\mathcal{X}(\boldsymbol{F}_{2^m})) = |1 - \omega_1^m|^2 |1 - \omega_2^m|^2.$$

Table 11.2 *Counts on a curve and its jacobian*

r	$\#\mathcal{X}(\boldsymbol{F}_{q^r})$	$\#\text{jac}(\mathcal{X}(\boldsymbol{F}_{q^r}))$
1	3	7
2	9	49
3	9	49
4	9	441
5	33	1057
6	97	2401

First use a bit of algebra to eliminate ω_1 and ω_2 in order to simplify the expression. Use the observation that $\omega_1^6 = 8 = \omega_2^6$ to simplify the expression and to write

$$\#\text{jac}(\mathcal{X}(\boldsymbol{F}_{2^m})) = \begin{cases} 2^{2m} + 2^m + 1 & \text{if} \quad m = 1, 5 \pmod 6 \\ (2^m + 2^{m/2} + 1)^2 & \text{if} \quad m = 2, 4 \pmod 6 \\ (2^m - 1)^2 & \text{if} \quad m = 3 \pmod 6 \\ (2^{m/2} - 1)^4 & \text{if} \quad m = 0 \pmod 6. \end{cases}$$

It is clear from this simplification that for this hyperelliptic curve, one cannot hope to find a large prime factor of $\#\text{jac}(\mathcal{X}(\boldsymbol{F}_{2^m}))$ that is comparable to $\#\text{jac}(\mathcal{X}(\boldsymbol{F}_{2^m}))$ unless $m = 1$ or $5 \pmod 6$. For any other m, every prime factor of $\#\text{jac}(\mathcal{X}(\boldsymbol{F}_{2^m}))$ cannot be larger than $\sqrt{\#\text{jac}(\mathcal{X}(\boldsymbol{F}_{2^m}))}$.

For example, with $m = 101$, which is equal to $5 \pmod 6$, one can compute for this curve that

$$\#\text{jac}(\mathcal{X}(\boldsymbol{F}_{2^{101}})) = 7 \cdot 607 \cdot p,$$

where p is a 58-digit prime. Because $\text{jac}(\mathcal{X}(\boldsymbol{F}_{2^{101}}))$ is a group, this means that there is a cyclic subgroup of $\text{jac}(\mathcal{X}(\boldsymbol{F}_{2^{101}}))$ whose order is given by a 58-digit prime.

For applications to cryptography, however, this subgroup might be regarded as having a potential weakness. This is because

$$(2^{2m} + 2^m + 1)(2^m - 1) = 2^{3m} - 1,$$

so this group embeds into $\boldsymbol{F}^*_{2^{303}}$. A discrete-log problem in $\text{jac}(\mathcal{X}(\boldsymbol{F}_{2^{101}}))$ can be mapped, in principle, onto a discrete-log problem in $\boldsymbol{F}^*_{2^{303}}$, where it may be easier to solve by using an attack along the lines of index calculus. Because this mapping into $\boldsymbol{F}^*_{2^{303}}$ is tractable, the discrete-log problem in $\text{jac}(\mathcal{X}(\boldsymbol{F}_{2^{101}}))$ may be vulnerable to such an attack in $\boldsymbol{F}^*_{2^{303}}$, which is known as the *Frey–Rück attack*. Because of the Frey–Rück attack, this particular curve is regarded as potentially vulnerable and, therefore, potentially insecure for cryptography. In fact, for any value of m, computing the discrete log in $\boldsymbol{F}_{q^{3m}}$ is asymptotically less complex than is computing the discrete log in $\text{jac}(\mathcal{X}(\boldsymbol{F}_{2^m}))$ directly.

Because of the fact that $\#\text{jac}(\mathcal{X}(\boldsymbol{F}_{2^m}))$ divides $2^{3m} - 1$, this particular hyperelliptic curve is regarded as potentially vulnerable for any choice of m.

The situation here should be compared with the fact that a supersingular elliptic curve is potentially vulnerable when used for cryptographic purposes because of the Weil pairing and Tate pairing (to be studied in Chapter 12) which give embeddings of the jacobian into a small multiplicative group of a finite field.

For our second example of a group extension of a hyperelliptic curve in the field $\boldsymbol{F}_{2^m}$, we study the curve

$$\mathcal{X} : y^2 + y = x^{383} + 1$$

which is easily seen to be nonsingular. Recalling that $\deg f = 2g + 1$, this curve has genus 191. We do not want to calculate $\omega_1, \ldots, \omega_{191}$ the hard way because it would use 191 equations in 191 unknowns, and would require counting the points of the curve in each subfield F_{2^i} for $i = 1, \ldots, 191$. The set of 191 equations would then need to be solved either algebraically or numerically with very high-precision arithmetic. An alternative method is to evaluate the zeta function using Jacobi sums, a topic that we do not describe. Using the method of Jacobi sums, one could conclude that

$$\begin{aligned}\#\text{jac}(\mathcal{X}(\boldsymbol{F}_q)) &= 1 - 711 \cdot 2^{87} + 2^{191}\\ &= ap,\end{aligned}$$

where p is a 58-digit prime and a is an integer. It can be verified by computation that p does not divide $2^k - 1$ for any positive k smaller than 2000. Therefore no group of order p can embed in $\boldsymbol{F}_{2^k}$ for any k smaller than 2000, and perhaps not even for k much larger. Thus because of the immense size of the field $\boldsymbol{F}_{2^{2000}}$, the discrete-log problem in this field is evidently intractable, so the Frey–Rück attack will fail.

However, a cryptosystem based on this hyperelliptic curve may be vulnerable to a different attack known as the *Adleman–deMarrais–Huang* (ADH) attack. The ADH attack has been developed for curves over fields of odd characteristic. For a fixed p, this attack has a complexity $\exp O\left(\sqrt{g \log g}\right)$ for curves of large genus.

For a third example, consider the curve

$$\mathcal{X}(\boldsymbol{F}_2) : y^2 + y = x^7$$

over $\boldsymbol{F}_2$. This curve has genus three. It is feasible to find ω_1, ω_2, and ω_3 directly by counting points in small extension fields. Then the order of the jacobian can be found to be

$$\#\text{jac}(\mathcal{X}(\boldsymbol{F}_{2^{47}})) = 7p,$$

where p is a 42-digit prime. Moreover, p does not divide $2^{47k} - 1$ for any positive k smaller than 2000. Therefore this cryptosystem is not afraid of a Frey–Rück attack because the group does not embed in $\boldsymbol{F}^*_{2^m}$ for any m smaller than 2000. Nor is the cryptosystem vulnerable to an ADH attack because $\log q$ is much larger than g. Perhaps this cryptosystem is vulnerable to some other attack, but none is known.

Two more examples are based on the curves over $\boldsymbol{F}_2$,

$$\mathcal{X}: y^2 + xy = x^5 + x^2 + 1$$

and

$$\mathcal{X}: y^2 + xy = x^5 + 1,$$

each of which has genus $g = 2$. For the first curve,

$$\#\text{jac}(\mathcal{X}(\boldsymbol{F}_{2^{61}})) = 2p,$$

where p is a 37-digit prime. For the second curve,

$$\#\text{jac}(\mathcal{X}(\boldsymbol{F}_{2^{67}})) = 2p,$$

where p is a 40-digit prime. Both of these curves are secure against both a Frey–Rück attack and an ADH attack. A cryptosystem based on either of these curves is secure against every currently known attack that is based on the mathematics of the curve. Of course, these examples may be considered vulnerable to a direct attack because the keyspace, as limited by the size of the prime, is small.

Our final example is the curve of genus two

$$\mathcal{X}: y^2 + xy = x^5 + x^2 + 1$$

for which

$$\#\text{jac}(\mathcal{X}(\boldsymbol{F}_{2^{113}})) = 2p,$$

where p is a 68-digit prime, given in decimal notation by

$$p = 53919893334301278715823297673841230760642802715019043549764193368381.$$

An element of the jacobian can be represented in the Mumford transform domain as $(a(x), b(x))$. One element of the jacobian of order p is represented

$$\begin{aligned} a(x) &= x^2 + \text{08B44E44B14ADACC86D8762982405}x \\ &\quad + \text{134B113A6992FECFC7D878550F4E3} \\ b(x) &= \text{08F7B4B8D16067C3561E600C51AA7}x \\ &\quad + \text{1D2057F41205A1701420E0C6F159D}, \end{aligned}$$

where the coefficients in $a(x)$ and $b(x)$ are elements of $\boldsymbol{F}_{2^{113}}$ expressed using the notation of the hexadecimal alphabet. Each hexadecimal symbol takes values in the alphabet $\{0, 1, 2, \ldots, F\}$, and represents a 4-bit binary number. Each coefficient of $a(x)$ or $b(x)$ is a hexadecimal number of length 29, and so corresponds to a 116-bit binary number. (Because this binary number specifies an element of $\boldsymbol{F}_{2^{113}}$, the three leading bits are always zero.) In a polynomial basis, the two coefficients of $a(x)$ as elements of the binary field $F_{2^{113}}$, can be represented as

$$a_1 = a_{1,112}z^{112} + \cdots + a_{1,1}z + a_{1,0}$$
$$a_2 = a_{2,112}z^{112} + \cdots + a_{2,1}z + a_{2,0},$$

now with coefficients in F_2.

Problems for Chapter 11

11.1 Show that if the field characteristic is not two and $h(x) = 0$, the elliptic curve

$$y^2 + h(x)y = f(x)$$

is nonsingular if, and only if, $f(x)$ is a square-free polynomial.

11.2 Verify that if the line $x - a = 0$ crosses an ordinary point (a, b) of the hyperelliptic curve $\mathcal{X} : y^2 + h(x)y = f(x)$, then it also crosses the ordinary point $(a, -b - h(a))$ of $\mathcal{X}$.

11.3 Show that the opposite of a point of a hyperelliptic curve is always itself a point of that hyperelliptic curve.

11.4 Consider the hyperelliptic curve $\mathcal{X} : y^2 + y = x^7$ of genus three over F_2. Find an expression for $\#\text{jac}(\mathcal{X})(F_{2^{47}})$ by finding the complex numbers ω_1, ω_2, and ω_3.

11.5 Prove the following statements for the hyperelliptic curve $\mathcal{X} : y^2 + h(x)y = f(x)$ over the finite field F_q.

a If q is even, then $h(x)$ is nonzero if the curve has no singular affine points.

b If q is odd, then $h(x)$ can be made to be zero by an appropriate change of variables. Then $\mathcal{X}$ is a hyperelliptic curve if, and only if, $f'(x)$ has no repeated zeros.

11.6 The point (6, 4) is a special point of the hyperelliptic curve

$$\mathcal{X} : y^2 + xy = x^5 + 5x^4 + 6x^2 + x + 3$$

over the field F_7. Show that this special point is a double zero. Is this an instance of a general principle?

11.7 The integer $p = 2^{31} - 1$ is a kind of prime known as a *Mersenne prime*. Estimate how many hyperelliptic curves of genus two exist over F_p where $p = 2^{31} - 1$.

11.8 The elliptic curve $\mathcal{X}(F_q)$ is a hyperelliptic curve of genus one, so Cantor's algorithm can be used for point addition. Show this correspondence by showing that Cantor's algorithm gives $(x - a_1, b_1) + (x - a_2, b_2) = (x - a_3, b_3)$, where the three points of the elliptic curve (a_1, b_1), (a_2, b_2), and (a_3, b_3) satisfy $(a_1, b_1) + (a_2, b_2) = (a_3, b_3)$.

11.9 Let

$$f(x, y) = \frac{y^4 + 1}{(x^2 + 1)^3}$$

and

$$g(x, y) = \frac{y^4}{(x^2 + 1)^3}$$

be two rational functions on the elliptic curve $\mathcal{X}(\boldsymbol{Q}) : y^2 = x^3 - x$ over the rational field $\boldsymbol{Q}$.

a Show that $f(x, y)$ has no poles or zeros on $\mathcal{X}(\boldsymbol{Q})$.

b Show that $g(x, y)$ has no poles on $\mathcal{X}(\boldsymbol{Q})$. Does $g(x, y)$ have any zeros on $\mathcal{X}(\boldsymbol{Q})$?

c Find the divisors of $f(x, y)$ and $g(x, y)$ over $\mathcal{X}(\boldsymbol{Q})$.

11.10 Specialize the Hasse–Weil interval for the jacobian to the elliptic curves. How does it relate to the Hasse–Weil bound for the elliptic curves?

11.11 Prove that for any two bivariate rational functions $f(x, y)$ and $g(x, y)$ on the elliptic curve $\mathcal{X}(\boldsymbol{F})$, if $g(x, y) = cf(x, y)$ for some nonzero constant c, then $f(D) = g(D)$ for any zero-degree divisor D. The multiplying constant c is irrelevant.

11.12 **a** Show that a hyperelliptic curve whose characteristic is an odd prime can always be written

$$\mathcal{X}(\boldsymbol{F}_p) : y^2 = (x - \alpha_1) \cdots (x - \alpha_{2g+1}),$$

where the α are distinct elements of some extension field.

b Prove that the *hyperelliptic involution* $(x, y) \to (x, -y)$ has $2g + 1$ fixed points.

11.13 Let $G(x, y) = a(x) - b(x)y$ and $H(x, y) = c(x) - d(x)y$ be two elements of the coordinate ring $\boldsymbol{F}[x, y]/\langle y^2 + h(x)y - f(x)\rangle$. Prove that $\text{norm}(G(x, y)H(x, y)) = (\text{norm}G(x, y))(\text{norm}H(x, y))$.

11.14 Generalize the Cantor reduction to the case in which $h(x)$ is not zero.

11.15 Can the proof of Weil reciprocity be generalized to hyperelliptic curves?

Notes for Chapter 11

In contrast to the vast and usually accessible literature of the theory of elliptic curves, the theory of hyperelliptic curves has a somewhat limited literature. Much of what does exist is scattered within advanced and sometimes arcane publications accessible only to a specialist. One accessible article is the survey by Menezes, Wu, and Zuccherato (1998), which presents the material at a level more suitable to the needs of the cryptographer rather than to the tastes of the algebraic geometer. A more recent survey is the article by Boston and Darnall (2009).

Point counting on a hyperelliptic curve of small characteristic depends, in part, on the remarkable properties of the zeta function, many of which were discovered by Weil (1948). Using the theory of the zeta function, the number of points of a hyperelliptic curve in an extension field can be computed based on the point count in several small subfields. This method is undesirable for curves of very large genus. Koblitz (1991) shows how to determine the zeta function in fields of large genus by using Jacobi sums. Generalization of the Schoof algorithm to hyperelliptic curves over very large ground fields is not immediate. Point counting for hyperelliptic curves defined in large fields has been studied by Kedlaya (2001) and Vercauteren (2006).

The application of hyperelliptic curves to cryptography was suggested by Koblitz (1989) as a generalization of the application of elliptic curves to cryptography. The algorithm for adding two elements of the jacobian of a hyperelliptic curve on a nonbinary field in the Mumford representation is due to Cantor (1987). This algorithm was described for curves in the binary fields by Koblitz (1987), and was formally proved by Menezes, Wu, and Zuccherato (1998).

Frey and Rück (1994), generalizing the MOV attack, show how to replace the discrete-log problem in the jacobian to the discrete-log problem in an extension field, thereby providing an attack on hyperelliptic curve cryptography.

The ADH attack on the discrete-log problem on the jacobian of a hyperelliptic curve, which is based on the ideas of index calculus, was devised by Adleman, DeMarrais, and Huang (1994). Curves with large genus, perhaps equal to ten or more, may be vulnerable to the ADH attack. More recently, Thériault (2003) and Gaudry and others (2000, 2007) described an improved attack on hyperelliptic-curve cryptography similar to the index-calculus attack. These papers study the complexity of such attacks on large-genus hyperelliptic curves asymptotically in the size of the finite field. The literature, with a few exceptions, is largely silent regarding the issue of the complexity of attacking hyperelliptic curves of practical size.

12 Cryptography based on bilinear pairings

A larger mathematical structure always can be built on top of a smaller mathematical structure. For example, a pair of sets, together with a function relating those two sets, becomes a larger mathematical package when the ensemble is viewed collectively. Thus, a large elliptic curve can be mapped into a large finite field by mapping each point of the elliptic curve into one point of the finite field. But we want to go beyond this: we want to map a pair of r-torsion points of an elliptic curve into one point of a finite field. More precisely, we want to map a pair of subgroups, each of the same prime order r of an elliptic curve, into a subgroup, also of prime order r, of the finite field. This is the structure that comprises this chapter's subject. A pair of points – one point from each of the two additive subgroups of order r, denoted G_1 and G_2, of a large elliptic curve under the operation of point addition – is mapped into one point of a subgroup, denoted G_T or $G_\times$, of the multiplicative group of a finite field. The mapping with the pair of groups as the domain and the single group as the range, taken as a package, becomes the new mathematical structure that we will want to explore.

We will study a special class of such mappings, called bilinear pairings, and the application of pairings in cryptography. We will show how the intractable problems in the two underlying groups passed up to the other group remain (apparently) intractable problems in most instances, and also how seemingly intractable problems in the two underlying groups may sometimes be mapped into a group where tractable attacks may be known. At the heart of this chapter is the fact that, even though the discrete-log problem on groups of interest appears to be intractable, a bilinear pairing enables a discrete-log problem on an elliptic curve to be changed into a discrete-log problem in a finite field.

We will study the use of bilinear pairings to provide additional methods for secure information systems. One may even regard the introduction of the mathematical objects discussed in this chapter as an attempt to move some of the protocol issues, in appropriate situations, inside the cryptography function where they may be less vulnerable to attack.

12.1 Bilinear pairings

A *bilinear pairing*, or *pairing*, will be defined abstractly in terms of three groups. The first two groups may be the same group, namely a subgroup of prime order of the group of an elliptic curve $\mathcal{X}(F_q)$ under point addition. Instead, the two groups may be different, though still closely related, namely two distinct cyclic subgroups of the same prime order of the group of an elliptic curve $\mathcal{X}(F_q)$ under point addition. When it is necessary to make this distinction, the first instance is called a *symmetric bilinear map*, which will be our exemplar case, and the second instance is called an *asymmetric bilinear map*. Each of the first two groups, G_1 and G_2, has a group operation that is referred to as addition and the group operation is denoted $+$. When the two groups under addition are the same group, that group may be denoted G_+. The third group is a cyclic subgroup of the same order of the multiplicative group $F_{q^k}^*$ of an extension F_{q^k} of the finite field F_q. This group has a group operation that is referred to as field multiplication and is denoted $\times$. The third group itself is denoted G_T or $G_\times$. A symmetric mapping from pairs of elements (Q, R) of G_+ into G_T is denoted Φ. Thus a pair of elements (Q, R) from G_+ is mapped into the element $\Phi(Q, R)$ of G_T. The symmetric bilinear map $\Phi(Q, R)$ is a bilinear pairing if it is linear in each variable. In particular, this requires that

$$\Phi(aQ + bQ', R) = \Phi(Q, R)^a \Phi(Q', R)^b,$$
$$\Phi(Q, aR + bR') = \Phi(Q, R)^a \Phi(Q, R')^b,$$

and

$$\Phi(aQ, bR) = \Phi(Q, R)^{ab}$$

for all integers a and b and for all Q, Q', R, and R' of the group G_+.

More concretely, a bilinear pairing – either a symmetric bilinear pairing or an asymmetric bilinear pairing – is a bilinear mapping that is nondegenerate, and is computationally tractable. The symmetric bilinear mapping is *nondegenerate* if for every $P \neq \mathcal{O}$, there is at least one Q such that $\Phi(P, Q) \neq 1$, and for every $Q \neq \mathcal{O}$, there is at least one P such that $\Phi(P, Q) \neq 1$. This condition is imposed to prevent triviality. The notion of computationally tractable has not been precisely defined, so the notion of a bilinear pairing is not precisely defined either. From a practical point of view, however, the meaning of the statement is clear.

We require that the cardinalities of the three groups, G_1, G_2, and $G_\times$, are all equal to the same prime number. That is, $\#G_1 = \#G_2 = \#G_\times$, and equal to a prime. Because the identity element is the only element of a group that has order one, and a group of prime order has more than one element, there must exist at least one element that is not the identity element. The order of any other element is larger than one and Lagrange's

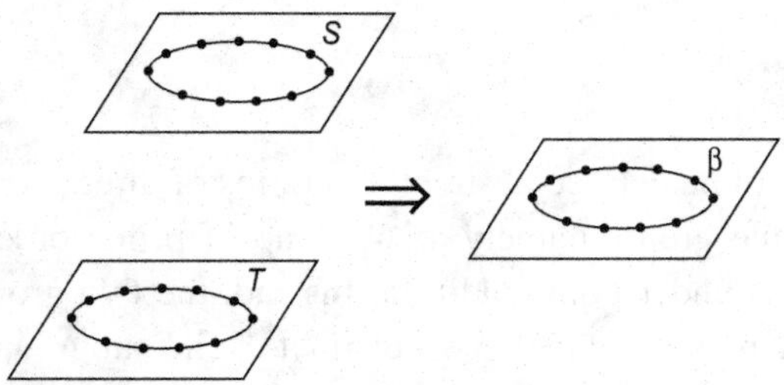

Figure 12.1 Mapping $\mathcal{X}[11] \times \mathcal{X}[11]$ to $\boldsymbol{F}_{q^k}[11]$

theorem states that its order must divide the order of the group. Because the order of the group has been specified to be a prime, the order of every nonidentity element must be equal to the order of the group. Therefore each of the three groups is cyclic, and, for each group, any nonidentity element g can be the generator of that group. We express a group generated by g as $G = \langle g \rangle$.

The group G_+ is generated by any element $S \in G_+$ that is not the identity element. Thus $G_+ = \langle S \rangle$. Because a bilinear pairing is nondegenerate, there is an R such that $\Phi(S, R)$ is not the identity element of $G_\times$, and so $\Phi(S, R)$ generates the group $G_\times$. That is, $G_\times = \langle \Phi(S, R) \rangle$. Therefore Φ takes a generator of group G_+ to a generator of group $G_\times$.

An illustration of a pairing for cyclic groups in $\mathcal{X}(\boldsymbol{F}_q)$ of order eleven is shown in Figure 12.1. The two input groups of order eleven are shown on the left, and the output cyclic group of order eleven is shown on the right. The points in the cyclic groups on the left are denoted aS and bT, respectively, with respect to reference points S and T used as generators. The pairing is $\Phi(aS, bT) = \beta^{ab}$, where β is an element of order eleven in an appropriate extension field $\boldsymbol{F}_{q^k}$. For this small example, the mapping is easy to compute by first computing a and b from aS and bT, and then computing β^{ab}. For a cryptographically large cyclic group of prime order, however, the computation of a and b from aS and bT is (apparently) intractable because the discrete-log problem on an elliptic curve is (apparently) intractable. The bilinear pairing bypasses the need for this discrete logarithm. The important structure of the function Φ is that one argument aS visits all points by taking b steps around the cycle with steps of size a, while the other takes a steps around the cycle with steps of size b. When defining a pairing, it does not matter which generators are chosen as the reference points, S and T. It is not even necessary when defining a pairing to make S and T explicit. We need not know them. It need not even be tractable to compute S and T, or a and b, as long as $\Phi(aS, bT)$ itself can be computed from aS and bT. Later, we construct specific pairings by devising explicit computational formulas. Then we must verify that the formulas lead to all of the pairing properties. Therefore in such cases, the indicated S and T must exist, though we will not need to know what they are, nor in cases of cryptographic interest, can we know.

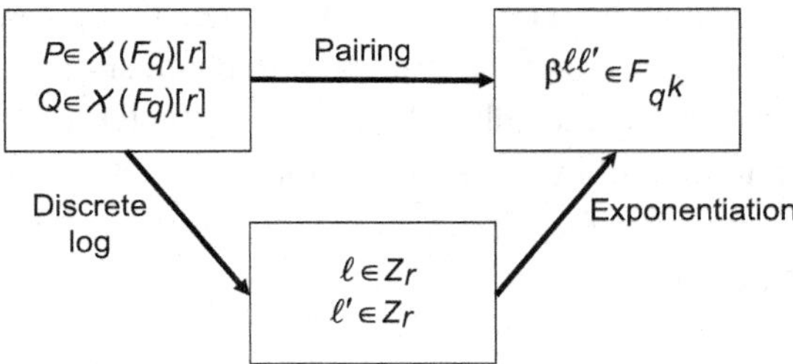

Figure 12.2 Skirting the intractable with pairing.

Without some strong reason to the contrary, one would be inclined to choose the groups G_+ and $G_\times$ such that the discrete-log problem in $G_\times$ is at least as hard as the discrete-log problem in G_+. For a pairing-based cryptosystem to be practical, the discrete-log problem in the subgroup of $\mathcal{X}(\boldsymbol{F}_q)$ should be intractable, and the discrete-log problem in $\boldsymbol{F}^*_{q^k}$ should also be intractable.

12.2 Pairing-based cryptography

Pairing-based cryptography is that branch of cryptography that uses the formal mathematical structure of a bilinear pairing, taking the points of a pair of elliptic curves into a finite field, thereby enriching a variety of topics of secure communication. Pairing-based methods are used for tripartite key exchange, for short signatures, and for identity-based encryption. The underlying structure is based on a bilinear mapping from a pair of like cyclic subgroups of an elliptic curve on a large finite field into an extension of that field. There are several important instances of such a bilinear pairing. Two of these are the *Weil pairing* and the *Tate pairing*. These, and variations, will be described in later sections of this chapter. These pairings can be rather indigestible topics of mathematics. However, for now, it is not necessary to explain or describe the specific bilinear pairings. It is only necessary to describe the formal properties that we require to be satisfied by any such bilinear pairing. These properties describe the external behavior, leaving the pairing function itself as a black box that can be used without looking inside. The first part of this chapter deals with how these bilinear pairings can be used for various cryptographic purposes. The pairings themselves will be discussed at the end of the chapter in a way that lets one look inside the black box to whatever extent is desired.

The purpose of a bilinear pairing is explained by the conceptual diagram in Figure 12.2. The bottom path shows an alternative procedure in which each copy of G_+ is mapped first to $\boldsymbol{Z}_r$, and then each pair of elements of $\boldsymbol{Z}_r$ is mapped to $G_\times$. While this procedure does give the form of a bilinear pairing, it is intractable as written because it involves a discrete logarithm. The pairing computation, shown at the top of

Figure 12.2, takes $G_+ \times G_+$ immediately to $G_\times$, bypassing the need for a discrete-log computation in the group of the elliptic curve. Surprisingly, in contrast to the indirect path, the direct path is computationally tractable. The pairing – though conceptually difficult compared to the discrete logarithm – is computationally much simpler when the group order is extremely large. This fact lies at the heart of this chapter. Without this fact, the chapter would be pointless.

12.3 Pairing-based key exchange

The Diffie–Hellman key exchange on an elliptic curve is based on a public point P of the curve, two randomly chosen secret integers a and b, and two computed points aP and bP that are made public. The Diffie–Hellman key exchange can be reformulated in terms of a bilinear pairing. Let $\Phi(P, Q)$ be a bilinear pairing known to the entire community, and let s be a large integer also publicly known. Parties A and B each randomly choose a point of the curve, denoted P_A and P_B, respectively. Party A computes $Q_A = sP_A$ and makes it public. Party B computes $Q_B = sP_B$ and makes it public. Party A secretly computes $\Phi(P_A, Q_B) = \Phi(P_A, P_B)^s$. Party B secretly computes $\Phi(P_B, Q_A) = \Phi(P_B, P_A)^s$. Thus, both parties now have the same field element, $\Phi(P_B, P_A)^s$ which is to be used as the key. However, although the cryptanalyst knows s, Q_A, and Q_B, it cannot compute $\Phi(P_B, P_A)^s$.

The use of a bilinear pairing can also provide a one-pass tripartite (three-party) key exchange that has the character of a Diffie–Hellman key exchange, whereas the conventional Diffie–Hellman protocol must transmit three messages for a tripartite key exchange. This means that each party must transmit three times. By using a bilinear pairing, each party needs to transmit only once.

Suppose that three parties, denoted A, B, and C, want to create a common key over a public network. This can be done by the standard Diffie–Hellman key exchange by using a protocol with multiple transmissions, as shown in Figure 12.3. In a symmetric configuration, each party must send three messages. Each party first chooses a random integer, denoted a, b, or c, respectively, and then each computes either α^a, α^b, or α^c, respectively. In the first transmission, each party broadcasts α^a, α^b, or α^c, respectively. Party A receives α^b and α^c, computes $(\alpha^b)^a$ and $(\alpha^c)^a$, then transmits α^{ab} to party C, and α^{ac} to party B. The other two parties do the equivalent. As a result, party A receives α^{bc} and computes α^{abc}, party B receives α^{ac} and computes α^{abc}, and party C receives α^{ab} and computes α^{abc}. After three transmissions by each party, all three parties have α^{abc}. Although three of the transmissions are redundant, they can only be eliminated by making the procedure asymmetric.

The cryptanalyst is able to observe the multiple transmissions and attempts to find α^{abc} from knowledge of α, α^a, α^b, α^c, α^{ab}, α^{ac}, and α^{bc}. This is a more general form

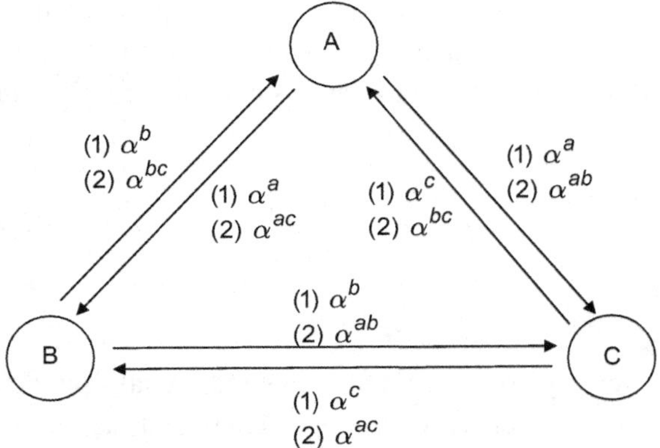

Figure 12.3 A three-pass key exchange

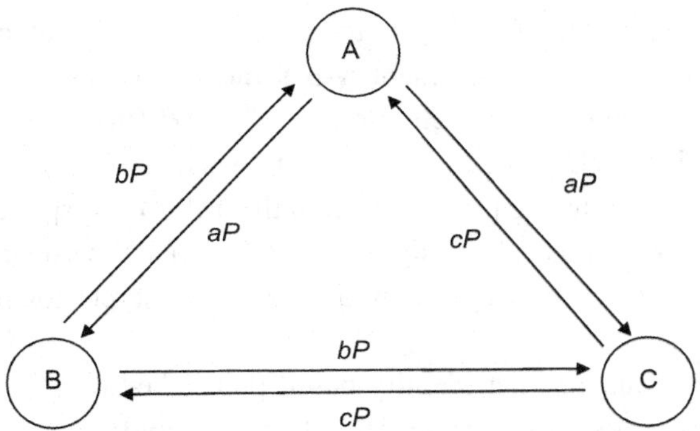

Figure 12.4 A one-pass key exchange

of the Diffie–Hellman problem. Clearly, this problem is not harder than the Diffie–Hellman problem. It may be easier, though it is thought not to be.

Instead, any bilinear pairing can be used for an alternative tripartite key exchange, shown in Figure 12.4. This alternative key exchange is known as the *Joux key exchange*. It does this with only one transmission from each node. The Joux key exchange requires a bilinear pairing $\Phi(P_1, P_2)$, such as one taking points P_1 and P_2 of an elliptic curve into a finite field $\boldsymbol{F}_{q^k}$. Each of the three parties A, B, and C chooses a random integer a, b, or c, respectively, and computes and transmits its point aP, bP, or cP accordingly. Party A receives bP and cP, and computes $\Phi(bP, cP)^a = \Phi(P, P)^{abc}$. Party B receives aP and cP, and computes $\Phi(aP, cP)^b = \Phi(P, P)^{abc}$. Party C receives aP and bP, and computes $\Phi(aP, bP)^c = \Phi(P, P)^{abc}$. When this is finished, all three share knowledge of $\Phi(P, P)^{abc}$, which is an element of the field $\boldsymbol{F}_{q^k}$. As such, it is used as a common key by the three parties.

The crystanalyst can observe aP, bP, and cP and can attack this protocol by attempting to compute $\Phi(P, P)^{abc}$ from aP, bP, and cP. It is obvious that, if

$\alpha = \Phi(P, P)$, the field elements α^{ab} and α^{ca} can be readily computed by the cryptanalyst. Thus the attack can be regarded as an attempt to compute the field element α^{abc} from α^{ab}, α^{bc}, and α^{ca}. This is believed to be as hard as computing α^{ab} from α^a and α^b.

12.4 Identity-based encryption

An attractive goal for public-key cryptography is to use an individual's public identity, as codified by a name, a network address, or a longer document, as an encryption key. When using the usual Diffie–Hellman key exchange, however, this goal of an identity-based encryption is thwarted by the intractability of the discrete-log problem.

To this point, recall that the Diffie–Hellman key exchange requires a user to choose a random integer a, depending on the group, then to compute the group element aP or α^a, which then is used to form the key. An identity-based encryption scheme based on the Diffie–Hellman key exchange would choose a so that aP (or α^a) is equal to the individual's public identity, such as the user's name and network address. But this requires the user to solve the discrete-log problem to find the a that corresponds to aP (or α^a). If the underlying group is chosen to be simple enough so that the user can solve the discrete-log problem, then the cryptanalyst also can solve the discrete-log problem, so the system is not secure.

Thus at first glance, it may appear that identity-based public-key cryptography is not possible. In fact, and perhaps surprisingly, a kind of identity-based public-key encryption is possible by using a bilinear pairing, though it requires a certification service to set up.

Identity-based encryption is a technique by which a user encrypts or verifies a signature without the need to have a shared key, or to look up a public key, or to refer to a trusted third party for each exchange. Instead, the user's public key consists of a common identification label such as a name and an address. The only constraints are that the user's public key must be sufficiently long, and, for signatures, that this identification could not be repudiated later by the owner. Secret keys that shadow the public key are generated from a "master" secret key by a certification service and are not public. A need for such a certification service is unavoidable, because if, by computing a discrete log, the user could derive a secret key from that user's identity, then so could anyone else.

Instead, a bilinear pairing $\Phi(Q, R)$ is used to establish an identity-based cryptosystem. To this purpose, a representation of the field element $\Phi(Q, R)$ that is compatible with the representation of the public key is required. For example, if the public key is represented as a binary string, then the field element $\Phi(Q, R)$ must be expressed in the form of a binary string.

	Secret	Public
Certifier	t	$tP = T$
Decryptor	tQ	Q

Figure 12.5 The key structure in an identity-based cryptosystem

Two hash functions H_1 and H_2 are required. One function is used to digest the user's identity in a standard representation, and one function is used to digest the user's messages. The identity hash function H_1 reduces the public identity to a point Q of the elliptic curve. The message hash function H_2 reduces the message to a binary string of a standard length. Thus, we can regard the identity to be a point of the elliptic curve. We define these two functions as:

$$H_1 : \{0, 1\}^* \Rightarrow G_+ \backslash \{0\}$$
$$H_2 : G_\times \Rightarrow \{0, 1\}^\ell.$$

These functions must be publicly known and collision-free. Each user's public encryption key is actually a digest of standard length of that user's public identity documents, as hashed by the public function H_1.

To certify users, there is a trusted certification authority with a secret certification key t, known only to the trusted certification authority, and a corresponding public certification key $T = tP$, where P is a generator of the group G_+. The point P is public, and is a standard known to all users, as in the certifier's public key T. The trusted certification authority certifies each identity-based public key Q in the system. The trusted certification authority can then withdraw from the situation by destroying the secret certification key t.

To certify, the certification authority observes a user's public key Q, and replies, over a protected channel, to that user's request for a permanent personal secret decryption key. The response is tQ, where t is the secret certification key and Q is the user's public key. For this function, the trusted certification authority is not required to verify the identity of the user. Verification of identity is a separate function and can be performed independently. The only task of this certification authority is to return, over a secure channel and without question, a private decryption key to anyone who presents a public encryption key.

The set of keys used by an identity-based cryptosystem is shown in Figure 12.5. This should be compared to the key structure in a Diffie–Hellman cryptosystem, as shown in Figure 12.6.

	Secret	Public
User A	a	aP
User B	b	bP

Figure 12.6 The key structure in a Diffie–Hellman cryptosystem

Notice that, for identity-based encryption, the user's keys have a relationship that reverses the relationship used in elliptic-curve cryptography. In that case, the public key was obtained as a point-multiple of the secret key. Now, the secret key is obtained as a point-multiple of the public key. The attraction of identity-based encryption comes down to the fact that multiplication of a point of the curve by t appears in different columns of Figure 12.5.

After certification, the key is valid indefinitely with no recertification or other action required from the certification authority. After all parties have been certified, the trusted certification authority can be dissolved because it is no longer needed by that network. The network is then deemed to be permanently secure. Even if a member of the network compromises its own key, as by a careless action, the remainder of the network remains secure.

To encrypt a message x for the user whose public identity key is Q, the encryptor randomly chooses an integer r and computes rP and $\Phi(Q, T)^r$, where Q is the identity-based public key and T is the public certification key. This field element $\Phi(Q, T)^r$ is then expressed as a binary sequence. The encrypted message is the pair

$$\{rP, x + \Phi(Q, T)^r\}.$$

Because r is randomly chosen, $(\Phi(Q, T)^r)$ is random as well, and the message is appropriately concealed.

Before describing the decryption, we provide the following reduction

$$\begin{aligned}\Phi(Q, T)^r &= \Phi(Q, tP)^r \\ &= \Phi(Q, P)^{rt} \\ &= \Phi(tQ, rP),\end{aligned}$$

and recall that rP is part of the encrypted message. Thus the decryptor – knowing tQ – can compute the final term on the right, and so can compute $\Phi(tQ, rP)$ thereby obtaining $\Phi(Q, T)^r$. The message is then easily recovered because

$$x + \Phi(Q, T)^r - \Phi(tQ, rP) = x.$$

Such a system presents a variety of practical security issues. The production and distribution of the secret key tQ to a user must be secure, and efforts must be made to prevent loss, duplication, or unauthorized access to each user's secret key. Compromise of any user's secret key, however, threatens only the messages to that user, past or present, not the entire network.

Computationally, identity-based encryption encounters many of the same issues as standard public-key cryptography. That is, when the secret key is known, computation is relatively easy. To recover the message without the secret key is intractable, requiring solution of the discrete-log problem.

Critics of identity-based encryption state that even if the public identity-based key is something as simple as a name and address, it must still be looked up in many common situations. In that case, it is a small matter, in the critics' view, to also look up a separate encryption key. Accordingly, there is no compelling advantage in equating the key to the identity.

12.5 Pairing-based signatures

A bilinear pairing can be used as a means to form a digital signature. This method is regarded as attractive in applications for which a short signature is required. The method makes use of a publicly known cryptographic hash function that maps a binary string of arbitrary length to a digest consisting of a nonzero element of the additive group G_+, such as a point of an elliptic curve. The group G_+ is chosen large enough so that the discrete-log problem is intractable for that group.

As is usual in signature protocols, the cryptographically secure hash function is used to reduce a binary string of arbitrary length into a digest, which has a fixed length. In this case, the digest must be a nonidentity element of the additive group G_+ that provides an input to the bilinear pairing. Presumably the group G_+ is a subgroup of the group of an elliptic curve. Thus the hash function has the form

$$H : \{0, 1\}^* \Rightarrow G_+ \backslash \{0\}.$$

The signer randomly chooses any nonzero integer a smaller than $\#G_+$ and publishes that user's public signature verification key aP. To sign the message x, the signer computes the digest $H(x) = \text{hash}(x)$, and the point

$$aH(x) \in G_+.$$

The signed message is the pair $(x, aH(x))$. The length of the signature is determined by $\#G_+$. To compute a from the signed message involves computing a discrete logarithm, and so this inverse computation is intractable.

The signature can be verified effectively by using the properties of the bilinear pairing and the signer's public signature verification key aP. Simply compute $\Phi(aH(x), P)$, which only involves the known signed message $aH(x)$ and the public point P. Then compute $\Phi(H(x), aP)$ by hashing the message x and using the public signature verification key aP. If $\Phi(aH(x), P) = \Phi(H(x), aP)$, the signature is declared to be valid.

The advantage of pairing-based signatures is that for a given level of presumed security, the signature is short compared to other signature methods. This advantage is a widely accepted opinion, but supported only by anecdotal evidence.

12.6 Attacks on the bilinear Diffie–Hellman protocol

Much of this chapter is written under the premise that the bilinear Diffie–Hellman problem is intractable. This is widely believed to be the case though, again, apparently no proof supporting this belief is known. The original Diffie–Hellman problem for the adversary, when stated in the group of an elliptic curve, is to find abP when given P, aP, and bP, where P is a point of an elliptic curve that generates an appropriate cyclic group, and where a and b are integers. The same Diffie–Hellman problem can be stated in any large cyclic group. In the multiplicative notation appropriate to $\boldsymbol{F}_q^*$, the problem is to find α^{ab} when given α, α^a, and α^b.

The Diffie–Hellman problem is believed to be an intractable problem in groups of cryptographic interest, but again the evidence for this belief is anecdotal, since no formal proof of this statement is known. A solution of the discrete-log problem provides a solution of the Diffie–Hellman problem, but a solution of the Diffie–Hellman problem need not solve the discrete-log problem. Moreover, it is not known whether the original Diffie–Hellman problem is easier than the discrete-log problem, but it clearly is not harder.

An enlarged version of the Diffie–Hellman problem, which arises in tripartite key exchange, is to find $abcP$ when given P, aP, bP, cP, abP, bcP, and caP. That this problem is clearly not harder than the original Diffie–Hellman problem is evident because one can always ignore some of the terms. For example: find $abcP$ when given P, aP, and bcP. It is generally believed that the tripartite version of the Diffie–Hellman problem is not easier than the original Diffie–Hellman problem, though apparently no proof is known for this statement.

Another variant is the so-called bilinear Diffie–Hellman problem. The bilinear Diffie–Hellman problem, when stated in the group of an elliptic curve, considers a bilinear pairing Φ between the additive group G_+ and the multiplicative group $G_\times$. The problem is to find $\Phi(P, P)^{abc}$ when given Φ, P, aP, bP, and cP. This problem is not harder than the discrete-log problem because, if the discrete-log problem were solved, one could compute a, b, and c. Moreover, this problem is not known to be easier.

The last variant we will discuss is known as the decisional Diffie–Hellman problem. In the case of an elliptic curve, the task is to determine whether a given point Q is equal to abP when given Q, P, aP, and bP. In the case of a finite field, the task is to determine whether a given element β is equal to α^{ab} when given β, α, α^a, and α^b. This problem has a deceptive resemblance to the original Diffie–Hellman problem, but is actually quite different. A solution to the decisional Diffie–Hellman problem need not be a solution to the original Diffie–Hellman problem. Indeed, the decisional Diffie–Hellman problem for an elliptic curve is trivial to solve if a bilinear pairing is available. Simply compute $\Phi(P, Q)$ and $\Phi(aP, bP)$. The second of these is equal to $\Phi(P, abP)$. This means that if $\Phi(P, Q) = \Phi(aP, bP)$, then $Q = abP$. Thus the complexity of the decisional Diffie–Hellman problem is the same as the complexity of computing the bilinear pairing.

We emphasize the point that although the decisional Diffie–Hellman problem may superficially appear to be only a minor variation of the original Diffie–Hellman problem, and to the casual eye may seem to be not much different, it is, in fact, actually much easier and, for the group of an elliptic curve, tractable algorithms are known. The lesson here is that the complexity of a problem can be a delicate question, and depends strongly on the precise statement of the problem.

12.7 Torsion points and embedding degree

Under the operation of point addition, an elliptic curve over the finite field $\boldsymbol{F}_q$ forms an abelian group. The Hasse–Weil bound states that the number of points in this group is approximately equal to the field size q. If the field size is very large, then the number of points on the elliptic curve is very large as well, so the points of the curve form a very large group. If the size of the field requires a 100-digit integer to state, then the order of the group also requires a 100-digit integer to state. While the sheer massiveness of this group may at first be quite intimidating, much about the group can be understood by patient reasoning. Indeed, we will study not only the group structure of the curve $\mathcal{X}(\boldsymbol{F}_q)$, but also the group structure of the curve $\mathcal{X}(\boldsymbol{F}_{q^k})$ in the extension field $\boldsymbol{F}_{q^k}$ for certain values of k.

Recall that any finite abelian group is isomorphic to a direct sum of cyclic groups as follows:

$$G \simeq \boldsymbol{Z}_{k_1} \oplus \boldsymbol{Z}_{k_2} \oplus \cdots \oplus \boldsymbol{Z}_{k_n}.$$

Each of the subgroups $\boldsymbol{Z}_i$ is cyclic with an order equal to a prime or a prime power. If the orders of the subgroups of this decomposition are pairwise coprime, then G itself is a cyclic group. Moreover, any subgroup of prime order must be cyclic. The group of the elliptic curve $\mathcal{X}(\boldsymbol{F}_q)$ under point addition is an abelian group, and so it has such a decomposition. In Chapter 10, we were interested in cyclic subgroups of this group of very large prime order.

Any abelian group of prime order r is isomorphic to $\boldsymbol{Z}_r$, so any subgroup G_+ of $\mathcal{X}(\boldsymbol{F}_q)$ of prime order r is isomorphic to $\boldsymbol{Z}_r$. This means that such a subgroup G_+ is cyclic. The theory of groups does assure us that the isomorphism exists, but to explicitly state the isomorphism is equivalent to solving the elliptic-curve discrete-log problem, which is computationally intractable. To learn more about the group G_+, we must study a subgroup of $\mathcal{X}(\boldsymbol{F}_q)$ as itself, not as an isomorphic copy of $\boldsymbol{Z}_r$, because from a computational point of view the isomorphism is hidden from us by the intractability of the discrete-log problem.

The decomposition of an elliptic curve into subgroups is captured by introducing the notion of a *torsion point*. For any integer r, an r-torsion point of $\mathcal{X}(\boldsymbol{F}_q)$ is a point P for which $[r]P = \mathcal{O}$. The elliptic curve $\mathcal{X}(\boldsymbol{F}_q)$ can also be considered in any extension field $\boldsymbol{F}_{q^k}$, or even in the union of all extension fields $\overline{\boldsymbol{F}}_q = \bigcup_{i=1}^{\infty} \boldsymbol{F}_{q^i}$. The set of r-torsion points of $\mathcal{X}(\boldsymbol{F}_q)$ in the ground field $\boldsymbol{F}_q$, denoted $\mathcal{X}(\boldsymbol{F}_q)[r]$, consists of all points of $\mathcal{X}(\boldsymbol{F}_q)$ of order r, all points of $\mathcal{X}(\boldsymbol{F}_q)$ whose order divides r, and the identity element $\mathcal{O}$. It is easy to see that $\mathcal{X}(\boldsymbol{F}_q)[r]$ is a subgroup of $\mathcal{X}(\overline{\boldsymbol{F}}_q)$. For most values of r, $\mathcal{X}(\boldsymbol{F}_q)[r]$ will consist only of the identity element. The set of rational r-torsion points of $\mathcal{X}(\boldsymbol{F}_q)$ will be nontrivial for the prime r only if r divides $\#\mathcal{X}(\boldsymbol{F}_q)$. The set of r-torsion points of $\mathcal{X}(\overline{\boldsymbol{F}}_q)$ is defined similarly.

In general, any group for which every point has finite torsion, even though the group itself may be infinite, is called a *torsion group*. Although the elliptic curve $\mathcal{X}(\overline{\boldsymbol{F}}_q)$ has an infinite number of elements, it is a torsion group because every point of $\mathcal{X}(\overline{\boldsymbol{F}}_q)$ is an element of $\mathcal{X}(\boldsymbol{F}_{q^k})$ for some k, and so has finite order. We will see that, for most values of r, if there are any r-torsion points in $\mathcal{X}(\overline{\boldsymbol{F}}_q)$, then there are exactly r^2 r-torsion points.

An elliptic curve over a finite field is a finite abelian group, so its structure is described by the representation theorem for finite groups. To analyze this structure, we first analyze the structure of the subgroup, $\mathcal{X}(\boldsymbol{F}_q)[r]$, consisting of all r-torsion points of $\mathcal{X}(\boldsymbol{F}_q)$. This is the set of all points, possibly in an extension field of $\boldsymbol{F}_q$, whose order divides r, together with the point at infinity $\mathcal{O}$. With the exception of the cases in which r is a multiple of the field characteristic p (left as an exercise), we will see that this subgroup always has a form given by the isomorphism[1]

$$\mathcal{X}(\overline{\boldsymbol{F}}_q)[r] \simeq \boldsymbol{Z}_r \oplus \boldsymbol{Z}_r.$$

This important isomorphism is the central statement of the next section. In particular, Theorem 12.8.2 of the next section tells us that there are exactly r^2 points of torsion

[1] It may be helpful to regard the left side of this group isomorphism as a "torus" defined on $\boldsymbol{Z}^2$. The torsion structure theorem describing $\mathcal{X}(\boldsymbol{F}_q)[r]$ then mimics the fact that an elliptic curve $\mathcal{X}(\boldsymbol{C})$ is isomorphic to a complex torus.

Table 12.1 *Two cycles in $\mathcal{X}(\overline{\boldsymbol{F}}_{67})$*

$P = (09, 01)$	$Q = (58, 01i)$
$2P = (59, 63)$	$2Q = (08, 63i)$
$3P = (64, 38)$	$3Q = (03, 38i)$
$4P = (17, 46)$	$4Q = (50, 46i)$
$5P = (36, 23)$	$5Q = (31, 23i)$
$6P = (62, 65)$	$6Q = (05, 65i)$
$7P = (21, 06)$	$7Q = (46, 06i)$
$8P = (06, 17)$	$8Q = (61, 17i)$
$9P = (06, 50)$	$9Q = (61, 50i)$
$10P = (21, 61)$	$10Q = (46, 61i)$
$11P = (62, 02)$	$11Q = (05, 02i)$
$12P = (36, 44)$	$12Q = (31, 44i)$
$13P = (17, 21)$	$13Q = (50, 21i)$
$14P = (64, 29)$	$14Q = (03, 29i)$
$15P = (59, 04)$	$15Q = (08, 04i)$
$16P = (09, 66)$	$16Q = (58, 66i)$
$17P = \mathcal{O}$	$17Q = \mathcal{O}$

r if there are any points of torsion r. If r, but not r^2, divides $\#\mathcal{X}(\boldsymbol{F}_q)$, then r of them are on the rational curve $\mathcal{X}(\boldsymbol{F}_q)$. Where then are the remaining $r^2 - r$ points of torsion r?

To preview the answer to this question by means of an example, note that the curve corresponding to the supersingular polynomial $y^2 = x^3 + x$ has 68 points in $\boldsymbol{F}_{67}$. Because 17 divides 68, the rational curve must have a subgroup of order seventeen. The point $(9, 1) \in \mathcal{X}(\boldsymbol{F}_{67})$ has order seventeen. Its orbit is given on the left side of Table 12.1. Thus, there are seventeen rational seventeen-torsion points of $\mathcal{X}(\boldsymbol{F}_{67})$.

Because the prime 67 is of the form $4k + 3$, the element $\sqrt{-1}$ does not exist in $\boldsymbol{F}_{67}$, so $\boldsymbol{F}_{67^2}$ can be defined with $i^2 = -1$. Moreover, the mapping $(x, y) \to (-x, iy)$ takes a point of $\mathcal{X}(\boldsymbol{F}_{67})$ to a point of $\boldsymbol{F}_{67^2}$. The orbit of $(-9, i)$ in $\boldsymbol{F}_{67^2}$ can be written by reference to the orbit of $(9, 1)$ in $\boldsymbol{F}_{67}$. The orbit generated by the point $(-9, i) \in \mathcal{X}(\boldsymbol{F}_{67^2})$ has order seventeen. Its orbit is shown on the right side of Figure 12.1. These points of $\mathcal{X}(\boldsymbol{F}_{67^2})$ are also seventeen-torsion points. In general, $[a]P + [b]Q$ is also a seventeen-torsion point for each a and b from 0 to 16. Thus the full set of seventeen-torsion points is given by

$$\mathcal{X}(\overline{\boldsymbol{F}}_{67})[17] = \mathcal{X}(\boldsymbol{F}_{67^2})[17] = \{[a]P + [b]Q \mid a = 0, \ldots, 16, b = 0, \ldots, 16\}.$$

These points can be displayed in an array as shown in Figure 12.7. There are 289 points in the array corresponding to the 17^2 seventeen-torsion points of $\mathcal{X}(\overline{\boldsymbol{F}}_{67})$, and if P and Q are specified, then any other point can be designated by giving only the integer coefficients a and b. An arbitrary point R of $\mathcal{X}(\boldsymbol{F}_{67^2})[17]$ can always be written as

b \ a	0	1	2	3	4	⋯	16
0	$\mathcal{O}$	P	$2P$	$3P$	$4P$	⋯	$16P$
1	Q	$P+Q$	⋯			⋯	$16P+Q$
2	$2Q$	$P+2Q$	⋯			⋯	$16P+2Q$
3	$3Q$	$P+3Q$	⋯			⋯	$16P+3Q$
4	$4Q$	$P+4Q$	⋯			⋯	$16P+4Q$
⋮	⋮	⋮					⋮
16	$16Q$	$P+16Q$	⋯				$16P+16Q$

Figure 12.7 The 17^2 seventeen-torsion points of $\mathcal{X}(\overline{\boldsymbol{F}}_{67})$

$[a]P + [b]Q$, because the values a and b can be found by first computing the elliptic curve trace[2]

$$[a]P = \text{trace}_q(R).$$

The trace projects the point R to a point of the group in the ground field. We can designate this point $[a]P$ for some a, even though a is not computable from P and $[a]P$. Then because $R = [a]P + [b]Q$, we can write $[b]Q = R - [a]P$ for some b, again even though b is not computable. Of course, Q can be replaced by any nonidentity point in its orbit, writing $R = [a]P + [c]Q'$, where $Q' = [c/b]Q$. One may also express the set of r-torsion points as

$$\mathcal{X}(\overline{\boldsymbol{F}}_{67})[17] = \langle P \rangle \oplus \langle Q \rangle.$$

where $\langle P \rangle$ and $\langle Q \rangle$ denote the orbits of P and Q. This is a concrete and natural way to refer to the array in Figure 12.7. Only two cycles and their elements are referred to explicitly. One cycle is given by the first row and one cycle is given by the first column. Instead, an alternative abstract and formal way to refer to the array is to partition the points of the curve into equivalence classes, and then refer to cycles of the equivalence classes. With respect to the array in Figure 12.7, regard the elements of the first row as one cycle, and regard the rows themselves as the elements of the second cycle. This interpretation is written

$$\mathcal{X}(\overline{\boldsymbol{F}}_{67})[17] = \mathcal{X}(\boldsymbol{F}_{67})[17] \oplus \mathcal{X}(\boldsymbol{F}_{67^2})/17\mathcal{X}(\boldsymbol{F}_{67^2}).$$

[2] The q-ary trace of a point $R = (x, y)$ of an elliptic curve $\mathcal{X}(\boldsymbol{F}_{q^r})$ is defined as

$$\text{trace}_q(x, y) = (x, y) + (x^q, y^q) + (x^{q^2}, y^{q^2}) + \cdots + (x^{q^r}, y^{q^r}).$$

The first term on the right is the cyclic group in the first row of the array. The second term on the right is the quotient group consisting of the set of equivalence classes corresponding to the rows of the array.

The example of $\mathcal{X}(\boldsymbol{F}_{67})[17]$ sets up an important definition.

Definition 12.7.1 *Let* GCD$(r, p) = 1$. *The embedding degree of r with respect to the prime p is the smallest integer k such that r divides $p^k - 1$. The same definition applies with the prime power q replacing p.*

Each prime factor r that divides the order of an elliptic curve $\mathcal{X}(\boldsymbol{F}_q)$ has its own embedding degree. (Usually one requires that r^2 does not divide $\#\mathcal{X}(\boldsymbol{F}_q)$.) These embedding degrees may be different for different r. We are usually interested in elliptic curves for which $\#\mathcal{X}(\boldsymbol{F}_q)$ itself is a prime or has a dominant prime factor. Then we may simply refer to the embedding degree of the elliptic curve, meaning the embedding degree of the dominant prime factor. It may be that the embedding degree is extremely large. Such curves are not useful for constructions that involve the embedding degree. For other curves, the embedding degree can be quite small. The example exhibited in Figure 12.1 shows an elliptic curve with an embedding degree of two.[3]

To see that the embedding degree always exists, use the division algorithm for integers to write $p^i - 1 = Q_i r + R_i$ for $i = 1, 2, 3, \ldots$. For each i, the remainder R_i is nonnegative and smaller than r. Therefore, it must repeat. Let $R_\ell = R_j$ be the first repetition. Then subtracting gives $p^j - p^\ell = (Q_j - Q_\ell)r$. Thus $p^\ell(p^{j-\ell} - 1) = (Q_j - Q_\ell)r$. But r does not divide p^ℓ, so it divides $p^k - 1$ with $k = j - \ell$.

For example, eleven divides $131^2 - 1 = (131 - 1)(131 + 1)$. Therefore the embedding degree of eleven with respect to 131 is two. We have seen that $\mathcal{X}(\boldsymbol{F}_{131}) : y^2 = x^3 - 3x + 8$ has 11 eleven-torsion points in $\boldsymbol{F}_{131}$, and 11^2 eleven-torsion points in $\boldsymbol{F}_{131^2}$. In fact, there are no other eleven-torsion points in any extension of $\boldsymbol{F}_{131}$, so all eleven-torsion points are in $\boldsymbol{F}_{131^2}$ because two is the embedding degree of eleven with respect to 131.

Let $P \neq \mathcal{O}$ be a nonidentity point of $\mathcal{X}(\boldsymbol{F}_{131})[11]$. Then because eleven is a prime, P generates $\mathcal{X}(\boldsymbol{F}_{131})[11]$. Let Q be an element of $\mathcal{X}(\boldsymbol{F}_{131^2})[11]$ that is not in $\mathcal{X}(\boldsymbol{F}_{131})$. Then Q generates an eleven-point cycle in $\mathcal{X}(\boldsymbol{F}_{131^2})$ with no points in $\mathcal{X}(\boldsymbol{F}_{131})$ other than $\mathcal{O}$. Finally, we can write

$$\mathcal{X}(\boldsymbol{F}_{131^2})[11] = \{[a]P + [b]Q \mid a = 0, \ldots, 10, b = 0, \ldots, 10\}.$$

By holding either a or b to a constant and varying the other, a cyclic subgroup of $\mathcal{X}(\boldsymbol{F}_{131^2})[11]$ is formed. In this way the 121 eleven-torsion points can be organized

[3] This is an instance of the general statement that for a supersingular elliptic curve over a prime field, the embedding degree is always two for any value of r. More generally, for any supersingular elliptic curve the embedding degree is at most six.

into twelve cyclic subgroups, each with eleven points. The identity element $\mathcal{O}$ is shared by all cycles, so the cycles are not disjoint. Only one of the twelve cycles lies entirely in $\boldsymbol{F}_{131}$. The other eleven cycles can have no elements in $\boldsymbol{F}_{131}$ except for the identity element because every element of a cycle of prime order is a generator of that cycle, and $\mathcal{X}(\boldsymbol{F}_{131})$ is closed under point addition.

Our second example is a supersingular elliptic curve with embedding degree k equal to four. The curve $\mathcal{X}(\boldsymbol{F}_{2^m}) : y^2 + y = x^3 + x + 1$ is supersingular for any m. It is easy to check that, over the ground field $\boldsymbol{F}_2$, the order of the curve is $\#\mathcal{X}(\boldsymbol{F}_2) = 1$. Then, by Theorem 10.11.2, for any m, $\#\mathcal{X}(\boldsymbol{F}_{2^m}) = 2^m + 1 - (1+i)^m - (1-i)^m$, where $i = \sqrt{-1} \in \boldsymbol{C}$. This expression for $\#\mathcal{X}(\boldsymbol{F}_{2^m})$ holds for any value of m, but it takes on a convenient form if m is odd. Then

$$\#\mathcal{X}(\boldsymbol{F}_{2^m}) = \begin{cases} 2^m - 2^{(m+1)/2} + 1 & \text{if} \quad m \pm 1 \pmod 8 \\ 2^m + 2^{(m+1)/2} + 1 & \text{if} \quad m \pm 3 \pmod 8 \end{cases}$$

and

$$2^{2m} + 1 = (2^m - 2^{(m+1)/2} + 1)(2^m + 2^{(m+1)/2} + 1).$$

Let $q = 2^m$, and let r divide $\#\mathcal{X}(\boldsymbol{F}_{2^m})$. Because r divides one of the two terms on the right, r divides $q^2 + 1$. Therefore, r divides $q^4 - 1 = (q^2 + 1)(q^2 - 1)$. We conclude that the embedding degree is at most four. It is easy to see that r does not divide $q^2 - 1$ or $q^3 - 1$. Therefore we conclude that the embedding degree is four for every r that divides the order of $\mathcal{X}(\boldsymbol{F}_{2^m})$, provided m is odd.

12.8 The torsion structure theorem

So far, other than referring to the canonical theory of abelian groups, we have done little more than recognize the cyclic subgroups in the group $\mathcal{X}(\boldsymbol{F}_q)$ by way of several examples. However, we are now prepared to step up to a higher level where we can see that a cyclic subgroup of $\mathcal{X}(\boldsymbol{F}_q)$ of prime order is embedded in the larger structure of a discrete torus. In this section, we will develop a deep and important theorem to this point, which has many implications. Our method of proof of this theorem uses elementary methods and involves extensive algebraic manipulation.

To preview the line of the proof, consider the case in which the prime r divides $\#\mathcal{X}(\boldsymbol{F}_q)$ and the embedding degree k of r is equal to two. These conditions are the statements that r divides $q + 1 - t$ and r divides $q^2 - 1$. Our goal is to show that r^2 then divides $\#\mathcal{X}(\boldsymbol{F}_{q^2})$, which means that $\mathcal{X}(\boldsymbol{F}_{q^2})$ has a subgroup of order r^2. The seventeen-torsion points introduced in Figure 12.7 are an instance of this fact. To this purpose, recall that Corollary 10.11.3 states that

$$\#\mathcal{X}(\boldsymbol{F}_{q^2}) = (q + 1 - t)(q + 1 + t)$$

and, by assumption, the embedding degree of r is two so r divides both $q+1-t$ and q^2-1, but does not divide $q-1$. Therefore r divides $q+1$. Because r divides both $q+1-t$ and $q+1$, it must divide $q+1+t$, so r^2 divides $\#\mathcal{X}(\boldsymbol{F}_{q^2})$. Conversely, if the embedding degree is not two, r does not divide $q+1$, so r does not divide $q+1+t$. Therefore r^2 does not divide $\#\mathcal{X}(\boldsymbol{F}_{q^2})$ if the embedding degree is not two.

Our featured theorem of this section states more generally that if k is the embedding degree of the prime r and if $\mathcal{X}(\boldsymbol{F}_q)$ has r r-torsion points, then $\mathcal{X}(\boldsymbol{F}_{q^k})$ has r^2 r-torsion points. The stronger statement that $\mathcal{X}(\overline{\boldsymbol{F}}_q)$ has no other r-torsion points will not be proved, though it is true.

Lemma 12.8.1 *Let r divide $\#\mathcal{X}(\boldsymbol{F}_q)$, then for every m,*

$$\frac{\#\mathcal{X}(\boldsymbol{F}_{q^m})}{\#\mathcal{X}(\boldsymbol{F}_q)} = m\frac{q^m-1}{q-1} \pmod r$$

which is nonzero if m is smaller than the embedding degree k and is zero if m is equal to k.

Proof The left side is zero modulo r when $m=k$ if $\#\mathcal{X}(\boldsymbol{F}_{q^m})$ is divisible by r^2 when $m=k$. To show this define

$$R_m = \frac{\#\mathcal{X}(\boldsymbol{F}_{q^m})}{\#\mathcal{X}(\boldsymbol{F}_q)} \pmod r.$$

That is, $R_m = \#\mathcal{X}(\boldsymbol{F}_{q^m})/(q+1-t)$ where $q+1-t$ has r as a factor. Lagrange's theorem states that every $\#\mathcal{X}(\boldsymbol{F}_{q^m})$ is divisible by $q+1-t$. Therefore, to show that $\#\mathcal{X}(\boldsymbol{F}_{q^m})$ is divisible by r^2, we must show that R_m itself is divisible by r.

The proof is by induction. The statement is clearly true for $m=1$. Suppose that the statement is true for all ℓ smaller than m. The cases of even m and odd m will be treated separately.

If m is even, then Corollary 10.11.3 can be written as

$$\#\mathcal{X}(\boldsymbol{F}_{q^{2m}}) = [\#\mathcal{X}(\boldsymbol{F}_{q^m})][2(q^m+1) - \#\mathcal{X}(\boldsymbol{F}_{q^m})].$$

Moreover, Lagrange's theorem says that $\#\mathcal{X}(\boldsymbol{F}_{q^m})$ is a multiple of $\#\mathcal{X}(\boldsymbol{F}_q)$ and so is a multiple of r. Then

$$\begin{aligned} R_{2m} &= R_m[2(q^m+1) - \#\mathcal{X}(\boldsymbol{F}_{q^m})] \pmod r \\ &= m\frac{q^m-1}{q-1}2(q^m+1) \pmod r \\ &= 2m\frac{q^{2m}-1}{q-1} \pmod r \end{aligned}$$

as was to be proved for m even.

If m is odd, then use Theorem 10.11.6 which states that for odd m,

$$\#\mathcal{X}(\boldsymbol{F}_{q^m}) = ((q+1)^m - t^m) - \sum_{i=1}^{(m-1)/2} \binom{m}{i} q^i \left[\#\mathcal{X}(\boldsymbol{F}_{q^{m-2i}})\right].$$

Therefore

$$\begin{aligned} R_m &= \sum_{i=0}^{m-1} (q+1)^i t^{m-1-i} - \sum_{i=1}^{(m-1)/2} \binom{m}{i} q^i R_{m-2i} \pmod r \\ &= m(q+1)^{m-1} - \sum_{i=1}^{(m-1)/2} \binom{m}{i} q^i R_{m-2i} \pmod r, \end{aligned}$$

where the second line follows because $q+1-t = 0 \pmod r$ so $t = q+1 \pmod r$. The task is to show that R_m is nonzero when m is less than k, and $R_m = 0$ when m is equal to k. Because $q^m - 1 \neq 0$ for $m < k$ and because $q^k - 1 = 0 \pmod r$ by the definition of k, we can write

$$\begin{aligned} \sum_{i=0}^{m-1} q^i \pmod r &= \frac{q^m - 1}{q-1} \pmod r \\ &\begin{cases} \neq 0 & m < k \\ = 0 & m = k \end{cases}, \end{aligned}$$

noting that $q - 1 \neq 0 \pmod r$. Accordingly, to complete the proof, we will prove by induction that

$$R_m = m \sum_{i=0}^{m-1} q^i \pmod r = m \frac{q^m - 1}{q-1} \pmod r.$$

Suppose that $R_\ell = \ell \sum_{i=0}^{m-1} q^i \pmod r$ for $\ell < m$. This does hold for $\ell = 1$. Then

$$\begin{aligned} R_m &= m(q+1)^{m-1} - \sum_{i=1}^{(m-1)/2} \binom{m}{i} q^i (m-2i) \frac{q^{m-2i} - 1}{q-1} \pmod r \\ &= m(q+1)^{m-1} - \sum_{i=1}^{(m-1)/2} \binom{m}{i} \frac{q^{m-i}(m-2i)}{q-1} - \sum_{i'=m-1}^{(m+1)/2} \binom{m}{m-i'} \frac{q^{m-i'}(m-2i')}{q-1}, \end{aligned}$$

where i has been replaced by $m - i'$ in the second summation. Now replace i' by i in the second summation and recall that $\binom{m}{m-i} = \binom{m}{i}$ to write

$$\begin{aligned} R_m &= m(q+1)^{m-1} - \sum_{i=1}^{(m-1)/2} \binom{m}{i} \frac{(m-2i)q^{m-i}}{q-1} - \sum_{i=(m+1)/2}^{m-1} \binom{m}{i} \frac{(m-2i)q^{m-i}}{q-1} \\ &= m(q+1)^{m-1} - \frac{m}{q-1} \sum_{i=1}^{m-1} \binom{m}{i} q^{m-i} + \frac{2}{q-1} \sum_{i=1}^{m-1} i \binom{m}{i} q^{m-i}. \end{aligned}$$

The two sums can be evaluated as

$$\sum_{i=1}^{m-1}\binom{m}{i}q^{m-i} = (q+1)^m - (q^m+1)$$

and

$$\sum_{i=1}^{m-1} i\binom{m}{i}q^{m-i} = m[(q+1)^{m-1} - 1].$$

Therefore, combining terms gives

$$R_m = \frac{m}{q-1}\big[(q^2-1)(q+1)^{m-2} - \big[(q+1)^m - (q^m+1)\big] + 2[(q+1)^{m-1} - 1]\big]$$

which can be simplified as

$$\begin{aligned} R_m &= \frac{m}{q-1}\big[(q^2-1-q^2-2q-1+2q+2)(q+1)^{m-2} + (q^m+1-2)\big] \\ &= m\frac{q^m-1}{q-1} \end{aligned}$$

as was to be proved for m odd. Thus the assertion holds for both odd m and even m, so the proof is complete. □

Because r divides $q^m - 1$ for $m = k$, but not for any m smaller than k, the lemma shows that $\#\mathcal{X}(\boldsymbol{F}_{q^k})$ has r^2 as a factor for embedding degree k. Therefore, $\mathcal{X}(\boldsymbol{F}_{q^k})$ contains r^2 points with torsion either r^2 or r. These r^2 points form an abelian subgroup, which for now is denoted G. Because r is prime, the finite group G is isomorphic either to $\boldsymbol{Z}_r \oplus \boldsymbol{Z}_r$ or to $\boldsymbol{Z}_{r^2}$. Our remaining task is to determine which it is. The first case, $\boldsymbol{Z}_r \oplus \boldsymbol{Z}_r$, has the form of a discrete torus. This option might be anticipated because in the complex field, an elliptic curve can be viewed as a complex torus. The real part of the complex torus is a real torus, and that real torus suggests the discrete torus $\boldsymbol{Z}_r \oplus \boldsymbol{Z}_r$. We will not pursue this abstract approach.

Our task is to show that the finite group G cannot be isomorphic to $\boldsymbol{Z}_{r^2}$. Observe that if this were true, there would be only r points of order r and they would all be in the ground field $\boldsymbol{F}_q$. To eliminate this case, we need to show that there is at least one point of order r not in the ground field. The example shown in Table 12.1 is one instance of this claim. We must show that this example is not an isolated instance of this fact.

We will first take a moment to motivate the forthcoming theorem by looking at curves with embedding degree two. Let r be a prime factor of $\#\mathcal{X}(\boldsymbol{F}_q)$ with embedding degree two. Then r is a divisor of $q + 1 - t$, which means that $q + 1 - t + ar = 0$ for some integer a, so q can be replaced by $t - 1 - ar$.

There are r^2 points in G and only r of them are rational points. Choose any point P that is not a rational point. Because the embedding degree is two, we know that $\pi_q^2(P) = \pi_{q^2}(P) = P$. Then, with these substitutions, the Frobenius endomorphism

$$\pi_q^2(P) - [t]\pi_q(P) + [q]P = \mathcal{O}$$

is reduced to

$$(\pi_q - 1)[t]P = [a][r]P = R$$

where $R = [a][r]P$ is a rational point of the curve because $[r]P$ is a rational point. But then, because $\pi_q^2 = 1$ and $\pi_q R = R$, multiplying both sides of this equation by π_q gives

$$(1 - \pi_q)[t]P = R.$$

This means that $R = -R$, so we conclude that $R = \mathcal{O}$ (or possibly, for some curves, $R = (0, 0)$). Finally, the equality $\pi_q[t]P = [t]P$ implies, by Theorem 10.13.1, that $[t]P$ is a rational point of the curve. Then, unless $t = 0 \pmod r$, we can conclude that the nonrational point P is a point of order r. Therefore, G must be isomorphic to $\mathbf{Z}_r \oplus \mathbf{Z}_r$. All points of G are r-torsion points.

We now want to show that a similar statement holds for a curve with an arbitrary embedding degree. This is the content of the following theorem. The proof of the theorem will introduce a more elegant line of reasoning. The formal statement of the group structure to be proved is

$$\mathcal{X}(\overline{\mathbf{F}}_q)[r] = \mathcal{X}(\mathbf{F}_q)[r] \oplus \mathcal{X}(\mathbf{F}_{q^k})/r\mathcal{X}(\mathbf{F}_{q^k})$$

where the second term on the right designates a set of equivalence classes such as those indicated by the rows in the example of Figure 12.7. The operational statement of this same group structure is

$$\mathcal{X}(\overline{\mathbf{F}}_q)[r] = \mathcal{X}(\mathbf{F}_q)[r] \oplus \mathcal{X}(\mathbf{F}_{q^r})[r]$$

where the second term on the right now refers to the set of canonical representatives of the equivalence classes comprising a single cycle of the nonrational r-torsion points. The canonical representatives will be described in the proof of the theorem.

Theorem 12.8.2 (Torsion structure theorem) *Let $\mathbf{F}_q$ be a finite field of characteristic p, and let $\mathcal{X}(\mathbf{F}_q)$ be an elliptic curve over $\mathbf{F}_q$. If r and p are coprime and k is the embedding degree of r, a prime divisor of $\#\mathcal{X}(\mathbf{F}_q)$, then*

$$\mathcal{X}(\overline{\mathbf{F}}_q)[r] = \mathcal{X}(\mathbf{F}_q)[r] \oplus \mathcal{X}(\mathbf{F}_{q^k})/r\mathcal{X}(\mathbf{F}_{q^k})$$

and so $\mathcal{X}(\mathbf{F}_{q^k})[r]$ is isomorphic to $\mathbf{Z}_r \oplus \mathbf{Z}_r$.

Proof Every point P of the curve $\mathcal{X}(\overline{\boldsymbol{F}}_q)$, satisfies the Frobenius endomorphism

$$\pi_q^2(P) - [t]\pi_q(P) + [q]P = \mathcal{O}.$$

In particular every point of $\mathcal{X}(\boldsymbol{F}_{q^k})[r]$, where k is the embedding degree, satisfies this endomorphism.

Because $\#\mathcal{X}(\boldsymbol{F}_q) = q + 1 - t$, we know that $t = q + 1 \pmod r$ whenever r divides $\#\mathcal{X}(\boldsymbol{F}_q)$. Therefore,

$$\begin{aligned} x^2 - tx + q &= x^2 - (q+1)x + q \pmod r \\ &= (x-1)(x-q) \pmod r \\ &= (x-q)(x-1) \pmod r. \end{aligned}$$

With this factorization, the Frobenius endomorphism becomes

$$\begin{aligned} (\pi_q^2 - [t]\pi_q + [q])P &= (\pi_q - [1])(\pi_q - [q])P = (\pi_q - [1])P' = \mathcal{O} \\ &= (\pi_q - [q])(\pi_q - [1])P = (\pi_q - [q])P'' = \mathcal{O}, \end{aligned}$$

where $P' = (\pi_q - [q])P$, and $P'' = (\pi_q - [1])P$.

We now require P to be any nonrational point of the subgroup G. By this choice of P, the point P'' is not equal to $\mathcal{O}$, and $\pi_q P'' = [q]P''$. Therefore $\pi_{q^k} P'' = \pi_q^k P'' = [q]^k P'' = [q^k]P'' = P''$ because $q^k = 1 \pmod r$. Therefore we now know that the order of the nonrational point P'' divides $q^k - 1$, and we knew before that the order of P'' divides r^2. Therefore, the order of P'' divides $\mathrm{GCD}[q^k - 1, r^2] = r$. Thus there is at least one nonrational point of order r. This means that G is not isomorphic to $\boldsymbol{Z}_{r^2}$. Therefore G is isomorphic to $\boldsymbol{Z}_r \oplus \boldsymbol{Z}_r$ and the proof is complete. □

The two cycles of r-torsion points $\{P \in \mathcal{X}(\overline{\boldsymbol{F}}_q)[r] \mid \pi_q(P) = [1]P\}$ and $\{P \in \mathcal{X}(\overline{\boldsymbol{F}}_q)[r] \mid \pi_q(P) = [q]P\}$ described in the proof are called the *Frobenius eigenspaces* of $\mathcal{X}(\overline{\boldsymbol{F}}_q)[r]$. They are often written as

$$\begin{aligned} G_1 &= \mathcal{X}(\overline{\boldsymbol{F}}_q)[r] \cap \ker(\pi_q - [1]) \\ G_2 &= \mathcal{X}(\overline{\boldsymbol{F}}_q)[r] \cap \ker(\pi_q - [q]), \end{aligned}$$

where the kernels are defined as

$$\begin{aligned} \ker(\pi_q - [1]) &= \{P \mid \pi_q(P) - [1]P = \mathcal{O}\} \\ \ker(\pi_q - [q]) &= \{P \mid \pi_q(P) - [q]P = \mathcal{O}\}. \end{aligned}$$

The eigenvalues corresponding to these Frobenius eigenspaces are 1 and q, respectively. The elements of these eigenvalues are the canonical representatives of the r-torsion points of the elliptic curve.

Corollary 12.8.3 *Let r be a prime that divides the order of $\mathcal{X}(\boldsymbol{F}_q)$, but does not divide $q-1$ or q. Then $\mathcal{X}(\boldsymbol{F}_{q^k})$ contains all r^2 points of order r if, and only if, r divides q^k-1.*

Proof Theorem 12.8.2 says that r^2 divides $\#\mathcal{X}(\boldsymbol{F}_{q^k})$ if k is the smallest integer for which k divides q^k-1. The statement of the corollary now follows. □

In Chapter 10, we discussed the two curves given by $\mathcal{X}(\boldsymbol{F}_{131}) : y^2 = x^3 - 3x + 8$ and its twist $\mathcal{X}'(\boldsymbol{F}_{131}) : y^2 = x^3 - 3x - 8$. The first curve has 110 points, and the second has 132 points. Both of these curves are groups under point addition, and both have subgroups of order eleven, because 11 divides both 110 and 132. What can we say about the eleven-torsion points on these two curves? By Theorem 12.8.2 we know that there are 11^2 points in $\mathcal{X}(\overline{\boldsymbol{F}}_{131})[11]$, but where are they? Only eleven of these points are in $\mathcal{X}(\boldsymbol{F}_{131})$. Moreover, there are 11^2 such torsion points in $\mathcal{X}'(\overline{\boldsymbol{F}}_{131})[11]$ as well.

We saw that the elliptic curve $\mathcal{X}(\boldsymbol{F}_{131^2}) : y^2 = x^3 - 3x + 8$ has 110×132 points, and 11^2 divides this number. Moreover, we can also see that $\mathcal{X}'(\boldsymbol{F}_{131^2}) : y^2 = x^3 - 3 - 8$ also has 110×132 points. Furthermore, 131 has the form $4k+3$, so we know that elements of $\boldsymbol{F}_{131^2}$ can be written as $a + ib$ with $i = \sqrt{-1}$. Thus we may conclude that, in the extension field, $\mathcal{X}$ and $\mathcal{X}'$ are somehow more directly related, perhaps as something resembling the real and imaginary parts of a curve with complex coordinates.

The r-torsion points of an elliptic curve over a finite field $\boldsymbol{F}_q$ comprise the set of points of the elliptic curve in any extension of $\boldsymbol{F}_q$ that lies in an orbit of order r. We have seen that, with mild exceptions, the set of r-torsion points of $\mathcal{X}(\overline{\boldsymbol{F}}_q)$ satisfies

$$\mathcal{X}(\overline{\boldsymbol{F}}_q)[r] \simeq \boldsymbol{Z}_r \oplus \boldsymbol{Z}_r.$$

By selecting one orbit in $\boldsymbol{F}_{q^k}[r]$ not in $\boldsymbol{F}_q[r]$, we can write this as

$$\mathcal{X}(\overline{\boldsymbol{F}}_q)[r] = \mathcal{X}(\boldsymbol{F}_q)[r] \oplus \mathcal{X}(\boldsymbol{F}_{q^k})[r].$$

An attractive choice for the canonical orbit is the orbit consisting of all r-torsion points satisfying $\pi_q(P) = [q]P$. This orbit does exist as is described further in Section 13.2. This description is sketched symbolically in Figure 12.8. Of course, there is only one identity $\mathcal{O}$, which is common to both orbits, although the illustration shows that identity point twice. The elliptic curve also has many other points that are not shown. Only those r-torsion points in the two orbits are shown.

If r is a prime, then any nonzero element of the cyclic group generated by a nonzero element is also a generator of that cyclic group. Two distinct cyclic subgroups only have the identity element in common. There are a total of $r^2 - 1$ nonidentity elements and $r-1$ of them appear in each cyclic subgroup. Therefore in such cases, there are $r+1$ cyclic subgroups in $\mathcal{X}(\boldsymbol{F}_q)[r]$.

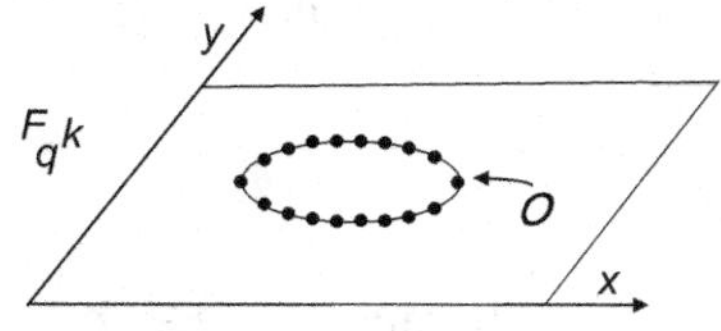

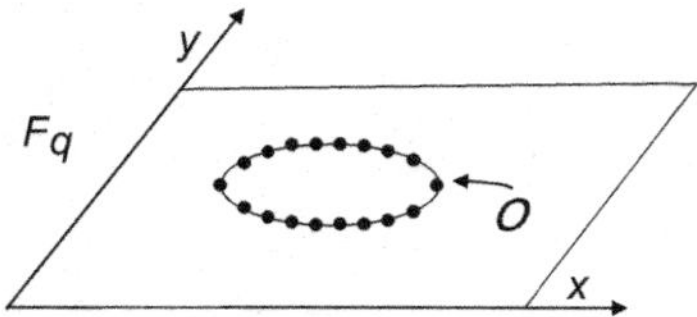

Figure 12.8 The r-torsion points of an elliptic curve

Theorem 12.8.4 *Let r be a prime that divides $\#\mathcal{X}(\boldsymbol{F}_q)$, but does not divide $q - 1$ or q. Then $\mathcal{X}(\boldsymbol{F}_{q^k})$ contains the r^2 elements of the torsion group $\mathcal{X}(\boldsymbol{F}_q)[r]$ if, and only if, r divides $\Phi_k(q)$ and does not divide $\Phi_\ell(q)$ for any ℓ less than k.*

Proof Let $n = q + 1 - t$, and let

$$\begin{aligned} N &= q^k + 1 - \alpha^k - \beta^k \\ &= (\alpha^k - 1)(\beta^k - 1), \end{aligned}$$

where $\alpha + \beta = t$ and $\alpha\beta = q$. Both r and n are integers. We are given that r divides n, and that r divides $\Phi_k(q)$. The task is to show that r also divides N.

As was discussed in Section 9.9, the cyclotomic polynomial $\Phi_k(x)$ always can be factored in a suitable extension of the prime field $\boldsymbol{F}_r$ as

$$\Phi_k(x) = \prod_{\mathrm{GCD}(\ell,k)=1} (x - \omega^\ell),$$

where ω is an element of order k in the extension field $\boldsymbol{F}_{r^\ell}$, and arithmetic in the field $\boldsymbol{F}_{r^\ell}$ is modulo r. Therefore

$$\Phi_k(q) = \prod_{\mathrm{GCD}(\ell,k)=1} (q - \omega^\ell).$$

We know that this equals zero modulo the prime r, so one of the factors on the right must equal zero modulo r. Thus for some value of ℓ, say ℓ', $q - \omega^{\ell'} = 0 \pmod r$, so $q = \omega^{\ell'} \pmod r$. In particular, now with the complex numbers α and β, we can write

$$\Phi_k(\alpha)\Phi_k(\beta) = \prod_{\mathrm{GCD}(\ell,k)=1} (\alpha - \omega^\ell)(\beta - \omega^\ell).$$

The two terms with $\ell = \ell'$ can be combined to give

$$\begin{aligned} (\alpha - \omega^{\ell'})(\beta - \omega^{\ell'}) &= \alpha\beta - (\alpha + \beta)\omega^{\ell'} + (\omega^{\ell'})^2 \\ &= q - t\omega^{\ell'} + (\omega^{\ell'})^2. \end{aligned}$$

The right side involves only integers and can be reduced modulo r to give

$$\begin{aligned}(\alpha - \omega^{\ell'})(\beta - \omega^{\ell''}) &= q^2 - tq + q \pmod r \\ &= q(q - t + 1) \pmod r.\end{aligned}$$

The next step is to observe that the right side is zero because $q - t + 1 = 0 \pmod r$. Finally, we conclude that

$$\begin{aligned}N &= \Phi_k(\alpha)\Phi_k(\beta)\prod \Phi_d(\alpha)\prod \Phi_d(\beta) \\ &= 0 \pmod r.\end{aligned}$$

This completes the proof of the theorem. □

It is now easy to anticipate what should happen. We have already said that for any prime r, the set $\mathcal{X}(\boldsymbol{F}_q)[r]$ is a nontrivial abelian group with order r or a power of r whenever r divides $\#\mathcal{X}(\boldsymbol{F}_q)$. Therefore for any prime r, $\mathcal{X}(\boldsymbol{F}_{q^k})[r]$ is a nontrivial abelian group with order equal to a power of r whenever r divides $\#\mathcal{X}(\boldsymbol{F}_{q^k})$. As k is increased one by one, starting from $k = 1$, all the r-torsion points, initially, are only in the ground field. Thus

$$\mathcal{X}(\boldsymbol{F}_{q^i})[r] = \mathcal{X}(\boldsymbol{F}_q)[r]$$

for $i = 1, 2, \ldots, k - 1$. Then when i is equal to k, a burst of new r-torsion points appears. These new points, of course, are themselves part of a larger group structure that can be described in that extension field. The original group of r-torsion points is a subgroup of this larger group. The group of all r-torsion points is isomorphic to $\boldsymbol{Z}_r \oplus \boldsymbol{Z}_r$. It contains r^2 points. Of these, r points are in $\mathcal{X}(\boldsymbol{F}_q)$, and the other $r^2 - r$ points are in $\mathcal{X}(\boldsymbol{F}_{q^k})$, but not in $\mathcal{X}(\boldsymbol{F}_q)$.

12.9 The structure of a pairing

From an abstract point of view, one may take the view that there is essentially only one bilinear pairing relating any two cyclic source groups G_1 and G_2 to a cyclic object group G_T, all of the same prime order r. This is because of the required bilinearity property. To specify $\Phi(P, Q)$ abstractly, it is only necessary to specify the image in G_T of any generator of the first group when Q is the identity of the second group, and to specify the image in G_T of any generator of the second group when P is the identity of the first group. Any other pairing has a simple relationship to this pairing and, because the three groups G_1, G_2, and G_T are all cyclic groups of prime order, the mapping is of the form $\boldsymbol{Z}_r \oplus \boldsymbol{Z}_r \to \boldsymbol{Z}_r$. However, this equivalence at the abstract level does not hold up at the structural level or at the computational level. The abstract isomorphism relating a pairing to $\boldsymbol{Z}_r \oplus \boldsymbol{Z}_r$ is hidden behind the computational intractability of the discrete-log problem.[4] In fact, the equation defining a bilinear pairing can take many different forms, as can the structure of the computation implementing it. It is in the detail of the definition and in the structure of the computational algorithms that the real differences between pairings are seen.

A bilinear pairing is a function on certain pairs of points of the plane, and can be described using bivariate rational functions. To construct such a pairing for our needs, our method will be to define one or more suitable bivariate rational functions that lead to the required bilinear property. We will specify a suitable rational function

$$f(x, y) = \frac{a(x, y)}{b(x, y)}$$

indirectly by specifying its poles and zeros on the elliptic curve $\mathcal{X}(\boldsymbol{F}_q)$ in a way that provides the required property. For this purpose, $b(x, y)$ must not be divisible by $p(x, y)$, the polynomial defining the elliptic curve.

To construct a pairing, we will construct an appropriate bivariate rational function, denoted $f(x, y)$ on the curve, $\mathcal{X}(\boldsymbol{F}_q)$, or perhaps two such rational functions, denoted $f(x, y)$ and $g(x, y)$. By specifying only the poles and zeros on the curve, a rational function $f(x, y)$ will be defined up to an unspecified multiplicative constant, as well as an additive multiple of $p(x, y)$. The additive multiple of $p(x, y)$ is irrelevant because it evaluates to zero on the curve. The unspecified multiplicative constants in numerator and denominator may cancel under the right circumstance, or these constants must be removed in some other way. The bilinear pairing is simply written as

$$\Phi(P, P') = \frac{f(P)}{g(P')},$$

where the notation $f(P)$ means $f(x, y)$ with $(x, y) = P$.

We are now ready to construct bilinear pairings, and the task will occupy most of the remainder of this chapter. We do so by defining the rational functions that are at the heart of such pairings. These rational functions will be massive and impossible to actually write as such. They are described only by giving a rule defining their poles and zeros. The poles and zeros must be specified respecting conjugacy conditions so that the polynomial evaluates as an element of the appropriate field. Moreover, the defining divisor must be a principal divisor. Otherwise, no rational function will exist with that divisor. Theorem 11.6.2 gives a necessary and sufficient condition for a divisor on an elliptic curve to be a principal divisor.

We are interested in bilinear pairings described as functions $\Phi(P, Q)$ from $\mathcal{X}(\boldsymbol{F}_q)[r] \oplus \mathcal{X}(\boldsymbol{F}_q)[r]$ into an orbit μ_r in $\boldsymbol{F}^*_{q^k}$ of order r. However, some attractive bilinear pairings will actually be constructed as functions from $\mathcal{X}(\boldsymbol{F}_q)[r] \oplus \mathcal{X}(\boldsymbol{F}_{q^k})[r]$

[4] Except for small instances of this structure, the isomorphism is (apparently) an impassable portal protected by the demon of computational complexity.

to $\mu_r \subset F^*_{q^k}$, and so are called asymmetric bilinear pairings. Accordingly, this alternative construction of a pairing must be augmented by a suitable map Ψ, conventionally called a *distortion map*, from $\mathcal{X}(F_q)[r]$ to $\mathcal{X}(F_{q^k})[r]\backslash\mathcal{X}(F_q)$ if such a map is available. We may then continue to speak of a pairing from $\mathcal{X}(F_q)[r] \oplus \mathcal{X}(F_q)[r]$, even though the construction actually gives a pairing from $\mathcal{X}(F_q)[r] \oplus \mathcal{X}(F_{q^k})[r]$. In such cases, it is understood that a distortion map Ψ is needed to go from the ground field to the extension field. Then the pairing should be written as $\Phi(P, \Psi(Q))$, but this may be informally written as $\Phi(P, Q)$. Although not always explicitly mentioned, the distortion map is understood to be present whenever it is necessary.

A distortion map Ψ from $\mathcal{X}(F_q)$ into $\mathcal{X}(F_{q^k})$ is defined as any endomorphism[5] such that the image of every nonidentity rational point of $\mathcal{X}(F_q)$ is not in $\mathcal{X}(F_q)$. In general, distortion maps exist only for supersingular curves.

For an example of a distortion map, recall the supersingular curve $\mathcal{X}(F_{2^m}) : y^2 + y = x^3 + x + 1$ discussed in Section 12.7. It was shown there that for any odd m and any prime r that divides $\#\mathcal{X}(F_{2^m})$, the embedding degree k is equal to four. Therefore, for every r dividing $\#\mathcal{X}(F_{2^m})$, all r-torsion points are in $\mathcal{X}(F_{2^{4m}})$. Now define the map $Q = \Psi(P)$ for $P = (x, y)$ by $\Psi(x, y) = (x + \xi^2, y + \xi x + \eta)$, where ξ and η are any nonzero elements of $F_{2^{4m}}\backslash F_{2^m}$ satisfying $\xi^3 = 1$ and $\eta^2 + \eta = \xi^2 + 1$. Then Q is an element of $\mathcal{X}(F_{2^{4m}})\backslash\mathcal{X}(F_{2^m})$, as can be easily checked. Moreover, it is easy to show that $\Psi(P_1 + P_2) = \Psi(P_1) + \Psi(P_1)$. This means that Q is an r-torsion point of $\mathcal{X}(F_q)$ if P is an r-torsion point of $\mathcal{X}(F_{q^m})$. Therefore $\Psi(P)$ is a distortion map. This distortion map allows an asymmetric bilinear pairing to be converted to a symmetric bilinear pairing.

If a distortion map does not exist, then the trace function can be used in the opposite direction to convert r-torsion points in the extension field into r-torsion points in the ground field. This again converts an asymmetric bilinear pairing into a symmetric bilinear pairing by $\widehat{\Phi}(S, T) = \Phi(\mathrm{Tr}(S), T)$. This, however, is less desirable because now S and T are no longer points in F_q^2. They are points in $F_{q^k}^2$, which is more complicated.

12.10 Attacks using bilinear pairings

The dominant theme in this chapter is the study of cryptographic methods that use a bilinear pairing to provide better cryptographic systems. However, the topic of bilinear pairings first entered the subject of cryptology not to provide improved security, but for the purpose of devising attacks against elliptic-curve cryptography. Accordingly, in this section we shall turn to this contrary, adversarial role of bilinear pairings. We

[5] The requirement that a distortion map be an endomorphism ensures that the map is analytically and computationally tractable. This restriction is not considered to be an important issue, but it does restrict the generality.

shall study attacks on the discrete-log problem based on the use of bilinear pairings. This method of attack on the discrete-log problem on an elliptic curve using a pairing such as the Weil pairing is called an *MOV attack*, and the analogous attack on the discrete-log problem on a hyperelliptic curve using a pairing such as the Weil pairing is called a *Frey–Rück attack*.

The central idea of a pairing-based attack is to use a bilinear pairing in some way to map a discrete-log problem in the additive group of an elliptic curve into a discrete-log problem in the multiplicative group of a finite field. The motivation for this line of attack is that direct attacks on the discrete-log problem in the multiplicative group of a finite field are known, while satisfactory direct attacks on the discrete-log problem on an elliptic curve are not known. Indeed, this indirect attack using a bilinear pairing is the only noteworthy attack known on the discrete-log problem on an elliptic curve.

The input to the Weil pairing is a pair of r-torsion points of the elliptic curve $\mathcal{X}(\boldsymbol{F}_q)$, and the output is an element of the field $\boldsymbol{F}_{q^k}$ contained within an orbit of order r. The Weil pairing can be used to transform the discrete-log problem on an elliptic curve in $\boldsymbol{F}_q$ into a discrete-log problem in the finite field $\boldsymbol{F}_{q^k}$, where k is the embedding degree of r. Because attacks are available on the discrete-log problem in the group $\boldsymbol{F}^*_{q^k}$ that are not directly available in the group of an elliptic curve, perhaps a bilinear pairing can provide a way to change the discrete-log problem on an elliptic curve into a different and easier discrete-log problem. On the other hand, because the embedding degree k is not equal to one, we know that the field elements of $\boldsymbol{F}_{q^k}$ will be in an arithmetic system larger than the field elements representing the points of $\mathcal{X}(\boldsymbol{F}_q)$. The larger size of the field elements may make the discrete-log problem in $\boldsymbol{F}_{q^k}$ harder than the discrete-log problem it replaces, which perhaps will offset the fact that the problem is now in a group that is in principle more vulnerable to attack. To assure this protection, the embedding degree should be made large enough, and this can be done by careful choice of the elliptic curve.

In order to illustrate the enormity of these operations with an example, choose the prime

$$p = 2^{130} + 169.$$

This prime has the 40-digit decimal representation

$$p = 1361129467683753853853498429727072845993.$$

The elliptic curve over $\boldsymbol{F}_p$ defined by the polynomial

$$y^2 = x^3 + x + 1230929586093851880935564157041535079194$$

has cardinality

$$r = \#\mathcal{X}(\boldsymbol{F}_p) = 1361129467683753853846060531160085896483,$$

which is a prime number. Notice that only the last 20 digits of r differ from the 40 digits of p, as is promised by the Hasse–Weil bound. The embedding degree is the smallest integer k for which r divides $p^k - 1$. Elements of $\boldsymbol{F}_{p^k}$ are k-tuples of 40-digit integers, and within $\boldsymbol{F}^*_{p^k}$, there is a cyclic subgroup of order r.

For a second example of an MOV attack, consider the polynomial

$$y^2 + y = x^3 + x$$

over the field $\boldsymbol{F}_{2^{173}}$. The order of this elliptic curve is

$$\#\mathcal{X}(\boldsymbol{F}_{2^{173}}) = 5 \cdot 13625405977 \cdot r$$

where r is a 42-digit prime. The embedding degree for this curve is four. Therefore an MOV attack will use a bilinear pairing to map $\mathcal{X}(\boldsymbol{F}_{2^{173}})[r]$ into $\boldsymbol{F}_{(2^{173})^4}$ and then solve the discrete-log problem. One analysis for the discrete-log problem on this curve using the MOV attack concludes that it requires 1.4×10^{18} operations in $\boldsymbol{F}_{2^{692}}$, the field of 692-bit integers. This would be a formidable calculation, though perhaps not intractable.

To protect against an MOV attack one uses an elliptic curve with a large embedding degree. In this way, the potential vulnerability to this attack is offset by forcing the adversary into a discrete-log problem of the same order, but within the large finite field $\boldsymbol{F}_{q^k}$. Neither the MOV attack nor the Frey–Rück attack is considered a threat unless the curve is chosen carelessly. Moreover, the pairing computation itself may be difficult if the embedding degree is large. In particular, because the embedding degree of a supersingular curve is small, a supersingular elliptic curve may be avoided because of a perceived vulnerability to an MOV attack. Recall that a supersingular elliptic curve over $\boldsymbol{F}_{p^r}$ is an elliptic curve for which the Frobenius trace t is divisible by the field characteristic p. In particular, for elliptic curves over the prime field $\boldsymbol{F}_p$, a supersingular elliptic curve is one for which t is zero. A cryptosystem based on a supersingular elliptic curve is potentially vulnerable to an MOV attack because the embedding degree of a supersingular elliptic curve is small. Specifically, the embedding degree k of a supersingular elliptic curve is not larger than six and if q is a prime, then the embedding degree is two. This raises the question of whether the discrete-log problem in $\boldsymbol{F}_{q^k}$ for k at most six is easier than a discrete-log problem on the group of an elliptic curve of order q. Rather than answer this difficult question, some simply avoid using supersingular elliptic curves in cryptography, preferring to use a curve with an embedding degree larger than six. This caution, however, may be offset by a deeper analysis of a specific choice of curve.

To find the embedding degree of a supersingular curve over $\boldsymbol{F}_p$, where p is a prime, other than two or three, we note that $\#\mathcal{X}(\boldsymbol{F}_p) = p + 1$. But $p^2 - 1$ is divisible by $p + 1$, and so is divisible by any factor of $p + 1$. Therefore the embedding degree is two for any r dividing $p + 1$. Consequently, the elliptic curve we want to study is the elliptic curve $\mathcal{X}(\boldsymbol{F}_{p^2})$ in the extension field $\boldsymbol{F}_{p^2}$. Suppose that p is a prime of the form

$4k + 3$. Because -1 is a nonsquare when p has the form $4k + 3$, we can write the field elements as $a + ib$ where $i^2 = -1$, and where a and b are elements of $\boldsymbol{F}_p$. The plane $\boldsymbol{F}_p^2$ can be mapped into $\boldsymbol{F}_{p^2}^2$ in the natural way by $(x, y) \to (x, y)$, which trivially maps $\mathcal{X}(\boldsymbol{F}_p)$ onto $\mathcal{X}(\boldsymbol{F}_{p^2})$. The plane $\boldsymbol{F}_p^2$ also can be mapped into the plane $\boldsymbol{F}_{p^2}^2$ by $(x, y) \to (-x, iy)$. But then if $(x, y) \in \boldsymbol{F}_p$, we show that $(-x, iy) \in \boldsymbol{F}_{p^2}$ by writing

$$(iy)^2 = (-x)^3 + a(-x),$$

which reduces to

$$y^2 = x^3 + ax.$$

Thus in this sense, a second copy of the elliptic curve $\mathcal{X}(\boldsymbol{F}_{p^2})$ appears in the larger plane $\boldsymbol{F}_{p^2}^2$.

12.11 The Tate pairing

The *Tate pairing* $\Phi(P, Q)$ is an asymmetric bilinear pairing from $\mathcal{X}(\boldsymbol{F}_q)[r] \oplus \mathcal{X}(\boldsymbol{F}_{q^k})/r\mathcal{X}(\boldsymbol{F}_{q^k})$ into $\boldsymbol{F}_{q^k}^*/(\boldsymbol{F}_{q^k}^*)^r$, where r, but not r^2, is a divisor of $\#\mathcal{X}(\boldsymbol{F}_q)$. The definition of these sets will be elaborated below. The Tate pairing can be stated more compactly in terms of the canonical representatives of these sets as a pairing from $\mathcal{X}(\boldsymbol{F}_q)[r] \oplus \mathcal{X}(\boldsymbol{F}_{q^m})[r]$ into $\mu_r \subset \boldsymbol{F}_{q^k}^*$. The Tate pairing is computationally simpler than the older Weil pairing, which will be described in Section 12.13. The Tate pairing can be evaluated by the Miller algorithm, which will be described in Section 12.12.

An elliptic curve over $\boldsymbol{F}_q$ can also be regarded over any extension of $\boldsymbol{F}_q$. Then for each m, this is an elliptic curve[6] $\mathcal{X}(\boldsymbol{F}_{q^m})$. This larger curve is itself an abelian group under point addition, and the order r of any subgroup of that larger curve must divide $\#\mathcal{X}(\boldsymbol{F}_{q^m})$. If one would calculate the subgroup structure of $\mathcal{X}(\boldsymbol{F}_{q^m})[r]$ for each value of m, in turn, one would come to the first value of m at which additional r-torsion points are found. This value, denoted k, is the embedding degree corresponding to r. The embedding degree k depends on the order r of the cyclic subgroup $\mathcal{X}(\boldsymbol{F}_q)[r]$ of the elliptic curve. Different cyclic subgroups may have different embedding degrees.

The Tate pairing can be described more concisely as a bilinear pairing from $G_1 \times G_2$ to G_T with $G_1 = \mathcal{X}(\boldsymbol{F}_q)[r]$, $G_2 = \mathcal{X}(\boldsymbol{F}_{q^k})/r\mathcal{X}(\boldsymbol{F}_{q^k})$, and $G_T = \boldsymbol{F}_{q^k}^*/(\boldsymbol{F}_{q^k}^*)^r$. To describe the quotient group $G_T = \boldsymbol{F}_{q^k}^*/(\boldsymbol{F}_{q^k}^*)^r$ in more detail, let β and γ be elements of $\boldsymbol{F}_{q^k}^*$ of order r and $(q^k - 1)/r$, respectively. Because r and $(q^k - 1)/r$ are coprime, we can write

$$\boldsymbol{F}_{q^k}^* = \{\beta^{\ell'}\gamma^{\ell''} \mid 0 \le \ell' < r\,;\, 0 \le \ell'' < (q^k - 1)/r\}.$$

[6] More generally, one would say that there is an elliptic curve in the algebraic closure $\overline{\boldsymbol{F}}_q$ of $\boldsymbol{F}_q$, where $\overline{\boldsymbol{F}}_q = \bigcup_{i=1}^{\infty} \boldsymbol{F}_{q^i}$ and $\mathcal{X}(\overline{\boldsymbol{F}}_q)$ has intersections with each of the extension fields of $\boldsymbol{F}_q$.

The ℓ'th equivalence class, which contains the element $\beta^{\ell'}$, is $\{(\beta^{\ell'}\gamma^{\ell''}) \mid 0 \le \ell'' < (q^k - 1)/r\}$. Then $\boldsymbol{F}^*_{q^k}/(\boldsymbol{F}^*_{q^k})^r$ is the set of such equivalence classes:

$$\boldsymbol{F}^*_{q^k}/(\boldsymbol{F}^*_{q^k})^r = \{\{(\beta^{\ell'}\gamma^{\ell''}) \mid 0 \le \ell'' < (q^k - 1)/r\} \mid 0 \le \ell' < r\}.$$

This set of sets can be represented by a cycle of order r, denoted μ_r consisting of those elements for which $\ell'' = 0$. These are the canonical representatives of the range G_T of the pairing. They comprise the orbit μ_r generated by an element β of μ_r with order r.

The Tate pairing will be defined in terms of a Miller function. A *Miller function* for the point P of the elliptic curve $\mathcal{X}(\boldsymbol{F}_q)$ is a bivariate rational function $f_{i,P}(x, y)$ with a zero of multiplicity i at the point P of the curve and a pole of multiplicity one at the point $[i]P$. Accordingly, for each P, the Miller function $f_{i,P}(x, y)$ is defined by its divisor

$$\text{div}(f_{i,P}) = i(P) - ([i]P) - (i - 1)(\mathcal{O})$$

up to an arbitrary multiplier by an element of the field $\boldsymbol{F}_{q^k}$. This divisor satisfies the conditions of Theorem 11.6.2, so this divisor is a principal divisor. Therefore the Miller function is well defined. In the special case that $i = r$ where r is the order of the point P, we have that $[r]P = \mathcal{O}$, so then the divisor of the Miller function can be written, more simply, as

$$\text{div}(f_{r,P}) = r(P) - r(\mathcal{O}),$$

although it can still be written in the more general form.

For any point $P \in \mathcal{X}(\boldsymbol{F}_q)[r]$ and any $Q \in \mathcal{X}(\boldsymbol{F}_{q^k})/r\mathcal{X}(\boldsymbol{F}^*_{q^k})$, the Tate pairing is a bilinear pairing denoted $f_{r,P}(Q)$. All elements of an equivalence class have the same value when raised to the $(q^k - 1)/r$ power. For our purposes, the Tate pairing is defined as

$$\Phi(P, Q) = f_{r,P}(Q)^{(q^k-1)/r} \in \mu_r,$$

where $f_{r,P}(Q)$ is the Miller function defined earlier. As written, this is sometimes referred to as the *reduced Tate pairing* to call attention to the supplementary exponentiation. The purpose of the supplementary exponentiation is to ensure that $\Phi(P, Q)$, if nonzero, is a canonical representative of $\boldsymbol{F}^*_{q^k}$ of order r. This is because the function $f_{r,P}(x, y)$ is defined only up to a multiplying constant. This constant is any nonzero element of $\boldsymbol{F}^*_{q^k}$ which might not be in the desired subgroup of order r.

The input to the Tate pairing is a pair of r-torsion points. We treat only prime r. Point P is from the cyclic group $\mathcal{X}(\boldsymbol{F}_q)[r]$ and point Q is from a cyclic subgroup of $\mathcal{X}(\boldsymbol{F}_{q^k})[r]$ other than $\mathcal{X}(\boldsymbol{F}_q)[r]$. For these points, the canonical representatives from the Frobenius eigenspaces can be chosen. The output of the Tate pairing is an element in the cycle of order r in the extension field $\boldsymbol{F}_{q^k}$, where the integer k is the embedding degree of r. Because r divides $q^k - 1$, $F^*_{q^k}$ does have the cyclic subgroup of order r denoted μ_r. This cyclic subgroup is not in the ground field F^*_q.

It remains to show that the Tate pairing is a bilinear pairing, and to provide a method of computation. The first task occupies the remainder of the section. The second task occupies the next section.

The study of the Tate pairing is based on the following three expressions for divisors:

$$\begin{aligned}
\text{div}(f_{i,P}) &= i(P) - ([i]P) - (n-1)(\mathcal{O}) \\
\text{div}(v_{P,R}) &= (P) + (R) + ([-1](P+R)) - 3(\mathcal{O}) \\
\text{div}(v_{P,\mathcal{O}}) &= (P) + ([-1]P) - 2(\mathcal{O}).
\end{aligned}$$

The first divisor is the divisor of the function we seek. The second divisor is the divisor of a straight line through the three points P, R, and Q with $P + R + Q = \mathcal{O}$. If $P = R$, this straight line is the straight line tangent to the curve at P and through the point $[2]P$. The third divisor is the divisor of the straight line through the points P, $-P$, and $\mathcal{O}$.

Setting n equal to one, the first divisor reduces immediately to the trivial divisor $\text{div}(f_{1,P}) = 0$, which corresponds to a function with no zeros or poles. This function is a constant which we will take to be the function $f_{1,P}(x, y) = 1$.

Setting n equal to two gives the divisor

$$\text{div}(f_{2,P}) = 2(P) - ([2]P) - (\mathcal{O})$$

which, because $2(P) = (P) + (P)$ and $[2]P = P + P$, can be written as[7]

$$\begin{aligned}
\text{div}(f_{2,P}) &= (P) + (P) - ([2]P) - (\mathcal{O}) \\
&= ((P) + (P) + ([-1](P+P)) - 3\mathcal{O}) - ([2]P + [-2](P) + [-2](\mathcal{O})) \\
&= \text{div}(v_{P,P}) - \text{div}(v_{[2]P,\mathcal{O}}).
\end{aligned}$$

This leads to the function

$$f_{2,P}(x, y) = \frac{v_{P,P}(x, y)}{v_{[2]P,\mathcal{O}}(x, y)},$$

where $P = (x_P, y_P)$, where $[2]P = (x_{[2]P}, y_{[2]P})$, and where m_P is the slope of the curve $\mathcal{X}(F_q)$ at the point P. Therefore, this divisor specifies the function

$$f_{2,P}(x, y) = \frac{(y - y_P) + m_P(x - x_P)}{x - x_{[2]P}},$$

where the numerator is the equation of a line of slope m_P passing through the point (x_P, y_P), and the denominator is the equation of a vertical line passing through the

[7] Because the symbol $+$ has two different meanings here, the meaning must be conveyed by the parentheses. Thus, $(P) + (Q)$ is a divisor, while $P + Q$ is a point addition. Furthermore $2(P)$ is a divisor, while $[2]P$ is a point multiple. Thus $(P) + (P) = 2(P)$, while $P + P = [2]P$.

point x_P. Continuing, we have

$$\begin{aligned}\operatorname{div}(f_{3,P}) &= 3(P) - ([3]P) - 2(\mathcal{O})\\ &((P) + ([2]P) + ([-1](P+2P)) - 4(\mathcal{O})) - ([3]P) + ([-3]P) - 2(\mathcal{O})\\ &= \operatorname{div}(v_{P,[2]P}) - \operatorname{div}(v_{[3]P,\mathcal{O}}).\end{aligned}$$

Therefore

$$\begin{aligned}f_{3,P}(x,y) &= f_{2,P}(x,y)\frac{v_{P,[2]P}(x,y)}{v_{[3]P,\mathcal{O}}(x,y)}\\ &= \left(\frac{(y-y_P)+m_P(x-x_P)}{x-x_{[2]P}}\right)\left(\frac{(y-y_{[2]P})+m_{[2]P}(x-x_{[2]P})}{x-x_{[3]P}}\right).\end{aligned}$$

Continuing in this way, the general case is the function

$$f_{r,P}(x,y) = \prod_{\ell=1}^{r}\frac{(y-y_{[\ell]P})+m_{[\ell]P}(x-x_{[\ell]P})}{x-x_{([\ell+1])P}}.$$

Because the product is over every term of the cycle, and the denominator takes on every value of ℓ, this can be rewritten

$$\begin{aligned}f_{r,P}(x,y) &= \prod_{\ell=1}^{r}\frac{(y-y_{[\ell]P})+m_{[\ell]P}(x-x_{[\ell]P})}{x-x_{[\ell]P}}\\ &= \prod_{\ell=1}^{r}\left(\frac{y-y_{[\ell]P}}{x-x_{[\ell]P}}+\frac{y_P-y_{[\ell]P}}{x_P-x_{[\ell]P}}\right).\end{aligned}$$

Then

$$f_{r,P}(Q) = \prod_{\ell=1}^{r}\left(\frac{y_Q-y_{[\ell]P}}{x_Q-x_{[\ell]P}}+\frac{y_P-y_{[\ell]P}}{x_P-x_{[\ell]P}}\right)$$

and

$$\Phi(P,Q) = f_{r,P}(Q)^{(q^k-1)/r}.$$

The function $f_{r,P}(x,y)$ has x degree equal to r. Because, in applications, r may be larger than a 50-digit integer, we do not ever want to see $f_{r,P}(x,y)$ explicitly, nor is it possible to do so even if the function is reduced by the equation of the elliptic curve. Instead, we deal only with its evaluation $f_{r,P}(Q)$ at the specific point Q, which we will see is possible to compute without explicitly computing the function $f_{r,P}(x,y)$. The function $f_{r,P}(x,y)$ will be implicit in an algorithm for computing $f_{r,P}(Q)$.

Written in this way as a simple product of r terms, the Tate pairing appears to be an intractable computation for very large values of r. However, it is in a form that is amenable to a doubling algorithm for computation. In this application, the doubling algorithm is known as the *Miller algorithm* and is described in the next section. Indeed, the MOV attack now can be regarded as a deflection of a discrete-log problem on an elliptic curve into a doubling algorithm on that curve, thereby pushing the problem into a domain where the discrete-log problem is easier.

Theorem 12.11.1 *The Tate pairing is a bilinear pairing.*

Proof To prove the theorem, we must show that $\Phi([a]P, Q) = \Phi(P, Q)^a$ and that $\Phi(P, [b]Q) = \Phi(P, Q)^b$. We prove these two statements in two steps using elementary manipulations

Step 1 To show that $\Phi([a]P, Q) = \Phi(P, Q)^a$, we will show that $f_{r,P}(x, y)^a = g(x, y)^r f_{r,[a]P}(x, y)$ where $g(x, y)$ is an unspecified rational function. This then implies that $[g(Q)^r]^{(q^k-1)/r} = 1$, and so $[f_{r,P}(Q)^a]^{(q^k-1)/r} = f_{r,aP}(Q)^{(q^k-1)/r}$. To this purpose, write the following divisors

$$\begin{aligned} \text{div}(f_{r,P}(x, y)) &= r(P) - r(\mathcal{O}) \\ \text{div}([f_{r,P}(x, y)]^a) &= ar(P) - ar(\mathcal{O}) \end{aligned}$$

and

$$\text{div}(f_{r,[a]P}(x, y)) = r([a]P) - r(\mathcal{O}).$$

Thus

$$\begin{aligned} \text{div}\left(\frac{[f_{r,P}(x, y)]^a}{f_{r,[a]P}(x, y)}\right) &= ra(P) - r([a]P) - r(a-1)(\mathcal{O}) \\ &= r\{a(P) - [a]P - (a-1)(\mathcal{O})\}. \end{aligned}$$

This means that, except for a multiplying constant, the function $[f_{r,P}(x, y)]^a / f_{r,aP}(x, y)$ is a rational function raised to the rth power, which we denote as $g(x, y)^r$. Therefore with the multiplying constant equal to one, $f_{r,[a]P}(x, y) = [f_{r,P}(x, y)]^a$. Because $g(Q)$ is an element of $\boldsymbol{F}_{q^k}$, $[g(Q)^r]^{(q^k-1)/r} = 1$ and $f_{r,[a]P}(x, y) = [f_{r,P}(x, y)]^a$, as was to be proved.

Step 2 To show that $\Phi(P, [b]Q) = \Phi(P, Q)^b$, recall that the evaluation $f_{r,P}(Q)$ can be written

$$f_{r,P}(Q) = \prod_{\ell=1}^{r} \left(\frac{y_Q - y_{[\ell]P}}{x_Q - x_{[\ell]P}} + \frac{y_P - y_{[\ell]P}}{x_P - x_{[\ell]P}}\right).$$

Accordingly, to focus attention on Q, temporarily define the function

$$g_{r,P,Q}(x, y) = \prod_{\ell=1}^{r} \left(\frac{y_Q - y_{[\ell]P}}{x_Q - x_{[\ell]P}} + \frac{y - y_{[\ell]P}}{x - x_{[\ell]P}}\right).$$

Then $f_{r,P}(Q) = g_{r,P,Q}(P)$. By identifying the equation of a line through the points Q and $[\ell]P$, this can be rewritten as

$$\begin{aligned} g_{r,P,Q}(x,y) &= \prod_{\ell=1}^{r}\left(\frac{(y - y_{[\ell]P}) + m_{Q,[\ell]P}(x - x_{[\ell]P})}{x - x_{[\ell]P}}\right) \\ &= \prod_{\ell=1}^{r}\left(\frac{v_{Q,[\ell]P}(x,y)}{v_{[\ell]P,\mathcal{O}}(x,y)}\right). \end{aligned}$$

The numerator of each factor of $g_{r,P,Q}(x, y)$ is the equation of a line passing through Q and $[\ell]P$. The numerator of the ℓth factor has the divisor

$$\operatorname{div}(v_{Q,[\ell]P}(x,y)) = ((Q) + ([\ell]P) + ([-1](Q + [\ell]P)) - 3(\mathcal{O}).$$

The denominator of each factor of $g_{r,P,Q}(x, y)$ is the equation of a line passing through $[\ell]P$ and $[-\ell]P$, and has divisor $([\ell]P) + ([-\ell]P) - 2(\mathcal{O})$. Therefore

$$\begin{aligned} \operatorname{div}(g_{r,P,Q}(x,y)) &= \sum_{\ell=1}^{r}\{(Q) + ([\ell]P) + ([-1](Q + [\ell]P)) - (([\ell]P) \\ &\quad + ([-\ell]P) - 2(\mathcal{O}))\} \\ &= \sum_{\ell=1}^{r}\{(Q) - ([-\ell]P) + ([-1](Q + [\ell]P)) + (\mathcal{O})\}. \end{aligned}$$

Thus the ratio $g^a_{r,P,Q}(x, y)/g_{r,P,[a]Q}(x, y)$, with (x, y) suppressed from the notation, has divisor

$$\begin{aligned} \operatorname{div}\frac{(g_{r,P,Q})^a}{g_{r,P,[a]Q}} &= \operatorname{div}(g_{r,P,Q})^a - \operatorname{div}(g_{r,P,aQ}) \\ &= \sum_{\ell=1}^{r}\{a[Q] - a[-\ell P] + a[-(Q + \ell P)] \\ &\quad - [aQ] + [\ell P] - [-(aQ + \ell P)] - (a-1)(\mathcal{O})\} \\ &= r(a[Q] - [aQ]) \\ &\quad + \sum_{\ell=1}^{r}\{a[\ell P] - [a\ell P] + a[-(Q + \ell P)] - [-a(Q + \ell P)] \\ &\quad - (a-1)(\mathcal{O})\}. \end{aligned}$$

Because the final summation on ℓ vanishes, this means that, up to a multiplying constant which can be taken to be one, the ratio $(g_{r,P,Q})^a/g_{r,P,[a]Q}$ is a rational function, say $g(x, y)$, raised to the rth power. Because $[g(Q)^r]^{(q^k-1)/r} = 1$, this shows that $f_{r,P,[a]Q}(x, y) = [f_{r,P,Q}(x, y)]^a$, as was to be proved. □

The proof does make it clear that the function $f_{r,aP}(x, y)$ must be computed with attention to the appropriate normalization. The description of the Tate pairing also makes it clear that the Tate pairing $\Phi(P, Q)$, as such, is not defined if $Q = [\ell]P$ for any nonzero value of ℓ. This is the reason that P and Q must be chosen from different cyclic subsets of r-torsion points.

Corollary 12.11.2 *The Tate pairing has the following properties*

(i) $\Phi(P_1 + P_2, Q) = \Phi(P_1, Q)\Phi(P_2, Q)$.
(ii) $\Phi(P, Q_1 + Q_2) = \Phi(P, Q_1)\Phi(P, Q_2)$.

Proof Write $P_1 = [a]P$ and $P_2 = [b]P$ for some P. Property (i) then follows. Property (ii) is proved in a similar manner. □

12.12 The Miller algorithm

The Miller algorithm is an efficient algorithm for computing the evaluation of certain bivariate rational functions at a point P. It is essential to the practical applications of pairing-based cryptography. The algorithm is constructed in terms of a particular rational function called a *Miller function*. A Miller function, denoted $f_{n,P}$, is any rational function on an elliptic curve with a divisor of the form

$$\text{div}(f_{n,P}) = n(P) - ([n]P) - (n-1)(\mathcal{O}),$$

where n is any positive integer, P is a point of the elliptic curve, and $[n]P$ is also a point of the elliptic curve. A rational function $f_{n,P}(x, y)$ with this divisor has a zero of order n at the point P of the curve, a pole of order one at the point $[n]P$ of the curve, and a pole of order $(n-1)$ at the point $\mathcal{O}$. It is obvious that this divisor satisfies the conditions of Theorem 11.6.2, so such a rational function does exist. The function $f_{n,P}(x, y)$ is uniquely determined by its divisor up to a multiplying constant. It is also clear that for a huge integer n, we do not ever want to see the function $f_{n,P}(x, y)$ because it will have a huge degree. We will only want a way to compute the evaluation $f_{n,P}(Q)$ for any specific point Q. Thus we must compute the evaluation of $f_{n,P}(x, y)$ at the point Q without actually computing the function itself.

The Miller algorithm is a recursive algorithm for computing an evaluation of the rational function $f_{n,P}(x, y)$ at a point Q. It expresses the rational function as a recursion of divisors. In particular, it relates the divisor $(f_{n+m,P})$ to divisors $(f_{n,P})$ and $(f_{m,P})$, as well as divisors for straight lines in the plane. Straight lines correspond to polynomials of degree one. After this recursion of divisors is defined, it can become a recursion on the functions themselves, or, more to our purpose, as a recursion on the evaluation of the function at a single point.

To understand the Miller algorithm, make the following observation. The divisor of the Miller function $f_{m+n,P}$ is

$$\text{div}(f_{m+n,P}) = (m+n)(P) - ([m+n]P) - (m+n-1)(\mathcal{O}),$$

and if one multiplies the functions $f_{m,P}(x, y)$ and $f_{n,P}(x, y)$, then their divisors add because the product has the poles and zeros of both, possibly with cancellation. That is,

$$\begin{aligned}\operatorname{div}(f_{m,P} f_{n,P}) &= m(P) - ([m]P) - (m-1)(\mathcal{O}) + n(P) - ([n]P) - (n-1)(\mathcal{O}) \\ &= ([m+n]P) - m(P) - n(P) - (m+n-2)(\mathcal{O}).\end{aligned}$$

This obviously is not the divisor of the function $f_{m+n,P}$. It has surplus poles at the points $[m]P$ and $[n]P$, a surplus zero at $\mathcal{O}$, and is lacking a needed pole at $[m+n]P$. However, we can conclude that the desired divisor satisfies

$$f_{m+n,P}(x, y) = f_{m,P}(x, y) f_{n,P}(x, y) g_P(x, y),$$

where $g_P(x, y)$ is a rational function, yet to be constructed, with zeros at $[m]P$ and $[n]P$, and poles at $[m+n]P$ and $\mathcal{O}$.

From a computational point of view, this expression allows us to write the evaluation at point Q as

$$f_{m+n,P}(Q) = f_{m,P}(Q) f_{n,P}(Q) g_P(Q).$$

This equation is fundamental to the development of computational algorithms. It can be used in various ways to recursively compute an evaluation $f_{n,P}(Q)$, even when n is a huge integer.

To construct the function $g_p(x, y)$, we will make use of the divisor of a straight line in the plane. The line $\ell_{U,V}$ through points U and V also goes through the point W, where $W = -(U + V)$, and has divisor

$$\operatorname{div}(\ell_{U,V}) = U + V + (-(U + V)) - 3\mathcal{O}.$$

This is the divisor of a rational function that has three zeros; one zero at point U, one zero at point V, and one zero at point $-(U + V)$. It satisfies the requirements of Theorem 11.6.2, so there does exist such a rational function, namely the straight line that we started with. Similarly, the vertical line v_P through a point P has divisor

$$\operatorname{div}(v_P) = P + (-P) - 2\mathcal{O}.$$

Each of these divisors defines a corresponding rational function up to a scalar multiplier.

Accordingly, the Miller algorithm is based on the *Miller reduction* contained in the following observation

$$f_{m+n,P}(x, y) = f_{m,P}(x, y) f_{n,P}(x, y) \frac{\ell_{mP,nP}(x, y)}{v_{(m+n)P}(x, y)},$$

which can be abbreviated

$$f_{m+n,P} = f_{m,P} f_{n,P} \frac{\ell_{mP,nP}}{v_{(m+n)P}},$$

where $f_{s,P}(x, y)$ is abbreviated simply as $f_{s,P}$, and where $\ell_{[m]P,[n]P}$ is the equation of the line through $[m]P$ and $[n]P$. If $[m]P = [n]P$, then it is the equation of the tangent line. Finally $v_{(m+n)P}$ is the equation of the vertical line through $[m+n]P$.

For incrementing n by one, the Miller reduction can be rewritten as

$$f_{n+1,P} = f_{n,P}\frac{\ell_{P,nP}}{v_{(n+1)P}}$$

by setting m equal to one.

For doubling n, the Miller reduction can be rewritten with $m = n$ as

$$f_{2n,P} = f_{n,P} f_{n,P}\frac{\ell_{nP,nP}}{v_{(2n)P}}.$$

In this way, $f_{2^\ell,P}$ can be expressed as a recursion for the first N powers of two, where $N = \lceil \log_2 n \rceil$. This doubling formula gives a practical method of computation of the first N doublings even if n is a very large number. Then with n expressed in binary notation as $n = \sum_{\ell=0}^{N} n_i 2^\ell$, the terms $f_{2^\ell,P}$ corresponding to the bits with $n_\ell = 1$ are combined using the Miller reduction to produce the desired $f_{n,P}$.

With the background reasoning now established, it is appropriate to summarize the main point by stating the following theorem.

Theorem 12.12.1 *The Miller reduction*

$$f_{m+n,P}(Q) = f_{m,P}(Q) f_{n,P}(Q)\frac{\ell_{mP,nP}(Q)}{v_{(m+n)P}(Q)}$$

computes the evaluation at Q of a function whose divisor has the form

$$\text{div}(f_{m+n,P}) = [m+n]P - (m+n)P - (m+n-1)\mathcal{O}$$

corresponding to a Miller function.

Proof The divisor of the function on the right will be regarded as a result of the process of appending new poles and zeros, and discarding old poles and zeros

$$\begin{aligned}
\text{div}(f_{m+n,P}) &= [m+n]P - (m+n)P - (m+n-1)\mathcal{O} \\
&= [m]P - mP - (m-1)\mathcal{O} \\
&\quad + [n]P - nP - (n-1)\mathcal{O} \\
&\quad + mP + nP - (m+n)P - 3\mathcal{O} \\
&\quad - (m+n)P + (m+n)P + 2\mathcal{O} \\
&= \text{div}(f_{m,P}) + \text{div}(f_{n,P}) + \text{div}(\ell_{[m]P,[n]P}) - \text{div}(v_{[m+n]P}).
\end{aligned}$$

Replacing the divisors with the corresponding functions completes the proof of the theorem. □

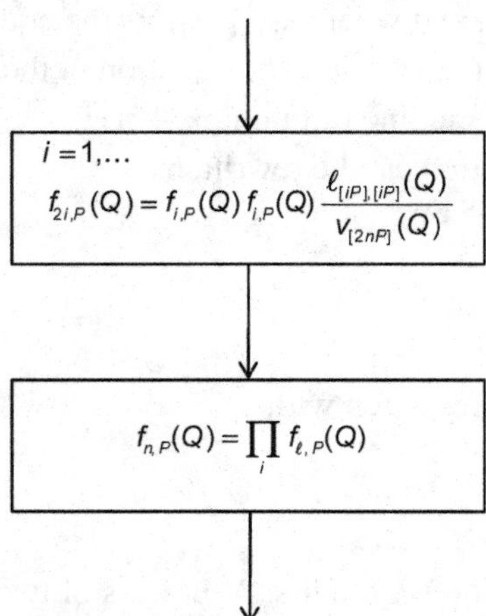

Figure 12.9 The Miller algorithm

To evaluate the rational function $f_{s,P}(x, y)$ for nonnegative integer s at the point Q, one can use the Miller algorithm, which is shown in Figure 12.9. The Miller algorithm computes a Miller function by recursive doubling, followed by combining a subset of the doublings. First use Theorem 12.12.1 to compute $f_{2^\ell,P}(Q)$ for $\ell = 1, 2, \ldots, \lfloor \log_2 s \rfloor$. Then, with the binary expansion of s as

$$s = \sum_\ell s_\ell 2^\ell,$$

repeatedly use the Miller reduction to combine all terms for which $s_\ell = 1$, so as to compute $f_{s,P}(Q)$. This requires no more than $2\lfloor \log_2 s \rfloor$ applications of the Miller reduction.

12.13 The Weil pairing

The Weil pairing is a symmetric bilinear pairing that predates the Tate pairing and, evidently, is in the shadow of the Tate pairing for practical applications. The Weil pairing is of secondary importance as a computational algorithm in cryptography. However, it does play a role in theoretical developments. The statement and structure of the Weil pairing are considerably more abstract. The final exponentiation found in the Tate pairing is not used in the Weil pairing. To eliminate the unknown multiplying constant, the Miller algorithm is executed twice with somewhat different divisors and the ratio of the two is computed in such a way as to cancel the unknown constant. We provide only a brief introduction to the Weil pairing.

The Weil pairing is a bilinear map from $\mathcal{X}(\boldsymbol{F}_q)[r] \oplus \mathcal{X}(\boldsymbol{F}_q)[r]$ into an r-point orbit μ_r in the extension field $\boldsymbol{F}_{q^k}$, where k is the embedding degree of the r-torsion points of the curve $\mathcal{X}(\boldsymbol{F}_q)$. We will consider only r-torsion points for prime r. The traditional form of the Weil pairing is given in this section, but it is very technical and not suitable for computations. An alternative computational form, which is not given, can be formulated.[8]

Recall that an r-torsion point of the elliptic curve $\mathcal{X}(\boldsymbol{F}_q)$ is defined as any point P for which $[r]P = \mathcal{O}$. The set of r-torsion points is denoted

$$\mathcal{X}[r] = \{P \in \mathcal{X}(\overline{\boldsymbol{F}}_q) \mid [r]P = \mathcal{O}\}.$$

The set $\mathcal{X}(\boldsymbol{F}_q)[r] = \mathcal{X}[r]$ consists of all points of $\mathcal{X}(\boldsymbol{F}_q)$ of order r, all points whose order divides r, as well as the identity element $\mathcal{O}$. Also recall that a principal divisor of $\mathcal{X}$ is a divisor for which there exists a rational function on $\mathcal{X}(\boldsymbol{F}_q)$ with that divisor.

Any r-torsion point of $\mathcal{X}(\boldsymbol{F}_q)$ lies within an orbit of order r. There is an orbit of order r in $\boldsymbol{F}^*_{q^k}$ as well, where k is the embedding degree. Because r is a prime, any element $P \in \mathcal{X}(\boldsymbol{F}_q)[r]$ except for $\mathcal{O}$ generates an orbit of order r within $\mathcal{X}(\boldsymbol{F}_q)$, and any element β of order r of $\boldsymbol{F}_{q^k}$ generates an orbit of order r in $\boldsymbol{F}^*_{q^k}$. Thus there is a mapping from one orbit to the other, sending P to β and $[\ell]P$ to β^ℓ. Of course, because it is intractable to find ℓ when given $[\ell]P$, it is also intractable to map $[\ell]P$ into β^ℓ by first finding ℓ. Continuing with this line of thought, if $[\ell]P$ and $[\ell']P'$ are two points on the two orbits of order r, then the pair $([\ell]P, [\ell']P')$ can be mapped to $\beta^{\ell\ell'}$ as was shown in Figure 12.2. Thus one can say that $[\ell]P$ is mapped to an orbit in $\boldsymbol{F}_{q^k}$ with a generator $\beta^{\ell'}$ determined by $[\ell']P'$.

To enable the definition of the Weil pairing, definitions of the bivariate rational functions, $f(x, y)$ and $g(x, y)$, are needed. These functions must be nontrivial and defined so that $f(P)$ and $g(P')$ are elements of $\boldsymbol{F}_{q^k}$ whenever P and P' are elements of $\mathcal{X}(\boldsymbol{F}_q)[r]$. The polynomials $f(x, y)$ and $g(x, y)$ are constructed by careful specification of their bivariate zeros and poles on the elliptic curve in an appropriate extension field.

There are several ways to approach the definition of the Weil pairing. We shall give a definition that is traditional but impractical and useless from a computational point of view. We will not give a practical reformulation based on an alternative and equivalent definition.

Let P be any point of $\mathcal{X}(\boldsymbol{F}_{q^k})$ of order r, and let T satisfy $[r]T = P$. Because the point P has order r, the point T has order r^2. Although such a point T does exist somewhere in $\mathcal{X}(\overline{\boldsymbol{F}}_q)$, it will be in a huge, massively large, extension field and hence not possible to explicitly compute. We require, for every r-torsion point $R \in \mathcal{X}[r]$, that $f(x, y)$ has a simple pole at every R and a simple zero at every $T + R$. In the formal language of algebraic geometry, this is written as the divisor

$$\mathrm{div}(f(x, y)) = \sum_{R \in \mathcal{X}[r]} ((T + R)) - \sum_{R \in \mathcal{X}[r]} (R),$$

[8] The original form of the Weil pairing is outlined here so that the literature is more accessible.

where $(T + R)$ is a point addition. This divisor $\operatorname{div}(f(x, y))$ has a plus one at every point of the form $T + R$ and a minus one at every point R, where R is a point in the set of r-torsion points of $\mathcal{X}(\boldsymbol{F}_q)$. This is just an alternative way of saying that the function $f(x, y)$ has a zero at every point of the form $T + R$ where $R \in \mathcal{X}[r]$, and a pole at every point of the form R. It is now clear that any choice of T such that $[r]T$ has order r will do because for any other such point T', the point $[r](T - T')$ has order r. To verify that this expression defines a principal divisor, refer to Theorem 11.6.2. Part (i) of that theorem is satisfied because there are r^2 zeros of order one, and r^2 poles of order one. Part (ii) is satisfied because the point addition

$$\sum_{R \in \mathcal{X}[r]} (T + R) - \sum_{R \in \mathcal{X}[r]} R = [r^2]T + \sum_{R \in \mathcal{X}[r]} (R - R) = \mathcal{O},$$

since there are r^2 terms in $\mathcal{X}[r]$, using the fact that T has order r^2. Therefore we are assured by Theorem 11.6.2 that the rational function $f(x, y)$ does exist. The function $f(x, y)$ is defined by its divisor up to multiplication by an unspecified element of $\boldsymbol{F}_{q^k}$.

To complete the definition of the Weil pairing, choose any point X at which f is nonzero, and for any point P define

$$g(P) = f(X + P).$$

The Weil pairing is defined as

$$\Phi(P, S) = \frac{f(X + P)}{f(X)}.$$

The Weil pairing is clearly an element of $\boldsymbol{F}_{q^k}$ if all coefficients of $f(x, y)$ and $g(x, y)$ are in $\boldsymbol{F}_{q^k}$. The definition is constructed so that the domain lies on an orbit of order r in $\boldsymbol{F}_{q^k}$. It is easy to conclude from this that the domain of the Weil pairing is equal to an orbit of order r in $\boldsymbol{F}_{q^k}$.

Theorem 12.13.1 *Suppose that the function $f(x, y)$ has the divisor*

$$\operatorname{div}(f(x, y)) = \sum_{R \in \mathcal{X}[r]} ((T + R)) - \sum_{R \in \mathcal{X}[r]} (R),$$

where $[r]T = P$ for any $P \in \mathcal{X}(\boldsymbol{F}_q)[r]$. Then $f(P)$ is an element of $\boldsymbol{F}_{q^k}$ of order r.

Proof By the nature of its construction, $f(P)$ is an element of $\overline{\boldsymbol{F}}_q$. It is enough to show that $f(P)$ has order r, because all points of $\overline{\boldsymbol{F}}_q$ of order r are zeros of $x^k - 1$, and so are in $\boldsymbol{F}_{q^k}$. This follows from an observation that $f(X + P)^n = f(X)^n$. □

The Weil pairing has been carefully constructed in such a way as to provide the properties tabulated in the following theorem, for which only an outline of a proof is given.

Theorem 12.13.2 *The Weil pairing has the following properties:*

(i) $\Phi(P_1 + P_2, T) = \Phi(P_1, T)\Phi(P_2, T)$*,*
(ii) $\Phi(P, T_1 + T_2) = \Phi(P, T_1)\Phi(P, T_2)$*,*
(iii) $\Phi(P, T) = \Phi(T, P)^{-1}$*,*
(iv) $\Phi(P, T) = 1$ *for all* P *if, and only if,* $T = \mathcal{O}$*,*
(v) $\Phi(P^a, T^a) = \Phi(P, T)^a$ *where* P^a *means* (x^a, y^a)*,*
(vi) $\Phi(P, P) = 1$ *for all* P.

Proof Part (i) is proved by writing

$$\begin{aligned}\Phi(P_1 + P_2, T) &= \frac{g(X + P_1 + P_2)}{g(X)} \\ &= \frac{g(X + P_1 + P_2)}{g(X + P_1)} \frac{g(X + P_1)}{g(X)} \\ &= \Phi(P_2, T)\Phi(P_1, T).\end{aligned}$$

Part (ii) is proved – pending the definition of appropriate terms – by writing

$$\begin{aligned}\Phi(P, T_1 + T_2) &= \frac{g_3(X + P)}{g_3(X)} \\ &= \frac{g_1(X + P)g_2(X + P)h([r]X + [r]P)}{g_1(X)g_2(X)h([r]X)} \\ &= \Phi(P, T_1)\Phi(P, T_2).\end{aligned}$$

To this end, define f_1, f_2, f_3, g_1, g_2, and g_3 as the appropriate rational functions for the points T_1, T_2, and $T_3 = T_1 + T_2$. Define a rational function $h(x, y)$ by the divisor

$$\mathrm{div}(h(x, y)) = (T_1 + T_2) - (T_1) - (T_2) + (\mathcal{O}).$$

Accordingly, because the divisors satisfy

$$\mathrm{div}\left(\frac{f_3(x, y)}{f_1(x, y)f_2(x, y)}\right) = r\,\mathrm{div}(h(x, y)),$$

we can conclude that the rational functions satisfy

$$f_3(x, y) = cf_1(x, y)f_2(x, y)h(x, y)^r$$

for some constant c. The equation for $\Phi(P, T_1 + T_2)$ now follows.

Part (iii) is proved by writing

$$\Phi(P + T, P + T) = \Phi(P, P)\Phi(P, T)\Phi(T, P)\Phi(T, T),$$

and then proving that $\Phi(T, T) = 1$ for all $T \in \mathcal{X}[r]$ so that the equation reduces to $1 = \Phi(P, T)\Phi(T, P)$. Thus it is enough to prove that $\Phi(T, T) = 1$ for all $T \in \mathcal{X}[r]$.

Part (iv) is proved by noting that if $\Phi(P, T) = 1$ for all P, then $g(X + P) = g(X)$ for all P.

Part (v) for any $R \in \mathcal{X}[r]$, let $R^a = (x^a, y^a)$ where $R = (x, y)$. Then $R^a \in \mathcal{X}[r]$ as well. Thus $\Phi(P^a, T^a)$ is well defined as an element of $\boldsymbol{F}_{q^k}$.

Part (vi) follows immediately from Part (iii). □

12.14 Pairing-friendly curves

For practical applications of pairing-based cryptography, it is important for cryptographic security that the embedding degree not be too small, but it is also important for the implementation complexity that the embedding degree not be too large. Thus it is necessary to have elliptic curves with a small embedding degree, though not too small. Because the embedding degree of a randomly chosen curve could be very large, which is not unusual, or could even be too small, it is desirable to have a more structured way to approach the search.

To be useful for pairing, an elliptic curve should have a large subgroup of prime order and a moderately small embedding degree (in the range of six to twelve might be suitable). Such elliptic curves, called *pairing-friendly elliptic curves*, are rare and so are hard to find by unstructured trial and error. Although we prefer to leave the term a little vague, a loose definition is that a pairing-friendly elliptic curve should be such that $\#\mathcal{X}(\boldsymbol{F}_q)$ has a prime factor r that is large compared to $\sqrt{q}$, and preferably on the order of q, and with an embedding degree small compared to $\log_2 r$. Although such pairing-friendly curves are sparse in the set of all elliptic curves, they are still plentiful because the set of elliptic curves over finite fields is so very large. Nevertheless, it requires well-considered methods to find the pairing-friendly curves.

One class of elliptic curves with small embedding degree that we have already encountered in Theorem 10.6.4 consists of those supersingular curves over a prime field $\boldsymbol{F}_p$ for which p has the form $p = 4\ell + 3$ and for which the defining polynomial has the form $y^2 = x^3 + ax$. Such elliptic curves always have an embedding degree of two. Because this embedding degree is small, these curves are commonly regarded as too friendly, and not really suitable for pairing-based cryptography.

A supersingular curve on a prime field always has an embedding degree of two. A supersingular curve on a prime-power field always has an embedding degree not larger than six, and not larger than three if the field characteristic is larger than three. An embedding degree of two is usually considered small, and even an embedding degree of six may be considered too small.

The Barreto–Naehrig curves, studied next in Section 12.15, are an interesting class of good pairing-friendly curves, always giving an elliptic curve with embedding degree of twelve. Other classes are the MNT curves, with embedding degree three, four, or five, which are not described here, and the Freeman curves, with embedding degree ten, which are touched on in Section 12.16. These classes usually have subgroups with prime orders comparable to the field size p. Other classes of curves, such as the Brezing–Weng curves, do not satisfy this condition.

12.15 Barreto–Naehrig elliptic curves

An attractive family of pairing-friendly elliptic curves is the family of *Barreto–Naehrig elliptic curves*. These curves constitute a family of elliptic curves that have reasonable embedding degrees. Each Barreto–Naehrig elliptic curve is an elliptic curve over a prime field with a prime order and an embedding degree of twelve.

To design a class of curves, the obvious approach is to first specify the field $\boldsymbol{F}_p$ and the curve $\mathcal{X}(\boldsymbol{F}_p)$, then to compute the order r of a subgroup of prime order and the embedding degree k of r. In contrast, the Barreto–Naehrig construction inverts this procedure. It designs a class of codes by first specifying the embedding degree k, then computing a combination of p, r, and t that are a consistent set of curve parameters. Then if p and r are both primes, these parameters admit a curve consistent with this procedure. With this approach, the curve itself is the last item to be defined.

A Barreto–Naehrig elliptic curve is defined over the prime field $\boldsymbol{F}_p$, with a defining polynomial of the form $y^2 = x^3 + b$, where b is a nonzero element of the field. We will see later that the constant b is arbitrary other than that $b + 1$ must be a square of $\boldsymbol{F}_p$. We will further show that for these curves, the prime p, the order r, and the trace t are described very neatly by the three simultaneous polynomials

$$p = 36z^4 - 36z^3 + 24z^2 - 6z + 1$$
$$r = 36z^4 - 36z^3 + 18z^2 - 6z + 1$$

and

$$t = 6z^2 + 1,$$

where z is an element of $\boldsymbol{Z}$ that can be chosen freely, provided only that p and r as specified by these polynomials in z are primes for that choice of z. The embedding degree of a curve defined in this way is always twelve.

Definition 12.15.1 *A Barreto–Naehrig elliptic curve is an elliptic curve $\mathcal{X}(\boldsymbol{F}_p) : y^2 = x^3 + b$ with order r where $b + 1$ is a square in $\boldsymbol{F}_p$, and p and r are both primes that can be expressed as*

$$p = 36z^4 - 36z^3 + 24z^2 - 64z + 1$$
$$r = 36z^4 - 36z^3 + 18z^2 - 64z + 1$$

for some integer z.

An attractive feature of a Barreto–Naehrig curve is the fact that the parameters p, r, and t of the curve are determined by the integer value of z and they need not be separately stored as integers. Each can be computed when needed.

For a simple example, take $z = 7$. Then $p = 100003$ and $r = 99709$, which are both prime, and $t = 295$. An appropriate Barreto–Naehrig elliptic curve with these parameters with embedding degree twelve is $\mathcal{X}(\boldsymbol{F}_{100003}) : y^2 = x^3 + 37$.

For a more elaborate example, the 160-bit Barreto–Naehrig curve, using the polynomial $y^2 = x^3 + 3$, has the following 48-digit primes as parameters:

$$p = 1461501624496790265145448589920785493717258890819$$

$$r = 1461501624496790265145447380994971188499300027613$$

and, because $t = p + 1 - r$,

$$t = 1208925843052179588632 07,$$

where z is a 12-digit integer. Only the polynomial and the value of z need to be stored because p and r can be computed when needed.

By changing the value of z, larger (or smaller) Barreto–Naehrig curves can also be constructed. For each integer value of z, one computes both p and r from the defining polynomials and tests both of them for primality. If both are prime, one has found an elliptic curve with an embedding degree of twelve. If either is not prime, that value of z does not produce a Barreto–Naehrig curve.

The number of Barreto–Naehrig curves is apparently very large. There are about 10^{12} choices of 12-digit integers for z, each leading to integers p and r with 49 digits. Because the density of primes among integers with 49 digits is about one in 100, we may conjecture that about one in 10^4 12-digit integers z will yield primes for both r and p. This conjecture suggests that there should be about 10^8 12-digit integers z that lead to Barreto–Naehrig curves. Of course, this is only an informal estimate obtained by treating the primality of p and r as random and independent events. It need not be true.

The formal development of the Barreto–Naehrig curves starts by recalling that $x^n - 1$ factors into a product of the cyclotomic polynomials $\Phi_d(x)$ for all d that divide n. To develop the construction, we will have a particular use for the twelfth cyclotomic polynomial, which is $\Phi_{12}(x) = x^4 - x^2 + 1$.

Recall that the embedding degree of r is the smallest k such that r divides $q^k - 1$. But $q^k - 1$ can be factored as a product of cyclotomic polynomials. This leads to the following theorem.

Theorem 12.15.2 *The order r of an elliptic curve of prime order and Frobenius trace t over a prime field $\boldsymbol{F}_p$ divides $\Phi_k(t-1)$, where $\Phi_k(x)$ is the kth cyclotomic polynomial* *and k is the embedding degree of r, but does not divide any $\Phi_i(t-1)$ with i less than k.*

Proof Because

$$q^i - 1 = \Phi_i(q) \prod_{\substack{d|i \\ d \neq i}} \Phi_d(q)$$

and k is the smallest value of i for which the prime r divides $q^i - 1$, then k is the smallest value of i for which r divides $\Phi_i(q)$. Because r divides $p^k - 1$, r also divides $(p-r)^k - 1$. By the definition of the Frobenius trace, $t - 1 = p - r$, so r divides $(t-1)^k - 1$. Therefore k is the smallest integer for which r divides $(t-1)^k - 1$. Because k is the smallest integer such that r divides $(t-1)^k - 1$, it must also be true that k is the smallest integer such that r divides $\Phi_k(t-1)$. □

Now specify that $k = 12$. To continue the development, consider only those values of the trace t that can be expressed as a polynomial in an integer parameter z as $t = 6z^2 + 1$. This restriction excludes the many elliptic curves for which t cannot be expressed in this way, but evidently still leaves an immense number of curves.

Theorem 12.15.3 *A Barreto–Naehrig elliptic curve has embedding degree equal to twelve.*

Proof The Frobenius trace is given by $t = p - r + 1 = 6z^2 + 1$. By Theorem 12.15.2, we see that r divides the integer $\Phi_{12}(6z^2)$, which can be written as

$$\begin{aligned} \Phi_{12}(6z^2) &= 6^4 z^8 - 6^2 z^4 + 1 \\ &= (36z^4 + 36z^3 + 18z^2 + 6z + 1)(36z^4 - 36z^3 + 18z^2 - 6z + 1). \end{aligned}$$

Thus with $r(z)$ defined as $36z^4 + 26z^3 + 18z^2 + 6z + 1$, we have

$$r(z)r(-z) = 6^4 z^8 - 6^2 z^4 + 1.$$

The polynomial $r(z)$ is irreducible over $\mathbf{Z}$, so except for the sign of z, $r(z)$ is the only factor of $6^4z^8 - 6^2z^4 + 1$. Thus $r(z)$ divides $\Phi_{12}(t-1)$. It does not matter whether we choose to use $r(z)$ or $r(-z)$ as the order r because z is allowed to take on both positive and negative integer values.

Finally, by the division algorithm, we can write

$$\Phi_d(z) = Q(z)r(z) + R(z)$$

for any $\Phi_d(z)$ with d smaller than k. Because $\Phi_d(z)$ and $r(z)$ are monic polynomials, $R(z)$ must be a polynomial with only integer coefficients. If $r(z)$ divides the polynomial $\Phi_d(z)$, then it must also divide $R(z)$. But the remainder polynomial $R(z)$ has a degree smaller than the degree of $r(z)$. This means that $r(z)$, which is a prime, divides $R(z)$ which is a nonnegative integer smaller than $r(z)$. Therefore $R(z) = 0$ and $r(z)$ cannot divide $\Phi_d(z)$ for any d smaller than k. □

To find a Barreto–Naehrig curve, recall that $r = p + 1 - t$, so $p = r + t - 1$. If we can find a z for which both r and $r + t - 1$ are primes, then we have a set of parameters p, r, and t for an elliptic curve with embedding degree twelve. That curve will exist if we can choose a value of b so that the polynomial $y^2 = x^3 + b$ gives such an elliptic curve. This is straightforward for prime r because then we only need to find one point of order r on the curve to generate all of the points of the curve. Simply find a nonzero integer b such that $b + 1$ is a square in $\boldsymbol{F}_p$, and observe that the point $(1, \sqrt{b+1})$ is a nonidentity point on the curve $\mathcal{X}(\boldsymbol{F}_p) : y^2 = x^3 + b$. Because r is a prime, the point $(1, \sqrt{b+1})$, then, has order r, which verifies that the curve defined by the polynomial $y^2 = x^3 + b$ has the required parameters. Because four is always a square modulo p, three is always a suitable choice for b.

12.16 More pairing-friendly curves

The constructions of this section, in general, construct elliptic curves using the theory of complex multiplication, as described in Section 10.17, and are based on integer solutions t and p to the diophantine equation

$$Dy^2 = 4p - t^2,$$

where D is the (complex multiplication) discriminant, often chosen to be a small integer such as one, two, or three.

In this section, to create a larger environment in which to work, t and p are each set equal to a polynomial in an indeterminant x, which will later be set to an integer. Thus

$$Dy^2 = 4p(x) - t(x)^2.$$

The general technique is to choose polynomials $p(x)$ and $t(x)$ with integer coefficients, then study whether there are integer values of x that solve the equation for an integer y, and for which p is a prime. Thus the prime p is not specified beforehand. It is a consequence of the method.

It can happen that the equation

$$Dy^2 = 4p(x) - t(x)^2$$

has an integer solution y for every integer value of x. Then the equation can be regarded as a polynomial equation in the integer parameter x for which there is a solution for every value of the integer x. Then, for any choice of x, one only needs to test $p(x)$ for primality. In other cases, only a sparse set of integer values of x correspond to integer solutions for y.

Recall that the Barreto–Naehrig curves were constructed with reference to the cyclotomic polynomial $\Phi_{12}(x)$. Inspired by this method, let $\Phi_k(x)$ be the kth cyclotomic polynomial. Let $p(x)$ be a polynomial with integer coefficients, and let $r(x)$ be an irreducible factor of the polynomial $\Phi_k(t(x) - 1)$. Let $p(x) = n(x) + t(x) - 1$. For any positive square-free integer D, suppose that (x^*, y^*) is an integer solution to the equation $Dy^2 = 4p(x) - t(x)^2$ and that $p(x^*)$ and $n(x^*)$ are both prime. Then provided D is not too large to manage, the method of complex multiplication is an efficient algorithm to construct an elliptic curve $\mathcal{X}(\boldsymbol{F}_{p(x_0)})$ that has prime order $n(x^*)$ and an embedding degree k.

In summary, for a given positive integer k and positive square-free integer D, an infinite family of elliptic curves with embedding degree k exists if three polynomials $t(x)$, $n(x)$, and $p(x)$ exist over $\mathbf{Z}$ such that the following proposition is satisfied.

Proposition 12.16.1 *Let $r(x)$, $p(x)$, and $t(x)$ be elements of the ring $\mathbf{Z}[x]$ satisfying*

(1) $n(x) = p(x) + 1 - t(x)$.
(2) $n(x)$ and $p(x)$ are irreducible polynomials over $\mathbf{Z}$.
(3) $n(x)$ divides $\Phi_k(t(x) - 1)$, where $\Phi_k(z)$ is the kth cyclotomic polynomial.
(4) For some square-free integer D, the equation $Dy^2 = 4p(x) - t(x)^2$ has infinitely many integer solutions (x, y) for which $p(x)$ is a prime integer.

Then the triple $(t(x), n(x), p(x))$ specifies an infinite family of elliptic curves of embedding degree k.

Proof Immediate from the preceding discussion. □

Of course, it may be possible that the final equation has only a finite number of solutions. This then gives a *sparse* family of pairing-friendly curves. It may also be that there are no solutions.

The Barreto–Naehring curves form a family of curves of embedding degree twelve satisfying Proposition 12.16.1. By devising factorizations of $\Phi_k(t(x) - 1)$ for other values of k, one may construct other families of pairing-friendly elliptic curves. If the square-free integer D is small enough, the curves of this family can be computed by the method of complex multiplication, as discussed in Section 10.17.

If, for any x satisfying Proposition 12.16.1, $p(x)$ is not a prime, then that value of x does not give an elliptic curve and is rejected. Moreover, if $n(x)$ is not a prime (or at least has a large prime factor), then that value of x is rejected as an elliptic curve that is not suitable for cryptography. The prime number theorem says that the density of primes near n is $\log n$. Therefore if there are sufficiently many x satisfying the proposition, then we may expect that about $(\log n)^2$ of them must be examined in order to find both a prime p and a large prime factor r of the integer n. Of course, there may be some unknown and unsuspected connection between the distribution of primes and the distribution of solutions to Proposition 12.16.1 that invalidates this plausible, but incidental, statistical premise.

Proposition 12.16.1 is an implicit statement of a set of conditions under which a pairing-friendly elliptic curve exists. One may attempt to satisfy these conditions by proceeding in any of a variety of ways. The *Freeman elliptic curves* are a family of curves with embedding degree ten based on the irreducible factorization

$$\Phi_{10}(10x^2 + 5 + 2) = (25x^4 + 25x^3 + 15x^2 + 5x + 1)s(x),$$

where $s(x)$ is a polynomial that is not of interest. Then with

$$t(x) = 10x^2 + 5x + 3$$

and

$$n(x) = 25x^4 + 25x^3 + 15x^2 + 5x + 1$$
$$p(x) = r(x) + t(x) - 1.$$

Then $Dy^2 = 4q(x) - t(x)^2$ results in the quadratic equation

$$Dy^2 = 15x^2 + 10x + 3.$$

This has the form of a well-studied diophantine equation known as the generalized Pell equation.[9] The Freeman elliptic curves correspond to those solutions of this equation for which $p(x)$ and $n(x)$ are prime. It is known that this can occur only if D is equal to 43 or 67 (mod 20). A 234-bit Freeman curve $y^2 = x^3 + Ax + B$ computed by the method of complex multiplication with $D = 1227652867$ has the parameters

$$q = 18211650803969472064493264347375950045934254696657090420726230043203803$$
$$n = 18211650803969472064493264347375949776033155743952030750450033782306651$$
$$A = -3$$

and

$$B = 15748668094913401184777964473522859086900831274922948973320684995903275.$$

Because this curve has embedding degree ten, n divides $q^{10} - 1$ and n does not divide $q^i - 1$ for i less than ten.

[9] The *Pell equation*

$$ny^2 - x^2 = 1$$

is a classical diophantine equation used early on to form approximations to $\sqrt{n}$.

Problems for Chapter 12

12.1 Can the inner product $\langle \boldsymbol{v}_1, \boldsymbol{v}_2 \rangle$ be used by means of the expression

$$\Phi(\boldsymbol{v}_1, \boldsymbol{v}_2) = e^{\langle \boldsymbol{v}_1, \boldsymbol{v}_2 \rangle}$$

to define a bilinear pairing from $\boldsymbol{R}^n \times \boldsymbol{R}^n$ into $\boldsymbol{R}$?

12.2 Let $\mathcal{X}(\boldsymbol{F}_q)$ be an elliptic curve over the finite field $\boldsymbol{F}_q$, where q is not a power of two. Find the form of all the two-torsion points of $\mathcal{X}(\boldsymbol{F}_q)$. Show that $\mathcal{X}(\boldsymbol{F}_q)[2] \simeq \boldsymbol{Z}_2 \oplus \boldsymbol{Z}_2$.

12.3 Can the Joux key exchange protocol be generalized to four parties?

12.4 A bilinear pairing satisfies the property that $\Phi(aP, bP) = \Phi(P, P)^{ab}$, where $\Phi(P, P)$ is an element of a finite field. What does $\Phi(aP, bP)\Phi(cP, dP)$ equal? How does this compare to $\Phi(aP, cP)\Phi(bP, dP)$?

12.5 Show that the bilinear-pairing approach to short signatures allows for multiple sequential signatures by parties $i = 1, \ldots, m$ by writing the message with m signatures as the recursion

$$\text{sign}_n(h(m)) = \text{sign}_{n-1}(h(m)) + a_n h(m),$$

where a_i is the key of the ith signature. How are the multiple signatures verified?

12.6 A twist of the elliptic curve

$$\mathcal{X}(\boldsymbol{F}_q) : y^2 = x^3 + ax + b$$

is the elliptic curve

$$\mathcal{X}'(\boldsymbol{F}_q) : y^2 = x^3 + d^2ax + d^3b,$$

where d is any nonsquare of $\boldsymbol{F}_q$. Are any two twisted curves equal if they use two distinct nonsquares d_1 and d_2 in the definition? Are the two twisted curves isomorphic as groups?

12.7 Let $\mathcal{X}(\boldsymbol{F}_p)$: $y^3 = x^3 + ax + b$ be an elliptic curve over $\boldsymbol{F}_p$ and let P be a point of $\boldsymbol{F}_{p^m}$.

a Does the Frobenius map π_q with $q = p$ take P to P?

b Does the Frobenius map π_q take $\mathcal{X}(\boldsymbol{F}_{p^m})$ to $\mathcal{X}(\boldsymbol{F}_{p^m})$?

c Is point addition preserved by π_q?

12.8 **a** Noting that $13 = 8 + 4 + 1$ and $13 = 2(4 + 2) + 1$, compare the complexity of several sequences of operations for computing the point multiple $13P$. With reference to Horner's rule, state the general case.

b Can Horner's rule be used with the Miller algorithm to compute the Tate pairing?

12.9 The Barreto–Naehrig elliptic curves form a class of curves with embedding degree twelve, each over its own prime field $\boldsymbol{F}_p$. What is the smallest prime p for which a Barreto–Naehrig curve exists? What is the order of this curve? Find all its points. What is the second smallest prime p for which a Barreto–Naehrig curve exists?

12.10 Let r divide $\#\mathcal{X}(\boldsymbol{F}_q)$, and let $\mathcal{X}(\overline{\boldsymbol{F}}_q)[r]$ denote the r-torsion points of $\mathcal{X}(\overline{\boldsymbol{F}}_q)$. Define a *Tate tower* as $\bigcup_{\ell=1}^{\infty} \mathcal{X}(\overline{\boldsymbol{F}}_q)[r^\ell]$. Assuming that r^ℓ-torsion points exist for all ℓ, what is the image of the Frobenius map applied to the Tate tower (that is, applied to each element of the Tate tower)?

12.11 Use Fermat's little theorem to prove that the embedding degree always exists.

12.12 Let $\boldsymbol{F}_q$ be a finite field of characteristic p and let $\mathcal{X}(\boldsymbol{F}_q)$ be an elliptic curve over $\boldsymbol{F}_q$. Show that if r and p are not coprime, then either

$$\mathcal{X}(\overline{\boldsymbol{F}}_q)[r] \simeq \boldsymbol{Z}_{r'} \oplus \boldsymbol{Z}_{r'}$$

or

$$\mathcal{X}(\overline{\boldsymbol{F}}_q)[r] \simeq \boldsymbol{Z}_{r} \oplus \boldsymbol{Z}_{r'},$$

where r' is the largest factor of r that is coprime to p.

12.13 Every polynomial over $\boldsymbol{F}_3$ of the form $y^2 = x^3 - x \pm 1$ is supersingular. Let m be coprime with 6. Show that if $P = (\alpha, \beta)$ is a point of the curve $\mathcal{X}(\boldsymbol{F}_3)$: $y^2 = x^3 - x \pm 1$, then $[3]P = (\alpha^9 \pm 2, -\beta^9)$, where the sign agrees with the sign in the defining polynomial.

12.14 Does Theorem 12.8.2 imply that every r^2-torsion point of $\mathcal{X}(\boldsymbol{F}_q)$ is also an r-torsion point of $\mathcal{X}(\boldsymbol{F}_q)$?

12.15 Show that every polynomial over $\boldsymbol{F}_{2^m}$ of either the form $y^2 = x^3 + x$ or the form $y^2 = x^3 + x + 1$ is supersingular. Show that an elliptic curve based on such a polynomial has order $2^m + 1 \pm 2^{(m+1)/2}$ and embedding degree four.

12.16 Show that

$$\Phi(x, y) = \left(-\frac{y^2}{2x^2}, \frac{-y(x^2 - 2)\sqrt{-2}}{4x^2} \right).$$

is an endomorphism for the elliptic curve $\mathcal{X}(\boldsymbol{Q}) : y^2 = x^3 + 4x^2 + 2x$. Describe $\Phi^4(x, y)$.

12.17 Is the Duursma–Lee polynomial supersingular?

Notes for Chapter 12

The notion of a bilinear pairing, introduced by Weil in 1940, predates its use in cryptology. The Tate pairing is due to Tate (1958, 1963) and developed further by Lichtenbaum (1969). The topic of bilinear pairings first entered the subject of cryptology not as a method of cryptography, but as a method of cryptanalysis. This first appearance of the Weil pairing was the introduction by Menezes, Okamoto, and Vanstone (1993) as a way of approaching the discrete-log problem on elliptic curves, a method of attack now known as an MOV attack. The MOV attack appears to expose a possible vulnerability in any elliptic-curve cryptosystem that uses a supersingular curve, but as shown by Balasubramanian and Koblitz (1998), the MOV attack almost never results in a subexponential attack algorithm for other elliptic curves. Soon after, Frey and Rück (1994), and Frey, Miller, and Rück (1999) proposed a more general attack on hyperelliptic-curve cryptography, an attack that includes elliptic-curve cryptography as a special case.

Perhaps in retrospect, it is not surprising that once the notion of bilinear pairing found its way into cryptology in 1993, it eventually morphed into a positive tool of cryptographers. However, that role reversal was quite surprising and unexpected at the time, as discussed in the tutorial by Koblitz, Koblitz, and Menezes (2011). Sakai, Ohgishi, and Kasahara (2000) recast the Diffie–Hellman key exchange using the Weil pairing, and Joux (2000) published a protocol that provides a one-round tripartite key exchange by using pairing, thereby drawing increased attention to bilinear pairing in the cryptography community. As a result, it can be said that these papers stimulated the development of other applications of bilinear pairing. In particular, the use of bilinear pairing to produce short, secure signatures was proposed by Boneh, Lynn, and Shacham (2001). Pairing-based cryptography is discussed by Boneh and Franklin (2001). Pairing-based methods are secure only if pairing inversion is intractable, so this is a fundamental topic of study. The difficulty of the pairing inversion problem was discussed by Galbraith, Hess, and Vercauteren (2008).

Cryptography based on bilinear pairing is further discussed by Boneh and Franklin (2001), introducing a method, based on pairing, of using identity as a public key. The notion of identity-based cryptography had been described earlier as a goal by Shamir (1985), but without a means to achieve it. It had been an unfilled and illusive dream for many years because no practical scheme had been found.

A noteworthy and popular class of pairing-friendly elliptic curves was introduced by Barreto and Naehrig (2006). The development of the popular Barreto–Naehrig curves rests on the deep insight of the important earlier work of Miyaji, Nakabayashi, and Takano (2001) which constructed curves with embedding degrees of three, four, and six. These were the first nonsupersingular elliptic curves known that have practical embedding degrees. Their methods led to further work in this direction by Galbraith, McKee, and Valença (2007). Such methods are needed because, as shown by Balasubramanian (1998) and Luca, Mireles, and Shparlinski (2004), a randomly chosen elliptic curve over a finite field almost surely has an embedding degree larger than the logarithm of the field size squared. These observations led to increased interest in directed methods to construct pairing-friendly elliptic curves. Such methods include the unpublished work of Cocks and Pinch (2001), the work of Miyaji, Nakabayashi, and Takano (2003), Dupont, Enge, and Moran (2005), Freeman (2006), Rubin and Silverberg (2010), Nogami and Morikawa (2010), and Izuta, Nogami, and Morikawa (2010). The known methods of constructing pairing-friendly curves were surveyed by Freeman, Scott, and Teske (2010).

The development of algorithms to compute bilinear pairings has its own history which, although clearly motivated by applications, runs on its own separate track. The widely popular Miller algorithm for computing a bilinear pairing was developed by Victor Miller in 1986, and known to the community, but was not published at the time. In addition to its value as a computational tool, the Miller algorithm has also stimulated other perspectives on the Weil and Tate pairings. The Duursma–Lee algorithm (2003) shifted the focus of pairing computations by introducing an alternative and faster algorithm that only applies to particular curves of a special form. This fruitful insight has led, in turn, to the introduction of other pairings to further reduce the computational burden.

13 Implementation

An implementation of a cryptographic algorithm is a hardware or software device that performs the function of encryption or decryption. A discussion of implementation extends the discussion of encryption and decryption algorithms down into the next level of detail. The coordination of the detail at this level of a hardware or software implementation is called an *architecture*. An architecture is a general plan or framework for a hardware or software implementation that provides a structure for the final details of the implementation but does not fully describe those details. The notion of an architecture is midway between the notion of an algorithm and the notion of an implementation.

At the level of an implementation, the computational equations are expanded into a configuration of adders, multipliers, and inverters, with data paths between them; or into a collection of software subroutines, along with the sequence of instructions linking and executing the subroutines. The designer of a hardware implementation must deal with details of data movement and must make decisions regarding whether various computational resources, such as an adder or a multiplier, should be shared by subsections of the algorithm, or should be resources dedicated to a specific computation. The many multiplications required by an algorithm, for example, can be provided by one multiplier shared in time, or by many distinct multipliers used concurrently. Each binary multiplier could be a serial multiplier, computing one bit of the product at a time, or a parallel multiplier, computing all bits of a product simultaneously, or even a hybrid serial/parallel multiplier.

Similar comments apply to the operations needed by the computations of a software implementation. Such decisions regarding the architecture of a software implementation are based on a trade between computational cost and computational speed. Representations of field elements must be chosen consistent with the constraints of a computer instruction set, which is not normally designed to suit the needs of finite-field arithmetic.

There are many layers of mathematics used in the theory of the algorithms that have been described in this book, and there are many layers in the implementation. Starting with the first layer and working up, the layers of the mathematics might be called integer arithmetic, modular integer arithmetic and prime fields, extension fields and binary fields, elliptic curves, and bilinear pairings. Within an implementation, each of the higher layers in this stack is built on top of the more basic layers. This chapter will examine a few topics of implementation, starting with the larger topics and ending with the details. To this purpose, the chapter begins at the higher level with a discussion of various methods of making the computations of bilinear pairing more efficient and ends at the lower level with the computations of finite-field arithmetic.

13.1 Pairing enhancements

The Tate pairing is preferred to the Weil pairing because it is less complex, but it is still computationally expensive. Therefore the structure and computations of the Tate pairing have been carefully scrutinized and reconstructed in order to accelerate those computations, at least for some curves with suitable properties. The study of such methods of acceleration blurs the usual distinction between theory and implementation. This section introduces the study of enhancements to accelerate the computation of a pairing. The next section examines this topic in depth.

Several straightforward pairing enhancements of the reduced Tate pairing that are suitable for some elliptic curves are referred to as the *Duursma–Lee enhancements*. The Duursma–Lee enhancements lead to a version of the Tate pairing that allows the Miller algorithm to be accelerated, notably for certain elliptic curves in fields of characteristic three.

Recall that the reduced Tate pairing

$$\Phi(P, Q) = f_{r,P}(Q)^{(q^k-1)/r}$$

is defined in terms of the Miller function $f_{r,P}(Q)$. The first of the Duursma–Lee enhancements of the Tate pairing is motivated by the observation that the final exponentiation by $(q^k - 1)/r$ that completes the reduced Tate pairing is computationally expensive. This enhancement replaces r by $q^{k/2} + 1$ (provided k is even) which then is a divisor of $q^k - 1$. Because, as asserted by the Hasse–Weil bound, the order r of an appropriate elliptic curve over $\boldsymbol{F}_q$ will be approximately $q + 1$, the integer $q^{k/2} + 1$ is much larger than r unless k equals two. This means that the final exponentiation requires only the smaller exponent of $q^{k/2} - 1$ instead of $(q^k - 1)/r$. However, this reduction in the exponent comes only in exchange for additional iterations of the Miller algorithm.

This useful technique can be made more general. Let m be any multiple of r that divides $q^k - 1$. This is written as $r|m|q^k - 1$. Then the pairing can be modified as $\Phi(P, Q) = f_{m,P}(Q)^{(q^k-1)/m}$, with m appearing both in the Miller function and in the exponentiation. By the choice of m, one can trade between the number of iterations of the Miller algorithm and the size of the final exponentiation. One may prefer an m for which the binary representations of both m and $(q^k - 1)/m$ have low Hamming weight (if such exist), so as to gain additional computational reductions.

The second Duursma–Lee enhancement is designed specifically for the elliptic curve $\mathcal{X}(\boldsymbol{F}_{3^m}) : y^2 = x^3 - x \pm 1$, where $\boldsymbol{F}_{3^m}$ is a field of characteristic three and $\mathrm{GCD}(m, 6) = 1$. The last condition means that m is not even and it is not divisible by three. It is easy to see that for $m = 1$, this curve has order equal to either one or seven depending on the choice of sign in the polynomial. Therefore, for $m = 1$, the Frobenius trace t, which is defined as $q + 1 - \#\mathcal{X}(\boldsymbol{F}_3)$, is equal to ± 3, which is a trivial multiple of the field characteristic. We conclude that the curve is supersingular in $\boldsymbol{F}_3$ and so is supersingular in $\boldsymbol{F}_{3^m}$ for all m.

The complex numbers α and β, as zeros of $z^2 - tz + q$, are $(3 \pm i\sqrt{3})/2$. These complex numbers are also written $\alpha, \beta = \sqrt{3}e^{\pm i\pi/6}$. For a general m, the order r of the curve $\mathcal{X}(\boldsymbol{F}_{3^m})$, as determined by Theorem 10.11.2, is

$$\begin{aligned}\#\mathcal{X}(\boldsymbol{F}_{3^m}) &= 3^m + 1 \pm 3^{m/2}(e^{\pi m/6} + e^{-\pi m/6}) \\ &= 3^m \pm 3^{(m+1)/2} + 1.\end{aligned}$$

The conclusion holds because m has been restricted to be ± 1 modulo 6. The choice of sign in the final line agrees with the choice of sign in the polynomial defining the curve.

The embedding degree for such a supersingular curve is shown to always equal six as follows. Observe that

$$(3^m + 3^{(m+1)/2} + 1)(3^m - 3^{(m+1)/2} + 1) = 3^{2m} - 3^m + 1.$$

The right side has the form $x^2 - x + 1$ with $x = 3^m$. But

$$x^6 - 1 = (x - 1)(x + 1)(x^2 + x + 1)(x^2 - x + 1)$$

so $x^k - 1$ is divisible by $x^2 - x + 1$ for $k = 6$, but for no smaller k. Then the order r divides 3^{mk} for $k = 6$ but for no smaller k. Thus the supersingular elliptic curve $\mathcal{X}(\boldsymbol{F}_{3^m})$ based on the polynomial $y^2 = x^3 - x \pm 1$ always has embedding degree six.

This curve also has the unusual property that with point P represented as the pair of field elements $(\alpha, \beta) \in \boldsymbol{F}_{3^m}^2$, the formula for point tripling is $3P = (\alpha^9 + 2, -\beta^9)$, which follows by an elementary computation using the usual formula for point addition. The particular special property to be noticed here is that for this particular curve, the x and y coordinates do not interact during point tripling. Therefore point multiples can be computed conveniently and rapidly using radix-three arithmetic, as will be described in Section 13.3.

The distortion map for this family of elliptic curves is $\Psi(P) = \Psi(x, y) = (\alpha - x, iy)$ where $i^2 = -1$ and α satisfies $\alpha^3 = \alpha \pm 1$, the sign agreeing with the sign in the polynomial defining the curve. To verify that the distortion map takes r-torsion points of $\mathcal{X}(\boldsymbol{F}_{3^m}) : y^2 = x^3 - x \pm 1$ to $\mathcal{X}(\boldsymbol{F}_{3^{6m}}) : y^2 = x^3 - x \pm 1$, in characteristic three, write

$$\begin{aligned}(\alpha - x)^3 - (\alpha - x) \pm 1 &= \alpha^3 - x^3 - \alpha + x \pm 1 \\ &= -x^3 + x \mp 1 + \alpha^3 - \alpha \mp 1 \\ &= -(x^3 - x \pm 1) = (iy)^2.\end{aligned}$$

Thus $\Psi(P)$ takes points of $\mathcal{X}(\boldsymbol{F}_{3^m})$ to points of $\mathcal{X}(\boldsymbol{F}_{3^{6m}})$. Because the coordinates of $\Psi(x, y)$ are (trivially) rational functions, it is a distortion map.

13.2 Accelerated pairings

Motivated by the Duursma–Lee enhancement, the bilinear pairings have been carefully redefined and remodeled so that the various kinds of accelerated computations apply more generally. The reduced complexity of the resulting variations of the bilinear pairings on the group of an elliptic curve accelerates the Tate pairing computations. Two such accelerations, known as the *eta pairing* and the *ate pairing*[1] play the same functional role as the Tate pairing from the point of view of the user, but the internal computational structure is somewhat different, and more efficient. They may be viewed as efficient versions of the Tate pairing, or as different pairings, according to preference.

Before describing the eta and ate pairing accelerations, we provide an alternative and more elegant description of the cyclic groups G_1 and G_2 of r-torsion points that will make the presentation cleaner.

This conclusion can also be understood in an elementary way. The first statement is immediate. The second statement follows easily when r is a prime by observing that the Frobenius operator π_q is a linear function and the r-torsion points form a cycle of prime order. Therefore, the linear function $\pi_q(P)$ has the form $\pi_q(P) = [a]P$ for some integer a. But for $P \in \mathcal{X}(\boldsymbol{F}_{q^k})[r]$,

$$P = \pi_{q^k}(P) = \pi_q^k(P) = [a]^k P = [a^k]P,$$

which means that $a^k = 1 \pmod r$. Because $q^k = 1 \pmod r$, we conclude that $a = q$.

Accordingly, the groups G_1 and G_2 are redefined. The group G_1 is defined as the set of r-torsion points P for which $\pi_q(P) = P$. This means that the points of G_1 are in the ground field, and so $G_1 = \mathcal{X}(\boldsymbol{F}_p)[r]$. The group G_2 is then defined as the set of r-torsion points P for which $\pi_q(P) = [q]P$. This is expressed as the two *Frobenius eigenspaces*

$$\begin{aligned}G_1 &= \mathcal{X}(\overline{\boldsymbol{F}}_q)[r] \cap \ker(\pi_q - [1]) \\ G_2 &= \mathcal{X}(\overline{\boldsymbol{F}}_q)[r] \cap \ker(\pi_q - [q]),\end{aligned}$$

[1] This outrageously cute term is a word play on both the Tate pairing and the eta pairing.

where the kernels are defined as

$$\ker(\pi_q - [1]) = \{P \mid \pi_q(P) - [1]P = \mathcal{O}\}$$
$$\ker(\pi_q - [q]) = \{P \mid \pi_q(P) - [q]P = \mathcal{O}\}.$$

It is clear that G_1 is simply the set $\mathcal{X}(\boldsymbol{F}_q)[r]$. The reason that $\mathcal{X}(\boldsymbol{F}_q)[r]$ is written in this unconventional way is to emulate the definition of G_2. The cyclic subgroup G_2 is a preferred representation of the quotient group for the remainder of this section because it allows the Frobenius map to be embedded within the equations of the Tate pairing.

Our next observation is that because $\#\mathcal{X}(\boldsymbol{F}_q) = q + 1 - t$ and r divides $\#\mathcal{X}(\boldsymbol{F}_q)$, then $q = t - 1 \pmod r$. Accordingly, let $T = t - 1$ and note that $T \leq 2\sqrt{q}$ by the Hasse–Weil bound. Thus, modulo r, the integer q can be replaced by $T \leq 2\sqrt{q}$. This is because multiplication of $P \in G_2$ by q goes around the modulo r cycle approximately an integer number of times, differing from a multiple of r only by T, so it is enough to multiply P by T rather than by q. That is $[q]P = [T]P$ if P is an r-torsion point. The development of the accelerated forms of the Tate pairing, to follow next, is based on these facts.

The development of the eta and ate pairings below will also make use of the fact that the Miller function $f_{\ell,P}(x, y)$ satisfies an expansion identity given by

$$f_{T^k,P} = f_{T,P}^{T^{k-1}} f_{T,TP}^{T^{k-2}} \cdots f_{T,T^{k-1}P}.$$

This equation is easy to verify by equating the divisors of the functions on both sides of the equal sign. It is closely related to the structure of the Miller algorithm. This identity is used only to verify that the eta pairing and the ate pairing are each not degenerate.

The eta pairing

The eta pairing is an accelerated form of the Tate pairing that can be used for appropriate curves. The eta pairing requires a distortion map, so it can be applied only for supersingular curves. The eta pairing is defined on the two nonidentity r-torsion points P and Q of the supersingular curve $\mathcal{X}(\overline{\boldsymbol{F}}_q)$. The Frobenius map is embedded into the structure of the eta pairing because that map is used to specify the group G_2. Recall that the Frobenius map $\pi_q(P)$ takes the point (x, y) of $\mathcal{X}(\overline{\boldsymbol{F}}_q)$ to the point (x^q, y^q) of $\mathcal{X}(\overline{\boldsymbol{F}}_q)$.

The discussion of the eta pairing begins with the supersingular elliptic curve $\mathcal{X}(\boldsymbol{F}_q)$ over $\boldsymbol{F}_q$ and with the formula for the reduced Tate pairing, which is given by

$$\Phi(P, Q) = f_{r,P}(Q)^{(q^k-1)/r},$$

where r is prime factor of $\#\mathcal{X}(\boldsymbol{F}_q)$, presumably the largest such prime factor, and k is the embedding degree of r. The r-torsion points P and Q are in the sets G_1 and G_2.

Points in these sets satisfy $\pi_q(P) = P$ and $\pi_q(Q) = [q]Q$, respectively. Accordingly, the Tate pairing requires a distortion map in a situation in which both P and Q originate as r-torsion points in the ground field. The eta pairing shares this requirement for a distortion map.

Because r divides $q^k - 1$ and $q = t - 1 \pmod r$, the Frobenius trace t is equal to $q + 1 - \#\mathcal{X}(\boldsymbol{F}_q)$. It follows that r divides $(t-1)^k - 1$. Thus r divides $T^k - 1$ where $T = t - 1$. Moreover r, but not r^2, divides $q^k - 1$. Define $N = \text{GCD}(T^k - 1, q^k - 1)$, which will be a multiple of r. Then there is an integer L not divisible by r, such that $T^k - 1 = LN$.

We can now define the eta pairing. Let P and Q be r-torsion points of $\mathcal{X}(\boldsymbol{F}_q)$ other than $\mathcal{O}$. The order of both P and Q divides N because N is a multiple of r. The eta pairing is defined by

$$\eta_T(P, Q) = f_{T,P}(\psi(Q)),$$

where ψ is a distortion map.

The importance of the eta pairing is due to the fact that r is replaced by T, which is a much smaller integer. The reason this matters is as follows. The loop length of the standard (radix-two) Miller algorithm for a Tate pairing of order r is $\log_2 r$. For a well-chosen curve, r is approximately equal to $\#\mathcal{X}(\boldsymbol{F}_q)$ and, by the Hasse–Weil bound, $\#\mathcal{X}(\boldsymbol{F}_q)$ is approximately equal to q. This means that the loop length of the Tate pairing is about $\log_2 q$. In contrast, again by the Hasse–Weil bound, $t = T + 1$ is not larger than $2\sqrt{q}$. Therefore, the eta pairing has a loop length of about $\log_2 \sqrt{q}$, which is a reduction in loop length of about a factor of two.

There remains one important and nontrivial detail that must be addressed. It is necessary to verify that the eta pairing is not degenerate. It would be degenerate if $\eta_T(P, Q) = 1$ for all allowable P and Q, and so useless. To show that the pairing is not degenerate, it suffices to find an integer c for which some pair of points, P and Q, satisfies

$$\eta_T(P, Q)^{c(q^k-1)/N} = \Phi(P, Q)^L.$$

The Lth power of the Tate pairing on the right side is an element of $\mu_r \subset \boldsymbol{F}_{q^k}^{\times}$, which is a cyclic group of order r. Therefore, the right side is equal to one for all P and Q if, and only if, r does not divide L. Because r does not divide L, $\eta_T(P, Q)$ is not equal to one for some P and Q. It then follows that $\eta_T(P, Q) \neq 1$ as well, so the pairing is not degenerate. It only remains to find such a constant c. Let

$$c = \sum_{i=0}^{k-1} T^{k-1-i} q^i = k q^{k-1} \pmod r$$

and so $c = (T^k - 1)/N$ is such a constant. This analysis is similar, though inverted, to the analysis given next for the ate pairing.

The ate pairing

The *ate pairing* is also an accelerated form of the Tate pairing. It extends the notions of the eta pairing to ordinary elliptic curves as well as to supersingular elliptic curves. The ate pairing eliminates the need for a distortion map Ψ. To do so, the ate pairing reverses the roles of G_1 and G_2, and operates on $G_2 \times G_1$ where, as before, $G_1 = \mathcal{X}(\overline{\boldsymbol{F}}_q)[r] \cap \ker(\pi_q - [1])$ and $G_2 = \mathcal{X}(\overline{\boldsymbol{F}}_q)[r] \cap \ker(\pi_q - [q])$. A variation of the ate pairing, called the *twisted ate pairing*, operates instead on $G_1 \times G_2$, thereby retaining the need for a distortion map.

Because r divides $\#\mathcal{X}(\boldsymbol{F}_q)$, we know that $q = t - 1 \pmod r$, where $t = q + 1 - \#\mathcal{X}(\boldsymbol{F}_q)$ is the Frobenius trace. Because r, but not r^2, divides $q^k - 1$, we can conclude that r divides $(t-1)^k - 1$. Thus r divides $T^k - 1$ where $T = t - 1$. Define $N = \text{GCD}[T^k - 1, q^k - 1]$. The integer N must be divisible by r. Moreover, $NL = T^k - 1$ for some integer L that is not divisible by r.

The ate pairing is then defined as a function from $G_2 \times G_1$ into μ_r given by

$$a_T(Q, P) = f_{T,Q}(P)^{(q^k-1)/N}.$$

The ate pairing is an accelerated form of the Tate pairing because $|T|$ is at most $2\sqrt{q} + 1$, whereas r is usually a large prime nearly equal to q. This means that, as for the eta pairing, not more than about half as many Miller iterations are needed to compute the ate pairing. However, for some elliptic curves, the ate pairing can be faster by as much as $\phi(k)$, where k is the embedding degree, and $\phi(k)$ is the totient function. Offsetting this advantage is the disadvantage that coefficients of the Miller polynomial are in $\boldsymbol{F}_{q^k}$ rather than $\boldsymbol{F}_q$.

The ate pairing satisfies the linearity properties required of a pairing. However, as for the eta pairing, we must ensure that the ate pairing is not degenerate. It is enough to show that there is at least one pair of points P, Q such that $a_T(Q, P)$ is nonzero. To show that the ate pairing is not degenerate, it is enough to show that

$$a_T(Q, P)^c = \Phi(Q, P)^{L(q^k-1)/r}$$

for some constant c. This is because r does not divide L, so the exponent $L(q^k - 1)/r$ is not a multiple of r. Therefore the right side cannot be equal to one for every P and Q. The method of proof is the same in spirit as the proof of nondegeneracy for the eta pairing. We will require the following string of equalities applied to the Tate pairing $\Phi(Q, P)$

$$\begin{aligned}[\Phi(Q, P)^{(q^k-1)/r}]^L &= \Phi(Q, P)^{L(q^k-1)/r} = f_{LN,Q}(P)^{(q^k-1)/N} \\ &= f_{t^k-1,Q}(P)^{(q^k-1)/N} \\ &= f_{T^k,Q}(P)^{(q^k-1)/N}.\end{aligned}$$

Only the last equality requires explanation. Note that the divisors satisfy

$$\begin{aligned}\operatorname{div}(f_{T^k,Q}) &= T^k(Q) - ([T^k]Q) - (T^k - 1)(\mathcal{O}) \\ &= T^k(Q) - (Q) - (T^k - 1)(\mathcal{O}) \\ &= (T^k - 1)(Q) - (T^k - 1)(\mathcal{O}) \\ &= \operatorname{div}(f_{T^k-1,Q}).\end{aligned}$$

Because $Q \in G_2$, we know that

$$\pi_q(Q) = [q]Q = [T]Q.$$

Therefore

$$f_{T,[T]Q}(P) = f_{T,\pi_q(Q)}(P) = f^q_{T,Q}(P).$$

Then using the expansion identity introduced above,

$$\begin{aligned}f_{T^k,Q}(P) &= f_{T,Q}(P)^{T^{k-1}} f_{T,[T]Q}(P)^{T^{k-2}} \cdots f_{T,[T^{k-1}]Q}(P) \\ &= f_{T,Q}(P)^{qT^{k-1}} f_{T,Q}(P)^{q^2T^{k-2}} \cdots f_{T,Q}(P)^{q^{k-1}T} \\ &= f_{T,Q}(P)^c,\end{aligned}$$

where

$$\begin{aligned}c &= qT^{k-1} + q^2T^{k-2} + \cdots + q^{k-1}T \\ &= (k-1)q^k \pmod{r}.\end{aligned}$$

Therefore, the ate pairing is nondegenerate.

13.3 Doubling and tripling

An efficient method to compute a point multiple $[r]P$ on an elliptic curve, where r is a large integer, is by the double-and-add method. This method is based on the binary expansion of the integer r as $r = \sum_i r_i 2^i$ and the binary representation $(r_\ell, r_{\ell-1}, \ldots, r_1, r_0)$, where $r_i \in \{0, 1\}$. Such a representation holds for every nonnegative integer r. The double-and-add method[2] is based on the innocent-looking identity

$$\begin{aligned}[r]P &= \left[\sum_{i=0}^{\ell} r_i 2^i\right] P = \sum_{i=0}^{\ell} [r_i][2^i]P \\ &= \sum_{i:r_i=1} [2^i]P.\end{aligned}$$

[2] This method can be used in any group. For example in $\boldsymbol{F}_p^*$, this becomes $\alpha^r = \alpha^{\sum_i r_i 2^i} = \Pi_{i:r_1=1}\alpha^{2^i}$.

This simple equality does have immense consequences, and can be considered as a foundation for much of modern elliptic-curve cryptography. This is because the right side can be computed by executing ℓ point doublings and at most $\ell - 1$ additions. Because this expression allows point multiples to be computed easily, cryptography based on elliptic curves is practical. Because a simple method of computation apparently does not exist in the other direction, computing a from $[a]P$, cryptanalysis is evidently intractable.

The point-doubling method requires that all doublings be computed, but not all additions are needed in general. Often, if one is willing to subtract points as well as add them, the computations can be reduced. For example, using only additions, $63 = 2^5 + 2^4 + 2^3 + 2^2 + 2^1 + 2^0$, but with subtraction allowed, $63 = 2^6 - 2^0$. This savings is more striking when written as

$$\begin{aligned}[63]P &= [2^5]P + [2^4]P + [2^3]P + [2^2]P + [2]P + P \\ &= [2^6]P - P.\end{aligned}$$

To compute the first expression, as written, requires five point doublings and five point additions. To compute the second expression requires six point doublings and one point subtraction.

An alternative method of computing a point multiple that may be even faster is the triple-and-add method. This method requires an efficient method of point tripling. The signed ternary expression of r is $r = \sum_i r_i 3^i$ where $r_i \in \{-1, 0, +1\}$ and expressed as $(r_\ell, \ldots, r_1, r_0)$ and $\ell = \lceil \log_3 r \rceil$. The successive triples of P are first computed recursively by $[3^i]P = [3][3^{i-1}]P$. The triple-and-add method then invokes the identity

$$[r]P = \sum_{i=0}^{\ell} [r_i][3^i]P = \sum_{i:r_i=1} [3^i]P - \sum_{i:r_i=-1} [3^i]P.$$

The method involves ℓ point triplings, at most ℓ point additions, and one point subtraction. Cubing in a field of characteristic three is a linear operation just as squaring in a field of characteristic two is a linear operation. That is, $(a + b)^3 = a^3 + b^3$ in a field of characteristic three.

The method of point tripling can be used for any curve, but it is especially attractive on a curve for which point tripling is simple. Therefore, to further reduce the computational burden, one may choose a curve based on a particular polynomial, such as the polynomial $y^2 = x^3 - x \pm 1$ over a field of characteristic three, for which the computation of a point triple is simplified. For example, as was shown in Section 13.1 for the elliptic curve $\mathcal{X}(\boldsymbol{F}_{3^m}) : y^2 = x^3 - x \pm 1$, point tripling is given by the simple expression $3(\alpha, \beta) = (\alpha^9 + 2, -\beta^9)$.

The methods of doubling and tripling hold in any abelian group. Because the jacobian of a hyperelliptic curve is an abelian group, these methods hold in the group of the jacobian. The Miller algorithm, which was introduced as a recursive doubling algorithm for divisors, defines a rational function by conceptualizing a table of its poles and zeros, recursively computing not the function itself, but instead computing an evaluation of the function at a chosen point. The Miller algorithm is based on the arithmetic of divisors, and on finite-field arithmetic. It can be executed by emulating the methods of point doubling and point tripling to formulate the methods of divisor doubling and divisor tripling. The resulting method lies at the core of the Miller algorithm.

To compute the divisor multiple aD by divisor doubling expand a in binary notation as $a = \Sigma_i a_i 2^i$ where $a_i \in [0, 1]$. Then

$$aD = (\Sigma_i a_i 2^i)D = \Sigma_i a_i (2^i D).$$

Although this equality is elementary, the difference in the computational burden of the two sides of the second equality is enormous. The left side is defined as the sum of a copies of the divisor D. The right side computes a table of the divisors $2^i D$, then adds a subset of the terms $2^i D$ as specified by the binary representation of a. The right side only requires computational work proportional to $\log_2 a$.

Similarly, point tripling computes aD by writing a in signed ternary notation as $a = \Sigma_i a_i 3^i$, where $a_i \in [-1, 0, +1]$. Then

$$\begin{aligned} aD &= (\Sigma_i a_i 3^i)D = \Sigma_i a_i (3^i D) \\ &= \sum_{i:a_i=1} (3^i D) - \sum_{i:a_i=-1} (3^i D). \end{aligned}$$

The right side first computes a table of $3^i D$ for $i = 1, \ldots, \lceil \log_3 a \rceil$, then adds or subtracts a subset of these entries as specified by the ternary representation of a. Point subtraction consists of adding the negative of a point, and computing the negative of a point is trivial. Therefore point subtraction has the same complexity as point addition.

13.4 Point representations

An affine point of an elliptic curve or a hyperelliptic curve over a finite field $\boldsymbol{F}_q$ is a point of the affine plane $\boldsymbol{F}_q^2$. The affine point is represented by a pair of elements of the finite field $\boldsymbol{F}_q$ and denoted $P = (x, y)$. The point at infinity is not an affine point. An examination of the equations for point addition on an elliptic curve will reveal that an operation of field division is required when adding two affine points of an elliptic curve. Division in a finite field, which is discussed in a later section, is a considerably harder operation than is multiplication. It can be worth making an effort to minimize the number of divisions in the field.

To simplify the process of point addition, other representations of the points of an elliptic curve can be used. Such representations are used primarily to minimize the need for field divisions as a part of the operation of point addition. An informal approach is simply to postpone execution of the divisions until after a series of point additions is complete so that the field divisions can be merged and executed simultaneously as a single field division. A formal way to do this is by choosing an alternative representation of the field elements.

The *projective representation* of the point P consists of three scalars (X, Y, Z) with the understanding that the usual affine coordinates can be recovered by the expressions $x = X/Z$ and $y = Y/Z$. The point at infinity is then any triple with the value of Z equal to zero. With the projective representation, the divisions needed to compute x and y can be replaced by multiplications needed to update Z. A series of point operations using the projective representation can be made without field divisions, only field multiplications. A division may be needed later to compute the affine coordinates from the projective coordinates, but this can be deferred to some future time, if it is needed at all.

The *jacobian representation* of an affine point is yet another alternative representation of that point. This representation again involves three variables (X, Y, Z) but now with the understanding that $x = X/Z$ and $y = Y/Z^2$. This means that Y in the projective representation and Y in the jacobian representation differ by a factor of Z. Again, the purpose of this representation is to insert flexibility into the computations so that divisions can be suppressed for as long as possible.

Either the projective representation or the jacobian representation can be used for any elliptic curve $\mathcal{X}(\boldsymbol{F}_q)$. The next point representation, called the *Edwards representation*, is a more delicate reformulation, and can be used only if the Weierstrass form of the elliptic curve can be written in the specific form

$$\mathcal{X} : y^2 = (x - c^4 d - 1)(x^2 - 4c^2 d)$$

for some constants c and d. Of course, the Weierstrass form always can be expressed in this way by lifting the setting to an appropriate extension field so that the right side can be factored as shown. However, the Edwards representation would then be expressed in the extension field and the disadvantage of a larger field would offset any anticipated advantage of this representation.

For elliptic curves of the stated form, the Edwards representation results in simplified expressions for point addition. The Edwards representation of the curve in the variables (u, v) is given by

$$\mathcal{X} : u^2 + v^2 = c^2(1 + du^2v^2).$$

This representation follows by the substitution

$$x = \frac{-2c(w - c)}{u^2} \qquad y = \frac{4c^2(w - c) + 2c(c^4 d + 1)u^2}{u^3}$$

into the previous expression for $\mathcal{X}$, where $w = (c^2du^2 - 1)$. The point $(u, v) = (0, c)$ is now the identity element for the group law. It corresponds to the point at infinity in the Weierstrass representation.

The addition law in the Edwards representation is

$$(u_1, v_1) + (u_2, v_2) = \left(\frac{u_1 v_2 + u_2 v_1}{c(1 + d u_1 u_2 v_1 v_2)}, \frac{v_1 v_2 - u_1 u_2}{c(1 - d u_1 u_2 v_1 v_2)}\right)$$

for all points of the curve. The advantage is that there is no need for a special formula for the case in which the two points, (u_1, v_1) and (u_2, v_2), are equal. The negative of a point in the Edwards representation is $-(u, v) = (-u, v)$.

13.5 Algorithms for elliptic-curve arithmetic

The primary computations of elliptic-curve cryptography are point addition and point doubling. Both operations are used in the computation of point multiples. The point multiple $[a]P$ is defined as the sum of a copies of the point P. The doubling of the point P is defined as $P + P$. A general procedure for the computation of a point multiple on an elliptic curve was developed in Chapter 10 for use on an arbitrary elliptic curve expressed in the short Weierstrass form. The procedure for the computation of a point multiple can instead be tailored to a specific elliptic curve by taking advantage of any special properties that curve may have. In either case, the computation of a point multiple rests on the operation of point addition.

The elliptic-curve operations of point addition and point doubling are described by sets of equations involving the more basic operations of addition, multiplication, and division in the underlying field of the curve. Division in a field can be executed by multiplying by the multiplicative inverse, and the multiplicative inverse in a finite field is computed by using the extended euclidean algorithm. In this way, division in a finite field is usually replaced by two operations: inversion and multiplication.

At the highest level, a careful implementation requires that the structure of the equations be scrutinized in order to arrange them in the most efficient form. The field operations themselves are studied at a lower level.

For an elliptic curve $\mathcal{X}(\boldsymbol{F}_p) : y^2 = x^3 + ax + b$ over a field of characteristic larger than three, the equations for point addition and point doubling are

$$x_3 = m^2 - x_1 - x_2$$
$$y_3 = m(x_1 - x_3) - y_1,$$

where

$$m = \frac{y_2 - y_1}{x_2 - x_1} \quad \text{or} \quad m = \frac{3x_1^2 + a}{2y_1}.$$

The first expression for m is to be used for the case of point addition and the second expression for point doubling. The indicated operations are operations in the field of the curve, with division implemented as field inversion followed by multiplication. A point addition requires one field inversion, three field multiplications, and six field additions. A point doubling requires one field inversion, four field multiplications, and five field additions, as well as one doubling of a field element and one tripling of a field element.

The form of these equations suggests that the denominators might be cleared into the numerators so as to postpone all divisions. In this case, the computation gives instead, a known scalar multiple of x_3 and known scalar multiple of y_3. In a long sequence of calculations to compute a point multiple, these factors continue to accumulate. They can be collected into a single term and divided out all at once. In fact, this strategy is suggestive of the alternative jacobian representation of a point as described in the previous section.

To specify a point P of an elliptic curve $\mathcal{X}(\boldsymbol{F}_p) : y^2 = x^3 + ax + b$, one only needs to give the x coordinate together with one additional bit that gives the sign of y. To recover y, one computes the square root of $x^3 + ax + b$ in the field $\boldsymbol{F}_p$ using the extra bit to resolve the sign ambiguity.

It is not completely straightforward to represent an arbitrary binary number by a point of an elliptic curve because not every value of x corresponds to a point of the curve.

13.6 Modular addition in an integer ring

Integer arithmetic is a topic that is both familiar and elementary, and although it underlies the other topics of this chapter, most of this topic is standard and need not be discussed here. Integer arithmetic in a modular ring is also standard, but the operation of modular reduction requires extra consideration. The elementary operations of modular addition and modular subtraction are discussed in this section. Efficient modular multiplication is more intricate, and is described in the next section. Division in a modular integer ring is executed by first using the extended euclidean algorithm to compute an inverse followed by modular multiplication. The extended euclidean algorithm consists of multiple applications of the operations of integer addition, integer subtraction, and integer multiplication.

Modular integer addition, which is written

$$z = x + y \text{ (modulo } N),$$

is computed by

$$z = \begin{cases} x + y & \text{if} \quad x + y < N \\ x + y - N & \text{if} \quad x + y \geq N. \end{cases}$$

Modular subtraction, which is written

$$z = x - y \text{ (modulo } N),$$

is computed by

$$z = \begin{cases} x - y & \text{if} \quad x - y \geq 0 \\ x - y + N & \text{if} \quad x - y < 0. \end{cases}$$

Both of these operations, addition and subtraction, are straightforward in implementations, either in software or in hardware.

13.7 Modular multiplication in an integer ring

Integer multiplication modulo N, where N is a large but arbitrary integer, is usually much more difficult than ordinary integer multiplication. This is because of the reduction modulo N, which is an operation defined as a remainder under the operation of integer division. Thus its complexity appears to depend on the complexity of division. However, much of this complexity can be tamed by a method known as the *Montgomery multiplication algorithm*, provided the amount of modulo N arithmetic is large enough to justify the preparation needed for this algorithm. The integer N may be a prime or may be a composite – though if it is a composite with known factors, other reductions, based on the chinese remainder theorem, are possible prior to and in conjunction with the Montgomery algorithm.

Select any integer R coprime to N of the form b^k, where b is the arithmetic base that is being used in the computations, and such that R is larger than N. For example, if integers are represented in binary notation, then $b = 2$ and $R = 2^k$ (with the smallest k such that R is larger than N) and R is coprime to N. If numbers are represented in decimal notation, then $b = 10$ and $R = 10^k$ is the suitable choice (with the smallest k such that R is larger than N) and R is coprime to N. The trick of Montgomery multiplication is to exchange integer multiplication modulo N for integer multiplication modulo R at the expense of a few side calculations. In base-b arithmetic, modulo b^k reduction is trivial. In base-10 arithmetic, modulo 10^k reduction is trivial.

Because R and N are coprime, integers r and n exist such that

$$Rr + Nn = 1.$$

Moreover, because

$$R(r + \ell N) + N(n - \ell R) = 1$$

must also be true, it is clear that r and n can be chosen so that either $0 < r < N$, or $0 < n < R$. The chosen values of r and n will be called R^{-1} and N^{-1} because $RR^{-1} = 1 \text{ (mod } N)$ and $NN^{-1} = 1 \text{ (mod } R)$.

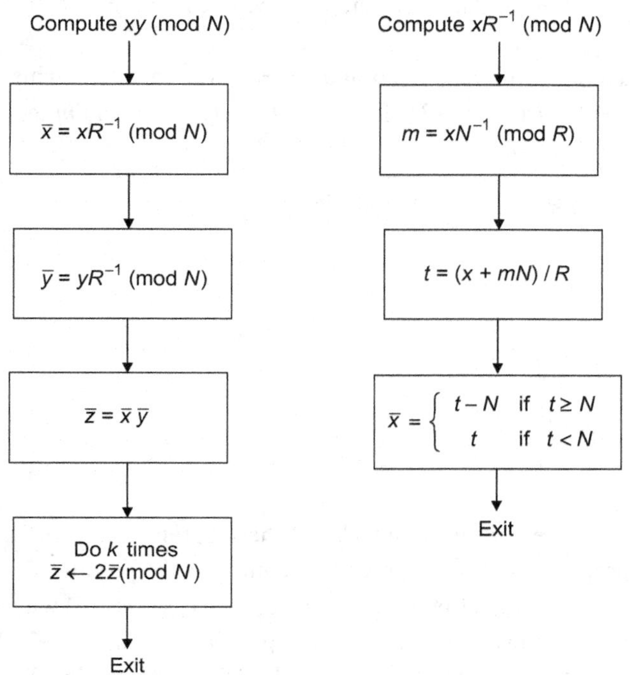

Figure 13.1 Montgomery multiplication

The task of modular multiplication is to compute $z = xy \pmod N$ where $0 \leq x < N$, $0 \leq y < N$, and $0 \leq z < N$. To this end, define the modified multiplicands by $\overline{x} = xR^{-1} \pmod N$ and $\overline{y} = yR^{-1} \pmod N$. These are called the *Montgomery representations* of the integers x and y. The maps from x to $\overline{x}$ and from y to $\overline{y}$ are clearly permutations of $\{0, 1, \ldots, N-1\}$ because x and y can be recovered by $x = \overline{x}R \pmod N$ and by $y = \overline{y}R \pmod N$. Suppose that we can find a method to compute $\overline{x}$ and $\overline{y}$ from x and y that avoids the modulo N reduction. Then we compute, as integers

$$\overline{z} = \overline{x}\ \overline{y}.$$

It is easy to recover the product xy from $\overline{z}$ because $xy = R^2\overline{z} \pmod N$, which, by assumption about the choice of R, is easy to compute. From $\overline{z}$, we compute

$$\begin{aligned} xy &= R^2\overline{z} \pmod N \\ &= b^{2k}\overline{z} \pmod N. \end{aligned}$$

Because R is a power of the arithmetic base, b, it is easy to compute $R^2\overline{z} \pmod N$ by an iterative procedure. This is shown on the left side in Figure 13.1, for the case where the arithmetic base is 2. Each step of the final iteration computes $2\overline{z} \pmod N$, and the iteration stops after $2k$ steps. For any other logarithm base b, the computation is the same, except that the doubling in the final iteration becomes a multiplication by the logarithm base.

It only remains to give a simple way to compute $\overline{x} = xR^{-1} \pmod{N}$. This method is depicted on the right side of Figure 13.1 in terms of the two integers m and t given as intermediate variables and defined by

$$m = xN^{-1} \pmod{R} \qquad (0 \leq m < R)$$

and

$$t = (x + mN)/R.$$

Then $\overline{x}$ is given by

$$\overline{x} = \begin{cases} t - N & \text{if} \quad t \geq N \\ t & \text{if} \quad t < N. \end{cases}$$

Figure 13.1 gives the complete algorithm in terms of these parameters.

Before proving the proposition that the procedure computes the product xy, we will first work through a simple example. Let $N = 13$ and let $z = xy \pmod{13}$ where $x = 5$ and $y = 3$. Of course, it is obvious in this simple example that $z = 2 \pmod{13}$. We will calculate this product, instead, using Montgomery multiplication. First, let $R = 16$ and notice that

$$9 \cdot 16 - 11 \cdot 13 = 1$$

so $R^{-1} = 9 \pmod{13}$, and $N^{-1} = 11 \pmod{16}$. To multiply 3 by 5, first write

$$m = xN^{-1} \pmod{R} = 55 \pmod{16} = 7$$

and $t = (x + mN)/R = (5 + 7 \cdot 13)/16 = 6$. Thus $\overline{x} = 6$.

Next, write

$$m = yN^{-1} \pmod{R} = 33 \pmod{16} = 1,$$

and $t = (y + mN)/R = (3 + 1 \cdot 13)/16 = 1$. Thus $\overline{y} = 1$.

Finally $\overline{z} = 16 \cdot 6 \cdot 1 \pmod{13} = 5$ and $z = 16 \cdot 5 \pmod{13} = 2$, which agrees with the obvious calculation.

Proposition 13.7.1 *Suppose that R satisfies* GCD$(R, N) = 1$ *and R^{-1} is the inverse of $R \pmod{N}$. Then*

$$xR^{-1} \pmod{N} = \begin{cases} t - N & \textit{if} \quad t \geq N \\ t & \textit{if} \quad t < N, \end{cases}$$

where $m = xn \pmod{R}$ and $t = (x + mN)/R$.

Proof The proof of this consists of the following three steps.

Step 1 $mN = xnN \pmod R = x(-1) \pmod R$, so R divides $x + mN$, which means that t is an integer.

Step 2 $tR = x + mN = x \pmod N$, so $t = xR^{-1} \pmod N$.

Step 3 $0 \le x + mN < RN + RN = 2RN$, so $(x + mN)/R < 2N$. □

In some cases, Montgomery multiplication can be embedded into a sequence of equations being solved in order to obtain greater efficiency. Thus the expressions $xy + uv$ and xyz can be converted to the Montgomery representation $\overline{x}\,\overline{y} + \overline{u}\,\overline{v}$ and $\overline{x}\,\overline{y}\,\overline{z}$, then executed as integer arithmetic. The R^2 multiplication modulo N of the Montgomery algorithm can be deferred until the end, and performed only once. Similarly, a sequence of repeated squares $x^2, x^4, x^8, x^{16}, \ldots$ can be executed as $\overline{x}^2, (R\overline{x}^2)(R(R\overline{x}^2)^2)^2, \ldots,$ with additional multiplications by R later. Conversion of the variables to the Montgomery representation occurs once at the beginning. Conversion back to the conventional representation occurs once at the end.

13.8 Representations of binary fields

Fields of characteristic two are commonly used in systems for cryptography. One reason for this popularity may be that binary arithmetic is convenient to implement in hardware using standard binary logic components. This is because addition and multiplication in the ground field $\boldsymbol{F}_2$ are trivial binary operations, and addition and multiplication in the extension field $\boldsymbol{F}_{2^m}$ can be decomposed into additions and multiplications in the ground field $\boldsymbol{F}_2$. There are a number of ways to do this.

The field $\boldsymbol{F}_{2^m}$ forms a vector space over $\boldsymbol{F}_2$ of dimension m, and so the elements of the field $\boldsymbol{F}_{2^m}$ can be expressed in terms of a basis for this vector space. A *basis* for $\boldsymbol{F}_{2^m}$, regarded as a vector space over $\boldsymbol{F}_2$, is any set $\{\boldsymbol{e}_0, \boldsymbol{e}_1, \ldots, \boldsymbol{e}_{m-1}\}$ of m linearly independent vectors in $\boldsymbol{F}_{2^m}$. Each basis vector is an m-bit binary number when expressed in any other basis. There are many sets of m linearly independent equations, so there are many bases.

Using the basis $\{\boldsymbol{e}_0, \boldsymbol{e}_1, \ldots, \boldsymbol{e}_{m-1}\}$, an m-bit binary number $(b_0, b_1, \ldots, b_{m-1})$ specifies the field element

$$\beta = b_0\boldsymbol{e}_0 + b_1\boldsymbol{e}_1 + \cdots + b_{m-1}\boldsymbol{e}_{m-1},$$

which is a linear combination of basis vectors. In this way, with a basis of $\boldsymbol{F}_{2^m}$ specified over the field $\boldsymbol{F}_2$, the elements of the extension field $\boldsymbol{F}_{2^m}$ can be represented as binary

m-tuples and stored in a binary register of length m. Addition of field elements is then simple. Addition is componentwise modulo-two arithmetic (or bitwise exclusive-or), and the execution is the same for any basis. Subtraction is also simple in $\boldsymbol{F}_{2^m}$ because in any field of characteristic two, subtraction is the same as addition. Multiplication and division, however, are not simple. The structure of multiplication and division depend on the choice of basis. It is in the implementation of multiplication and division that the basis becomes relevant, and should be chosen to simplify the computations.

The most common basis for $\boldsymbol{F}_{2^m}$ is the *polynomial basis*

$$\{1, \alpha, \alpha^2, \ldots, \alpha^{m-1}\},$$

where α is a zero of an irreducible polynomial of degree m over $\boldsymbol{F}_2$. A different choice of an irreducible polynomial of degree m results is a different basis. It is common to choose a primitive polynomial $p(x)$ as this irreducible polynomial so that α is a primitive element. A primitive polynomial over $\boldsymbol{F}_2$ does exist of every degree m. Then the set of elements of the field is given by the set

$$\boldsymbol{F}_{2^m} = \{1, \alpha, \alpha^2, \ldots, \alpha^{m-1}, \alpha^m, \ldots, \alpha^{2^m-1}\}.$$

Each element is a linear combination of basis vectors $\{1, \alpha, \alpha^2, \ldots, \alpha^{m-1}\}$. More specifically, these basic vectors can be represented as the binary words

$$\begin{gathered}(1, 0, 0, 0, \ldots, 0)\\ (0, 1, 0, 0, \ldots, 0)\\ (0, 0, 1, 0, \ldots, 0)\\ \vdots\\ (0, 0, 0, 0, \ldots, 1).\end{gathered}$$

The corresponding linear combination of basis vectors corresponding to α^i is found for each α^i written as α^i by repeated application of the reduction $p(\alpha) = 0$.

For an example of a field representation in a polynomial basis, use the primitive polynomial $p(x) = x^4 + x + 1$. The extension field $\boldsymbol{F}_{16}$ is then constructed with this irreducible polynomial, and is shown in Table 13.1. The polynomial basis for $\boldsymbol{F}_{16}$ (in this representation) is $\{\alpha^0, \alpha^1, \alpha^2, \alpha^3\}$, which correspond to the polynomials 1, x, x^2, and x^3.

Theorem 9.13.5 states that there are $\phi(2^m - 1)/m$ primitive polynomials of degree m over $\boldsymbol{F}_2$. Of these, only one primitive polynomial is needed to construct the field. That polynomial should be chosen to simplify the implementation. This is usually interpreted to mean that the polynomial with the fewest nonzero coefficients should be chosen.

The polynomial $x^4 + x + 1$ has three nonzero coefficients. Every irreducible polynomial $p(x)$ over $\boldsymbol{F}_2$ must have an odd number of nonzero monomials. Otherwise, $p(1)$ would equal zero, so $x - 1$ would be a factor of $p(x)$. This means that an irreducible

Table 13.1 *The binary field F_{16} in a polynomial basis*

0		
α^0	=	1
α^1	=	x
α^2	=	x^2
α^3	=	x^3
α^4	=	$x+1$
α^5	=	x^2+x
α^6	=	x^3+x^2
α^7	=	x^3+x+1
α^8	=	x^2+1
α^9	=	x^3+x
α^{10}	=	x^2+x+1
α^{11}	=	x^3+x^2+x
α^{12}	=	x^3+x^2+x+1
α^{13}	=	x^3+x^2+1
α^{14}	=	x^3+1

Table 13.2 *A short list of primitive polynomials*

Degree		
2	=	x^2+x+1
3	=	x^3+x+1
4	=	x^4+x+1
5	=	x^5+x^2+1
6	=	x^6+x+1
7	=	x^7+x^3+1
8	=	$x^8+x^4+x^3+x^2+1$
9	=	x^9+x^4+1
10	=	$x^{10}+x^3+1$
11	=	$x^{11}+x^2+1$
12	=	$x^{12}+x^6+x^4+x+1$

polynomial must have at least three nonzero coefficients, which is the case for the primitive polynomial x^4+x+1. Such a polynomial is called a *trinomial.* Irreducible trinomials of degree m exist for some values of m, but not for every value of m. A list of irreducible polynomials is shown in Table 13.2. Each primitive polynomial in that table is a trinomial whenever a primitive trinomial exists for that m.

Another popular basis for F_{2^m} is the *normal basis*, which has the form

$$\{\alpha^{2^0}, \alpha^{2^1}, \alpha^{2^2}, \alpha^{2^3}, \ldots, \alpha^{2^{m-1}}\},$$

provided that these elements are linearly independent over $\boldsymbol{F}_2$. In general, such a set need not be linearly independent, and in such a case it does not form a basis. Of course, when a set of this form does not form a basis, it is not called a normal basis for $\boldsymbol{F}_{2^m}$.

A field element β expressed in a normal basis has the form

$$\beta = \sum_{i=0}^{m-1} b_i \alpha^{2^i},$$

where $b_i \in \boldsymbol{F}_2$. As for any basis of $\boldsymbol{F}_{2^m}$, it is easy to add field elements in a normal-basis representation because addition is componentwise modulo-two addition. It is also easy to square field elements in a normal basis representation. First note that

$$\beta^2 = \sum_{i=0}^{m-1} b_i^2 (\alpha^{2^i})^2.$$

But $b_i^2 = b_i$ because b_i is either zero or one, therefore

$$\beta^2 = \sum_{i=0}^{m-1} b_i \alpha^{2^{i+1}}$$
$$= \sum_{i=0}^{m-1} b_{i-1} \alpha^{2^i}$$

because $\alpha^{2^m} = \alpha$. Thus with a normal basis, squaring can be implemented as a simple cyclic shift of the binary representation of β. Squaring, however, is not as important as multiplication. Multiplication of field elements in the normal basis does not share the simple cyclic structure of squaring of field elements in the normal basis, but the normal basis does facilitate a serial form of multiplication described in the next section.

To form a normal basis corresponding to the field representation of $\boldsymbol{F}_{16}$ given in Table 13.1, we may first consider $\{\alpha^1, \alpha^2, \alpha^4, \alpha^8\}$. But this set is not a basis because, for instance, x^3 is not a linear combination of these four elements. However, if $\beta = \alpha^3$, then $\{\beta^1, \beta^2, \beta^4, \beta^8\}$ is a normal basis, as can easily be verified from the list

$$\begin{aligned}
\beta^1 &= \alpha^3 &&= x^3 \\
\beta^2 &= \alpha^6 &&= x^3 + x^2 \\
\beta^4 &= \alpha^{12} &&= x^3 + x^2 + x + 1 \\
\beta^8 &= \alpha^9 &&= x^3 \qquad\quad + x.
\end{aligned}$$

These four elements, viewed as vectors, are linearly independent and span the space, so they form a basis. A representation of $\boldsymbol{F}_{16}$ in this basis is shown in Table 13.3. However, this β is not a primitive element. In the field $\boldsymbol{F}_{16}$, it is not possible to use powers of a primitive element to form a normal basis. Thus in some finite fields of characteristic two, it is possible to use powers of a primitive element to form a normal basis, in others this is not possible.

Table 13.3 *The binary field in a normal basis*

0	$=$	0
β^1	$=$	x^3
β^2	$=$	$x^3 + x^2$
$\beta^2 + \beta^1$	$=$	x^3
β^4	$=$	$x + 1$
$\beta^4 + \beta^2 + \beta^1$	$=$	$x^2 + x$
β^8	$=$	$x^3 + x^2$
β^8	$=$	$x^3 + x + 1$
β^8	$=$	$x^2 + 1$
β^8	$=$	x^3
β^8	$=$	$x^2 + x + 1$
β^8	$=$	$x^3 + x^2 + x$
β^8	$=$	$x^3 + x^2 + x + 1$
$\beta^8 + \beta^4 + \beta^2$	$=$	$x^3 + x^2 + 1$
$\beta^8 + \beta^4 + \beta^1$	$=$	$x^3 + 1$

For general multiplication in F_{2^m}, it is sometimes useful to express numbers in a certain particularly nice form of a normal basis. Let $\{e_i \mid i = 0, \ldots, m-1\}$ be a basis for F_{2^m}. Because e_i and e_j are themselves field elements, the product $e_i e_j$ is again a field element and so can be expressed in terms of the same basis. For each i and j, let M_{ijk}, by its value as a one or a zero, express whether basis vector e_k occurs in the product of e_i and e_j. Thus

$$e_i e_j = \sum_l M_{ijk} e_k$$

where $M_{ijk} \in F_2$. A multiplication of field element a by field element b then follows directly. Thus, if $a \in F_{2^m}$ is represented by $(a_0, \ldots, a_{m-1})$ and $b \in F_{2^m}$ is represented by $(b_0, \ldots, b_{m-1})$, then the product $c = ab$ is an element of F_{2^m} represented by $(c_0, \ldots, c_{m-1})$, where $c_k = \sum_{i,j} a_i b_j M_{ijk}$. For each k, let $\boldsymbol{M}_k = [M_{ijk}]$, which is an n-by-n matrix over F_2. Then $\boldsymbol{c}$ can be written $\boldsymbol{c}_k = \boldsymbol{a}\boldsymbol{M}_k\boldsymbol{b}^t$. The $\boldsymbol{M}_k$ give multiplication tables for F_{2^m} over F_2. For the simplest matrix, we should choose that normal basis for which the number of nonzero entries in each $\boldsymbol{M}_k$ is as small as possible. The number of nonzero entries cannot be smaller than $2n-1$. The normal basis is called an *optimal normal basis* if this number is equal to $2n-1$. Such an optimal normal basis does exist for some values of m, but it does not exist for every value of m. Other applications of a normal basis are in the next section.

13.9 Multiplication and squaring in a binary field

Multiplication in a binary field is easy to describe if the field elements are expressed as powers of the primitive element α, because then the multiplicative group $F^*_{2^m}$ is displayed as a cyclic group. The nonzero field elements β and γ, represented as powers of α, are

$$\beta = \alpha^i \qquad \gamma = \alpha^j,$$

and these can be represented as the integers i and j that appear in the exponents. Representation by the exponent is the discrete-logarithm representation. This representation requires a special symbol and special handling for the zero field element because zero is not a power of α and so has no logarithm.

Because $\alpha^i\alpha^j = \alpha^{i+j}$ multiplication in the discrete-log representation becomes integer addition of the exponents modulo $q - 1$. However, in this representation, although multiplication is simpler, addition becomes more difficult because it involves the deeper structure of the field. To add the field elements represented by i and j, the elements α^i and α^j must be expressed in terms of a basis, then added to obtain α^k, from which k is obtained by taking the finite-field logarithm to the base α. Clearly the use of logarithms becomes intractable if the field is large, and is impractical even if the field is moderate in size. This method is useful only if the field is small enough so that it is reasonable to store finite-field log and antilog tables.

Efficient hardware circuits for multiplication in a large field must work more closely with the deeper structure of the field. This is in contrast to addition, which does not need to refer to the field structure. When elements of the field $\boldsymbol{F}_{2^m}$ are represented in terms of any basis, addition is simply bit-by-bit modulo-two addition, which is easily implemented in digital logic by using exclusive-or gates. All m bits of the sum can be computed in parallel. Circuits that compute the m bits of a multiplication in parallel are much more complicated. There are many ways of designing such circuits. Some multipliers may be designed to reduce the topological (wiring) complexity, some may be designed to reduce arithmetic complexity, and some may be designed to reduce computation time.

To compute the finite-field product $c = ab$, with the field elements represented in the polynomial basis, we can regard the field elements as polynomials in x with coefficients in $\boldsymbol{F}_2$. Then

$$c(x) = a(x)b(x) \pmod{p(x)}.$$

Recall that the modulo-$p(x)$ operation can be distributed across both addition and multiplication by x at any convenient point in the computation. The polynomial $b(x)$ can be rearranged using *Horner's rule* as follows:

$$b(x) = (\cdots((b_{n-1}x + b_{n-2})x + b_{n-3})x + \cdots + b_1)x + b_0.$$

By using the distributivity property, $b(x)$ can be multiplied by $a(x)$ as follows:

$$a(x)b(x) = (\cdots((b_{n-1}a(x)x + b_{n-2})a(x)x + b_{n-3})a(x)x + \cdots + b_1)a(x)x + b_0a(x).$$

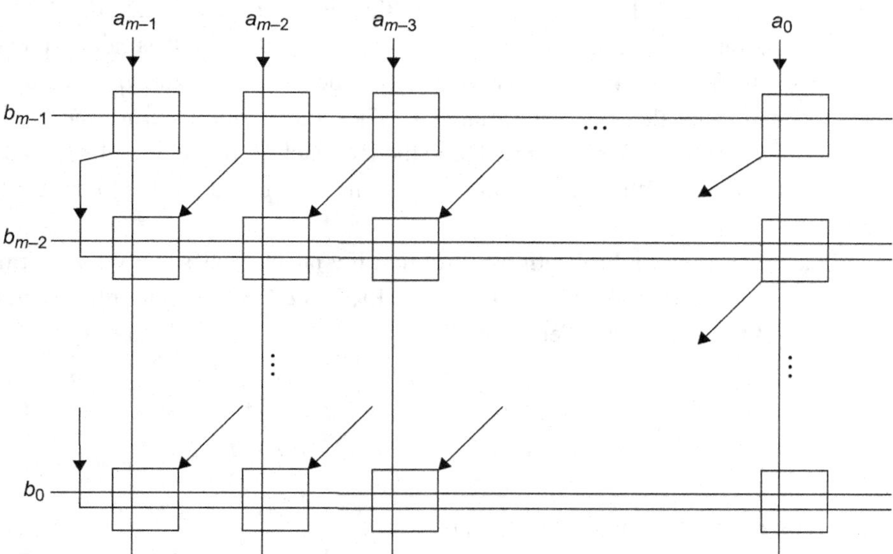

Figure 13.2 A cellular array for F_{2^m} multiplication

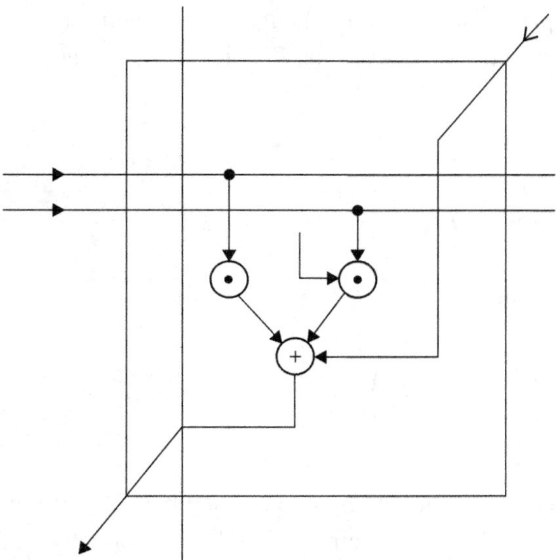

Figure 13.3 A single cell of the cellular array

Inserting a modulo-$p(x)$ operation at each step leads to

$$c(x) = R_{p(x)}[\ldots x R_{p(x)}[x R_{p(x)}[x b_{m-1} a(x)] + b_{m-2} a(x)] + \ldots] + b_0 a(x).$$

This suggests a multiplier with a cellular structure as shown in Figure 13.2. It consists of m^2 standard cells, one of which is shown in Figure 13.3, arranged in an array and interconnected in order to execute the multiplication. In one standard cell, one bit of a and one bit of b are multiplied ("and" gate), and the result is added ("exclusive-or" gate) to other terms and passed out the line on the diagonal, corresponding to multiplication by the position variable x. The other terms entering the modulo-two sums are the output of an earlier cell and the feedback from the modulo-$p(x)$ overflow. The coefficient p_i of the polynomial $p(x)$ determines whether the feedback is used in the ith cell.

The structure of multiplication is different if a normal basis is used. A normal basis $e_1, \ldots, e_m$ has the property $e_i = e_{i-1}^2$ for all i, and $e_1 = e_m^2$. Squaring is then a cyclic rotation of the vector representation because

$$\begin{aligned}(a_1 e_1 + \cdots + a_m e_m)^2 &= a_1^2 e_1^2 + \cdots + a_m^2 e_m^2 \\ &= a_1 e_2 + a_2 e_3 + \cdots + a_{m-1} e_m + a_m e_1\end{aligned}$$

(because $a_i = 0$ or 1, and so $a_i^2 = a_i$).

To compute the finite-field product $c = ab$ with the field elements represented in the normal basis, a bit-serial multiplier may be appropriate. The *Omura–Massey multiplier* is a bit-serial multiplier that computes every bit in the same way, but not at the same time. It exchanges circuit complexity for time complexity. Only a cyclic shift of the two multiplicands is required to compute successive bits. The field multiplication $c = a \cdot b$ can be written as

$$\sum_{k=0}^{m-1} c_k \alpha^{2^k} = \left(\sum_{i=0}^{m-1} a_i \alpha^{2^i}\right)\left(\sum_{j=0}^{m-1} b_j \alpha^{2^j}\right).$$

Squaring both sides and reindexing leads to

$$\sum_{k=0}^{m-1} c_{k-1} \alpha^{2^k} = \left(\sum_{i=0}^{m-1} a_{i-1} \alpha^{2^i}\right)\left(\sum_{j=0}^{m-1} b_{j-1} \alpha^{2^j}\right).$$

Therefore if a circuit is designed to compute any one bit, say bit c_{m-1} of the product $a \cdot b$, other bits can be computed by that same circuit simply by cyclically shifting the m bits in a and in b.

Bit c_{m-1} is linear in the coefficients of a, and is linear in the coefficients of b. Therefore c_{m-1} can be written as the bilinear form

$$c_{m-1} = \sum_{i=0}^{m-1} \sum_{j=0}^{m-1} a_i \boldsymbol{M}_{ij} b_j,$$

where the matrix $\boldsymbol{M}$ is symmetric and its elements are in $\boldsymbol{F}_2$. The matrix can be deduced by writing out the expansions in detail and identifying terms. For example, if $\boldsymbol{F}_{256}$ is constructed by using the prime polynomial $p(x) = x^8 + x^7 + x^5 + x^3 + 1$, then

$$\boldsymbol{M} = \begin{bmatrix} 0 & 0 & 0 & 1 & 0 & 0 & 0 & 0 \\ 0 & 0 & 0 & 0 & 0 & 0 & 1 & 1 \\ 0 & 0 & 0 & 0 & 1 & 0 & 1 & 0 \\ 1 & 0 & 0 & 0 & 0 & 1 & 1 & 1 \\ 0 & 0 & 1 & 0 & 0 & 1 & 0 & 0 \\ 0 & 0 & 0 & 1 & 1 & 0 & 0 & 0 \\ 0 & 1 & 1 & 1 & 0 & 0 & 0 & 1 \\ 0 & 1 & 0 & 1 & 0 & 0 & 1 & 1 \end{bmatrix}.$$

Then bit seven of the product $c = ab$ is computed by

$$c_7 = \boldsymbol{a}^T \boldsymbol{M} \boldsymbol{b}.$$

Each element of $\boldsymbol{M}$ that is equal to one corresponds to the $\boldsymbol{F}_2$ product of one bit of a and one bit of b. Because there are twenty-one ones in $\boldsymbol{M}$, there are twenty-one instances where a bit of a multiplies a bit of b. Each of these instances is an "and" operation. The result is a one only if both bits in the product are a one. All of these twenty-one product terms are then added together in $\boldsymbol{F}_2$. To compute the other components of c, we use

$$c_{7-\ell} = (S^\ell \boldsymbol{a})^T \boldsymbol{M} (S^\ell \boldsymbol{b}),$$

where S is the cyclic shift operator.

To form the matrix $\boldsymbol{M}$ is straightforward, though lengthy. To illustrate this process, the matrix $\boldsymbol{M}$ for multiplication in $\boldsymbol{F}_{2^4}$ will be developed as a simpler example. Using the field representation in Table 13.1 and $\beta = \alpha^3$, the product is

$$\begin{aligned} & c_0\beta^1 + c_1\beta^2 + c_2\beta^4 + c_3\beta^8 \\ & \quad = (a_0\beta^1 + a_1\beta^2 + a_2\beta^4 + a_3\beta^8)(b_0\beta^1 + b_1\beta^2 + b_2\beta^4 + b_3\beta^8). \end{aligned}$$

Elementary manipulations lead to

$$c_3 = a_2b_2 + (a_0b_1 + a_1b_0) + (a_0b_2 + a_2b_0) + (a_1b_3 + a_3b_1)$$

so that

$$\boldsymbol{M} = \begin{bmatrix} 0 & 1 & 1 & 0 \\ 1 & 0 & 0 & 1 \\ 1 & 0 & 1 & 0 \\ 0 & 1 & 0 & 0 \end{bmatrix}.$$

Then bit three is computed by $c_3 = \boldsymbol{a}^T \boldsymbol{M} \boldsymbol{b}$, and other bits are computed by cyclically shifting a and b, then using the same expression as described above.

13.10 Complementary bases

An m-bit binary number can be represented as an element of the field $\boldsymbol{F}_{2^m}$ in many ways depending on the basis. Let $\{\alpha_0, \alpha_1, \ldots, \alpha_{m-1}\}$ be a basis for $\boldsymbol{F}_{2^m}$. Then the binary number $(b_0, b_1, \ldots, b_{m-1})$ is represented by the field element $b = \sum_{k=0}^{m-1} b_k \alpha_k$. The element of $\boldsymbol{F}_{2^m}$ that corresponds to an m-bit binary number depends on the choice of basis. A specified element of $\boldsymbol{F}_{2^m}$ can be represented by an m-bit binary number by expressing the field element as a linear combination of basis elements. The binary number that results depends on which basis is used. It is possible to use two different bases for the variables in a finite field to execute a finite-field multiplication. Thus when two or more bases are used at the same time, care must be taken to distinguish which binary number is represented in which basis.

Two bases for $\boldsymbol{F}_{2^m}$, denoted $\{\mu_0, \mu_1, \ldots, \mu_{m-1}\}$ and $\{\lambda_0, \lambda_1, \ldots, \lambda_{m-1}\}$ are called *complementary bases* (or *dual bases*) if they satisfy the property

$$\text{trace}(\mu_i \lambda_k) = \delta_{ik},$$

where

$$\delta_{ik} = \begin{cases} 1 & \text{if} \quad i = k \\ 0 & \text{if} \quad i \neq k \end{cases}$$

is the Kronecker delta function. The binary trace of a element β of a binary field $\boldsymbol{F}_{2^m}$ is defined as $\text{trace}(\beta) = \sum_{i=0}^{m-1} \beta^{2^i}$. Every basis does have a unique complementary basis. A basis may be its own complementary basis. The complementary basis of the polynomial basis $\{1, \alpha, \alpha^2, \ldots, \alpha^{m-1}\}$ is called the *complementary polynomial basis*.

Theorem 13.10.1 (Projection property) *Let $\{\lambda_k\}$ and $\{\mu_k\}$ be a pair of complementary bases of $\boldsymbol{F}_{2^m}$. Then the field element β has the representation*

$$\beta = \sum_{k=0}^{m-1} b_k \mu_k,$$

where the coefficients are either zero or one, as given by

$$b_k = \text{trace}(\beta \lambda_k).$$

Proof Let b be any field element to be expressed in the $\{\mu_k\}$ basis. Because b_k is either zero or one, the binary trace of $b\mu_i$ can be evaluated as follows:

$$\begin{aligned}\text{trace}(b\mu_i) &= \text{trace}\left(\lambda_i \sum_{k=0}^{m-1} b_k \mu_k\right) \\ &= \sum_{k=0}^{m-1} b_k \text{trace}(\mu_k \lambda_i) \\ &= \sum_{k=0}^{m-1} b_k \delta_{ik} \\ &= b_i\end{aligned}$$

as was to be proved. □

The *Berlekamp multiplier* is a bit-serial multiplier that uses the polynomial basis $\{1, \alpha, \alpha^2, \ldots, \alpha^{m-1}\}$ and the complementary polynomial basis $\{\lambda_0, \lambda_1, \ldots, \lambda_{m-1}\}$. The algorithm starts with the two field elements b and c, one expressed in the polynomial basis and one expressed in the complementary polynomial basis, and it computes the product bc expressed in the complementary polynomial basis $\{\lambda_0, \lambda_1, \ldots, \lambda_{m-1}\}$. Let

$$\begin{aligned}b &= \sum_{i=0}^{m-1} b_i \alpha^i \\ c &= \sum_{k=0}^{m-1} c_k \lambda_k.\end{aligned}$$

Then the product $s = bc$ can be expanded as

$$\begin{aligned}s &= \sum_{i=0}^{m-1} \sum_{k=0}^{m-1} b_i c_k \alpha^i \lambda_k \\ &= \sum_{k=0}^{m-1} s_k \lambda_k\end{aligned}$$

where s has been expanded in the complementary polynomial basis $\{\lambda_0, \lambda_1, \ldots, \lambda_{m-1}\}$. Then by the projection property,

$$s_k = \text{trace}(s\alpha^k).$$

In particular, because $\alpha^0 = 1$,

$$\begin{aligned}s_0 &= \text{trace}(s) = \text{trace}(bc) \\ &= \text{trace}\left(\sum_{i=0}^{m-1} \sum_{k=0}^{m-1} b_i c_k \alpha^i \lambda_k\right).\end{aligned}$$

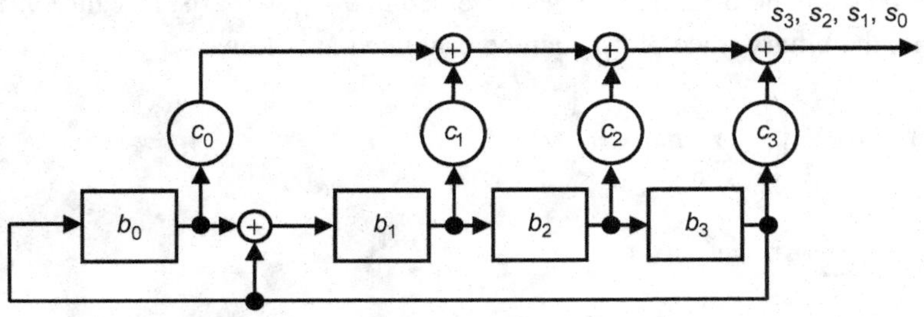

Figure 13.4 A Berlekamp bit-serial multiplier

But b_i and c_k are each either zero or one. Hence

$$s_0 = \sum_{i=0}^{m-1}\sum_{k=0}^{m-1} b_i c_k \text{trace}(\alpha^i \lambda_k)$$
$$= \sum_{i=0}^{m-1} b_i c_i,$$

which is a simple sum of products. Moreover, for each k, s_k can be obtained simply by first multiplying β by α^k, then repeating the same sum of products.

For example, let $p(z) = z^4 + z + 1$ be the primitive polynomial used to construct $\boldsymbol{F}_{16}$. Then over $\boldsymbol{F}_2$, $\{1, \alpha, \alpha^2, \alpha^3\}$ is the polynomial basis of $\boldsymbol{F}_{16}$, and $\{\alpha^{14}, \alpha^2, \alpha, 1\}$ is the complementary polynomial basis, as can be verified by straightforward computation. Let

$$b = b_0\alpha^0 + b_1\alpha^1 + b_2\alpha^2 + b_3\alpha^3$$
$$c = c_0\alpha^{14} + c_1\alpha^2 + c_2\alpha^1 + c_3\alpha^0$$

and $s = bc$ be written as

$$s = s_0\alpha^{14} + s_1\alpha^2 + s_2\alpha^1 + s_3\alpha^0.$$

Then, as can be easily verified,

$$s_0 = b_0c_0 + b_1c_1 + b_2c_2 + b_3c_3.$$

To obtain s_1 by the same expression, β is replaced by $\beta\alpha$. In turn, to obtain s_k, β is replaced by $\beta\alpha^k$.

A circuit for implementing the Berlekamp bit-serial multiplier is shown in Figure 13.4. The register is initialized with b_0, b_1, b_3, and b_4. Each clock forms one value s_k and multiplies $\beta 2^{k-1}$ by α to obtain $\beta\alpha^k$. The bit-serial multiplier has on the order of m fewer gates and on the order of m more clock cycles than the usual parallel multiplier. A set of m bit-serial multipliers operating concurrently has about the same number of gates and the same throughput as the parallel multiplier. The advantage of the circuit is the great reduction in the number of wires and the option of distributing multipliers among other circuit elements. The disadvantage is the need to use more than one representation of field elements. This may entail the need for conversion from one basis to the other. It can be attractive when the application allows the bases of the multiplicands to be preplanned to avoid basis conversion.

13.11 Division in a finite field

In most arithmetic systems, division is considerably harder than multiplication. Division is defined to be the inverse of multiplication. In a prime field, $\boldsymbol{F}_p$, multiplication is modulo p multiplication. Division in the prime field $\boldsymbol{F}_p$ is defined in terms of integer division with remainder and the euclidean algorithm, and so is closely related to division in $\boldsymbol{Z}$, but its execution is really quite different. To divide by a in $\boldsymbol{F}_p$, one multiplies by a^{-1}, so the task of division consists of computing a^{-1} modulo p, followed by multiplication.

In the prime field $\boldsymbol{F}_p$, the inverse of a nonzero element a can be determined in terms of the value A that satisfies the Bézout identity

$$aA + pP = 1.$$

The usual way to find A and P satisfying this equation is to compute A using the extended euclidean algorithm. Then $a^{-1} = A \pmod{p}$, so A is the inverse of a in $\boldsymbol{F}_p$. Division by a can be executed as a multiplication by A.

In the general finite field $\boldsymbol{F}_{p^m}$, division has a corresponding structure, but now the field elements are polynomials over $\boldsymbol{F}_p$ rather than integers. Every element of $\boldsymbol{F}_{p^m}$ can be represented as a polynomial over $\boldsymbol{F}_p$ of degree at most $m - 1$. The usual way to find $A(x)$ is to use the extended euclidean algorithm for polynomials to find polynomials $A(x)$ and $P(x)$ that satisfy the Bézout identity for polynomials

$$a(x)A(x) + p(x)P(x) = 1.$$

Division by the field element $a(x)$ is multiplication by $A(x)$, where $A(x)a(x) = 1$ modulo $p(x)$. Then $a(x)A(x) = 1 \pmod{p(x)}$, so $A(x) = a^{-1}(x)$.

The computation of $A(x)$ uses the extended euclidean algorithm in $\boldsymbol{F}_{p^m}$, which itself requires the computation of inverses in $\boldsymbol{F}_p$. The full computation can be structured as shown in Figure 13.5, with the computation of the inverse in $\boldsymbol{F}_{p^m}$ calling on the computations of the inverse in $\boldsymbol{F}_p$. One can even add further layers, if desired computing inverses in $\boldsymbol{F}_{p^m}$ by calling inverses in $\boldsymbol{F}_{p^{m'}}$, provided m' divides m. This requires $\boldsymbol{F}_{p^{m'}}$ to be represented as an extension of $\boldsymbol{F}_p$, and $\boldsymbol{F}_{p^m}$ to be represented as an extension of $\boldsymbol{F}_{p^{m'}}$.

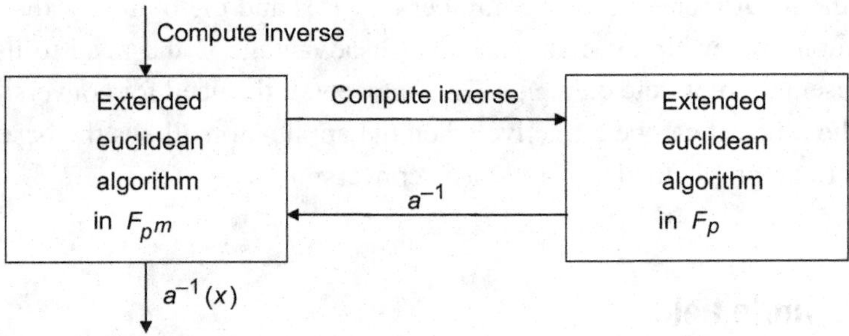

Figure 13.5 Computing inverses in a finite field

Problems for Chapter 13

13.1 The supersingular elliptic curve $\mathcal{X}(\boldsymbol{F}_2) : y^2 + y = x^3 + x + 1$ consists of a single point at infinity.

a Determine $\#\mathcal{X}(\boldsymbol{F}_{2^m})$ as a function of m (mod 8).

b Show that for odd m, any prime factor of $\#\mathcal{X}(\boldsymbol{F}_{2^m})$ has embedding degree four.

c Show that $\Psi(x, y) = (x + s^2, y + s, y + sx + t)$ is a distortion map from $\mathcal{X}(\boldsymbol{F}_{2^m})$ to $\mathcal{X}(\boldsymbol{F}_{2^{4m}})$ provided s and t are elements of $\boldsymbol{F}_{2^{4m}}$ satisfying $s^4 = s$ and $t^2 + t = s^6 + s^2$.

13.2 **a** Describe a procedure for converting a base-10 representation of a large integer to a base-2 representation.

b Describe a procedure for converting a base-10 representation of a large integer to a signed base-3 representation. That is, $a = \Sigma_i a_i 3^i$ where $a_i \in \{-1, 0, +1\}$.

13.3 Sometimes it is possible to bury the change of the basis of a finite field within the structure of a dual-basis, bit-serial multiplier. Let α be a zero of polynomial $p(x) = x^5 + x^2 + 1$ in the field $\boldsymbol{F}_{32}$.

a Prove that

$$\{1, \alpha, \alpha^2, \alpha^3, \alpha^4\}$$

and

$$\{\alpha^{26}, \alpha^{25}, \alpha^{29}, \alpha^{28}, \alpha^{27}\}$$

are complementary bases. Find a five-by-five matrix over $\boldsymbol{F}_2$ that provides a transformation between these complementary bases. Sketch a bit-serial multiplier in $\boldsymbol{F}_{32}$ using these bases.

b Let

$$\beta = b_0 1 + b_1\alpha + b_2\alpha^2 + b_3\alpha^3 + b_4\alpha^4$$
$$\gamma = c_1 1 + c_0\alpha + c_4\alpha^2 + c_3\alpha^3 + c_2\alpha^4$$

and

$$\sigma = s_1 1 + s_0\alpha + s_4\alpha^2 + s_3\alpha^3 + s_2\alpha^4.$$

Show that

$$s_0 = \text{trace}(\alpha^{25}\beta\gamma).$$

Sketch a bit-serial multiplier based on this expression. Why might this multiplier be preferred to the multiplier given in part (a)?

13.4 Prove that any sequence of additions, subtractions, multiplications, and squarings in $\boldsymbol{Z}_n$ can be executed entirely in the Montgomery domain.

13.5 Give a flow diagram for point subtraction in the curve $\mathcal{X}(\boldsymbol{F}_p)$, $p \neq 2$.

13.6 Give a flow diagram for computing the sequence $x^2, x^4, x^8, x^{16}, \ldots \pmod{N}$, for any integer x, using the Montgomery multiplication algorithm.

13.7 A "quarter-square" multiplier eliminates the need for general multiplication by using the identity

$$xy = \frac{1}{4}\left[(x+y)^2 - (x-y)^2\right]$$

to replace a multiplication by two squarings. Can this method be used for multiplication in a binary field?

13.8 The q-ary trace of an element β of $\boldsymbol{F}_{q^m}$ is defined as

$$\text{trace}(\beta) = \sum_{i=0}^{m-1} \beta^{q^i}.$$

Prove that over the elements of $\boldsymbol{F}_{q^m}$, the q-ary trace takes on each value of $\boldsymbol{F}_q$ equally often.

13.9 Write out explicit equations to compute the point $[3]P$ where P is the point (x, y) of an elliptic curve.

13.10 By repeated doubling of P, Q, and $P+Q$, develop a procedure for computing $aP + bQ$ where a and b are integers and P and Q are points of an elliptic curve. How does your procedure compare to first computing aP and bQ, then adding them?

13.11 Any point of a elliptic curve based on the polynomial $y^2 = x^3 - x \pm 1$ can be tripled by the formula $[3](\beta, \gamma) = (\beta^9 \pm 2, -\gamma^9)$. Design an algorithm for computing a point multiple $[r]P$ on such an elliptic curve based on the formula for point tripling. Compare the computational work of your algorithm with the computational work of the more conventional algorithm based on point doubling.

13.12 Find all points of the affine plane $\boldsymbol{F}_{11}^2$ satisfying the Edwards polynomial

$$x^2 + y^2 = 1 - x^2y^2.$$

Find all points of the affine plane $\boldsymbol{F}_{11}^2$ satisfying the Weierstrass polynomial

$$y^2 = x^3 + 4x.$$

13.13 **a** Construct and sketch logic circuits for addition and multiplication in $\boldsymbol{F}_8$.
b Find an irreducible polynomial of degree two in $\boldsymbol{F}_8$.
c Describe addition and multiplication circuits in $\boldsymbol{F}_{64}$ that use the arithmetic of $\boldsymbol{F}_8$ as components.
d Compare this with a more direct implementation.

13.14 Describe a detailed procedure for adding two points expressed in projective coordinates of an elliptic curve on a field of characteristic three. Can you give a procedure that uses only nine multiplications in the underlying field?

13.15 Show that the Miller function satisfies

$$f_{T^kP} = f_{T,P}^{T^{k-1}} f_{T,TP}^{T^{k-2}} \cdots f_{T,T^{k-1}P}$$

by showing that the divisors of the functions on the two sides are equal and referring to the Miller algorithm.

Notes for Chapter 13

Between the theory and the implementation of cryptography lie the topics of algorithms and architectures. In the case of pairing-based cryptography, theory and implementation often lie close together. Each feeds the other and both grow. The work of Barreto, Kim, Lynn, and Scott (2002) spans these viewpoints. Fast implementations of pairings were discussed by Aranha, Karabina, Longa, Gebotys, and López (2011).

Because of the recently discovered practical importance of pairing in cryptography, many versions of pairing have now been proposed and explored. In their 1994 paper, Frey and Rück introduced the Tate pairing as an alternative to Weil pairing. Later, Duursma and Lee (2003) modified the Tate pairing for certain special elliptic curves to reduce its complexity. The methods of Duursma and Lee were refined by Barreto, Galbraith, ÓhÉigeartaigh, and Scott (2007), leading to the introduction of the *eta pairing*. This was followed by the *ate pairing* by Hess, Smart, and Vercauteren (2006), then by other pairings by Lee, Lee, and Park (2009) and by Vercauteren (2010).

Implementation of hyperelliptic-curve cryptosystems has been studied by Wollinger (2001) and by Lange (2002). Algorithms for computation are in widespread use for many purposes and have a diverse history in many fields. Algorithms for modular integer multiplication include the Karatsuba algorithm (1962), the Montgomery algorithm (1985), and the Barrett algorithm (1987). The Edwards polynomial (2007) was introduced to simplify elliptic-curve calculations. Double-base methods were introduced by Dimitrov and Cooklev (1995). Addition–subtraction chains for exponentiating or computing point multiples were surveyed by Gordon (1998).

The advantage of using a normal basis for finite fields was noticed by Hensel (1888). The Omura–Massey multiplier (1986) works with any normal basis, but satisfies a certain optimality property if an optimal normal basis is used. Mullin, Onyszchuk, Vanstone, and Wilson (1989) showed that a normal basis must always use at least $2n - 1$ bitwise multiplications. It was proved by Gao and Lenstra (1992) that an optimal normal basis exists for F_{2^m} if, and only if, either $m + 1$ or $2m + 1$ is prime. Itoh and Tsujii (1988) gave a fast algorithm for computing multiplicative inverses in a finite field by using normal bases.

The search for primitive binary trinomials has an interesting history of steady advances well beyond the current practical needs by those with a taste for such activities, and extensive tables are available, as by Seroussi (1998). More recently, Brent and Zimmermann (2009) have reported three primitive trinomials over $\boldsymbol{F}_2$ of degree 32582657, this integer being chosen because it is a Mersenne prime; it has the form $2^m - 1$. The Brent–Zimmermann polynomials are $x^r + x^s + 1$, where $r = 32582657$ and s can be either 5110722, 5552421, or 7545455.

14 Cryptographic protocols for security and identification

The methods of cryptography can be successful only if they are properly incorporated into an application. A cryptographic technique can be unbreakable as such, and yet could still be insecure if it is used unwisely. This observation motivates the subject of security protocols. A security protocol consists of a formal method of using a cryptographic technique in a practical application. Without a well-defined security protocol, a perfectly secure cryptographic technique could be compromised by its use.

We do not draw a sharp distinction between the topic of secrecy and the topic of security. The two are closely related, and it is not helpful to attempt to draw a sharp line between them. In this context, a *primitive* is a basic method of obtaining secrecy and a *protocol* is a body of rules for using that primitive.

In this chapter we will also study other forms of information protection such as identification and secret sharing that can be regarded as topics closely related to security protocols. The topic of identification deals with methods for establishing the identity of the source of a transmission. This is a subtle and difficult topic because the very concept of identity is vague and subjective. The identity of an isolated individual or device may be self-defined and subject to no outside verification. That identity is useful only to establish that all messages in a sequence of messages are from the same entity. A more extensive notion of identity requires that the entity be associated with known documentation or with a history that can be verified. Ultimately the task of identification consists of binding an individual, or a device, to a community. To do so, the entity can establish a secret that is not directly divulged, but its existence is made public. That secret enables the entity to solve a certain class of mathematical problems that could not otherwise be solved. If, when challenged, the entity can display this ability to solve such problems, the entity thereby establishes its identity whenever required. In effect, the identity of that entity consists only of the fact that it is the owner of the secret that enables such problems to be solved.

The topic of secret sharing, also studied in this chapter, deals with methods for sharing a secret or sharing access to a secret among n participants in such a way that any t of the participants must collaborate in order to access the secret. It is required that if fewer than t participants collaborate, then the secret must be completely inaccessible, while if t participants collaborate, the secret is fully revealed. A collaboration of $t - 1$ participants cannot retrieve any information about the secret, however meager. The method of secret sharing can also be used to distribute a data set across multiple storage systems such that any t of them are sufficient to recover the entire data set. In this application, it becomes a method of data protection against storage failures, rather than a method of secret sharing.

14.1 Protocols for cryptographic security

A protocol for cryptographic security is a detailed procedure for using a chosen cryptographic technique so as to ensure that its security is not compromised by careless use. Accordingly, it is usually regarded as good practice to precisely write out in formalized detail a procedure for using a cryptographic method, including appropriate formal statements about data structure and message format. It is possible that seemingly reasonable procedures that are fully inside the crytographic protection and apparently completely isolated by the cryptographic function can actually leak information about the plaintext or key. For example, an application that is protected by a cryptosystem may innocently require that all received and decrypted plaintexts be in a standard format with a designated header. To enforce this requirement, the receiver of an encrypted message returns an error message to the sender whenever the decrypted plaintext message is not in the required format. It could be possible that in some cryptosystems this practice introduces a vulnerability based on traffic analysis, even if the communication in both directions is encrypted. Although it may appear unlikely that such a weakness would exist, such possibilities must be thought through and formally dismissed. For such reasons, a system of security protocols should be used to protect a cryptosystem from potential harmful practices.

The functions of identification and authentication also create the need for protocols. The widely accepted method of using a password for identification, no matter how complicated, is vulnerable to an eavesdropper anywhere in the path of the transmitted message. Indeed, even the intended receiver can later become an adversary and use that password for the purpose of deception. This shows that there is a need for a more sophisticated identification protocol. For this reason, identification schemes have been developed that do not reveal the secret password. Remarkably, such schemes verify that the password is known to the owner, yet reveal nothing about the password to the verifier.

A formal security protocol will normally contain multiple requirements integrated into a coherent package. These functions may include: key agreement or key establishment, identification, authentication, symmetric encryption, and nonrepudiation. More advanced protocols may include such topics as blind signatures and deniable encryption. A blind signature is one used to notarize a message that is already encrypted so that the signer cannot and does not read the message, but the message can be decrypted without invalidating the signature. This allows the message to be decrypted without involving the destination of the message. Deniable encryption refers to the formation of an encrypted message that appears to be a nonencrypted message. Deniable encryption is closely related to steganography.

An example of a formal cryptographic security protocol suite in wide use is one known as Kerboras. This protocol suite combines the functions of security and authentication. It protects both parties of a communication against eavesdropping and replay attacks. It uses symmetric-key cryptography and requires a trusted third party. It also uses public-key cryptography and asymmetric-key cryptography, as well as identification protocols, which are described in the next section.

14.2 Identification protocols

An *identification protocol* is a scheme to verify the identity of the entity that transmitted a message on a public channel. A trusted authority validates the identity of any user that intends to transmit messages, or perhaps creates a new identity and certifies that the user has the indicated identity. The only purpose of the identification protocol is to verify that the trusted authority is convinced of the identity of that user, by whatever standard the trusted authority uses to define identity. The identification protocol must satisfy the requirement that no adversary can impersonate a user's assigned identity. An ordinary password is a simple, but weak, form of protection. The common method of a simple password or a personal identification number is not satisfactory, in general, because an adversary can eavesdrop and steal the password. Even a legitimate receiver could become an adversary in the future and then use the password adversarially. Instead, to prevent such attacks, an identification code must be computed from a secret identity key that is not revealed. An entity then establishes its identity when so requested by proving that it possesses the identity key. It must be computationally intractable to extract the identity key from the identification code, yet the identification code must allow verification that the entity possesses the identity key. In effect, the identity key plays the role of a password that is retained in secret by the entity and is never transmitted or revealed. Yet its ownership can be verified.

An identification protocol verifies the identity of an individual, here called the *claimant*, by verifying that the individual does have possession of a secret key or password that only that individual can have. The claimant identifies itself to the *verifier* by proving that it knows the asserted secret without actually revealing the secret. A *zero-knowledge* protocol is one in which the claimant proves that it possesses the secret without revealing other information of any kind, either to the verifier or to a third party intercepting the messages.

The difference between a signature and an identification is that a signature validates a message, while an identification validates a participant in a communication. A properly signed document could be in the legitimate possession of and transmitted by a third party, or it may be in the possession of an adversary, and so transmitted for some purpose, possibly one of deception. The function of an identification protocol is to verify that this latter possibility is not the case. In other situations, one may care only about the authenticity of the document, and not care about the identification or authenticity of the transmitter of that document. An alternative and weaker requirement is consistency. Consistency is required when one wants only to know that every message in a sequence of messages is coming from the same entity. The identity is then contained in the first message. Indeed, the identity can be regarded as the sender of the first message, nothing more.

An identification scheme is secure if it does not reveal any information that subsequently enables another participant to falsely identify itself as the possessor of the identity secret. An identification scheme is sound and complete if knowledge of the secret is necessary and sufficient for identification. The broader community must, in some way, bind the possession of the secret to a general consensus regarding the identity of that individual. This requires the existence of a trusted certification authority, validated by the community. The trusted certification authority publicly publishes an announcement linking an identity to the possessor of the secret.

An identification protocol is executed on a public channel, but should not reveal any information that will allow a subsequent impersonation by an adversary that has intercepted the messages. An identification protocol will use time-variant parameters in order to make each instance of its use unique. This is to prevent replay attacks and interleaving attacks, to prevent certain chosen-text attacks, and to guarantee uniqueness.

An identification scheme usually performs its task by using a so-called *challenge–response sequence* (or a *claim–challenge–response sequence*). This sequence begins with a *claim message* from the claimant to the verifier, whereby the claimant requests verification. The verifier then responds with a *challenge message* to the claimant. In turn, the claimant sends a *response message* to the verifier, and the verifier checks for consistency of the claim message with the response message. If the claimant is a device, then the procedure only verifies the identity of the device, not the user of the device. There still remains the possibility that the rightful owner of the device is not the one actually using it. For this reason, the owner of the device should also participate in the identification protocol.

An identification protocol must not be vulnerable to a man-in-the-middle attack. A man in the middle is an unseen relay that intercepts, understands, alters, and retransmits a message for the purpose of deception. The goal of the man in the middle is to steal the identity of the claimant by deceiving the protocol rather than by breaking the cryptographic system.

14.3 Zero-knowledge protocols

Zero-knowledge techniques solve the seemingly conflicting goals of proving a statement without revealing information of any kind, including information about how to prove the statement. Perhaps surprisingly, zero-knowledge techniques do exist based on the premise that certain trapdoor functions are computationally intractable. Zero-knowledge techniques are commonly used in the form of protocols for identification and authentication.

To understand the nature of a zero-knowledge protocol in an intuitive setting, consider two long corridors, side by side, connected at the far end by a self-closing door that is out of sight of the two entrances to the two corridors. To prove that it possesses the key to the self-closing door, the claimant randomly enters either corridor without being observed. The verifier then chooses a corridor from which the claimant is to exit. The valid claimant has the key and can unlock the door between corridors, and so the valid claimant exits by the requested corridor with certainty. A false claimant does not have the key, and so exits from the requested corridor with probability one-half. By repeating this process m times, the verifier ensures that the claimant possesses the key with a probability of error of 2^{-m}.

A zero-knowledge protocol for secure identification consists of some secret knowledge that enables challenges in the form of queries to be answered, though the query could not be answered with certainty without the secret knowledge. Moreover, the query must be answered without revealing the secret knowledge.

A zero-knowledge proof must satisfy three properties restated as follows:

(1) **Completeness:** If the statement is true, then the honest verifier will be convinced by an honest claimant within some probability of failure that can be specified as small as desired.

(2) **Soundness:** If the statement is false, then the honest verifier cannot be deceived, except with some small probability that can be specified as small as desired.

(3) **Zero-knowledge:** If the statement is true, the verifier can obtain no information other than that the statement is true. This property must hold even if the verifier chooses to alter its use of the protocol with an intention to deceive.

A zero-knowledge protocol is based on a secret key that is known only to the claimant. The claimant must prove its ownership of the key without divulging the key or any other information. This is done by setting up a computational problem with a trapdoor. The problem is intractable as such, but easy if one has the key to the trapdoor. Such problems usually rest on either the biprime factoring problem or on the discrete-log problem, both of which are deemed to be intractable.

14.4 Methods of secure identification

Some methods of secure identification that will be discussed in this section are the *Schnorr identification protocol*; the *Okamoto identification protocol*; the *Feige–Fiat–Shamir identification protocol*; and the *Guillou–Quisquater identification protocol*. Such protocols are usually based on a *claim–challenge–response sequence* for identification. In such a protocol, the *claimant* to an identity, when challenged by a *verifier*, makes an appropriate response that only that claimant could make. Such a sequence, more simply called a *challenge–response sequence*, may be repeated multiple times, if necessary, to reduce the probability of a false identification. The *Fiat–Shamir identification protocol* is also discussed in this section, but only for pedagogical purposes because the Feige–Fiat–Shamir protocol is superior.

The performance of an identification protocol can be judged by a variety of considerations. The primary considerations are security, computational requirements, communication requirements, and probability of false identification, but these are not the only considerations. Some protocols need the participation of a trusted certification authority and the method of satisfying this need is also important, as is the need for the storage and protection of the secret. The tasks of the trusted certification authority are to validate and certify the identity of the claimant, and to issue an appropriate certificate enabling a challenge to the identification secret held by the claimant. Some protocols can be expanded to embed this certificate into the protocol. These then are called *identity-based* identification protocols.

The many identification protocols differ from each other in the level of security, the amount of computation, the amount of communication, and the amount of memory required. The needs of an application will determine which protocol should be preferred. For instance, the Schnorr identification protocol is attractive because there is little on-line calculation, and then only in the third step. The Schnorr identification protocol does have the largest amount of computation, but most of this computation can be done off-line. The Guillou–Quisquater identification protocol uses fewer rounds, less memory, but more computations than does the Fiat–Shamir identification protocol.

Schnorr identification protocol

The only goal of the Schnorr identification protocol is to verify that the claimant knows the finite-field discrete logarithm of an openly published field element. Only the claimant knows the discrete logarithm of this field element. The Schnorr protocol is based on the fact that the computation of a discrete logarithm in $\boldsymbol{F}_p^*$ is intractable if p is sufficiently large. Therefore, knowledge of the discrete logarithm of a field element is a secret held only by that entity, and can be used as a means of identification. A trusted certification authority must determine the identity of the claimant by some means and assert that the claimant does uniquely know the discrete logarithm to the base α of an element v of the field $\boldsymbol{F}_p$. A certificate to this effect is issued and is publicly available. The signature of the trusted authority on the certificate is the guarantee that the certificate was issued by the trusted certification authority after establishing the identity of that claimant.

Let p be a prime large enough and for which $p-1$ has a large prime factor, denoted r, so that the discrete-log problem in $\boldsymbol{F}_p^*$ to the base r is not tractable. Let α be a publicly known generator of the cycle μ_r of order r (mod p). The integers p, r, and α are fixed and public parameters of the identification protocol. Each potential claimant chooses an integer a satisfying $0 \leq a \leq r-1$ and keeps it secret. Although the integer a is kept secret, the integer $v = \alpha^a \pmod p$ is computed and made public and is validated by the trusted certification authority as the identity of that user of the system. Ensuring the validity of the certificate provided by the trusted authority is not addressed by the Schnorr protocol. This is a separate task. The goal of the verifier using the Schnorr identification protocol is to verify that the claimant does know the discrete logarithm of v.

The challenge–response sequence for the Schnorr protocol, which may be repeated many times, requires that the claimant chooses a random integer s and computes $x = \alpha^s \pmod p$ at the start of each round, and also requires that the verifier then chooses a random integer e. The verifier already knows that the true claimant knows the discrete logarithm of v but the verifier cannot itself compute that discrete logarithm. The sequence of messages between the claimant and the verifier is a sequence of integers given by

$$\begin{aligned}
&\text{claimant to verifier:} \quad x = \alpha^s \pmod p \\
&\text{verifier to claimant:} \quad e \qquad 0 < e \leq r-1 \\
&\text{claimant to verifier:} \quad y = s + ae \pmod r,
\end{aligned}$$

where $a = \log_r v$ is known only to the claimant. To compute the response y to the challenge e, the claimant must know that $a = \log_\alpha v$. It is not tractable to compute a from v because this discrete log is intractable.

The verifier then tests whether

$$\alpha^y v^e = x \pmod p$$

holds. Because $y = s + ae + rQ$ for some Q, and α has order r, the computation becomes the test $\alpha^{s+ae+rQ}\alpha^{-ae} = \alpha^s \pmod p$. This can be true only for the correct response y, which can be computed only if the claimant knows $\log_\alpha v$. Because s is random in the interval from 0 to $r-1$, y is random as well and gives no new information regarding the secret to an adversarial eavesdropper. A false verification occurs only if the random choice of s and e satisfy the verification equation, which happens with probability $1/r$. If r is a 50-bit integer, this probability is 2^{-50}, which is about 10^{-15}. This is also the probability of correctly guessing $\log_\alpha v$.

Okamoto identification protocol

The Okamoto identification protocol is a modification of the Schnorr identification protocol. It is also based on the intractability of computing the discrete log in $\boldsymbol{F}_p^*$ for large values of p. Let r be a large prime factor of $\#\boldsymbol{F}_p^*$, and, as for the Schnorr protocol, let μ_r be a subgroup of order r. Then any two nonzero elements α_1 and α_2 of μ_r are generators of the subgroup μ_r.

Let $\alpha_2 = \alpha_1^c$, meaning that $c = \log_{\alpha_1} \alpha_2$ where p is a publicly known element of $\boldsymbol{F}_p$. The value of c is chosen randomly by the trusted authority. The trusted authority computes α_2, and the integers α_1 and α_2 are made public. Finally, the verifier randomly chooses a pair of integers (a_1, a_2) as a private key and computes a public key $v = \alpha_1^{-a_1}\alpha_2^{-a_2}$, which is certified by the trusted certification authority.

To use the Okamoto protocol, the claimant randomly chooses two nonnegative integers k_1 and k_2 not larger than r and computes $x = \alpha_1^{k_1}\alpha_2^{k_2} \pmod p$. Although the two exponents could be merged, in principle, as $k_1 + k_2 \log_{\alpha_1} \alpha_2$, it is not computationally tractable to do so without knowing $\log_{\alpha_1} \alpha_2$. It does help the explanation to write x this way.

The sequence of messages between the claimant and the verifier is as follows:

claimant to verifier: $x = \alpha_1^{k_1}\alpha_2^{k_2} \pmod p$

verifier to claimant: $e \quad 1 \le e \le r$

claimant to verifier: $y_1 = k_1 + \alpha_1 e \pmod r$

$y_2 = k_2 + \alpha_2 e \pmod r$.

The verifier then computes $\alpha_1^{y_1}\alpha_2^{y_2}v^e \pmod p$ and compares it to x. If these agree, the claim is accepted. Otherwise the claim is rejected. To reduce the probability of misidentification, the protocol can be repeated several times.

For an example of the Okamoto identification protocol, choose the prime $p = 88667$ and note that the prime 1031 is a divisor of $\#\boldsymbol{F}_{88667}^*$, so $\mu_{1031} \subset \boldsymbol{F}_{88667}^*$ is a cyclic subgroup. The trusted authority chooses $\alpha_1 = 73611$ and $\alpha_2 = 58902$, both elements of μ_{1031}, but does not divulge $\log_{\alpha_1} \alpha_2$.

For the private key, the claimant chooses $(a_1, a_2) = (515, 846)$ and computes the public key $v = 13078$, which is certified by the trusted certification authority.

To present a claim to the verifier, the claimant randomly chooses $(k_1, k_2) = (16, 899)$ and computes $x = \alpha_1^{k_1}\alpha_2^{k_2} \pmod p = 73611^{16}58902^{899} \pmod{88667} = 14574$. The verifier randomly chooses $e = 489$ as the challenge. The response to the challenge is $(y_1, y_2) = (287, 131)$. The verifier determines that

$$73611^{287}58902^{131}13078^{489} = 14574 \pmod{88667}.$$

Because this equality holds, the claim is accepted. Otherwise, if the equality did not hold, the claim would be rejected.

Fiat–Shamir identification protocol

The goal of the Fiat–Shamir identification protocol is to verify only that the claimant knows the square root of an integer v modulo a biprime n, which must be a square root that is not tractable to compute. The claimant does not need to reveal the square root of v, and does not. The Fiat–Shamir identification protocol is based on biprime cryptography and on the fact that the task of computing a square root modulo a biprime is equivalent to the task of factoring that biprime, and so is intractable. The identification secret is the knowledge of the square root of an integer that is made public. The secret is protected by the intractability of biprime factoring and the intractability of computing square roots modulo a biprime. It is not tractable to compute the square root in such a case by any known means.

The Fiat–Shamir protocol requires the existence of a trusted certification authority. The trusted authority validates the identity of the claimant by some means, then issues an identity certificate which includes a statement that the claimant knows the square root of a large integer v. Of course, there remains the possibility that at the time it published the certificate, the trusted authority was duped regarding the identity of that claimant. This concern must be dealt with at a higher level. Those methods of the trusted authority are not addressed by the Fiat–Shamir identification protocol.

Let p and q be large primes such that it is not tractable to factor the biprime $n = pq$. The primes p and q are selected by the trusted certification authority, but are not divulged, not to the claimant and not to the verifier, but the biprime n is made public. The claimant chooses as a secret an integer s, that is large and comparable to n, but smaller than n and coprime to n. Then the claimant computes $v = s^2 \pmod{n}$, and registers the integer v with the trusted certification authority. The trusted authority attaches v to a certificate of identification and publishes that certificate. In this way, the trusted authority assures the community that the square root of v is known only to the entity described in the identification certificate. Thus, for our present purpose, the knowledge of the square root itself becomes the identity.

The challenge–response sequence for the Fiat–Shamir protocol, which may be repeated many times, requires that, to begin each round, the claimant first chooses a random integer r and squares it modulo n. Then the sequence of communications between claimant and verifier is

$$
\begin{aligned}
&\text{claimant to verifier:} \quad x = r^2 \pmod{n} \\
&\text{verifier to claimant:} \quad e \in \{0, 1\} \\
&\text{claimant to verifier:} \quad y = r \cdot s^e \pmod{n}.
\end{aligned}
$$

Honest

	$x=r^2$	$x=r^2$
e = 0	?	?
e = 1	Maybe	Maybe

Dishonest

	$x=r^2$	$x=r^2/v$
e = 0	?	No
e = 1	No	Maybe

Figure 14.1 The Fiat–Shamir identification game

The challenge sent by the verifier is e, a single bit. The verifier tests whether

$$y^2 = r^2 v^e \pmod{n}.$$

The protocol is designed to verify a claim, but also to detect an imposter. If $y = 0$, the result of the protocol is inconclusive and a new round will be necessary. If the inequality is not satisfied, then the claim is rejected. If the equality is satisfied, the claim may or may not be valid. If $e = 0$, then r, which is a number that is known to the claimant, is indeed the square root of x. This only ensures, however, that the claimant did not attempt to subvert the protocol by improperly sending $x = r^2/v$ as the initial claim, then sending $y = r$ as the response to the challenge. It does not ensure that the claimant knows the square root of x. If $e = 1$, then either the claim is valid or the claimant has attempted to game the system and the equality is deceptive. The equality can hold only if the claimant knows that r is the square root of v. Otherwise, the test fails and the claimant is an imposter. To increase the probability of a correct decision, the process will be repeated a large number of times, each time with a new valve of r.

The Fiat–Shamir protocol can be treated as a game between the verifier and the imposter. The honest claimant does know v and does know the square root of x. Its response to either challenge is valid. The dishonest claimant does not know v. If it sends $x = r^2/v$, whose square root it does not know and cannot compute, then this deception is detected whenever the challenge is $e = 0$, but is not detected whenever the challenge $e = 1$. This is illustrated in Figure 14.1. The question-mark in a box indicates that the challenge was $e = 0$, and so the challenge did not ask for verification of the secret. The challenge asked for verification of an honest claimant. The word "maybe" in a box indicates an apparently successful verification, but possibly a deception by an imposter.

Suppose that the imposter chooses from its two options randomly and equiprobably, and the challenger also chooses from its two options randomly and equiprobably. Then the imposter will be detected with probability one-half, and the process will be inconclusive with probability one-half. After m repetitions of the protocol, the probability of failing to detect an imposter would be 2^{-m}.

The Fiat–Shamir protocol, as such, is not a zero-knowledge protocol. This is because a cryptanalyst does obtain some slight information about the secret value of s. Specifically, one bit of information is leaked during each use of that protocol. The claimant can avoid this one bit of information leakage by randomly replacing $x = r^2$ by $x = \pm r^2$, and replacing the test by $y^2 = \pm r^2 v^3 \pmod{n}$.

Feige–Fiat–Shamir identification protocol

The Feige–Fiat–Shamir identification protocol is a zero-knowledge protocol based on biprime cryptography and on the fact that computing a square root modulo a biprime is equivalent to factoring a biprime. The protocol is a stronger form of the Fiat–Shamir identification protocol, and is motivated by the repetition required by that protocol.

The trusted authority chooses two large integers, p and q, both primes of the form $4k + 3$, and sets $n = pq$. The primes p and q are not divulged by the trusted authority. The claimant forms k secret numbers, $s_1, \ldots, s_k$, such that $\mathrm{GCD}(s_i, n) = 1$, and k secret control bits $b_i \in \{0, 1\}$, then computes $v_i = (-1)^{b_i}/s_i^2 \pmod{n}$ for $i = 1, \ldots, k$. The control bits are used to ensure a random distribution of signs, thereby preventing information leakage. The vector $\boldsymbol{v} = \{v_1, v_2, \ldots, v_k\}$ forms the public key.

claimant to verifier: $x = \pm r^2$

verifier to claimant: $(e_1, e_2, \ldots, e_k)$ where $e_i \in \{0, 1\}$

claimant to verifier: $y = r s_1^{e_1} s_2^{e_2} \cdots s_k^{e_k}$.

The verifier tests whether

$$y^2 = \pm x v_1^{e_1} v_2^{e_2} \cdots v_k^{e_k}.$$

It is straightforward to demonstrate the validity of this test. In this way, one repetition of the Feige–Fiat–Shamir identification protocol is a replacement for k repetitions of the Fiat–Shamir identification protocol.

For an example of the Feige–Fiat–Shamir identification protocol, select the biprime $n = 553913 = 683 \cdot 811$. This biprime is much too small to be secure, but is large enough for an example. The claimant randomly chooses three integers $(s_1, s_2, s_3) = (157, 4646, 43215)$ and three control bits $(b_1, b_2, b_3) = (1, 1, 0)$, then using $v_i = (-1)^{b_2}/s_i^2$, computes the claimant's public key

$$(v_1, v_2, v_3) = (441845, 124423, 338402) \pmod{553913},$$

which is entered by the trusted certification authority on the claimant's public certificate. To execute the protocol, the claimant randomly selects $r = 1279$, then computes $x = r^2 \pmod{553913} = 25898$. This is the claim sent to the verifier. The verifier sends the challenge $(0, 1, 0)$ to the claimant. The claimant computes $y = rs_2 \pmod{n} = 403104$ and returns this to the verifier. In turn, the verifier computes $z = y^2 v_2 \pmod{n} = 25898$, which is equal to x, thereby verifying the claim.

Guillou–Quisquater identification protocol

The Guillou–Quisquater identification protocol is also based on the intractability of biprime factorization. It can be regarded as a variation of the Fiat–Shamir identification protocol replacing the squaring operation with the calculation of a larger power. The identification secret is that the claimant knows the eth root of an integer x. The Guillou–Quisquater protocol allows the claimant to verify that it knows this secret. This protocol reduces the amount of communication needed during the witness–challenge–response sequence, but increases the amount of computation.

The Guillou–Quisquater identification protocol requires the existence of a trusted certification authority. The trusted authority validates the identity of the claimant by some means separate from this protocol, then issues a public identity certification which includes a statement that the claimant knows the eth root, modulo a biprime pq, of an integer x. Computing the eth root, modulo a biprime, is also intractable. This is an immediate observation if e is an even number because computing a square root is intractable. It also holds if e is odd.

Let p and q be large primes such that factoring the biprime $n = pq$ is not tractable. Let s and v be two integers satisfying

$$sv = 1 \pmod{\phi(n)}.$$

A trusted authority chooses the integers p, q, and s. These integers are kept secret, known only to the trusted authority. The trusted authority computes $n = pq$ and $v = s^{-1} \pmod{\phi(n)}$, then provides n and v to all users by making them public. A positive integer e, larger than two, is also a public and permanent part of the protocol.

The claimant picks a random positive integer r smaller than n and computes $x = r^e \pmod{n}$. The challenge–response sequence is

$$\begin{aligned} &\text{claimant to verifier:} \quad && x = r^e \pmod{n} \\ &\text{verifier to claimant:} \quad && c \quad 1 \le c \le e-1 \\ &\text{claimant to verifier:} \quad && y = r \cdot s^e. \end{aligned}$$

Although this protocol can be repeated several times, there is little reason to do so because it is very unlikely that a false claimant will guess the correct e. For example, if n is a 50-bit integer, then the probability of false identification is 2^{-50}, or about 10^{-15}.

14.5 Signature protocols

We have already studied several signature schemes that are closely related to popular methods of public-key cryptography. In Section 3.1, a signature scheme based on biprime cryptography similar to RSA was introduced. The RSA signature scheme and the Elgamal signature scheme were discussed in detail in Section 8.3. These keyed signatures for authentication are reviewed in this section in order to provide contrast with the other signature protocols. The keyed signature schemes provide authentication, but they do not in themselves provide identification. The public-key signatures can be described as authentication signatures, but they are not identification signatures. The authentication signatures allow a potential signer to establish a method proving that the signer is, indeed, the source of the signed message. The authentication signature does not validate the identity of the signer or the message source, only the consistency of the message source. The signer does not participate in the signature verification, and no communication is needed for verification.

A keyed signature that includes identification is the Schnorr signature. This identification signature, which is described below, may be regarded as a variant of the Elgamal signature because it also is based on the intractability of the discrete-log problem.

To extend the topic of signatures to the task of identification, signature schemes can be based on a challenge–response protocol. Every identification protocol based on a challenge–response sequence, as described in Section 14.4, can be turned into a signature protocol. The direct way to do this is to replace the challenge by a digest of the message, as computed by an appropriate hash function. This may require increasing the blocklength of the challenge because, although a blocklength of 40 bits is often considered adequate for an identification challenge, a blocklength of 160 bits is considered necessary for a message digest. To illustrate this procedure, we will describe the Guillou–Quisquater signature protocol below, which is obtained by appropriately modifying the corresponding identification protocol of Section 14.4.

Asymmetric keyed signatures

Asymmetric signature protocols based on intractable inverse problems were described in Sections 3.1 and 8.3. Those signature protocols allow any user to establish a personal signature so that all future messages from that user can be signed to authenticate the user as the source of these messages. These signature protocols share the property that the signer does not participate in the verification of the signature. The signer could be inaccessible or, indeed, no longer exist. The basic asymmetric keyed signature protocols do not include the function of identification, although the function would normally be incorporated into the system by another means.

To summarize the biprime signature protocol, let $n = pq$ be a biprime that is large enough so that factoring is intractable. Let a and b be two integers for which $ab = 1 \pmod{\phi(n)}$. The integers n and b comprise the public key, while a, p, and q comprise the private key, known only to the signer. The digital signature on the message x is $y = \text{sign}_a(x) = x^a \pmod{n}$. The signed message is the pair $(x, y) = (x, \text{sign}_a(x))$. To verify the signature, compute $\text{ver}_b(y) = y^b$, and compare it to the message x. If $y^b = x$, then the signature is accepted as valid. Of course, in practice, the signature is applied to a message digest rather than to the message itself.

An asymmetric signature protocol based on the discrete log, known as the Elgamal signature, is also described in Section 8.3. The Elgamal signature scheme, in itself, does not include the function of identification, although that function would normally be adjoined by a separate means. The Elgamal signature protocol is based on the intractability of the discrete-log problem. The Elgamal signature uses two secret keys. One key, denoted k, is unique to the message and that key is randomly chosen anew for every message. The other key, denoted i, is a permanent secret key belonging to only that signer. The signer chooses the secret individual key once and computes $I = \alpha^i$, which is the permanent authentication code of the signer. To sign the message x, the signer computes $\Delta = k^{-1}(x - iK) \pmod{p-1}$. The signature is the pair $\text{sign}_k(x) = (K, \Delta)$, and the signed message is $(x, \text{sign}_k(x)) = (x, (K, \Delta))$. The verification procedure consists of testing the equality $x = \Delta k + iK \pmod{p-1}$. If the equality holds, the signature is valid. Otherwise the signature is not valid.

Schnorr identification signature

The Schnorr identification signature is another signature protocol based on the intractability of the discrete-log problem in the group $\boldsymbol{F}_p$, where p is a sufficiently large integer. Let p be a prime for which $p - 1$ has a large prime factor, denoted r, that is large enough so that the discrete-log problem in $\boldsymbol{F}_p^*$ is not tractable. The Schnorr protocol uses the cyclic subgroup $\mu_r \subset \boldsymbol{F}_p^*$, with order r and generator α. The security of the signature is protected by the intractability of the discrete-log problem.

The Schnorr signature protocol uses a hash function, denoted hash:$\{0, 1\}^* \to \{0, \ldots, r-1\}$, on the set of binary strings, and embeds the hash function within the signature. To set up the signature protocol, the signer randomly chooses a large nonnegative integer i smaller than r and computes $\alpha^i \pmod{r}$, where α is a generator of μ_r. The integer i is the signer's permanent secret key and the integer I is the signer's permanent public key. The public key I can be validated by a trusted certification authority as an identification key belonging to that user.

To sign a message digest, the signer randomly chooses a nonnegative integer k smaller than r as the secret message key and computes $K = \alpha^k \pmod{p}$ as the public message key. The secret message key k is chosen anew and independently for each message. The signer's public message key K is concatenated with the message x that is to be signed, and the concatenation $x \| K$ is hashed to produce the message digest $h = \text{hash}(x \| K)$. It is not tractable to invert the message digest to recover x or K.

The signer then sets $s = k - hi \pmod{p}$. The signature is the pair (h, s). Thus $\text{sign}(x) = (h, s)$.

To verify the signature (h, s) on message x, the verifier multiplies α^s by I^k, noting that both α and I are publicly known, while s and k are given by the signature. The verifier then tests the equality

$$\text{hash}(x\|\alpha^s I^k) = h$$

which will hold if the signature is valid. The signature is deemed valid if the equality holds. Otherwise the signature is declared not valid. The probability of falsely claiming a signature to be valid is governed by the probability of a hash collision.

The Schnorr signature protocol is widely believed to be secure. It is clear that the security of the Schnorr protocol is limited by the intractability of the discrete-log problem. To argue further that it is not vulnerable to any other kind of direct attack, one argues that any successful attack on the Schnorr signature can be used to successfully attack the discrete-log problem, though only in a probabilistic way. Then, if one accepts the premise that the discrete-log problem is intractable, it follows that the Schnorr protocol is secure. Such arguments are considered compelling but not crisp. We do not present such an argument.

Lamport authentication signature

The Lamport signature scheme is a simple authentication protocol based on an iterated password. It is useful in some applications. The Lamport signature protocol only validates a message as originating at a specific source. It does not assert an identity for the source. The main attribute of the Lamport protocol is its extreme simplicity. The Lamport protocol does not require the verifier to transmit. It does not make use of a trusted certification authority. It may be considered a weakness of this protocol that the verifier, and any potential adversary, learns each key in the sequence when it is used, although each key is used only once. Each key is learned only after it has expired.

Let h be a secure one-way function that is used to iteratively compute a new password from the current password. The claimant randomly chooses an initial password w, such as a long binary sequence, and computes and stores $h^\ell(w)$ for $\ell = 1, \ldots, t$, where h^ℓ denotes ℓ repeated applications of h as $h^\ell(w) = h(h^{\ell-1}(w))$ and t is some convenient stopping integer larger than the number of keys to be used. The passwords are used in reverse order, starting with the last password on the list.

At the ith identification, for $i = 1, 2, \ldots$, the claimant uses $h^{t-i}(w)$ as the password. The verifier stores the most recent past password used by the claimant. To verify the ith password, the verifier checks that $h(h^{t-i}(w)) = h^{t-(i-1)}(w)$ is equal to the password that was used by the claimant on the previous round $i - 1$. It is not possible, however, to predict the new password from the old password because h is a one-way function.

A suitable choice for the one-way function h is exponentiation in a prime field. Then starting with any integer $k = k_t$, the sequence of keys in reverse order is given by $k_{i-1} = \alpha^{k_i}$, where α is a primitive element in $\boldsymbol{F}_p$. The sequence of keys is $k_1, k_2, \ldots, k_{t-1}, k_t$. To initialize the protocol, the verifier is sent key k_1. Thereafter, when given the ith message with identification key k_i appended, the verifier computes α^{k_i}, which must equal k_{i-1}. Otherwise, the verification fails.

The verifier, and any adversarial eavesdropper, can observe and record all past passwords, but cannot predict the password attached to the next message. The index i is incremented by the verifier only after the ith password is properly received and recognized. The Lamport protocol can be used only in applications in which an old password has no value after a new password is given.

Guillou–Quisquater identification signature

The Guillou–Quisquater identification signature is based on the intractability of factoring large biprimes. The Guillou–Quisquater identification signature can be regarded as a modification of the Guillou-Quisquater identification protocol in which a challenge is replaced by a hash function. To set up the signature process, the signer randomly chooses large secret primes p and q and computes the product $n = pq$, and also chooses a positive integer e smaller than n that is coprime to the totient function $\phi(n)$. Both n and e are made public. The signer also selects any integer I smaller than n and coprime to n. The signer's public identity key is I. The signer's corresponding secret key is the integer i satisfying $I\alpha^i = 1 \pmod{n}$. To compute the secret key i requires knowledge of the factors p and q. Use the extended euclidean algorithm to compute $I^{-1} \pmod{n}$, $d_1 = e^{-1} \pmod{p-1}$, and $d_2 = e^{-1} \pmod{q-1}$. Then compute $a_1 = (I^{-1})^{d_1} \pmod{p}$ and $a_2 = (I^{-1})^{d_2} \pmod{q}$. Finally, compute a satisfying $a_1 = a \pmod{p}$ and $a_2 = a \pmod{q}$.

Let $h : \{0, 1\}^* \to \mathbf{Z}_n$ for some positive integer n be a hash function that provides a digest, denoted hash(x), of the message x to be signed. To produce a signed message, the signer randomly chooses an integer k as a private message key and computes $K = k^e \pmod{n}$ as a public message key. The signer then computes the digest of the concatenate of x and K, denoted $h = \text{hash}(x \| K)$ and then computes $s = ki^h \pmod{n}$.

To verify that the signed message (x, s, h) is valid, the verifier computes that $s^e I^h = x$.

For a simple example of the Guillou–Quisquater signature protocol, let $n = pq$ with $p = 20849$, $q = 27457$, and $n = 572450993$. To generate the key, the encryptor selects an integer $e = 47$, an identifier $I = 1091522$, and computes the private key $i = 214611724$. Then

$$(n, e, I) = (572450993, 47, 1091522)$$

in the public information.

To sign the message $x = 1101110001$, the signer randomly selects an integer $k = 42134$ and computes $k^e \pmod{n}$ to obtain $r = 297543350$. The concatenate of message x and message key is hashed to compute the digest $h = \text{hash}(x \| K)$. For this example, we will suppose that the message is such that the digest h is $\text{hash}(x \| r) = 2713833$.

Then

$$(x, s, h) = (1101110001, 252000854, 2713833)$$

in the signed document.

To verify the signature, the verifier computes $s^e \pmod n = 398641962$, $I^h \pmod n = 110523867$, and $r' = s^e I^h \pmod n = 297543350$. Finally, because $u = r$, and $h' = h(x\|u) = h(x\|r) = h$, the signature is declared valid.

14.6 Protocols for secret sharing

The task of sharing access to a secret by a group of individuals can be described as the task of distributing access subkeys, or *shadow keys*, according to a stated purpose. The standard version of this protocol consists of a protected secret message, with access subkeys distributed to a group of participants of size n in such a way that no subset of fewer than t participants of the group can read the message, but any subset of t members of the group can read the message if they cooperate. Thus any set of t out of n subkeys constitutes a complete key and can be used to unlock the secret message. Fewer than t subkeys will not unlock the secret, either in whole or in part.

An elementary example of a protocol for secret sharing with $n = 2$ and $t = 2$ is as follows. The key k is a random positive integer smaller than some fixed prime p. Subkey k_1 is a randomly chosen integer smaller than p, and subkey k_2 is equal to $k - k_1 \pmod p$. The subkeys k_1 and k_2 should each be on the order of $p/2$. Then the two users must each provide their subkey in order to compute the key as $k = k_1 + k_2 \pmod p$. Even though the prime p may be known, the knowledge of either k_1 or k_2 divulges nothing about the key k.

In general, let t and n be positive integers with $t \le n$. A (t, n) secret-sharing scheme is a method to share the binary key k among a set of n participants such that at least t of them are needed to reconstruct the secret key. An unconditionally secure secret-sharing scheme is one for which any set of $t - 1$ shares reveals no information about the secret. Each share of an unconditionally secure secret-sharing scheme must be at least as large as the secret itself. This is because, even when $t - 1$ shares are known, the tth share must still provide all of the information. The known $t - 1$ shares taken together do not contain any information.

A trivial instance of an (n, n) perfect secret-sharing system consists of $n - 1$ randomly chosen elements $a_1, a_2, \ldots, a_{n-1}$ of the vector space $\boldsymbol{F}_2^r$. Given the binary message m of blocklength r, the shares are $a_1, a_2, \ldots, a_{n-1}, m + \sum_\ell a_\ell$. In this trivial system, all n subkeys are necessary to unlock the message. The (2, 2) secret-sharing system described earlier is a special case of this trivial (n, n) secret-sharing system.

A more interesting secret-sharing protocol is a method based on dividing a key (or a message) into n shadow keys (or subkeys), each of which gives no information about the key itself, yet any t parts are sufficient to recover the key. A secret-sharing scheme based on interpolation, referred to as the *Shamir secret-sharing scheme*, is based on the evaluation of polynomials, usually polynomials in a finite field. The method uses the fact that a unique polynomial of degree $t-1$ is defined by any t points in the plane F^2. For example, two points of the plane define a line, three points define a quadratic polynomial, four points define a cubic polynomial, and so forth. The same method, when used for the task of error control, is called a *Reed–Solomon code.*

A specific finite field F_q underlies the secret-sharing scheme, and is public. The secret consists of one specific element of the field, denoted a_0. The secret can take on any of q values, and so consists of $\lfloor \log_2 q \rfloor$ bits. The encryptor randomly chooses $t-1$ elements of F_q, denoted $a_1, a_2, \ldots, a_{t-1}$. These field elements are publicly known. The encryptor then defines the polynomial of degree t

$$a(x) = a_0 + \sum_{i=1}^{t} a_i x^i$$

and evaluates this polynomial at the n chosen points of the field to give

$$A_j = a(x_j) \qquad j = 1, \ldots, n.$$

Any t values of the set of A_j determine the polynomial $a(x)$ of degree t by Lagrange interpolation and so determine a_0. However any $t-1$ values of A_j determine only a set of q polynomials of degree t, one such polynomial for each value of a_0. Therefore, $t-1$ values of A_i give no information about a_0.

Polynomial interpolation is closely related to the subject of error-control codes as described in Chapter 15. In particular, these methods are related to Reed–Solomon codes. More generally, the secret-sharing scheme can reject a subversive member of the group of participants intending to deceptively prevent the recovery of the correct secret by falsifying its shadow key. To detect this false key requires that $t+2$ shadow keys be known, which allows the incorrect subkey to be detected and rejected.

Problems for Chapter 14

14.1 Suppose that a given password consists of an ordered sequence of twenty characters. Five characters of this password are given to each of four persons, so all four must collaborate to recover the password. Is this a secure secret-sharing scheme?

14.2 Develop the Schnorr signature protocol from the Schnorr identification protocol by taking inspiration from the Guillou–Quisquater signature scheme.

14.3 Use the Lamport identification protocol to design a procedure whereby the door lock on a hotel room self-updates its access key if, and only if, the room changes its occupant. (The lock has no source of new information except the sequence of keys presented to it.)

14.4 **a** Let $s \in \mathbf{Z}$ be an integer that represents a secret. (For example, its binary representation denotes a binary message.) Let r_i for $i = 1, \ldots, t-1$ be random integers smaller than s, and let $r_t = s - \sum_{i=1}^{t-1} r_i$. Does the set $\{r_1, \ldots, r_t\}$ constitute a proper t-user secret-sharing scheme?

b Let p be an arbitrary prime and s an integer smaller than p representing a secret. Can the above scheme be restated in $\mathbf{Z}_p$?

14.5 Develop an identity-based version of the Feige–Fiat–Shamir identification protocol by taking inspiration from the Guillou–Quisquater identity-based identification protocol.

14.6 Show that if computing square roots modulo a large biprime is not tractable, then computing eth roots modulo that biprime is not tractable either if e is even. What can be said for a general e?

14.7 A fraudulent claimant using the Fiat–Shamir protocol either claims that $x = r^2$ or that $x = r^2/v$ with probabilities q_0 and $q_1 = 1 - q_0$, respectively. The verifier makes challenges either $e = 0$ or $e = 1$ with probability p_0 and $p_1 = 1 - p_0$, respectively. According to the game-theoretic min–max principle, how should the probabilities q_0 and p_0 be chosen? Should the imposter and the claimant choose their options independently from repetition to repetition or should there be dependence? Why?

Notes for Chapter 14

Some well-known identification protocols are those introduced by Fiat and Shamir (1987), Feige, Fiat, and Shamir (1988), Guillou and Quisquater (1988), Schnorr (1991a and b), and Okamoto (1992). Most of these identification protocols are zero-knowledge protocols, and challenge–response sequences. The formal study of zero-knowledge methods has evolved out of interactive proof systems, that were first described formally by Goldwasser, Micali, and Rackoff (1989) and by Babai and Moran (1988). The general technique of constructing a signature scheme from an identification protocol was published by Fiat and Shamir (1987).

The topic of secret sharing was introduced independently by several people simultaneously, including Blakley (1979) and Shamir (1979). The connection to Reed–Solomon codes was recognized by McEliece and Sarwate (1981). The proof that every share of a perfect secret-sharing scheme must be at least as large as the secret itself is due to Karnin, Greene, and Hellman (1983).

15 More public-key cryptography

Underlying each popular public-key cryptosystem is a difficult mathematical problem for which no tractable computational algorithm is known. For many of these problems, however, no proof is known that a tractable computational algorithm does not exist. Accordingly, it is wise to develop alternative cryptographic procedures based on other apparently intractable problems. Some computational problems that have to do with lattices or codes provide alternative apparently intractable problems. Lattices were first introduced into the theory of cryptography, especially public-key cryptography, for the contrary purpose of demonstrating or exploiting weaknesses in some cryptosystems. This, in turn, soon was turned to the study of lattices for the positive purpose of developing alternative forms of public-key cryptography.

A lattice can be used to formulate a kind of public-key cryptosystem that is not based on biprime factorization and is not based on a discrete-log problem. The primary problem of interest for these purposes is the so-called shortest-vector problem in lattices of very large dimension. In contrast to the biprime factoring and the discrete-log problems, the shortened-vector problem in a lattice is formally intractable in the sense that definite statements within the formalism of complexity theory can be made about the general case of the shortest vector problem.

An n-dimensional lattice is a periodic arrangement of points in an n-dimensional space, typically euclidean space $\boldsymbol{R}^n$. The most familiar example of a lattice is $\boldsymbol{Z}^n$, the n-dimensional integer lattice, which can be embedded in $\boldsymbol{R}^n$. Nowadays, lattice-based methods of cryptography are well studied though their security is still debated. Lattice-based public-key cryptosystems, at this time, are not nearly as popular as cryptosystems based on integer factorization or on the discrete-logarithm problem. This is perhaps due in part to the complexity and computational overhead of these methods, but also because the supporting historical and anecdotal evidence is not as great as for other methods. However, if the integer-factorization problem and the discrete-log problem are both solved at some future time, or if the cryptographic systems based on these problems are broken in some other way, then lattice-based cryptosystems may become very important. At the present time, lattice-based cryptography remains in the shadow of these other popular methods.

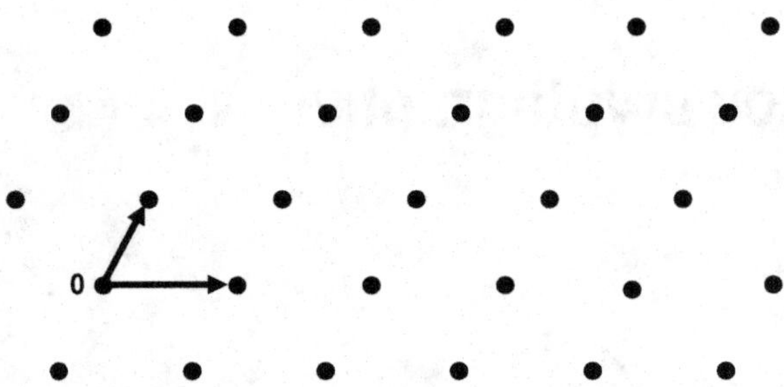

Figure 15.1 A lattice in $\boldsymbol{R}^2$

15.1 Introduction to lattices

Let $\boldsymbol{B}$ be a real n-by-n square matrix over the real field $\boldsymbol{R}$. The set of all real-valued vectors $\boldsymbol{v} \in \boldsymbol{R}^n$ that can be written as $\boldsymbol{v} = \boldsymbol{B}\boldsymbol{a}$, where $\boldsymbol{a}$ is a vector of integers of length n, is called a *lattice* and is denoted $\boldsymbol{\Lambda}$. Thus

$$\boldsymbol{\Lambda} = \left\{ \boldsymbol{v} = \sum_{i=1}^{n} a_i \boldsymbol{b}_i \mid a_i \in \mathbf{Z} \right\},$$

where the vectors $\boldsymbol{b}_i$, with components b_{ij}, are the columns of the matrix $\boldsymbol{B}$. The lattice $\boldsymbol{\Lambda}$ can be expressed more concisely as

$$\boldsymbol{\Lambda} = \{\boldsymbol{v} = \boldsymbol{B}\boldsymbol{a} \mid \boldsymbol{a} \in \boldsymbol{Z}^n\}.$$

The lattice $\boldsymbol{\Lambda}$ has full rank if the matrix $\boldsymbol{B}$ has full rank.

Every nonsingular real square matrix $\boldsymbol{B}$ corresponds to a full-rank lattice $\boldsymbol{\Lambda}$ in $\boldsymbol{R}^n$, and that square matrix $\boldsymbol{B}$ is called a *generator matrix* of the lattice $\boldsymbol{\Lambda}$. Every nonsingular real square matrix $\boldsymbol{B}$ is a generator matrix for some lattice. Some such $\boldsymbol{B}$ may have additional structure. An n-by-n square matrix for which the n columns are the distinct cyclic shifts of each other, in order, is known as a *circulant matrix.* Requiring that $\boldsymbol{B}$ be a circulant matrix imposes considerable structure on the resulting lattice.

A typical lattice in $\boldsymbol{R}^2$ is shown in Figure 15.1. The lattice has an *origin*, denoted $\mathbf{0}$, as the point $\boldsymbol{v}$ at which all components of $\boldsymbol{a}$, or of $\boldsymbol{v}$, are equal to zero.

If the matrix $\boldsymbol{B}$ is nonsingular, then the columns of $\boldsymbol{B}$ are linearly independent. In this case, the set of columns of $\boldsymbol{B}$ is called a *basis* of the lattice.

For example, let

$$\boldsymbol{b}_1 = \begin{bmatrix} 1 \\ 2 \end{bmatrix}, \qquad \boldsymbol{b}_2 = \begin{bmatrix} 1 \\ -1 \end{bmatrix}$$

be two basis vectors. Then the generator matrix is

$$\boldsymbol{B} = \begin{bmatrix} 1 & 1 \\ 2 & -1 \end{bmatrix}$$

and the lattice consists of all $\boldsymbol{Ba}$, where $\boldsymbol{a}$ is an integer vector of dimension two.

A basis of a lattice is not unique. Indeed, every lattice whose dimension is at least two has an infinite number of bases. For example, $\{(\boldsymbol{b}_1, \boldsymbol{b}_2) = (1, 0), (0, 1)\}$ is perhaps the most reasonable basis for $\boldsymbol{Z}^2$, while $\{(\boldsymbol{b}_1, \boldsymbol{b}_2) = (1, 0), (97, 1)\}$ is also a basis, but evidently an unreasonable one.

A basis is a set of linearly independent points of the lattice, regarded as vectors. This means that a new generator matrix can be expressed in terms of an old basis by the matrix equation $\boldsymbol{B}' = \boldsymbol{BA}$ where $\boldsymbol{A}$ is an invertible matrix of integers, and called a *unimodular matrix*. This is because $\{\boldsymbol{Ba} \mid \boldsymbol{a} \in \boldsymbol{Z}^n\} = \{\boldsymbol{BAa} \mid \boldsymbol{a} \in \boldsymbol{Z}^n\}$ if $\boldsymbol{A}$ is an invertible matrix of integers. The following theorem further characterizes such $\boldsymbol{A}$.

Theorem 15.1.1 *Two generator matrices* $\boldsymbol{B}$ *and* $\boldsymbol{B}'$ *generate the same lattice* $\boldsymbol{\Lambda}$ *if, and only if, they are related by*

$$\boldsymbol{B}' = \boldsymbol{BA},$$

where $\boldsymbol{A}$ *is a matrix with integer elements that satisfies* $\det \boldsymbol{A} = \pm 1$.

Proof Because $\boldsymbol{B} = \boldsymbol{B}'\boldsymbol{A}^{-1}$, we can write $1 = \det(\boldsymbol{I}) = \det(\boldsymbol{AA}^{-1}) = (\det \boldsymbol{A})$ $(\det \boldsymbol{A}^{-1})$. But $\det \boldsymbol{A}$ and $\det \boldsymbol{A}^{-1}$ are both integers, so they are both $+1$ or -1. □

The theorem implies that two generator matrices for the lattice $\boldsymbol{\Lambda}$ have the same determinant but for sign. Accordingly, the absolute value of the determinant of any generator matrix of the lattice $\boldsymbol{\Lambda}$ is called the *determinant* of that lattice.

If $\boldsymbol{B}$ is the n-dimensional identity matrix, then the lattice $\boldsymbol{\Lambda}$ is the integer lattice $\boldsymbol{Z}^n$. Thus every n-dimensional lattice is just a copy of $\boldsymbol{Z}^n$ that is skewed by the matrix $\boldsymbol{B}$. Any subset of $\boldsymbol{\Lambda}$ that is itself a lattice is called a *sublattice* of $\boldsymbol{\Lambda}$.

Because any point $\boldsymbol{x}$ of a lattice is an element of $\boldsymbol{R}^n$, the *euclidean norm* $\|\cdot\|$ is defined in the usual way on the points of a lattice as

$$\|\boldsymbol{x}\| = \sqrt{x_1^2 + x_2^2 + \cdots + x_n^2}.$$

Then the *euclidean distance* between any two points $\boldsymbol{x}$ and $\boldsymbol{y}$ of the lattice $\boldsymbol{\Lambda}$ is $d(\boldsymbol{x}, \boldsymbol{y}) = \|\boldsymbol{x} - \boldsymbol{y}\|$. The *minimum distance* of the lattice $\boldsymbol{\Lambda}$ is defined as $d_{\min} = \min_{\boldsymbol{x},\boldsymbol{y} \in \boldsymbol{\Lambda}} d(\boldsymbol{x}, \boldsymbol{y})$. The *inner product* of two elements, $\boldsymbol{x}$ and $\boldsymbol{y}$, of the lattice $\boldsymbol{\Lambda}$ is defined in the usual way as

$$\langle \boldsymbol{x}, \boldsymbol{y} \rangle = x_1 y_1 + x_2 y_2 + \cdots + x_n y_n.$$

If $\langle \boldsymbol{x}, \boldsymbol{y} \rangle = 0$, then lattice points $\boldsymbol{x}$ and $\boldsymbol{y}$ are said to be *orthogonal*.

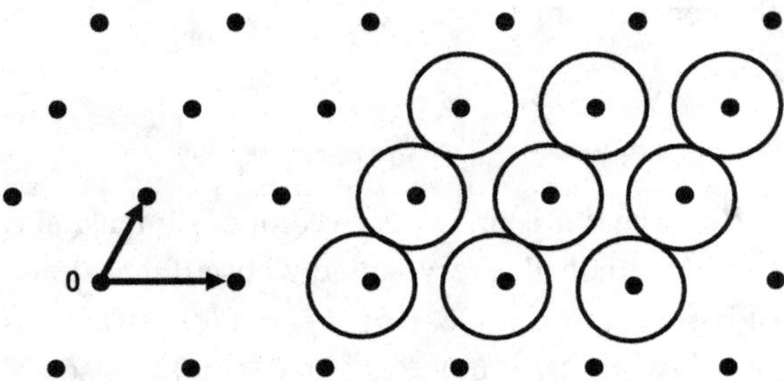

Figure 15.2 Sphere packing in $\boldsymbol{R}^3$

If the generator matrix $\boldsymbol{B}$ is required to be nonsingular, then every lattice of dimension n, as a set of vectors, spans $\boldsymbol{R}^n$. A lattice induces a certain kind of *tessellation* of $\boldsymbol{R}^n$, and each tessellate is called a *cell* of the lattice. The set of points of $\boldsymbol{R}^n$ bounded by the set of lattice points of the form $\boldsymbol{Bi}$ for which every component of $\boldsymbol{i}$ is either a zero or a one is called the *standard cell* of the lattice, and is denoted $\mathcal{F}$.

The standard cell is a well-behaved subset of $\boldsymbol{R}^n$, and so it has a volume in the usual sense. The volume of a standard cell is denoted vol($\mathcal{F}$) or, when more convenient, vol($\boldsymbol{\Lambda}$). Every cell of the lattice has the same volume vol($\boldsymbol{\Lambda}$) as the standard cell, which is equal to det$\boldsymbol{B}$. The *Hadamard inequality* on the volume of a lattice cell is

$$\text{vol}(\boldsymbol{\Lambda}) \leq \|\boldsymbol{b}_1\| \, \|\boldsymbol{b}_2\| \cdots \|\boldsymbol{b}_n\|,$$

where $\{\boldsymbol{b}_1, \boldsymbol{b}_2, \ldots, \boldsymbol{b}_n\}$ is a basis consisting of all nonzero vertices of the standard cell. Hadamard's inequality follows from the fact that the volume of a parallelepiped is never greater than the product of the lengths of its sides.

Two basis vectors are orthogonal if $\boldsymbol{b}_i \cdot \boldsymbol{b}_j = 0$. A basis is an *orthogonal basis* if all pairs of basis vectors are orthogonal. In general, because of its discrete structure, a given lattice does not have an orthogonal basis. The *orthogonality defect* of a basis is defined as $\prod_{i=1}^{n} \|\boldsymbol{b}_i\|/\text{det}\boldsymbol{B}$. The orthogonality defect measures how far the basis vectors are from being orthogonal. It compares the product of the length of the basis vectors with the volume of the parallelepiped that they define. The orthogonality defect is always larger than or equal to one, and is equal to one only if the basis vectors are orthogonal.

Every lattice is associated with both a *sphere packing*, as shown in Figure 15.2, and with a *bilinear form*, given by

$$b(x_1, \ldots, x_n) = \sum_{i=1}^{n} \sum_{j=1}^{n} b_{ij} x_i x_j.$$

The sphere packing illustrated in Figure 15.2 clearly is not an efficient sphere packing of $\boldsymbol{R}^2$. Apparently, in two dimensions, the hexagonal lattice should lead to the most efficient sphere packing. This is indeed the case.

A circle of radius $2/\sqrt{\pi}$ centered at the origin of $\boldsymbol{R}^2$ will enclose at least one lattice point other than the origin for any lattice in $\boldsymbol{R}^2$ whose cells have area one. This statement holds as well for any horizontal or vertical ellipse centered at the origin, provided the ellipse has area equal to at least four. Each case is an instance of the following general theorem.

Let $\mathcal{R}$ be a convex region of $\boldsymbol{R}^n$ that is symmetric with respect to the origin, meaning that if any real vector $\boldsymbol{x}$ is in $\mathcal{R}$, then $-\boldsymbol{x}$ is in $\mathcal{R}$ where $-\boldsymbol{x}$ replaces each component of the vector $\boldsymbol{x}$ by its negative. A circle is an example of a convex region in $\boldsymbol{R}^2$. A hypersphere is an example of a convex region in $\boldsymbol{R}^n$.

Theorem 15.1.2 (Minkowski convex body theorem)[1] *Let $\boldsymbol{\Lambda}$ be a lattice of dimension n. Then every compact, convex, symmetric region $\mathcal{R}$ of $\boldsymbol{R}^n$ with volume at least $2^n \text{vol}(\boldsymbol{\Lambda})$ contains at least one nonzero lattice point.*

Proof We treat the case of inequality in Step 1. The case of equality will be established by a subsequent limiting argument in Step 2.

Step 1 (Inequality) Let $\mathcal{R} \subset \boldsymbol{R}^n$ be a compact, convex, symmetric region of $\boldsymbol{R}^n$ that satisfies

$$\text{vol}(\mathcal{R}) > 2^n \text{vol}(\mathcal{F}),$$

where $\mathcal{F}$ is the standard cell of $\boldsymbol{\Lambda}$. Define the set $\mathcal{R}'$ by shrinking $\mathcal{R}$ by a factor of two. That is

$$\mathcal{R}' = \{\boldsymbol{x}' \in \boldsymbol{R}^n | 2\boldsymbol{x}' \in \mathcal{R}\},$$

where $2\boldsymbol{x}'$ is the vector obtained from $\boldsymbol{x}'$ by multiplying each component of $\boldsymbol{x}'$ by two. Because shrinking $\mathcal{R}$ by a factor of two reduces its volume by a factor of 2^n, we know by assumption that

$$\text{vol}(\mathcal{R}') > \text{vol}(\mathcal{F}).$$

Let $\phi(\boldsymbol{x})$ be the indicator function of $\mathcal{R}'$. That is

$$\phi(\boldsymbol{x}) = \begin{cases} 1 & \text{if} \quad \boldsymbol{x} \in \mathcal{R}' \\ 0 & \text{if} \quad \boldsymbol{x} \notin \mathcal{R}'. \end{cases}$$

[1] The important Minkowski theorem can be regarded as a consequence of the Blichfeldt theorem, which states that any closed set $S \in \boldsymbol{R}^n$, whose volume is larger than the determinant of the full-rank lattice $\boldsymbol{\Lambda} \in \boldsymbol{R}^n$, contains two distinct points, s_1 and s_2, such that $s_1 - s_2 \in \boldsymbol{\Lambda}$.

Next translate $\phi(\boldsymbol{x})$ to each lattice point. We will show that these translates must overlap. To this purpose, add the translates and form the function

$$\Phi(\boldsymbol{x}) = \sum_{\lambda \in \boldsymbol{\Lambda}} \phi(\boldsymbol{x} + \lambda).$$

We will show that for some $\boldsymbol{x}$, $\Phi(\boldsymbol{x}) \geq 2$.

The space $\boldsymbol{R}^n$ is the union of all cells of the lattice. That is, the equality $\boldsymbol{R}^n = \cup_{\boldsymbol{v} \in \Lambda}(\mathcal{F} + \boldsymbol{v})$ holds. Therefore

$$\begin{aligned}\int_{x \in \mathcal{F}} \Phi(\boldsymbol{x}) d\boldsymbol{x} &= \int_{\boldsymbol{x} \in \mathcal{F}} \sum_{\lambda \in \boldsymbol{\Lambda}} \phi(\boldsymbol{x} + \lambda) d\boldsymbol{x} = \int_{\boldsymbol{x} \in \boldsymbol{R}^n} \phi(\boldsymbol{x}) d\boldsymbol{x} \\ &= \mathrm{vol}(\mathcal{R}') > \mathrm{vol}(\mathcal{F}).\end{aligned}$$

This says that the integral of $\Phi(\boldsymbol{x})$ over the hypervolume of region $\mathcal{F}$ is strictly larger than $\mathcal{F}$. Therefore $\Phi(\boldsymbol{x})$ is larger than one at some $\boldsymbol{x} \in \mathcal{F}$ and so, because it has only integer values, it is equal to two or greater at some $\boldsymbol{x}$. Therefore some $\phi(\boldsymbol{x} + \lambda)$ overlap.

We conclude that the compact set $\mathcal{R}'$ is not contained in the standard cell $\mathcal{F}$, so some equivalence classes of the quotient group $\mathcal{R}'/\Lambda$ contain more than one $\boldsymbol{x}$. That is, in at least one equivalence class there are two points $\boldsymbol{x}_1'$ and $\boldsymbol{x}_2'$ of $\mathcal{R}'$, corresponding to the points $\boldsymbol{x}_1$ and $\boldsymbol{x}_2$ of $\mathcal{R}$, that can be written

$$\begin{aligned}\boldsymbol{x}_1' &= \boldsymbol{f} + \boldsymbol{v}_1 \\ \boldsymbol{x}_2' &= \boldsymbol{f} + \boldsymbol{v}_2,\end{aligned}$$

where $\boldsymbol{f}$ is a point of $\mathcal{F}$, and $\boldsymbol{v}_1$ and $\boldsymbol{v}_2$ are distinct elements of $\boldsymbol{\Lambda}$. Therefore $\boldsymbol{x}_1 - \boldsymbol{x}_2 = \boldsymbol{v}_1 - \boldsymbol{v}_2$, where $\boldsymbol{v}_1$ and $\boldsymbol{v}_2$ are different elements of $\boldsymbol{\Lambda}$, and so the point $\boldsymbol{v} = \boldsymbol{v}_1 - \boldsymbol{v}_2$ is a nonzero element of the lattice $\boldsymbol{\Lambda}$. To complete the proof, we must show that the lattice element $\boldsymbol{v}$ is an element of the specified region $\mathcal{R}$.

To this end, observe that $\mathcal{R}'$ is symmetric, so because $\boldsymbol{x}_2'$ is in $\mathcal{R}'$, $\boldsymbol{x}_3' = -\boldsymbol{x}_2'$ is also in $\mathcal{R}'$. This means that both $\boldsymbol{x}_1 = 2\boldsymbol{x}_1'$ and $\boldsymbol{x}_3 = 2\boldsymbol{x}_3'$ are in $\mathcal{R}$. Because $\mathcal{R}$ is convex, $\frac{1}{2}(\boldsymbol{x}_1 + \boldsymbol{x}_3) = \boldsymbol{x}_1' + \boldsymbol{x}_3'$ is in $\mathcal{R}$. This means that $\boldsymbol{x}_1' - \boldsymbol{x}_2'$ is in $\mathcal{R}$. Therefore $\boldsymbol{v} = \boldsymbol{v}_1 - \boldsymbol{v}_2$ is the required nonzero lattice point that is contained in $\mathcal{R}$.

Step 2 (Equality) Let $\mathcal{R} \subset \boldsymbol{R}^2$ be a compact, convex, symmetric region of $\boldsymbol{R}^2$ that satisfies

$$\mathrm{vol}(\mathcal{R}) = 2^n \mathrm{vol}(\mathcal{F}).$$

Let $\mathcal{R}_k$ denote the set of points obtained from $\mathcal{R}$ by multiplying each point of $\mathcal{R}$ componentwise by $1 + \frac{1}{k}$. That is,

$$\mathcal{R}_k = \{(1 + \tfrac{1}{k})\boldsymbol{x} \in \boldsymbol{R}^n | \boldsymbol{x} \in \mathcal{R}\}.$$

Then for any k,

$$\mathcal{R}' \subset \mathcal{R} \subset \mathcal{R}_k \subset \mathcal{R}_{k-1} \subset \cdots \subset \mathcal{R}_1.$$

By Step 1, every $\mathcal{R}_k$ contains at least one nonzero lattice point. Because $\mathcal{R}_1$ is bounded, it contains only a finite number of lattice points. No other lattice points can appear in any $\mathcal{R}_k$ for larger k. Moreover, $\mathcal{R}_k \subseteq \mathcal{R}_1$ for all k. Hence, there is at least one nonzero lattice point that is common to all $\mathcal{R}_k$. Thus because $\mathcal{R}$ is compact,

$$\bigcap_{k>1}^{\infty} \mathcal{R}_k = \mathcal{R}.$$

This means that $\mathcal{R}$ contains at least one nonzero lattice point, which proves the theorem. □

For an example of the Minkowski theorem, note that the volume of the lattice $\mathbf{\Lambda} = \mathbf{Z}^2$ is given by $\text{vol}(\mathbf{Z}^2) = 1$. The theorem says that every compact, convex, symmetric region of $\mathbf{R}^2$ of area at least four must include at least one point of $\mathbf{Z}^2$ other than the origin. Let $\mathcal{R}$ be a convex set bounded by the ellipse $ax^2 + bxy + cy^2 = 1$ where a, b, and c are such that this set has an area equal to at least four. Because the volume of the $\mathbf{Z}^2$ lattice is one, the Minkowski theorem asserts that $\mathcal{R}$ contains at least one point of $\mathbf{Z}^2$ other than the origin.

Now recall that the Hadamard inequality states that any basis $\{\boldsymbol{b}_n\}$ satisfies

$$\text{vol}(\mathbf{\Lambda}) \leq \|\boldsymbol{b}_1\| \cdot \|\boldsymbol{b}_2\| \cdots \|\boldsymbol{b}_n\|.$$

The extent of slackness in this inequality describes the extent to which the basis is nonorthogonal. Another statement is the following.

Theorem 15.1.3 (Hermite) *Any lattice* $\mathbf{\Lambda}$ *of dimension* n *has a nonzero element* $\boldsymbol{v}$ *that satisfies*

$$\|\boldsymbol{v}\| \leq \sqrt{\frac{2n}{\pi e}} [\text{vol}(\mathbf{\Lambda})]^{1/n}.$$

Proof The proof is based on the Minkowski inequality given in Theorem 15.1.2. Let $\mathcal{B}_r \subset \mathbf{R}^n$ be a hypersphere of radius r in n-dimensional euclidean space and let $k = \lfloor n/2 \rfloor$. The volume of an n-dimensional hypersphere of radius r is given by

$$\text{vol}_{2k}(\mathcal{B}_r) = \frac{\pi^k}{k!} r^{2k}$$

$$\text{vol}_{2k+1}(\mathcal{B}_r) = \frac{k!}{(2k+1)!} 2^{2k+1} \pi^k r^{2k+1}$$

according to whether the dimension is even ($n = 2k$) or odd ($n = 2k + 1$). In either case, if n is large, then the hypersphere $\mathcal{B}_r$ in n-dimensional euclidean space has approximate volume

$$\mathrm{vol}(\mathcal{B}_r) \approx \left(\frac{2\pi e}{n}\right)^{n/2} r^n$$

as can be shown using Stirling's approximation.[2] Setting

$$r = \sqrt{\frac{2n}{\pi e}}\,\mathrm{vol}(\mathbf{\Lambda})^{1/n}$$

gives

$$\mathrm{vol}(\mathcal{B}_r) \gtrsim 2^n \mathrm{vol}(\mathbf{\Lambda}).$$

Theorem 15.1.2 says that the ball $\mathcal{B}_r$ of radius r contains a lattice point. Therefore there is a nonzero element of $\mathbf{\Lambda}$, such that $\|\boldsymbol{v}\| \leq \sqrt{\frac{2n}{\pi e}}\mathrm{vol}(\mathbf{\Lambda})^{1/n}$, as was to be proved. □

Theorem 15.1.3 gives an upper bound on the shortest vector $\boldsymbol{v}$ of every n-dimensional lattice, but this bound is not the tightest possible such bound. Accordingly, for each n, define γ_n as the smallest number such that the inequality

$$\|\boldsymbol{v}\| \leq \gamma_n[\mathrm{vol}(\mathbf{\Lambda})]^{1/n}$$

holds for the shortest vector $\boldsymbol{v}$ of *every* lattice $\mathbf{\Lambda}$ of dimension n. This γ_n is called a *Hermite constant.* The Hermite constant can be restated as

$$\gamma_n = \max_{\mathbf{\Lambda}:\mathrm{vol}\mathbf{\Lambda}=1} \min_{\substack{\boldsymbol{v}\in\mathbf{\Lambda} \\ \boldsymbol{v}\neq 0}} \|\boldsymbol{v}\|.$$

Accordingly, if each n-dimensional lattice is scaled so that it has volume equal to one, then γ_n is the squared length of the largest possible shortest vector to a nonzero lattice point.

It follows from the definition of the Hermite constant γ_n that for any lattice $\mathbf{\Lambda}$ of dimension n, there is a basis $\{\boldsymbol{b}_1, \boldsymbol{b}_1, \ldots, \boldsymbol{b}_n\}$, such that

$$\|\boldsymbol{b}_1\| \cdot \|\boldsymbol{b}_2\| \cdots \|\boldsymbol{b}_n\| \leq \gamma_n^{n/2}\mathrm{vol}(\mathbf{\Lambda}).$$

The Hermite constants are not known in general, but are known to satisfy

$$n/2\pi e \leq \gamma_n \leq (1.744\ldots)n/2\pi e.$$

For example, the constant $\gamma_n = (4/3)^{(n-1)/4}$ satisfies the Hermite inequality for every n, but it is not necessarily the smallest such constant. For $n = 2$, the Hermite constant is known and is $\gamma_2 = 2/\sqrt{3}$. This means that the distance between any two vertices of a (planar) polygon with four sides cannot be larger than $2/\sqrt{3}$ times the square root of

[2] Sterling's approximation to $n!$ states that

$$\sqrt{2n\pi}\left(\frac{n}{e}\right)^n < n! < \sqrt{2n\pi}\left(\frac{n}{e}\right)^n\left(1 + \frac{1}{12n-1}\right).$$

the area of that polygon. The hexagonal lattice has the largest possible shortest vector of any two-dimensional lattice, and so satisfies this bound with equality.

The known values of γ_n, up to two decimal places, are $\gamma_2 = 1.16$, $\gamma_3 = 1.26$, $\gamma_4 = 1.41$, $\gamma_5 = 1.52$, $\gamma_6 = 1.67$, $\gamma_7 = 1.81$, $\gamma_8 = 2$, and $\gamma_{24} = 4$. These can be written more precisely as $\gamma_2^2 = 4/3$, $\gamma_3^3 = 2$, $\gamma_4^4 = 4$, $\gamma_5^5 = 8$, $\gamma_6^6 = 64/3$, $\gamma_7^7 = 64$, $\gamma_8^8 = 256$, $\gamma_{24}^{24} = 4^{24}$. The value of γ_n is unknown for other values of n.

The Hermite inequality is a special case of the Mordell inequality, which is the statement

$$\gamma_d^{k-1} \leq \gamma_k^{d-1} \qquad 2 \leq k \leq d,$$

where γ_n is the Hermite constant. This inequality is satisfied with equality for $(k, d) = (3, 4)$. We will not prove the Mordell inequality.

15.2 Elementary problems in lattice theory

The fundamental problems in lattice theory may superficially appear to be simple, but they are actually hard problems when n is large. Indeed, these problems are formally intractable, known to be in the class of problems designated as NP-hard problems. The fundamental intractable problems in lattice theory are as follows:

Shortest-vector problem: Find a shortest nonzero vector in a given lattice $\mathbf{\Lambda}$. That is, find a nonzero lattice point that is closest to the origin in euclidean distance.

The shortest vector need not be unique. For example, in $\mathbf{Z}^2$, there are four shortest vectors, namely $(0, 1)$, $(0, -1)$, $(1, 0)$, and $(-1, 0)$.

Closest-vector problem: Given any arbitrary $\boldsymbol{w} \in \boldsymbol{R}^n$, find an element $\boldsymbol{v}$ of the given lattice $\mathbf{\Lambda}$ closest to $\boldsymbol{w}$ in euclidean distance.

The closest vector need not be unique. There may be several elements of $\mathbf{\Lambda}$ at the same distance from $\boldsymbol{w}$. For example, the point $\boldsymbol{w} = (0, 5, 0.5)$ has four nearest points in $\mathbf{Z}^2$.

Other problems in lattice theory are the following.

Approximate shortest-vector problem: Find a nonzero vector in the lattice whose norm is not larger than γ times the norm of the shortest vector, where γ is any given number larger than one.

Covering-radius problem: Find r such that $\boldsymbol{R}^n$ is covered by the union of all the hyperspheres of radius r centered at all lattice points.

The shortest-vector problem and the closest-vector problem are the primary computational problems associated with lattices. These problems become computationally intractable for an arbitrary lattice as the dimension of the lattice increases, and may be beyond current computational resources even for n as small as 100. One reason that the shortest-vector problem is so difficult is that there are so many bases for a given lattice in $\boldsymbol{R}^n$. Of course, the shortest vector problem is not computationally intractable for every lattice of large n. For example, it is trivial for $\boldsymbol{Z}^n$ for every n.

The shortest-vector problem and the closest-vector problem are different problems, but are closely related. A variation of the closest-vector problem is the *approximate closest-vector problem*, which may be acceptable in some applications in which the closest vector cannot be found, as is the usual situation.

Yet another problem, but one that we will not study, is the shortest-basis problem.

Shortest-basis problem: Find the shortest set of linearly independent vectors $\{\boldsymbol{b}_i\}$ that span the given lattice.

The shortest-basis problem has many versions, depending on the definition of the shortest set of vectors. There are several reasonable ways to define a shortest basis. The shortest basis can refer to the basis that minimizes the maximum length of any basis vector, or the basis that minimizes the product of the lengths of the basis vectors thereby minimizing the orthogonal defect. Another definition of a shortest basis, but not the only other definition, is described by the following procedure of successive minima. Let $\boldsymbol{b}_1$ be a shortest nonzero vector in $\boldsymbol{\Lambda}$. Then at step i, let $\boldsymbol{b}_i$ be a shortest vector in $\boldsymbol{\Lambda}$ that is linearly independent of $\boldsymbol{b}_1, \ldots, \boldsymbol{b}_{i-1}$. These are the nonzero lattice points contained in the smallest ball about the origin that contains i nonzero lattice points (with appropriate regard for ties). This is an appealing notion of a shortest basis, but we have little hope of discovering a tractable algorithm to find such a basis. For an arbitrary lattice of large dimension, we cannot even find $\boldsymbol{b}_1$. At best, we can hope to find a coarse approximation to $\boldsymbol{b}_1$.

Each of these elementary problems in lattice theory, in general, does not have a tractable computational algorithm to find a solution. Instead, these exact problems are replaced with approximate lattice problems. These surrogate problems are defined by replacing the requirement for an exact solution with a requirement for an approximate solution, even a very crude approximate solution.

15.3 Reduction of a lattice basis

The study of practical methods for the solution of the shortest-vector problem and the closest-vector problem, either exactly or approximately, comprises the topic of lattice reduction. This topic includes the study of practical methods of computing a shortest basis or even a basis that is reasonably short and approximately orthogonal. It is apparently hopeless, in general, to find an exact solution to a lattice-reduction problem for large n because such problems are known to be formally intractable. Even to find an approximate solution is formally intractable. At best, one can hope to find solutions under sufficiently weak notions of approximation, or to find algorithms that work only in special cases[3] or for small n.

The *Lenstra–Lenstra–Lovász algorithm* (or the LLL algorithm) is a widely used basis-reduction algorithm. Starting from any basis for a lattice, the LLL algorithm computes a basis that consists of moderately short lattice vectors, though not the shortest possible. The new basis is called a *reduced basis*, meaning not much more than the basis that the algorithm computes. More specifically, the LLL algorithm approximately solves the shortest-vector problem, finding a vector with length not larger than $(2/\sqrt{3})^n$ times the length of the shortest vector. Notice that this tolerance is rather large when n is on the order of 100, so the algorithm does not promise an answer that is tight. Thus, for example, $(2/\sqrt{3})^{100} \approx 10^6$, which is only a weak guarantee. Normally, however, the algorithm will far outperform this weak guarantee.

In an *orthogonal basis*, the basis vectors are pairwise orthogonal. For any subspace of real euclidean space $\boldsymbol{R}^n$, we can always require that a basis be an orthogonal basis. This is because any linearly independent set of vectors $\boldsymbol{v}_1, \ldots, \boldsymbol{v}_n \in \boldsymbol{R}^n$ can be turned into another such set in which elements are pairwise orthogonal and span the same subspace. An efficient process, called the *Gram–Schmidt orthogonalization algorithm*, can compute such a basis. In contrast, an orthogonal basis need not exist for a lattice, so this notion of orthogonalization does not provide the solution to the shortest-vector problem or to the closest-vector problem. This is because a lattice is a discrete set of points as shown in Figure 15.1, and the basis vectors must be elements of the lattice. Nevertheless, the notion of orthogonalization does provide a useful starting point for this discussion.

The reason that the Gram–Schmidt orthogonalization algorithm is important to this section is that the notions that are described in that algorithm provide the inspiration and guidance that underlies the development of the LLL algorithm. The LLL algorithm attempts to emulate the Gram–Schmidt algorithm within the constraint that the lattice is a discrete set of points. This attempt does not find the best basis, however, because the greedy structure of the Gram–Schmidt algorithm fails to find the optimal solution when the space is discrete. A greedy decision during one iteration of the algorithm will preempt possible decisions during later iterations of the algorithm, even though these preempted decisions might be the better decisions.

The task of lattice reduction differs from the task of basis orthogonalization in that the continuous nature of euclidean space has been replaced by the discrete nature of a lattice. At each iteration, the algorithm will select a lattice point as a basis vector. This lattice point need not be the best choice, but it is the best that the algorithm can find at that time, and the decision, once made, is not revisited. For this reason,

[3] As in the case of the Merkle–Hellman trapdoor knapsack.

the LLL algorithm is unable to solve the shortest-vector problem. Instead, it finds an approximate solution. Specifically, it finds moderately short lattice vectors. Finding a basis that is optimal in some sense is believed to be exponentially hard in the dimension of the lattice.

The Gram–Schmidt algorithm in $\boldsymbol{R}^n$ uses the notion of a projection of a vector $\boldsymbol{v} \in \boldsymbol{R}^n$ onto a vector $\boldsymbol{u}$ defined as $\mu \boldsymbol{u}$ where the projection coefficient is

$$\mu = \frac{\boldsymbol{v} \cdot \boldsymbol{u}}{\|\boldsymbol{u}\|^2},$$

and where $\boldsymbol{v} \cdot \boldsymbol{u}$ is the inner product between vector $\boldsymbol{u}$ and vector $\boldsymbol{v}$ and $\|\boldsymbol{u}\|^2 = \boldsymbol{u} \cdot \boldsymbol{u}$ is the norm squared of $\boldsymbol{u}$. The algorithm deals with two sets of basis vectors. The given nonorthogonal basis is $\{\boldsymbol{v}_1, \ldots, \boldsymbol{v}_n\}$ and the orthogonal basis to be computed is $\{\boldsymbol{u}_1, \ldots, \boldsymbol{u}_n\}$. The projection coefficients are expressed as $\mu_{ij} = \frac{\boldsymbol{v}_i \cdot \boldsymbol{u}_j}{\|\boldsymbol{u}_i\|^2}$. The recursion of the Gram–Schmidt algorithm then uses $\mu_{ij} \boldsymbol{u}_j$, which is the projection of $\boldsymbol{v}_i$ onto $\boldsymbol{u}_j$. The core operation

$$\boldsymbol{u}_i = \boldsymbol{v}_i - \mu_{ij} \boldsymbol{u}_j$$

removes from $\boldsymbol{v}_i$ the projection of $\boldsymbol{v}_j$ onto $\boldsymbol{u}_i$.

Stated explicitly, the Gram–Schmidt algorithm starts with any basis $\{\boldsymbol{v}_1, \boldsymbol{v}_2, \ldots, \boldsymbol{v}_n\}$ of an n-dimensional subspace of $\boldsymbol{R}^m$, and computes an orthogonal basis of the subspace by the following equations

$$\begin{aligned}
\boldsymbol{u}_1 &= \boldsymbol{v}_1 \\
\boldsymbol{u}_2 &= \boldsymbol{v}_2 - \frac{\boldsymbol{v}_2 \cdot \boldsymbol{u}_1}{\|\boldsymbol{u}_1\|^2} \boldsymbol{u}_1 \\
\boldsymbol{u}_3 &= \boldsymbol{v}_3 - \frac{\boldsymbol{v}_3 \cdot \boldsymbol{u}_2}{\|\boldsymbol{u}_2\|^2} \boldsymbol{u}_2 - \frac{\boldsymbol{v}_3 \cdot \boldsymbol{u}_1}{\|\boldsymbol{u}_1\|^2} \boldsymbol{u}_1 \\
&\vdots \\
\boldsymbol{u}_n &= \boldsymbol{v}_n - \frac{\boldsymbol{v}_n \cdot \boldsymbol{u}_{n-1}}{\|\boldsymbol{u}_{n-1}\|^2} \boldsymbol{u}_{n-1} - \frac{\boldsymbol{v}_n \cdot \boldsymbol{u}_{n-2}}{\|\boldsymbol{u}_{n-2}\|^2} \boldsymbol{u}_{n-2} - \cdots - \frac{\boldsymbol{v}_n \cdot \boldsymbol{u}_1}{\|\boldsymbol{u}_1\|^2} \boldsymbol{u}_1.
\end{aligned}$$

At each step, $\boldsymbol{u}_i$ is computed by projecting $\boldsymbol{v}_i$ onto the orthogonal complement of the vector subspace spanned by $\boldsymbol{u}_1, \ldots, \boldsymbol{u}_{i-1}$, which is denoted $\text{span}^{\perp}(\boldsymbol{u}_1, \ldots, \boldsymbol{u}_{i-1})$. This greedy process is then a correct and tractable algorithm for computing the desired orthogonal basis in $\boldsymbol{R}^m$.

The result of the Gram–Schmidt orthogonalization algorithm is an orthogonal basis in $\boldsymbol{R}^m$, but the basis vectors need not be short vectors, nor is shortness a meaningful goal because a basis vector in euclidean space can be of any length. If desired, the resulting basis vectors can be normalized to each have length one.

In contrast, to find the shortest vector in a lattice is quite difficult and indeed is formally intractable. We must be satisfied to find only an approximation. The LLL algorithm refers to the output of the Gram–Schmidt algorithm, but alters that output in order to accommodate the new goal and the discrete lattice constraints, then executes the Gram–Schmidt algorithm again. The goal is to obtain short vectors, though not necessarily the shortest. The LLL algorithm uses and updates two bases: the lattice basis $\{\boldsymbol{b}_1, \ldots, \boldsymbol{b}_n\}$ and the orthogonal shadow basis $\{\boldsymbol{b}_1^*, \ldots, \boldsymbol{b}_n^*\}$ in $\boldsymbol{R}^n$ computed by the Gram–Schmidt algorithm. The lattice basis always consists of nonzero points of the lattice. The shadow basis consists of orthogonal vectors that need not be – and usually will not be – points of the lattice. Both bases are iterated during each iteration. Based on examination of the shadow basis, the lattice basis vectors are updated by rounding and swapping. Then a new shadow basis is computed from the new lattice basis, again by using the Gram–Schmidt algorithm. This alternating process is repeated until it forms the required basis.

The key observation underlying the LLL algorithm is that if some projection coefficient in the Gram–Schmidt procedure satisfies

$$\mu_{ij} = \frac{\boldsymbol{b}_i \cdot \boldsymbol{b}_j^*}{\|\boldsymbol{b}_j^*\|^2} > \tfrac{1}{2},$$

then for an appropriate integer a, the vector $\boldsymbol{b}_i$ can be replaced by $\boldsymbol{b}_i - a\boldsymbol{b}_j$, thereby reducing the magnitude of the Gram–Schmidt projection coefficient. This is because the new projection coefficient μ'_{ij} is

$$\begin{aligned}\mu'_{ij} &= \frac{(\boldsymbol{b}_i - a\boldsymbol{b}_j) \cdot \boldsymbol{b}_j^*}{\|\boldsymbol{b}_j^*\|^2} \\ &= \frac{\boldsymbol{b}_i \cdot \boldsymbol{b}_j^*}{\|\boldsymbol{b}_j^*\|^2} - a > \tfrac{1}{2} - a \in [-\tfrac{1}{2}, \tfrac{1}{2}]\end{aligned}$$

by choice of the integer a.

In general, the elements of a vector-space basis do not have a preferred order. A basis remains a basis under any permutation of its basis vectors. The Gram–Schmidt algorithm, however, treats the basis vectors one at a time and so infers an order, although the algorithm has no prior preference for any particular order. If the input basis vectors are permuted, the output of the Gram–Schmidt algorithm will be a different basis although it will still be an orthogonal basis.

In contrast, the LLL algorithm does provide an ordered basis at its output, but it imposes no requirement for the order of the basis vectors at the algorithm input.

Our discussion of the LLL algorithm will be composed of three parts. First, we state the form of the output that is to be computed by the algorithm. Then we state the algorithm and prove that it halts with the required output. Finally, we bound the length of the basis vectors that the algorithm computes.

Definition 15.3.1 *An LLL-reduced basis for the lattice* $\boldsymbol{\Lambda}$ *is an ordered basis* $\{\boldsymbol{b}_1, \boldsymbol{b}_2, \ldots, \boldsymbol{b}_n\}$ *of lattice elements whose Gram–Schmidt orthogonalized shadow* *basis* $\{b_1^*, \ldots, b_n^*\}$ *and Gram–Schmidt projection coefficients* $\mu_{ij} = (b_i \cdot b_j^*)/(b_j^* \cdot b_j^*)$ *satisfy*

(i) $|\mu_{ij}| \leq \frac{1}{2}$ *for* $1 \leq i < j \leq n$.
(ii) $\|b_i^*\|^2 \geq \left(\frac{3}{4} - \mu_{i,i-1}^2\right) \|b_{i-1}^*\|^2$ *for* $1 < i \leq n$.

A reduced basis is an ordered basis. Its elements have a specific order that cannot be changed without violating the definition. A reduced basis is not unique. There can be many reduced bases for a given lattice. The LLL algorithm finds one of them.

The elements of the Gram–Schmidt orthogonalized shadow basis $\{b_1^*, \ldots, b_n^*\}$ in the definition are not, in general, elements of the lattice. They are elements of R^n that guide the development of the LLL algorithm. The constant $\frac{3}{4}$ in the definition of an LLL-reduced basis can be replaced by any number strictly between $\frac{1}{4}$ and 1 and the LLL algorithm will still work. These alternative choices for the value $\frac{3}{4}$ will not be considered.

The second condition of the definition, called the *Lovász condition*, is equivalent to the inequality

$$\|b_i^* + \mu_{i,i-1} b_{i-1}^*\|^2 \geq \frac{3}{4}\|b_{i-1}^*\|^2$$

because $b_i^* \cdot b_j^* = 0$. The Lovász condition is structured so as to determine whether the projection of b_{i+1} onto $\text{span}^{\perp}(b_1, \ldots, b_{i+1})$ is larger than three-fourths of the projection of b_i onto $\text{span}^{\perp}(b_1, \ldots, b_{i+1})$. This condition ensures that b_{i+1} is not much shorter than b_i.

The LLL algorithm computes a reduced basis using the following procedure.

(1) Set $i = 1$.
(2) For each $j = 1, \ldots, i-1$, replace b_i by $b_i - ab_j$ (possibly with $a = 0$) as needed so that each new Gram–Schmidt coefficient μ_{ij} satisfies $|\mu_{ij}| \leq \frac{1}{2}$.
(3) Next, do the following:
    (a) If the Lovász condition is satisfied and $i = n$, halt.
    (b) If the Lovász condition is satisfied and $i < n$, increment i and go to Step 2.
    (c) If the Lovász condition is not satisfied, swap b_{i-1} with b_i, decrement i and go to Step 2.

The LLL algorithm is illustrated in Figure 15.3. The LLL algorithm is guaranteed to find a lattice vector $b \in \Lambda$ as a basis vector satisfying

$$0 < \|b\| \leq 2^{(n-1)/2}\|v_{\min}\|,$$

as shown below in Theorem 15.3.5 (or, better, as $(2/\sqrt{3})^n v_{\min}$), where $v_{\min}$ is the lattice point with the shortest length. To appreciate the weakness of the algorithm, observe that if $n = 100$, then the multiplying coefficient is 2^{49}, which is larger than 10^{16}. Thus in response to a request for the smallest lattice point, the LLL algorithm can only assert that the magnitude of its selected lattice point is not more than 10^{16} times larger than the magnitude of the smallest lattice point. Clearly, this is a very weak guarantee.

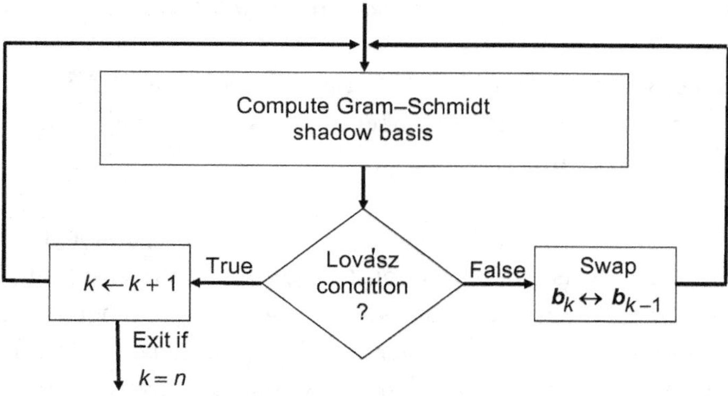

Figure 15.3 Overview of the LLL algorithm

As bad as this might seem, the LLL algorithm is somehow surprisingly useful. It is easy to see that the swap step of the LLL algorithm takes care of the second condition required of a reduced basis. Indeed if the LLL algorithm terminates, then its output must satisfy the second property. The reduction step takes care of the first property. A later proposition shows that it does terminate.

Proposition 15.3.2 *The LLL algorithm cannot terminate on a basis that is not a reduced basis.*

Proof The second property of a reduced basis is enforced by the test triggering the swap step. □

The key to understanding the LLL algorithm lies in proving that the swap step can only be visited a finite number of times, as is shown next. This implies that the algorithm must halt.

Proposition 15.3.3 *The swap step of the LLL algorithm can be visited only a finite number of times.*

Proof Let $D = \Pi_{i=1}^{n} d_i$ where $d_i = \Pi_{j=1}^{i} \|\boldsymbol{b}_j^*\|^2$. Because for every j, $\|\boldsymbol{b}_j^*\| \leq \|\boldsymbol{b}_{\min}\|$ at every iteration, D cannot be smaller than $\|\boldsymbol{b}_{\min}\|^{n^2}$. Thus the value of D is bounded away from zero. The value of D changes only in Step 3, where it changes only when $i = j - 1$. Then d_j is replaced by $d'_{j-1} \leq \sqrt{3/4}d_j$ and D is replaced by $D' \leq \sqrt{3/4}D$. But $(3/4)^{\ell/2}$ goes to zero as ℓ goes to infinity. However, D is bounded away from zero, which means that ℓ cannot go to infinity. This means that the swap step can be executed only a finite number of times, as was to be proved. □

Theorem 15.3.4 *The LLL algorithm computes a reduced basis.*

Proof Follows directly from Proposition 15.3.2 and Proposition 15.3.3. Because the swap step can only be executed a finite number of times, it must halt. But it can only halt on a reduced basis. The proof is now complete. □

We now know that the LLL algorithm halts on a reduced basis, although it need not be the best such reduced basis, and it is almost certainly not the best reduced basis. Even if it were to halt on the best reduced basis, the algorithm does not assert this.

It now remains to prove a statement regarding the performance of the algorithm. To end the section, we show that the reduced basis computed by the LLL algorithm satisfies $|\boldsymbol{b}_1| \leq 2^{(n-1)/2}\|\boldsymbol{v}\|$ for all $\boldsymbol{v} \in \boldsymbol{\Lambda}$. In particular, this holds for that nonzero $\boldsymbol{v}$ for which the magnitude $\|\boldsymbol{v}\|$ is the smallest.

Theorem 15.3.5 *Let* $\{\boldsymbol{b}_1, \ldots, \boldsymbol{b}_n\}$ *be an* LLL*-reduced basis for a lattice* $\boldsymbol{\Lambda} \in \boldsymbol{R}^n$*. Then*

$$\|\boldsymbol{b}_1\| \leq 2^{(n-1)/2}\|\boldsymbol{v}\|$$

for every $\boldsymbol{v} \in \boldsymbol{\Lambda}$.

Proof The smallest length $\|\boldsymbol{v}_{\min}\|$ of any point of the lattice satisfies

$$\|\boldsymbol{v}_{\min}\| \geq \min_i |\boldsymbol{b}_i^*|,$$

where $\{\boldsymbol{b}_1^*, \ldots, \boldsymbol{b}_n^*\}$ is the Gram–Schmidt orthogonalized shadow basis of the reduced basis $\{\boldsymbol{b}_1, \ldots, \boldsymbol{b}_n\}$. But

$$\|\boldsymbol{b}_n^*\|^2 \geq \tfrac{1}{2}\|\boldsymbol{b}_{n-1}^*\|^2 \geq \cdots \geq \tfrac{1}{2}\|\boldsymbol{b}_1^*\|^2 = \tfrac{1}{2}\|\boldsymbol{b}_1\|^2,$$

where the last equality, $\boldsymbol{b}_1^* = \boldsymbol{b}_1$, follows from the definition. Hence, for any i,

$$\|\boldsymbol{b}_1\| = \|\boldsymbol{b}_1^*\| \leq \cdots \leq \left(\tfrac{1}{2}\right)^{(i-1)/2}\|\boldsymbol{b}_i^*\| \leq \left(\tfrac{1}{2}\right)^{-(n-1)/2}\|\boldsymbol{b}_i^*\|.$$

Hence

$$\|\boldsymbol{b}_1\| \leq \left(\tfrac{1}{2}\right)^{(n-1)/2}\min_i \|\boldsymbol{b}_i^*\| \leq \left(\tfrac{1}{2}\right)^{(n-1)/2}\|\boldsymbol{v}\|$$

for all $\|\boldsymbol{v}\| \in \boldsymbol{\Lambda}$, as was to be proved. □

The statement of the theorem can be expanded to the stronger statement that an LLL-reduced basis as is computed by the LLL algorithm satisfies

$$\|\boldsymbol{b}_n\| \leq 2^{(n-1)/2}\max_{1\leq j\leq t}\|\boldsymbol{v}_j\|$$

for any set $\{\boldsymbol{v}_1, \boldsymbol{v}_2, \ldots, \boldsymbol{v}_n\}$ of nonzero lattice points. We will not prove this stronger statement.

15.4 Lattice-based cryptography

Lattice-based cryptography is an alternative form of public-key cryptography. A lattice-based cryptosystem is based on a problem in lattice theory that is deemed to be hard. Specifically, lattice-based cryptography rests on the intractability of the shortest-vector problem or the closest-vector problem. A lattice-based cryptosystem does not depend on the intractability of the discrete-log problem nor on the intractability of factoring biprimes. Thus if lattice-based cryptography is secure, then it will still be secure even if the discrete-log problem and the integer-factoring problem were to be successfully solved. Because of this, lattice-based cryptography is considered to be important as an alternative to existing methods of public-key cryptography, should those methods be compromised. Moreover, unlike the factoring of biprimes and the computation of discrete logs, which are believed to be hard problems but without a formal statement of this belief known, the shortest-vector problem is known to be hard in the sense of formal intractablity. The goal of lattice-based cryptography is to devise a cryptosystem whose breaking is equivalent to solving the shortest-vector problem or the closest-vector problem in a general lattice.

A practical lattice-based cryptosystem employs a lattice $\mathbf{\Lambda}$ that is isomorphic to $\mathbf{Z}^n$, but this isomorphism is hidden by computational complexity. The lattice $\mathbf{\Lambda}$ has an infinite number of points, as does $\mathbf{Z}^n$. The design of a cryptosystem must associate a key with the trellis in some way. The keyspace $\mathcal{K}$ has $\#\mathcal{K}$ keys, which are addressed by $\log_2 \#\mathcal{K}$ bits. It is required that $\log_2 \#\mathcal{K}$ not be too large, preferably a few hundred bits, that it is easy to compute the plaintext when given the key k and the ciphertext, and that it is computationally intractable to compute the plaintext when given the keyspace $\mathcal{K}$ and the ciphertext. The main point here is that the keyspace $\mathcal{K}$ and its use are known. The known relationship between the lattice and the keyspace must not obviate the intractability of the unconstrained lattice-reduction problem.

Lattice-based cryptography is closely related to ring-based cryptography. This is because a ring constructed from the ring of integers $\mathbf{Z}$ is likely to have a representation as a lattice. The ring-based cryptosystem that we describe below is a public-key cryptosystem that is closely related to the theory of lattices. It is an asymmetric-key cryptosystem with two different keys, an encryption key and a decryption key. The cryptographic secrecy is based on the intractability of the shortest-vector problem in a lattice.

The elements of the lattice $\mathbf{Z}^n$ can be represented by the elements of a ring $\mathbf{Z}[x]/\langle x^n - 1\rangle$. Each element of the lattice $\mathbf{Z}[x]/\langle x^n - 1\rangle$ is a polynomial of degree at most n with integer coefficients. Thus each element of the ring $\mathbf{Z}[x]/\langle x^n - 1\rangle$ can be identified with an element of $\mathbf{Z}^n$. This ring is, however, a richer structure than a lattice because both addition and multiplication are defined in the polynomial ring $\mathbf{Z}[x]/\langle x^n - 1\rangle$. This additional structure enables a practical method of encryption to be devised, but may also introduce a vulnerability that could be exploited by a cryptanalyst to avoid the intractability of general lattice reduction.

An element of the polynomial ring $\boldsymbol{Z}[x]/\langle x^n - 1\rangle$ is

$$a = a_{n-1}x^{n-1} + a_{n-2}x^{n-2} + \cdots + a_1 x + a_0,$$

where all coefficients of this polynomial are integers. Multiplication of ring elements is multiplication as polynomials modulo $x^n - 1$. Under the operation of addition, $\boldsymbol{Z}[x]/\langle x^n - 1\rangle$ is isomorphic to the lattice $\boldsymbol{Z}^n$. For this reason a cryptosystem based on $\boldsymbol{Z}[x]/\langle x^n - 1\rangle$ is regarded as a lattice-based cryptosystem.

The Ajtai–Dwork public-key cryptosystem

To introduce the subject of lattices as used in cryptography, we briefly describe the concept of the Ajtai–Dwork cryptosystem, which was an early attempt to develop what is now seen as a lattice-based cryptosystem. The Ajtai–Dwork cryptosystem is an important development in the history of cryptography because it is an approach that rests on the formal intractability of the shortest-vector problem, whereas the earlier and widely used methods of cryptography are based on integer factoring or on the discrete logarithm, which are not known to be formally intractable. Indeed, the Ajtai–Dwork cryptosystem is as secure as the worst case of the shortest-vector problem. The Ajtai–Dwork cryptosystem is unsuccessful because it is completely impractical, requiring an enormous key. A brief summary is useful, however, as a historical and introductory pedagogical example because the lattice-based cryptosystems that came later evolved from it. We provide only a general sense of the ideas.

The Ajtai–Dwork keys are randomly chosen points in the n-dimensional hypersphere of radius one in $\boldsymbol{R}^n$. Because the key space must be finite, the n-dimensional hypersphere is covered with a fine grid in the form of the lattice $n^{-2}\boldsymbol{Z}$. This lattice $n^{-2}\boldsymbol{Z}$ is formed by the trivial generator matrix equal to a diagonal matrix with n^{-2} in all elements of the diagonal. The secret key for the Ajtai–Dwork cryptosystem is a randomly chosen point in the n-dimensional unit ball of $\boldsymbol{R}^n$. The components of the chosen point are restricted to lie in the lattice $n^{-2}\boldsymbol{Z}$ and inside the unit ball. The general notion is to construct hyperplanes in this lattice, and to specify a point slightly displaced from a hyperplane that cannot be returned to the hyperplane without knowing the secret key. The message is encrypted one bit at a time. Clearly this is an impractical cryptosystem because of the immense key, and further details are not interesting and are not given.

The GGH public-key cryptosystem

The Goldreich–Goldwasser–Halevi cryptosystem, abbreviated GGH, is an asymmetric cryptosystem based on lattices. The GGH cryptosystem is based on the fact that the closest-vector problem in a lattice is intractable for an arbitrary basis, but it is tractable for a nearly orthogonal basis. As defined, the GGH cryptosystem is not practical because the key is too large, but it does introduce a satisfying concept that has been refined in subsequent improvements.

The GGH private and public keys are two representations of the same lattice. To create these keys, a lattice $\mathbf{\Lambda}$ with a short and nearly orthogonal basis is chosen. The short and nearly orthogonal basis $\{\boldsymbol{b}_1, \boldsymbol{b}_2, \ldots, \boldsymbol{b}_n\}$ for the lattice $\mathbf{\Lambda}$ is the private key: and a long and highly nonorthogonal basis $\{\boldsymbol{w}_1, \boldsymbol{w}_2, \ldots, \boldsymbol{w}_n\}$ of the same lattice is chosen as the public key.

For the encryption of the plaintext message $\boldsymbol{m}$, represented in the binary form $(m_1, m_2, \ldots, m_n)$, the binary components m_i are used as coefficients to form a linear combination of the public basis vectors $\boldsymbol{w}_i$. Thus the plaintext m corresponds to a unique lattice point $\boldsymbol{w}$ given by

$$\begin{aligned}\boldsymbol{w} &= m_1\boldsymbol{w}_1 + m_2\boldsymbol{w}_2 + \cdots + m_n\boldsymbol{w}_n\\ &= \boldsymbol{m}\boldsymbol{W},\end{aligned}$$

where $\boldsymbol{W}$ is the matrix with the basis vectors $\boldsymbol{w}_i$ as columns. Indeed, the lattice point $\boldsymbol{w}$ is the message. Then a small random offset vector $\boldsymbol{e}$ is generated by the encryptor, and added to the lattice point, but otherwise the offset vector is kept secret by the encryptor. The encryptor has no further use for the offset vector, and it can be destroyed. In this way, the ciphertext $\boldsymbol{c}$ is computed as

$$\begin{aligned}\boldsymbol{c} &= m_1\boldsymbol{w}_1 + m_2\boldsymbol{w}_2 + \cdots + m_n\boldsymbol{w}_n + \boldsymbol{e}\\ &= \boldsymbol{m}\boldsymbol{W} + \boldsymbol{e}.\end{aligned}$$

Because of the offset vector $\boldsymbol{e}$, the ciphertext $\boldsymbol{c}$ is not a point of the lattice $\mathbf{\Lambda}$. Instead, it is slightly displaced from the lattice point that holds the message.

When the offset vector $\boldsymbol{e}$ is small enough, and the good basis is known, the message can be decrypted by finding the closest lattice point to the ciphertext $\boldsymbol{c}$. That is, for decryption, we need to solve the closest-vector problem using a good basis. First, the representation of $\boldsymbol{c}$ in the short and nearly orthogonal basis is computed as $\boldsymbol{m}' = \boldsymbol{c}\boldsymbol{V}^{-1}$ so that $\boldsymbol{c} = m_1\boldsymbol{v}_1 + \cdots + m_n\boldsymbol{v}_n + \boldsymbol{e}$. Then the basis coefficients are rounded to the nearest integers, and rounded coefficients are used in the representation as $\boldsymbol{u} = [m_1]\boldsymbol{v}_1 + \cdots + [m_n]\boldsymbol{v}_n$. Hence, $\boldsymbol{u}$ is the closest lattice vector to $\boldsymbol{c}$. In turn, we cannot decrypt the message $\boldsymbol{m}$ as the representation of $\boldsymbol{u}$ in the bad basis, i.e., $\boldsymbol{m} = \boldsymbol{u}\boldsymbol{W}^{-1}$. The security of the GGH cryptosystem rests on the difficulty of the closest vector problem starting from a highly nonorthogonal basis.

The key in this scheme contains n^2 numbers because the basis consists of n vectors, each an element of $\boldsymbol{Z}^n$. Possibly n may be chosen as large as 500 for a secure system that encrypts 500 bits at a time. The private key then is a 500-by-500 matrix, which has 250,000 elements. The public key is the same. Consequently the GGH system requires an impractically large key size. It is not a practical cryptosystem. It is a conceptual cryptosystem whose place in history was to stimulate the development of more practical lattice-based cryptosystems.

The generation of the keys for the GGH cryptosystem first requires the construction of the random private key $\{\boldsymbol{b}_1, \boldsymbol{b}_2, \ldots, \boldsymbol{b}_n\}$. The vectors of the public key should be nearly pairwise orthogonal and not too long. A simple way to do this is to choose the n components for each of the n elements randomly and independently from some symmetric set of positive and negative numbers of small magnitude. The law of large numbers will ensure that these basis vectors are nearly orthogonal with high probability.

Next a unitary matrix must be chosen to convert this private basis to the public basis. The unitary matrix must be random and secret, it must be unitary and known to be unitary, and it must be computable in practice. One way to form this matrix $\boldsymbol{A}$ is to write $\boldsymbol{A} = \Pi_i \boldsymbol{A}_i$ where the $\boldsymbol{A}_i$ are a large number of simple unitary matrices. These can be multiplied onto $\boldsymbol{B}$ one by one until $\boldsymbol{W}$ is formed. This method of forming $\boldsymbol{A}$ will be known to the cryptanalyst. To protect against an attack, a large number of such $\boldsymbol{A}_i$ will be needed.

Ring-based public-key cryptosystems

The large key size required to state an n-by-n matrix in the GGH cryptosystem can be avoided by using a cyclic basis. This is a basis in which every basis vector is defined as the cyclic translate of the previous basis vector. The basis is

$$\begin{aligned}
\boldsymbol{b}_1 &= (b_1, b_2, b_3, \ldots, b_{n-1}, b_n)\\
\boldsymbol{b}_2 &= (b_2, b_3, \ldots, b_{n-1}, b_n, b_1)\\
&\vdots\\
\boldsymbol{b}_n &= (b_n, b_1, b_2, b_3, \ldots, b_{n-1})
\end{aligned}$$

and the corresponding generator matrix of the lattice is the circulant matrix

$$\boldsymbol{G} = \begin{bmatrix} b_1 & b_2 & b_3 & \cdots & b_n \\ b_2 & b_3 & b_4 & \cdots & b_1 \\ \vdots & & & & \\ b_n & b_1 & b_2 & \cdots & b_{n-1} \end{bmatrix}.$$

With this choice of matrix as the generator matrix (or as a subblock of a more general block-partitioned generator matrix), the key size is reduced to only n numbers (or so) because only one column needs to be stated. This yields a practical key size for the cryptosystem. A lattice of this special form can be related to a basic integer ring, and so a mathematical system with many strong properties is now in play. This not only reduces the size of the key and provides computational advantages to the encryptor, but it also presents new adversarial opportunities to the cryptanalyst.

An early cryptosystem of this kind is one known as the NTRU cryptosystem. The term NTRU is a longstanding acronym with an uninteresting history. It originated as an unconventional and uninteresting private designation, or jargon, for the polynomial ring $R[x]/\langle x^N - 1\rangle$ over the ring R. Usually R is the ring $\mathbf{Z}_q$. The term now designates a particular method of cryptography based on the ring $\mathbf{Z}_q[x]/\langle x^N - 1\rangle$. The elements of this ring are polynomials of degree at most $N - 1$ with integer coefficients, and multiplication of polynomials uses the reduction $x^N = 1$. This ring has the structure of a lattice. It is worth noting that NTRU really is a ring-based public-key cryptosystem in that it uses both addition and multiplication of the ring. Most do not. The security of the NTRU cryptosystem is hard to analyze because of the many parameters involved. Indeed, flaws were found in early versions, and have been since repaired.

Even though the shortest-vector problem for an arbitrary lattice, when it is given in the worst-case basis, is formally intractable, it need not be formally intractable for a lattice of a special form. It is not known whether the intractability of the shortened-vector problem holds for the NTRU lattice. A lattice that has a special form is not an arbitrary lattice, so the general statement regarding intractability might not hold.

15.5 Attacks on lattice cryptosystems

A direct attack on a lattice-based cryptosystem usually comes down to an attempt to solve the closest-vector problem or the shortest-vector problem. For example, the Merkle–Hellman knapsack cryptosystem can be mentioned as an early form of lattice-based cryptography. Because the integer packing problem can be described as a lattice problem, the Lenstra–Lenstra–Lovász algorithm can be used to attack the Merkle–Hellman knapsack cryptosystem. The shortest-vector problem and the closest-vector problem are believed to be formally intractable for a general lattice, not just for some lattices, but for most lattices. Nevertheless, these problems are quite easy for some lattices, such as $\mathbf{Z}^n$ expressed using the conventional basis. The knapsack cryptosystem builds a trapdoor into the packing problem, which means that the corresponding lattices are special lattices with side conditions that create exceptions to the general statement regarding the complexity of lattice reduction. This is an instance of a general concern. Although, in principle, lattice-based cryptosystems are protected by the intractability of the shortest-vector problem, a practical lattice-based cryptosystem does not use a random lattice. The cryptosystem does introduce constraints on the lattice structure to reduce key size or to facilitate decryption and those constraints may prove to be a vulnerability. In general, known lattice-based cryptosystems have not been proved to be protected by this intractability of the general problem.

The goal of a cyberattack on a typical lattice-based cybersystem is to determine the message $m_1, m_2, \ldots, m_n$ when given the public key $\boldsymbol{w}_1, \boldsymbol{w}_2, \ldots, \boldsymbol{w}_n$ and the perturbed subset sum $\boldsymbol{c} = \sum_i m_i \boldsymbol{w}_i + \boldsymbol{e}$. The attack is based on a lattice interpretation, observing that with the perturbation $\boldsymbol{e}$ removed, the set of all such messages is a lattice. Thus, the most direct attack is to first find the closest lattice point to $\boldsymbol{c}$, then to project $\boldsymbol{c}$ onto the set of basis vectors $\{\boldsymbol{w}_1, \boldsymbol{w}_2, \ldots, \boldsymbol{w}_n\}$. This direct approach is, on its face, intractable because it requires solving the closest-vector problem.

A deeper study, however, reveals other potential attacks that must be anticipated. These possibilities can be illustrated in the context of the GGH cryptosystem. A practical realization of the GGH cryptosystem must choose a private key and a public key. These are created by first randomly choosing the private key, then randomly choosing a unitary matrix so that the public key can be computed from the private key. It would not be possible to go in the other direction, randomly choosing a public key, then computing the private key from it, because this would be intractable.

One way to compute the private key is to randomly choose the n components of each of n basis vectors from the set of integers $\{-\ell, -(\ell-1), \ldots, (\ell-1), \ell\}$ for some small integer ℓ. These basis vectors will, with large probability, be nearly orthogonal. This gives the private key. Then choose the unitary matrix. One way to do this is to define $\boldsymbol{A} = \Pi_i \boldsymbol{A}_i$ where each $\boldsymbol{A}_i$ is a randomly chosen but simple unitary matrix, say a matrix having mostly zeros and a few randomly chosen components arranged to ensure the unitary property. This matrix $\boldsymbol{A}$ now gives the public key. Multiplication of the private generator matrix by $\boldsymbol{A}$ can be done by multiplying by each of the $\boldsymbol{A}_i$ in order.

The cryptanalyst does know this process of computing the public key, but not the specific random selection of the generator matrix and the unitary matrix. The public lattice is not arbitrary. Perhaps this knowledge of the process can be used by the cryptanalyst to circumvent the intractability of the closest-vector problem.

Similar concerns carry over to other formulations of lattice-based cryptography. All such concerns have evidently been put to rest by the community, but it is important to always be aware of the possibility of vulnerability. A general statement about intractability must always be inspected in the context of a specific application.

15.6 Introduction to codes

The vector space $\boldsymbol{F}_q^n$ consists of all n-tuples of elements of the finite field $\boldsymbol{F}_q$. A *code* of size M is a set of M vectors, or points, of $\boldsymbol{F}_q^n$. A code will be denoted $\mathcal{C}$ and the elements of $\mathcal{C}$ will be denoted $\boldsymbol{c}$ and called *codewords*. A *linear code* is a set of points of $\boldsymbol{F}_q^n$ that forms a vector subspace of $\boldsymbol{F}_q^n$. An (n, k) linear code is a subspace of $\boldsymbol{F}_q^n$ of dimension k. Then $M = q^k$. We will be interested only in linear codes.

The *Hamming weight* of any vector $\boldsymbol{v} \in \boldsymbol{F}_q^n$ is the number of nonzero components of $\boldsymbol{v}$. The Hamming weight of $\boldsymbol{v}$ is denoted $w_H(\boldsymbol{v})$. The *Hamming distance* between two vectors $\boldsymbol{u}$ and $\boldsymbol{v}$ of the vector space $\boldsymbol{F}_q^n$ is the Hamming weight of their componentwise difference. The Hamming distance between $\boldsymbol{u}$ and $\boldsymbol{v}$ is denoted $d_H(\boldsymbol{u}, \boldsymbol{v})$.

The *minimum distance* $d_{\min}$ of the code $\mathcal{C}$ is given by the smallest Hamming distance between any two distinct codewords. Thus

$$d_{\min} = \min_{\substack{\boldsymbol{c},\boldsymbol{c}' \\ \boldsymbol{c}\neq\boldsymbol{c}'}} d_H(\boldsymbol{c}, \boldsymbol{c}').$$

For a linear code, the componentwise difference of any two codewords is itself a codeword. Because $d_H(\boldsymbol{c}, \boldsymbol{c}') = w_H(\boldsymbol{c} - \boldsymbol{c}')$, the minimum distance $d_{\min}$ of a linear code is equal to the weight of the minimum-weight nonzero codeword. The integers n, k, and $d_{\min}$ are the standard external parameters describing a code. Two codes with the same n, k, and $d_{\min}$ can have a vastly different internal structure. One prefers a linear code $\mathcal{C}$ for which n is large, and both k and $d_{\min}$ are as large as possible for this n. One also prefers a code in which computations relevant to an application, described later, are tractable.

Let the vectors $\boldsymbol{g}_1, \boldsymbol{g}_2, \ldots, \boldsymbol{g}_k$ be a basis for the k-dimensional subspace comprising $\mathcal{C}$. Codewords are linear combinations of the basis vectors. Let $\boldsymbol{G}$ be a matrix, called a *generator matrix*, with the basis vectors $\boldsymbol{g}_1, \boldsymbol{g}_2, \ldots, \boldsymbol{g}_k$ as rows. Then every codeword of $\mathcal{C}$ can be written as

$$\boldsymbol{c} = \boldsymbol{a}\boldsymbol{G},$$

where $\boldsymbol{a}$ is a vector of length k called the *dataword*. Let the vectors $\boldsymbol{h}_1, \boldsymbol{h}_2, \ldots, \boldsymbol{h}_{n-k}$ be a basis for the $(n-k)$-dimensional subspace comprising the dual space $\mathcal{C}^\perp$, which is the null space of $\boldsymbol{G}$. Let $\boldsymbol{H}$ be a matrix, called the *check matrix*, whose rows are basis vectors of the dual space $\mathcal{C}^\perp$. Clearly, rank $\boldsymbol{H} = n - k$. Then $\boldsymbol{H}^T\boldsymbol{c} = \boldsymbol{0}$ if, and only if, $\boldsymbol{c}$ is an element of $\mathcal{C}$. Moreover, the smallest weight of a nonzero codeword $\boldsymbol{c}$ is equal to the cardinality of the smallest set of dependent columns of $\boldsymbol{H}$. Therefore, recalling Definition 9.11.3,

$$d_{\min} - 1 = \text{heft}\, \boldsymbol{H}$$

for a linear code. Moreover,

$$\text{heft}\, \boldsymbol{H} \leq \text{rank}\, \boldsymbol{H} = n - k.$$

Combining these statements gives $d_{\min} \leq n - k + 1$. This inequality is known as the *Singleton bound*. If the Singleton bound is satisfied with equality for the linear code $\mathcal{C}$, then code $\mathcal{C}$ is called a *maximal-distance code*. In general, maximal-distance codes exist over $\boldsymbol{F}_q$ only if $n \leq q + 1$. The Singleton bound is not satisfied with equality for any nontrivial code over $\boldsymbol{F}_2$. Binary codes always fall far short of equality in the Singleton bound.

It is not known how large $d_{\min}$ can be for any linear binary code with a specified n and k. Equivalently, it is not known how large the heft of a large binary (or q-ary) k-by-n matrix can be, not even approximately, even for rather small codes. The search for linear (n, k) codes over $\boldsymbol{F}_q$ with large minimum distance comes down to a search for k by n matrices over $\boldsymbol{F}_q$ with large heft.

A (truncated) vandermonde matrix with $k \leq n$, given by

$$\boldsymbol{H} = \begin{bmatrix} \beta_0^0 & \beta_0^1 & \beta_0^2 & \cdots & \beta_0^{n-1} \\ \beta_1^0 & \beta_1^1 & \beta_1^2 & \cdots & \beta_1^{n-1} \\ \vdots & \vdots & \vdots & & \vdots \\ \beta_{r-1}^0 & \beta_{r-1}^1 & \beta_{r-1}^2 & \cdots & \beta_{r-1}^{n-1} \end{bmatrix},$$

in general does not have its heft equal to its rank, but if the β_j are chosen to be successive powers of a field element ω of order n, the heft will be equal to the rank. That is, the elements must satisfy $\beta_j = \omega^j$ for $j = j_0, \ldots, j_0 + r - 1$. The check matrix then has the form

$$\boldsymbol{H} = \begin{bmatrix} 1 & \omega^{j_0} & \omega^{2j_0} & \cdots & \omega^{(n-1)j_0} \\ 1 & \omega^{j_0+1} & \omega^{2(j_0+1)} & \cdots & \omega^{(n-1)(j_0+1)} \\ \vdots & \vdots & \vdots & & \vdots \\ 1 & \omega^{j_0+r-1} & \omega^{2(j_0+r-1)} & \cdots & \omega^{(n-1)(j_0+r-1)} \end{bmatrix}.$$

By factoring out the term $\omega^{\ell j_0}$ from the ℓth column, every k-by-k submatrix of $\boldsymbol{H}$ then can be put in the form of a k-by-k vandermonde matrix (transposed) with nonzero determinant. Hence such an $\boldsymbol{H}$ matrix has a heft equal to k and so gives a code over $\boldsymbol{F}_q$ with $d_{\min} = n - k + 1$, which is known as a *Reed–Solomon code*. Such a check matrix exists, however, only if $n \leq q - 1$. Thus a Reed–Solomon code exists only for n less than $q - 1$. There is a way to extend the construction of a Reed–Solomon code to exploit all $q + 1$ points of the projective line $\boldsymbol{F}_q$, thereby extending the blocklength to $n = q + 1$, but not larger.

To obtain a code with a blocklength n larger than $q + 1$, one uses elements in the extension field $\boldsymbol{F}_{q^m}$. In this way, one can obtain a blocklength as large as $q^m + 1$, but the matrix $\boldsymbol{H}$ will have elements in the extension field $\boldsymbol{F}_{q^m}$. The expression $\boldsymbol{H}^T\boldsymbol{c} = \boldsymbol{0}$ now describes a code not in $\boldsymbol{F}_q^n$ but in $\boldsymbol{F}_{q^m}^n$. However, some $\boldsymbol{c}$ that satisfy $\boldsymbol{H}^T\boldsymbol{c} = \boldsymbol{0}$ will have all of their components in the small field $\boldsymbol{F}_q$, and the set of such codewords is a linear subspace of $\boldsymbol{F}_q^n$. It is called a *subfield-subcode*.

One defines a subfield-subcode directly by

$$\mathcal{C} = \{\boldsymbol{c} \in \boldsymbol{F}_q^n | \boldsymbol{H}^T\boldsymbol{c} = \boldsymbol{0}\},$$

where the elements of $\boldsymbol{H}$ are in the extension field $\boldsymbol{F}_{q^m}$. This, however, is an implicit definition, useful only to test a vector $\boldsymbol{c}$ to ensure that it is a codeword. It is not an explicit statement that is useful for finding the codewords. To generate codewords, one must use the dual of $\boldsymbol{H}$ regarded as a check matrix over the ground field $\boldsymbol{F}_q$.

The vandermonde matrix formulation of $\boldsymbol{H}$ can be restated in terms of polynomials. Each row of $\boldsymbol{H}^T\boldsymbol{c} = \boldsymbol{0}$ has the form $\sum_{i=0}^{n-1} \beta_j^i c_i = 0$. This is the statement that the polynomial $c(x) = \sum_{i=0}^{n-1} c_i x^i$ has a zero at β_j. Thus, $\boldsymbol{H}^T\boldsymbol{c} = \boldsymbol{0}$ is the statement that the polynomial $c(x)$ has zeros at $\beta_1, \beta_2, \ldots, \beta_{n-k}$. The set of all $c(x)$ with a specified set of zeros is an ideal of $\boldsymbol{F}_q[x]$, which is generated by a single polynomial $g(x)$, called a *generator polynomial* of the code $\mathcal{C}$. The polynomial $g(x)$ then divides $x^n - 1$ which means that $x^n - 1 = h(x)g(x)$ for some polynomial $h(x)$. The polynomial $h(x)$ is called the *check polynomial*. Every codeword satisfies $c(x)h(x) = 0 \,(\mathrm{mod}\, x^n - 1)$. If $c(x)h(x) = 0 \,(\mathrm{mod}\, x^n - 1)$, then $xc(x)h(x) = 0 \,(\mathrm{mod}\, x^n - 1)$ as well, so $xc(x) \,(\mathrm{mod}\, x^n - 1)$ is an element of the code $\mathcal{C}$ whenever $c(x)$ is an element of the code $\mathcal{C}$. For this reason, such codes are called *cyclic codes*. Because of the form of $\boldsymbol{H}$, the minimum distance of $\mathcal{C}$ must be larger than the number of consecutive zeros of $g(x)$. Thus if $g(x)$ has r consecutive zeros, the cyclic code $\mathcal{C}$ has minimum distance at least $r + 1$, a statement known as the *BCH bound* and $r + 1$ is known as the *BCH distance*. With this formulation, it is now clear how to ensure that the codeword is in the small field $\boldsymbol{F}_q$. Because every conjugate of a zero of $c(x)$ must also be a zero of $c(x)$, every conjugate of β in the *defining set* is a zero of $c(x)$. We can conclude that $\mathcal{C}$ can be described as the set of polynomials $c(x)$ of degree at most $n - 1$ with a specified defining set of $d_{\min} - 1$ consecutive zeros, as well as zeros at all q-ary conjugates of the zeros in the defining set. Thus

$$\mathcal{C} = \{c(x) \,|\, c(\omega^j) = 0 \text{ if } j^\ell \,(\mathrm{mod}\, q^m - 1) \in \{j_0, j_0 + 1, \ldots, j_0 + d - 2\} \text{ for } \ell \in \boldsymbol{Z}\}.$$

The defining set is also written as the field elements

$$\omega^{j_0}, \omega^{j_0+1}, \ldots, \omega^{j_0+d-2}.$$

This set of zeros gives a subfield-subcode over $\boldsymbol{F}_q$ of a Reed–Solomon code over $\boldsymbol{F}_{q^m}$, which is called a *BCH code*. A BCH code has a minimum distance at least as large as the BCH bound, but sometimes the minimum distance of the code may be larger, so $d_{\min} \geq d$. Because it is necessary to insert zeros into $g(x)$ for all conjugates of a defining zero, a BCH code almost always has a disappointing dimension k for a given value of $d_{\min}$.

A cyclic code can also be described in the language of the Fourier transform. To this end, use the coefficients of the polynomial $c(x)$ to form the vector $\boldsymbol{c}$, which has Fourier transform $C_j = n^{-1} \sum_{i=0}^{n-1} \omega^{ij} c_i$. The zeros of $c(x)$ are at the values of j such that $c(\omega^j) = nC_j = 0$. Thus the cyclic code consists of all vectors such that a prescribed set of components of the Fourier transform are zero. If these zeros of the Fourier transform are at $d - 1$ consecutive values of j, then the code has minimum distance at least d. This statement is known as the BCH bound. If ω is in the field of the code $\boldsymbol{F}_q$, then the code is a Reed–Solomon code. If ω is in an extension field of $\boldsymbol{F}_q$, then the code is a BCH code.

The class of *Goppa codes* is another very large class of codes that are obtained as subfield-subcodes of modified Reed–Solomon codes. It is known that the class of Goppa codes contains some very good codes as measured by n, k, and $d_{\min}$, but we have no idea of how to identify which of the Goppa codes are good. Nor do we have satisfactory algorithms for computations related to large Goppa codes.

The Goppa codes can be defined most directly as subfield-subcodes given by the $\boldsymbol{F}_q$ null space of the matrix $\boldsymbol{H} = \boldsymbol{H}_2\boldsymbol{H}_1$ where $\boldsymbol{H}_2$ is an n-by-n diagonal matrix with nonzero elements in $\boldsymbol{F}_{q^m}$ and $\boldsymbol{H}_1$ is a k-by-n (truncated) vandermonde matrix formed from r consecutive powers of some element of $\boldsymbol{F}_{q^m}$ of order n, as described earlier. There are $(q^m - 1)^n$ such diagonal matrices $\boldsymbol{H}_2$ over $\boldsymbol{F}_{q^m}$. Each such matrix gives a subfield-subcode with minimum distance at least as large as $r + 1$ and dimension not larger than $n - r$. It is known that some of these Goppa codes have large dimension and minimum distance and some do not. No formal procedure is known for specifying the matrix $\boldsymbol{H}_2$ so that both the dimension and the minimum distance of the Goppa code are large.

15.7 Subspace projection

A linear code $\mathcal{C}$ is a k-dimensional vector subspace of $\boldsymbol{F}_q^n$. Any vector $\boldsymbol{v} \in \boldsymbol{F}_q^n$ not in $\mathcal{C}$ can be "projected" to the subspace $\mathcal{C}$ by finding a codeword $\boldsymbol{c} \in \mathcal{C}$ that is closest to $\boldsymbol{v}$ in Hamming distance. This operation of projection into the subspace $\mathcal{C}$ is commonly known as the task of *decoding*. In general, the task of projection into a subspace of $\boldsymbol{F}_q^n$ is formally intractable for large n. It is, however, tractable if the subspace satisfies certain conditions, and if other restrictions are imposed.

Let $d_{\min}$ be the minimum distance of code $\mathcal{C}$, and let $t = \lfloor (d_{\min} - 1)/2 \rfloor$. A *Hamming sphere* of radius t about any element $\boldsymbol{v}$ of $\boldsymbol{F}_q^n$ consists of all elements of $\boldsymbol{F}_q^n$ that are within Hamming distance t of $\boldsymbol{v}$. Two Hamming spheres of Hamming radius t about two distinct codewords of a code $\mathcal{C}$ of minimum distance $d_{\min}$ are disjoint. Thus the set of all Hamming spheres of radius t about the codewords of $\mathcal{C}$ is a set of disjoint sets. If $\boldsymbol{v}$ lies in (or even near) such a Hamming sphere, then depending on the structure of the code, it may be computationally easy to find the codeword $\boldsymbol{c}$ at the center of the sphere. Said another way, if it is known that there is exactly one codeword $\boldsymbol{c}$ in a Hamming sphere of radius t about a given $\boldsymbol{v}$, then for some codes satisfying a linear complexity property, it may be easy to determine $\boldsymbol{c}$. This statement holds for many popular codes

$\mathcal{C}$, but it does not hold in general. However, even for codes of this form, if $\boldsymbol{v}$ does not lie in (or near) a Hamming sphere, it can be computationally intractable to find the closest $\boldsymbol{c} \in \mathcal{C}$.

An arbitrary point $\boldsymbol{v}$ of $\boldsymbol{F}_q^n$ is at Hamming distance $d_H(\boldsymbol{v}, \boldsymbol{c})$ from codeword $\boldsymbol{c}$. A codeword that is nearest to $\boldsymbol{v}$ is at distance $\min_{c \in \mathcal{C}} d_H(\boldsymbol{v}, \boldsymbol{c})$. The codeword closest to $\boldsymbol{v}$ need not be unique, and if this is the situation, the task is much more difficult. Accordingly, there are several formulations of the task of decoding. The various formulations of this task can be made precise as follows:

Closest-codeword problem: Given $\boldsymbol{v} \in \boldsymbol{F}_q^n$, find any (or all) $\boldsymbol{c} \in \mathcal{C}$ for which $d_H(\boldsymbol{v}, \boldsymbol{c})$ is minimum.

Restricted-codeword problem: Given $\boldsymbol{v} \in \boldsymbol{F}_q^n$, find $\boldsymbol{c} \in \mathcal{C}$ provided a codeword $\boldsymbol{c}$ satisfying $d_H(\boldsymbol{v}, \boldsymbol{c}) \leq \tau$ exists.

Bounded-codeword problem: Given $\boldsymbol{v} \in \boldsymbol{F}_q^n$, find $\boldsymbol{c} \in \mathcal{C}$ provided a codeword $\boldsymbol{c}$ satisfying $d_H(\boldsymbol{v}, \boldsymbol{c}) < d_{\min}/2$ exists.

In general, it is not computationally tractable to find even one closest codeword to a given vector $\boldsymbol{v}$. The closest-codeword problem is intractable. Even the simpler bounded-codeword problem can be difficult and perhaps intractable for an arbitrary linear code. For a Reed–Solomon code or a BCH code, however, the bounded-codeword problem is easy to solve. Although the task of projecting an arbitrary $\boldsymbol{v} \in \boldsymbol{F}_q$ into the subspace $\mathcal{C}$ is intractable in general, for these codes one can satisfy this task by using the methods of linear algebra augmented by the task of finding the zeros of a polynomial, called the *locator polynomial*, over $\boldsymbol{F}_q$. This method holds as well for Goppa codes, but only up to the BCH distance as specified by the BCH bound. The actual minimum distance of the Goppa code, in general, is larger and, for large codes, much larger.

In general, the task of projection of $\boldsymbol{v} \in \boldsymbol{F}_q$ into a subspace of $\boldsymbol{F}_q$ is formally intractable, meaning that any polynomial–time algorithm for this task can be converted to a polynomial–time algorithm to solve any problem in a large class of problems that are believed to be intractable.

15.8 Code-based cryptography

A code-based cryptosystem is based on the fact that, in general, it is not tractable to find a codeword that is closest to a given $\boldsymbol{v} \in \boldsymbol{F}_q^n$. This is because the task of projection into a subspace is formally intractable. This is the closest codeword problem. To use this fact to construct a cryptosystem, one must devise a way by which a secret key enables a method of projecting a point of the vector space $\boldsymbol{F}_q^n$ to a codeword $\boldsymbol{c} \in \mathcal{C}$.

As is also the case for lattice-based cryptography, the task of code-based cryptography is to secretly constrain this intractable problem by introducing a keyspace $\mathcal{K}$ in such a way that the problem remains intractable when given only the ciphertext and the keyspace $\mathcal{K}$, but becomes easy when given a specific key k from the keyspace.

The *McEliece cryptosystem* is an asymmetric public-key cryptosystem that neither depends on the discrete-log problem, nor on the factoring of integers. In fact, it is known that the McEliece cryptosystem is formally intractable. Specifically, any subexponential algorithm for breaking the McEliece cryptosystem can be used to solve a great many other computational problems that are generally believed to be exponentially hard. As with other applications involving intractability, this application uses a specific instance – namely a specific code – whereas the intractability statement refers to the general case.

The McEliece cryptosystem has some significant disadvantages, which is why it is not considered attractive. One disadvantage is that the length of the ciphertext is expanded by a factor of n/k. Moreover, the public key is large, perhaps as large as 500,000 bits. However, because it does not depend on the intractability of the discrete-log problem, the McEliece cryptosystem is an alternative cryptosystem that might become important if systems depending on the discrete-log problem or the factoring of integers would ever be broken.

The McEliece cryptosystem uses a Goppa code for the code $\mathcal{C}$. Although the same construction can be used to design a cryptosystem based on a code other than a Goppa code, the resistance against attacks must be demonstrated in each instance.

The code $\mathcal{C}$ is an (n, k) Goppa code with minimum distance $d_{\min}$ at least $2t + 1$, and for which a bounded-distance decoding algorithm up to radius τ is known. The parameters n, k, and t are public. The encryptor uses three matrices, a k-by-n generator matrix $\boldsymbol{G}$, a k-by-k nonsingular binary matrix $\boldsymbol{S}$, and an n-by-n permutation matrix $\boldsymbol{P}$, then computes as the generator matrix $\widehat{\boldsymbol{G}} = \boldsymbol{SGP}$. The public key is the n-by-k binary generator matrix $\widehat{\boldsymbol{G}}$.

A plaintext message $\boldsymbol{m}$, consisting of a binary block of blocklength n, is encrypted as $\widehat{\boldsymbol{c}} = \boldsymbol{m}\widehat{\boldsymbol{G}} + \widehat{\boldsymbol{e}}$, where the offset $\widehat{\boldsymbol{e}}$ is a randomly chosen binary word of blocklength n and weight τ. This can be regarded as $\widehat{\boldsymbol{c}} = \boldsymbol{aGP} + \widehat{\boldsymbol{e}}$, where $\boldsymbol{a} = \boldsymbol{mS}$ is a surrogate message and $\boldsymbol{P}$ is the permutation known to the decryptor. This can also be regarded as $\boldsymbol{c} = \boldsymbol{aG} + \widehat{\boldsymbol{e}}\boldsymbol{P}^{-1}$, where $\boldsymbol{c} = \widehat{\boldsymbol{c}}\boldsymbol{P}^{-1}$.

To decrypt the ciphertext $\widehat{\boldsymbol{c}}$, first undo the permutation by $\boldsymbol{cP}^{-1} = \widehat{\boldsymbol{c}}$. Then, by using the decoding algorithm to project $\widehat{\boldsymbol{c}}$ onto the subspace $\mathcal{C}$, remove $\widehat{\boldsymbol{e}}$ from $\widehat{\boldsymbol{c}}$ to recover $\boldsymbol{c}$. The modified message $\boldsymbol{a} = \boldsymbol{mS}$ is recovered from the codeword $\boldsymbol{c}$. Finally, the plaintext message is given by $\boldsymbol{m} = \boldsymbol{aS}^{-1}$.

Problems for Chapter 15

15.1 Prove that a lattice is a group under the addition of points.

15.2 A two-dimensional lattice is defined by the generators

$$\boldsymbol{m}_1 = (12, 5)$$
$$\boldsymbol{m}_2 = (14, 5).$$

Find the lattice point closest to (9, 3).

15.3 Prove that every discrete additive subgroup of $\boldsymbol{R}^n$ is a lattice either of $\boldsymbol{R}^n$ or of a subspace of $\boldsymbol{R}^n$. What is the meaning of the word "discrete" in this problem?

15.4 The dual lattice to the full-rank lattice $\boldsymbol{\Lambda} = \{\boldsymbol{v} = \boldsymbol{M}\boldsymbol{a} \mid \boldsymbol{a} \in \boldsymbol{Z}^n\}$ is $\boldsymbol{\Lambda}^\perp = \{\boldsymbol{v} \in \boldsymbol{Z}^n \mid \langle \boldsymbol{v} \cdot \boldsymbol{u} \rangle \in \boldsymbol{Z}$ for all $\boldsymbol{v} \in \boldsymbol{\Lambda}\}$. Prove that $\boldsymbol{\Lambda}^\perp$ is a full-rank lattice. Prove that if $\boldsymbol{B}$ is a generator matrix for $\boldsymbol{\Lambda}$, then $(\boldsymbol{B}^{-1})^T$ is a generator matrix for Λ^T, and that $\det \boldsymbol{\Lambda}^\perp = 1/\det \boldsymbol{\Lambda}$.

15.5 An ellipse in the plane $\boldsymbol{R}^2$ is given by

$$\left(\frac{u}{a}\right)^2 + \left(\frac{v}{b}\right)^2 = \frac{4}{\pi ab},$$

where

$$u = x\cos\theta - y\sin\theta$$
$$v = x\sin\theta + y\cos\theta.$$

This ellipse encloses a region of area four. Is there a choice of the three parameters (a, b, θ) to give an ellipse with area four, centered at the origin of the (x, y) plane, that does not enclose a nonzero point of the lattice $\boldsymbol{Z}^2$? Can you show this directly, without explicit reference to the Minkowski theorem?

15.6 Does Minkowski's theorem hold if the word "compact" is deleted?

15.7 Illustrate the proof of the Minkowski theorem for $n = 2$ by a series of sketches.

15.8 Describe the relationship between the generator polynomial and the generator matrix of a cyclic code. Describe the relationship between the check polynomial and the check matrix of a cyclic code.

15.9 Relate the minimum distance of a BCH code to the number of consecutive zeros of the generator polynomial $g(x)$.

Notes for Chapter 15

Lattices have been studied by mathematicians for many years, including early studies by Lagrange (1736–1813) and Gauss (1777–1855). The formal mathematical theory of lattices is now extensive. The Hadamard inequality was first published by Jacques Hadamard (1865–1963). The Hermite constant is named for Charles Hermite (1822–1901). Minkowski's convex body theorem, which is important to several branches of mathematics, is due to Hermann Minkowski (1864–1909). It was generalized by Blichfeldt (1873–1945). The LLL algorithm was published by Lenstra, Lenstra, and Lovász (1982), and later modified by Pohst (1987), Schnorr (1987), and others.

Certain lattices play a role in the efficient packing of spheres. It is well known that the most efficient packing of two-dimensional spheres (circles) places the sphere centers on the points of a hexagonal lattice. Gauss determined that the most efficient regular (lattice) packing of three-dimensional spheres is the familiar cannonball packing, formally known as the face-centered cubic packing. The *Kepler conjecture* (1611) opined that no nonregular packing in three dimensions is more efficient than the cannonball packing. This was settled by Hales (2005), by massive computer calculations that are still being verified. This anecdote, which relates to lattice theory, does give some indication that lattice problems are extremely hard.

Codes have been studied rather recently, first by engineers, and usually as subspaces of vector spaces over finite fields. Perhaps the most noteworthy early work was by Reed and Solomon (1960) as inspired by Shannon (1948). The intractability of the closest-codeword problem was demonstrated by Berlekamp, McEliece, and van Tilborg (1978). Public-key cryptography based on codes was introduced by McEliece (1978). The McEliece cryptosystem provides a form of public-key cryptography that is quite different from methods based on the intractability of biprime factoring or the discrete-log problem.

Public-key cryptography systems that are based on discrete geometrical structures such as codes or lattices have been under development starting with the code-based cryptography of McEliece (1978). Attention later turned to lattices based on the groundbreaking work of Ajtai (1996) on the complexity of lattice problems. In this paper, Ajtai presented a family of one-way functions whose worst-case hardness is protected by the hardness of certain shortest-vector lattice problems. Micciancio (2007) argued that this protection holds even for lattices with a circulant generator matrix. Motivated by the Ajtai paper, Ajtai and Dwork (1997) proposed the first lattice-based cryptosystem. Lattices had been introduced earlier into cryptology as a tool for cryptanalysis, as by Adleman (1983) and Shamir (1984) in their successful attacks on the trapdoor knapsack cryptosystem. This attack continues to raise cautions and provide a challenge with respect to lattice-based cryptography. Ajtai and Dwork turned this tool of attack into a tool of defense. Although Ajtai and Dwork proved that breaking their system in the *average* case is as hard as the shortest-vector problem, their method is impractical. An improved and practical lattice-based cryptosystem, similar to the McEliece code-based cryptosystem, was proposed later by Goldreich, Goldwasser, and Halevi (1997). The Goldreich–Goldwasser–Halevi cryptosystem is a trapdoor one-way function based on the fact that the *closest*-vector problem can be a hard problem. This encryption is fast, but the key size is impractically large. A specific and well-studied instance of ring-based cryptography is a cryptography system known as the NTRU cryptosystem. The NTRU cryptosystem of Hoffstein, Pipher, and Silverman (1998) is based on the fact that the *shortest*-vector problem can be a hard problem. Their work was not originally formulated using lattices. Its relationship to cyclic lattices and to the Goldreich, Goldwasser, and Halevi notions of lattice cryptography was only realized later as by Coppersmith and Shamir (1997). The vulnerability of lattice-based cryptosystems to cryptanalysis was studied by Nguyen and Stern (1998) and by Nguyen (1999). Yoshida (2003) suggested a particular family of elliptic curves to be used for the discrete vector decomposition problem. Duursma and Kiyavash (2005) showed that this family of curves is not secure.

Bibliography

L. M. Adleman, A Subexponential Algorithm for the Discrete Logarithm Problem with Applications to Cryptography, *Proceedings of the 20th Annual Symposium on the Foundations of Computer Science*, 55–60, 1979.

L. M. Adleman, On Breaking Generalized Knapsack Public Key Cryptosystems, *Proceedings of the 15th Annual ACM Symposium on the Theory of Computing*, 402–412, 1983.

L. M. Adleman, The Function Field Sieve, *Algorithmic Number Theory, Lecture Notes in Computer Science*, vol. 877, L. M. Adleman and M.-D. Huang, editors, pp. 108–121, New York, Springer, 1994.

L. M. Adleman and J. DeMarrais, A Subexponential Algorithm for Discrete Logarithms over All Finite Fields, *Mathematics of Computation*, **61**, 1–15, 1993.

L. M. Adleman and M.-D. Huang, Counting Rational Points on Curves and Abelian Varieties over Finite Fields, *Algorithmic Number Theory*, *Lecture Notes in Computer Science*, vol. 1122, H. Cohen, editor, pp. 1–16, Springer, 1996.

L. M. Adleman, J. DeMarrais, and M.-D. Huang, A Subexponential Algorithm for Discrete Logarithms over the Rational Subgroup of the Jacobians of Large Genus Hyperelliptic Curves over Finite Fields, *Algorithmic Number Theory*, *Lecture Notes in Computer Science*, vol. 877, pp. 28–40, New York, Springer, 1994.

M. Ajtai, Generating Hard Instances of Lattice Problems, *Proceedings of the 28th Annual ACM Symposium on the Theory of Computing*, 99–108, 1996.

M. Ajtai, The Shortest Vector Problem in NP-Hard for Randomized Reductions, *Proceedings of the 30th Annual ACM Symposium on Theory of Computing*, 10–19, 1998.

M. Ajtai and C. Dwork, A Public-Key Cryptosystem with Worst-Case/Average-Case Equivalence, *Proceedings of the 29th Annual ACM Symposium on the Theory of Computing*, 284–293, 1997.

W. Alford, A. Granville, and C. Pomerance, There Are Infinitely Many Carmichael Numbers, *Annals of Mathematics*, **140**, 703–722, 1994.

T. M. Apostol, *Introduction to Analytic Number Theory*, New York, Springer, 1976.

D. F. Aranha, K. Karabina, P. Longa, C. H. Gebotys, and J. López, Faster Explicit Formulas for Computing Pairings over Ordinary Curves, *Advances in Cryptology*, *EUROCRYPT11*, T. Rabin, editor, pp. 48–68, New York, Springer, 2011.

M. Artin, *Algebra*, Englewood Cliffs, NJ, Prentice Hall, 1991.

D. W. Ash, I. F. Blake, and S. A. Vanstone, Low Complexity Normal Bases, *Discrete Applied Mathematics*, **25**, 191–210, 1989.

C. Asmuth and J. Bloom, A Modular Approach to Key Safeguarding, *IEEE Transactions on Information Theory*, **IT-28**, 208–210, 1983.

A. O. L. Atkin, The Number of Points on an Elliptic Curve Modulo a Prime, (unpublished) 1988.

A. O. L. Atkin and F. Morain, Elliptic Curves and Primality Testing, *Mathematics of Computation*, **61**, 29–68, 1993.

L. Babai and S. Moran, Arthur–Merlin Games: A Randomized Proof System, and a Hierarchy of Complexity Class, *Journal of Computer and System Sciences*, **36**, 254–276, 1988.

E. Bach, Explicit Bounds for Primality Testing and Related Problems, *Mathematics of Computation*, **55**, 355–380, 1990.

E. Bach and K. Huber, Note on Taking Square-Roots Modulo N, *IEEE Transactions on Information Theory*, **IT-45**, 807–808, 1999.

R. Balasubramanian and N. Koblitz, The Improbability that an Elliptic Curve has a Subexponential Discrete Log Problem Using the Menezes–Okamoto–Vanstone Algorithm, *Journal of Cryptology*, **11**, 141–145, 1998.

T. H. Barr, *Invitation to Cryptology*, Upper Saddle River, NJ, Prentice Hall, 2002.

P. S. L. M. Barreto and M. Naehrig, Pairing-Friendly Elliptic Curves of Prime Order, *Selected Areas of Cryptography 05, Lecture Notes in Computer Science*, vol. 3897, B. Prennel and S. Tavares, editors, pp. 319–331, New York, Springer, 2006.

P. S. L. M. Barreto and J. F. Voloch, Efficient Computation of Roots in Finite Fields, *Design, Codes, and Cryptography*, **39**, 275–280, 2006.

P. S. L. M. Barreto, S. Galbraith, C. ÓhÉigeartaigh, and M. Scott, Efficient Pairing Computation on Supersingular Abelian Varieties, *Designs, Codes, and Cryptography*, **42**, 239–271, 2007.

P. S. L. M. Barreto, H. Kim, B. Lynn, and M. Scott, Efficient Algorithms for Pairing-Based Cryptosystems, *Advances in Cryptology, CRYPTO02*, M. Yung, editor, pp. 354–368, New York, Springer, 2002.

P. S. L. M. Barreto, B. Lynn, and M. Scott, Efficient Implementation of Pairing-Based Cryptosystems, *Journal of Cryptology*, **17**, 321–334, 2004.

P. Barrett, Implementing the Rivest, Shamir, and Adleman Public Key Encryption Algorithm on a Standard Digital Processor, *Advances in Cryptology, CRYPTO86*, A. M. Odlyzko, editor, pp. 311–323, New York, Springer, 1986.

M. Bauer, A Subexponential Algorithm for Solving the Discrete Logarithm Problem in the Jacobian of High Genus Hyperelliptic Curves over Arbitrary Finite Fields, (preprint) 1999.

I. Ben-Aroya and E. Biham, Differential Cryptanalysis of Lucifer, *Journal of Cryptology*, **9**, 21–34, 1996.

C. H. Bennett, G. Brassard, and A. K. Ekert, Quantum Cryptography, *Scientific American*, **10**, 132–134, 1992.

N. Berger and M. Scott, Constructing Tower Extensions of Finite Fields for Implementation of Pairing-Based Cryptography, *Arithmetic of Finite Fields, WAIFI 2010, Lecture Notes on Computer Science*, vol. 6087, M. A. Hasan and T. Helleseth, editors, pp. 180–195, 2010.

E. R. Berlekamp, *Algebraic Coding Theory*, New York, McGraw-Hill, 1968.

E. R. Berlekamp, Bit Serial Reed–Solomon Encoders, *IEEE Transactions on Information Theory*, **IT-28**, 869–874, 1982.

E. R. Berlekamp, Factoring Polynomials over Finite Fields, *Bell System Technical Journal*, **46**, 1853–1859, 1967.

E. R. Berlekamp, Factoring Polynomials over Large Finite Fields, *Mathematics of Computation*, **24**, 713–735, 1970.

E. R. Berlekamp, R. J. McEliece, and H. C. A. van Tilborg, On the Inherent Intractability of Certain Coding Problems, *IEEE Transactions on Information Theory*, **IT-24**, 203–207, 1978.

D. Bernstein and T. Lange, Faster Addition and Doubling on Elliptic Curves, *Advances in Cryptology, Asiacrypt07*, K. Kurosawa, editor, pp. 29–50, New York, Springer, 2007.

T. Beth and F. Piper, The Stop and Go Generator, *Advances in Cryptology, EUROCRYPT84*, T. Beth, N. Cot, and I. Ingemarsson, editors, pp. 88–92, New York, Springer, 1984.

E. Biham and A. Shamir, Differential Cryptanalysis of DES-Like Cryptosystems, *Journal of Cryptology*, **4**, 3–72, 1991.

E. Biham and A. Shamir, *Differential Cryptanalysis of the Data Encryption Standard*, New York, Springer, 1993.

R. E. Blahut, *Algebraic Codes on Lines, Planes, and Curves*, Cambridge University Press, 2008.

R. E. Blahut, Transform Techniques for Error Control Codes, *IBM Journal of Research and Development*, **23**, 299–315, 1979.

I. F. Blake, Curves, Codes and Cryptography, *Codes, Curves, and Signals*, A. Vardy, editor, pp. 63–75, Boston, MA, Kluwer, 1998.

I. F. Blake, Lattices and Cryptography, *Codes, Graphs, and Systems*, R. E. Blahut and R. Koetter, editors, pp. 317–332, Boston, MA, Kluwer, 2002.

I. F. Blake, R. Fuji-Hara, R. C. Mullin, and S. A. Vanstone, Computing Logarithms in Finite Fields of Characteristic Two, *SIAM Journal on Algebraic and Discrete Methods*, **5**, 276–285, 1984.

I. F. Blake, V. K. Murty, and G. Xu, Refinements of Miller's Algorithm for Computing the Tate/Weil Pairing, *Journal of Algorithms*, **58**, 134–149, 2006.

I. F. Blake, G. Seroussi, and N. Smart, *Elliptic Curves in Cryptography*, Cambridge University Press, 1999.

G. R. Blakley, Safeguarding Cryptographic Keys, *Proceedings of the National Computer Conference*, **48**, 313–317, 1979.

D. Bleichenbacher, A Chosen Ciphertext Attack against Protocols Based on the RSA Encryption Standard PKCS #1, *Advances in Cryptology, CRYPTO98*, H. Krawczyk, editor, pp. 1–12, New York, Springer, 1998.

M. Blum and S. Micali, How to Generate Cryptographically Strong Sequences of Pseudo-Random Bits, *SIAM Journal of Computing*, **13**, 850–864, 1984.

D. Boneh, The Decision Diffie–Hellman Problem, *Proceedings of the 3rd Algorithmic Number Theory Symposium, Lecture Notes in Computer Science*, vol. 1423, pp. 48–63, New York, Springer, 1998.

D. Boneh, Twenty Years of Attacks on the RSA Cryptosystem, *Notices of the American Mathematics Society*, **46**, 203–213, 1999.

D. Boneh and M. Franklin, Identity-Based Encryption from the Weil Pairing, *Advances in Cryptology, CRYPTO01*, J. Kilian, editor, pp. 213–229, New York, Springer, 2001.

D. Boneh and M. Franklin, Identity-Based Encryption from the Weil Pairing, *SIAM Journal of Computing*, **32**, 586-615, 2003.

D. Boneh and R. Venkatesan, Breaking RSA May Not Be Equivalent to Factoring, *Advances in Cryptology, EUROCRYPT98*, K. Nyberg, editor, pp. 59–71, New York, Springer, 1998.

D. Boneh, R. DeMillo, and R. Lipton, On the Importance of Checking Cryptographic Protocols for Faults, *Advances in Cryptology, EUROCRYPT97*, W. Fumy, editor, pp. 37–51, New York, Springer, 1997.

D. Boneh, B. Lynn, and H. Shacham, Short Signatures from the Weil Pairing, *Advances in Cryptology, Asiacrypt01*, C. Boyd, editor, pp. 514–532, New York, Springer, 2001.

N. Boston and M. Darnall, Elliptic and Hyperelliptic Curve Cryptography, *Cryptographic Engineering*, C. K. Koc, editor, pp. 171–189, New York, Springer, 2009.

L. Breiman, The Individual Ergodic Theorem of Information Theory, *Annals of Mathematical Statistics*, **28** (correction in vol. 31), pp. 809–811, 1957.

R. P. Brent and P. Zimmermann, Ten New Primitive Binary Trinomials, *Mathematics of Computation*, **78**, 1197–1199, 2009.

F. Brezing and A. Weng, Elliptic Curves Suitable for Pairing-Based Cryptography, *Designs, Codes, and Cryptography*, **37**, 133–141, 2005.

E. F. Brickell, Breaking Iterated Knapsacks, *Advances in Cryptology, CRYPTO84*, G. R. Blakley and D. Chaum, editors, pp. 342–358, New York, Springer, 1984.

E. F. Brickell and A. M. Odlyzko, Cryptanalysis, a Survey of Recent Results, *Proceedings of the IEEE*, **76**, 578–593, 1988.

M. E. Briggs, *An Introduction to the General Number Field Sieve*, M.S. Thesis, Virginia Polytechnic Institute, 1998.

L. Brynielsson, On the Linear Complexity of Combined Shift Register Sequences, *Advances in Cryptology, EUROCRYPT85*, H. C. Williams, editor, pp. 156–160, New York, Springer, 1985.

D. G. Cantor, Computing in the Jacobian of a Hyperelliptic Curve, *Mathematics of Computation*, **48**, 95–101, 1987.

D. G. Cantor and H. Zassenhaus, A New Algorithm for Factoring Polynomials over Finite Fields, *Mathematics of Computation*, **36**, 587–592, 1981.

C. Carlet, On Cryptographic Complexity of Boolean Functions, *Proceedings of the 6th Conference on Finite Fields and Applications to Coding Theory, Cryptography and Related Articles*, G. L. Mullen, H. Stichtenoth, and H. Tapia-Recillas, editors, pp. 53–69, New York, Springer, 2002.

R. D. Carmichael, Note on a New Number Theory Function, *Bulletin of the American Mathematical Society*, **16**, 232–238, 1910.

R. D. Carmichael, On Composite Numbers which Satisfy the Fermat Congruence, *American Mathematical Monthly*, **19**, 22–27, 1912.

R. D. Carmichael, On Sequences of Integers Defined by Recurrence Relations, *Quarterly Journal of Pure and Applied Mathematics*, **48**, 343–372, 1920.

A. Chan and R. A. Games, On the Quadratic Spans of Periodic Sequences, *Advances in Cryptology, CRYPTO89*, G. Brassard, editor, pp. 82–89, New York, Springer, 1989.

A. H. Chan, R. A. Games, and E. L. Key, On the Complexities of deBruijn Sequences, *Journal of Combinational Theory, Series A*, **33**, 233–246, 1982.

A. H. Chan, M. Goresky, and A. Klapper, On the Linear Complexity of Feedback Registers, *IEEE Transactions on Information Theory*, **IT-36**, 640–644, 1990.

D. Chaum, E. van Heijst, and B. Pfitzmann, Cryptographically Strong Undeniable Signatures, Unconditionally Secure for the Signer, *Advances in Cryptology, CRYPTO91*, J. Feigenbaum, editor, pp. 470–484, New York, Springer, 1991.

C. Cocks, Split Knowledge Generation of RSA Parameters, *Proceedings of the 6th IMA International Conference*, M. Darnell, editor, pp. 89–95, New York, Springer, 1997.

C. Cocks and R. G. E. Pinch, Identity-Based Cryptosystems Based on Weil Pairing, (unpublished) 2001.

D. Coppersmith, Fast Evaluation of Logarithms in Fields of Characteristic Two, *IEEE Transactions on Information Theory*, **IT-30**, 587–594, 1984.

D. Coppersmith, Modifications to the Number Field Sieve, *Journal of Cryptology*, **6**, 169–180, 1993a.

D. Coppersmith, Solving Linear Equations over $GF(2)$: Block Lanczos Algorithms, *Linear Algebra and its Applications*, **192**, 33–60, 1993b.

D. Coppersmith, Solving Homogeneous Linear Equations over $GF(2)$ via Block Wiedemann Algorithms, *Mathematics of Computation*, **62**, 333–350, 1994a.

D. Coppersmith, The Data Encryption Standard (DES) and Its Strength against Attacks, *IBM Journal of Research and Development*, **38**, 243–250, 1994b.

D. Coppersmith, Small Solutions to Polynomial Equations, and Low Exponent RSA Vulnerabilities, *Journal of Cryptology*, **10**, 233–260, 1997.

D. Coppersmith and A. Shamir, Lattice Attacks on NTRU, *Advances in Cryptology, EUROCRYPT97*, W. Fumy, editor, pp. 52–61, New York, Springer, 1997.

D. Coppersmith, M. Franklin, J. Patarin, and M. Reiter, Low-Exponent RSA with Related Messages, *Advances in Cryptology, EUROCRYPT96*, U. Maurer, editor, pp. 1–9, New York, Springer, 1996.

D. Coppersmith, H. Krawczyk, and Y. Mansour, The Shrinking Generator, *Advances in Cryptology, CRYPTO93*, D. R. Stinson, editor, pp. 22–39, 1993.

D. Coppersmith, A. M. Odlyzko, and R. Schroeppel, Discrete Logarithms in $GF(p)$, *Algorithmica*, **1**, 1–15, 1986.

T. H. Cormen, C. E. Leiserson, and R. L. Rivest, *Introduction to Algorithms*, Cambridge, MA, Massachusetts Institute of Technology Press, 1990.

G. Cornacchia, Su di un Metodo per la Risoluzione in Numeri Interi dell Equazione $\sum_{k=0}^{n} C_k x^{n-k} y^k = P$, *Giomale di Mathematiche di Battaglini*, **46**, 33–90, 1908.

D. Cox, J. Little, and D. O'Shea, *Ideals, Varieties, and Algorithms*, New York, Springer, 1992.

R. Cramer and V. Shoup, A Practical Public Key Cryptosystem Provably Secure against Adaptive Chosen Ciphertext Attack, *Advances in Cryptology, CRYPTO98*, H. Krawczyk, editor, pp. 13–25, New York, Springer, 1998.

I. Csiszár and J. Körner, Broadcast Channels with Confidential Messages, *IEEE Transactions on Information Theory*, **IT-24**, 339–348, 1978.

J. Daemen and V. Rijmen, The Block Cipher Rijndael, *Smart Card Research and Applications*, J.-J. Quisquater and B. Schneier, editors, pp. 288-296, New York, Springer, 2000.

J. Daemen and V. Rijmen, Rijndael, the Advanced Encryption Standard, *Dr. Dobb's Journal*, **26**(3), 137–139, 2001.

I. B. Damgaard, A Design Principle for Hash Functions, *Advances in Cryptology, EUROCRYPT89*, J.-J. Quisquater and J. Vandewalle, editors, pp. 416–427, New York, Springer, 1989.

Data Encryption Standard (DES), National Bureau of Standards FIPS Publication 46, 1977.

N. G. deBruijn, A Combinatorial Problem, *Indagationes Mathematicae*, **8**, 461–467, 1946.

N. G. deBruijn, On the Number of Positive Integers $\leq x$ and Free of Prime Factors $> y$, *Indagationes Mathematicae*, **13**, 50–60, 1951.

J. M. DeLaurentis, A Further Weakness in the Common Modulus Protocol for the RSA Cryptoalgorithm, *Cryptologia*, **8**, 253–259, 1984.

K. Dickman, On the Frequency of Numbers Containing Prime Factors of a Certain Relative Magnitude, *Arkiv for Matematik Astronomi och Fysic*, **10**, 1–14, 1930.

W. Diffie and M. E. Hellman, New Directions in Cryptography, *IEEE Transactions on Information Theory*, **IT-22**, 644–654, 1976a.

W. Diffie and M. E. Hellman, Multiuser Crytographic Techniques, *Federal Information Processing Standard Conference Proceedings*, **45**, 109–112, 1976b.

W. Diffie and M. E. Hellman, Privacy and Authentication: An Introduction to Cryptography, *Proceedings of the IEEE*, **67**, 397–427, 1979.

V. S. Dimitrov and T. Cooklev, Hybrid Algorithm for the Computation of the Matrix Polynomial $I + A + \cdots + A^{N-1}$, *IEEE Transactions on Circuits and Systems*, **CS-42**, 377–380, 1995.

J. D. Dixon, Asymptotically Fast Factorization of Integers, *Mathematics of Computation*, **36**, 255–260, 1981.

E. Dubrova, Finding Matching Initial States for Equivalent NLFSRs in the Fibonacci and the Galois Configurations, *IEEE Transactions on Information Theory*, **IT-56**, 2961–2966, 2010.

R. Dupont, A. Enge, and F. Moran, Building Curves with Arbitrary Small MOV Degree over Finite Prime Fields, *Journal of Cryptology*, **18**, 79–89, 2005.

I. M. Duursma and N. Kiyavash, The Vector Decomposition Problem for Elliptic and Hyperelliptic Curves, *Journal of the Ramanujan Mathematics Society*, **20**, 59–76, 2005.

I. M. Duursma and H.-S. Lee, Tate Pairing Implementation for Hyperelliptic Curves $y^2 = x^p - x + d$, *Advances in Cryptology, Asiacrypt03*, C.-S. Laih, editor, pp. 111–123, New York, Springer, 2003.

H. M. Edwards, A Normal Form for Elliptic Curves, *Bulletin of the American Mathematics Society*, **44**, 393–422, 2007.

K. Eisenträger, K. Lauter, and P. L. Montgomery, Improved Weil and Tate Pairings for Elliptic and Hyperelliptic Curves, *Algebraic Number Theory, Lecture Notes in Computer Science*, vol. 3076, pp. 169–183, 2004.

P. Ekdahl and T. Johansson, Another Attack on A5/1, *IEEE Transactions on Information Theory*, **IT-49**, 284–289, 2003.

A. K. Ekert, Quantum Cryptography Based on Bell's Theorem, *Physics Review Letters*, **67**(6), 661–663, 1991.

T. Elgamal, A Public Key Cryptosystem and a Signature Scheme Based on Discrete Logarithms, *IEEE Transactions on Information Theory*, **IT-31**, 469–472, 1985a.

T. Elgamal, A Subexponential-Time Algorithm for Computing Discrete Logarithms over $GF(p^2)$, *IEEE Transactions on Information Theory*, **IT-31**, 473–481, 1985b.

N. D. Elkies, Explicit Isogenics, (unpublished) 1991.

N. D. Elkies, Elliptic and Modular Curves over Finite Fields and Related Computational Issues, *Advances in Cryptology, Asiacrypt98*, K. Ohta and D. Pei, editors, pp. 21–76, New York, Springer, 1998.

A. Enge, *Elliptic Curves and Their Application to Cryptography: An Introduction*, Dordrecht, Kluwer, 1999.

T. Etzion, Linear Complexity of deBruijn Sequences: Old and New Results, *IEEE Transactions on Information Theory*, **IT-45**, 693–698, 1999.

U. Feige, A. Fiat, and A. Shamir, Zero-Knowledge Proofs of Identity, *Journal of Cryptology*, **1**, 77–94, 1988.

H. Feistel, *Cryptographic Coding for Data-Bank Privacy*, RC2827, Yorktown Heights, NY, IBM Research, 1970.

H. Feistel, Block Cipher Cryptographic System, US Patent #3,798,359 (filed June 1971) March 1974.

H. Feistel, Cryptography and Computer Privacy, *Scientific American*, **228**, 15–23, 1973.

A. Fiat and M. Naor, Rigorous Time/Space Trade-Offs for Inverting Functions, *Proceedings of the 23rd Annual ACM Symposium on the Theory of Computing*, 534–541, 1991.

A. Fiat and A. Shamir, How to Prove Yourself: Practical Solutions to Identification and Signature Problems, *Advances in Cryptology, CRYPTO86*, A. M. Odlyzko, editor, pp. 186–194, New York, Springer, 1986.

C. Flye Sainte-Marie, Solution to Question No. 48, *Intermédiare des Mathématiciens*, **1**, 107–110, 1894.

K. Fong, D. Hankerson, J. López, and A. Menezes, Field Inversion and Point Halving Revisited, *IEEE Transactions on Computers*, **C-53**, 1047–1059, 2004.

D. Freeman, Constructing Pairing-Friendly Elliptic Curves with Embedding Degree 10, *Algorithmic Number Theory, Lecture Notes in Computer Science*, vol. 4076, 2006.

D. Freeman, M. Scott, and E. Teske, A Taxonomy of Pairing-Friendly Curves, *Journal of Cryptology*, **23**, 224–280, 2010.

G. Frey and H. G. Rück, A Remark Concerning m-divisibility and the Discrete Logarithm Problem in the Divisor Class Group of Curves, *Mathematics of Computation*, **62**, 865–874, 1994.

G. Frey, M. Miller, and H. G. Rück, The Tate Pairing and the Discrete Logarithm Applied to Elliptic Curve Cryptosystems, *IEEE Transactions on Information Theory*, **IT-45**, 1717–1719, 1999.

S. D. Galbraith, Supersingular Curves in Cryptography, *Advances in Cryptology, Asiacrypt01*, C. Boyd, editor, pp. 495–513, New York, Springer, 2001.

S. D. Galbraith, *Mathematics of Public-Key Cryptography*, Cambridge University Press, 2012.

S. D. Galbraith, K. Harrison, and D. Soldera, Implementing the Tate Pairing, *Algorithmic Number Theory, Lecture Notes in Computer Science*, vol. 2369, C. Fieker and D. Kohel, editors, pp. 324–337, Springer, 2002.

S. D. Galbraith, F. Hess, and F. Vercauteren, Aspects of Pairing Inversion, *IEEE Transactions on Information Theory*, **IT-54**, 5719–5728, 2008.

S. D. Galbraith, F. McKee, and P. C. Valença, Ordinary Abelian Varieties Having Small Embedding Degrees, *Finite Fields and Their Applications*, **13**, 800–814, 2007.

S. D. Galbraith, K. Paterson, and N. Smart, Pairing for Cryptographers, *Discrete Applied Mathematics*, **156**, 3113–3121, 2008.

S. Gao and H. W. Lenstra, Jr., Optimal Normal Bases, *Designs, Codes, and Cryptography*, **2**, 315–323, 1992.

S. Gao and J. von zur Gathen, Berlekamp's and Niederreiter's Polynomial Factorization Algorithms, *Contemporary Mathematics*, **168**, 101–116, 1994.

P. Garrett, *Making, Breaking Codes: An Introduction to Cryptology*, Upper Saddle River, NJ, Prentice Hall, 2001.

P. Gaudry, An Algorithm for Solving the Discrete Log Problem on Hyperelliptic Curves, *Advances in Cryptology, EUROCRYPT00*, B. Preneel, editor, pp. 19–34, New York, Springer, 2000.

P. Gaudry, E. Thome, N. Thériault, and C. Diem, A Double Large Prime Variation for Small Genus Hyperelliptic Index Calculus, *Mathematics of Computation*, **76**, 475–492, 2007.

P. R. Geffe, How to Protect Data with Ciphers that Are Really Hard to Break, *Electronics*, **46**, 99–101, 1973.

C. Gentry, *A Fully Homomorphic Encryption System*, Ph.D. Thesis, Stanford University, 2009.

C. Gentry, Fully Homomorphic Encryption Using Ideal Lattices, *Proceedings of the 41st Annual ACM Symposium on the Theory of Computing*, 169–178, 2009.

J. K. Gibson, Discrete Logarithm Hash Function that Is Collision Free and One Way, *IEEE Proceedings*, **138**, 407–410, 1991.

R. Gold, Optimal Binary Sequences for Spread Spectrum Multiplexing, *IEEE Transactions on Information Theory*, **IT-13**, 619–621, 1967.

O. Goldreich, S. Goldwasser, and S. Halevi, Public-Key Cryptosystems from Lattice Reduction Problems, *Advances in Cryptology, CRYPTO97*, B. S. Kaliski, Jr., editor, pp. 112–131, New York, Springer, 1997.

S. Goldwasser, The Search for Provably Secure Cryptosystems, *Proceedings of Symposia in Applied Mathematics*, vol. 42, *Cryptology and Computational Number Theory*, pp. 89–113, Providence, RI, American Mathematical Society, 1990.

S. Goldwasser and J. Kilian, Primality Testing Using Elliptic Curves, *Journal of the Association for Computing Machinery*, **46**, 450–452, 1999.

S. Goldwasser, S. Micali, and C. Rackoff, The Knowledge Complexity of Interactive Proof Systems, *SIAM Journal of Computing*, **18**, 186–208, 1989.

J. Dj. Golic and R. Menicocci, Statistical Distinguishers for Irregularly Decimated Linear Recurring Sequences, *IEEE Transactions on Information Theory*, **IT-52**, 1153–1159, 2006.

S. W. Golomb, *Digital Communications with Space Applications*, Englewood Cliffs, NJ, Prentice-Hall, 1964.

S. W. Golomb, *Shift Register Sequences*, San Francisco, CA, Holden-Day, 1967, 2nd edition, Walnut Creek, CA, Aegean Park Press, 1982.

S. W. Golomb and G. Gong, *Signal Design for Good Correlation*, Cambridge University Press, 2005.

S. W. Golomb and L. R. Welch, *Nonlinear Shift-Register Sequences*, JPL Memo No. 20–149, Pasadena, CA, Jet Propulsion Laboratory, 1957.

G. Gong and S. W. Golomb, Transform Domain Analysis of DES, *IEEE Transactions on Information Theory*, **IT-45**, 2065–2073, 1999.

D. M. Gordon, Discrete Logarithms in $GF(p)$ Using the Number Field Sieve, *SIAM Journal of Discrete Mathematics*, **6**, 124–138, 1993.

D. M. Gordon, A Survey of Fast Exponentiation Methods, *Journal of Algorithms*, **27**, 129–146, 1998.

M. Goresky and A. Klapper, *Algebraic Shift Register Sequences*, Cambridge University Press, 2012.

M. Goresky and A. Klapper, Fibonacci and Galois Representations of Feedback-with-Carry Shift Registers, *IEEE Transactions on Information Theory*, **IT-48**, 2826–2836, 2002.

R. Granger and F. Vercauteren, On the Discrete Logarithm Problem on Algebraic Tori, *Advances in Cryptology, CRYPTO05*, V. Shoup, editor, pp. 66–85, New York, Springer, 2005.

R. Granger, D. Page, and M. Stam, Hardware and Software Normal Basis Arithmetic for Pairing-Based Cryptography in Characteristic Three, *IEEE Transactions on Computers*, **C-54**, 852–860, 2005.

E. J. Groth, Generation of Binary Sequences with Controllable Complexity, *IEEE Transactions on Information Theory*, **IT-17**, 288–296, 1971.

G. Guanella, Means for and Method for Secret Signaling, US Patent #2,405,500, 1946.

L. C. Guillou and J. J. Quisquater, A Practical Zero-Knowledge Protocol Fitted to Security Microprocessor Minimizing Both Transmission and Memory, *Advances in Cryptology, EUROCRYPT87*, D. Chaum and W. L. Price, editors, pp. 123–128, New York, Springer, 1987.

L. C. Guillou and J. J. Quisquater, Method and Apparatus for Authenticating Accreditations and for Authenticating and Signing Messages, US Patent #5, 140, 634, 1992.

C. G. Günther, Alternating Step Generators Controlled by deBruijn Sequences, *Advances in Cryptology, EUROCRYPT87*, D. Chaum and W. L. Price, editors, pp. 5–14, New York, Springer, 1987.

T. C. Hales, A Proof of the Kepler Conjecture, *Annals of Mathematics*, **162**, 1065–1185, 2005.

M. Hall, An Isomorphism between Linear Recurring Sequences and Algebraic Rings, *Transactions of the American Mathematics Society*, **44**, 196–218, 1938.

D. Hankerson, A. Menezes, and S. Vanstone, *Guide to Elliptic Curve Cryptography*, New York, Springer, 2004.

R. Hartshorne, *Algebraic Geometry*, New York, Springer, 1977.

H. Hasse, Theorie der höheren Differentiale in einem algebraischen Funktionenkörper mit vollkommenen Konstantenkörper bei beliebiger Charakteristik, *Journal für die Reine and Angewandte Mathematik*, **175**, 50–54, 1936.

M. E. Hellman, An Extension of the Shannon Theory Approach to Cryptography, *IEEE Transactions on Information Theory*, **IT-23**, 289–294, 1977.

M. E. Hellman, A Cryptanalytic Time–Memory Tradeoff, *IEEE Transactions on Information Theory*, **IT-26**, 401–406, 1980.

M. E. Hellman and J. M. Reyneri, Fast Computation of Discrete Logarithms in $GF(q)$, *Advances in Cryptology, CRYPTO83*, D. Chaum, editor, pp. 3–13, New York, Plenum Press, 1983.

K. Hensel, Über die Darstellung der Zahlen eines Gattungsbereiches für einen Beliebigen Primdivisor, *Journel für die Reine und Angewandte Mathematik*, **103**, 230–237, 1888.

T. Herlestam, On Functions of Linear Shift Register Sequences, *Advances in Cryptology, EUROCRYPT85*, F. Pichler, editor, pp. 119–129, New York, Springer, 1985.

F. Hess, N. P. Smart, and F. Vercauteren, The Eta Pairing Revisited, *IEEE Transactions on Information Theory*, **IT-52**, 4595–4602, 2006.

L. S. Hill, Cryptography in an Algebraic Alphabet, *American Mathematical Monthly*, **36**, 306–312, 1929.

L. S. Hill, Concerning Certain Linear Transformation Apparatus of Cryptography, *American Mathematical Monthly*, **38**, 135–154, 1931.

J. Hoffstein, J. Pipher, and J. H. Silverman, NTRU: A Ring-Based Public Key Cryptosystem, *Algorithmic Number Theory, Lecture Notes in Computer Science*, vol. 1423, J. P. Buhler, editor, pp. 267–288, New York, Springer, 1998.

J. Hoffstein, J. Pipher, and J. H. Silverman, *An Introduction to Mathematical Cryptography*, New York, Springer, 2008.

D. A. Huffman, A Method for the Construction of Minimum Redundancy Codes, *Proceedings of the IRE*, **40**, 1091–1101, 1952.

T. Itoh and S. Tsujii, A Fast Algorithm for Computing Multiplicative Inverses in $GF(2^m)$ Using Normal Bases, *Information and Computation*, **78**, 171–177, 1988.

M. J. Jacobson, N. Koblitz, J. H. Silverman, A. Stein, and E. Teske, Analysis of the Xedni Calculus Attack, *Designs, Codes, and Cryptography*, **20**, 41–64, 2000.

M. Jacobson, Jr., A. Menezes, and A. Stein, Hyperelliptic Curves and Cryptography, *High Primes and Misdemeanors: Lectures in Honour of the 60th Birthday of Hugh Cowie Williams*, A. van der Poorten and C. M. Ringel, editors, pp. 255–282, Toronto, Fields Institute Communications, 2004.

W. S. Jevons, *The Principles of Science*, London, Macmillan, 1874.

A. Joux, A One-Round Protocol for Tripartite Diffie–Hellman, *Proceedings of the 4th International Symposium on Algorithmic Number Theory*, 385–394, New York, Springer, 2000.

A. Juels and M. Sudan, A Fuzzy Vault Scheme, *Design, Codes, and Cryptography*, **38**, 237–257, 2006.

D. Kahn, *The Codebreakers: The Story of Secret Writing*, London, Macmillan, 1967. Revised edition, New York, Scribner, 1996.

B. S. Kaliski, R. L. Rivest, and A. T. Sherman, Is the Data Encryption Standard a Group?, *Advances in Cryptology, EUROCRYPT85*, F. Pichler, editors, pp. 81–92, New York, Springer, 1985.

A. Karatsuba and Y. Ofman, Multiplication of Many-Digital Numbers by Automatic Computers, *Proceedings of the USSR Academy of Science*, **145**, 293–294, 1962.

E. Karnin, J. Greene, and M. Hellman, On Secret Sharing Systems, *IEEE Transactions on Information Theory*, **IT-29**, 35–41, 1983.

T. Kasami, *Weight Distribution Formula for Some Class of Cyclic Code*, Technical Report No. R-285, Urbana–Champaign, IL, University of Illinois, 1966.

J. Katz and Y. Lindell, *Introduction to Modern Cryptography*, Boca Raton, FL, CRC Press, 2007.

K. S. Kedlaya, Counting Points on Hyperelliptic Curves Using Monsky–Washnitzer Cohomology, *Journal of the Ramanujan Mathematical Society*, **16**, 323–338, 2001.

J. Kelsey and T. Kohno, Herding Hash Functions and the Nostradamus Attack, *Advances in Cryptology, EUROCRYPT06*, S. Vaudenay, editor, pp. 183–200, New York, Springer, 2006.

E. L. Key, An Analysis of the Structure and Complexity of Nonlinear Binary Sequence Generators, *IEEE Transactions on Information Theory*, **IT-22**, 732–736, 1976.

A. Klapper, The Vulnerability of Geometric Sequences Based on Fields of Odd Characteristic, *Journal of Cryptology*, **7**, 33–51, 1994.

T. Kleinjung, On Polynomial Selection for the General Number Field Sieve, *Mathematics of Computation*, **75**, 2037–2047, 2006.

E. Knudsen, Elliptic Scalar Multiplication Using Point Halving, *Advances in Cryptology, Asiacrypt99*, K.-Y. Lam, E. Okamoto, and C. Xing, editors, pp. 1351–1491, New York, Springer, 1999.

A. H. Koblitz, N. Koblitz, and A. Menezes, Elliptic Curve Cryptography: The Serpentine Course of a Paradigm Shift, *Journal of Number Theory*, **131**, 781–814, 2011.

N. Koblitz, Elliptic Curve Cryptosystems, *Mathematics of Computation*, **48**, 203–209, 1987.

N. Koblitz, A Family of Jacobians Suitable for Discrete Log Cryptosystems, *Advances in Cryptology, CRYPTO88*, S. Goldwasser, editor, pp. 94–99, New York, Springer, 1988.

N. Koblitz, Hyperelliptic Cryptosystems, *Journal of Cryptology*, **1**, 139–150, 1989.

N. Koblitz, Jacobi Sums, Irreducible Zeta Polynomials, and Cryptography, *Canadian Mathematical Bulletin*, **34**, 229–235, 1991.

N. Koblitz, *Algebraic Aspects and Cryptography*, Berlin, Springer,1998.

N. Koblitz, A. Menezes, and S. Vanstone, The State of Elliptic Curve Cryptography: Towards a Quarter-Century of Public Key Cryptography, *Designs, Codes, and Cryptography*, **19**, 173–193, 2000.

P. C. Kocher, Timing Attacks on Implementations of Diffie–Hellman, RSA, DSS, and Other Systems, *Advances in Cryptology, CRYPTO96*, N. Koblitz, editor, pp. 104–113, New York, Springer, 1996.

A. R. Korselt, Probléme chinois, *L'Intermédiaire des Mathématiciens*, **6**, 142–143, 1899.

K. Koyama, U. Maurer, T. Okamoto, and S. A. Vanstone, New Public-Key Schemes Based on Elliptic Curves over the Ring Z_n, *Advances in Cryptology, CRYPTO91*, J. Feigenbaum, editor, pp. 252–266, New York, Springer, 1991.

M. Kraitchik, *Théorie des Nombres*, vol. 1, Paris, Gauthier-Villars, 1922.

J. C. Lagarias and A. M. Odlyzko, Solving Low-Density Subset Sum Problems, *Journal of the Association of Computing Machinery*, **32**, 229–246, 1985.

X. Lai, J. L. Massey, and S. Murphy, Markov Ciphers and Differential Cryptanalysis, *Advances in Cryptology, EUROCRYPT91*, D. W. Davies, editor, pp. 17–38, New York, Springer, 1991.

B. A. LaMacchia and A. M. Odlyzko, Computation of Discrete Logarithms in Prime Fields, *Design, Codes, and Cryptography*, **1**, 47–62, 1991.

T. Lange, *Fast Arithmetic on Hyperelliptic Curves*, Ph.D. Thesis, Institute for Information Security and Cryptography, Ruhr-Universität Bochum, 2002.

S. K. Langford and M. E. Hellman, Differential-Linear Cryptanalysis, *Advances in Cryptology, CRYPTO94*, Y. Desmedt, editor, pp. 17–25, New York, Springer, 1994.

E. Lee, H.-S. Lee, and C.-M. Park, Efficient and Generalized Pairing Computation on Abelian Varieties, *IEEE Transactions on Information Theory*, **IT-55**, 1793–1803, 2009.

A. K. Lenstra and H. W. Lenstra, Jr., Algorithms in Number Theory, *Handbook of Theoretical Computer Science*, vol. A, *Algorithms and Complexity*, J. van Leeuwen, editor, pp. 673–715, New York, Elsevier, 1990.

A. K. Lenstra and H. W. Lenstra, Jr., *The Development of the Number Field Sieve, Lecture Notes in Mathematics*, vol. 1554, New York, Springer, 1993.

A. K. Lenstra, H. W. Lenstra, Jr., and L. Lovász, Factoring Polynomials with Rational Coefficients, *Mathematische Annalen*, **261**, 515–534, 1982.

A. K. Lenstra, H. W. Lenstra, Jr., M. S. Manasse, and J. M. Pollard, The Number Field Sieve, *The Development of the Number Field Sieve, Lecture Notes in Mathematics*, A. K. Lenstra and H. W. Lenstra, Jr., editors, vol. 1554, pp. 11–42, New York, Springer, 1993.

H. W. Lenstra, Jr., Primality and Factorization, *Proceedings of the 4th Symposium on Information Theory in the Benelux*, Acco, Leuven, Belgium, pp. 13–15, 1983.

H. W. Lenstra, Jr., Factoring Integers with Elliptic Curves, *Annals of Mathematics*, **126**, 649–673, 1987.

H. W. Lenstra, Jr., Rijndael for Algebraists, (unpublished) 2002.

R. Lercier, Computing Isogenies in F_2^n, *Algorithmic Number Theory, Proceedings of the 2nd International Symposium ANTS-II*, pp. 197–212, New York, Springer, 1996.

R. Lercier and F. Morain, Counting the Number of Points on Elliptic Curves over Finite Fields: Strategies and Performances, *Advances in Cryptology, EUROCRYPT95*, L. C. Guillou and J.-J. Quisquater, editors, pp. 79–94, New York, Springer, 1995.

S. K. Leung-Yan-Cheong and M. E. Hellman, The Gaussian Wire-Tap Channel, *IEEE Transactions on Information Theory*, **IT-24**, 451–456, 1978.

W. J. LeVeque, *Fundamentals of Number Theory*, Reading, MA, Addison-Wesley, 1977; republished by Dover, Mineola, NY, 1996.

S. Levy, *Crypto: How the Code Rebels Beat the Government – Saving Privacy in the Digital Age*, New York, Penguin Books, 2001.

S. Lichtenbaum, Duality Theorems for Curves over P-adic Fields, *Inventiones Mathematicae*, **7**, 120–136, 1969.

R. Lidl and H. Niederreiter, *Finite Fields*, vol. 20 of *The Encyclopedia of Mathematics*, Cambridge University Press, 1983.

L. Lovasz, *An Algorithmic Theory of Number, Graphs, and Convexity*, Philadelphia, PA, SIAM Publications, 1986.

M. Luby and C. Rackoff, How to Construct Pseudorandom Permutations and Pseudorandom Functions, *SIAM Journal of Computing*, **17**, 373–386, 1988.

F. Luca, D. J. Mireles, and I. E. Shparlinski, MOV Attack in Various Subgroups on Elliptic Curves, *Illinois Journal of Mathematics*, **48**, 1041–1052, 2004.

J. L. Massey, Shift Register Synthesis and BCH Decoding, *IEEE Transactions on Information Theory*, **IT-15**, 122–127, 1969.

J. L. Massey and R.-W. Liu, Equivalence of Nonlinear Shift Registers, *IEEE Transactions on Information Theory*, **IT-10**, 378–379, 1964.

J. L. Massey and S. Serconek, A Fourier Transform Approach to the Linear Complexity of Nonlinearly Filtered Sequences, *Advances in Cryptology, CRYPTO94*, Y. Oesmedt, editor, pp. 332-340, New York, Springer, 1994.

J. L. Massey and S. Serconek, Linear Complexity of Periodic Sequences, *Advances in Cryptology, CRYPTO96*, N. Koblitz, editor, pp. 358–371, New York, Springer 1996.

M. Matsui, Linear Cryptanalysis Method for the DES Cipher, *Advances in Cryptology, EUROCRYPT93*, T. Helleseth, editor, pp. 386–397, New York, Springer, 1993.

M. Matsui, The First Experimental Cryptanalysis of the Data Encryption Standard, *Advances in Cryptology, EUROCRYPT94*, A. De Santis, editor, pp. 1–11, New York, Springer, 1994.

U. M. Maurer, Conditionally Perfect Secrecy and a Provably Secure Randomized Cipher, *Journal of Cryptology*, **5**, 53–66, 1992.

U. M. Maurer, Secret Key Agreement by Public Discussion from Common Information, *IEEE Transactions on Information Theory*, **IT-39**, 733–742, 1993.

U. M. Maurer and J. L. Massey, Cascade Ciphers: The Importance of Being First, *Journal of Cryptology*, **6**, 55–61, 1993.

K. McCurley, The Discrete Logarithm Problem, *Proceedings of Symposia in Applied Mathematics*, vol. 42, *Cryptology and Computational Number Theory*, pp. 49–74, Providence, RI, American Mathematical Society, 1990.

R. J. McEliece, *A Public-Key Cryptosystem Based on Algebraic Coding Theory*, DSN Progress Report No. 42–44, pp. 114–116, Pasadena, CA, Jet Propulsion Laboratory, 1978.

R. J. McEliece and D. V. Sarwate, On Sharing Secrets and Reed-Solomon Codes, *Communications of the Association for Computing Machinery*, **24**, 583–584, 1981.

B. McMillan, The Basic Theorems of Information Theory, *Annals of Mathematical Statistics*, **24**, 196–219, 1953.

B. McMillan, Two Inequalities Implied by Unique Decipherability, *IRE Transactions in Information Theory*, **IT-2**, 115–116, 1956.

W. Meier and O. Staffelbach, Fast Correlation Attacks on Stream Ciphers, *Advances in Cryptology, EUROCRYPT88*, C. Günther, editor, pp. 301–314, New York, Springer, 1988.

W. Meier and O. Staffelbach, Fast Correlation Attacks on Certain Stream Ciphers, *Journal of Cryptography*, **1**, 159–176, 1989.

A. J. Menezes, Hyperelliptic Cryptosystems, *Journal of Cryptology*, **1**, 139–150, 1989.

A. J. Menezes, *Applications of Finite Fields*, Dordrecht, Kluwer, 1993.

A. J. Menezes, *Elliptic Curve Cryptosystems*, Dordrecht, Kluwer, 1997.

A. J. Menezes, An Introduction to Pairing-Based Cryptography, *Recent Trends in Cryptography Summer School*, I. Luengo, editor, vol. 477, Providence, RI, American Mathematical Society, 2005.

A. J. Menezes and S. A. Vanstone, The Implementation of Elliptic Curve Cryptography, *Advances in Cryptology, AUSCRYPT90*, J. Seberry and J. Pieprzyk, editors, pp. 2–13, New York, Springer, 1990.

A. J. Menezes, T. Okamoto, and S. A. Vanstone, Reducing Elliptic Curve Logarithms to Logarithms in a Finite Field, *IEEE Transactions on Information Theory*, **IT-39**, 1639–1646, 1993.

A. J. Menezes, P. van Oorschot, and S. A. Vanstone, *Handbook of Applied Cryptography*, New York, CRC Press, 1997.

A. J. Menezes, S. A. Vanstone, and R. J. Zuccherato, Counting Points on Elliptic Curves over F_2^m, *Mathematics of Computation*, **60**, 407–420, 1993.

A. J. Menezes, Y.-H. Wu, and R. J. Zuccherato, An Elementary Introduction to Hyperelliptic Curves, Appendix in *Algebraic Aspects of Cryptography*, N. Koblitz, editor, pp. 155–178, New York, Springer, Berlin, 1998.

R. C. Merkle, Secure Communications over Insecure Channels, *Communications of the Association for Computing Machinery*, **21**, 294–299, 1978.

R. C. Merkle, *Secrecy, Authentication, and Public-Key Systems*, Ph.D. Dissertation, Department of Electrical Engineering, Stanford University, 1979.

R. C. Merkle, One Way Hash Functions and DES, *Advances in Cryptology, CRYPTO89*, G. Brassard, editor, pp. 428–446, New York, Springer, 1989.

R. C. Merkle, A Fast Software One-Way Hash Function, *Journal of Cryptology*, **3**, 43–58, 1990.

R. C. Merkle and M. E. Hellman, Hiding Information and Signatures in Trapdoor Knapsacks, *IEEE Transactions on Information Theory*, **IT-24**, 525–530, 1978.

R. C. Merkle and M. E. Hellman, On the Security of Multiple Encryption, *Communications of the Association for Computing Machinery*, **24**, 465–466, 1981.

D. Micciancio, The Hardness of the Closest Vector Problem with Preprocessing, *IEEE Transactions on Information Theory*, **IT-47**, 1212–1215, 2001a.

D. Micciancio, The Shortest Vector Problem is NP-Hard to Approximate to Within Some Constant, *SIAM Journal on Computing*, **30**, 2008–2035, 2001b.

D. Micciancio, Generalized Compact Knapsacks, Cyclic Lattices, and Efficient One-Way Functions from Worst-Case Complexity Assumptions, *Computational Complexity*, **16**, 365–411, 2007.

D. Micciancio and O. Regev, Lattice-Based Cryptography, *Post-Quantum Cryptography*, D. J. Bernstein, J. Buchmann, and E. Dahmen, editors, pp. 147–191, New York, Springer, 2009.

F. Miller, *Telegraphic Code to Insure Privacy and Secrecy in the Transmission of Telegrams*, New York, C. M. Cornwell, 1882.

G. L. Miller, Riemann's Hypothesis and Tests for Primality, *Journal of Computer and Systems Science*, **13**, 300–317, 1976.

V. S. Miller, Short Programs for Functions on Curves, (unpublished) 1986.

V. S. Miller, The Weil Pairing, and its Efficient Calculation, *Journal of Cryptology*, **17**, 235–261, 2004.

V. S. Miller, Uses of Elliptic Curves in Cryptography, *Advances in Cryptology, CRYPTO85*, H. C. Williams, editors, pp. 417–426, New York, Springer, 1985.

A. Miyaji, M. Nakabayashi, and S. Takano, New Explicit Conditions of Elliptic Curve Traces for FR-Reduction, *IEICE Transactions on Fundamentals*, **E84-A**, 1234–1243, 2003.

P. L. Montgomery, Modular Multiplication without Trial Division, *Mathematics of Computation*, **44**, 519–521, 1985.

J. H. Moore, Protocol Failures in Cryptosystems, *Contemporary Cryptology: The Science of Information Integrity*, G. J. Simmons, editor, pp. 541–548, New York, IEEE Press, 1992.

L. J. Mordell, Observation on the Minimum of a Positive Quadratic Form in Eight Variables, *Journal of the London Mathematical Society*, **19**, 3–6, 1944.

M. A. Morrison and J. Brillhart, A Method of Factoring and the Factorization of F_7, *Mathematics of Computation*, **29**, 183–205, 1975.

P. Moulin and J. A. O'Sullivan, Information-Theoretic Analysis of Information Hiding, *IEEE Transactions on Information Theory*, **IT-49**, 563–593, 2003.

R. C. Mullin, I. M. Onyszchuk, S. A. Vanstone, and R. M. Wilson, Optimal Normal Bases in $GF(p^n)$, *Discrete Applied Mathematics*, **22**, 149–161, 1989.

D. Mumford, *Tate Lectures on Theta II*, Boston, MA, Birkhauser, 1984.

B. Murphy and R. P. Brent, On Quadratic Polynomials for the Number Field Sieve, *Australian Computer Science Communications*, **20**, 199–213, 1998.

National Bureau of Standards, *Secure Hash Standard*, FIBS Publication No. 180, Gaithersburg, MD, NBS, 1993.

P. Q. Nguyen, Cryptanalysis of the Goldreich–Goldwasser–Halevi Cryptosystem from *CRYPTO97, Advances in Cryptology, CRYPTO99*. M. J. Wiener, editor, pp. 288–304, New York, Springer, 1999.

P. Q. Nguyen and J. Stern, Cryptanalysis of the Ajtai–Dwork Cryptosystem, *Advances in Cryptology, CRYPTO98*, B. S. Kaliski, Jr., editor, pp. 223–242, New York, Springer, 1998.

P. Q. Nguyen and J. Stern, The Two Faces of Lattices in Cryptography, *Cryptography and Lattices, Lecture Notes on Computer Science*, vol. 2146, pp. 146–180, New York, Springer, 2001.

I. Niven, H. S. Zuckerman, and H. L. Montgomery, *An Introduction to the Theory of Numbers*, 5th edition, New York, Wiley, 1991.

Y. Nogami and Y. Morikawa, Ordinary Pairing Friendly Elliptic Curve of Embedding Degree 3 whose Order Has Two Large Prime Factors, *Memoirs of the Faculty of Engineering, Okayama University*, **44**, 60–68, 2010.

K. Nyberg and R. A. Rueppel, Message Recovery for Signature Schemes Based on the Discrete Logarithm Problem, *Designs, Codes, and Cryptography*, **7**, 61–81, 1996.

A. M. Odlyzko, Discrete Logarithms in Finite Fields and their Cryptographic Significance, *Advances in Cryptology, EUROCRYPT84*, T. Beth, N. Cot, and I. Ingemarsson, editors, pp. 224–314, New York, Springer, 1984.

A. M. Odlyzko, The Rise and Fall of Knapsack Cryptosystems, *Proceedings of the Symposia on Applied Mathematics*, vol. 42, *Cryptology and Computational Number Theory*, pp. 75–88, Providence, RI, American Mathematical Society, 1990.

T. Okamoto, Provably Secure and Practical Identification and Corresponding Signature Schemes, *Advances in Cryptology, CRYPTO92*, E. F. Brickell, editor, pp. 31–53, New York, Springer, 1992.

J. K. Omura and J. L. Massey, Computational Method and Apparatus for Finite Field Arithmetic, US Patent 4,587,627, May 6, 1986 (filed September 14, 1982).

C. Paar and C. Pelzl, *Understanding Cryptography: A Textbook for Students and Practitioners*, New York, Springer, 2009.

S. K. Park, *Applications of Algebraic Curves to Cryptography*, Ph.D. Dissertation, University of Illinois, 2007.

N. J. Patterson, The Algebraic Decoding of Goppa Codes, *IEEE Transactions on Information Theory*, **IT-21**, 384–386, 1975.

J. Pila, Frobenius Maps of Abelian Varieties and Finding Roots of Unity in Finite Fields, *Mathematics of Computation*, **55**, 745–763, 1996.

R. G. E. Pinch, The Carmichael Numbers up to 10^{15}, *Mathematics of Computation*, **61**, 381–391, 1993.

J. B. Plumstead, Inferring a Sequence Generated by a Linear Congruence, *Proceedings of the 23rd IEEE Symposium on the Foundations of Computer Science*, 153–159, 1982.

H. C. Pocklington, The Determination of the Prime or Composite Nature of Large Numbers by Fermat's Theorem, *Proceedings of the Cambridge Philosophical Society*, **18**, 29–30, 1914–16.

S. C. Pohlig and M. E. Hellman, An Improved Algorithm for Computing Logarithms in $GF(p)$ and its Cryptographic Significance, *IEEE Transactions on Information Theory*, **IT-24**, 106–110, 1978.

M. Pohst, A Modification of the LLL Algorithm, *Journal of Symbolic Computation*, **4**, 123–128, 1987.

J. M. Pollard, Theorems on Factorization and Primality Testing, *Proceedings of the Cambridge Philosophical Society*, **76**, 521–528, 1974.

J. M. Pollard, A Monte Carlo Method for Factorization, *BIT Numerical Mathematics*, **15**, 331–334, 1975.

J. M. Pollard, Monte Carlo Methods for Index Computation mod p, *Mathematics of Computation*, **32**, 918–924, 1978.

J. M. Pollard, Factoring with Cubic Integers, *The Development of the Number Field Sieve, Lecture Notes in Mathematics*, A. K. Lenstra and H. W. Lenstra, Jr., editors, vol. 1554, pp. 50–94, New York, Springer, 1993.

J. M. Pollard and C. P. Schnorr, An Efficient Solution of the Congruence $x^2 + Ky^2 = m \pmod{n}$, *IEEE Transactions on Information Theory*, **IT-33**, 702–709, 1987.

C. Pomerance, Recent Developments in Primality Testing, *The Mathematical Intelligencer*, **3**(3), 97–105, 1981.

C. Pomerance, The Quadratic Sieve Factoring Algorithms, *Advances in Cryptology, EUROCRYPT84*, T. Beth, N. Cot, and I. Ingemarsson, editors, pp. 169–182, New York, Springer, 1984.

C. Pomerance, Fast, Rigorous Factorization and Discrete Logarithm Algorithms, *Discrete Algorithms and Complexity*, D. S. Johnson, editor, pp. 119–143, New York, Academic Press, 1987.

C. Pomerance, A Tale of Two Sieves, *Notices of the American Mathematical Society*, **43**, 1473–1485, 1996.

C. Pomerance, J. W. Smith, and R. Tuler, A Pipe-line Architecture for Factoring Large Integers with the Quadratic Sieve Algorithm, *SIAM Journal on Computing*, **17**, pp. 387–403, 1988.

B. Preneel, R. Govaerts, and J. Vandewalle, Information Authentication: Hash Functions and Digital Signatures, *Computer Security and Industrial Cryptography: State of the Art and Evolution, Lecture Notes in Computer Science*, B. Preneel, R. Govaerts, and J. Vandewalle, editors, vol. 741, pp. 87–131, New York, Springer, 1993.

B. Preneel, R. Govaerts, and J. Vandewalle, Hash Functions Based on Block Ciphers: A Synthetic Approach, *Advances in Cryptology, CRYPTO93*, D. R. Stinson, editor, pp. 368–378, New York, Springer, 1993.

G. Purdy, A High-Security Log-in Procedure, *Communications of the Association for Computing Machinery*, **17**, 442–445, 1974.

M. O. Rabin, *Digital Signatures and Public-Key Functions as Intractable as Factorization*, Technical Report No. LCS-TR-212, Cambridge, MA, Massachusetts Institute of Technology Laboratory for Computer Science, 1979.

M. O. Rabin, Probabilistic Algorithm for Testing Primality, *Journal of Number Theory*, **12**, 128–138, 1980.

I. S. Reed and G. Solomon, Polynomial Codes over Certain Finite Fields, *Journal of the Society of Industrial and Applied Mathematics*, vol. 8, pp. 300–304, 1960.

O. Regev, Lattice-Based Cryptography, *Advances in Cryptology, CRYPTO06*, N. Koblitz, editor, pp. 131–141, New York, Springer, 2006.

B. Riemann, On the Number of Primes Less Than a Given Quantity, *Monatsberichte der Berliner Akademie*, 1859.

R. L. Rivest, The MD4 Message Digest Algorithm, *Advances in Cryptology, CRYPTO90*, A. Menezes and S. A. Vanstone, editors, pp. 303–311, New York, Springer, 1990.

R. L. Rivest, A. Shamir, and L. Adleman, A Method for Obtaining Digital Signatures and Public-Key Cryptosystems, *Communications of the Association for Computing Machinery*, **21**, 120–126, 1978.

J. Rosenthal, A Polynomial Description of the Rijndael Advanced Encryption Standard, *Journal of Algebra and its Applications*, **2**, 223–236, 2003.

K. Rubin and A. Silverberg, Torus-Based Cryptography, *Advances in Cryptology, CRYPTO03*, D. Boneh, editor, pp. 349–365, New York, Springer, 2003.

K. Rubin and A. Silverberg, Choosing the Correct Elliptic Curve in the CM Method, *Mathematics of Computation*, **79**, 545–561, 2010.

H. G. Rück, On the Discrete Logarithms in the Divisor Class Group of Curves, *Mathematics of Computation*, **68**, 805–806, 1999.

R. A. Rueppel, *Analysis and Design of Stream Ciphers*, New York, Springer, 1986.

R. A. Rueppel and O. Staffelbach, Products of Linear Recurring Sequences with Maximum Complexity, *IEEE Transactions on Information Theory*, **IT-33**, 124–131, 1987.

R. Sakai, K. Ohgishi, and M. Kasahara, Cryptosystems Based on Pairing, *Proceedings of the Symposium on Cryptography and Information Security*, Okinawa, Japan, 2000.

D. V. Sarwate and M. B. Pursley, Crosscorrelation Properties of Pseudorandom and Related Sequences, *Proceedings of the IEEE*, **68**, 593–619, 1980.

T. Satoh, On p-adic Point Counting Algorithms for Elliptic Curves over Finite Fields, *Journal of the Ramanujan Mathematics Society*, **15**, 247–270, 2000.

T. Satoh, On p-adic Point Counting Algorithms for Elliptic Curves over Finite Fields, *5th International Symposium on Algorithmic Number Theory V, Lecture Notes in Computer Science*, C. Fieker and D. R. Kohel, editors, vol. 2369, pp. 43–66, New York, Springer, 2002.

E. Savaş, T. A. Schmidt, and Ç. K. Koç, Generating Elliptic Curves of Prime Order, *Cryptographic Hardware and Embedded Systems, Lecture Notes in Computer Science*, G. Goos, J. Hartmanis, and J. van Leeuwen, editors, vol. 2162, pp. 142–158, 2001.

B. Schneier, *Applied Cryptography*, New York, Wiley, 1996.

C. P. Schnorr, A Hierarchy of Polynomial Time Lattice Basis Reduction Algorithms, *Theoretical Computer Science*, **53**, 201–224, 1987.

C. P. Schnorr, Efficient Signature Generation by Smart Cards, *Journal of Cryptology*, **4**, 161–174, 1991a.

C. P. Schnorr, Method for Identifying Subscribers and for Generating and Verifying Electronic Signatures in a Data Exchange Signature, US Patent 4,995,082A, February 19, 1996.

R. J. Schoof, Elliptic Curves over Finite Fields and the Computation of Square Roots Mod p, *Mathematics of Computation*, **44**, 483–494, 1985.

R. J. Schoof, Counting Points on Elliptic Curves over Finite Fields, *Journal de Théorie des Nombres de Bordeaux*, **7**, 219–254, 1995.

R. Schroeppel, Elliptic Curve Point Halving Wins Big, *Proceedings of 2nd Midwest Arithmetical Geometry in Cryptography Workshop*, Urbana, IL, 2000.

E. S. Selmer, *Linear Recurrence Relations over Finite Fields*, Department of Mathematics, University of Bergen, Norway, 1966.

G. Seroussi, *Table of Low-Weight Irreducible Polynomials over F_2*, Technical Report No. HPL-98-135, Palo Alto, CA, Hewlett-Packard Laboratories, 1998.

A. Shamir, How to Share a Secret, *Communications of the Association for Computing Machinery*, **22**, 612–613, 1979.

A. Shamir, A Polynomial-Time Algorithm for Breaking the Basic Merkle–Hellman Cryptosystem, *IEEE Transactions on Information Theory*, **IT-30**, 699–704, 1984a.

A. Shamir, Identity-Based Cryptosystems and Signature Schemes, *Advances in Cryptology, CRYPTO84*, G. R. Blakly and D. Chaum, editors, pp. 47–53, New York, Springer, 1984b.

D. Shanks, Class Number, a Theory of Factorization, and Genera, *Proceedings of the Symposia on Pure Mathematics*, **20**, 415–440, Providence, RI, American Mathematical Society, 1971.

D. Shanks, Five Number-Theoretic Algorithms, *Proceedings of the 2nd Manitoba Conference on Numerical Mathematics*, 51–70, 1972.

C. E. Shannon, A Mathematical Theory of Communication, *Bell System Technical Journal*, **27**, 379–423 and 623–656, 1948 (Part I) pp. 623–656 (Part II). Reprinted in book form with postscript by W. Weaver, University of Illinois Press, Urbana, IL, 1949, Anniversary edition 1998.

C. E. Shannon, The Communication Theory of Secrecy Systems, *Bell System Technical Journal*, **28**, 656–715, 1949.

P. Shor, Polynomial-Time Algorithms for Prime Factorization and Discrete Logarithms on a Quantum Computer, *SIAM Review*, **41**, 303-332, 1999.

V. Shoup, Lower Bounds for Discrete Logarithms and Related Problems, *Advances in Cryptology, EUROCRYPT97*, W. Fumy, editor, pp. 256–266, New York, Springer, 1997.

T. Siegenthaler, Correlation-Immunity of Nonlinear Combining Functions for Cryptographic Applications, *IEEE Transactions on Information Theory*, **IT-30**, 776–780, 1984.

J. Silverman, *Arithmetic of Elliptic Curves*, New York, Springer, 1986.

J. Silverman, The Xedni Calculus and the Elliptic Curve Discrete Logarithm Problem, *Designs, Codes, and Cryptography*, **20**, 5–40, 2000.

J. Silverman and J. Tate, *Rational Points on Elliptic Curves*, New York, Springer, 1992.

D. R. Simon, Finding Collisions on a One-Way Street: Can Secure Hash Functions Be Based on General Assumptions? *Advances in Cryptology, EUROCRYPT98*, K. Nyberg, editor, pp. 334–345, New York, Springer, 1998.

S. Singh, *The Code Book: The Science of Secrecy from Ancient Egypt to Quantum Cryptography*, New York, Anchor Books, 1999.

B. Smeets, *Some Results on Linear Recurring Sequences*, Ph.D. Dissertation, University of Lund, Sweden, 1987.

M. E. Smid and D. K. Branstad, The Data Encryption Standard: Past and Future, *Contemporary Cryptology: The Science of Information Integrity*, G. J. Simmons, editor, pp. 43–64, New York, IEEE Press, 1992.

J. Solinas, Efficient Arithmetic on Koblitz Curves, *Designs, Codes, and Cryptography*, **19**, 195–249, 2000.

R. Solovay and V. Strassen, A Fast Monte Carlo Test for Primality, *SIAM Journal on Computing*, **6**, 84–85, 1977.

A. Sorkin, LUCIFER, A Cryptographic Algorithm, *Cryptologia*, **8**, 22–35, 1984.

M. Steiner, G. Tsudik, and M. Waidner, Key Agreement in Dynamic Peer Groups, *IEEE Transactions on Parallel and Distributed Systems*, **PDS-11**, 769–780, 2000.

I. N. Stewart and D. O. Tall, *Algebraic Number Theory*, London, Chapman and Hall, 1979.

H. Stichtenoth, *Algebraic Function Fields and Codes*, Berlin, Springer, 1993.

D. R. Stinson, *Cryptography: Theory and Practice*, 3rd edition, Boca Raton, FL, CRC Press, 2006.

J. Tate, WC-Group over p-adic Fields, *Séminaire Bourbaki 10e Année*, Paris, Sécretariat Mathématique, 1958.

J. Tate, Duality Theorems in Galois Cohomology over Number Fields, *Proceedings of the International Congress on Mathematics*, Stockholm, 1962.

J. Tate, Duality Theorems in Galois Cohomology over Number Fields, *Proceedings of the International Congress of Mathematicians*, pp. 288–295, Djursholm, Sweden, Institut Mittag-Leffler, 1963.

N. Thériault, Index Calculus Attack for Hyperelliptic Curves of Small Genus, *Advances in Cryptology, Asiacrypt03*, C.-S. Laih, editor, pp. 75–92, New York, Springer, 2003.

M. Tompa and H. Woll, How to Share a Secret with Cheaters, *Journal of Cryptology*, **1**, 133–138, 1988.

A. Tonelli, Bemerkung über die Auflösung quadratischer Congruenzen, *Universität zu Göttingen Nachrichen*, pp. 344–346, 1891.

W. Trappe and L. Washington, *Introduction to Cryptography with Coding Theory*, New York, Prentice Hall, 2006.

S. M. Turner, Square Roots mod p, *American Mathematical Monthly*, **101**, 443–449, 1994.

B. van der Waerden, *Modern Algebra*, vol. 2, New York, Frederick Ungar, 1950.

F. Vercauteren, An Extension of Kedlaya's Algorithm to Hyperelliptic Curves in Characteristic 2, *Journal of Cryptology*, **19**, 1–25, 2006.

F. Vercauteren, Pairings on Elliptic Curves, *Identity-Based Encryption*, M. Joye and G. Neven, editors, Amsterdam, IOS Press, 2009.

F. Vercauteren, Optimal Pairings, *IEEE Transactions on Information Theory*, **IT-56**, 455–461, 2010.

G. S. Vernam, Cipher Printing Telegraph Systems for Secret Wise and Radio Telegraphic Communications, *Journal of the American Institute of Electrical Engineering*, **55**, 109–115, 1926.

M. Walker, Information-Theoretic Bounds for Authentication Systems, *Journal of Cryptology*, **2**, 131–143, 1990.

X. Wang and H. Yu, How to Break MD5 and Other Hash Functions, *Advances in Cryptology, EUROCRYPT05*, R. Cramer, editor, pp. 19–35, New York, Springer, 2005.

X. Wang, X. Lai, D. Feng, H. Chen, and X. Yu, Cryptanalysis of the Hash Functions MD4 and RIPEMD, *Advances in Cryptology, EUROCRYPT05*, R. Cramer, editor, pp. 1–18, New York, Springer, 2005.

M. Ward, An Arithmetical Theory of Linear Recurring Sequences, *Transactions of the American Mathematics Society*, **35**, 600–628, 1933.

L. C. Washington, *Elliptic Curves: Number Theory and Cryptography*, 2nd edition, Boca Raton, FL, CRC Press, 2008.

M. N. Wegman and J. L. Carter, Universal Classes of Hash Functions, *Journal of Computer and System Sciences*, **10**, 143–154, 1979.

A. Weil, *Courbes Algébraiques et les Variétés Abéliennes*, Paris, Hermann, 1948.

A. Weil, Numbers of Solutions of Equations in Finite Fields, *Bulletin of the American Mathematics Society*, **55**, 497–508, 1949.

R. Wernsdorf, The One-Round Functions of the DES Generate the Alternating Group, *Advances in Cryptology, EUROCRYPT92*, R. A. Rueppel, editors, pp. 99–112, New York, Springer, 1992.

D. Wiedemann, Solving Sparse Linear Equations over Finite Fields, *IEEE Transactions on Information Theory*, **IT-32**, 54–62, 1986.

M. Wiener, Cryptanalysis of Short RSA Secret Exponents, *IEEE Transactions on Information Theory*, **IT-36**, 553–559, 1990.

M. V. Wilkes, *Time-Sharing Computer Systems*, Amsterdam, Elsevier, 1968.

T. Wollinger, *Computer Architectures for Cryptosystems Based on Hyperelliptic Curves*, MSc Thesis, Worcester Polytechnic Institute, 2001.

P. W. Wong and N. Memon, Secret and Public Key Image Watermarking Schemes for Image Authentication and Ownership Verification, *IEEE Transactions on Image Processing*, **IP-10**, 1593–1601, 2001.

A. Wyner, The Wire-Tap Channel, *Bell System Technical Journal*, **54**, 1355–1387, 1975.

G.-Z. Xiao and J. L. Massey, A Spectral Approach to Correlation-Immune Combining Functions, *IEEE Transactions on Information Theory*, **IT-34**, 569–571, 1988.

M. Yoshida, Inseparable Multiplex Transmission Using the Pairing on Elliptic Curves and its Application in Watermarking, *Proceedings of the 5th Conference on Algebraic Geometry, Number Theory, Coding Theory, and Cryptography*, University of Tokyo, 2003.

C.-A. Zhao, F. Zhang, and J. Huang, A Note on the Ate Pairing, *International Journal of Information Security*, **7**, 379–382, 2008.

N. Zierler, Linear Recurring Sequences, *Journal of the Society of Industrial and Applied Mathematics*, **7**, 31–48, 1959.

N. Zierler and J. Brillhart, On Primitive Trinomials, *Information and Control*, **13**, 541–544, 1968.

Index

Printed in the United States
by Baker & Taylor Publisher Services